EARTH SCIENCES

Edited by **Imran Ahmad Dar**

INTECHOPEN.COM

Earth Sciences
Edited by Imran Ahmad Dar

CBS, Edition 2016

Published by InTech

Janeza Trdine 9, 51000 Rijeka, Croatia

Publishing Process Manager Danijela Duric

Technical Editor Teodora Smiljanic

Cover Designer InTech Design Team

Additional hard copies can be obtained from orders@intechopen.com

Earth Sciences, Edited by Imran Ahmad Dar
p. cm.
ISBN 978-953-307-861-8

INTECH

open science | open minds

Contents

Preface

The studies of Earth's history, and of the physical and chemical properties of the substances that make up our planet, are of great significance to our understanding both of its past and its future. The geological and other environmental processes on Earth and the composition of the planet are of vital importance in locating and harnessing its resources.

Earth Sciences deploy an interdisciplinary mix of physics, chemistry, biology, math, and natural science to understand the mysteries of nature. They focus on the solid Earth (rocks, minerals, mountain belts, volcanoes, earthquakes, sedimentary basins, oil and gas deposits, etc.) as well as on the history of life (paleontology) and its impact on the Earth. In recent decades, geologists have become increasingly concerned with the history of the Earth's climate, how the physical and chemical behavior of the oceans has changed over time, and how drifting continents and evolving life have interacted to control the composition of the atmosphere and oceans and hence to control global climate. Environmentalists also examine how human activities affect our environment, including the quality of air, water, and soil.

This book is primarily written for research scholars, geologists, civil engineers, mining engineers, and environmentalists. Hopefully the text will be used by students, and it will continue to be of value to them throughout their subsequent professional careers. This does not mean to infer that the book was written solely or mainly with the student in mind. Indeed from the point of view of the student in environmental science it could be argued that this text contains more detail than he will require in his initial studies.

This book offers one of the most comprehensive and up-to-date treatments of topics from fundamental geologic principles, to the specifics of environmental and geologic hazards, all from a truly environmental perspective. It fully discusses both processes and environmental issues and details, where appropriate, of the quantification of geologic processes. The book has been divided into 11 sections; geology, geochemistry, seismology, petroleum geology, hydrology, hydrogeology, mineralogy, mining, volcanology, remote sensing, and environmental sciences.

A great many people have contributed to the development of this book. The editor is very thankful to all the authors. Portions of the manuscript of the Book *Earth Sciences* were contributed by Mr. Clemente Jude, Prof. Ceryan Sener, Dr Clary Renee, Dr Wang Dayong, Dr Ercilla Gemma, Dr Ivanyuk Gregory, Dr Zhang Haijiang, Prof. Jia Ailin, Dr Ge Jiandang, Prof. Muto Tetsuji, Dr Li Zhonghai, Dr Adey Walter, Prof. Ariffin Kamar Shah, Prof. Erarslan Kaan, Mr. Istadi Bambang, MSc Yuangao Zhang, Dr Tim Webster, Dr Jeff Harris, Mr. Oudijk Gil, Dr Garcia-Aguirre, Dr Calligaris, Dr Imran Ahmad Dar, Dr Matsumaru Kuniteru, Dr Ben Fadhel Moez and Dr Jiaojiao Geng. The Editor wishes to offer his thanks to all those involved. The Editor is also very thankful to InTech President, Aleksandar Lazinica. In particular, the Editor would like to record his grateful appreciation to Ana Pantar, editorial consultant at InTech for inviting him to edit this prestigious book. Lastly, but by no means least, one must thank Ms. Danijela Duric, publishing process manager, for her timely and kind communication with the Editor during the entire process.

Imran Ahmad Dar
Department of Industries and Earth Sciences, The Tamil University, Ocean and
Atmospheric Sciences & Technology Cell,
India

Part 1

Geology

1

The Expanding International Coal Market

Jude Clemente
Department of Homeland Security
JTC Energy Research Associates, LLC
San Diego State University, California
USA

1. Introduction

"Coal is the single largest source of primary energy in the world, and in absolute terms coal use has grown faster than use of any other fuel for most of this decade. As the developing world industrializes and struggles to meet seemingly insatiable demand for power, coal has become the fuel of choice. An expanding seaborne trade in coal has started to connect previously isolated regional markets......[and]*.....reflects the internationalization and commoditization of the coal market, in a transformation reminiscent of that of the oil market in the 1970s and 1980s,"* (Stanford University, Program on Energy and Sustainable Development, 2009)

Coal provides nearly 30% of the world's primary energy and generates over 40% of global electricity (IEA, 2010a). Historically a domestically produced and consumed resource, the international coal market meets just 16% of total consumption (World Coal Association, 2011a). Global coal trade patterns can be parsed into two distinct regions, the Atlantic region and the Pacific region. The Americas and Europe are the primary demand centers in the Atlantic, while Asia, dominated by Japan, South Korea, India, and China, is the hub in the Pacific. Geography has been critical when deciding which producers serve which markets, with suppliers generally serving only one region. The United States, holding roughly 30% of the world's coal endowment, is considered the "swing supplier," balancing markets when supply is short (BP, 2010). There are also two parallel coal markets; thermal (steam) coal is used to generate electricity, and metallurgical (coking) coal is used to manufacture steel and iron. In the past 20 years:

- The seaborne trade in thermal and coking coal has increased by 7% and 1.6% per year respectively (World Coal Association, 2011a).
- Global coal consumption has increased by more than 50% - from 2,235 million tonnes of oil equivalent (Mtoe) to 3,430 Mtoe (IEA, 2010a).
- Asia Pacific has extended its share of world coal consumption from 37% to 66% and will account for basically all of the incremental demand increase over the next 20 years (IEA, 2010a).

Today, coal is the fuel of choice in the developing world, and a widening array of suppliers has the international coal market in the midst of a rapid and dramatic transformation. The relatively high cost supply source of the United States will give lower cost producers the edge in Asia, where the dependence on coal for incremental electricity is stunning. Looking forward, "Chindia" will play the central role in determining global trade flows and prices.

China and India, with 36% of humanity (The World Bank, 2010), will increasingly rely upon coal to feed their flourishing economies - coal already generates 80% and 70% of their power respectively (IEA, 2010a). As a result, these nations will dominate the growth in coal-based electricity over the next several decades. And, importantly, a parallel expansion of coal consumption in Asia will occur in the industrial sector where the demand for coal to manufacture steel, cement, and liquid fuel is soaring. From 2008 to 2030, the International Energy Agency (IEA, 2010a) projects that global coal consumption will expand by over 1,600 Mtoe - roughly as much incremental energy as natural gas, nuclear, wind, and solar will provide together. The IEA's Coal Industry Advisory Board (2009) succinctly states the reality of the world's coal situation:

- "The future use of increasing quantities of coal worldwide is inevitable if the world is to avoid a damaging energy crunch and support the development needs of poorer nations."
- "All major studies that have examined the outlook for world energy demand indicate that the world will remain dependent on the continued use of coal for many decades to come."

Developing nations realize that breaking the pattern of systematic poverty depends upon available and affordable energy in general and access to reliable electricity in particular. On a per capita basis, The World Bank (2010) reports that China consumes 17% as much power as the United States, and India uses 8% as much as the European Union (EU). With the devastating consequences of electricity deprivation well documented, the international coal market will continually expand as China and India strive to raise the quality of life of their people (see Morse & He, 2010). For example, Platts (2005) cites the Asia Pacific region specifically as the driving force behind the now thriving coal derivatives market, whether in futures or over-the-counter contracts. In addition to having the fastest growing consumers, Asia Pacific also retains the two largest coal exporters, Australia and Indonesia (see Figure 1). The present chapter examines the three main factors that are now converging to expand the international coal market: 1) the need for more electricity, 2) the need for more coal, and 3) the need for more coal imports in China and India. Developing nations have made clear that poverty reduction goals will not be sacrificed for climate change mitigation. India, for instance, has 400 million citizens without access to electricity - more people than the United States and Germany have combined.

Top Exporters (million tonnes)

Country	Thermal	Coking	Total
Australia	134	125	259
Indonesia	200	30	230
Russia	105	11	116
Colombia	69	0	69
S. Africa	66	1	57
USA	20	33	53
Canada	7	21	28

Top Importers (million tonnes)

Country	Thermal	Coking	Total
Japan	113	52	165
China	102	35	137
S. Korea	82	21	103
India	44	23	67
Chinese Taipei	57	3	60
Germany	32	6	38
UK	33	5	38

Fig. 1. The International Coal Market, 2009
Source: developed from World Coal Association, 2011a

2. The need for more electricity

"Electrification greatly improves the quality of life. Lighting alone brings benefits such as increased study time and improved study environment for schoolchildren, extended hours for small businesses, and greater security," (The World Bank, 2008)

Electricity is the *sine qua non* of modern society. Electricity is essential to gains in quality of life, economic well-being, and a cleaner environment. The U.S. National Academy of Engineering (2004) identified societal electrification as the "most significant engineering achievement" of the 20th Century - a century that saw a population swell of over four billion people, the rise of the metropolis, a transportation revolution, historic improvements in medical care, and the emergence of a vast system of electronic communication. As recognized by the Global Energy Network Institute (2002), "Every single one of the United Nations Millennium Development Goals requires access to electricity as a necessary prerequisite." The socioeconomic benefits of the U.S. Rural Electrification Act of 1936 alone demonstrate the scope of electricity's importance to living a longer and better life. Access to electricity brought about a sea change to the American quality of life, ranging from childhood survival to clean drinking water to literacy. Arguments that some states, such as California, have grown their economy and flattened electricity consumption through efficiency policies are largely rhetorical. Regression analyses confirm that approximately 80% of California's lower per capita use of electricity is due to unique characteristics like higher prices, milder weather, and smaller homes with more people (see Mitchell et al., 2009). Electricity is unique and employed in ways that no other energy form can be:

- High quality and convertible to virtually any energy service - light, motion, heat, electronics, and chemical potential
- Permits previously unattainable precision, control, and speed
- Provides temperature and energy density far greater than those attainable from standard fuels
- Has no inertia - instantaneous access and 100% convertible to work

Electricity has wide ranging environmental benefits. Electro-technologies are more efficient than their fuel-burning counterparts and, unlike standard fuels, have no waste products at the point of use. No smoke, ash, combustion gas, noise, or odor. Electrification increases the efficiency of a society's primary energy consumption and decreases emissions of pollutants and greenhouse gases (GHGs). Electro-technologies produce less carbon dioxide (CO_2) per unit of Gross Domestic Product (GDP), leading to findings by the Electric Power Research Institute (2003) that "technology innovation in electricity use is a cornerstone of global economic progress" and "deploying the technology of an enhanced electricity infrastructure would include.....a 13-25% reduction in carbon dioxide emissions" and "a 10% increase in real GDP." Environmental management depends upon electricity for the movement of water and waste. A sustainable environment requires clean water and sanitation facilities. Electricity is the key to providing these services and pollution controls, and power consumption is directly related to their corresponding environmental benefits. Given the extraordinary virtues of electricity, it is no surprise that demand for power will remain a steady drumbeat across the globe (see Figure 2).[1]

[1]Even under the IEA's (2010a) 450 Scenario, which optimistically assumes that "collective policy action is taken to limit the long-term concentration of greenhouse gases in the atmosphere to 450 parts per

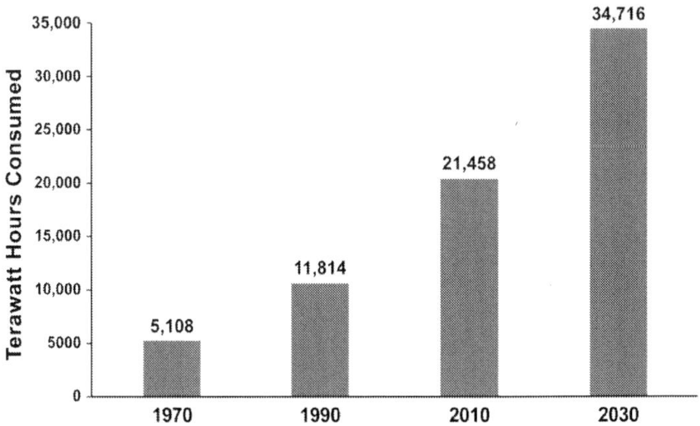

Fig. 2. The Constant Increase in Global Electricity Consumption, 1970-2030
Source: developed from International Energy Agency, 1999; 2010a

The depth of global poverty and energy deprivation is difficult for most Westerners to comprehend. The world is confronted with a human crisis at a horrific scale - 2,600 million people live on less than \$2 (all \$ in USD) a day, 2,000 million have minimal access to electricity, and 1,400 million have no electricity at all (The World Bank, 2010). In addition, another 2,000 million people will be born in the next several decades. In developing countries, household tasks requiring energy - gathering wood, carrying water, and cooking - are typically delegated to women and their children. Chores made all the more easier, safer, and healthier with the availability of electricity. The lack of electricity perpetuates the cycle of poverty because it not only blocks access to electronic communication but also means inadequate illumination for reading and studying at night. Access to electricity is a necessary condition for economic and human progress. Societies with more access to electricity survive childhood, eat better, drink cleaner water, and learn to read. Women and children are among the greatest beneficiaries of electrification as new doors of opportunity open for these particularly vulnerable segments of the population. At least 70% of all people living in poverty are female (The Global Poverty Project, n.d.). This problem of "feminization of poverty" will be impossible to resolve unless households have adequate access to electricity and other forms of modern energy. Thus, the major challenge of our time is not merely to reach 2050 with a significant reduction in GHGs emissions but to also create electricity access for the vast multitude of men, women, and children who toil grimly in the dark. Access to electricity should be a human right: every 10-fold increase in electricity use is linked to a 10-year increase in life expectancy (Boyce, 2010).

As noted by the IEA (2002), a "lack of electricity exacerbates poverty and contributes to its perpetuation, as it precludes most industrial activities and the jobs they create." The consequences of abject poverty and energy deprivation continue to devastate around the world. UNICEF (2009) reports that 24,000 children die each day from preventable causes. To put this amount in perspective, consider the cumulative effect of child deaths: 8,760,000

million of CO_2-equivalent," the world is sill projected to require 30,170 terawatt hours of electricity in 2030, or 40% more than it did in 2010.

children die each year from causes that electricity could help eliminate, a sum equal to the urban population of New York City. Pasternak (2000) found that a per capita annual consumption rate of at least 4,000 kilowatt hours (kWh) of electricity is required for a nation to reach a significant Human Development Index of 0.9. Electricity deprivation is thus a global blight (see Figure 3). Well over four billion people, at least 60% of the world's population, use fewer than 2,350 kWh per year, or only one-third as many as a typical resident of the EU (The World Bank, 2010). The challenges put forth transcend national borders and are calamities for humanity at large. While the *Copenhagen Accord 2009* stated that the eradication of poverty should be the "first and overriding" priority of developing nations, there is strong reason to argue that it did not go far enough (United Nations, 2009). The elimination of poverty and energy deprivation is a global responsibility. Based on the Accord, the path forward for nations is clear:

- Eliminate energy poverty as a first-order priority
- Create access to energy for everyone, everywhere by 2050
- Advance all energy forms for long-term access
- Promote advanced generation technologies to reach near zero emissions from coal and natural gas power plants

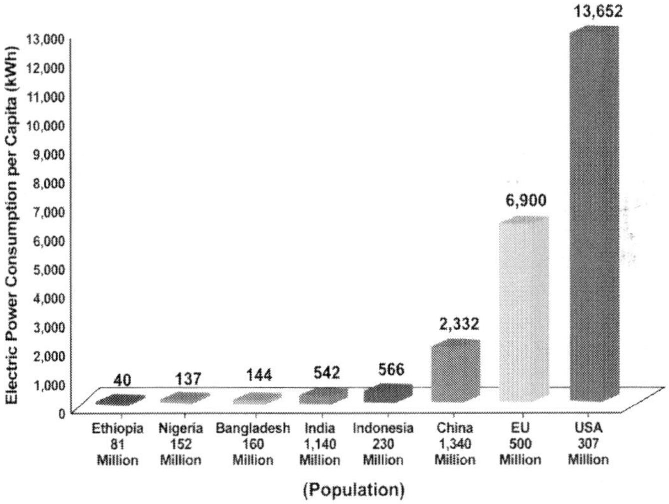

Fig. 3. The Scale of Global Electricity Deprivation, 2008
Source: developed from The World Bank, 2010

3. The need for more coal

"In the past quarter of a century, China has created wealth for many of its people, lifted many out of poverty, and helped drive and sustain global economic growth. Coal has underpinned China's massive and unprecedented growth in output, fuelling an economic miracle that has helped to improve the standard of living in many countries," (IEA, 2009)

If the goal of eliminating abject poverty and energy deprivation is ever to be attained, the supply and affordability of energy, particularly electricity, must improve dramatically.

Although the scale that will be required to meet these goals cannot be met by just one fuel, coal will stay the strategic choice since it generally has the lowest cost on a heat equivalency basis. The provision of adequate and affordable electricity to the 8.2 billion people who will inhabit the planet in 2030 will depend upon the increased availability, production, and consumption of coal-based electricity. Developing nations will increasingly lean on coal because it has the abundance, technology, and scalability to meet their enormous power generation challenges. From 2008 to 2030, the IEA (2010a) projects that coal will provide an additional 6,500 terawatt hours of electricity, nearly twice as much as the current total generation of the EU. This is more incremental power than natural gas, nuclear, wind, and solar will generate over that span combined. Indeed, for coal, the past is prologue. Coal has been, is now, and will continue to be a fundamental building block of socioeconomic development throughout the world. Over the half century from 1980-2030, despite a population increase of 3.7 billion people, economic growth of $121 trillion, and energy demand increase equivalent to more than 10,000 million tonnes of oil, coal will have actually extended its contribution to global energy production from 25% to 29% (IEA, 2010a; EIA, 2010a). The IEA (1999; 2010a) reports that coal has maintained roughly a 40% share of world electricity generation since 1970 and is on pace to provide 43% in 2030.

The importance of the low cost electricity that can be derived from coal cannot be overstated.[2] For the industrialized nations, high electricity prices disrupt family budgets and erode the ability of domestic firms to compete in their increasingly competitive global industries. It is in the developing world, however, where high electricity prices *wreck the most havoc since the people* have almost no capacity to absorb them. Data gathered from the U.S. Energy Information Administration's (EIA) *International Energy Outlook 2010* indicate that China's per capita GDP (2005 US$) in 2010 was $6,027 and India's was $2,963. By comparison, the average American made over $45,100, and the average EU citizen made over $30,000. Electric lighting is far less expensive and consumes less fuel than the kerosene lamps that are commonplace in the developing world. And light bulbs cut indoor air pollution. According to the World Health Organization (2005), the burning of solid biomass fuels is responsible for 1.6 million deaths a year and 2.7% of the global burden of disease. Low cost electricity powers water pumps, allowing the distribution of potable water and reducing waterborne parasitic diseases. Cost-efficient power also promotes the use of modern computers, information systems, and electric motors in manufacturing, thereby substantially improving productivity. Today, virtually all societies seek to enhance their "Three Es" - energy, economy, and environment. Coal is the fuel of choice for measurable reasons:

Abundance and Accessibility - BP's *Statistical Review of World Energy 2010* reports that coal is the most prevalent and widely distributed fossil fuel, accounting for 64% of global economically recoverable fossil resources, compared to 19% for oil and 17% for natural gas. The amount of proven recoverable coal reserves is enormous and exceeds 820 billion tonnes. Coal is distributed across every continent and every region of the world. For example, the Western Hemisphere and Asia Pacific each have about 260 billion tonnes of coal, Russia has 157 billion, Europe has 73 billion, and South Africa has 31 billion. The world consumed a total of 6.8 billion tonnes of coal in 2009 (World Coal Association, 2011b).

[2]In January 1930, when he was Governor of the state of New York, Franklin D. Roosevelt wrote an article in *The New York Times*, stating that: ".....high rates, of course, bear hard on the individual. But from a social standpoint they are chiefly to be regretted because they restrict the use of electricity. Rate schedules should be so adjusted as to induce the freest possible use of electricity both in the home and on the farm."

Secure Energy - As stated by the IEA (2008), "It is widely acknowledged that the oil and natural gas markets provide risks that undermine security of supply in the medium and long term." The widespread physical distribution of coal, on the other hand, readily enhances energy security across broad political arenas by being a buffer against supply disruptions. For example, the three largest nations, China, India, and the United States, have 40% of the population and 50% of the coal but only 4% of the oil and 5% of the natural gas (The World Bank, 2010; BP, 2010). By comparison, the Middle East (including Egypt) and Russia have just 6% of the population but control 62% of the oil and 65% of the natural gas (The World Bank, 2010; BP, 2010).

Reliability - Coal's abundance and even distribution, added to its low and stable price pattern, set the stage for a prolonged and reliable supply of energy. In many countries, coal-based generation is one of the first sources to be dispatched throughout the electric grid, as predictability makes coal a very attractive baseload fuel. Compared to other sources, the amount of electricity that can be generated from coal significantly exceeds its relative capacity. In 2008, for instance, coal accounted for 31% of total generation capacity but produced 41% of the world's electricity (IEA, 2010a). The EIA (2010b) concludes that in 2016 all three types of coal-fired plants (conventional, advanced, and advanced with carbon capture and sequestration, CCS) will have capacity factors of 85%, compared to just 34% for wind and 25% for solar.

Affordability - For example, based on IEA (2010b) analyses of levelized costs of electricity, supercritical coal-based plants are some of the most affordable sources of power in China, \$33 per megawatt hour, versus \$39 for natural gas (combined cycle combustion turbines), \$50 for hydro, \$53 for nuclear (Westinghouse AP1000), and \$71 for wind. Both China (Large Substituting for Small, LSS) and India (Ultra Mega Power Plants, UMPP) have implemented national strategies to deploy larger and more efficient supercritical and ultrasupercritical coal plants to capitalize on economies of scale. From 2008 to 2030, the great bulk of the combined 1,050 gigawatts of new coal capacity that "Chindia" is projected to add will be these more CCS-ready advanced units (IEA, 2010a).

Versatility - Countries around the world have been initiating an increasing number of coal projects converting coal-to-liquids (CTL), substitute natural gas, or chemicals. The scale of China's coal conversion plans is especially informative since various conversion efforts could utilize an additional few billion tonnes of coal over the next decade. CTL projects in particular will become more important with the approach of global peak conventional oil production - new petroleum finds are getting more complex, deeper, and smaller. China wants a 50 million tonne per year CTL industry by 2020 (Royal Society of Chemistry, 2007). The need for more oil from the destabilizing Middle East is a concern for both China and India.

Steel - Coal is vital to the production of steel, accounting for almost 70% of global output (World Coal Association, 2011c). And steel is a core component of our rapidly urbanizing world. In 2050, the global urban population is projected to be 70%, up from 50% today (World Health Organization, 2010). Urbanization will make huge demands on infrastructure - more buildings, roads, pipes, and machines. This equals more steel, which in turn means more coal. In addition, Dargay et al. (2007) project that the world will have 2,080 million vehicles in 2030, up from 960 million in 2007 - the average vehicle contains over 1,600 pounds of steel (Automotive News, 2007). Constant industrial development quadrupled the price of export coking coal from \$44 in 2000 to over \$176 a tonne in 2010 (Metals Consulting International, 2011).

4. Rising coal imports in China and India

"The seaborne thermal coal market is experiencing a transformation which may be as significant as that which occurred for the iron ore market over the past decade. In a similar way, we believe China and India together could transform the demand landscape for thermal coal over the next decade, displacing current western importers and evolving to dominate the industry," (Deutsche Bank, 2010)

Led by China and India, the EIA (2010a) projects that Asia's share of world coal imports will increase from 59% to 70% in the next 25 years. Although the region imports a growing majority of the world's thermal coal used for power generation, there is also immense potential for coking coal. The urbanization process continues apace in China and India, with more steel and cement translating into more coal. Simply put, the overwhelming reliance of these two nations on their domestic resources is unsustainable (see Figure 4). China accounts for 14% of global coal reserves but 47% of both production and consumption (BP, 2010). The central government therefore plans to cap coal output during the 12th Five-Year Plan (2011-2015) at between 3.6 and 3.8 billion tonnes, compared to 3.2 billion tonnes mined in 2009 (Reuters, 2010). India, meanwhile, in equilibrium holding and producing 7% of the world's coal (BP, 2010), must rely upon inefficient Coal India for over 80% of national output (Coal Explorer, 2011). India's coal minister, Sriprakash Jaiswal, claims coal is India's "most notorious" sector, consistently under-producing due to mismanagement, bloated bureaucratic operations, corruption, and theft (Daily News & Analysis, 2010). Many factors suggest that "Chindia" will increasingly require the international coal market to feed its insatiable appetite for energy.

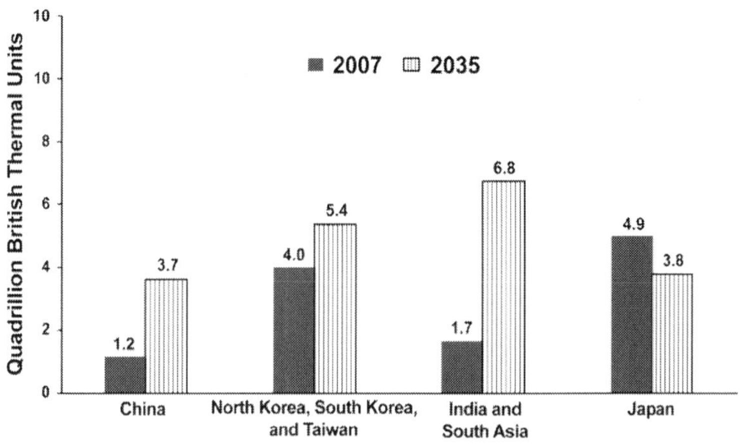

Fig. 4. Asia and the Growing International Coal Market, Coal Imports, 2007-2035
Source: developed from U.S. Energy Information Administration, 2010a

4.1 China

China's coal imports began to increase at the close of 2008, while accelerating the closure of smaller, inefficient, and less safe mines in Shanxi, historically the country's most productive province. In 2008, China installed a $586 billion economic stimulus package for infrastructure projects and other stimulus measures to bolster domestic demand (The

Economist, 2008).[3] Coal imports tripled to over 130 million tonnes in 2009 (Xinhua News Agency, 2010) and grew to 150 million in 2010 (Jin, 2011). The EIA (2010a) conservatively projects that China's imports will triple in the next 25 years. Wood Mackenzie (2010) expects that the first major shift in the international coal market will occur sometime in 2011 or 2012 when China overtakes Japan to become the world's largest thermal coal importer. Looking forward, there are at least three reasons why China will need more foreign coal:

- **Infrastructure Issues -** Perhaps the greatest challenge facing China's coal industry today is the transportation of coal from the distant producing regions, mostly in the North and West, to the high demand areas, located in the South and East. Escalating demand has rail lines heavily overburdened by coal transport, so more trucks are being used to take coal from Inner Mongolia and Xinjiang, energy-rich autonomous regions in North China, to population centers. In 2010, over 10,000 trucks moving coal from Inner Mongolia to Chinese cities got stuck in a 120 kilometer long traffic jam for nine days (Yahoo Finance, 2010). Gridlock is common in China and roads need repaired from the heavy coal traffic. Now, China's largest coal-producing region, the remote Inner Mongolia is becoming more crucial as an energy base and holds a staggering 1,080 billion tonnes of prospective coal reserves (Sinoc, 2005). China will need to invest some $150 billion in coal infrastructure by 2020 (China Business News, 2010). Importantly, the cost of coal transport in China can constitute as much as 60% of the fuel's final delivery price (Stanway & Wong, 2011).
- **Higher Costs -** China has a rather complex distribution chain, and non-coal costs can have an unusually large impact. Improving safety and environmental standards, for instance, have helped double China's coal production costs to $47 a tonne in the last five years (Stanway & Wong, 2011). Recently, the decommissioning of small mines in Shanxi, rising labor costs, and a greater reliance on road transport have all exacerbated the issue - coal trucks are twice as expensive as rail. Further, China's shallow-lying, cheaper reserves are depleting. Miners are now burrowing as deep as 1,500 meters in search of coal. In stark contrast to India, coal over-production has been the concern in China, where the central government wants a 3-5% resource tax to slow companies - China could reach peak coal production by as early as 2020 (Peng, 2010). As noted, foreign coking coal costs quadrupled in the 2000s, and in 2009 JPMorgan Chase estimated that higher global demand for thermal coal would lift prices from $77 to $94 per tonne in 2011 (Mining Exploration News, 2011). By March 2011, however, Sify Finance (2011) was reporting that only the catastrophic Japanese earthquake (i.e., weakened demand) was preventing thermal coal prices from reaching the $145 level.
- **Rising Steel Needs -** China is in the midst of the largest and fasted infrastructure build out in world history. In 2009, China imported about 35 million tonnes of coking coal, a whopping 400% increase from 2008 (Metal First, 2011). Industry consolidation led to new facilities strategically located in coastal areas, so steel plants now have greater flexibility to utilize the international coal market. Steel from coal is China's fundamental material to build up, out, and down. An unprecedented urban influx that will see 325 million more Chinese living in cities in less than a generation will use coal to produce

---

[3]The impact that the economic stimulus will have on energy demand in China is particularly noteworthy. As stated by Keidel (2008), China's consumption should increase accordingly because the money is "going into the real economy, not into the balance sheets of troubled financial institutions," as occurred in the United States.

steel for skyscrapers and other urban accoutrements (Moody & Lan, 2010). By 2025, McKinsey & Company (2008) report that China will have 219 cities with a population in excess of one million people (Europe now has just 35). Rio Tinto says that China could build 50,000 skyscrapers before 2030, the equivalent of 10 New York Citys (Teather, 2008). With 390 million vehicles, Dargay et al. (2007) project that China will have the world's largest vehicle fleet by 2030 - a 10-fold leap from today. China already accounts for nearly half of world steel *production.*

4.2 India

India's coal supplies and transportation systems are struggling to keep pace with surging domestic demand, and foreign coal will be needed to fill the gap. Electricity and steel comprise the vast majority of India's total consumption. India's long-term demand for thermal coal stems from a massive coal-fired power station build out and continued cost inflation pressures on the domestic coal industry. The demand for coking coal will arise from India's quickly urbanizing population; McKinsey & Company (2010) project that India's urban population will increase from 340 million in 2008 to 590 million by 2030. Steel pipemaker Jindal SAW estimates that India will need 200 million tonnes of steel by 2020, a 210% increase from the existing base (Metal First, 2010). According to India Coal Market Watch (ICMW), the country's overall coal imports in 2010 increased by 14% from 2009 to 78 million tonnes, 70% of which were of the thermal variety (Sethuraman, 2011). ICMW projects that India's total coal imports will rise to more than 90 million tonnes in 2011, with an additional 4,000 megawatts of coal-fired electricity capacity coming online (Sethuraman, 2011). Wood Mackenzie (2010) believes that India's demand for imported thermal coal will eclipse that of China before 2020. According to the EIA (2010a), India will surpass Japan as the world's largest importer of coal by 2025 and be importing four times 2008 levels in 2035. India's unmatched latent demand for coal-based electricity is simply overwhelming (see Figure 5).[4]

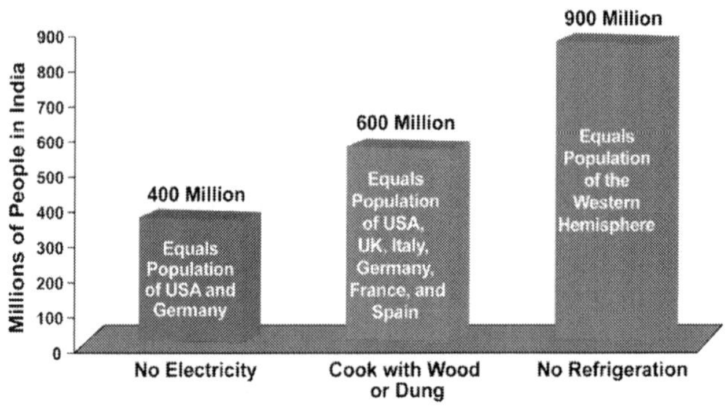

Fig. 5. The Scale of Latent Demand for Coal-Based Electricity in India
Source: developed from International Energy Agency, 2010a; The World Bank, 2010

[4]The IEA (2010a) projects that electricity from nuclear energy (10%/year) and natural gas (6%/year) will grow at impressive rates from 2008-2035, but coal will still easily be the dominant source of India's power under any foreseeable scenario.

Gearing up for the vast expansion of coal imports, India's ongoing improvements in infrastructure include the extended use of smaller ports and the construction of deeper ports. In fact, UMPP focuses specifically on coal imports because more efficient plants require higher quality coal to achieve full load. Larger plants can obtain the economies of scale needed to compensate for the increased costs associated with foreign coal. Most boilers in India are oversized (because of the elevated ash content of the country's coals), and these units cannot use better quality imported coals without blending in the domestic varieties. And because India's domestic thermal coal is just half the price of imports, Chikkatur & Sagar (2009) recommend a "bifurcated" approach: "wherein using imported coal with high-efficiency global technologies.....will serve as a complement to the existing pathway of adapting such technologies for Indian coals." According to the Harvard researchers, the popular belief that India has ample reserves of coal should be tempered by a number of factors:

- With the need to move toward more expensive underground mining, a significant expansion in India's coal mining capacity faces technical/environmental hurdles and social/environmental constraints. Problems related to displacement (rehabilitation and resettlement) have been persistent in India. Too much opencast mining, due to its lower costs and reduced losses, has stagnated underground mining. Technical and institutional problems have limited mechanized longwall technology. India's underground mining output requires a "quantum jump" from 0.45 to 2.7 tonnes per man year (Bucyrus, 2010).
- Despite India's substantial coal endowment, the real size is likely smaller than generally assumed. India's resources are often assessed geologically, not techno-economically. It is the latter form of assessment, however, that indicates the amount of coal that can be feasibly extracted under prevailing technical and economic conditions. Technical terms like "resources" and "reserves" have been routinely misused in India, so it is by no means clear exactly how much coal the country really has. Some more recent Indian coal resource inventories have even included coal that had already been mined (see Chikkatur & Sagar, 2009).
- As noted, the quality of Indian coals is poor. Chikkatur & Sagar (2009) report coals typically have high ash content (40-50%), high moisture content (4-20%), and low calorific values (2,500-5,000 kcal/kg). The ash content has been increasing in recent decades due to more opencast mining and a greater reliance on inherently inferior grades. India's coal does have low sulfur content (0.2-0.7%) but the low calorific value leads to lower boiler efficiency, meaning more coal is needed for the same amount of electricity generated. With India's own coking coal supply on the decline, domestic reserves constitute just 18% of the country's total coal endowment (Steel Exchange India, 2002). India's demand for coking coal is expected to double from 2011-2014 (Kumar, 2010).

5. Concluding remarks

"The spike in Chinese demand for imported coal.....requires careful examination.....the unique structure of the Chinese coal market creates a series of key arbitrage relationships.....Developments in China's domestic coal market will be a dominant factor determining global coal prices and trade flows.....and by implication power prices in many regions.....The nature of Chinese demand for international coal is.....fundamentally different from India, the other source of dramatic demand growth in international coal markets.....In other words.....the unique politics and economics of the

Chinese coal market are now therefore by necessity the politics and economics of the global market," (Richard Morse & Gang He, Stanford University, 2010)

It is no coincidence that the world's electric power system will remain built upon coal. Consumers prefer low cost, reliable energy, and the producers that can provide these services continue to prosper. Coal was the backbone of the Industrial Revolution in England during the 18th Century, America's emergence as a major economic power during the late 19th and 20th Centuries, Germany's manufacturing prowess during the early 20th Century, and coal is now fueling the 21st Century economic miracles rapidly unfolding in both China and India. Following the overall trend of globalization, the expanding international coal market is now in a state of transition. In the years ahead, geological realties suggest that "Chindia" will be scrambling to meet soaring needs with domestic reserves, as more electricity and steel will steadily translate into more imports. Specifically, as the world's largest coal arbitrageur, the international price of coal will be increasingly linked to the domestic price in China. Higher prices will be a problem for importers. With rising prices, exporters are unlikely to accommodate long-term supply contracts and block off reserves to ensure supply for the entire lifetime of a coal plant - which can be up to 50 years.

For the world's top suppliers, significant challenges pose the question of whether or not coal infrastructure expansions will keep pace with higher import demand. All three leading exporters, Australia, Indonesia, and Russia, require major investments to increase exports. In fact, with an emerging economy and a population of 240 million people, Indonesia wants to preserve its coal for future needs - and South Africa's Mineral Resources Minister wants laws to impose the same strategy. Colombia has been mostly focused on the lower cost European market, and the country's low sulfur coal is not much of an advantage for fast growing India, for instance, because India already has a substantial low sulfur reserve base. In the United States, higher costs and a heightened risk aversion to coal export projects could limit the availability of the world's leading coal endowment. For example, the largest U.S. coal producer, Peabody Energy, faces resistance in its attempt to use Wyoming's coal-rich Powder River Basin to serve Asia through an export terminal in the state of Washington.

Analysis of the dangers posed by a society's use of fossil fuels and the emission of CO_2 generally focuses on the potential for climate change impacts. It is important, however, in the context of assessing the risk of anthropogenic CO_2 emissions, to also examine the reasons why CO_2 is emitted in the first place. CO_2 is not released in a socioeconomic vacuum. CO_2 is emitted because it is the inevitable by-product of combusting fossil fuels. The generation of electricity from coal, for instance, results in CO_2 emissions but it also yields significant benefits for the health and welfare of the people. Thus, it is important to strike a balance in the equation - both an assessment of the dangers posed to the atmosphere by CO_2 emissions and the powerful benefits created by the energy usage that results in CO_2 emissions. Electricity produced from fossil fuels - namely coal - has been, is, and will continue to be a cornerstone of global development. Going forward, the developing nations will need full access to the very same diverse range of fuels that has empowered their industrialized counterparts to raise the living standards for, and extend the lives of, billions of people.

The material reduction of CO_2 is a global challenge but is tractable through advancing technologies. The commercialization and deployment of clean coal technologies as they develop will accelerate environmental improvement and progress toward the goal of near zero emissions. In the area of power generation, the two processes of (a) higher efficiency through supercritical and ultrasupercritical coal power plants and (b) CCS presents a unique opportunity to not only meet the climate change goals delineated at Copenhagen in 2009 but to also propel the world toward the Accord's objective of eradicating poverty and energy deprivation. The synergy between these twin forces can help solve constraints in electricity, natural gas, and liquid fuel supplies. Developed nations need to work within the global context to deploy clean coal technologies as quickly as possible: current IEA (2010a) projections indicate that 1.2 billion people will still be without electricity in 2030. Indeed, coal's story is far from told.

6. References

Automotive News. (March 12, 2007). Carmakers still choose steel over aluminum; New high-strength alloys cut weight, are cheaper to use, In: *Goliath*, April 1, 2011, Available from: http://goliath.ecnext.com/coms2/gi_0199-6343230/Carmakers-still-choose-steel-over.html

Boyce, G. (2010). The Peabody Plan, *Proceedings of the World Energy Council*, Montreal, Canada, September 14, 2010, Retrieved from www.peabodyenergy.com/ pdfs/WorldEnergyCongressremarks.pdf

BP. (2010). *Statistical Review of World Energy 2010*. Retrieved from http://www.bp.com/ productlanding.do?categoryId=6929&contentId=7044622

Bucyrus. (2010). The Future of Overground Lies Belowground. *Proceedings of the National Seminar on Underground Coal Mining*, Hyderabad, India, August 28, 2010, Retrieved from http://scclmines.com/downloads/NationalSeminarSOUVENIR.pdf

Chikkatur, A. & Sagar, A. (2009). Rethinking India's Coal Power Technology Trajectory. *Economic & Political Weekly*, Vol. 44, No. 46, (November 14, 2009), pp. (53-58)

China Business News. (March 11, 2010). China coal and mining equipment industry, March 16, 2011, Available from: http://cnbusinessnews.com/china-coal-and-mining-equipment-industry/

Coal Explorer. (February 12, 2011). Coal Pricing in India is Controlled by the Government, March 18, 2011, Available from: http://www.coalexplorer.com/20111202/coal-pricing-india-controlled-government-coal-india-contributes-about-82-countrys-total-pro

Coal Industry Advisory Board. (September 4, 2009). The Role of Coal in the Post-2012 Greenhouse Gas Reduction Agreement, International Energy Agency, Paris, France, February 27, 2011, Available from: http://www.iea.org/ciab/Post-2012_GHG_Red_Statement.pdf

Daily News & Analysis. (May, 25, 2010). End of coal mafia, pilferage and theft top priority: Sriprakash Jaiswal, April 19, 2011, Available from: http://www.dnaindia.com/ india/report_end-of-coal-mafia-pilferage-and-theft-top-priority-sriprakash-jaiswal_1387638

Dargay, J., Gately, D., & Sommer, M. (January 2007). *Vehicle Ownership and Income Growth, Worldwide: 1960-2030*, New York University, New York, New York, USA, February 13, 2011, Available from: http://www.econ.nyu.edu/dept/courses/gately/ DGS_Vehicle%20Ownership_2007.pdf

Deutsche Bank. (June 6, 2010). Steam Coal May Become Next Booming Commodity in Australia - DB, In: *Coal Spot*, March 15, 2011, Available from: http://www.coalspot.com/news-detail.php?nid=1153&page=79

Electric Power Research Institute. (November 2003). *Electricity Technology Roadmap: 2003 Summary and Synthesis,* Retrieved from http://www.iea.org/work/2004/distribution/presentations/Gellings_bkgd_%20paper.pdf

Global Energy Network Institute. (October 2002). Electricity for all: Targets, timetables, instruments, March 20, 2011, Available from: http://www.geni.org/globalenergy/library/media_coverage/electricite-de-france/electricity-for-all--targets-timetables-instuments.shtml

International Energy Agency (IEA). (1999). *World Energy Outlook 1999,* Paris, France

International Energy Agency (IEA). (2002). Chapter 13: Energy and Poverty, In: *World Energy Outlook 2002,* pp. (365), Paris, France

International Energy Agency (IEA). (April 2008). Energy security and the role of coal, April 12, 2011, Available from: http://www.iea-coal.org.uk/publishor/system/component_view.asp?LogDocId=81977

International Energy Agency (IEA). (2009). *Cleaner Coal in China,* Paris, France

International Energy Agency (IEA). (2010a). *World Energy Outlook 2010,* Paris, France

International Energy Agency (IEA). (2010b) *Projected Costs of Generating Electricity,* Paris, France

Jin, T. (January 26, 2011). China Commodities & Energy Wrap: Refined Copper, Nickel, Coal, Gold, 2010 Imports, In: *The China Perspective,* April 12, 2011, Available from: http://www.thechinaperspective.com/articles/chinacommoditiesampenergywrap refinedcoppernickelcoalgold2010imports8140/index.html

Keidel, A. (November 2008).*China's Stimulus Lesson for America,* Carnegie Endowment for International Peace, Washington, D.C., USA

Kumar, A. (April 10, 2010). I expect coking coal prices to touch $300 per tonne, In: *Business Standard,* April 3, 2011, Available from: http://www.business-standard.com/india/news/%5Ci-expect-coking-coal-prices-to-touch-300-per-tonne%5C/391446/

McKinsey & Company. (March 2008). Preparing for China's urban billion, March 20, 2011, Available from:
http://www.mckinsey.com/mgi/publications/china_urban_summary_of_findings.asp

McKinsey & Company. (March 2009). Preparing for China's urban billion Impacts of Urbanization: Implications for building construction, March 29, 2011, Available from:
http://www.mckinsey.com/mgi/reports/pdfs/china_urban_billion/CUB_2_3_3_Implications_for_building_construction.pdf

McKinsey & Company. (April 2010). India's urban awakening: Building inclusive cities, sustaining economic growth, March 23, 2011, Available from:
http://www.mckinsey.com/mgi/publications/india_urbanization/index.asp

Metal First. (August 12, 2010). India needs 208% more steel by 2020, April 9, 2011, Available from: http://jennifer.metalfirst.com/news_detail.php?id=2992

Metal First. (February 17, 2011). Compete in the global coal resources in India and strong demand is expected to push up prices, March 3, 2011, Available from: http://ferroalloy.metalfirst.com/subsites/news_detail.php?id=4583

Metals Consulting International. (2011). Metallurgical Coal Prices, March 28, 2011, Available from: http://www.steelonthenet.com/files/metallurgical_coal.html

Mining Exploration News. (March 12, 2011). U.S. China Mining Group Expected Full Coal Production at Xing An Coal Mine, March 29, 2011, Available from:
http://paguntaka.org/tag/china-coal-demand

Mitchell, C., Deumling, R., & Court, G. (March 2009). Stabilizing California's Demand: The Real Reasons Behind the State's Energy Savings, In: *Public Utilities Fortnightly*, March 27, 2011, Available from: http://www.fortnightly.com/exclusive.cfm?o_id=159

Moody, A. & Lan, L. (March 22, 2010). Focusing on future urbanization, In: *China Daily*, March 25, 2011, Available from: http://www.chinadaily.com.cn/bizchina/2010-03/22/content_9621374.htm

Morse, R. & He, G. (August 2010). *The World's Greatest Coal Arbitrage: China's Coal Import Behavior and Implications for the Global Coal Market*, Stanford University, Stanford, California, USA, March 29, 2011, Available from: http://pesd.stanford.edu/publications/the_worlds_greatest_coal_arbitrage_chinas_coal_import_behavior_and_implications_for_the_global_coal_market/

Pasternak, A. (October 2000). *Global Energy Futures and Human Development: A Framework for Analysis*, Lawrence Livermore National Laboratory, U.S. Department of Energy, Washington, D.C., USA, Available from: https://e-reports-ext.llnl.gov/pdf/ 239193.pdf

Peng, N. (November 1, 2010). Report: China may start 3-5% sales tax on coal, In: *China Daily*, March 24, 2011, Available from: http://www.chinadaily.com.cn/bizchina/2010-11/01/content_11485276.htm

Platts. (2005). The Rise of Futures in International Coal Markets - The Asian Challenge, April 2, 2011, Available from: http://www.platts.com/IM.Platts.Content/InsightAnalysis/IndustrySolutionPapers/coalfutures0709.pdf

Reuters. (November 2, 2010). UPDATE 1-China to limit coal production output - media, March 20, 2011, Available from: http://af.reuters.com/article/ energyOilNews/idAFTOE6A101420101102

Roosevelt, F. (January 12, 1930). The Governor States the Power Issue, In: *The New York Times*. p. 127, June 1, 2011, Retrieved from Proquest Historical Newspapers (1851-2007) from Penn State University's Pattee Library

Royal Society of Chemistry. (August 31, 2007). Coal chemical projects flourish, April 1, 2011, Available from: http://www.rsc.org/chemistryworld/Issues/2007/September/CoalChemicalProjectsFlourish.asp

Sethuraman, D. (January 12, 2011). India's Coal Imports Rose by 14 Percent in 2010, ICMW Says, In: *Bloomberg*, April 1, 2011, Available from: http://www.bloomberg.com/news/2011-01-12/india-s-coal-imports-rose-by-14-percent-in-2010-icmw-says.html

Sify Finance. (March 25, 2011). Japan quake to depress thermal coal prices, March 28, 2011, Available from: http://www.sify.com/finance/japan-quake-to-depress-thermal-coal-prices-news-default-ldzbbVcebhd.html

Sinoc. (February 20, 2005). Energy Development Strategy for Inner Mongolia Autonomous Region (Asian Development Bank Project), April 17, 2011, Available from: http://www.sinoc.com.cn/news/view_en.asp?id=118&cataid=3

Stanford University, Program on Energy and Sustainable Development. (2009). *Global Coal Markets*, Stanford University, Stanford, California, USA, March 19, 2011, Available from: http://sprie.stanford.edu/research/coal_markets/

Stanway, D. & Wong, F. (January 15, 2011). China tries to give coal a merrier soul, In: *The Peninsula*, March 19, 2011, Available from:

http://www.thepeninsulaqatar.com/saturday-editions/139068-china-tries-to-give-coal-a-merrier-soul.html
Steel Exchange India. (2002). Coal and Coke, April 2, 2011, Available from: http://www.steelexchangeindia.com/insidesteel/utilities/cokingcoal.htm

Teather, D. (August 27, 2008). Chinese skyscraper builders to put up equivalent of 10 New Yorks, says Rio Tinto, In: *Guardian UK*, May 3, 2011, Available from: http://www.guardian.co.uk/business/2008/aug/27/riotinto.commodities

The Economist. (November 10, 2008). China Seeks Stimulation, April, 3, 2011, Available from: http://www.economist.com/node/12585407

The Global Poverty Project. (n.d.) Global Poverty Info Bank: Women, April 23, 2011, Available from: www.globalpovertyproject.com/infobank/women

The World Bank. (2008). The Welfare Impact of Rural Electrfication: A Reassessment of the Cost and Benefits, Washington, D.C., USA

The World Bank. (2010). World Development Indicators 2010, Washington, D.C., USA

UNICEF. (October 14, 2009). UNICEF and WHO launch report on diarrhea, the second greatest killer of children, April 13, 2011, Available from: http://www.unicefusa.org/news/releases/unicef-and-who-launch-report.html

United Nations. (December 18, 2009). Framework Convention on Climate Change, Bonn, Germany, March 26, 2011, Available from: http://unfccc.int/resource/docs/2009/cop15/eng/107.pdf

U.S. Energy Information Administration (EIA). (2010a). *International Energy Outlook*, Washington, D.C., USA

U.S. Energy Information Administration (EIA). (2010b). Levelized Cost of New Generation Resources in the Annual Energy Outlook 2011, April 13, 2011, Available from: http://www.eia.doe.gov/oiaf/aeo/electricity_generation.html

U.S. National Academy of Engineering. (2004). Greatest Engineering Achievements of the 20th Century, April 10, 2011, Available from: http://www.greatachievements.org

Wood Mackenzie. (October 19, 2010). China and India's Unabated Demand Will Shift Global Thermal Coal Trade says Wood Mackenzie, April 23, 2011, Available from: http://www.woodmacresearch.com/cgi-bin/corp/portal/corp/corpPressDetail.jsp?oid=2273475

World Coal Association. (2011a). Coal Market & Transportation, April 1, 2011, Available from: http://www.worldcoal.org/coal/market-amp-transportation/

World Coal Association. (2011b). Frequently Asked Questions, April 11, 2011, Available from: http://www.worldcoal.org/resources/frequently-asked-questions/

World Coal Association. (2011c). Uses of Coal, April 16, 2011, Available from: http://www.worldcoal.org/coal/uses-of-coal/

World Health Organization. (June 2005). Indoor air pollution and health, March 16, 2011, Available from: http://www.who.int/mediacentre/factsheets/fs292/en/index.html

World Health Organization. (2010). Urbanization and health. *Bulletin of the World Health Organization*, Vol. 88, No. 4, (April 2010), pp. (241-320)

Xinhua News Agency. (February 10, 2010). China's 2009 coal imports triple after stimulus increases demand, March 29, 2011, Available from: http://news.xinhuanet.com/english2010/china/2010-02/10/c_13171366.htm

Yahoo Finance. (September 3, 2010). Thousands of trucks stuck in China traffic jam, April 17, 2011, Available from: http://finance.yahoo.com/news/Thousands-of-trucks-stuck-in-apf-2254555704.html?x=0&.v=1

2

Weathering Indices for Assessment of Weathering Effect and Classification of Weathered Rocks: A Case Study from NE Turkey

Sener Ceryan
Geology Engineering Department,
Balikesir University, Balikesir
Turkey

1. Introduction

Weathering state and weatherability of rocks are highly important for engineering geology projects and the use of rocks as building. The state of weathering resulting physical and chemical processes may be reflected by changes in index properties such as dry density, void ratio, clay content and seismic velocity. Thus, it is important for geotechnical engineers to estimate weatherability of rocks, quantitatively the changes during weathering and classification of the weathered rocks. In this study, first these topics which are the classification of weathered rocks and the indices for definition of the effects of the weathering is discussed and then a case study from NE Turkey is given. The research reported here was carried out in a 40 km^2 area of Upper Cretaceous Eocene granitic rocks along the River Harsit around Dogankent (Giresun) in the North eastern part of Turkey (approximately 41^0 N, 39 ^{0}E). In the case study, the definition of mineralogical and chemical changes created by the weathering of the Harsit granitic rocks and the classification of weathered rock materials and masses from the granitic rocks were investigated

2. Weathering indices

Several weathering indices have been devised for quantifying the changes in the intrinsic properties of rocks from different points of view, some of which can be related to the engineering properties of weathered rocks (Tecer 1999, Gupta and Rao 2001, Tecer and Cerit 2002, Ceryan et al. 2005). The most commonly used methods can be broadly categorized as chemical, mineralogical-petrographical and engineering indices (Gupta and Rao 2001). Several mineralogical and micropetrographical parameters have been proposed as the basis for weathering indices in view of their variation with weathering (e.g. Lumb 1962, Weinert 1964, Mendes et al. 1966, Dixon 1969, Onodera et al. 1974, Irfan and Dearman 1978a, 1978b, Cole and Sandy 1980, Howarth and Rowlands 1987, Tugrul and Gurpinat 1997, Rigopoulos et al 2010).
Chemical change during weathering and hydrothermal alteration are quantified in several ways including the normalized value of element (or oxide) using their parent rock concentrations or immobile element concentrations in the samples (Krauskopf 1967, Minarik

et al. 1983), standard cell calculation (Colman 1982), ratio of elements to immobile elements (Chesworth et al. 1981, Colman 1982, Guan et al. 2001), measurement and calculation of loss or gain of weight (or volume) based on immobile element (Gresens 1967, Grant 1986, MacLean 1990, Huston 1993), using cation packing index (Ceryan et al 2008c, Ceryan 2011), cation exchange capacity (Arikan et al. 2007), modeling of compositional change due to chemical weathering (Eynatten et al. 2003), using an EC/pH meter (Shalkowski et al. 2009) and chemical weathering indices (Vogel 1973, Jayawardena and Izawa, 1994a,1994b, Düzgören-Aydın et al. 2002, Düzgören-Aydın and Aydin 2003, Price and Vebel 2003, Bozkurtoglu et al 2006, Ohta and Arai 2007, Ceryan et al. 2008a,Yildiz et al 2010, Ceryan 2011). Chemical weathering indices have also been proposed by numerous authors. These indices were summerized in Duzgoren-Aydın et al. (2002).

According to Ceryan (2008), no single weathering index given in the literature meets the modeling of the process involved in chemical weathering outlined above, and no weathering index would give unequivocal results when applied to the prediction models to assess the mechanical behavior of rocks materials. Thus, a theoretical model was developed by Ceryan (2008). The said model depends on isovolumetric approach and take into consideration of the definition of Loughnan (1969). In order to explain the change of the volumetric concentration of major oxides across a weathering profile, the following steps were applied (Ceryan 2008).

- Modal analysis and whole rock analyses of the sample taken across a weathering profile are performed.
- The weight percentage of the major oxide under interest from whole rock analyses of fresh samples are multiplied by dry density of the respective sample. Then Amob value in Figure 1a is obtained. By means of this Amos value, a parallel line (OA) was drawn.
- For each sample weathered to various degrees, the weight percentage of the major oxide is multiplied by dry density. By this way, upon plotting of the dry density as a function of the volumetric concentration of the major oxide, the OB line in Figure 1a is obtained.
- By microprobe analysis of the fresh minerals, the major oxide composition of the minerals in the sample is determined.
- To calculate total amount of the major oxide in unaltered portion, the following equation (Banfield 1985) is applied.

$$Wmo = \left[\sum_{i=1}^{n} Mv(i) \times Ow(i) \right] \quad (1)$$

  where Wmo is the weight percentage of the major oxide in weathered sample, i values represent minerals such as plagioclase (i= 1), orthoclase (i= 2), hornblende (i=3), biotite (i=3), pyroxene (i= 4), quartz (i= 5), opaque minerals (i= 6), Mv is the volume percentage of minerals found by the modal analysis, Ow is the concentration (in weight percentage) of the major oxide in minerals from the microprobe analysis.
- The total amount of any major oxide from the equation 1 is multiplied by its dry density and then the volumetric concentration of the major oxide in the unaltered portion of the sample is found
- The volumetric concentration of the major oxide in the unaltered parts of the samples versus the dry density of the samples weathered to various degrees are drawn and then the OC line in Figure 1a is obtained.

Ca, Na, Mg, and K are geochemically mobile elements. Chemical leaching results in a significant decrease of the oxides of these elements. The ratio of the volumetric concentration of ($CaO+MgO+Na_2O+K_2O$) in a weathered sample to those in the fresh sample taken from the same weathering profile gives the amount of leaching for the weathered sample. Therefore, this ratio given at the Equation 2 is defined as the *Chemical Leaching Index* (CLI) (Ceryan 2008). Al, Fe, and Ti are less affected by chemical leaching than alkali and alkali-earth elements, but tend to concentrate in weathering products (Loughnan 1969). If the drainage is well-developed, Si moves away, if not, it also tends to concentrate in the weathering products. The ratio of the total amount of these oxides in weathering product to those in the respective sample yields the amount of weathering products. Therefore, the *chemical weathering product index* (CWPI) is defined through the Equation 3 (Ceryan 2008).

$$CLI = \frac{100(A_{mob} - B_{mob})}{A_{mob}} \quad (2)$$

$$CWPI = \frac{100(B_{immob} - C_{immob})}{B_{immob}} \quad (3)$$

where A_{mob} and B_{mob} are the total volumetric concentration of mobile oxides in fresh sample and weathered sample, respectively. B_{immob} and C_{immob} are the total volumetric concentration of immobile oxides in the whole sample and unaltered portion of the sample, respectively. If y axis in the Figure 1a represents the volumetric concentration of the mobile elements, likewise CLI can be found for the weathered sample.If y axis in the Figure 1b represents the volumetric concentration of immobile elements, CWPI can be found for the weathered sample as defined above.Considering the definition of Loughman (1969), *Total Chemical Weathering Index* can be defined as the sum of CWPI and CLI. Since the rock material can be weathered 100% at most, TCWI value should be also at most 100. Therefore, (CWPI+CLI) value has been divided by 2 in order to get TCWI given by the following equation (Ceryan 2008).

$$TCWI = \frac{(CWPI + CLI)}{2} \quad (4)$$

According to Olier (1984), the weatherability of a rock depends on the number of cation replaceable with hydrogen in a mineral. When considering this definition of the k-value, it is possible to say that the k-value can be used for characterizing the weathering state and weatherability of a rock (Ceryan et al 2008b, 2008c). In addition;

a. By using the k-value, the amount of the removed minerals by chemical leaching can be estimated,
b. The amount of weathering products can be found by the k-value,
c. The petro-physical properties of a rock can be expressed depending on weathering degree by k-value,
d. Although the chemical weathering indices are calculated by results of the chemical analyses, the k-value is obtained from modal analyses (Ceryan et al 2008c)

The cation distribution is defined by the "Cation Packing Index", k-value, for each (stochiometric) mineral phase. k-value ($mole/cm^3$) is described as follows:

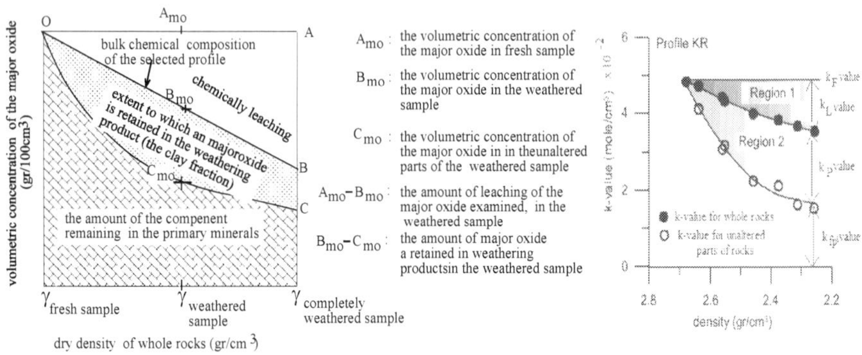

Fig. 1. A hypothetical model illustrating the behavior of major oxides during chemical weathering (modified Banfield 1985) (a), the graphs showing the relationhips between the density and the cation packing index, (k-value (Ceryan 2011) (b).

$$k = \frac{C}{N_L V_M} \tag{5}$$

where C is number of cations per mole, N_L is Avogadro's number and V_M is molar volume. For a certain rock, k-value can be calculated by using the following expression:

$$k = \sum x_i k_i \tag{6}$$

where k_i is the k value of the i mineral phase, x_i is is mod of the mineral in the rock determined by modal analysis of thin sections

Ceryan (2011) said that as decomposition of a mineral result in the formation of new minerals with lower k-value, the k-value of whole rocks is generally regarded as a measure of the degree of weathering (Table 1), therefore, it is possible to say that the k-value can be used for characterizing the weathering state and weatherability of a rock.

The region 1 on Figure 1b represents the chemical leaching ratio while the region 2 on Figure 1b gives the weathering product ratio. In the Figure 1b, k_F is cation packing index (k-value) of fresh sample, k_L is the amount of chemical leaching, k_P is k-value of weathered parts of samples (weathering products) and k_{pf} is k-value of unaltered parts of the samples (Ceryan 2011). The difference between the k-value of fresh sample (k_F) and the k-value (k_W) of weathered sample gives the amount of chemical leaching (k_L, Figure 1b). The ratio of the difference to k-value of the fresh sample taken from the same weathering profile is defined as k- leaching index (k^*_L). On the other hand The difference between the k-value of the whole sample and the k-value of the unaltered parts of the same samples (k_{fp}) gives the amount of weathering product. Therefore the ratio of the difference to k-value of the same sample is defined as k-product index (k^*_P) (Ceryan 2011);

$$k^*_L = \frac{k_L}{k_F} \tag{7}$$

$$k^*_P = \frac{k_P}{k_{fp}} \tag{8}$$

Minerals		PrI	Wm	Ks	k-value (10^{-2} mole/cm^3)	V_p (m//n)
Olivine	*Fayalite*	0.057	138.36	3.385	6.85	8400
	Forsterite	0.709	46.132	1.813	6.80	
Pyroxene	*Diopside*	0.362	24.434	1.829	6.05	7330-7200
	Enstatite	0.515	22.646	1.6431	6.37	
Amphibole	*Tremolite*	0.425	4.2954	1.004	5.54	6800
	Hornblende	0.849	2.8744	0.5296	5.309	
	Labradorite	0.389	0.542	0.144		
Plagioclase	*Andesine*	0.479	0.5875	0.088	4.99-4.97	7250-6250
	Oligooclase	0.529	0.5916	0.048		
Alkali-Feldspar	*Orthoclase*	0.719	0.7178	-0.0007	4.577	5800
Quartz		-	-		4.41	6050
Mica	*Moscovite*	-	-	-	4.98	5880
	Biotite	1.1613 *	4.3741*	0.5791	4,656	5360
Vermiculite		0.4439*	1.8176*	1.8176	3.95*	-
Chlorite (coronsite)		0.4996*	1.6406*	1.6406	4.10	5000
	Sericite	0.5547	0.2133	-0.4151	4.52	
Clay	*İllite*	0.1725	0.1907	0.0435	4.099	2400-1800
	Smectite	0.1044	0.1226	0.0697	3.997	
	Kaolinite	0.0462	0.0429	-0.0325	4.058	

Table 1. The chemical weathering indices values, k-value and P-wave velocity of selected rock-forming minerals and their weathering products (from Ceryan et al 2008b, 2008c)

From a geotechnical standpoint, indices based on key engineering properties generally have more applicability than those based on chemistry and mineralogy and are also usually more simple and less time consuming (Gupta and Rao 2001). Martin (1986) pointed out that in principle a simple quantitative degree of weathering scale can be established based on a reliable index of any rock property which changes unidi rectionally throughout the weathering process and whose value can be readily determined at any weathering stage. A simple and rapid test to obtain a quick absorption index (QAI) or void index has been proposed by Hamrol (1961) for the assessment of weathering of granite and schist. Water absorption by weight were used to determine weathering degree of marble by for marble Gulec (1973) and create weathering classification of rock materials by Kılıc (1995). Lee (1987) and Ceryan (1999) used it for predicting of mechanical properties of weathered granitic rocks. A different type of measure, the abrasion resistance hardness index (Ha), was devised by Conca and Cubba (1986) to study the abrasion hardness and extent of weathering in different rocks - sandstone, gabbro, tonalite and crystalline limestone (from Gupta and Rao 2001).

The slake durability index (Sd) was devised by Franklin and Chandra (1972) to assess the durability or weatherability of clastic sedimentary rocks such as mudstone, claystone and shale, particularly useful for rocks with significant clay content (Moon and Beattie 1995, Gokceoglu 1997, Koncagul and Santi 1999, Gokceoglu et al. 2000, Sadisun et al. 2005), but there are some limitations and weaknesses associated with this method (Erguler and Ulusay 2009). The other weathering indices are dry density (e.g.Banfield 1985, Turk and Dearman 1985, Eggleton et al. 1987, Irfan 1996, Ceryan 2008, Ceryan 2011), Schmidt hammer rebound value (e.g.Irfan and Dearman 1978a, Martin and Hencher 1982, İrfan and Powell 1985, Lee 1987, Guolin and Yushan 1990, Zhao et al. 1993, GCO 1994, Gokceoglu 1997, Ceryan et al 2008a, Basu et al 2009), the elastic wave velocity (e.g. Illev 1967, Kılıc 1995, 1999, Dearman and Irfan 1978, Krank and Watters 1983, Turk and Dearman 1985, Lee 1987, Dearman et al. 1987, Dobereiner et al. 1993, Weiss et al. 2002, Ceryan and Sen 2003, Kocbay 2003, Gurocak and Kilic 2005, Arikan et al. 2007,Ceryan et al. 2008a, 2008b, Basu et al. 2009, Korkmaz and Ceryan 2011), porosity, effective porosity and void ratio (e.g. Ondera et al. 1974, İrfan and Dearman 1978b, Türk ve Dearman 1985, Paşamehmetoğlu et al.1981, Lumb 1983, Lee, 1987, Esaki and Jiang 1999, Gupta and Rao 2001, Ceryan et al. 2008a, Gokceoglu et al. 2009, Rigopoulos et al. 2010, Marques et al. 2010). Ceryan et al (2008a) suggested new index base on porosity (n) and effective porosity (n_e).

$$Iefp = \frac{n_e - n}{n} x100 \tag{9}$$

Ceryan et al (2008b) suggested two new indices representing mineralogical and physical changes due to weathering. These indices, Mineralogical Change Parameter (Imp) and Physical Change Parameter (Ifp) were given following equations:

$$\mathrm{Im}\,p = \frac{100(V_{pf}^{*} - V_{pw}^{*})}{V_{pf}^{*}} \tag{10}$$

$$Ifp = \frac{100(V_p^{*} - V_p)}{V_p^{*}} \tag{11}$$

where Ifp is the Physical Change Parameter, V_p is P-wave velocity of the investigated dry sample and V_p^* is P wave velocity of the same samples which would have lacked pores and fissures (Foumaintraux 1976), w refers to weathered rocks, while f refers to fresh rocks. If the mineral composition of the samples is known, V_p^* can be calculated by employing the Equation 37 (Foumaintraux 1976).

$$\frac{1}{V_p^{*}} = \sum_{i=1}^{n} \frac{xi}{V_{pi}} \tag{12}$$

Where x_i is mod of the mineral in the rock and V_{pi} is P-wave velocity in the mineral constituent (i).

Aydın and Basu (2006) said that microstructural weakening accompanying this process is expected to be dramatic, especially in terms of tensile strength during the early stages of weathering and the behavior of rocks in tension may therefore be an effective indicator of their microstructure, and hence state of weathering. Considering the diffuculty the applicability of the index properties indicates obtained from their study, Gupta and Rao

(2001) sugested a new engineering index, strength ratio (Rs), based on unconfined compressive strength. This is expressed as

$$Rs = \sigma_{CA} / \sigma_{CFF} \times 100 \quad (13)$$

where; Rs is the strength ratio (%), σ_{CA} is the uniaxial compressive strength of altered rock,(MPa) and σ_{CFF} is the uniaxial compressive strength of fresh rock (MPa).

3. Classification of weathered rocks for engineering purpose

Although descriptions and classifications are related, their purpose is fundamentally different, the description of a rock being a record of what is present and the classification; of the rock being an assessment of its character in a form which permits a comparison to be made with other rocks of similar character. A classification is derived from descriptions whereas descriptions cannot be derived from a classifications (Lee 1987, Anon 1995, Ceryan 1999). Description and classification of weathered rocks are necessary to obtain the changes of its engineering properties. The first step in classification is to determine the parameters of rocks related to classification purpose and to define the rock according to these parameters and properties (Lee 1987, Anon 1995, Ceryan 1999). Defining the weathered rocks for the purpose of engineering goals is make sense to determine the degree of weathering effect, extend and characteristics in detail at that momen (Lee 1987, Anon 1995, Ceryan 1999). There are the disadvantages in using the classifications proposed for weathered rock (Table 2). Nevertheless, there are several good reasons for employing such classifications for certain rock types, particularly at higher degrees of weathering (Anon, 1995);

	Disadvantages		Advantages
a.	The possibility of finding rock mass properties not included in standard weathering scales in the field limits the use of these classifications.	a.	Without an appreciation of the degree of weathering as a process a far poorer understanding of the engineering performance would result.
b.	It is known that the classifications widely used are not handled identically and they are differently applied to all people in varying forms, and this shows that the said circumstance is treated bound to the thickness of weathering profile in that area and the knowledge and experience of the applicants.	b.	Grades will often provide a framework within which test results can be interpreted and linked to engineering performance.
c.	Classification includes interpretation and simplification, hence this constitutes missings in original data and descriptions.	c.	Because extremely weathered rocks are often sensitive to disturbance during sampling and testing, good quality geotechnical test data can be difficult to obtain. The framework of understanding provided by a workable classification based on index properties can ensure the optimum use of the available information.

Table 2. Disadvantages and advantages in using the classifications of weathered rocks (Anon 1995)

In the literature, there are different classifications systems for weathered rocks.These systems, qualitative classification of weathered rocks, are mainly based on the visual definition of the geological properties, the index properties and the basic mechanical test that can be applied also in the field. The common properties used for classification of weathered rock materials are presence of original texture, degree of discolouration of rock, degree of chemical decomposıtion of biotite and feldspar, degree of physical disintegration, disintegration of material in water, relative rock material strength, breakability of NX core in the hand, friability, relative hardness by hammer blow, Schmidt hammer value, method of hand excavation, degree of plucking of individual grains, degree of penetration of geological pick or knife, hand penetrometer, tests (Dearman 1976, BSI 1981,Martin and Hencer 1986, Lee 1987, GCO 1994, Anon, 1995, Ceryan 1999, Ceryan et al 2008b). The common rock mass properties used in the weathered rocks classification system are rock mass: degree of discolorations along joint plane, presence of original structure, rock to soil ratio, degree of weathering along joint plane, angularity of corestone, opening of joint, NX core recovery, relative rock mass permeability, RQD (BSI 1981;Martin and Hencer 1986, Lee 1987, GCO 1994, Anon 1995, Ceryan 1999, Ceryan et al 2008b). In the present, the the most widely used weathering classification system among quantitative classification system in the literature were suggested by Anon (1995). Price (1993) suggested a ratings (quantitative) system for the description of rock mass weathering. Ratings for rock materials and ratings for discontinuity are mainly parameter used in the said system (Table 3). The system is based on visual impression. However, an approach which seems to be successful for many engineering applications is the use of a rating system to place a rock mass within a classification. In Figure 2, the rating system in graphical form and comparison of the rating system with the qualitative system suggested by Anon (19995). Akgun and Ceryan (2010) obtained that meaningful relationship between the rock mass strength properties and the weathering rating (Rw). (Eqs 13-16) And they said that Geological Strength Index (GSI) value and shear strenth of rock mass decrease with the weathering rating (Figure 3)

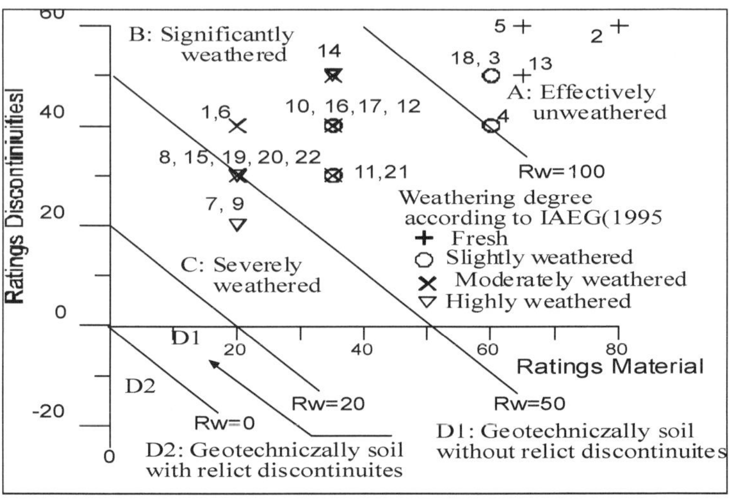

Fig. 2. The rating system in graphical form and showing the weathering degree of the geotechnical units selected from volcanic rocks from Giresun-Gumushane road, NE Turkey Number is representing geotecnical unit name (Akgun and Ceryan 2010).

Prp	Fresh	Discolored (some loss of strength)	Friable (and discolored) (considerable loss of strength, geotechnicaly an engineering soil, σ_{ci}<1.25 Mpa)
4/4	40	0	0
3 / 4	30	5	5
2 / 4	20	10	10
1 / 4	10	15	15
0	0	20	20

	Ignous rocks-joint only			All discontiniuties in all type of rocks
Prp	Unweathered	Surface stained	Rock material weathered to depth> joint wavines	Proportion of discotinuities present as relict in geotechnical soil
4 / 4	20	0	0	-20
3 / 4	15	5	5	-15
2 / 4	10	10	10	-10
1 / 4	5	15	15	-5
0	0	20	20	0

Sedimentary and metamorphic rocks (including limestone)- Ratings for joint and bedding or foliation planes

Prp	Unweathered	Surface staining or modified by solution	Rock material weathered to depth> joint waviness or open by solution
4 / 4	20	0	0
3 / 4	15	5	5
2 / 4	10	10	10
1 / 4	5	15	15
0	0	20	20

Table 3. Rating for all rocks materials and joint and relict discontinuities in all rocks (Price, 1993; Prp: Proportion)

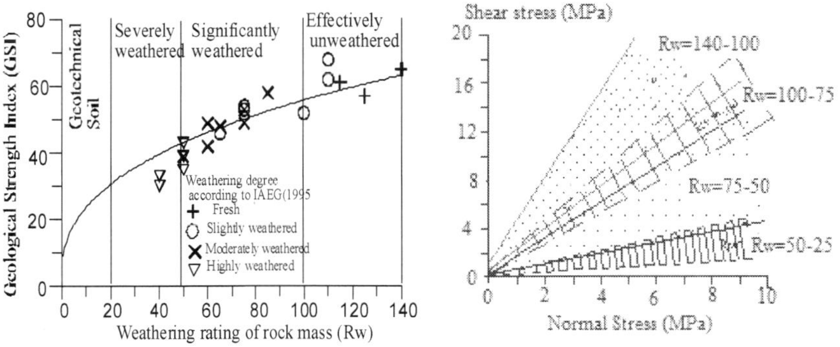

Fig. 3. Changing of the GSI and shear strength of the geotechnical units by weathering condition for volcanic rocks exposed Giresun-Gumushanr roads, NE Turkey (Akgun and Ceryan 2010).

$$GSI = 4.32Rw^{0.564} \quad (r=0.930) \tag{14}$$

$$\sigma_{cm} = 5.76Rw^{2.975} \ (r=0.945) \text{ and } \sigma_{tm} = -7.35Rw^{3.11} \ (r=0.923) \tag{15}$$

$$E_m = 5.09Rw^{1.60} \ (r=0.938) \tag{16}$$

$$c'_m = 6.54Rw^{1.982}10^{-5} \ (r=0.943) \text{ and } \phi'_m = 6.54Rw^{1.982} \ (r=0.928) \tag{17}$$

Where σ_{cm} is uniaxial compressive strength of rock mass (MPa), σ_{tm} is tensile strength of rock mass (MPa), E_m is the deformation modulus of rock mass (MPa), c/ and ø/ are cohesion (MPa) and frictional angle (degree) rock mass,

The quantitative weathering systems are the second one in creating the classification systems of the weathered rock materials. The approaches used for the creation of the quantitative weathering classifications are handled within 4 groups (Ceryan et al 2008b). First approach is that the weathering grades are defined numerically according to the only one index property (Hamrol 1961, Onodera et al. 1974, Zhao and Broms 1993, Gokceoglu and Aksoy 2000). However, weathering may not be expressed by the change in one index property used in the classification. For example, measured crack density and dry density from different locations in a granitic batholith may vary depending on its heterogeneity. Moreover, crack density varies depending on the tectonic activity on the location of the sample and the technique used when preparing a thin-section. Furthermore, using only one index property does not give enough information about all the weathering processes. In the second approach, the change amount of an index property measured on weathered sample to the value measured in the fresh sample is taken essentially. In this approach, the qualitative definition of the weathering grade has been shown in the following equation.

$$WD = \frac{100(Z_{fresh} - Z_{weathered})}{Z_{fresh}} \tag{18}$$

where, WD is weathered degree of the sample, Z_{fresh} is the measured value of the fresh rock property basically used. $Z_{weathered}$ is the value of the measured weathered rock property

Weathering classifications using elastic wave velocity (Illiev 1967), water absorption (Gulec 1973) and unconfined compressive strength (Gupta and Rao 2001) are some examples for this approach. The third approach is the use of empirical formulae which are commonly used in obtaining the quantitative weathering scales. In these formulae, two or more properties are used. The approaches proposed by Guolin and Yushan (1990), Kılıç (1995, 1999), Kocbay (2003) and Lan and et al. (2003) can be given as examples.The last one proposing quantitative weathering scale uses the statistical analyses such as hieracical cluster analysis (Wei and Lui 1990) and multiple regression and factor analysis (e.g. Arikan et al. 2007).

The changes caused by weathering processes in rock material may be mainly considered under the two topics; first is the mineralogical change (and directly chemical change) and second is the physical change. Each of these changes can be defined separately and measured (Ceryan et al 2008b). The width of micro-cracks (Onodera et al. 1974), crack density (Dixon 1969, Davis 1984), micro-fracture index (Irfan and Dearman 1978a, Al-Qudami et al. 1997) and linear crack density (Sousa et al. 2006) were used order to measure the physical change occurred by the weathering in rock material (from Ceryan et al 2008b). To define the mineralogical and chemical changes numerically due to weathering, it is possible to find various methods in the literature including chemical weathering indices and petrographical indices given above (from Ceryan et al 2008b). If we are able to measure the mineralogical and physical changes separately in the weathered rock material, we can show each of these changes in the distant axis in the Cartesian coordinate system (in the continual axis set that the other one admitted) (from Ceryan et al 2008b). Fig 4a, the definition of weathering degree based on mineralogical change and the physical change shown in thin section image. In Fig. 4b from Al-Qudami et al. 1997, the mineralogical change was defined by the secondary mineral content and the physical change was described by micro fissuring index. As a consequence of the demonstration of the physical and mineralogical changes in this way (on the different axes in the Cartesian coordinate system), the weathering state can be defined as the distance from the origin (from Ceryan et al 2008b).

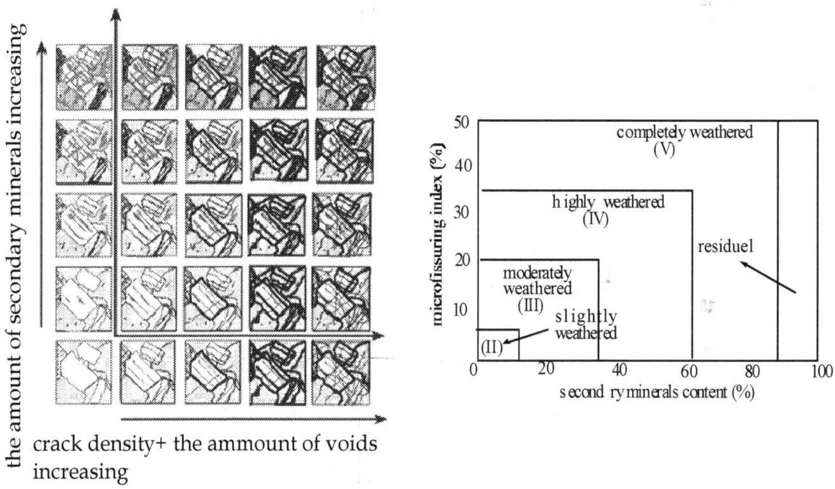

Fig. 4. The definition of weathering degree based on mineralogical change and the physical change caused by weathering processes(a) and the secondary mineral content and micro-fissuring index

Ceryan et al. (2008b) said that If Imp and Ifp are shown on the distance axis in the cartesian coordinate system, the distance from the origin will show the weathering condition of the sample. Therefore, "Weathering State Parameter ", Iad, showing the weathering condition has been defined by the following equation (Ceryan et al 2008b).

$$Iad = \sqrt{\frac{Ifp^2 + \mathrm{Im}\,p^2}{2}} \quad (19)$$

Iad together with slake-durability index value (Id), an indicator of the rock to soil ratio in the weathered rock material (Lee and Freitas 1988), should be considered in order to be able to define the weathering grades completely and significantly. As a result of this approach,

"Quantitative Weathering Index (Ia)" can be formulated following formula (Ceryan et al 2008b),

$$Ia = 100 - ((100 - Iad) * Id * 0.01) \tag{20}$$

Advantageous of the numerical weathering index proposed by Ceryan et al (2008b) were given as follows;

- During the P-wave velocity tests, the samples are not disturbed and hence, the test can produce many results using one sample.
- Descriptions of mineralogical and physical changes seperately allow to make a comparison between two. By using this comparison, it is possible to assess the weathering process.
- It is possible to construct some correlations between the numerical weathering index and the engineering properties of rocks. By using these correlations, the engineering properties of the weathered rock material can be predicted easily and reliably.
- It is possible to classify the weathering degrees of igneous rocks by using the numerical weathering index and the weathering classes provide important informations about engineering behaviour of igneous rocks.
- When determining the weathering degress of pyroclastic rocks such as tuff by using the numerical weathering index, the homogeneity of the rock to be employed should be checked, because the P-wave velocity shows a high variety depending on heterogeneity.
- It is evident that if the pre-existing voids and/or the voids created by weathering are oriented, V_p and mechanical properties also change depending on measurement orientation. To cope with this difficulty, the minimum V_p value can be used as stated by Weiss et al. (2002).
- The numerical weathering index should not be applied on the rocks showing the soluble type weathering.

4. A case study: Weathering and weatherability of Harsit granitic rock

4.1 Weathering of Harsit granitic rock

In the study area, the basement rocks are basalts, andesites and pyroclastic units (Figure 5). The Harsit granitoid was intruded in the Upper Cretaceous/Eocene period Towards the periphery of the pluton, it consists of lucocratic quartz diorite, quartz monzonite and quartz monzodiorite while towards the centre the rocks are granodiorite. Towards the NW and SE the granites are terminated by NE-SW faults (Ceryan 2008a). During these processes some elements are released and combine with other minerals to form new minerals, e.g., the development of smectite from plagioclase is a consequence of the addition of Ca, Na and Fe while the subsequent removal of these changes the clay mineral to kaolinite (Figure 6). Orthoclase minerals are more resistance to weathering than plagioclase and holes on their surfaces indicated acid attack at an early stage of the weathering (Figure 6).

Hydrothermal weathering products are sericite and allunite. The weathering of the minerals in the Harsit granitic rocks. The weathering and hydrothermal alteration of the minerals in the Harsit granitic rocks and the type of change which occurs are indicated in Figure 7. Chemical weathering causes variation in the composition of the rocks by leaching and the introduction/ development of new components. SiO_2 concentrations show a continuous decrease with increasing weathering. Some of the free silicon ions released during weathering are transported in solution but some combine to form new clay minerals. Al_2O_3, because of its low solubility, tends to be concentrated in residual weathering products. CaO and Na_2O, being soluble, either quickly move out of the system or combine with epidote and hornblende ± plagioclase. MgO is rapidly leached and removed in the early stage of chemical weathering, although a certain proportion is retained in the mineral structures of clays and chlorite. FeO remains relatively constant, although it may change from ferric to ferrous.

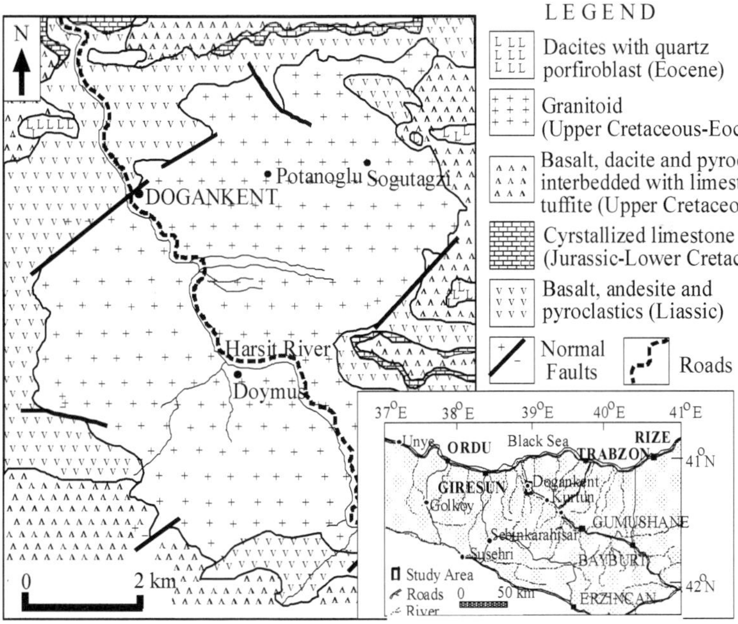

Fig. 5. Geological map of the study area

4.2 Weatheability of Harsit granitic rocks

Durability, i.e., the rock's ability to resist degradation during its working life, is depended on a number of important parameters; weatherability of rock material, degree of imposed during winning, production, placing and service, the climatic, topographic and hydrological environments in service (Fookes et al 1988). In the studies, which give assessment of durability of Harsit granitic rocks, performed Ceryan and Ceryan (2005) and Ceryan et al (2008), rock durability indicators, Static Rock Durability Indicator and Dynamic Durability Indicator purposed by Fookes et. al (1988) were used. Index properties, petrographic and chemical weathering indices and rock durability indices of rock materials from Harsit granitic rocks used these study were given in Table 4.

During chemical weathering; both chemical leaching of mobile elements (oxides) and forming weathering product occur on the rock materials. Thus, for the prediction of the engineering performance of the stone in service, Chemical Weathering Index and Chemical Leaching Index were used together in the study performed Ceryan and Ceryan. (2005). Simple and multiple regression analyses using chemical indexes, setting in this study, of the sample from the weathering profiles samples of the Harşit granitic rocks show that the rock durability indexes purposed by Fookse et al. (1988) can be obtained easily and cheaply. In the study performed by Ceryan et al (2008), an application of fuzzy modeling to the prediction of potential rock durability indexes from rock sample taken from Harşit Granitoid was given. Depending on cation packing index and micro-cracks plus voids ratio, important changes in Static Rock Durability Indicator (RDIs) were determined. However, weatherability of the building stone depends on both mineralogical properties and fabric. Therefore, cation packing index representing mineralogical and micro-cracks plus voids ratio representing fabric properties are considered together in the fuzzy model to estimate the durability of the sample from Harsit granitic rocks. In the fuzzy model described input-output relationships by fuzzy if-then rules, Cation Packing Index and micro-cracks plus voids ratio used such as input data. The fuzzy model constructed in this study exhibited higher performance and showed good generalization ability (Ceryan et al 2008)

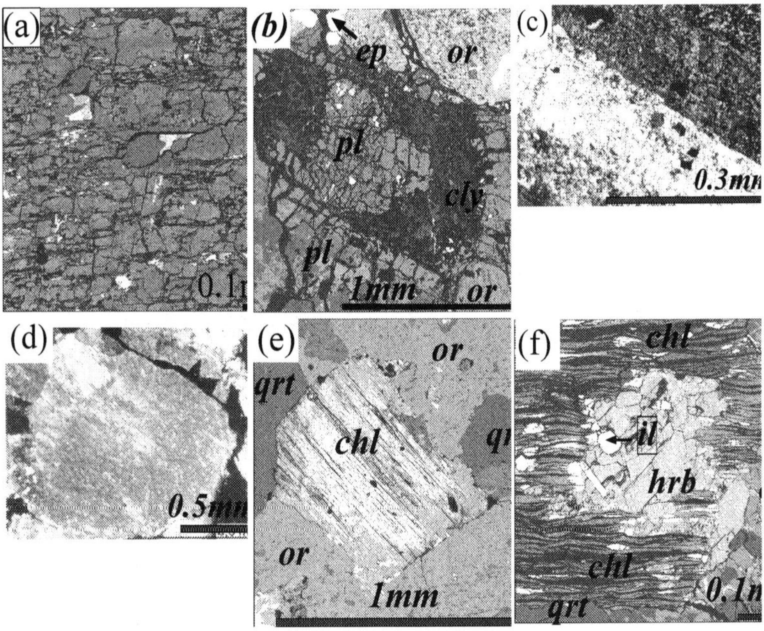

Fig. 6. Weathering products of plagioclase (a,b,c); clay minerals (dark phases)(a, b), epidote (bright phases)(a,b,c), the etching caused by the acid effect in orthoclase (b) the clays occured by the weathering of orthoclase (d), chlorite occurrence due to the hydrothermal alteration of the biotite(e), weathering products of hornblende, chlorite (fibrous phase) and titanite (bright phases)(f). (a,b,e and f are scaninig electron microskope images, c and d are optical microscope image; qrt: quartz; pl: plagioclase; or: orthoclase; ep: epidote; chl: chlorite; cly: clay ; bi: biotite, il:ilmenite; hrn: hornblende)

Sample	WG	Ip	SGssd	WA	Is(50)	SST	RDIs	CD (%)	k-value	CLI	CWPI
C-1A	F	8,158	2,706	0,4	6,65	0,3	2,37	0,25	4,621	0	0
C-23A	SW	2,882	2,698	0,5	3,59	8,7	0,92	1,84	4,449	21	23,92
C-3B	MD	1,521	2,63	1,2	0,785	34,6	-1,25	7,25	4,177	31,6	33,31
D-1A	F	28,387	2,68	0,4	6,865	0,3	2,48	0,16	4,682	0	0
D-1B	SW	3,129	2,674	0,5	3,915	5,8	1,15	0,18	4,61	-	-
D-2AB	SW	1,864	2,648	0,9	3,275	9,2	0,72	3,15	4,395	18,3	23,58
D-56	MW	1,159	2,615	1,4	0,725	42,6	-1,62	4,26	4,308	37,4	44,42
P-1A	F	7,149	2,719	0,6	7,845	0,2	2,77			0	
P-2A	F	5,502	2,705	0,5	5,935	0,3	2,09	1,17	4,654	0,69	9,397
P-2BC	MW	1,751	2,652	1,4	1,32	14,6	-0,32	3,85	4,417	-	-
P-3A	HW	1,18	2,629	1,5	0,305	43,1	-1,81	9,21	4,051	29	47,62
P-3B	HW	0,852	2,575	1,6		38,6				43,8	65,91
P-4	HW	1,089	2,605	1,3	0,28	45,6	-1,89	5,65	4,006	33	52,12
S-1A	F	14,198	2,693	0,5	6,58	0,5	2,33	0,61	4,728	3,79	0
S-3B	F	6,376	2,666	0,8	6,51	0,4	2,28	0,68	4,648	0	11,18
S-4A	SW	3,514	2,667	0,8	4,07	12,6	0,9	1,25	4,579	15,4	20,03
S-2A	MW	1,494	2,632	1,3	1,265	34,6	-1,08	3,42	4,427	26,2	34,74
S-3C	MW	1,697	2,628	1,4	2,68	24,7	-0,19	6,65	4,392	26,4	35,68
S-3A	MW	1,002	2,559	2	2,16	40,1	-1,11	2,48	4,407	27,6	41,48
S-5B	HW	0,84	2,611	1,5	0,905	58,6	-2,18	5,88	4,2619	45,4	51,74

(WG:weathering grade; SGssd=specific gravity (saturated and surface dry) $I_{s(50}$:point load index (I0,5 $I_{s(50)dry}$+0, 5 $I_{s(50)sat}$) (Mpa), WA: percentage water absorption (%) Ip:mikropetrographic index SST:MgSO4 soundness value (%); RDIs:static rock stability indicator, CD: micro-cracks plus voids ratio (%); k-value : cation packing index (10^{-2} mol/cm^3), *CLI:Chemical leaching Index; CWPI=Chemical Product Index)*

Table 4. Index properties, and rock durability indices of rock materials from Harsit granitic rocks (Ceryan and Ceryan 2005, Ceryan et al 2008)

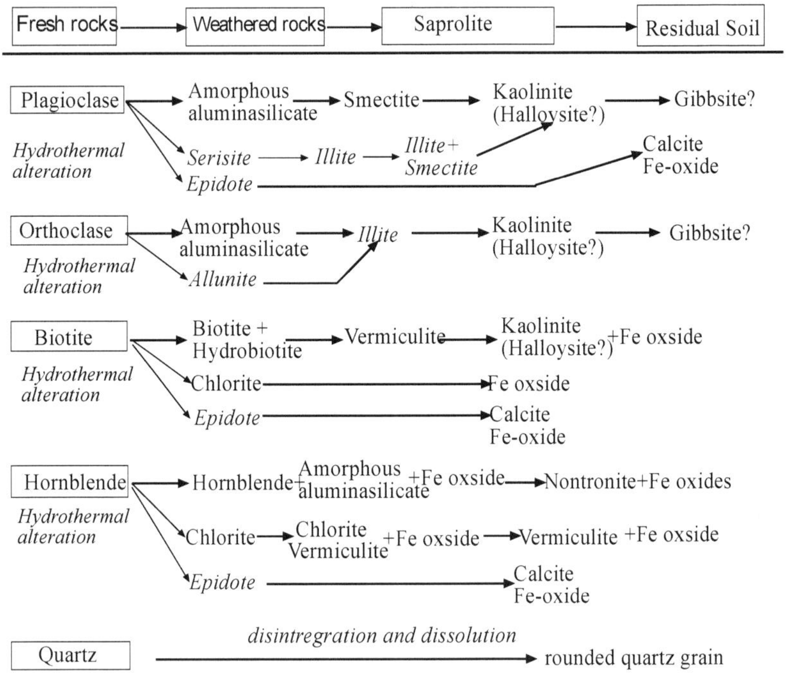

Fig. 7. Schematic representation of probable paths of rock-forming minerals transformation in the Harsit Granitoid

4.3 Classification of of Harsit Granitic Rocks for engineering purpose

Using petrographic techniques, the percentage of secondary minerals was established and relating this to the percentage microcracks and voids, a weathering classification from fresh to residual soils was established (Figure 8). While the engineering behaviors of the weathered rocks are assessed, the physical and mineralogical (directly chemical) changes caused by the weathering to be considered together will be significant (Table 5) The physical change is mainly in the direction of the ratio of effective porosity/total porosity increase and the ratio of micro-fracture + voids. Thus, this condition must be taken into consideration while the statistical relations between the weathering indices and the strength and deformation properties for weathered rocks. On the other hand, from the point of view of the geotechnical standpoint, indices based on the measurement of P-wave velocity generally have more applicability than those based on chemical, mineralogical and engineering properties (Ceryan et. al. 2008a). The classification of rock mass in the Harşit Granitoid is performed in accordance with the procedure suggested by Anon (1995) (Table 6). Transitions in the weathering zones of the Harsit Granitoid have graded. Because of changes even at the small scale, the same micro-region of the area in which exposed Harsit granitic rock masses in different weathering degrees (Figure 9).

Weathering Degree	Fresh	Slightly weathered	Moderately Weathered	Highly weathered	Completely weathered
			Weathering Indices		
γ (g/cm³)	2,664 (±0,042) (176)	2,634(±0,038) (99)	2,576(±0,053) (134)	2,553(±0,051) (87)	2,257 (±0,092) (127)
n (%)	1,92(±0,54) (176)	2,32(±0,59)(99)	4,43(±1,3)(134)	4,83(±1,6)(87)	15,32(±/3,7)(127)
ne (%)	1,50(±0,45(176)	1,92(±0,55)(99)	3,74(±1,2)(134)	4,11(±1,4)(87)	14,08(±3,6)(127)
Sa (%)	0,57(±0,18)(176)	0,73(±0,21)(99)	(1,46)(±0,5)(134)	1,55(±0,64)(87)	6,31((±1,9)(127)
FMC (%)	89,6(±4,1)(24)	73,2(±4,9)(18)	58,4(±5,5)(21)	47,5(±5,7)(15)	38,6(±4,1)(33)
SMC(%)	10,1(±4,7)(24)	25,2(±4,73)(18)	36,9(±6,3)(21)	42,3(±4,3)(15)	47,43(±4,9)(33)
CD (%)	0,66(±0,4)(24)	1,61(±1,3)(18)	4,65 (±1,9)(21)	7,79(±2,43)(15)	14,9(±3,7)(33)
Irms(%)	11,54(±5,9)(24)	34,9(±9,6)(18)	64,43(±17,7)(21)	88,9(±15)(15)	128,1(±29,4)(33)
WPI	12,8(±0,48)(12)	7,2(±1,7)(9)	3,7(±1,6)(13)	-0,1(±2,4)(11)	-5,61(±4,1)(11)
P	74,9(±6,4)(12)	64,1(±3,3)(9)	59,4(±2,4)(13)	60,2(±6,4)(11)	46,8(±8,0)(11)
PI	84,5(±0,8)(12)	83,4(±0,8)(9)	82,2(±1,5)(13)	82,1(±0,9)(11)	81,7(±1,29)(11)
Imob	0,0 (12)	0,137(±0,01)(9)	0,126(±0,01)(13)	0,118(±0,02)(11)	0,096(±0,02)(11)
CWPI(%)	4,1(±0,5)(12)	22,5(±1,8)(9)	37,9(±4,8)(13)	54,1(±6,9)(11)	60,2(±7,3)(11)
I_{QAB} (%)	0,21(±0,07) (8)	0,41(±0,21)(7)	1,20(±0,23)(7)	1,80(±0,68)(6)	7,36(±1,63)(11)
Id (%)	99,3(±0,40)(8)	98,4(±1,03)(7)	88,8(±10)(7)	56,8(±13)(6)	7,62(±(5,7)(11)
Iefp (%)	23,5 (±11)(176)	19,6(±6,9)(99)	16,7(±5,2)(88)	13,6(±6,1)(25)	13,2(±3,2)(66)
Vp (m/sn)	4111(±198)(176)	3553(±396)(99)	2769(±553)(134)	2158(±486)(87)	753(±184)(127)
Vpm (m/sn)	5732(±127)(176)	5109(±342)(99)	4507(±555)(134)	4079(±280)(87)	3617(±231)(127)
IQ(%)	75,6(±7,4)(176)	69,8(±7,2)(99)	59,7(±6,8)(134)	42,2(±14)(87)	20,0(±8,9)(127)
Ivp(%)	96,8(±1,8)(176)	83,0(±11,7)(134)	74,8(±6,8)(134)	65,6(±4,5)(87)	60,1(±0,06)(127)
PWD	0,5(±0,4)(12)	2,80(±1,9)(9)	10,4(±3,3)(13)	17,5(±13,7)(11)	17,5(±13,7)(11)
CWD	0,3 (±0,1)(1760)	10,2(±3,9)(99)	18,2(±5,9)(134)	27,3(±6,7)(87)	27,3(±6,7)(87)
			Mechanical Properties		
$I_{s(50)}$, MPa	7,3 (±1,8) (96)	5,0(±1,6)(68)	1,9(±0,9)(88)	1,0(±0,7)(61)	-
σ_c (MPa)	160,3(±34)(40)	128,9(±42)(37)	66,8(±33)(39)	39,3(±30)(27)	2,2(±1)(61)
σ_t (MPa	13,2(±1,4)(25)	8,4(±2,2)(20)	3,0(±1,8)(30)	1,8(±1,4)(14	-
Ed x10⁴ (MPa)	4,575(±0,5)(176)	3,407(±0,7)(134)	2,089(±0,8)(134)	1,269(±0,5)(87)	0,14(±0,06)(127)
Et x10⁴ (MPa)	5,483(±0,5)(8)	2,085(±0,4)(9)	0,869(±0,3)(15)	0,491(±0,17)(5)	-
Es x10⁴ (MPa	1,967(±0,48)(8)	1,499(±0,47)(9)	0,607(±0,26)(15)	0,355(±0,16)(5)	-

(γ (g/cm³): Dry density; n (%);Total prosity, ne (%)Effective porosity;, Sa (%) Water absorbtion, (atmospheric pressure); Vp: P-wave veloccity in dry samples; Vpm: : P-wave veloccity in solid part of the sample, I_{QAB} (%) Quick absorbsiyon, Id(%)Slake durability (second cycle), SHV: Schmidt rebound hardness ; $I_{s(50)}$ (MPa) Point load strength index , σ_c (MPa Unconfined compressive strength , σ_t (MPa) Indirect (Brazilian) tensile strength; Ed (MPa): Dynamic Elastisite modulu; Et (MPa) Tanjant Elastisite modulu, Es (MPa):Deformasyon Modulu, (N-type Schmid hammer is used),FMC: fresh mineral)

Table 5. Average (± standard deviation) and (number of data) weathering indices and mechanical properties of each weathering grade defined for granitic materials from Harsit Granotoid (Ceryan et. al. 2008a)

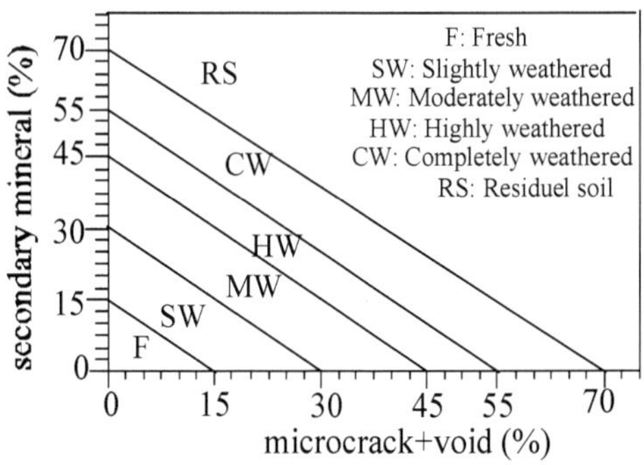

Fig. 8. The definition of the rock material weathering grade for Harsit Granitic Rocks

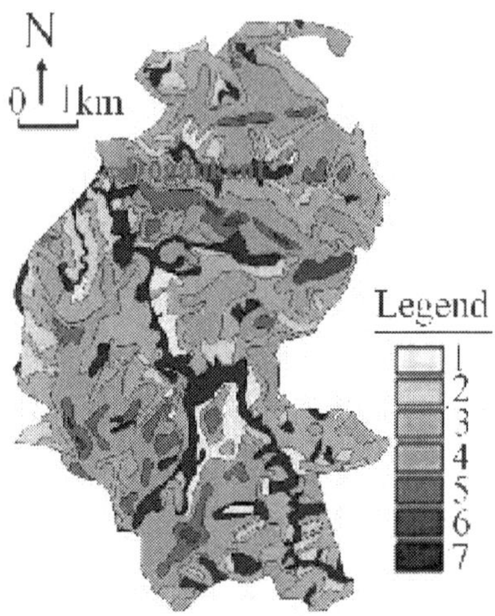

Micro-region adjusted to weathering

1. Un weathered rock masses % 80-100, Slightly weathering rock masses %0-20
2. Unweathered and Slightly weathering rock masses % 70-90, Moderately weathering rock masses %10-30,
3. Unweathered and Slightly weathering rock masses % 0-30, Moderately weathering rock masses %60-100, Saprolite % 0-40,
4. Moderately weathering rock masses %0-20, Saprolite % 0-80,
5. Saprolite % 10 0-90, Residual soils % 0-10,
6. Saprolite % 0-20, Residual soils % 100-80 and
7. Colluvium)

Fig. 9. Micro-region adjusted to weathering in urban area of Dogankent (NE Turkey (Ceryan and Ceryan (2008)

Zon	Weathering Profile	Description and typical characteristics
6	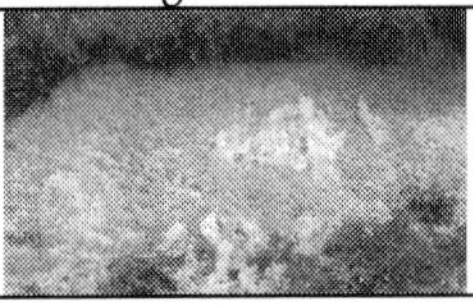	*Material:* % 100 RS-CW It is separated material including more sand, clay and silt.c=0.1-0.2 MPa, ϕ=36-40º.,*Mass structure* : Mass structure isn't preserved. The thickness varies between 0.5 and 5m.
5		*Material*: >%30 F-MW, >%70 HW-RS. The dispersion of the material is generally in order and the zone is homogeneousϕ=38-42º,c=0.2-0.3 MPa. *Discontinuities*: Systematic discontinuities and the fractures formed with weathering are preserved. Filling of the discontinuities usually consists of clay and iron oxide. ϕ_b=22-16º, JRC=2-6, Discontinuity frequency (Df) = 142.4(±1.5(discontiunity length/m^2). *Rock mass structure:* May behave as soil although relict fabric may be significant. Weak grade will control behavior of soil mass. For rock mass with relict structure ϕ_m =18-20, c_m =0.06-0.04 MPa The thickness varies between 2.5 and 11 meter.
4		*Material*: % 30-50 F-MW, % 50-70 HW-CW. *Discontinuities:* The thickness of the filling become lager amount. The filling generally consists of silt and sand. Rock bridge is completely removed. There are plenty of fractures formed by weathering in the corestones. ϕb =26-18º, JRC=2-4(±2), JCS=23(±12) MPa, Discontinuity frequency (Df)=11.6(±2.1) *Rock mass structure:* Corestones are beginning breaking and may be significant for investigation and construction Rock framework still locked and controls strength and stiffness, matrix control permeability. ϕm =21-24, cm=1.3-1.8 MPa The thickness varies 2.5 and 11 meter.
3		*Material*: % 50-90 G: I-III % 10-50 G: IV- VI. *Discontinuities*: Weathering deepness is usually bigger than the roughness. The fractures formed by weathering are seen in the blocs. The number of rock bridges gets less. Spacing of the discontinuities is more less than first zone. ϕb =32-25º, JRC=4-8(±2), JCS=53(±14) MPa, Df=2.4(±1.3)*Rock mass structure:* The edge of the corestones becomes circular. Mass structure is preserved. But the blocks tend to divide each other. Rock framework still locked and controls strength and stiffness, matrix control permeability. ϕm =22-25, cm=1.8-2.4 MPa The thickness varies 1.5 and 6 m.

2		*Material*: >%9U F-MW, <%10 HW-CW..*Discontinuities*: Weak materials along discontinuities. Shear strength stiffness and permeability affected. ϕb =28-26°, JRC=6-10(±4), JCS=112(±21) MPa, Df=2.4(±1.3). *Rock mass structure* : The blocks are cornered and interconnect each other.For closely jointed rock masses ϕm =26-28°, cm=2.3-4.4 MPa. The thickness varies 3 and 40 meter.
1		*Material*: %100 F-MW. *Discontinuities:* Discolour on discontinuities surface. ϕb =21-18°, JRC=8-12(±2),JCS=160(±34) MPa Df=1.6(±0.3). *Rock mass structure* : The blocs are interconnect firmly each other Properties of deformation and strength depend on direction and properties of the discontinuities. Behaves as rock apply rock mechanics principles to mass assessment and design

Table 6. The weathering classification of rock mass of Harsit granitic rocks (Ceryan and Ceryan 2008)

5. References

Akgun, A. And Ceryan, Ş. 2010.Estimating of the Mechanical Properties of Rock Masses by the Rating System Representing Rock Mass Weathering: A Case Study", 9th International Congress on Advances in Civil Engineering, 27-30 September 2010 Karadeniz Technical University, Trabzon, Turkey, p.1-6, 2010

Al-Qudami, Shehata WM, Al-Harthi AA, Sabtan A. 1997. On weathering of syenite under arid conditions. *Bulletin of the International Association of Engineering Geology*, 56, 3-8

Anon 1995, The description and classification of weathered for engineering purposes (Geological Society Engineering Group Working, Party Report). Quarterly Journal of Engineering Geology, 28, 207-242

Arıkan, F, Ulusay, R. And Aydın N. 2007. Characterization of weathered acidic volcanic rocks and a weathering classification based on a rating System . *Bulletin of Engineering Geology and the Environment* 66,415-430

Aydin, A. and Basu A. 2006. Technical Note :The use of Brazilian Test as a Quantitative Measure of Rock Weathering. *Rock Mech. Rock Engng.* 39 (1), 77-85 83

Banfield, J. F. 1985, *The Mineralogy and Chemistry of Granite Weathering*, Master of Science Thesis. Australian National University.

Basu, A., Celestino, T. B. and Bortolucci, A. A. 2009. Evaluation of rock mechanical behaviors under uniaxial compression with reference to assessed weathering grades. *Rock Mech Rock Engng,* 42: 73-93

Bozkurtoğlu, E., Vardar, M., Suner, F., and Zambak C. 2006. A new numerical approach to eathering and alteration in rock using a pilot area in the Tuzla geothermal area, Turkey. *Engineering Geology*, 87, 33-47

British Standards Institution (BSI), 1981. Code of Practice for Site Investigations. London, 147.

Ceryan, Ş. 2011. Estimating of the Durability and Weathering State of Rock Materials Using Chemical Weathering Indices based on Cation Packing Value: A Case Study, *Proceding of 11th International Multidsiplinary Sceintific Geoconferance 20-25 June 2011, SGEM 2011*, Bulgaria, Volume I, 437-444

Ceryan, S. 1999. *Weathering and Classification of Harsit Granitoid and the Effect of Weathering on Engineering Properties*. Ph.D. Thesis. Karadeniz Technical. University, Trabzon, Turkey, 300p, (in Turkish)

Ceryan, Ş. 2008. New Chemical Weathering Indices for Estimating Engineering Properties of Rocks: A Case Study from Kürtün Granodiorite, NE Turkey, *Turkish Journal of Earth Sciences*, 17, 187-207

Ceryan, Ş. and Ceryan, N. 2005. The relationship between the chemical weathering indices and potential durability of stone in service. *Proceeding of 1st International Symposium on Treavertine, September 21-25, Pamukkale University, Denizli, Turkey*, 291–296

Ceryan, Ş., Ceryan N. and Aydın, A. 2005. Determination of Weathering In Engineering Time Using Interaction Matrices. Proc. of 1st International Symposium on Travertine, September 21-25, Pamukkale University, Denizli, Turkey, 297-304, 2005.

Ceryan, N. and Ceryan, S. 2008. An application of interaction matrice method for slope failure susceptibility zoning: Dogankent settlement area (Giresun, NE Turkey). Bulletin of Engineering Geology and Environment, 67,375-385

Ceryan, S. and Sen, C.2003. Two Potential Methods: Chemical Weathering Index and Elastic Wave Velocity Defining Potential Rock Durability *Proceeding of industrial minerals and building stones. Intertanional Association for Engineering Geology and Environment The Turkish National Group of IAEG, İstanbul, Turkey*, 139–146

Ceryan, S., Tudes., S. and Ceryan, N. 2008a. Influence of weathering on the engineering properties of Harsit granitic rocks (NE Turkey)", Bulletin of Engineering Geology and Environment, 67, 97-10

Ceryan, S., Tudes, S. and Ceryan, N. 2008b. A new quantitative weathering classification for igneous rocks. *Environmental Geology*, 55, 1319-1336, 2008

Ceryan, S., Zorlu K., Gökçeoğlu, C. and Temel, A. 2008c. The use of cation packing index for characterizing the weathering degree of granitic rocks. *Engineering Geology*, 98, 2008, 60-74

Ceryan, Ş., Ceryan, N. and Bostancı, H. T. 2008d. An example for the prediction of the rock durability using fuzzy model. Proceedings of the 8th International Congress on Advances in Civil Engineering, ACE 2008, Famagusta, North Cyprus, 15-17, 2008, September, Vol 3: Engineering Materials/Structural& Earthquake Engineering, 41,50

Chesworth, W. Dejou, J.and Larroque, P. 1981.The weathering of basalt and relative mobilities of the major elements at Belbex. France, *Geochimica et Cosmochimica Acta* 45, 7, 1235–1243.

Cole, W.F and Sandy M.J. 1980. A Proposed Secondary Mineral Rating for Basalt Road Aggregate Durability. *Australian Road Research*, 10(3), 27-3

Colman, S.M. 1982. Chemical Weathering of Basalts and Andesites: Evidence from Weathering Rinds, U.S. Geological Survey Professional Paper No:1246, 51p.

Conca, J.L. and Cubba, R. 1986. Abrasion resistance hardness testing of rock materials. *Int J Rock Mech Min Sci Geomech Abstr* 23(2),141-149

Davis G.H. 1984. *Structural Geology of Rocks and Regions*. John Wiley and Sons, New York

Dearman W.R. and Irfan T.Y. 1978. Assessment of The Degree of Weathering in Granite Using Petrographic And Physical Index Tests. *International Symposium on Deterioration and Protection of Stone Monuments*, Unesco, Paris, 2.3,3-35

Dearman W.R., Turk, N, Irfan T.Y. and Rowshane, H. 1987. Detection of rock material variation by sonic velocity zoning. *Bulletin of the International Association of Engineering Geology*, 35, 3-8

Dearman, W. R. 1976. Weathering Classification in the Characterization of Rock: A Revision *Bulletin of the International Association of Engineering Geology*, 13, 123-127.

Dixon, H.W. 1969. Decomposition products of rock substances. Purposed engineering geological classification. *Rock Mechanics Symp.* Stephen Roberts Theatre, Univ. Sydney, 39-44

Dobereiner, L, Durville, J.L. and Restituito, J. 1993. Weathering of the Massiac Gneiss (Massif Central, France). *Bulletin of the International Association of Engineering Geology*, 47,79-97

Düzgören-Aydin, N.S. and Aydin A. 2003. Chemical heterogeneities of weathered ignous profiles: implications for chemical indices. *Environmental and Engineering Geoscience* 9, 4, 363–376.

Düzgören-Aydin, N.S., Aydin A. and Malpas, J. 2002. Re-assessment of chemical weathering indices: case study of pyroclastic rocks of Hong Kong. *Engineering Geology* 63, 99–119.

Erguler Z.A., and Ulusay, R. 2009. Assessment of physical disintegration characteristics of clay-bearing rocks: Disintegration index test and a new durability classification chart. *Engineering Geology*, 105(1-2), 11-19

Esaki, T and Jiang, K. 1999. Comprehensive study the weathered condition of welded tuff from a historic stone bridge in Kagoshima, Japan, *Engineering Geology* 55, 121–130.

Eynatten, H., Barcelo/-Vidal, C. and Pawlowsky-Glahn. 2003. Modelling Compositional Change: the Example of Chamical Weathering of Granitoid Rocks. *Mathematical Geology*, 35 (3), 211-251

Fookes, P.G., Gourley, C. S. and Ohikere, C. 1988. Rock weathering in engineering time. *Quaterly Journal of Engineering Geology* 21, 33–57

Fourmaintraux, D. 1976. Characterization of rocks; laboratory tests, Chapter IV in La Mecanique des roches appliquee aux ouvrages du genie civil by Marc Panet et all. Ecole Nationale des Ponts et Chaussees. Paris.

Franklin JA, Chandra R (1972) The slake durability test. *Int JRock Mech Min Sci* 9,325-341

GCO, 1994. *Guide to Rock and Soil Descriptions. Geoguide 3*, 2nd edition, Geotechnical Engineering Office Hong Kong.

Gokceoglu C., Zorlu ,K., Ceryan, S. and Nefeslioglu, H. A. 2009. comparative study on indirect determination of degree of eathering of granites from some physical and strength parameters by two soft computing techniques., *Mateials Characteization*, 60 1317-1327,

Gokceoglu, C. and Aksoy ,H . 2000. New approaches to the characterization of clay-bearing, densely jointed and weak rock masses. *Engineering Geology*, 58 1-23.

Gokceoglu, C., 1997. *The approaches to overcome the difficulties encountered in the engineering classification of clay-bearing, densely jointed and weak rock masses*. PhD Thesis, Hacettepe University, Geological Engineering Department (in Turkish).

Grant, J.A. 1986. The isocron diagrams-a simple solution the Gresens equations for metasomatic alterations. *Economic Geology* 87, 1976-1982.

Gresens, R. L. 1967. Composition-volume relationships of metasomatism, *Chemical Geology*, 2, 47-65.

Guan, P., Ng, C.W.W. Sun, M. and Tang W. 2001. Weathering indices for rhyolitic tuff and granite in Hong Kong, *Engineering Geology* 59, 147–159.

Gulec, K. 1973. *Engineering geology of Afyon Marbles and the relationships between its physico-mechanical properties and weathering*. . Ph.D. Thesis İstanbul Technical. University, Istanbul, Turkey, (in Turkish).

Guolin R, Yushan L (1990). Engineering geological zonation of Xiamen granitic weathered crust and bearing capacity of residual soil. *6 th International IAEG Congress*, 1989-1996, Balkema Rotterdam.

Gupta. A.S. and Rao K.S. 2001. Weathering indices and their applicability for crystalline rocks. *Bulletin of Engineering Geology and the Environment.*,60, 201-221

Gurocak Z, Kilic, R. 2005. Effect of weathering on the geomechanical properties of the Miocene basalts in Malatya, Eastern Turkey, *Bulletin of Engineering Geology and the Environment*, 64, 373-381

Hamrol, A. 1961. A quantitative classification of weathering and weatherability of rocks. *Proc.5th. Int. Conf. Soil Mechanics. Foundation Engineering*, Paris, 2, 771-773

Howarth, D.F. and Rowlands, J.C., 1987. Quantitative Assesment of Rock Texture and Correlation with Drillability and Strength Properties. *Rock Mechanics and Rock Engineering*, 20, 57-85.

Huston, D.L. 1993. The effect alteration and metamorphism on wall rocks to the Balcooma and Dry River South volcanic-hosted massive sulfide deposits, Queensland, Australia. *Journal of Geochemical Exploration* 48, 277-307.

Illev, I.G. 1967. An attempt to estimate the degree of weathering of intrusive rocks from their physico-mechanical properties. Proc. 1 st. Cong. Int. Soc. Rock Mechanics., Lisbon, 109-114

Irfan TY (1996). Mineralogy, fabric properties and classification of weathered granites in Hong Kong. Quarterly Journal of Engineering Geology, 1996, 29,5-35

Irfan, T.Y. and Dearman, W.R. 1978a The engineering petrography of a weathered granite in Cornwall, England. *Quarterly Journal of Engineering Geology*, 11,233-244

Irfan, T.Y. and Dearman, W.R. 1978b. Engineering classificational and index properties of a weathered granite, *Bull. Int. Assoc.Engng Geol*, 32, 67-80.

Irfan, T.Y. and Powell, G.E. 1985. Engineering geological investigations for pile foundation on a deeply weathered granitic rock in Hong Kong. Bull Int Assoc Eng Geol 32:67-80

Jayawardena U. de S. and Izawa, E., 1994a. Application of present indices of chemical waethering for precambrian metamorphic rocks in Sri Lanka. *Bulletin of the Intertanional Association of Engineering Geology* 49, 55–61.

Jayawardena U. de S. and Izawa, E. 1994b. A new chemical index of weathering for metamorphic silicate rocks in tropical regions: A study from Sri Lanka, *Engineering Geology* 36, 303–310.

Kilic, R. 1995. Geomechanical properties of the ophiolites (Çankırı, TURKEY) and alteration degree of diabase *Bulletin of the International Association of Engineering Geology*, 51, 63-69

Kilic, R. 1999. A unified alteration index(UAI) for mafic rocks. *Environmental and Engineering Geoscience*, 4, 475-483

Kocbay, A. 2003. *Investigation of characteristics and alteration degree of basalts outcropped in the vicinity of Osmancik-Corum*, Ph.D. Thesis Ankara University, Ankara, Turkey, 46p. (in Turkish)

Koncagul, E.C. and Santi, P.M. 1999. Predicting the unconfined compressive strength of the Breathitt shale using slake durability, Shore hardness and rock structural properties. *Int J Rock Mech Min* 36(2),139–153

Korkmaz N. and Ceryan N. 2011. An Example for The Prediction of the Uniaxial Compressive Strength of Weathered Rocks Using The Wethering Indices based on P-Wave Velocity. Proceding of *11th International Multidsiplinary Sceintific Geoconferance* 20-25 June 2011, SGEM 2011, Bulgaria, Volume I,.347-354

Krank, A.K.D. and Watters, R.J. 1983. Geotechnical properties of weathered Sierra Nevada granodiorite. *Bull. Association. Engineering. Geol.*, 20,173-184

Krauskopf, K.B., 1967. *Introduction to Geochemistry*. McGraw-Hill Book Company Inc., 176,1967, 518-527.

Lan, H.X., Hu, R.L., Yue, Z.Q. Lee, C.F. and Wang, S.J. 2003. Engineering and geological characteristics of granite weathering profiles in South China. *Journal of Asian Earth Science* 21, 353–364.

Lee, S.G. 1987. *Weathering and geotechnical characterization of Korean granites*, PhD Thesis, Imperial College, University of London.

Loughnan, F. C. 1969. *Chemical Weathering of the Silicate Minerals*, Elsevier, Amsterdam.

Lumb, P., 1983. Engineering properties of fresh and decomposed igneous rock from Hong Kong. Engineering Geology 19, 81- 94.

Lump, P. 1962. The properties decomposed granite. *Geotecnique*, 12, 226-243

Maclean, W. H. 1990. Mass change calculations in altered rock series. *Mineraliu Deposita* 25, 44–49.

Marques, E.A.G., Barroso E.V., Menezes Filho A.P. and Vargas Jr. E. do A. 2010. Weathering zones on metamorphic rocks from Rio de Janeiro—Physical, mineralogical and geomechanical characterization, Engineering Geology, 111 (1-4), 1-18

Martin, R.. 1986. Use Of Index Tests For Engineering Assessment Of Weathered Rocks. *Proceedings of 5 th Congress of International Association of Engineering Geologists*, Buenos Aires, 433-450.

Martin, R., P. and Hencher, S., R., 1986, Principles For Description And Classification Of Weathered Rocks For Engineering Purposes. In: Hawkins, A., B. (ed.), *Site Investigation Practice*, Geological Society, London, Engineering Geology Special Publication, 2, 299-308.

Mendes, F.M., Aires-Barros L and Rodriques, F.P. 1967. The use of modal analysis in the mechanical characterization of Rock Masses. Proc. 1 st. Int. Congress, In. Soc. Roc. Mech., Lizbon, 1, 217-233

Minarik, L., Absolon, K., Köllnerova, Z. and Klecka, M. 1983. Chemical change of granite during its weathering, *In*: *Leaching and diffusion in rocks and their weathering products*. Augustithis, S. S., (ed), Theophrastus Publications. S.A. Athens, 293-306.

Moon V.G. and Beattie, A.G. 1995. Textural and microstructural influences on the durability of Waikato Coal Measures mudrocks.*Quarterly Journal of Engineering Geology* 28, 303–312

Ohta, T. and Arai, H. 2007. Statistical empirical index of chemical weathering in igneous rocks: A new tool for evaluating the degree of weathering. *Chemical Geology* 240, 280–297

Ollier, C. 1984. *Weathering Geomorphology Text.* 2nd edition, Longman, London.

Onodera, T.F., Yoshinaka, R. and Oda, M. 1974. Weathering and its relation to mechanical properties of granite. Third International Conference of ISRM, Denver 2A, 71--78.

Pasamehmetoglu, A.G., Karpuz C. and Irfan, T.Y. 1981. The weathering classification of Ankara andesites form rock mechanics point of view, *Proc. Int. Symp. Weak Rocks, Tokyo* Vol.1, 185–190

Price, D. G. 1995. Weathering and Weathering Processes. *Quarterly Journal of Engineering Geology* ,28, 34-252.

Price, J. R. and Velbel, M. A. 2003. Chemical weathering indices applied to weathering profiles developed on heterogeneous felsic metamorphic parent rocks. *Chemical Geology* 202, 397–416.

Rigopoulos, I., Tsikouras, B., Pomonis, P. and Hatzipanagiotou, K. 2010. The influence of alteration on the engineering properties of dolerites: The examples from the Pindos and Vourinos ophiolites (northern Greece). *International Journal of Rock Mechanics & Mining Sciences* 47, 69–80

Sadisun, I.A., Shimada, H., Ichinose M. and Matsui, K. 2005. Study on the physical disintegration characteristics of Subang claystone subjected to a modified slaking index test, *Geotechnical and Geological Engineering* 23, 199–218

Shalkowski, A. Yoshinori Kodama, Y. and Nakano, S. 2009.The assessment of weathering stages in granites using an EC/pH meter. *Geomorphology* 105, 253–260

Sousa, L. M. O., Suarez del Rio L.M, Calleja L., Vicente, G., Argandofia V.G R and Rey A. R. 2005. Influence of microfractures and porosity on the physico-mechanical properties and weathering of ornamental granites.*Engineering Geology* 77, 2005, 153-168

Tecer L. 1999. Laboratory Experiments on the Investigation of the effects of Sulphuric Acid on the Deterioration of Carbonate Stones and Surface corrosion. Water, Air, and Soil Pollution, 114, 1-12.

Tecer L and Cerit O. 2002.The Effects Of Air Pollution On Carbonate Stone Monuments In Urban Areas (Sivas, Turkey), Fresennius Environmental Bulletin, 11:8, 505-509.

Tuğrul, A and Gurpınar, O. 1997. The effect of chemical weathering on the engineering properties of Eocene basalts in northeastern Turkey. *Environmental & Engineering Geosciense* 3,225–234.

Turk, N. and Dearman, W.R. 1985. Influence of water on engineering properties of weathered rocks. Proc. 21st Raq. Conf., Engng. Grouyp of the Geol. Soc. On Groundwater in Engineering Geology, Sheffield, UK. 109-121

Vogel, D. E. 1973. Precambrian Weathering in Acid Metavolkanic Rocks from The Superior Province, Villebond Township, Southcentral Quebec., *Canadian Journal Earth Science*,12, 2080-2085.

Wei K. and Liu D. 1990. The zoning system for assessment of weathering states of granites, *Proceedings 6th Int. IAEG Congress, Amsterdam,* 657–663.

Weinert, H.H. 1964. Engineering petrology for roads in South Africa. *Engineering Geology,* 2, 363-395

Weiss, T., Rasolofosaon. P. N. J. and Siegesmund, S. 2002. Ultrasonic wave velocities as a diagnostic tool for the qualityassessment of marble. *Natural Stone, Weathering Phenomena, Conservation Strategies and Case Studies* (edited by. Siegesmund S, Weiss T. and Vollbrecht A) Geological Society, London, Special Publications. 205,2002, 149-164

Yıldız A., Kuşcu, M., Dumlupunar, İ., Arıtan, A. E. And and Bağcı, M. 2010. The determination of the mineralogical alteration index and the investigation of the efficiency of the hydrothermal alteration on physico-mechanical properties in volcanic rocks from Köprülü, Afyonkarahisar, West Turkey. *Bulletın of Engıneerıng Geology and the Envıronment,* 69(1), 51-61,

Zhao, J and Broms B.B. 1993. Mechanical and physical properties of weathered Bukit Timah granite of Singapore. *Geotechnical Engineering of Hard Soils-Soft Rocks,* Anagnostopoulos et. all (eds). Balkema, Rotterdam, 883-890.

3

The Effectiveness of Petrified Wood as a Geobiological Portal to Increase Public Understanding of Geologic Time, Fossilization, and Evolution

Renee M. Clary[1] and James H. Wandersee[2]
EarthScholars Research Group
[1]Mississippi State University
[2]Louisiana State University
USA

1. Introduction

Petrified wood is ubiquitous, and is found in every US state and on every continent. Because of its abundance and intriguing nature, we hypothesized that fossilized wood might be an effective gateway through which important interdisciplinary scientific concepts could be taught. Our earliest investigations of *in situ* petrified wood at informal education sites revealed that petrified wood could spark public interest about its formation, and its seemingly paradoxical nature. Upon this original viewer interest, scientific content in chemical composition, fossilization processes, extinction events, evolutionary processes, and geologic time could then be scaffolded, in either informal or formal educational settings. In our first classroom investigations, we probed the effectiveness of petrified wood as a portal in college Earth History courses to address geologic age, fossilization processes, and fossil properties (Clary & Wandersee, 2007). Through classroom incorporation of petrified wood, instructors identified students' alternative conceptions, and significant student gains in some scientific content were evident at the end of the semester. In a subsequent research investigation, students in a junior level Landscape Architecture design class were assigned a project in which they developed an informal educational space that conceptualized geologic time (Clary, Brzuszek, & Wandersee, 2009). Petrified wood was used to measure student gains in the understanding of geologic time, and data revealed that a threshold petrified wood conceptual knowledge was present in all successful design solutions. Through our research, we identified petrified wood as a potential geobiological portal to address public understanding of geologic time, identified by Stephen Jay Gould (1987) as one of the major scientific constructs of all time, paralleling evolutionary theory in its importance.

Our latest petrified wood research focuses upon an earlier educational influence in our future citizens' geological literacy. In this current study, we investigate primary and secondary teachers' geobiological content knowledge of petrified wood, and probe potential investigative techniques for effective petrified wood study within K-12 classrooms. We also attempt to ascertain the role of science professional development programs for increasing teacher content knowledge in geologic time and fossilization processes.

2. Theoretical background

One goal of the EarthScholars Research Group™ is to improve public understanding of science through novel, integrated learning approaches. With researchers' scientific backgrounds in geology and biology (botany), we are inspired to identify, emphasize and promote the intersections between our two disciplines for an integrated, geobiological understanding of science. We research science education opportunities at informal science sites, as well as the potential for the improved articulation of integrated science instruction in formal science classrooms. Our research is guided by the learning theory of human constructivism, and we strive to promote meaningful learning in science classrooms by introducing new concepts that connect in substantive ways to our learners' prior knowledge. We also strive to provide context to scientific concepts and constructs in order to accurately depict the nature of science to our students, in contrast to the "final form science" (Duschl, 1990) that exists in many modern classrooms. Because our research agenda promotes integrated geobiological education, we ascribe to the Big Ideas in Earth Science as identified by the Earth Science Literacy Initiative (2010). These Big Ideas are interdisciplinary in content, and promote an authentic view of science.

2.1 Nature of science, integration of scientific concepts, and the history of science in science instruction

To promote a valid view of science, traditional classroom and informal science instruction must move beyond a collected set of seemingly unrelated facts, concepts, and theories. Without the context for how the scientific knowledge emerged and developed, students may leave the classroom with only the "rhetoric of conclusions" (Schwab, 1962), and fail to connect the content knowledge between one science course and the next. The non-contextual accumulation of knowledge is also present outside the classroom: Suzuki (2004) noted that world events are presented without context, leading to confusion and frustration in the general population. Integrated, interconnected science can help facilitate student understanding of our planet's complexity (Orr, 1994).

The context in which scientific concepts and theories developed also has value for the science classroom. The history of science can help students view science as an interesting practice, as it reveals the social, political, and cultural influences from which the constructs and theories emerged (Matthews, 1994). The nature of science is revealed when students understand how scientific knowledge is restructured as new data are collected and new research investigations modify previous conclusions (Duschl, 1994). Although some conclusions are later exposed as erroneous, the inclusion of this history is intellectually honest for our students, and reveals how scientific methodology ultimately overturns poor research studies and corrects false conclusions (Clary, Wandersee, & Carpinelli, 2008). The history and philosophy of science not only provides context, but can also humanize a science curriculum (Jenkins, 1989).

2.2 Learning theory of human constructivism

The learning theory guiding EarthScholars Research Group™ is that of human constructivism, as developed by Novak (1977) and elaborated by Mintzes, Wandersee, and Novak (1998, 2000). Human constructivism builds upon the work of cognitive psychologist David Ausubel (1963, 1968; Ausubul, Novak, & Hanesian, 1978) and Novak's (1963) fundamental principles for science education research. Theoretical principles arising from human constructivism posit that humans are meaning-makers, and that learning occurs when there is a change in the meaning of experience. Meaningful learning occurs when learners can connect new concepts to their existing knowledge frameworks in new, non-verbatim ways, leading to either strong or weak cognitive restructuring. Therefore, the goal of science education is to foster conceptual change in learners that can lead to more powerful knowledge structures and an understanding of our planet.

Science instruction that leads to meaningful learning will promote quality over quantity of material, meaning of concepts over memorization, and an understanding of scientific concepts and constructs over simple awareness (Mintzes et al., 1998). Therefore, a selected set of concepts should be taught in a meaningful way, as opposed to brief overview of a large number of seemingly unrelated constructs. This reduction of material presents one of the major dilemmas for Earth Science education (Bybee & Pratt, 1996). However, the core constructs of the geosciences can be identified as plate tectonics and geologic time, which also influence important constructs in other disciplines, including evolutionary theory in biology (Dodick & Orion, 2003). Because meaningful learning occurs when learners connect new concepts to their existing knowledge frameworks, teachers should assess students' prior knowledge before instruction begins, and then proceed accordingly (Ausubel et al., 1978; Mintzes et al., 1998). Some students harbor alternative or non-scientific conceptions about how science operates within the natural world (Novak & Musonda, 1991; Wandersee, Mintzes, & Novak, 1994). If these alternative conceptions (often called misconceptions in the science education literature) are not identified and addressed, science instruction can be ineffective.

2.3 Geoliteracy and the Big Ideas in Earth Science education

There is an undeniable need for the public to know and understand the planet on which we live (Clary & Wandersee, 2009). Pursuant to this understanding and compiled in congruence with the learning theory of human constructivism, the Earth Science Literacy Initiative (funded by the US National Science Foundation) identified and published the nine Big Ideas needed by the general population for geological literacy (2010). These nine big concepts are interdisciplinary in nature, and are aligned with the US National Science Education Standards (National Research Council, 1995).

The Big Ideas include the nature of repeatable and replicable scientific methodologies (Big Idea 1); geologic time (Big Idea 2); Earth System Science, in which complex and interacting "spheres" of the planet interact and influence each other (Big Idea 3); the dynamic nature of the planet (Big Idea 4); the unique role of water on our planet (Big Idea 5); the evolution of life forms and their influences on the planet (Big Idea 6); the dependence of humans on Earth's natural resources (Big Idea 7); the planet's natural hazards for humans (Big Idea 8); and the significant influence of humans on our planet (Big Idea 9). Although these Big Ideas represent the core concepts for Earth Science literacy, each concept also has connections and interactions with the other concepts.

For example, geologic time, or the 4.6 billion year history of our planet, provides the needed reference scale for tectonic evolution of the Earth, and the origin, evolution, and extinction of its life forms. Often referred to as "deep time" (Carlyle, 1832, McPhee, 1981, Rudwick, 1992), geologic time can inform us of the principles that determined our planet's history, and can help us to uncover the past trends and patterns that may help us to effectively predict Earth's future (Soreghan, 2005). An understanding of geologic time is crucial for the general geological knowledge needed by all citizens to make informed decisions about our planet (Clary & Wandersee, 2009). Notably, petrified wood study can address many of these Big Ideas, including scientific methodology (1), geologic time (2), Earth System Science (3), our changing planet (4), hydrologic influences on fossilization (5), evolution of life forms (6), and climatic changes over time, extending into the present time (9).

3. EarthScholars Research Group: Previous petrified wood investigations

The EarthScholars Research Group™ began its investigation into the role of petrified wood in science education in 2003. Because of fossil wood abundance, numerous locations exist where petrified wood is publicly displayed. While the biggest public display of petrified wood in the United States is at the Petrified Forest National Park in Arizona, other smaller sites across the US display petrified wood *in situ* as well. We visited and investigated two informal sites: the Mississippi Petrified Forest in Flora, Mississippi (Fig. 1) and the Petrified Forest, in Calistoga, California (Fig. 2). We observed and interviewed visitors and staff, analyzed each site for its effectiveness, and determined the potential of petrified wood in science learning.

We identified several opportunities to learn important scientific constructs in these informal learning environments, but noted that geologic time was not emphasized or was avoided on the self-guided trails. Likewise, scientific explanations of the formation of petrified wood were missing, partial, or incorrect. However, opportunities existed to help the public visualize how the current location differed from the past paleoenvironment in which the fossilized trees originally lived, and how scientists arrived at the ages for the fossilized wood specimens preserved at each site. A context of the displayed fossilized trees within a larger context of plant evolution would have been informative and helpful.

Fig. 1. The Mississippi Petrified Forest in Flora, MS has the largest public display of petrified wood east of the Mississippi River in the US. Signage is minimal at the site, but visitors are provided a brochure, which offers descriptions of the numbered waypoints. The fossilized wood on display is Oligocene in age (30 million years old).

The fact that both forests have drawn the public since 1871 (California) and 1854 (Mississippi) provides convincing documentation of the value of petrified wood and petrified forests for teaching geobiological literacy. Our investigation into the visitors' comments (Fig. 3) revealed that although visitors were awed by the display of the fossilized trees, several visitors invoked non-scientific views to explain their formation and presence at the site.

Fig. 2. The Petrified Forest in Calistoga, CA is noted for the druzy quartz crystals within the fossilized wood specimens. Brochures are used in lieu of signage at this site. A tourist site since 1871, the Petrified Forest displays fossilized wood from the Pliocene, approximately 3 million years in age.

3.1 Introductory college science classrooms, non-majors

As a result of our research of petrified wood in informal education sites, the EarthScholars Research Group™ embarked on a longitudinal study of the effectiveness of petrified wood in introductory college geology classrooms in 2004. We also chose petrified wood as a geobiological portal for our Earth Science classrooms because our earlier surveys indicated that petrified wood is a topic to which most students are introduced in the K-12 school environment. Furthermore, students in our Earth History class had previous exposure in their prerequisite physical geology course to famous plant fossils: *Glossopteris* specimens from Australia, South America, India, Africa, and Antarctica provided convincing evidence for the existence of the southern supercontinent Gondwana, whose collision with Laurasia resulted in the formation at Pangaea at the end of the Paleozoic Era. This fossil evidence, as part of the history of science in the development of the plate tectonic theory, is a typical topic of study in most physical geology courses.

We designed and tested the Petrified Wood Survey™ (Appendix A) to probe our students' understanding of geologic time, fossilization processes, and petrified wood's chemistry and geographic occurrences (Clary & Wandersee, 2007). Our investigation was conducted across

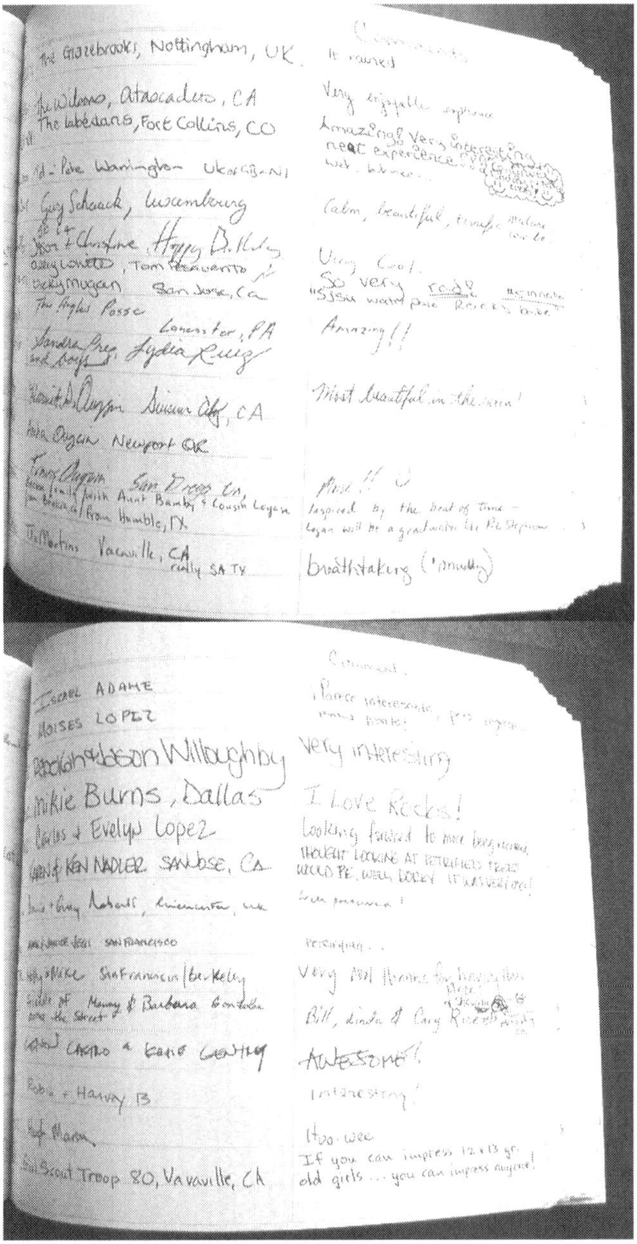

Fig. 3. Sample pages from the Visitors Comments, Petrified Forest, California, attest to visitor interest in the petrified wood display.

three semesters in large, non-science major courses at a research university in the southern US. Through our extended research program, the Petrified Wood Survey™ (PWS) was validated as a testing instrument, and we verified that its use as a pre-survey assessment of incoming student knowledge did not contribute to any testing effect at the end of the semester. The PWS effectively exposed student alternative conceptions about fossilization processes, geologic time, and the formation of petrified wood, as well as the fossils' composition, properties, and geographic occurrence. The PWS also measured significant student gains ($\alpha = 0.05$) between pre- and post-surveys across a semester of Earth History instruction, irregardless of whether petrified wood was included as a geobiological portal in the classroom.

However, the PWS measured even greater student gains in understanding of geologic time, fossilization, and evolution when petrified wood *was* included in the Earth History classroom. During the treatment semester, we integrated hands-on mini-laboratory sessions in which students compared and contrasted (unidentified) fossilized wood and modern log samples, and then discussed and debated the specimens within online discussion groups. Petrified wood specimens from Calistoga, California were incorporated in the classroom, and students also had the option of participating in a petrified wood research project, in which they investigated a unique site where petrified wood was displayed or could be collected.

Our research revealed that the areas of student understanding most affected by petrified wood integration were the abundance of petrified wood, its properties, and the nature of its formation. Other student gains were made in geologic time understanding. From our identification of initial student alternative conceptions, it appears that both weak and strong restructuring of students' conceptual frameworks occurred (Clary & Wandersee, 2007). However, some geochemical content areas were not as affected by petrified wood integration, particularly the role of oxygen (i.e., the lack of oxygen) in fossilization processes, and the role of dissolved minerals in petrified woods' colors.

3.2 Junior level landscape architecture design class

In 2008, we extended our integrated petrified wood study to a junior-level design class at a different US research university in the southern US. Students enrolled in Landscape Architecture Design I were instructed to design an informal educational garden that depicted geological time in front of the campus geosciences building. We predicted petrified wood would be an effective geobiological portal for this assignment, since the design site had an existing petrified log on display that had to be included in each student's final project solution. Students were also familiar with petrified wood: The largest public display of petrified wood east of the Mississippi River is located within the state, and petrified wood is also acknowledged as the state rock. In the university's geology museum, petrified wood is on display, with fairly large trunk segments available for hands-on investigation (Fig. 4.).

We used the Petrified Wood Survey™ to assess design students' incoming knowledge and perceptions, and then utilized it as a posttest to measure any content knowledge gains throughout the project. Although the sample size was small ($N = 25$), the successful design solutions that effectively depicted geologic time were developed by students who had achieved a minimum geology content knowledge of 75% as measured by the PWS (Clary, Brzuszek, & Wandersee, 2009). The mean gain in geologic content knowledge as measured by the PWS was 3.56, a significant gain that was also higher than the previously reported gains in undergraduate classes for non-science majors (Clary et al, 2009).

Fig. 4. The Dunn-Seiler Museum at Mississippi State University displays several specimens of petrified wood, including this fossilized log that is available for hands-on investigation.

4. Extending petrified wood research to inservice teachers: Methods

Our latest research investigation with petrified wood involves inservice K-12 teachers (N = 97) who self-elected to participate in either a state-funded Mathematics and Science Partnership (MSP) science professional development program (n = 28, 41, 18), or a state-funded professional development program for teachers targeted toward improving science instruction in high needs school districts (n = 10). Our previous research indicated that non-science majors in introductory college Earth History courses had some familiarity with petrified wood, with only 17-18% of students reporting that they never had been introduced to the topic previously (Clary & Wandersee, 2007). Likewise, our investigation with Landscape Architecture upperclassmen revealed that only 8.3% of students had not received any prior instruction in petrified wood (Clary et al., 2009). Therefore, this current investigation probes the geobiological content knowledge of the inservice teachers who introduce petrified wood in K-12 classrooms. We further implemented hands-on petrified wood and fossilization investigations with some teacher groups (N = 58) and sought teacher feedback as to the activities' effectiveness in the classroom.

In February 2011, the Petrified Wood Survey™ was administered to teachers participating in a MSP program in central Mississippi (n = 28). The group is identified in this study as MSP1 . Teachers self-identified that they were primarily K-8 classroom instructors, although one high school teacher (grades 9-12) was present. Following the Petrified Wood Survey™, whole group discussion was conducted on the topic of petrified wood, including some of the alternative conceptions held by teachers and/or students about fossilized wood specimens. Teachers were then divided into five groups (5-6 teachers/group) and provided hand samples of petrified wood and modern wood (Fig. 5).

Fig. 5. Teacher groups were provided hand samples of cut and polished petrified wood (right) and a cut modern log (left) for comparison.

Because not all teachers have direct access to petrified wood samples, we introduced and field-tested a simple activity to mimic the fossilization process that involved Epsom salts and a paper towel. We provided teachers with a simple laboratory procedure from TOPS Learning Systems (n.d.). On a small plate, 14.7 cm^3 of $MgSO_4$ (one level US tablespoon) is dissolved in 29.6 ml (2 level US tablespoons) water. A paper towel is folded into quarters, and then rolled up loosely to form the "log." This paper log is placed in the salt mixture, and allowed to dry over time. The salt solution is drawn into the paper towel log, and the $MgSO_4$ precipitates as the log dries. We previously enlisted middle school students (Fig. 6)

Fig. 6. A middle school student (US grade 7) uses the TOPS Learning Systems (n.d.) petrified paper activity to replicate the fossilization of wood.

to follow the activity sheet procedure, resulting in the "petrified paper logs" that the teachers investigated. Following the investigation of the petrified paper, teachers reflected as to whether this activity would be useful in their science classrooms.

In late February 2011, we investigated a second group of inservice teachers enrolled in a Mathematics and Science Partnership. However, this MSP targets middle school teachers of science, so all participating teachers taught at least one science class at grade 6, 7, or 8. Teachers enrolled in this professional development program, are subdivided into three content areas (chemistry, geosciences, and physics), and rotate through each discipline on an annual basis. We originally engaged these teachers at the conclusion of their first professional development year; this group is identified in this study as TANS1. Therefore, we used the Petrified Wood Survey™ to determine the geobiological content knowledge of the inservice teachers according to the groups they were assigned (chemistry n = 13; geosciences n = 11; physics n = 17). We hypothesized that the teachers who had received instruction in the geosciences may perform better than those teachers who had received instruction in chemistry or physics. Following the survey, teachers returned to their laboratories, and the geosciences teachers formed groups (n = 4) to compare hand samples of petrified wood and modern wood (Fig. 7). Because our original activity implementation with inservice teachers in the earlier group resulted in minimal reflection, we developed a hand-out that provided probing questions on the petrified wood and modern wood hand samples (Fig. 8), and utilized this during the activity. Following this activity, we initiated a whole group discussion on the properties and characteristics of fossilized wood. Teachers then returned to their groups, and investigated the petrified paper laboratory activity (Fig. 9). We developed and included a hand-out that included organizational charts and probing questions for this activity (Fig. 10).

Fig. 7. Teachers in the TANS1 geosciences group compare samples of modern and petrified wood in February 2011.

PW Mini Lab Comparison Questions
1. Which of the specimens is fossilized? How can you tell?
2. What is the composition of each specimen?
3. How are the two specimens alike?
4. How are they different?
5. What process (or processes) was (were) responsible for making the specimens different? How did the fossilized sample form?
6. What do you suspect is the age difference between the two specimens?

Fig. 8. Questions probing the differences between petrified wood and modern wood specimens (Fig. 5) were provided to inservice teachers in the TANS1 geosciences professional development program to facilitate reflection.

In June 2011, we surveyed teachers who participated in a different professional development program, which consisted of four weeks of instruction for the integration of mathematical and science content. Ten elementary teachers participated, and constitute the group ELE1. The Petrified Wood Survey™ was administered, and fossilized and modern wood samples were provided for comparison. Because of time limitations, the full hands-on investigative activities were not conducted.

Fig. 9. Inservice teachers in the TANS1 geosciences group investigate the petrified paper activity. In this photo, teachers are trying to determine why a sample "log" made from coarse brown paper did not dry properly when compared to the petrified paper log.

In June 2011, the TANS professional development group returned to the university campus for an intensive two-week summer academy. Teachers originally in the chemistry group were assigned to geosciences. However, nine of the 2011 TANS geosciences teachers were new recruits. Therefore, the Petrified Wood Survey™ was administered to the new

Petrification Activity Questions
1. *Does this activity effectively replicate the fossilization process? (chart provided with Accuracies/Inaccuracies)*
2. *Can you effectively incorporate this activity into your classroom? (chart provided with Advantages/Disadvantages)*
3. *Please list the topics/objectives you feel you can address with this activity:*
4. *Can you modify this activity to address additional topics? If yes, please explain:*
5. *What required MDE competencies* [Mississippi Department of Education science standards] *does this activity address?*
6. *What do you predict your student response will be to this activity?*

Fig. 10. Organizing charts (questions 1 and 2) and probing questions were provided to inservice teachers in the TANS1 geosciences group to facilitate reflection on the petrification activity.

geosciences group (n = 18), referred to in this study as TANS2. Teachers self-organized into four groups, and the petrified wood and modern wood comparison activity was administered using the organizing hand-out (Fig. 8). Following the comparison activity, we initiated whole group discussion on the properties and composition of petrified wood. We concluded the session with the petrified paper activity with small groups, providing the organizational charts and questions (Fig. 10) to facilitate teachers' reflections.

5. Data and analysis

Following our research investigation with our first group of teachers participating in a professional development program (MSP1), we refined the collection instruments for the petrified wood and modern wood comparison activities, and the petrification paper activity. Therefore, in the analyses that follow, we utilize the full data set (N = 97) when comparing the Petrified Wood Survey™ results, as this instrument was consistent throughout our four professional development program investigations (MSP1, TANS1, ELE1, TANS2). However, for the analyses of the petrified wood and modern wood comparisons, and the petrification paper activity, we only utilized the data from MSP1 to inform the development of subsequent investigations. In addition time factors prohibited sufficient data collection from ELE1. The discussion of the two investigative activities, therefore, relies on the data generated from TANS1 and TANS2.

5.1 Petrified Wood Survey™ results for inservice teachers

In our previous research investigations where the Petrified Wood Survey™ (PWS) was utilized, incoming geobiological knowledge on the 12 scored items (questions 3-14, Appendix A) averaged 3.34 (2003 Earth History course for non-science majors), 3.12 (2004 Earth History course for non-science majors), and 3.73 (2008 junior level Landscape Architecture Design I course). All teacher groups performed better than these student averages in our current research investigation. MSP1 (n =28) averaged 4.12, while TANS1 (n = 41) averaged 4.88 and TANS2 (n = 18) averaged 5.00. The smaller group of elementary teachers, ELE1, averaged 3.80 (Table 1).

Furthermore, we were able to separate the discipline averages of TANS1 teacher groups. Although the TANS teachers did not encounter previous instruction that specifically

Group Symbol	Group Description	Number	PWS average
MSP1	K-8 teachers, 1 high school teacher, MSP program	28	4.12
TANS1	6-8 teachers in MSP program	41	4.88
TANS2	6-8 teachers, MSP program 2nd year	18	5.00
ELE1	K-8 teachers, math/ science integrated PD program	10	3.80

Table 1. Average scores on the Petrified Wood Survey™ are organized according to each professional development teacher group.

Group Symbol	Group Description	Number	PWS average
TANS1 - chemistry	6-8 teachers in MSP program	13	4.31
TANS1 - geosciences	6-8 teachers in MSP program	11	5.46
TANS1 - physics	6-8 teachers in MSP program	17	4.94
TANS2 - geosciences	6-8 teachers, MSP program 2nd year	18	5.00

Table 2. Average scores on the Petrified Wood Survey™ are organized according to TANS subgroup.

targeted petrified wood, those TANS teachers who participated in the geosciences content program revealed superior performance with an average PWS score of 5.46. The TANS2 group, with approximately half of the teachers experiencing chemistry instruction the previous year, had the second highest PWS average (Table 2).

We utilized the full set of PWS scores (N = 97) when collecting and analyzing background knowledge for the teachers (Table 3). Most teachers regarded petrified wood as an interdisciplinary topic, although the majority did not think that they had ever had instruction in petrified wood in the K-12 environment before. Only 25% had high school Earth Science instruction, but 43.6% recalled a college course in geology or Earth Science. Few teachers could identify a paleontologist or scientist associated with the study of fossils. The majority of teachers left this item blank, or alternatively wrote the name of the instructor or a famous musician or athlete (e.g., Katy Perry, Lady Gaga). Of the 16 legitimate responses received, only one person was associated with fossils, Roy Chapman Andrews. The highest vote recipients were Charles Darwin and Alfred Wegener. Although Darwin proposed evolution by natural selection, he is not noted for his particular attention to the study of fossils. Wegener, who proposed continental drift, is associated with fossils only as they supported the hypothesis of moving continental masses.

For an in-depth analysis of the different content items of the Petrified Wood Survey™, we opted to analyze and compare the scores on content items of teachers who had previous

Within which school science subject does petrified wood and petrified forests seem to fit best?	
	Percentage
Biology	1.0%
Chemistry	3.1%
Earth Science/Geology	26.8%
Choices Biology/Geology	16.5%
All of the above choices	52.6%
Looking back across all your years in school, how many times were you taught about petrified wood?	
	Percentage
Never	39.4%
Once	33.0%
Twice	14.9%
Three or more times	12.8%
Did you have an Earth Science or Geology class in high school?	
	Percentage
yes	25.3%
no	74.7%
Did you have an Earth Science or Geology class in college?	
	Percentage
yes	43.6%
no	56.4%
*Which famous scientist do you most strongly associate with the study of fossils?**	
	Votes
Roy Chapman Andrews	1
Charles Darwin	5
Albert Einstein	1
Robert Hooke	1
Louis Leakey	1
Alfred Wegener	6

Table 3. Background information from the Petrified Wood Survey™ for all teacher groups (N = 97) revealed an interdisciplinary perception of the petrified wood topic, but little previous instruction. *Few teachers could recognize a famous scientist associated with fossil study. Only 16 legitimate votes were registered.

geoscience instruction (TANS1 geosciences, n = 11), teachers with chemistry instruction (TANS2, n = 18), and the remainder of teachers in professional development programs (MSP1, TANS1 chemistry, TANS1 physics, ELE1, n = 68). We made this decision in an attempt to ascertain whether the professional development program in the geosciences (TANS1 geosciences) had a potential effect on teacher performance. Earlier research investigations also revealed that the most problematic topic areas of petrified wood involved chemical processes (Clary & Wandersee, 2007), and we sought to determine if prior chemistry instruction had potential impact on these teachers' scores (TANS2).

Which of these methods do you think scientists are LEAST likely to use to estimate the age of petrified wood samples?			
	TANS1G	TANS2	Other
Analyzing assoc. fossils of other organisms found surrounding the petrified wood	0%	11%	18%
Analyzing sequence of rock layers in which petrified wood was found	18%	17%	10%
Counting tree rings visible in a stem section	82%	72%	72%
If you held a piece of South African black ironwood (today's heaviest wood) in one hand, and an identically sized piece of petrified wood in your other hand, which do you predict would happen?			
	TANS1G	TANS2	Other
Ironwood would feel heavier.	18%	28%	16%
Petrified wood feels heavier.	45%	56%	37%
Both feel the same.	36%	17%	46%
Which of the following do you think is NOT true of petrified wood?			
	TANS1G	TANS2	Other
It's a fossil.	0%	0%	5%
It's a rock.	45%	39%	36%
It mostly contains ancient wood, plus some minerals.	36%	44%	39%
Many species of petrified plants have been found.	18%	17%	20%

Table 4. Comparison among teacher response percentages for questions probing the properties of petrified wood. TANS1G (n = 11) received previous instruction in geosciences, while some participants in TANS2 (n = 18) received previous instruction in chemistry. Other teachers (n =68) are grouped together for comparative purposes.

Inservice teachers scored well on the methods by which scientists estimate the age of petrified wood samples (Table 4). The teachers with previous geoscience instruction (TANS1 geosciences) and the teachers with previous chemistry instruction (TANS2) performed slightly better on the density of fossilized wood. However, the difference is slight, and not significant. Ironically, although the majority of teachers in TANS1 geosciences acknowledged that petrified wood is of greater density than the world's heaviest wood, they failed to recognize that petrified wood is also a rock. The other teachers performed slightly better on this item. (Table 4).

Questions on the Petrified Wood Survey™ that probed geological content knowledge of the geological time scale and construct of uniformitarianism ("the present is the key to the past") were answered correctly by the majority of the inservice teachers (Table 5).

Which one of these three time periods is the one that occurred longest ago on the geologic time scale?			
	TANS1G	TANS2	Other
Cambrian	64%	56%	49%
Devonian	18%	28%	32%
Jurassic	18%	17%	19%

Someone says, "Today's physical processes, such as those mentioned in the previous question, continue to operate on the earth in the same rate and in the same way as they did in the past." Do you think this is true or false?			
	TANS1G	TANS2	Other
TRUE	64%	72%	51%
FALSE	36%	28%	49%

How long do you estimate it took for a tree from the past to become completely petrified?			
	TANS1G	TANS2	Other
Less than 100 yrs.	9%	11%	24%
Hundreds of yrs.	18%	33%	19%
Thousands of yrs.	36%	22%	26%
Millions of yrs.	36%	28%	28%
Billions of yrs.	0%	6%	3%

Table 5. Comparison among teacher response percentages for questions probing the geological concepts of petrified wood. TANS1G (n = 11) received previous instruction in geosciences, while some participants in TANS2 (n = 18) received previous instruction in chemistry. Other teachers (n =68) are grouped together for comparative purposes.

However, teachers with previous instruction in geosciences (TANS1 geosciences) or chemistry (TANS2) performed slightly better. More problematic was an understanding of the time required for fossilization. The TANS1 geosciences teachers scored slightly higher on this question, but only 36% realized that *millions* of years are typically involved in the fossilization process. However, many teachers in all professional development groups thought that thousands of years, or only hundreds of years, were required. A larger population of inservice teachers outside the geosciences group thought that little time (hundreds of years, or even less than a hundred years) was the required time for the complete fossilization of wood (Table 5).

Perhaps because petrified wood is abundant in the state of Mississippi, the majority of teachers thought that their home state was the site of the largest public display of petrified forests in the US (Table 6). Many inservice teachers also failed to recognize that petrified wood has been found on every continent, even Antarctica.

Teachers who had previous geosciences instruction (TANS1 geosciences) or chemistry instruction (TANS2) performed slightly better in recognizing the conditions necessary for fossilization, and acknowledged that oxygen was detrimental to preservation (Table 7).

To which state would you travel if you wanted to visit the largest public display of petrified forests?			
	TANS1G	TANS2	Other
Arizona	27%	6%	18%
Colorado	0%	6%	12%
Mississippi	73%	56%	51%
New Mexico	0%	6%	3%
Wyoming	0%	28%	16%

Petrified wood has been found			
	TANS1G	TANS2	Other
in a few US states	9%	0%	12%
in a few countries	0%	6%	6%
on a few continents	9%	11%	9%
every continent except Antarctica	55%	33%	37%
on every continent	27%	50%	37%

Table 6. Comparison among teacher response percentages for questions probing petrified woods' geographic locations. TANS1G (n = 11) received previous instruction in geosciences, while some participants in TANS2 (n = 18) received previous instruction in chemistry. Other teachers (n =68) are grouped together for comparative purposes.

However, the TANS2 group appeared to struggle with the role that dissolution (and permineralization) plays in the fossilization of wood.

Similar to our earlier investigations with students with petrified wood, the most difficult PWS questions for inservice teachers were those that probed the chemical composition of

Which of the following do you think is NOT required in order to form petrified wood?			
	TANS1G	TANS2	Other
Dead trees	9%	11%	13%
Rapid burial	36%	39%	53%
O2-rich environment	55%	50%	31%
Time	0%	0%	3%

Which of these natural processes do you think is NOT important to the formation of petrified wood?			
	TANS1G	TANS2	Other
Dissolving	36%	56%	28%
Permeating w/mineral-rich sol'n	9%	0%	9%
Replacing with minerals	0%	6%	12%
Decaying	55%	39%	51%

Table 7. Comparison among teacher response percentages for questions probing the fossilization processes. TANS1G (n = 11) received previous instruction in geosciences, while some participants in TANS2 (n = 18) received previous instruction in chemistry. Other teachers (n =68) are grouped together for comparative purposes.

the fossilized wood (Table 8). Teachers wanted to incorrectly attribute the colors of petrified wood to the original wood color and climate, in addition to the minerals associated with groundwater. Teachers also did not recognize the role of silica (SiO_2) in most petrified wood samples.

What gives petrified wood its colors?			
	TANS1G	TANS2	Other
Climate during fossilization	0%	0%	1%
Minerals assoc. with groundwater	27%	17%	19%
Natural color of original wood	9%	6%	3%
All of the above	64%	78%	76%

With a typical petrified wood sample, what type of mineral has replaced the original wood?			
	TANS1G	TANS2	Other
Carbonate, CaCO3	45%	56%	49%
Phosphate, Ca2(PO4)3	9%	22%	28%
Pyrite, FeS2	9%	11%	14%
Quartz, SiO2	36%	11%	9%

Table 8. Comparison among teacher response percentages for questions probing the chemical composition of petrified wood. TANS1G (n = 11) received previous instruction in geosciences, while some participants in TANS2 (n = 18) received previous instruction in chemistry. Other teachers (n =68) are grouped together for comparative purposes.

5.2 Petrified wood comparisons

Our original group of inservice teachers (MSP1) responded well to the petrified wood and modern wood comparisons. In whole group discussion, teachers identified similarities between samples, as well as contrasting features. Most teachers expressed surprise at our earlier research results, which revealed persistent student alternative conceptions about the age difference between the two samples, and confusion among some students as to which sample was fossilized (Clary & Wandersee, 2007).

With the TANS inservice teachers (TANS1 geosciences, TANS2), we utilized the data collecting hand-out (Fig. 8) to probe teacher perceptions more deeply. We assimilated these two data sets (n = 39), and analyzed the comments via Neundorf's (2002) content analysis guidelines. Persistent themes emerged from teachers' written observations, including 1) samples exhibited similarities with tree rings and preservation of "fibrous" structure; 2) differences included samples' density, composition, and texture; and 3) compositional differences resulted from rock (crystals) and cellulose. Some teachers observed that both specimens were once living plants, but the petrified wood sample had been subjected to additional fossilization processes.

The teachers' comments also revealed several alternative conceptions about fossilized wood. Some teachers did not commit to time differences, reverting instead of "lots of time" and "petrified wood is older," but some teachers (n = 9) persisted in thinking that fossilized wood was only hundreds or thousands of years older than the modern wood sample. Some teachers also exhibited alternative conceptions about the composition, suggesting that fossilized wood was "marble" or that both specimens were made of "carbon."

5.3 Fossilization replication

The first inservice teacher group (MSP1) came to the consensus that the petrified paper activity provided a good simulation of petrified wood formation, but the required drying time would force its use as an exploratory activity before an appropriate science unit was implemented. Several teacher groups suggested incorporating this as a group activity to reduce the space requirement. Another group suggested that measuring activities and comparisons could be incorporated to extend the lesson.

For the TANS inservice teachers (TANS1 geosciences and TANS2), we utilized the data collecting hand-out (Fig. 10) to probe teacher perceptions of this classroom activity. The data sets were combined (n = 39), and teachers' reflections were assimilated and analyzed with Neuendorf's (2002) content analysis guidelines.

Teachers thought that the petrified paper accurately depicted fossilization in that minerals filled in the porous areas of the paper towel similar to a permineralization process; the structure ended up hardened; and the resulting petrified paper did not burn. Teachers noted that the investigation inaccurately portrayed fossilization in that the petrified paper was "not as hard" as petrified wood samples; the minerals did not replace original structures of the paper towel; atmospheric oxygen was present; and there was no burial or pressure. Teachers also pointed out that this process took an extremely short amount of time when compared to natural fossilization processes, and the chemicals involved ($MgSO_4$) were not analogous to the silica of which most petrified wood is composed.

Several classroom advantages were observed, however, including the ease of the activity, the availability of materials, and the ability to modify this process to include different salts, or to extend the activity to include stalactite and stalagmite formation. Several teachers noted that this activity "covers many areas." The primary disadvantages to the petrified paper investigation were the long amount of classroom time needed for the petrified "logs" to dry (which, conversely, might confuse students as to the brevity of the fossilization process), and the use of lighters or matches in the classroom.

The majority of teachers listed scientific investigation, data collection and analysis, and inquiry as topics that were directly addressed by this activity. Many teachers suggested an investigation of other minerals and substrates for classroom-produced "petrified" products. In addition, some teachers noted that this activity could be extended into classroom study of chemical versus physical changes, solutions and solubility, mixtures, elements and compounds, and the Periodic Table. Some of the more unique suggestions included scaffolding into a discussion of the human skeleton and the mineral composition of bones, the progression of life forms (once-living organisms that died and were fossilized), and whether the soaked logs collected by the adventurers in the "Ax Men" show on the History Channel represented petrified samples. Teachers noted that this activity could address required standards in Scientific Inquiry, Earth and Space Science, Physical Science, and Life Science as outlined by the 2010 Mississippi Science Framework (Mississippi Department of Education, 2010).

Finally, teachers' comments were positive as to how their students would receive this activity. One teacher wrote, "Most students will enjoy literally getting their hands wet at the beginning. But to then watch the process over several days and not really 'see' the change occurring from 'wood' to 'rock' until they attempt to light the log. I think there will be some amazed students and some disbelieving students." Other teachers remarked that their students will "be curious as how to change the variables" and that any hands-on investigations were well-received by students.

6. Conclusions and implications

Our previous research confirmed that petrified wood could serve as an effective portal for interdisciplinary study in college science classrooms (Clary & Wandersee, 2007), and re-affirmed the use of the Petrified Wood Survey™ as an instrument to ascertain student understanding of fossilization processes, geologic time, and fossil wood properties (Clary et al., 2009). This latest research extends our previous petrified wood research in that it focuses upon inservice teachers who will, in all likelihood, provide K-12 students with an introduction to petrified wood and the associated scientific constructs that it addresses. Our current investigation probed the geobiological content knowledge of inservice teachers who participated in professional development programs, and sought teacher reflections on the effectiveness and usefulness of petrified wood investigations in the K-12 classroom.

The Petrified Wood Survey™ results revealed that inservice teachers who participated in this investigation ($N = 97$) performed better than the college students in our previous research studies. This is a positive finding, although we caution that our population is not random, and does not necessarily represent an average teacher performance in the state. Teachers in our investigation elected to participate in science professional development programs, and their geobiological content knowledge may be elevated as a result of this participation.

We found it interesting that, similar to our earlier researched student populations, the most difficult concepts for these inservice teachers were geochemical ones. Few teachers could identify the mineral that replaced the original wood during the fossilization process, and the process that was responsible for the various colors exhibited by petrified wood samples. Few teachers could identify the largest public display of petrified wood in the US, reverting to the largest public display of petrified wood east of the Mississippi River (their own state). Similar to the students we had formerly surveyed, teachers also had difficulties with the role of oxygen (or lack thereof) in the fossilization process. The concepts with which the teachers were most proficient were geologic time, methods by which scientists date fossils, geographic occurrence of petrified wood specimens, and the lack of decay in the fossilization process. Although these teachers performed significantly better than non-science introductory students and junior level design students in their knowledge of fossilization processes, evolution, and geologic time, the majority (56.4%) had never taken a college Earth Science or geology course, and only 25.3% had an Earth Science or geology course in high school! It is also important to note that although the teachers performed significantly better than the college students on the Petrified Wood Survey™, their collective averages—in all professional development groups—were still below the 50% content knowledge level as measured by the PWS. We propose that these findings support the need for science teacher professional development programs in the geosciences.

When we separated scores on the PWS from those teachers with a year of geoscience instruction (TANS1 geosciences, n = 11) from other teacher groups, we observed that the scores for these teachers averaged better than their colleagues. However, we caution that this sample size is small, and does not lend itself to a conclusion as to the effectiveness of the geosciences professional development instruction. Our further analysis of independent items on the PWS did not reveal a remarkable performance of these geosciences teachers on any particular topic. Likewise, an analysis of TANS2 teachers' PWS scores did not reveal superior performance on questions probing the chemical properties of petrified wood, although approximately half of these teachers received professional development in chemistry the previous year. We caution again that our sample size is small (n = 18), and we cannot make conclusions based on these data. However, we think that these data suggest that the chemical concepts involved with petrified wood formation may be best presented in a specialized context of fossilized wood study; if these topics were covered in the chemistry professional development instruction, there does not appear to be transfer of the content knowledge to petrified wood. More research is needed to investigate this possibility.

The comparison activity in which teachers investigated hand samples of petrified wood and modern wood appears to be an effective classroom technique. After our first professional development investigation (MSP1), we suspected that teachers understood the differences between the samples and the processes that had produced them. The use of the data-collecting hand-out in subsequent investigations (TANS1 geosciences and TANS2) was an effective device for recording teacher observations, as well as exposing teacher alternative conceptions. Through the data collected, we learned that some inservice teachers retain alternative conceptions about the length of the fossilization process (hundreds or thousands of years as opposed to millions of years), and the composition of petrified wood (marble, or carbon-based compounds).

Teachers acknowledged that the petrified paper investigation could be an effective tool for demonstrating petrification in the classroom. However, teachers pointed out potential problems with the activity, including classroom time and the use of a lighter and/or matches. These issues can be solved through classroom planning (e.g., initiating the fossil investigation with the petrified paper activity, and returning to the samples at the end of the unit), and/or teacher demonstration (e.g., teachers demonstrate whether or not the petrified paper will burn). Our inservice teachers wisely noted that this activity may foster some alternative conceptions in their students, including the short amount of time for the "fossilization" process, the presence of oxygen, and the lack of burial and pressure. However, the benefits of this activity, and teachers' perceptions that it will be well-received by students, seem to support its classroom incorporation. We suggest that teachers should thoroughly discuss the activity inaccuracies within the classroom, and reaffirm the time and physical factors associated with actual fossilization processes.

Thus far, our research results indicate that petrified wood can serve as an effective geobiological portal to address several important scientific constructs, and that the Petrified Wood Survey™ offers a convenient instrument to identify student and teacher alternative conceptions, and target them with classroom instruction. More research is needed, however. We are currently refining our petrified wood classroom activities so that they can be implemented in K-12 classrooms to effectively address geologic time, fossilization, and evolution, without conveying scientific inaccuracies to students.

7. Appendix A. Petrified Wood Survey™ (©Clary & Wandersee, 2005)

*Where appropriate, correct answers are marked with an asterisk**

1. Looking back across all your years in school, how many times were you taught about petrified wood and petrified forests?
   a. Never
   b. Once
   c. Twice
   d. Three or more times
2. Within which school science subject does the study of petrified wood and petrified forests seem to fit best?
   a. Biology
   b. Chemistry
   c. Earth Science/Geology
   d. Choices A & C
   e. All of the above
3. To which US state would you travel if you wanted to visit the largest public display of petrified forests?
   a. Arizona*
   b. Colorado
   c. Mississippi
   d. New Mexico
   e. Wyoming
4. Petrified wood has been found
   a. In a few US states
   b. In a few countries
   c. On a few continents
   d. On every continent except Antarctica
   e. On every continent*

5. If you held a piece of South African black ironwood (today's heaviest wood) in one hand, and an identically sized piece of petrified wood in your other hand, which do you predict would happen?
   a. The ironwood would feel heavier.
   b. The petrified wood would feel heavier.*
   c. They would seem to be about the same weight.
6. Which of the following do you think is NOT required in order to form petrified wood?
   a. Dead trees
   b. Rapid burial
   c. O_2-rich environment*
   d. Time
7. How long do you estimate it took for a tree from the past to become completely petrified?
   a. Less than 100 years
   b. Hundreds of years
   c. Thousands of years
   d. Millions of years*
   e. Billions of years
8. Which of these methods do you think scientists are LEAST likely to use to estimate the age of petrified wood samples?
   a. Analyzing associated fossils of other organisms found surrounding the petrified wood.
   b. Analyzing the sequence or rock layers in which the petrified wood was found.
   c. Counting tree rings visible in a stem section *
9. Which of these natural processes do you think is not important to the formation of petrified wood?
   a. Dissolving
   b. Permeating with mineral-rich solution
   c. Replacing with minerals
   d. Decaying*
10. Someone says, "Today's physical processes, such as those mentioned in the previous question, continue to operate on earth at the same rate and in the same way as they did in the past." Do you think this is true or false?
    a. True*
    b. False
11. With a typical petrified wood sample, what type of mineral has replaced the original wood?
    a. Carbonate, $CaCO_3$
    b. Phosphate, $Ca_2(PO_4)_3$
    c. Pyrite, FeS_2
    d. Silica/Quartz SiO_2*
12. What gives petrified wood its colors?
    a. Climate during fossilization
    b. Minerals associated with groundwater *
    c. Natural color of original wood
    d. All of the above

13. Which of the following do you think is NOT true of petrified wood?
    a. It's a fossil.
    b. It's a rock.
    c. It mostly contains ancient wood, plus some minerals.*
    d. Many species of petrified plants have been found.
14. Which one of these three time periods is the one that occurred longest ago on the geologic time scale?
    a. Cambrian*
    b. Devonian
    c. Jurassic
15. Did you have an Earth Science or Geology class in high school (grades 9-12)?
    a. Yes
    b. No
16. Did you have an Earth Science or Geology class in college?
    a. Yes
    b. No
17. Which famous scientist do you most strongly associate with the study of fossils?

    ____

8. References

Ausubel, D. P. 1963. *The psychology of meaningful verbal learning*. New York: Grune and Stratton.

Ausubel, D. P. 1968. *Educational psychology: A cognitive view*. New York: Holt, Rinehart and Winston.

Ausubel, D.P., Novak. J.D., & Hanesian, H. 1978. *Educational psychology: A cognitive view (2nd ed.)*. New York: Holt, Rinehart, and Winston.

Bybee, R. W., & Pratt, H. A. 1996. National standards: Challenges for earth science education. *Geotimes, 41*, 16-19.

Carlyle, T. 1832. Boswell's life of Johnson. *Fraser's Magazine, 5*, 379-413.

Clary, R.M., Brzuszek, R.F., & Wandersee, J.H. 2009. Students' geocognition of deep time, conceptualized in an informal educational setting. *Journal of Geoscience Education, 57*(4), 275-285.

Clary, R.M., & Wandersee, J.H. 2007. A mixed methods analysis of the effects of an integrative geobiological study of petrified wood in introductory college geology classrooms. *Journal of Research in Science Teaching, 44*(8),1011-1035.

Clary, R.M., & Wandersee, J.H. 2009. All are worthy to know the Earth: Henry De la Beche and the origin of geological literacy. *Science and Education, 18*(10), 1359-1376.

Clary, R.M., Wandersee, J.H., & Carpinelli, A. 2008. The great dinosaur feud: Science against all odds. *Science Scope 32* (2), 34-40.

Dodick, J., and Orion, N. 2003. Measuring student understanding of geological time. *Science Education, 87*, 708-731.

Duschl, R.C. 1990. *Restructuring science education: The importance of theories and their development.* New York: Teachers College Press.

Duschl, R. A. 1994. Research on the history and philosophy of science. In D.L. Gabel (Ed.), *Handbook of research on science teaching and learning* (pp. 443-465). New York: Macmillan.

Earth Science Literacy Initiative. 2010. Earth science literacy principles: The big ideas and supporting concepts of Earth science. Retrieved May 2, 2011 from http://www.earthscienceliteracy.org/es_literacy_6may10_.pdf

Gould, S.J. 1987. *Time's Arrow, Time's Cycle.* Cambridge, MA: Harvard University Press.

Jenkins, E. 1989. Why the history of science? In M. Shortland & A. Warwick (Eds.) *Teaching the history of science* (pp. 19-29). Oxford: Basil Blackwell.

Matthews, M.R. 1994. *Science teaching: The role of history and philosophy of science.* New York: Routledge.

McPhee, J. 1981. *Basin and Range.* New York: Farrar, Straus, and Giroux.

Mintzes, J.J., Wandersee, J.H., & Novak, J.D. (Eds.) 1998. *Teaching science for understanding: A human constructivist view.* San Diego, CA: Academic Press.

Mintzes, J.J., Wandersee, J.H., & Novak, J.D. (Eds.) 2000. *Assessing science understanding: A human constructivist view.* San Diego, CA: Academic Press.

Mississippi Department of Education. 2010. 2010 Mississippi Science Framework. Retrieved June 5, 2009 from http://www.mde.k12.ms.us/acad/id/curriculum/Science/Webpage%20links%207%2031%2008.htm

National Research Council (US). 1995. *National Science Education Standards.* Washington DC: National Academy Press.

Neuendorf, K. 2002. *The content analysis guidebook.* Thousand Oaks, CA: Sage.

Novak, J.D. 1963. What should we teach in biology? *NABT News and Views, 7*(2). Reprinted in *Journal of Research in Science Teaching, (1),* 241-243.

Novak, J.D. 1977. *A theory of education.* Ithaca, NY: Cornell University Press.

Novak, J.D., & Musonda, D. 1991. A twelve-year longitudinal study of science concept learning. *American Educational Research Journal, 28,* 117-153.

Orr, D. 1994. *Earth in mind: on education, environment, and the human prospect.* Washington, DC: National Academy Press.

Rudwick, M.J.S. 1992. *Scenes from deep time.* Chicago, IL: University of Chicago Press.

Schwab, J. 1962. The teaching of science as inquiry. In J. Schwab & P. Brandwein (Eds.), *The teaching of science* (pp. 1-104). Cambridge, MA: Harvard University Press.

Soreghan, G.S. 2005. Lessons from earth's deep time. *The Chronicle of Higher Education, 51,* B10.

Suzuki, D. (January 2, 2004). Media could help build better connections. *ScienceMatters.* Retrieved January 5, 2005 from http://www.davidsuzuki.org/about_us/dr_david_suzuki/article_archives/weekly02020401.asp

TOPS Learning Systems. n.d. Petrified paper. Retrieved January 30, 2011 from http://topscience.org/activities_print/FreeDownload23.pdf

Wandersee, J.H., Mintzes, J.J., & Novak, J.D. 1994. Research on alternative conceptions in science. In D. Gabel (Ed.), *Handbook of research on science teaching and learning* (Chapter 5, pp. 177-210). New York: Macmillan.

4

Debris Flow Phenomena: A Short Overview?

Chiara Calligaris and Luca Zini
University of Trieste, Geosciences Department
Italy

1. Introduction

Debris flows are one of the most dangerous and destructive processes affecting the second order streams in the mountain areas (Cavalli et al., 2005; Boniello et al., 2010; Santi P.M., 2008). This very common phenomenon in the Alpine environment is a type of landslide defined by several authors (Varnes, 1978; Hutchinson, 1988; Pierson, 2005; Pierson and Costa, 1987; Coussot and Meunier, 1996; Hungr et al., 2001) trough focusing on the involved material, on the water saturation and on the mass velocity. Debris flows usually consist of a complex mixture of fine (clay, silt and sand) and coarse (gravel, cobbles and boulders) materials with a variable water quantity (Nettleton et al., 2005). The outcoming mixture has a behaviour similar to a viscous "slurry" with a high density, 60% to 80% by weight solids (Varnes, 1978; Hutchinson, 1988; Pierson, 2005). The same Hutchinson (1988) is describing them as "wet concrete".

These phenomena are rapid mass movements, gravity induced able to transport large quantities of sediments and wood downslope, producing complex distribution of deposits and eroding surfaces along their flowpath (Remaitre et al., 2003).

Several other classifications try to define these processes. For example, Aulitzky in 1982 provided a classification focused on the typologies of the materials involved making a macroscopic distinction between the rocks and the engineering soils.

Pierson and Costa, in 1987, proposed their classification basing it on the sediment concentration and on the average flow velocity.

Paoluzzi, Coussot and Meunier, in 1996, described debris flow as a function of sediment concentration and material typology, between hyperconcentrated flows and landslides. Celerity, deposit nature and flow type are the parameters considered. Two of them are appropriate for a practical classification: solid fraction and material type (Paoluzzi et al., 1996).

Hungr in 2001 (Hungr et al., 2001) elaborated a classification having as main distinctive parameters the water content, the velocity and the material typology.

Seen that existing classifications for landslides were based on process, morphology, geometry, movement type and rate, type of material and activity, in 2005, Jakob (Jakob, 2005) proposed a different categorization based on a size classification. This classification is rarely used because it provides too little information on morphology or process characteristics of a landslide. It has been prevailing studied for regional studies along infrastructures corridors because it addresses variables that are part of a hazard evaluation.

Anyway, in the present work, a simple criterion of identification is proposed. Debris flows must be seen as intermediate phenomena between hyper concentrated flows (intense bed load transport) and landslides separated from them by sharp transitions of some characteristics (celerity, deposit nature and flow type). Two parameters, solid fraction and material type, thought to be appropriate for a sound and practical classification, are brought out, and the corresponding complete classification of flow and mass movements in mountain areas is presented. Two extreme debris flow types are thus distinguished: muddy debris flows and granular debris flows.

Regardless of classification, all are agreed that debris flow phenomena, throughout the world, cause considerable damages, but nowadays researchers are trying to better understand their behaviour in order to prevent them, to identify the warning signs and to build alert systems that allow to save many lives and properties. Even if they remain poorly understood, a basic knowledge is available concerning their recognition and propagation.

The knowledge of the possible inundation areas, the thickness of the deposits and the velocity expressed during the event are really useful to define, but especially to delineate the vulnerable areas in order to identify the structural and non-structural mitigation measures that have to be realized to protect the existing infrastructures (Boniello et al, 2010).

The volume and the composition of the mixture of a debris flow are the main factors that contribute to determine the hazards associated with such phenomena, since they govern the mobility and impact energy of the debris (Iverson, 1997; Jakob, 2005). In this regard, an adequate work must be carried out in the field of non-Newtonian fluid mechanics. In particular, one fundamental rheological property of debris flow materials is the yield stress, which explains thick deposits on steep slopes and can be inferred from field measurements. Furthermore it can be used to estimate viscous dissipation within the bulk during the flow. Relevant models predicting muddy/debris flow dynamics are already available whereas further progress is needed concerning granular flows. During the last years, several simulation models and approaches have been implemented (Cesco Bolla, 2008; O'Brien, 1998; Pirulli, 2005; Avolio et al.., 2011; Rickenmann, 1999) and created to reconstruct the path of a debris-flow phenomena, but a believable scenario can be obtained only by resorting to real parameters that are suitable to characterise the involved material (Sosio et al., 2006). Thus, it is necessary to calibrate those available computational codes through back-analysis simulations and laboratory analysis (Tecca et al., 2006).

In this chapter a fast overview will take the reader into the debris-flow world giving some fixed points on these particular events, how they trigger, which are the boundary conditions, how they develop along a slope. Than, an Italian severe damaged area will be described and used as test site for presenting the obtained results that could contribute to the knowledge of these dangerous phenomena.

2. How the debris flows occur? Predisposing and triggering factors

On steep slopes, in mountainous areas, could occur assorted types of flow or mass movement involving water and sediments. Among these events, debris flows are peculiar phenomena during which a large volume of a highly concentrated viscous water-debris mixture flows through a stream channel or on an open plain. For the occurrence of these types of landslide predisposing and triggering factors need to be present in the rough area.

Debris flows are geomorphological easy to recognize on the field. Mainly they are formed by a source area, a stream transport channel and a depositional area having a fan morphology (Figure 1).

Fig. 1. Identification of the three main parts of a debris flow phenomena: 1) source area, in red; 2) transport channel, in green; 3) depositional area, in blue.

The source area of a debris flow must have the following conditions to be defined a source area: 1) a very steep slope (>15°); 2) an abundant supply of loose debris (Bovis and Jakob, 1999); 3) a source of abundant moisture; 4) spare vegetation.

Among all the predisposing factors, the morphometric parameters play a very important role. They define the geometry of the catchments and their characteristics. Among all the possible parameters, the most important are: the area and the perimeter of the catchment, the average lenght, the maximum, minimum and average elevation, the average slope angle, the shape factor (intended as function of lenght and surface F=0.89L/S) the circularity rate and the Melton number (I_M).

But not only them have a huge impact on debris flow. The most important predisposing factor can be considered the debris avaliability (Bovis and Jakob, 1999). Concerning this sentence, the catchments can be divided in two main categories: one that have limited debris avaliability and the other that have an illimitated avaliability. In this case it is easy to understand when a debris flow can occurr along a torrent adding only the precipitations as triggering factor.

In this sense, a typical characteristic of the debris flow is their close connection with high intensity meteorological events. On one hand, it is possible to say that deep landslides are usually associated with structural causes (morphology, shear strength, etc. …) and triggered by long term weather events, able to saturate deep layers. In the case of debris flow, the slope equilibrium conditions are governed by effective stresses reduction, due to pore water pressure having a hydrostatic distribution. For these reasons, debris flows, but also the soil slips, are typically triggered by high intensity meteorological events occurred in a short time period, that can uplift the water table reaching a critical level (Skempton and Lory, 1957) or,

conversely, when the rainfall intensity exceeds the infiltration rate creating a saturated layer from the surface (Green and Ampt, 1911; Fredlund et al., 1978). Anyway, in both cases, infiltration phenomena create an additional system of forces increasing the destabilization.
For all these reasons, a study on the triggering factors of a debris flow should start from a multidisciplinary approach founded on hydrological, meteorological and geotechnical basis.
Debris flow can be triggered also from shallow landslides originating on steep slopes, from landslides in topographic swales or hollows, from the entrainment of materials within stream channels, from diffuse erosion, from rock glacier bursts (Mariis, 2006). Landslides that mobilize into debris flows often occur along topographic concavities, which concentrate groundwater flow and contain thicker accumulations of fine materials than surrounding ridges. Concentrated groundwater flow increases the wetness of clay and fine materials in hollows, making it particularly susceptible to destabilizing groundwater pressure increases during and immediately after rainstorms. Debris stops flowing when the internal kinetic energy drops below the level necessary to maintain the fluid to flow, commonly because slope of the channel through which the debris flows flattens or widens.
Debris flows can be triggered by many other different factors. Among the ones previously described, the addition of moisture can be considered the main one: without water, the debris has no possibilities to occur.
Another triggering factor can be considered the erosion of the material along the banks of the streams. This erosion can cut into thick deposits of saturated materials stacked high up the valley walls removing support from the base of the slope triggering a sudden flow of debris.
Talking about the possible triggering factors, wildfires can be considered one of them, not as the main factor, but as a help in creating boundary conditions. Some debris flows occur after wildfires have burned the vegetation from a steep slope or after logging operations have removed vegetation. Land use is one of the most important surroundings that needed to be taken into account when studying a landslide. The loss of support induced by the removed water from soil and the burning of the roots create the condition for a debris flow to occur: in this case, also a moderate amount of rain on a burn scar can trigger a large event.
Volcanic eruptions and earthquake have also to be considered as triggering factors in debris flow occurring.
Going back to rainfall heavy conditions, the scientific community is trying to define hydrological models on statistic base finalized to identify the critical amount of rain and the tresholds over which the triggering risk can be considered very high.
These tresholds are given by the following empirical equation:

$$I = a\,D^{-b} \tag{1}$$

where I is the rainfall intensity (mm/h) and D is the duration of a rainfall (hours). a and b are empirical coefficients (Bruschi, 2008). For the Friuli Venezia Giulia Region, the only values of a and b have been obtained by Paronuzzi et al.(1998) but they not take into account the recent alluvial events.
Once defined when and under which kind of rainfall conditions a phenomena can be triggered, it is important to quantify the magnitude (M) in order to extimate the flooded areas and to recognize the different hazard conditions.
It is possible to obtain data on magnitude trough empirical methods: among all of these, there are some really simple that correlate magnitude [m^3]with the catchment area S [km^2]

and the average fan slope i, expressed in percentage. Some of the most common methods are the following: Ceriani et al. (2000), Bianco & Franzi (2000), Hampel (1977), D'Agostino et al. (1996) and Marchi & Tecca (1996).
In addition to magnitude, a value that have a real important meanning, is the runoff determination. Its definition permits to identify the extension of the potentially hazardous flooded areas and could be extimated trought the relation proposed by Rickenmann (1999) based on the observation and analysis of 150 swiss catchments. The formula is a product between the magnitude (M) and the difference in elevation between the starting and end point of the triggered debris flow (H)

$$L_{tot} = 1.9\, M^{0.16}\, H^{0.83} \tag{2}$$

All the parameters previously described permit to widely characterize a debris flow, but one thing is still missing. An evaluation on the grain size distribution and the definition of its vertical depositional shape.
Debris flows deposits are characteristically poorly sorted, commonly contain large fragments resting unsupported in a finer-grained matrix, may be internally structureless and may contain elongate fragment strongly aligned approximately parallel to flow surfaces, that are indicative of laminar flow. They ara sometimes characterized by an inverse grading (Fischer, 1971).
Inverse grading can occur in two different type of deposits: distribution inverse grading or coarse-tail grading. The distribution inverse grading shows a steady increase of the grains' dimensions from the base to the top of the deposit and characterizes poor matrix deposits. This kind of flows move through the high rate of grain collisions; in this conditions the coarser clasts are pushed upward by dispersive pressure and/or the finer grains are pushed downward by kinetic sieving (Figure 2).

Fig. 2. Debris flow deposit in Gilgit region (north east Pakistan). The inverse gradation is present at the top of the debris.

The coarse-tail inverse grading shows a quite progressive increase of the size of the clasts in the basal layer, while the top contains the largest grains together with a chaotic mixture of sediment. The differential reduction of the matrix strength, caused by shear strain, produces selective setting of coarser clasts from the flow (Postma and Nemec, 1991). In the Friuli Venezia Giulia Region it is very difficult to find a depositional fan with a clear inverse grading. The reason is due to the short flow path and the presence, along the transport and depositional areas of a lot of obstacles as trees, houses or infrastructures. Figure 2 is showing a debris fan in the northern part of Pakistan, close to Gilgit. The dimension of the fan and the flow path permit to the debris mixture to become mature and to make the floating boulders to reach the top of the fan and the frontal area.

3. A case study: More than 300 debris flow in Val Canale valley

3.1 Val Canale, environmental settings

Val Canale valley, located in the extreme north eastern part of Italy, during the last century has been repeatedly affected by debris flow phenomena that generated serious economic and social damages. From a geological point of view, in the valley, outcrop continuously, in the hydrographic right of Fella River dolostones belonging to Sciliar and in the left, scists belonging to Werfen Formation. Fella River entered along one of the major regional thrust fault: the Fella-Sava line (Figure 3).

Val Canale valley, in 2003, during the occurred alluvial event has been severely affected by debris flow phenomena: the quite narrow valley, the steepness of the slopes and the high tectonic grade, created the conditions not only for the predisposing factors, but also for the triggering ones that permitted the developing of geostatic phenomena.

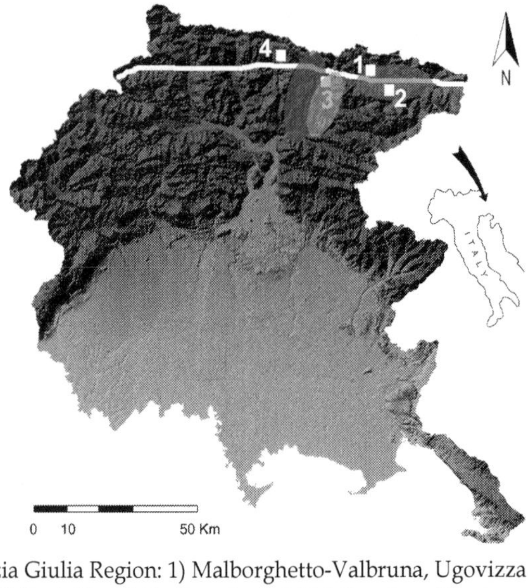

Fig. 3. Friuli Venezia Giulia Region: 1) Malborghetto-Valbruna, Ugovizza and Mount Cucco; 2) Mount Lussari; 3) Pontebba; 4) Paularo. In red: Val Canale valley; in green: Canal del Ferro valley; in blue: Val Aupa valley and Moggio Udinese municipality.

This very intense event, has meant that not only old rock falls or debris were reactivated, but occurred also new hyper concentrated flows that suddenly got into debris flow with a load of debris, mud, boulders and pieces of wood. Tropeano and Turconi (2004) estimated in about 1 million of cubic meters the total amount of debris and sediments mobilized and stored during the event.

The fluvial impact of 29th August produced important modifies on the morphology of the invested area causing severe damages and erosions, creating gullies and expanding the existing riverbeds (Borga et al., 2007). Debris flow invested houses and roads isolating, for days, the villages of Ugovizza, Valbruna, Malborghetto and Pontebba.

For the Val Canale valley, in 2003, was in process of adoption the Hydrogeological Basin Plan (P.A.I. Piano di Assetto Idrogeologico di Bacino) in which were defined areas at risk of debris flow. Its safeguards lines were suspended for the areas affected by the alluvial event of the 29th August due to the commissioner who established it under the occurring of such events. The phenomena occurred during the alluvial event, in some cases, exceeded the perimeters proposed in the Plan. The geostatic phenomena stored thousands of cubic meters also outside the known areas causing severe damages.

In the following years, Civil Defence of Friuli Venezia Giulia Region realized several mitigation measures in the hit areas, for this reason, was discerned the need to upgrade the perimeter areas using tools able to ensure their non-subjectivity. In this respect, are increasing the prospects of software development capable to provide modelling scenarios more and more responsive to realty.

As test sites for the whole area researchers of Geosciences Department studied 12 catchments that have been affected by debris flow phenomena. On every single basin has been realized a back analysis simulation trough commercial software called FLO-2D (O'Brien et al., 1993) this permitted to define physical and rheological parameters that better reproduce the occurred phenomena. For some of the basins different approaches have been used in order to define the runoff and the expansion areas: DF-SIM (for Rio Cucco basin) and Debris software (for Pontebba 01 basin) have been used (Di Gregorio et al., 1994; Segre at al., 1995; Bruschi, 2008).

For Fella sx catchment a rheological specific study has been realized. This permitted to go deeper into the rheology world and to try to better define characteristic values of viscosity and yield stress that heavy influence these so complex phenomena.

3.2 The alluvial event

The north eastern part of Friuli Venezia Giulia Region, especially Val Canale valley, Canal del Ferro and Aupa valleys have been interested, on 29th August 2003, by harsh weather conditions characterized by heavy rainfall since 12 o'clock. Rainfalls firstly affected high mountain areas, between Mount Cucco and Malborghetto-Ugovizza pastures, and then moved downstream with a gradually increasing intensity.

Pontebba's rain gauge, which is part of the network managed by Regional Directorate of Civil Defence, was the only instrument, close to the study area, that worked properly during the alluvial event. Data recorded by Pontebba's rain gauge, indicate the extreme gravity of the occurred phenomenon. Since 1928, when rainfall data recording started, had never occurred events of this entity. In the range between 1928 and 2010, the only comparable event was on 22nd June 1996 when occurred 78.4, 155, 345.6 and 465 mm of rain in 1, 3, 6, 12 and 24 hours respectively (Table 1).

Time (hours)	Height (mm)	
	Pontebba (1996)	Pontebba (2003)
1	78.4	88.6
3	155.0	233.4
6	199.6	343.0
12	345.6	389.6
24	465.0	396.2

Table 1. Heigh and duration time of rainfalls recorded by Pontebba's rain gauge (modified from Norbiato et al., 2007).

What is clear from data recorded on 2003, is that the event has reached remarkable precipitation values especially in the ranges between 3 and 12 hours. Specifically: have been observed maximum values of 50.8 mm in 30 minutes (between 17 and 17.30), of 88.6 mm for an hour (15.30 - 16.30), of 233.4 mm for three hours (14.30 - 17.30) and of 343.0 for six hours (12.0 - 18.0). The total amount of the event, which lasted about 12 hours, was equal to 389.6 mm. If compared with the series of heavy rainfall recorded by Pontebba's rain gauge and processed using Gumbel distribution, precipitations of 29th August 2003 are associated to a return time of over 100 years. Particularly impressive are the values corresponding to 3 and 6 hours. The strong detected intensities are in accordance with the great intensity of the morphodynamics actions induced by this event (Norbiato et al., 2007).

The most part of the landslides has been triggered between 14.00 and 18.00 when, at Pontebba's pluviometric station has been recorded a total rainfall value equal to 293.0 mm. On the northern side of the alignment Pontebba - Ugovizza occurred limited bursts over 400 mm (Borga et al., 2005).

Borga's researchers (2005) realized on signal probabilities rainfall lines, obtained trough linear moments method and GEV model (Generalized Estreme Value) for the north east Italian area, recognized the statistical rarity of the event that generated the 2003 flash flood in Val Canale.

2003 event, with its extraordinary features, is not an isolated one in the climatologic context of the Region: the event magnitude is in fact comparable to the one of other two events occurred in the previous 20 years and happened on 11st September 1983 with the center in Paularo and the second on 22nd June 1996 with the center on Moggio Udinese, Pontebba and Paularo areas.

These observations emphasise that extreme events are really rare if one refers to the specific site, while they occur with not negligible frequency when one considers the entire mountain areas of the Region. In Borga's paper were also estimated the return time of the heights of rain in August 2003 in Pontebba. Return times characterizing the event vary considerably with the duration: for duration between 1 and 24 hours, return time is calculated to be between 50 and 100 years; for 12 hours it is between 200 and 500 years, while for duration between 3 and 6 hours return time has been calculate to be in the range between 500 and 1000 years (Borga et al., 2005; Zanon, 2010).

3.3 Debris flow simulations in the 12 basins

12 catchments tributary of Fella River were chosen to realize debris flow event simulations (Calligaris et al., 2008). Everyone has been analyzed separately, but the methodological approach has been the same for all of them.

Rio Cucco basin has been modelled also with another software called DF-SIM, developed in an original way by O.U.C. Civil Defence and Soil Defence of Udine Province. Torrent Pontebba 01 has been analyzed instead, trough DEBRIS commercial software.

3.3.1 FLO-2D for simulating events in 12 basins

The two-dimensional numerical code FLO-2D is based on volume conservation. This code simulates a debris flow event along a defined topographical surface, using, as input data, an inflow hydrograph, the plastic viscosity of the material and the yield stress, being these a function of the Concentration by volume.

For the basins, the simulation has been realized on a computational domain made by a grid of 5m*5m obtained from the regional cartography CTRN at a scale 1 to 5.000 o, where possible, from laser scanner data.

Inflow hydrographs have been realized by the researchers of Padova University (Dipartimento Territorio e Sistemi Agro Forestali) that developed an hydrologic model spatially distributed (KLEM) setting it on the alluvial event of 29th August. The model uses rainfall data coming from rain gauges stations and from high resolution radar observations (Borga et al., 2007). For the back analysis, seen that rheological data were not available, have been used parameters described in literature and characteristics of the studied lithologies (O'Brien et al., 1988).

For every basin, at least 12 simulations have been realized (Figure 4), varying every time the input data and determining the physical and rheological couples of parameters that better approximate, as flooded area and thickness of deposits, the occurred event (Table 2).

η		τ		References	Studied basin
α_1	β_1	α_2	β_2		
0.036	22.1	0.181	25.7	Aspen Pit 1	Pontebba 2
0.0538	14.5	2.72	10.4	Aspen Pit 2	Rio Pirgler
0.00136	28.4	0.152	18.7	Aspen Natural Soil	Malborghetto Centro, Abitato Cucco
0.128	12	0.0473	21.1	Aspen Mine Fill	Malborghetto est, Studena bassa
0.000495	27.1	0.0383	19.6	Aspen Watershed	Fella sx
0.000201	33.1	0.291	14.3	Aspen Mine Source Area	Rio Cucco, Rio Ruscis
0.00283	23	0.0345	20.1	Glenwood 1	
0.0648	6.2	0.0765	16.9	Glenwood 2	
0.00632	19.9	0.000707	29.8	Glenwood 3	
0.000602	33.1	0.00172	29.5	Glenwood 4	Malborghetto nuovo, Pontebba 1
0.0075	14.39	2.6	17.48	Dai et al. (1980)	
0.0075	14.39	0.152	18.7	Tecca et al. (2006)	

Table 2. Couple of rheological parameters responding to the different hydrogeological context used for the back analysis simulations (from O'Brien et al., 1985). In the last column has been reported the correspondence with the studied basins.

Concerning Concentration by volume (Cv), values were varying in the range between 0.2 and 0.55. For the Manning coefficient has been used a value of 0.1, typical of soils made of debris deposits with no bushes on them. The specific weight of the mixture γ_m and the resistance parameter for laminar flow K, were assumed to be $26.5KN/m^3$ and 2085 respectively, values usually used in literature (Boniello et al., 2010; Calligaris et al., 2009; Tecca et al., 2006).

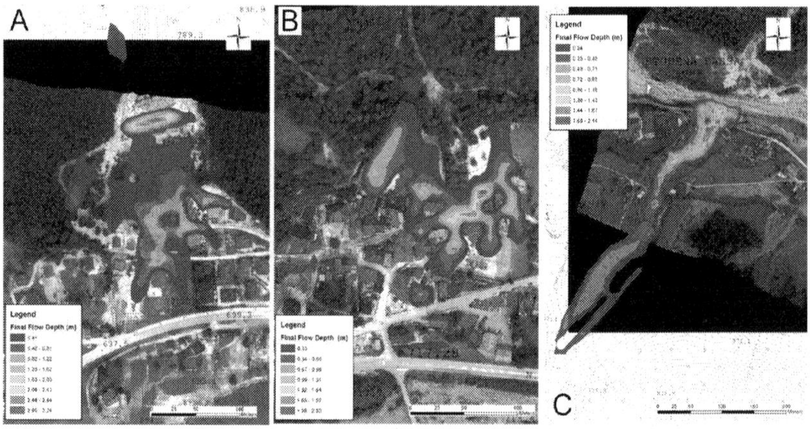

Fig. 4. Three of the twelve analyzed basins, flooded areas and heights of deposit: A) Abitato Cucco basin; B) Malborghetto centro basin; C) Rio Ruscis basin.

3.3.2 Rio Cucco simulation using DF-SIM

The calculation code has originally been developed by U.O.C. Soil Defence and Civil Protection of Udine Province. The program allows reproducing a debris flow phenomena adopting the cellular automata theory (CA) (Segre et al, 1995; Di Gregorio et al, 1994). A model for cellular automata is a complex system represented as simple composed of many parts; each of them, to evolve, has its own internal rule and interacts only with the parties close to it. Each automaton making system can take states and receive input according to a discrete time scale from where it is immersed and react to these challenges with a transition state or a response (output). The system evolution is performed through a transitional function used at each time step. In this way, it is possible to determine the new state of each cell starting from the current status and from cells state making up the neighbourhood of the cell itself.

A pretty flexible tool for debris flow simulation can be obtained through modelling based on Cellular Automata Theory; varying appropriately control parameters, the simulator fits the different rheological and fluid dynamics characteristics. With this software has been possible to simulate the event occurred on Rio Cucco and compare the obtained results with the one obtained from FLO-2D simulation (Figure 5).

Trough DF-SIM is possible to characterize the debris flow behaviour with the following parameters: solidification, dynamic friction angle and internal friction angle (characteristic parameters), critical height and humidity (critical parameters). Not only topographical files but also information about sediment sources areas, their location and debris thickness are

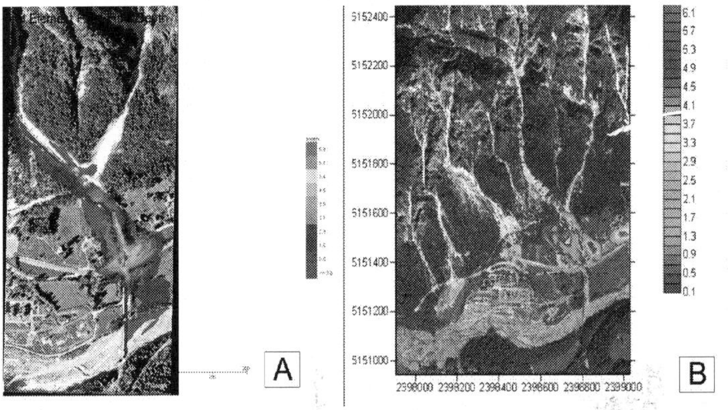

Fig. 5. A) Rio Cucco scenario event realized trough FLO-2D software; B) same scenario realized with DF-SIM software.

requested to run the simulation. For Rio Cucco, a total volume of about 100,000 m^3 was determined on the basis of the debris expansion areas outlined on field just after the event. Like for the volume, parameter values useful for the simulation were empirically defined. Several back-analysis simulations were performed with different parameters in order to obtain a good matching between flooded areas and thickness of deposits observed post-event and the simulated ones.

3.3.3 Pontebba 01 simulation using DEBRIS

Debris is a software that allows performing evaluations and inspections related to debris flow phenomena, analyzing their two-dimensional motion or evaluating the magnitude in the fan area. The software is organized with a vertical structure that makes obligatory to follow a precise sequence of commands to perform the analysis, in order to successfully insert the all data required. The code is composed by two sections: a one-dimensional one regarding the river auction until the outlet on the fan on which are calculated the discharge, the heights of debris at a decided section starting from a pure water hydrograph and a two-dimensional section regarding only the fan. To obtain good results, this last section needs as input data a detailed topography on which calculating the dynamic mixture on the fan. The two modules do not evaluate the triggering possibilities, but simulate the flow when it is already triggered. Output files are concerning the following variables: heights of flow, heights of stored debris, final topographic elevations, flow velocity and direction and a file regarding the phenomena intensity classes (intensity evaluated as pressure and heights of stored debris).
Made all the computes, it is possible to visualize the results, in a graph showing, for every path segment, the flow velocity and heights, indeed the point in which the debris is starting its deposition and the indicative runoff.
With the kinematic wave theory instead, the same parameters are calculated on the base of Arattano and Sauvage (1992) theories but visualized with the same mode as the ones obtained with the Takahaski theory (1991). A comparison between the obtained results is necessary to evaluate heights and velocity of the front of the debris while evolving the phenomena along the transport area. For the velocity others evaluations are available.

For the present work, a comparison has been realized between the results obtained trough FLO-2D software and Debris one (Figure 6). Results in plan are different due to the different approaches used by the two softwares, but heights of debris and flow velocity are comparable.

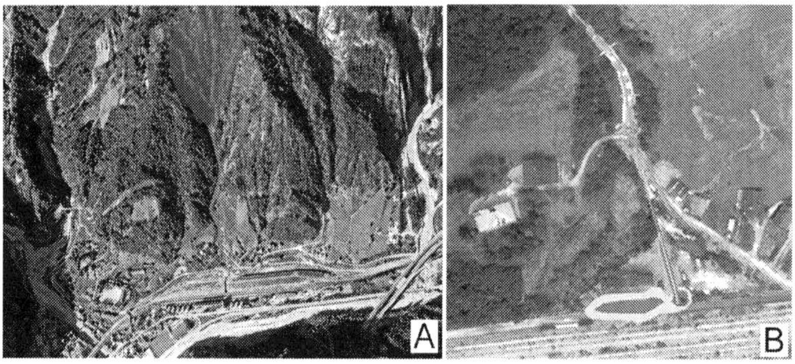

Fig. 6. Pontebba 01 basin: A) Location of the studied area; B) Comparison between obtained results (in the ranges of blue: Flo-2D simulation; in the ranges of red, mapped results with the software Debris).

3.4 Debris flow rheological characterization: an example on Fella sx basin

The debris flows mixture composition and the involved volumes are the main factors that must be defined to determine their hazard: from them, indeed, depend their energy and impact. To find an answer to the question, is fundamental to characterize the material involved in the detritical flow (debris flow or mud flow). According to some authors (Whipple, 1992; Calligaris et al., 2010), it is really difficult to establish a correlation between the rheological parameters identified trough laboratory analysis and the ones obtained trough empirical tests with the help of simulators.
In the examinated area, in Val Canale valley, some tests have been realized to try to better define viscosity and yield stress. An example will be shown on Fella sx basin (Figure 7).

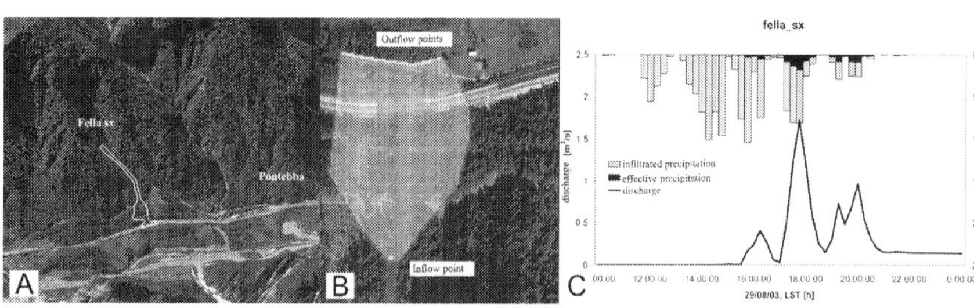

Fig. 7. A) Fella sx location; B) Computational domain; C) Inflow hydrograph.

For the studied sites, the flow can be considered monophasic and viscoplastic, and in a first approximation, governed by the clay that determine its plastic viscosity and the yield stress. Knowing the variation of these parameters with the solid concentration, is possible to proceed to a numeric simulation of the phenomenon trough the use of computer codes as

previously realized in order to compare the obtained values with the ones gained from the best fitting simulation (back analysis). Several limitations are recognizable in this approach due to the too driver schematization for a so complex phenomenon and for so many parameters, but anyway is a new frontier that will permit to the professionals to use and simulate run out with a good approximation being able to predict future scenario events.
Samples collected along the transport area, have been submitted to the grain size analysis, and on the finer fraction than <0.063mm, a rheological analysis has been realized.
Gran size analysis highlighted a 64.8% of gravel, a 14.8% of sand and a 20.4% of silt and clay defining the sample as a gravel with sandy silt. The rheological studies concerned the fine fraction (passing at 63 μm) obtained through the sieving of the collected samples. With this fraction have been prepared suspensions with a different water weight: 33, 36, 40, 44 and 48%.
The tests consisted of a sequence of segments at a constant stress with increasing values in a geometric progression that highlighted that all the examinated systems have a plastic behavior. Figure 8 explains, as example, the system answer at 40%, described in terms of viscosity and deformation.

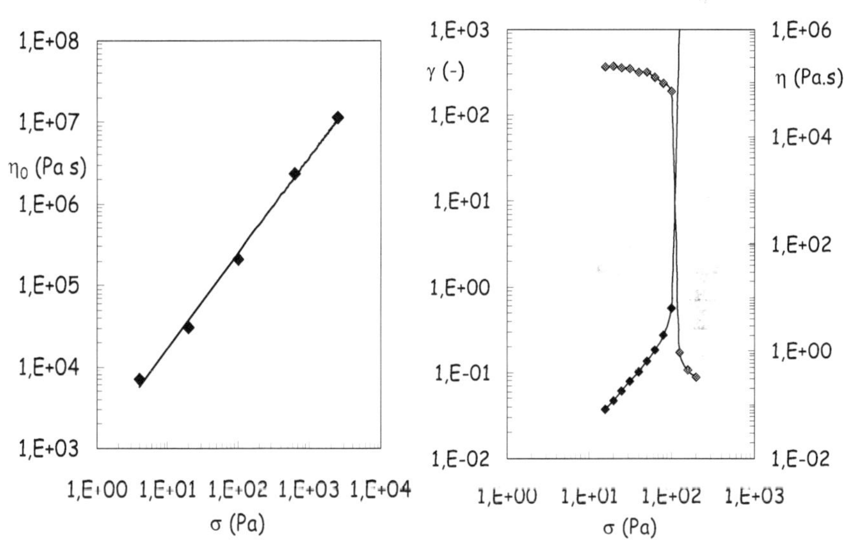

Fig. 8. A) plastic answer to the system at 40% of water content; B) correlation between parameters η_0 and σ_y of the examinated systems.

At low stresses, the answer of the system is viscoelastic: the deformation is firstly linearly increasing with the stress and it is possible to identify a Newtonian plateau of viscosity η_0. Later, in a short interval of stress values, the system goes to a regime of continuous deformation to a significant flow condition and, just after to the sample brake. The apparent yield stress σ_y can be placed in correspondence of the point that define the drop of viscosity, or the sharp increasing of the deformation. For all the systems the transitioning deformation/flow happen in the same range of deformation values (between 0.3 and 0.6), while η_0 and σ_y increase with the concentration of the solid phase (Figure 8B).

To characterize the shear-dependent behavior in the region of medium to high shear rates, is preferable the sequential procedure at controlled rate.
Data obtained can be correlated, with a sufficient approximation, with Bingham model; the correlation is similar to the one obtainable with the other plastic models at three parameters (Casson in the generalized version, Herschel-Bulkley). Table 3 shows Bingham parameter values (η_p,σ_B) and the experimental yield stress σ_y. Even if the values are correlated, their σ_y and σ_B recall the multiplicity of values that can be assigned to the yield stress, since they are dependent from the procedure adopted for their determination.

%		33	36	40	44	48
C_v	(-)	0.421	0.389	0.349	0.313	0.279
σ_B	(Pa)	1254	261	132	25.0	2.03
η_p	(Pa s)	30.5	1.40	0.12	0.10	0.06
σ_y	(Pa)	2500	630	100	20.0	4.0

Table 3. Parameters of Bingham model and experimental yield stress values.

To describe the dependence of η_p and σ_B from the concentration by volume C_v, the most frequently used exponential relations were adopted, although being the same inconsistent when extended at low and high solid content:

$$\eta_p = \alpha \exp(\beta c_v) \quad \sigma_B = \gamma \exp(\delta c_v) \tag{3}$$

3.4.1 FLO-2D computational code to simulate debris flow scenario event

To model the phenomenon on Fella sx has been used the FLO-2D software. The two-dimensional numerical code has been used for the back analysis trough the model of Aspen Watershed defining with the following rheological parameters: α = 4.95*10^{-5}, β = 27.1, γ = 3.8 *10^{-2} and δ = 19.6 that best approximate the event occurred for inundated area and thickness of the stored deposits.
Subsequently, several numerical simulations were carried out using the parameters obtained from correlation of experimental data and postulating different peak of concentration by volume (0.35, 0.42, 0.50). From the tests at controlled rate were obtain the following values: α= 2.49*10^{-13}, β=76.5, γ=3.44*10^{-3}, δ=29.55; from the one conducted at controlled stress dedicated to the measurement of the yield stress, the following results were obtained: γ'=1.67*10^{-5} and δ' =44.75. During all the simulation phases, the value of Manning coefficient was 0.1, the weight of volume γ_m and the resistance flow parameter K have been estimated at 26.5 kN/m^3 and 2085, respectively (O'Brien et al., 1988; Tecca et al., 2006)

3.4.2 Results and discussion

Simulations of the phenomenon were realized at different values of concentration by volume peak and produced 9 different scenarios, three of which are derived from back analysis (Figure 9, scenario A, B and C), the others come out from the two sets of experimental data ($\alpha - \beta - \gamma - \delta$, $\alpha - \beta - \gamma' - \delta'$). The comparison has given the way to verify that at small values of Cv (0.35 - scenarios A, D, G), the behavior of the debris mixture is very similar in all three cases. With the increase of the value of Cv (up to 0.5), we are witnesses of an increase in the flooded area and of a flow divagation that mainly tends to

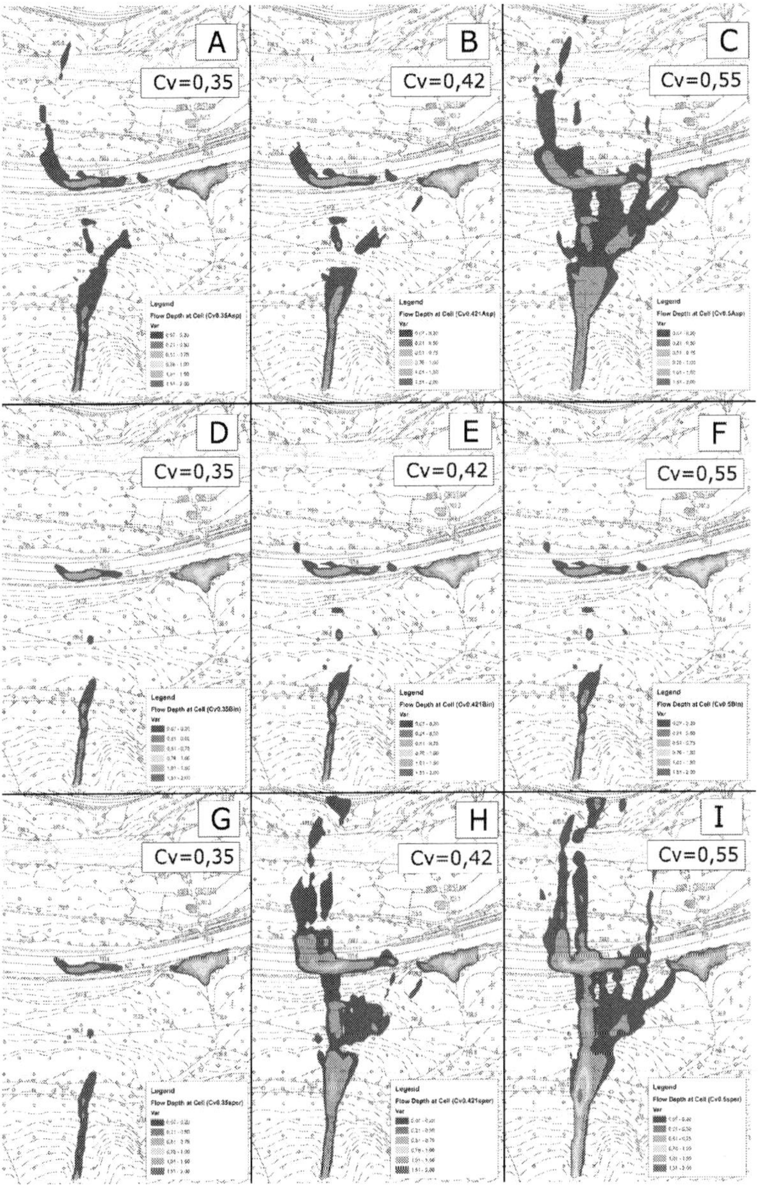

Fig. 9. Simulation scenario proposed (flooded area and thickness of the deposits). A, B, C: Aspen Watershed parameters; D, E, F: parameters obtained from Bingham α–β–γ–δ; G, H, I: parameters obtained from experimental data α–β–γ'–δ' for different values of C_V: 0.35 (A, D, G); 0.42 (B, E, H) and 0.50 (C, F, I).

the right side. At the Cv variation, the plastic viscosity and the yield stress vary considerably and with them also the scenarios produced by the numerical simulation. Nevertheless, they outline similar debris flow divagations of the mixture. The scenarios C and I proposed in Figure 9, point out that the rheological parameters obtained experimentally $\alpha - \beta - \gamma' - \delta'$, although differing from a numerical point of view from those extrapolated from back analysis, allow a good representation of the analyzed event whether the method of deposition are different: in scenario I we are witnessing in a debris accumulation at the transport channel which is not happening during the simulation of scenario C. Scenario I is more similar to the event occurred than the C one.

3.5 Areas at risk ridefinition

The procedure for hazard assessment has been realized using the reference of the Adriatic Basin Authority within the norms for the preparation of the Basin Plan (ADB, 2007). They include the evaluation of the hydrogeological risk by dividing the outlined areas into 4 hazard classes from P1 to P4, from moderate to very high hazard.

The methodological protocol proposed by the Basin Authority takes up the Swiss method (Bundesamt für Umwelt, Wald und Landschaft - BUWAL, 1997) and context it to its territorial jurisdiction (ADB, 2007).

The procedure provides that any landslide has to be characterized in according to three parameters: the geometric severity, speed and return time. Each parameter is divided into three classes.

These parameters have to be inserted in matrices at cascade defining the magnitude and hazard for every single phenomenon.

In all those situations where data on the geometric severity are not available and it is not possible to make an estimation of the damages, the hazard can be defined directly intersecting the velocity data with the frequency.

In the case of debris flows, the application of this method is simple for all those cases where informations about events that have already happened are available but it becomes very uncertain and subjective in all those situations in which there is no data, or in those basins where are present mitigation measures. Here the use of simulators can assist and provide more objective data as base for the application of the method and the definition of the areas at risk.

Here are the criteria by which areas at hydrogeological risk have been redefined in Val Canale valley.

The outline of the flooded areas and store heights of debris in the 12 analyzed basins were made, when possible, by entering directly into the matrix of the BUWAL Protocol, with the values recorded immediately after the event of 2003 (the last event occurred in the investigated area characterized by return time of 500-1000 years).

In all the basins in which no post-event surveys were carried out, the flooded areas and heights of deposit were derived by integrating the data with the ones obtained from the simulations.

In the basins in which have been realized, after the flood of 2003, mitigation measures, the flooded areas and heights of deposit have been refined by integrating the available data with the results of simulations that took into account the works done. Downstream of the mitigation works in which the simulations where no highlighting leakage of material, a review of hazard levels has been realized by decreasing the value of it of at least one class such as in the case of Malborghetto basin center (Figure 10).

Fig. 10. Area at risk for Rio Malborghetto centro basin: A) Outlined area just after the 2003 alluvial event, B) Simulated event, C) New perimeter.

4. Conclusion remarks and outlooks

The present contribute wanted to be an overview on the debris flow world, describing generally the predisposing and the triggering factors, its magnitude, morphometric and rheological parameters and some of the software that nowadays permit to contribute to the reconstruction of a scenario event. This because debris flows are still one of the most dangerous phenomena due to their velocity and quickly happening. At the present time researchers are trying to go deep into the different parameters that characterize a so complex phenomenon in order to try to better define it and to be able, in the future, to simulate a real phenomena. So, the future goal will be to define clearly the input variables in order to better understand the construction of debris flow fans and to predict, mitigate or control the hazard posed by these phenomena to communities situated into mountain areas. For this reason, rheological analysis and debris flow hydrograph will be the two most studied variables in the next years.

5. Acknowledgment

The present work was supported by the Geosciences Department of Trieste University and the Geological Survey of Friuli Venezia Giulia Region. The authors would like to acknowledge prof. F. Cucchi for his valuable comments and suggestions during the study and for the patience; furthermore dott. Chiara Boccali for helping with the images.

6. References

Arattano, M., Sauvage, W. (1992) A kinematic wave model for debris flow, *U.S. Geological Survey Open File Report*, pp.92-290

Aulitzky, H. (1982). Preliminary two-fold classification of torrents. *Mitteil. der Forst. Bundesversuchsanstalt*, Vol.144, pp. 243-256

Autorità di bacino dei fiumi dell'Alto Adriatico (2007). *Progetto di piano stralcio per l'assetto idrogeologico dei bacini dei fiumi Isonzo, Tagliamento, Piave, Brenta-Bacchiglione*, Variante 1, adottato con delibera del comitato istituzionale No.4, June 19, 2007

Avolio, M.V.; Bozzano F.; D'Ambrosio, D.; Di Gregorio, S.; Lupiano, V.; Mazzanti, P.; Rongo, R. & Spataro, W. (2011). Debris flow simulation by cellular automata: a short review of the SCIDDICA models. *Italian Journal of Engineering Geology and Environmental*, DOI 10.4408/IJEGE.2011-03.B-044, pp. 387-397

Bianco, G., Franzi, L. (2000). Estimation of debris flow volumesfrom storm events. *Debris Flow Hazard Mitigation: Mechanics, Prediction and Assesment.* Wieczorek & Naeser (eds), Balkema, Rotterdam, pp.441-448

Borga, M.; Vezzani, C. & Dalla Fontana, G. (2005). Regional rainfall depth-duration-frequency equations for an alpine region. *Natural Hazards*, Vol.36, pp. 221-235

Borga, M.; Boscolo, P.; Zanon, F. & Sangati, M. (2007). Hydrometeorological analysis of the 29 August 2003 flash flood in the eastern Italian Alps. *Journal of hydrometeorology*, Vol.8, pp. 1049-1067

Boniello, M.A.; Calligaris, C.; Lapasin, R. & Zini, L. (2010). Rheological investigation and simulation of a debris-flow event in the Fella watershed. *Nat. Hazards Earth Syst. Sci.*, Vol.10, pp. 989-997

Bovis, M.J. & Jakob, M. (1999). The role of debris supply conditions in predicting debris flow activity. *Earth surface and landforms and processes*, Vol.24, pp. 1039-1504

Bruschi, A. (2008). *Colate detritiche - Stima del percorso e della pericolosità,* Dario Flaccovio Editore, ISBN 978-88-7758-801-2, Palermo, Italy, 183 pp.

BUWAL (Bundesamt für Umwelt, Wald und Landschaft); BWW (Bundesamt für Wasserwirtschaft) & BRP (Bundesamt für Raumplanung) (1997). *Berücksichtigung von Massenbewegungsgefahren bei raumwirksamen Tätigkeiten*, Bern, Switzerland, 42 pp.

Calligaris, C.; Boniello, M.A. & Zini, L. (2008). Debris flow modelling in Julian Alps using FLO-2D, In: *Monitoring, simulation, prevention and remediation of dense and debris flows II*, D. De Wrachien, C.A. Brebbia & M.A. Lenzi, (Eds.), 81-88, Witpress, ISBN 978-1-84564-118-4, Southampton, UK

Calligaris, C.; Boniello, M.A.; Kranitz, F.; Manca, P. & Zini, L. (2009). Areas subjected to debris flow phenomena in Friuli Venezia Giulia Alps, *Proceedings of 6th Euregeo*, Vol.1, pp. 37-40, Münich, Germany, June 9-12, 2009

Calligaris, C.; Cucchi, F.; Nicola, G.; Zini, L.; Kranitz, F. & Manca, P. (2009). Modellazione della colata detritica del Rio Ruscis (Val Pontebbana). *Rassegna tecnica del Friuli Venezia Giulia*, Vol.4, pp. 25-29

Calligaris, C., Lapasin, R.and Zini, L. (2010). Caratterizzazione reologica e modellazione di colate detritiche nel bacino idrografico del Fella. *In proceedings of XI CONVEGNO NAZIONALE DI REOLOGIA*, pp.165-170

Calligaris C., Zini L., Kranitz F. and Manca P. (2011) Delimitazione delle aree a rischio da collate detritiche in Friuli Venezia Giulia. In proceedings of: dissesto idrogeologico-il pericolo geoidrologico e la gestione del territorio in Italia; Roma 10 giugno 2011. In press.

Cavalli, M. & Grisotto, S. (2005). GIS-based identification of debris flow dominated channels: application to the upper Avisio Basin (Trento). Servizio Sistemazione Montana della Provincia Autonoma di Trento, Interreg III Project Alpine Space, Work Package 7, Innovative Tools for Information Collection

Cerini. M., Crosta, G., Frattini, P. & Quattrini, S. (2000). Evaluation of hydrogeological hazard on alluvial fans. *Proc. Int. Symp. Interpraevent 2000*, Villach - Austria

Cesco Bolla, B. (2008). *Modellazione di colate detritiche: software a confronto.* M.S. Thesis, Geosciences Department, University of Trieste, Italy, 91 pp.

Coussot, P. & Meunier, M. (1996). Recognition, classification and mechanical description of debris flows. *Earth-Science Reviews*, Vol.40, No.3-4, pp. 209-227.

D'Agostino, V., Cerato, M. & Coali, R. (1996). Il trasporto solido di eventi estremi nei Torrenti del Trentino Orientale. *Int. Symp. Interprevent.*, Vol. 1, pp. 377-386

Di Gregorio, S.; Nicoletta, F.; Rongo, R.; Spezzano, G.; Talia, D. & Sorriso-Valvo, M. (1994). Landslide simulation by cellular automata in a parallel environment, *Proceedings of 2nd International Workshop on Massive Parallelism*, pp. 392-407, Capri, Italy, October 3-7, 1994

Fisher, R.V. (1971). Features of coarse-grained, high-concentration fluids and their deposits. *Journal of sedimentary petrology*, Vol.41, No.4, pp. 916-927

Fredlund, D.G.; Morgenstern, N.R. & Widger, R.A. (1978). Shear strength of unsaturated soils. *Canadian Geotechnical Journal*, Vol.15, No.13, pp. 313-321

Green, W.H. & Ampt, G.A. (1911). Studies on Soil Physics: 1. Flow of Air and Water Through Soils. *J. Agric. Sci*, Vol.4, pp. 1-24

Hampel, R. (1977). Geschiebewirtschaft in Widbachen. *Wildbach und Lawinenverbau*, Vol.41, pp.3-34

Hungr, O.; Evans, S.G.; Bovis, M.J. & Hutchinson, J.N. (2001). A review of the classification of landslides in the flow type. *Environmental and Engineering Geoscience*, Vol.VII, No.3, pp. 221-228

Jakob, M. (2005). A size classification for debris flows. *Engineering Geology*, Vol.79, pp. 151-161

Marchi, L. & Tecca, P.R. (1996). Magnitudo delle colate detritiche nelle Alpi Orientali italiane. *Geoingegneria Ambientale e Mineraria*, Vol. 33(2/3), pp.79-86

Mariis (2006). Debris flow, Available from: http://www.unicaen.fr/mountainrisks/

Nemec, W. & Postma, G. (1991) Inverse grading in gravel beds. *In: Abstracts IAS 12th Regional Meeting, International Association of Sedimentologists*, p. 38

Norbiato, D.; Borga, M.; Sangati, M. & Zanon, F. (2007). Regional frequency analysis of extreme precipitation in the eastern Italian Alps and the August 29, 2003 flash flood. *Journal of Hydrology*, Vol.345, pp. 149-166

O'Brien, J. & Julien, P.Y. (1985). Physical process of hyperconcentrated sediment flows, *Proceedings of the ASCE specialty conference on the delineation of landslides, flood and debris flow*, pp. 260-279, Logan, Utah, USA, June 14-15, 1984

O'Brien, J. & Julien, P.Y. (1988). Laboratory analysis of mudflow proprieties. *Journal of hydraulic engineering*, Vol.114, pp. 877-887

O'Brien, J.S.; Julien, P.Y. & Fullerton, W.T. (1993). Two dimensional water flood and mudflow simulation. *Journal of hydraulic engineering*, Vol.119, pp. 244-259

Paronuzzi, P., Coccolo, A., Garlatti, G: (1998). Eventi meteorici critici e debris flows nei bacini montani del Friuli. *ACQUA – Sezione I/Memorie Idrogeologia*, pp.39-50

Pierson, T.C. (2005). Hyperconcentrated flow-transitional process between water flow and debris flow, In: *Debris flow Hazard and Related Phenomena*, M. Jakob & O. Hungr, (Eds.), pp. 159-202, Springer, ISBN 3-540-20726-0, Chichester, UK

Pirulli, M. (2005). Numerical modelling of landslide runout, a continuum mechanics approach, Dissertation, Politecnico di Torino, Torino, Italy

Remaitre, A.; Malet, J.P. & Maquaire, O. (2003). Study of a debris flow event by coupling a geomorphological and rheological investigation, example of the Faucon stream (Alpes-de-Haute-Provence, France), In: *Debris flow Hazard Mitigation: Mechanics,*

Prediction and Assessment, D. Rickenmann & C.I. Chen, (Eds.), ISBN 978-90-77017-78-4, pp. 375-385, Rotterdam, Holland

Rickenmann, D. (1999). Empirical relationships for debris flows. *Natural Hazards*, Vol.19, pp. 47–77

Santi, P.M. (2008). Introduction to the special issue on debris flow initiated by runoff, erosion, and sediment entrainment in western North America. *Geomorphology*, Vol.96, pp. 247-249

Segre, E. & Deangeli, C. (1995). Cellular automaton for realistic modelling of landslides. *Nonlinear Processes in Geophysics*, Vol.2, pp. 1-15

Skempton, A. & De Lory, F. (1957). Stability of natural slope in London clay, *Proceedings of the 4th International Conference on Soil Mechanics and Foundation Engineering*, Vol. 2, pp. 378-381, London, UK, August 12-24, 1957

Sosio, R.; Crosta, G. B.; Frattini, P. & Valbuzzi, E. (2006). Caratterizzazione reologica e modellazione numerica di un debris flow in ambiente alpino. *Giornale di Geologia Applicata*, Vol.3, pp. 263–268

Takahashi, T. (1991). Debris flow, *Monograph IAHR*, A.A. Balkema, Rotterdam, pp.63-75

Tecca, P.R.; Armento, C. & Genevois, R. (2006). Debris flow hazard and mitigation works in Fiames slope (Dolomites, Italy). *Transaction on Ecology and Environment*, Vol.90, pp. 15-25

Tropeano, D.; Turconi, L. & Sanna, S. (2004). Debris flows triggered by the 29 August 2003 cloudburst in Val Canale, eastern Italian Alps, *Proceedings of the 10th Internationales Symposion InterPraevent 2004*, pp. 121-132, Riva del Garda, Italy

Varnes, D.J. (1978). Slope movement types and processes. In: *Landslides Analysis and Control (Special report - Transportation Research Board, National Research Council; 176)*, R.L. Schuster & R.J. Krizek, (Eds.), 11–33, Washington DC, USA

Zanon, F. (2010). *Radar hydrology and flash flood event analysis*. Phd. Thesis, Cycle XXII, TESAF Dept. Land, Environment, Agriculture and Forestry, University of Padova, Italy

Whipple, K.X. (1992). Predicting debris-flow runout and deposition on fans: the importance of the flow hydrograph. Erosion, Debris Flows and Environment in Mountain Regions, *Proceedings of the Chengdu Symposium*, pp. 337-345, IAHS Publications No. 209, Chengdu, China, July 5-9, 1992

Winter, M.G; Shackman, L.; Macgregor, F. & Nettleton, I.M. (2005). Background to Scottish landslides and debris flows, In: *Scottish road network landslide study*, M.G. Winter, F. Macgregor & L. Shackman, (Eds.), 12-24, ISBN 0 7559 4649 9, Scottish Executive, Edinburgh, UK

5

Heat-Transfer-Model Analysis of the Thermal Effect of Intrusive Sills on Organic-Rich Host Rocks in Sedimentary Basins

Dayong Wang, Minglong Zhao and Tian Qi
Key Laboratory of Ocean Energy Utilization and Energy Conservation of Ministry of Education, School of Energy and Power Engineering, Dalian University of Technology China

1. Introduction

Numerous geological explorations demonstrate that magmatic intrusions may increase the geothermal gradient in sedimentary basins, accelerating the thermal maturation of organic matter in strata, and promoting the hydrocarbon generation (Fjeldskaar et al., 2008; Jones et al., 2007). They may also be beneficial to the migration and accumulation of oil and gas by providing them with pathways, reservoirs, covering conditions and trapping constructions (*Feng and Tang*, 1997; *Li*, 2000; *Othman and Ward*, 2002; *Othman et al.*, 2001; *Wang et al.*, 1990). Therefore, it is of great significance to study the thermal effect of igneous intrusions on organic-rich host rocks.

A lot of the organic-rich host rocks are argillaceous rocks, e.g., shales in the DSDP 41-368 hole near Cape Verde Rise in eastern Atlantic and mudstones in Xia 38 well block in the Huimin Sag of Bohai Bay, which generally have the relatively low permeability (e.g., $<10^{-16}$ m^2). Under such circumstances, the hydrothermal convection in host rocks can be reasonably ignored, and heat conduction models can be used to approximately describe the heat transfer in host rocks (*Hanson*, 1995; *Hayba and Ingebritsen*, 1997). Thus, these models are often used as a geothermometer to indicate the temperature range in which the thermal metamorphism of these host rocks takes place (*Barker et al.*, 1998; *Santos et al.*, 2009; *Stewarta*, 2005; *Turcotte and Schubert*, 1982; *Wang et al.*, 2007, 2008). Several types of heat conduction models have been constructed and used in some geological researches (*Galushkin*, 1997; *Wang et al.*, 2007, 2011). However, only a small portion of these researches specially compare and quantify the difference in the prediction results of different heat conduction models (*Jeager*, 1959; *Galushkin*, 1997; *Wang et al.*, 2011). Due to the apparent importance of the accuracy of heat conduction models in these researches, it is still required to further explore and distinguish the applicable conditions of these models based on some geological cases.

In this study, we accordingly investigate the difference in the prediction results of three types of commonly used heat conduction models by taking an intrusive sill in the Bellata-1 Well in the Gunnedah Basin, Australia as an example. These models assume different intrusion mechanisms of magma and different evolution states of pore water during cooling of magma. By comparing the prediction results of these models with the measured vitrinite-reflectance (*Ro*) geothermometer, we also discuss the potential intrusion mechanism of the sill and the state of pore water in host rocks during cooling of the sill.

2. Geology of the Bellata-1 Well in the Gunnedah Basin, Australia

The Bellata-1 Well is located in the Gunnedah Basin of northern Australia and intersects Permian, Triassic, Jurassic and Cretaceous strata in turn (Fig 1). The total thickness of the Permian and Triassic strata reaches 451 m, overlain by the 640m thick Jurassic and Cretaceous sediments. A 15.68m thick mafic basaltic sill was found in the lower part of Triassic Napperby Formation. The intrusion of the sill took place between the Late Triassic and Early Jurassic with a current burial depth of 847.60 m (*Othman and Ward*, 2002; *Othman et al.*, 2001). The Triassic strata are mainly composed of organic-rich mudstones. The *R*o profile adjacent to the sill shows the effect of the significant local heating: the *R*o value can be as high as up to 2.43% within the contact aureole, whereas the *R*o value in the unaffected parts is only 0.57-0.74%. The oil generated by these organic-rich rocks due to the thermal effect of the intrusive sill is found in the Jurassic Pilliga Sandstone (Othman et al., 2001). Therefore, the igneous sill of the Bellata-1 Well and its host rocks constitute an ideal geological example for numerically investigating the thermal effects of igneous intrusions on organic-rich host rocks.

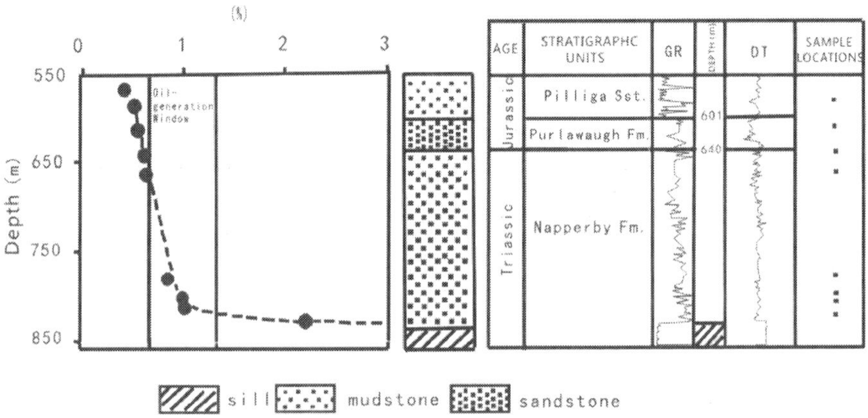

Fig. 1. Stratigraphic characteristics and vitrinite reflectance profile around an igneous sill in Well Bellata-1 (*Othman and Ward,* 2002; *Othman et al.,* 2001)

3. Method

3.1 Heat conduction models

Some general assumptions are required in constructing heat conduction models: 1) The shape of the intrusion is regular, dike-like or sill-like; 2) Convection motion in the intrusion is not considered; 3) Heat loss due to the escape of volatiles is neglected. Thus, the basic heat conduction equations in one dimension which can be used to describe the heat transfer

between intrusive magmas and host rocks are expressed (*Barker et al.*, 1998; *Shan et al.*, 1998; *Stewarta et al.*, 2005):
For magma intrusions：

$$\frac{\partial}{\partial Z}\left(K_{magma}\cdot\frac{\partial T}{\partial Z}\right)=\rho_{magma}\cdot\frac{H}{L_2-L_1}\cdot\frac{\partial T}{\partial t}+\frac{\partial(\rho_{magma}\cdot C_{magma}\cdot T)}{\partial t} \quad (1)$$

For host rocks：

$$\frac{\partial}{\partial Z}\left([1+[Nu]\cdot K_{host}]\cdot\frac{\partial T}{\partial Z}\right)=\frac{\partial}{\partial t}(\rho_{host}\cdot C_{host}\cdot T)+A_1+[A_2] \quad (2)$$

Where T is the temperature; t represents the time; K means the thermal conductivity; C is the specific heat; ρ denotes the density; A_1 and A_2 represent the latent heat consumed by the dehydration and decarbonation reactions and pore-water volatilization per unit volume of host rocks and per unit time; the subscripts, i.e., magma and host, denote magma and host rocks, respectively. H represents the latent crystallization heat of melted magma; L_1 - L_2 is the crystallization temperature range of intrusive magma. Nu represents the Nusselt number and can be used to implement approximately the hydrothermal convection in host rocks (*Galushkin*, 1997). In this study, a finite difference method is used to obtain the numerical solution of Eqns. (1) and (2).

3.2 Model parameters

As the host rocks were located on or near the ground surface at the intrusion moment of the sill (between Late Triassic and Early Jurassic), we approximately assume that the strata with the current depth of 640 m is located on the ground surface during the intruding of magma. The surface temperature and the geothermal gradient are assumed to be approximately equal to 25 °C and 30 °C/Km, respectively. The temperature of the melted mafic magma is usually about 1250 °C (*Barker et al.*, 1998; *Wohletz*, 1999). Its thermal conductivity and density are usually equal to 2.1 J m^{-1} s^{-1} $°C^{-1}$ and 2700 Kg/m^3 (*Barker et al.*, 1998; *Wohletz*, 1999), respectively. We specify the specific heat of the sill to be equal to 1200 J/Kg (*Galushkin* 1997; *Wang et al.*, 2010). The latent heat of crystallization of melted magma equals 400 KJ/Kg, and the corresponding crystallization temperature range is 1150 °C - 1250 °C. According to Wang et al. (2007), Organic-rich mudstones generally have the relatively low thermal conductivity. For example, the thermal conductivity of mudstones in the region of England is 1.4-1.6 W/mK (*MidttØmme et al.*, 1998), and the thermal conductivity of mudstones at the depth of 850 m in the Huimin Sag of Bohai Bay is about 1.4 W/mK (*Wang et al.*, 2007). The specific heat and density of mudstone matrix can be specified to be equal to 820 J/Kg and 2700 Kg/m^3, respectively. We specify 1.9 J m^{-1} s^{-1} $°C^{-1}$ as its thermal conductivity. This value is the same with that of the host rocks of the diabase sill of Well Xia38 in the Huimin Sag of Bohai Bay, as the latter has the same lithology and the similar intrusion depth with our example. At the buried depth of about 200 m, the boiling point of pore water may reach 200 °C, and its latent volatilization heat is about 1939.73 KJ/Kg. The porosity of the host rock is about 0.5 in terms of the depth - porosity relationship of mudstones (*Allen and Allen*, 2005). We calculate the total specific heat and total thermal conductivity of host rocks based on the computational equations of Galushlkin (1997), Travis et al. (1991), Wang et al. (2007) and Wohletz et al. (1999).

3.3 Simulated cases

Three types of one-dimensional heat conduction models are built to simulate the heat transfer between the sill and its host rocks (Table 1). We adopt the method of Galushkin (1997) to implement the finite-time intrusion mechanism of magma: The temperature at the axis of the sill is set as 300 °C when the sill begins to form; and the time of the pre-cooled shell formation is equal to 2.2 hours; the total time of the sill formation is about 4.4 hours.
In order to verify the applicability of these three heat conduction models to the modeled sill, we need to compare the prediction results of the models with the measured vitrinite-reflectance (*R*o) geothermometer. We adopt the vitrinite reflectance - the peak-temperature (T_{peak}) relation (i.e. T_{peak} = (ln*R*o + 1.19) / 0.00782) of *Barker et al.* (1998) to calculate the T_{peak} of the overlying host rocks based on the measured *R*o values and then compare it with the predictions of the models.

Case No.	Intrusion mechanism of magma	Pore-water volatilization	Hydrothermal convection in overlying host rocks	Dehydration and decarbonation of host rocks
1	instantaneous	not considered	not considered	considered
2	instantaneous	considered	not considered	considered
3	finite-time	not considered	not considered	considered

Table 1. Three cases for simulation

4. Results and discussion

The T_{peak} profiles of host rocks predicted by three types of heat conduction models are shown in Fig. 2. The contact temperature (T_c) predicted by Case 2 reaches 852 °C, and is higher than that predicted by the other cases. Actually, pore-water volatilization can decrease the thermal conductivity of host rocks. As a result, the diffusion of the heat of the sill in host rocks is depressed, and near the contact, heat from the sill congregates and rapidly increases the contact temperature. Comparably, the T_c predicted by Case 3 is lowest and only reaches 706 °C. This is apparently due to the heat loss caused by the pre-cooled shell of the sill compared to the instantaneous intrusion mechanism. In addition, the computation based on Case 1 deduces the highest degree of the thermal effect of the sill on its host rocks, whereas the prediction from Case 3 results in the lowest one. This indicates that the intrusion mechanism of magma may play a more important role in lowering the thermal effect of the intrusion than the heat sinks in host rocks.
By comparing the predicted T_{peak} with the measured *R*o geothermometer, it is obviously observed that the T_{peak} predicted by all of these three models is much lower than the *R*o geothemometer in the region where it is 75 m away from the margin of the sill (i.e. $X/D \approx 5$). This demonstrates that the increase in the temperature of strata due to the subsequent sedimentation after cooling of the sill have covered up the thermal influence of the intrusion on host rocks in this region. The heat conduction model assuming the instantaneous intrusion mechanism and ignoring pore-water volatilization matches well with the measured *R*o geothermometer. Othman et al. (2001, 2002) once reported that the Napperby Formation is mainly composed

of low-permeability mudstones (shale) and has the abnormal high pressure. Consequently, volatilization and escape of pore water can likely be restricted. This is consistent with the prediction of Case 1. All of these indicate that the instantaneous intrusion mechanism likely represents natural conditions and that the effect of pore-water volatilization is insignificant.

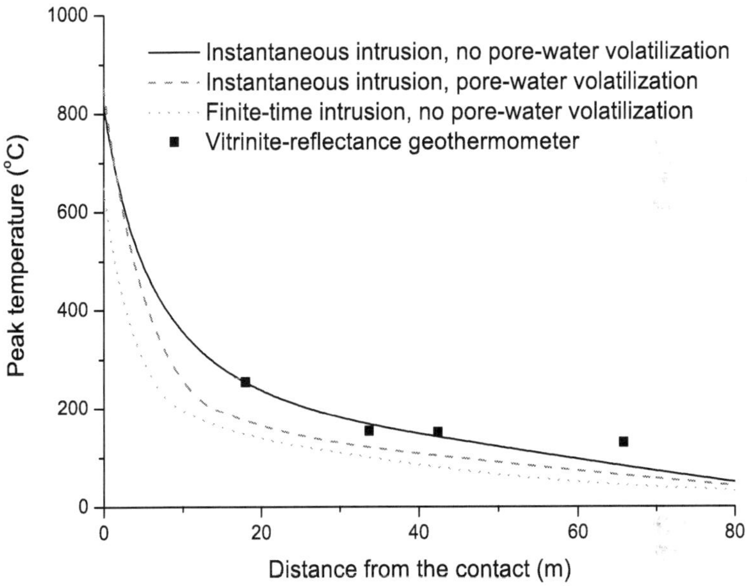

Fig. 2. Comparison between virtrinite-reflectance geothermometer of Baker et al. (1998) and peak temperature of host rocks predicted by three types of heat conduction models, assuming different intrusion mechanisms of magma and the state of pore water during cooling of magma

5. Conclusions

The following conclusions can be made based on the heat-conduction-model analysis of the T_{peak} of the host rocks of a mafic sill of the Bellata-1 Well from the Gunnedah Basin, Australia:

1. The consideration of pore-water volatilization can increase the T_c prediction, while it is converse for the finite-time intrusion mechanism. The computation based on the heat conduction model assuming the instantaneous intrusion mechanism and considering

pore-water volatilization deduces the highest T_c among three types of heat conduction models, whereas the computation based on the model assuming the finite-time intrusion mechanism and ignoring pore-water volatilization results in the lowest T_c.

2. The degree of thermal effect deduced by the heat conduction model assuming the instantaneous intrusion and ignoring pore-water volatilization is highest, while that deduced by the model assuming the finite-time intrusion and ignoring pore-water volatilization is lowest. This indicates that the intrusion mechanism of magma may play a more important role in lowering the thermal effect of the intrusion than the heat sinks in host rocks.
3. The heat conduction model assuming the instantaneous intrusion mechanism and ignoring pore-water volatilization matches well with the measured vitrinite-reflectance geothermometer. Considering the real geological characteristics of the host rocks, it can be concluded that the instantaneous intrusion mechanism likely represents natural conditions and that the effect of pore-water volatilization is insignificant.

6. Acknowledgement

This work was financially supported by the National Science Foundation of China (No. 41004031 and 50736001), the 863 program (No. 2008AA062303 and 2006AA09209), the 973 program (No. 2009CB219507), the Ph.D. Programs Foundation of Ministry of Education of China (No. 20100041120039), the Research Foundation of Dalian University of Technology (No. 852003, 893210, 893316) and the Fundamental Research Funds for the Central Universities.

7. References

Allen P A, Allen J R. Basin Analysis: Principles and Applications [M]. Oxford: Blackwell Scientific Publicatons, 2005, pp549.

Barker C E, Boney Y, Lewan M D. Fluid inclusion and vitrinite-reflectance geothermometry compared to heat-flow models of maximum paleotemperature next to dikes, western onshore Gippsland Basin, Australia [J]. International Journal of Coal Geology, 1998, 37: 73~111.

Feng Qiao, Tang Xiyuan. Magma activity's influence on conditions forming oil and gas pools[J]. Geological Science and Technology Information (In Chinese with English abstract), 1997, 16(4): 59~65.

Fjeldskaar W, Helset H M, Johansen H, et al. Thermal modeling of magmatic intrusions in the Gjallar Ridge, Norwegian Sea: implications for vitrinite reflectance and hydrocarbon maturation [J]. Basin Research, 2008, 20: 143-159.

Galushkin Y I. Thermal effects of igneous intrusions on maturity of organic matter: A possible mechanism of intrusion [J]. Organic Geochemistry, 1997, 26(11-12): 645-658.

Hanson R B. The hydrodynamics of contact metamorphism [J]. Geological Society of America Bulletin, 1995, 107: 595-611.

Hayba D O., Ingebritsen S E. Multiphase groundwater flow near cooling plutons [J]. Journal of Geophysical Research, 1997, 102: 2235-12252.

Jaeger J C. Temperatures outside a cooling intrusive sheet [J]. American Journal of Science, 1959, 257: 44~54.

Jones S F, Wielens H, Williamson M C, Zentilli M. Impact of magmatism on petroleum systems in the Sverdrup Basin, Canadian Arctic Islands, Nunavut: A numerical modeling study [J]. Journal of Petroleum Geology, 2007, 30(3): 237-255.

Li Yahui. Diabase and hydrocarbon reservoir formation on thc nor-thern slope of Gaoyou sag[J]. Journal of Geomechanics (In Chinese with English abstract), 2000, 6(2): 17~22.

MidttØmme K, Roaldset E, Aagaard P. Thermal conductivity of seleted claystones and mudstones from England [J]. Clay Minerals, 1998, 33: 131~145.

Othman R, Arouri K R, Ward C R, et al. Oil generation by igneous intrusions in the northern Gunnedah Basin, Australia[J]. Organic Geochemistry, 2001, 32: 1219~1232.

Othman R, Ward C R. Thermal maturation pattern in the southern Bowen, northern Gunnedah and Surat Basins, northern New South Wales, Australia[J]. International Journal of Coal Geology, 2002, 51: 145~167.

Santos R V , Dantas E L, de Oliveira C G, de Alvarenga C J, dos Anjos C W, Guimaraes E M, Oliveira F B. Geochemical and thermal effects of a basic sill on black shales and limestones of the Permian Irati Formation [J]. Journal of South American Earth Sciences, 2009, 28 (1): 14-24.

Shan Yehua, Yuan Ding, Li Zhian. Numerical simulation of magmatic metamorphism [J]. Coal Geology of China (In Chinese with English abstract), 1998, 10(2): 19~26.

Stewarta A K, Masseya M, Padgettb P L, et al. Influence of a basic intrusion on the vitrinite reflectance and chemistry of the Springfield (No. 5) coal, Harrisburg, Illinois[J]. International Journal of Coal Geology, 2005, 63: 58~67.

Travis B J, Janecky D R, Rosenberg N D. Three-dimensional simulation of hydrothermal circulation at mid-ocean ridges[J]. Geophysical Research Letters, 1991, 18(8): 1441~1444.

Turcotte D L, Schubert G.. Geodynamics Applications of Continuum Physics to Geological Problems [M]. New York: Wiley, 1982, 1~450.

Wang D., Lu X., Song Y., et al. Influence of the Temperature Dependence of Thermal Parameters of Heat Conduction Models on the Reconstruction of Thermal History of Igneous-intrusion-bearing Basins [J]. Computers & Geosciences, 2010, 36, 1339-1344, doi:10.1016/j.cageo.2010.03.014.

Wang D., Lu X., Xu S., Hu W. Comment on "Influence of a basic intrusion on the vitrinite reflectance and chemistry of the Springfield (No. 5) coal, Harrisburg, Illinois" by Stewart et al. (2005) [J]. International Journal of Coal Geology, 2008, 73: 196-199.

Wang D, Lu X, Zhang X, Xu S, Hu W, Wang L. Heat-model analysis of wall rocks below a diabase sill in Huimin Sag, China compared with thermal alteration of mudstone to carbargilite and hornfels and with increase of vitrinite reflectance [J]. Geophysical Research Letters, 2007, 34, L16312, doi:10.1029/2007GL030314.

Wang D., Song Y., Liu W., Zhao M., Qi T. Numerical investigation of the effect of volatilization and the supercritical state of pore water on maturation of organic matter in the vicinity of igneous intrusions[J]. International Journal of Coal Geology, 2011, 87: 33-40.

Wang Youxiao, Fan Pu, Cheng Xuehui, et al. Abnormal geothermal influence on hydrocarbon genesis of sedimentary organics[J], Oil & Gas Geology (In Chinese with English abstract), 1990, 11(1): 73~77.

Wohletz K H, Orsi G, Civetta L. Thermal evolution of the Phlegraean magmatic system [J]. Journal of Volcanology and Geothermal Research, 1999, 91: 381~414.

6

Submarine Mass Movements: Sedimentary Characterization and Controlling Factors

Gemma Ercilla[1] and David Casas[2]
[1]Instituto de Ciencias del Mar, CSIC. Grupo de Márgenes Continentales Passeig Marítim de la Barceloneta, Barcelona
[2]Actual Address: Instituto Geológico y Minero de España, IGME, Madrid, Spain

1. Introduction

Marine geology studies about submarine mass-movements, especially during the last 20 years, have demonstrated their relevance in building and evolution of continental margins. This is because mass-movements represent the main mechanism of sediment transport from continent to deep-sea areas, and one of the most common geological hazards in submarine environments (Masson et al., 2011). They occur in all the sea and oceans of the world, and may develop in all the physiographic environments, from the shelf, slope, continental rise to deep sea areas. Their resulting deposits have variable dimensions, from metric to several hundreds of km in length, and from centimetric to severel tens meters of thick. Their sedimentary record informs about variations of glacioeustacy and hinterland sediment sources, occurrence of meteoceanic processes that can affect to seafloor, active tectonism and seabed fluid flow related processes.

Likewise, the study of mass-movement deposits is nowadays important for applied scientific, mostly related to hydrocarbon exploration and geological hazards. These deposits may represent important accumulations of clastic sediments for hydrocarbon accumulation. In spite of their important role in submarine environments their occurrence has never been directly observed (Hueneke and Mulder, 2011). Their occurrence has been detected due to injuries caused on infrastructures resting on or fixed to the seafloor and/or subbottom, as cables and pipelines. In this sense, mass-movement processes in marine environments, both at shallow and deep sea areas can represent a hazard for any marine infrastructure. During recent years there has been an important growth in the level of study of these processes and their deposits. This is due to the increasing development of deep water exploration activity and the joined scientist efforts, from geologists and engineers, on working in specific objectives to resolve unknowns about mass-movements. These efforts have produced an important collection of scientific literature about the variability of failures and their resulting sedimentary products as well as potential hazards, all based on indirect (acoustic and seismic analysis) and direct (sedimentological, geotechnical and geochemical) approaches.

The present work goes into deep the knowledgement of submarine sedimentary instabilities combining multidisciplinary and multiscale approaches. Four main topics for mass-movements investigations are presented: 1) characterization of continental margin and historic register of the instabilities on a continental slope; 2) definition of the dynamics of slope failures; 3) study of physical and mechanical properties of sediments; and 4) definition of forces that may trigger submarine mass-movements and determinate their evolution. These topics are studied with several direct and indirect techniques of different degrees of resolution (millimetric to decametric) and in different geological contexts.

2. Selected case studies in the Mediterranean, Atlantic and Antarctic Sea

Six case studies will allow discussing, using as thread the topics above mentioned, the different results obtained as well as their interrelation. Based on indirect, morphologic and acoustic evidences three case studies are presented: 1) Mass-movement on the Ebro continental slope (NW Mediterranean); 2) Evidence of gas in the continental slope of the Gulf of Cadiz; and 3) The dynamics of the Baraza Slide in the northern Alboran Sea (SW Mediterranean). Based on the combining morphologic and seismic facies analysis and direct (sedimentological, geotechnical analysis) techniques, three other cases are presented: 4) Physical-geotechnical properties and texture of the Pliocene-Quaternary sediments in the Madeira Abyssal Plain; 5) Sedimentary processes in sediments recovered from mud volcanoes in the Anaximander Mountains (Eastern Mediterranean); and 6) Sedimentary stability of the continental slope and adjacent deep sea areas in the Bransfield Basin (Antarctic Peninsula).

2.1 Methodology

The bathy-morphologic, seismic and acoustic facies analysis will allow to observe the tecto-sedimentary framework of the margin on which mass-movements occur or may occur, as well as to characterize in detail the resulting sedimentary products from the different types of mass-movement processes. Bathymetric data include multibeam records obtained with SIMRAD EM12 multibeam system and processed with NEPTUNE and CARAIBES softwares, combined with and information provided from GEBCO gridded bathymetry data. Seismic and acoustic facies analyses have done through the analysis of different single-channel seismic profiles. Those include: 1) very high resolution TOPAS (Topographic PArametric Sonar) profiles, with a penetration of the acoustic signal between 30 and 200 milliseconds (two way travel time -twtt-) and a decimetric resolution; 2) high resolution Sparker profile which provides a penetration of about 1.5 s (twtt) and a metric resolution; and 3) medium resolution airgun profile (sleeve guns, 120 c.i.) with a penetration of 2-3 s (twtt) and providing a resolution of tens of meters.

The sedimentological and physical and geotechnical analysis allow to ground truth information about the properties of nearsurface sediments, both failured/deformed and non-failured. A great diversity of techniques are involved in these analysis which allow characterizing grain size, composition, grain density, water content, porosity, shear strength etc. Physical properties of marine sediments are important variables to understand geological processes and events of marine environments. Physical properties (density, magnetic susceptibility and P-wave velocity) depend to a large extent on lithology, grain size and composition of sediment (Hamilton et al., 1982; Nobes et al, 1991). The bulk density, for example, is related to porosity, grain density and is partially controlled by grain size (Johnson and Olhoeft, 1984). The P-wave velocity is controlled by porosity, carbonate and clay contents (Hamilton et al., 1982; Mienert, 1984; Nobes et al., 1986). Physical properties are also influenced by diagenetic effects, not only decrease of porosity with increasing compaction, but also cementation and carbonate dissolution (Nobes et al., 1992).

2.2 Variability of mass-movement on the Ebro continental slope (NW Mediterranean)

The study of the mass-movement on the Ebro continental slope (Fig. 1) is a good example showing the great variety in scale and distribution of mass-movement features, both erosive and depositional, that can affect to slopes of continental margins (Casas et al., 2003a). They affect to any part of the slope. The triggering factors for this slope are variable, being the background factor the high sediment supply during the regressive sea-level falls of high amplitude.

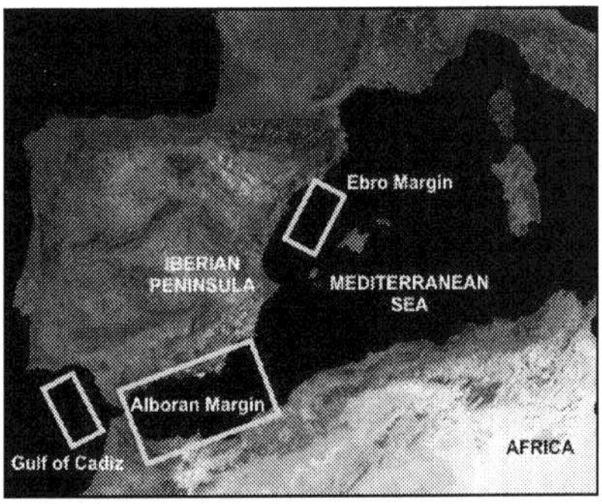

Fig. 1. Map of location of the study area for case studies 2.2 (Ebro Margin), 2.3 (Gulf of Cadiz) and 2.4 (Alboran Sea).

The Ebro continental slope locates in the Spanish NW Mediterranean continental passive margin, off the Ebro River, the second largest rivers in the Western Mediterranean Basin. It extends between 160 ± 20 m and 1000 ± 200 m water depth, its width decrease southward from 20-25 km to 10 km, and its gradients increases southward from 2.5-4° to 3.8-5.8°. The Ebro is a prograding margin since the late Oligocene, mainly fed by the Ebro River, whose tectonic adjustments have been active as late as to the Upper Pliocene and the Quaternary (Dañobeitia et al., 1990). Seismicity is fairly moderate; magnitudes between 4 and 5 have been measured during the last century (Surinach and Roca, 1982). The oceanography is manly characterized by permanent flows along the continental shelf and slope resulting from wind-driven surface currents from northwest that extend to water depths of 150-200 m, interacting with upper slope topography (Castellón et al. 1990; Arnau et al. 2004).

About 37 % of the Ebro slope (Fig. 2) is affected by a great variety of mass-movement features, both erosive and depositional (Casas et al., 2003a). Erosive features predominate along the entire slope, whereas depositional ones are dominant on the lower slope. Submarine canyons and gullies eroding their heads and walls, are indicating removal of sediments through the continental slope. Their formation and enlargement are mostly related to the increase of sediment supplied that slope receives during high amplitude sea-level falls and lowstand stages in the Quaternary (Alonso and Maldonado, 1990). Slide scars, isolated or associated downslope to slide deposits, occur between canyons in the open upper slope (Fig. 3). They probably formed by the decreases in shear strength due to high sediment supply during those mentioned regressive falls in sea-level. Small-scale slides (tens ms thick, hundreds of meters to few kilometers in length) also occur locally on the canyon walls, due to oversteeping of seafloor (> 10º).

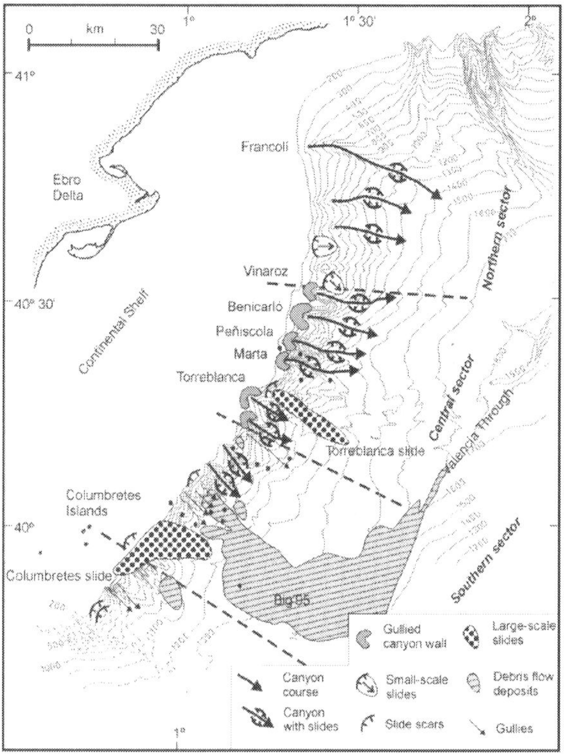

Fig. 2. Mass-movement features identified on the Ebro slope. From Casas et al. (2003a).

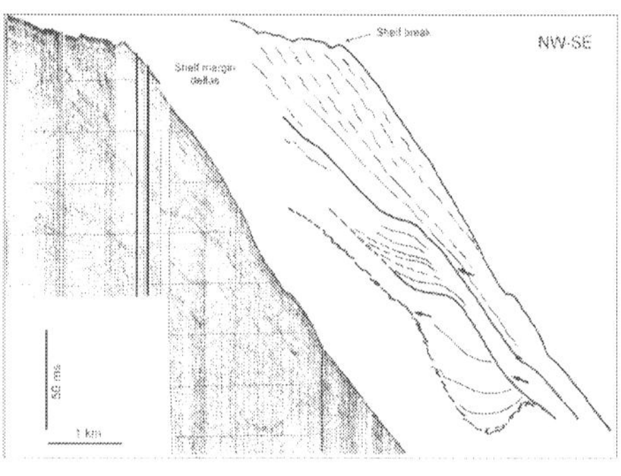

Fig. 3. Sparker profile showing the small-scale slides identified on the upper slope. The arrows indicate truncation of sediments. From Casas et al. (2003a).

The most prominent depositional mass-movement features occur on the open slope, and include three large slides, from north to south: Torreblanca (300 ms thick, > 40 km long), Big (< 135 m thick; 110 km long>; Lastras et al., 2002) and Columbretes (150 ms thick, 10 km long). They extend from the upper slope and enter into the base-of-slope where the Big Slide becomes to reach to the Valencia Channel, a mid-ocean valley crossing the Algero-Balear Basin. These slides represent slumps and associated mass-flows deposits with different post-failure behaviour. The main heads (unknown for the Big Slide) are defined by a main scar with an amphitheatre-like failure surface that evolves downslope to a depositional area defined by deformed stratified and chaotic facies with a rugged seafloor (Big Slide shows blocks metric in size). The Torreblanca and Columbretes slides represent frontal confined slides, whereas the Big Slide can be fit into the category of the unconfined frontally slides (Frey Martinez et al., 2005). In this sense, the Torreblanca Slide (Fig. 4), with a limited deformation and preservation of original stratification, and the Columbretes Slide (Fig. 5), more deformed and with a chaotic pattern, both display an increase of their thickness (few meters) at the distal end and do not rest upon the undisturbed sediments. Contrasting, the Big Slide, with a chaotic and transparent internal acoustic pattern, displays debris flow material over running the previous seabed. Addition to these large slides, on the lower slope of the south sector are also identified relatively small-scale debris flow deposits (< 8 km long, 0.26 s) interrupting the lateral continuity of the undisturbed stratified open slope deposits.

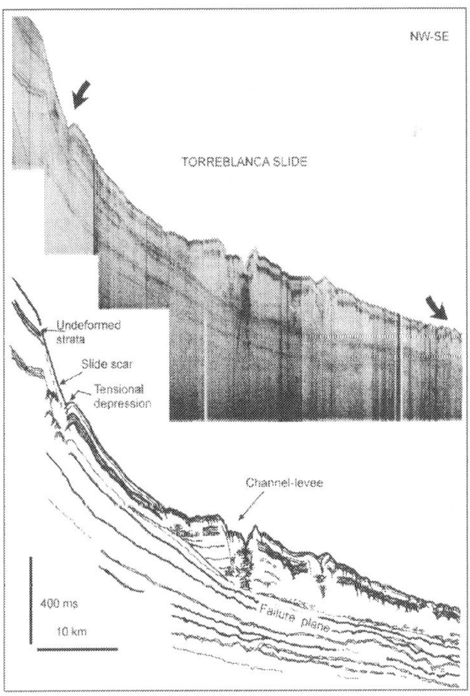

Fig. 4. Sparker profile along the Torreblanca Slide. From Casas et al. (2003a).

The distribution and variability of the mass movements in the Ebro margin are mainly conditioned by several factors: high sediment supply, oversteeping of the slope, sediment thickness involves in the failure, frequency of failures and structural location. The combination between high sediment supply and oversteeping of the sea surface favours the occurrence of small-scale slides and canyon incisions on the upper slope. The mass

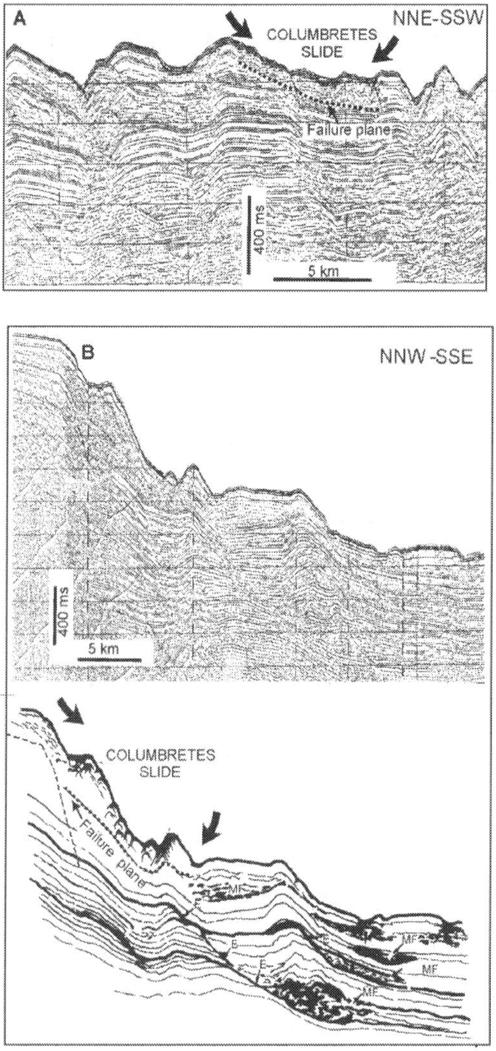

Fig. 5. Airgun profiles of Columbretes slide A) parallel to the slope; B) along the slide and line drawing. Legend: The thick arrows indicate the lateral extension of the slide; E, erosive surfaces; MF, mass-flow deposits. From Casas et al. (2003a).

movements forming both features, channelized and unchannelized (canyon vs. open slope), differ in the volume of sediment and occurrence frequency, being lower in volume and higher in frequency for the canyons and reverse for the small-scale slides. The failure style on the open slope involves short movements because those slides have not been transported over significant distances (few meters). On the other hand, the preferential location of the three large slides in the south sector of the Ebro slope, suggests a tectonic control in their triggering, specifically a seismic shacking related to volcanic activity (Field and Gardener, 1990). The timing occurrence of the two mass-movement groups seems to be different; during the regressive stages of sea level formed the small-scale slides and canyon incisions, being gravity loading plus storm and internal waves the possible triggering mechanisms; and whatever stage of sea-level for the three large slides, being the seismic activity their triggering. The distribution and variability of the mass movements on the Ebro slope also indicates that the depletive features of the mass movements are preferentially located on the upper slope whereas the accumulative features predominate in the lower slope.

2.3 A potential hazard for mass movement occurrence in the Gulf of Cadiz: Gas in slope sediments

Low to very high resolution seismic studies have revealed the presence of gas-charged sediments around the oceans and seas, from shallow to deep water depths (Judd and Hovland, 2007). Presence of gas in marine sediments, in its two versions as free gas and gas hydrate, is considered one of the main factors influencing on seafloor slope stability. Gas is considered a pre-disposition and triggering factor of mass movements. Although the direct evidence for the presence of gas in marine sediments can be only confirmed trough drilling and coring, it can be also identified indirectly by means of acoustic methods. The present study case has been chosen because the Gulf of Cadiz is a good example to show the great variety of acoustic evidences observed in marine sediments hosting gas, free and gas hydrate (Casas et al., 2003b). Its presence has to be always considered as potential geological hazard, of natural, induced or mixed type, and a predisposition factor to trigger mass movements (Judd and Hovland, 2007). Because of that the indirect interpretation of gas in marine sediments is essential in assessment of the factors controlling the occurrence of submarine mass-movements.

The Gulf of Cadiz (Fig. 1&6) is located in the southwestern Iberian, Atlantic Sea, and records a complex tectonic evolution because it occupies a focal position between the westernmost Mediterranean segment and the Iberian-African boundary (Bonnin et al., 1975; Dewey et al., 1989). This tectonic complexity is reflected by the seafloor morphostructure defined by a prograding margin with a continental slope (130 to > 900 m water depth) affected by numerous diapiric ridges and mud volcanoes separated by submarine valleys (Baraza et al., 1999; Maldonado et al., 1999). The Upper Miocene to Quaternary stratigraphic architecture of this margin corresponds to three main depositional systems (Maldonado et al., 1999). The main sedimentary systems comprise Plio-Quaternary shelf margin deltas on the proximal margin developed during the sea-level falls and rises and fed mainly from the two major rivers in the area (Guadiana and Guadalquivir); Upper-Pliocene to Quaternary mixed contourite-turbidite deposits on continental slope formed by the action of the Mediterranean outflow water after the opening of the Strait of Gibraltar (Hernández-Molina et al., 2003) ; and Upper Miocene fan lobe deposits in the distal continental slope fed by the two mentioned rivers. The above mentioned shelf margin deltas and fan lobe deposits contain substantial amount of gas, both in origin biogenic and thermogenic (Baraza et al., 1999; Leon et al., 2009).

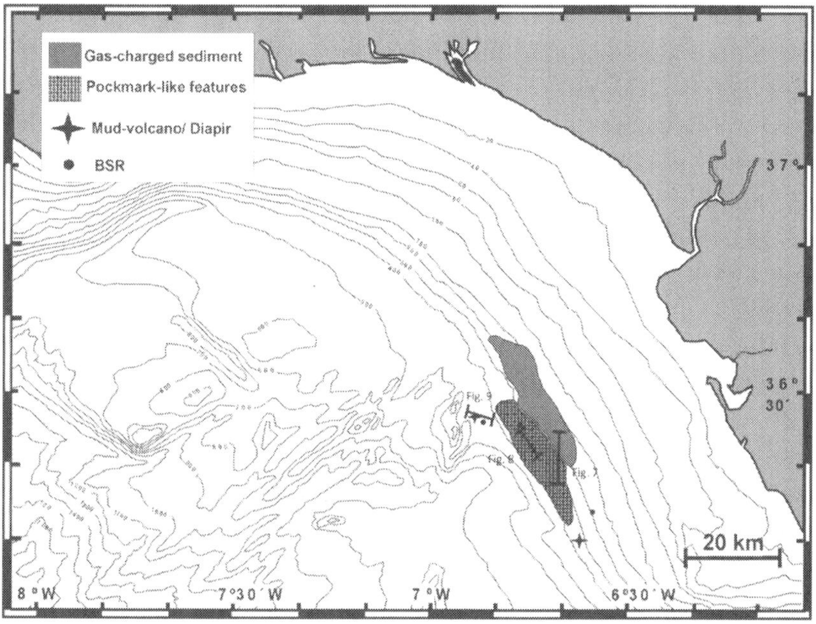

Fig. 6. Bathymetric map of the Gulf of Cadiz showing the location of the gas-related features observed in this study. Modified from Casas et al. (2003b). Location of Figs. 7, 8 & 9 are also displayed.

The presence of this gas has been defined in high resolution seismic profiles by a number of acoustic and morphological evidences: acoustic turbidity and blanking, bright spots, pockmarks, high amplitude refractions, acoustic plumes and turbidity in the water column, and bottom simulating reflectors (BSRs). These are the most typical gas related acoustic evidences defined in literature (Judd and Hovland, 2007). The acoustic turbidity and blanking features occur in the shelf margin muddy deposits on the uppermost slope (130 to 300 m water depth, area of 210 km^2), and reflects gas bubbles presence in the pore space on those deposits (Fig. 7). This gas is escaping to the immediately above water column, as it is suggested by the presence of acoustically reflective plumes. The mapped area with the free gas is locally affected by mass-movement deposits which tend to be rotational and are usually associated, forming multiple slides. They also occur immediately downslope of the area with free gas, and are underlined by acoustic turbidity and bights spots.

Modern and ancient pockmarks, hundreds meter in length and metric in relief, develop downslope from the gas-charged sediment area and surrounding diapirs (300 to 400 m water depth). These pockmarks are affected by high amplitude diffractions suggesting the escape of gas through them. Bright spots are also defined between pockmarks indicating gas concentrations trapped within sediment layers. The pockmarks shows a modern activity confirmed by the presence of acoustic reflective plumes above the pockmarks that indicate gas is ascending from the seafloor up to 50 m (Fig. 8). The expulsion processes implies a reworking of the subbottom and seafloor sediments that form dispersed craters whose steep walls can be affected by small scale mass movements. The pockmarks do not represent mass

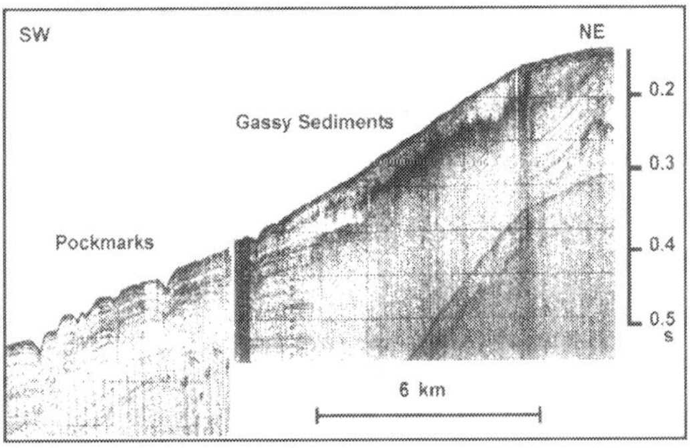

Fig. 7. GeoPulse seismic profile showing gas-charged sediment and pockmark-like features. From Casas et al. (2003b). See location in Fig. 6.

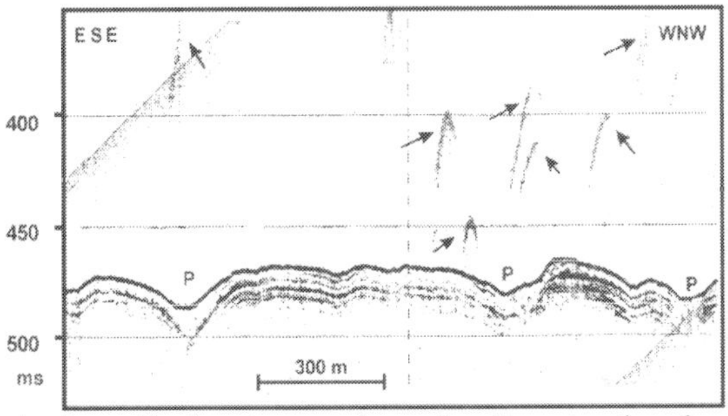

Fig. 8. GeoPulse seismic profile showing gas-charged sediment and pockmark-like features. From Casas et al. (2003b). See location in Fig. 6.

movements but they can be considered as geohazard because interrupt the lateral continuity of sediments and strata, becoming to rework sediment column "in situ". The BSR-like acoustic anomalies occur locally between 140 to 388 m water depth, and at 80 to 150 ms deep (Fig. 9). The gas saturated sediments that appear in the slope plus the diapiric activity have created the Pressure-Temperature conditions to form gas hydrate (Casas et al., 2003b).

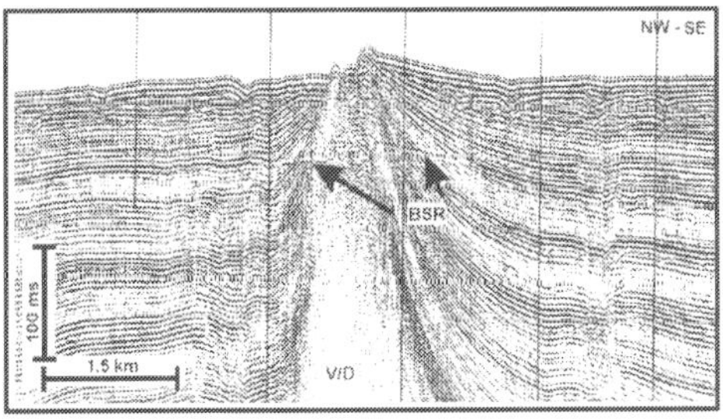

Fig. 9. BSR-like feature around the top of a mud volcanoe/diapir (V/D). From Casas et al. (2003b) Location of air-gun profile in Fig. 6.

It is evident a relationship between mass-movement occurrence and gas presence in the slope of the Gulf of Cadiz, but it is true that not elsewhere where gas exists there are mass-movements features. This means that gas alone cannot be responsible, but should be considered as a predisposition factor especially in tectonically active areas as the Gulf of Cadiz.

2.4 The dynamics of the Baraza Slide

The main objective of this case study is to show mass movement features imaged by multibeam systems on the seafloor should not be assumed to be recent alone and may have an instability history more complex than they seems. This study characterizes the Baraza Slide (Ercilla et al., 2009; Casas et al., 2011), located in the continental slope of the northern Alboran Sea (southwestern Mediterranean Sea; Fig. 1&10&11), between 590 and 830 m water depth, trough the analysis of their morphology, subbottom seismic facies and deformational features. The combined results indicate the Baraza Slide is a Late Pleistocene-Holocene sedimentary instability complex, that it has undergone repeated slope failures. The complex is formed by a mud flow system that changes to a slide system with time.

The mud flow system is characterized by the occurrence of a slope failure at the steepest (3 to 3.5°) sector of the open slope; the displaced mass moves down to the slope where gradients decrease sharply (0.5° to 1°). Mud flow is affected by a progressive dilution in flow concentration during its downslope movement. This dilution is suggested by the changes in acoustic facies of the mud-flow sediments, from chaotic to transparent, and also in the convexity of their cross-sections. It is inferred the occurrence of several mud flow events, at least 2 events based on multibeam mapping and seismic data, migrating from east to west and decreasing in their magnitude at the same time.

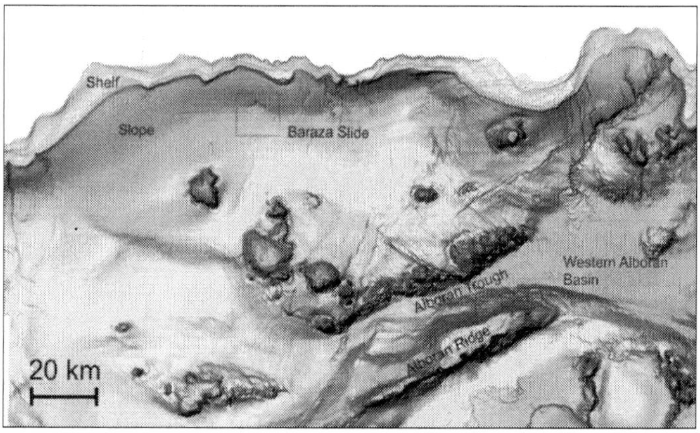

Fig. 10. Bathymetry of the Alboran Sea in the westernmost Mediterranean Sea showing the location of the Baraza Slide. The seafloor multibeam bathymetry, provided by the Spanish *Ministerio de Medio Ambiente y Medio Rural y Marino.*

The occurrence of mud-flow type movement ceases and the deformed resulting deposit is covered by a layer of late Pleistocene-Holocene sediments (Figs. 12&13). This level is affected by an unequal occurrence of structural (inverse faults, anticline folds) and outflowing

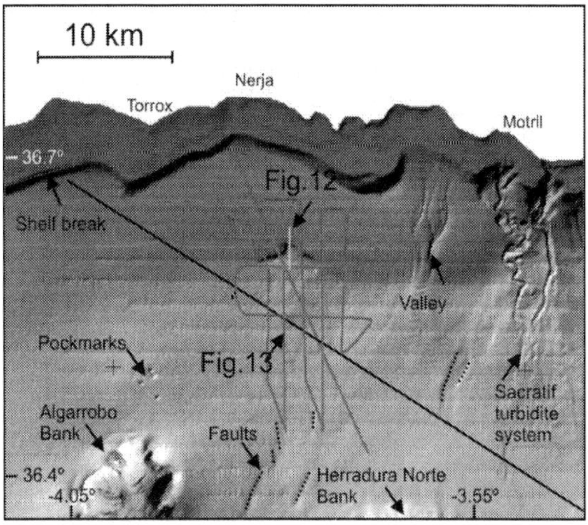

Fig. 11. Physiography of the present-day seafloor of the Baraza Slide area. Location of the Figs. 12 and 13 are also displayed.

features (water escape features). This fact suggest the occurrence of a slide-type movement at different time intervals up to recent times that affect different domains of the Baraza Slide although only the eastern sector of the slide still remains active. This type of movement affects to both the buried mud-flow deposits and to the overlying sediments. In fact, the slide scarp observed in the present day seafloor, results from the upslope propagation trough the overlying sediments of the ancient deeper scar that originated the mud-flow sediments. This

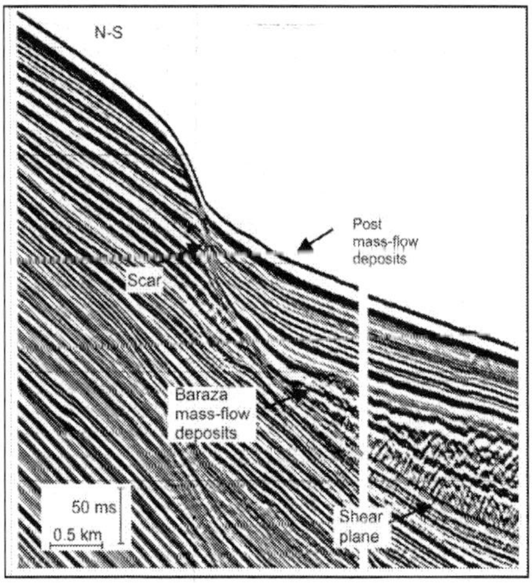

Fig. 12. Air-gun seismic profile illustrating how the scar plane extends downslope, going into subsurface sediments and joining with the shear plane. From Casas et al. (2011). See the location in Fig. 11.

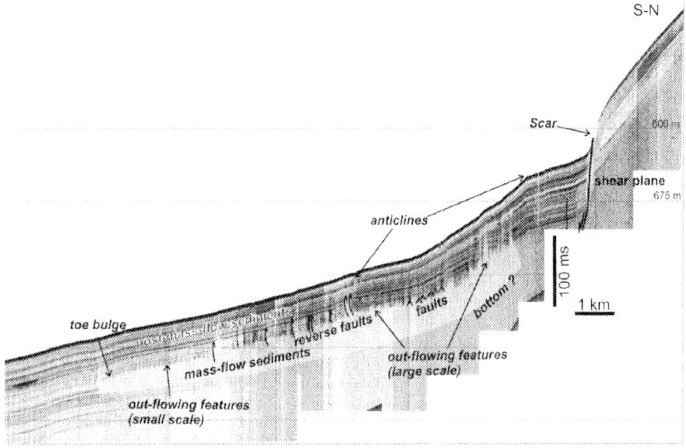

Fig. 13. TOPAS seismic profile which shows the main sedimentary structure and seismic facies of the Baraza Slide. From Ercilla et al. (2009). See the location on Fig. 11.

slide model combines extensional and compressional deformations. The extensional domain occurs at the slide scar where tensional failure occurs, and at foot of the scar where the post-mud-flow sediments are non-affected by outflowing features. The post-mud-flow level moves with a shear-dominated movement along the plane of the scar. Its stress originates the deformational an outflowing features; in fact, the anticline folds, water escape structures and their particular size distribution, the concomitant occurrence of slope breaks on the seafloor and upper surface of the mudflow deposits, all together would represent the structural criteria revealing the absorption of the compressional deformation.

2.5 Analysis of the physical-geotechnical properties and texture in the Pliocene-Quaternary sediments from the Madeira Abyssal Plain

This case study is centred on the analysis and the relationship of sedimentologic changes with the physical properties acquired on the upper 200 m of continuous coring at sites 950a to 952a of ODP Leg 157 (Baraza et al., 1996; Schmincke et al. 1995) which recovered more than 1000 m thick sediment sequence, from the deep floor of the Madeira Abyssal Plain (Fig. 14).

Four lithologic units define the Eocene to Quaternary sedimentary stratigraphy of the Madeira Abyssal Plain (Schmincke et al. 1995). The upper 200 m of cores studied corresponds to the Unit I (Pleistocene to middle Miocene). It consists of turbidite layers interbedded with pelagic nannafossil oozes. There are three primary types of turbidites (volcaniclastic, organic-rich and calcareous), originated from volcanic islands, the northwestern African margin and seamounts respectively. Results of grain-size analysis show homogeneity of grain-size distributions among all three sites. Attending to the percentages of the three main size-fractions, most of the analyzed samples are classified as silty-clays with only small amounts of sand (less than 17%).

Regarding the physical properties, they are controlled by the degree of compaction, rather than by changes in lithology. Differences in magnetic susceptibility (Fig. 15) appear to be related to changes in the mineralogical assemblage. Several high-amplitude peaks of magnetic susceptibility clearly differentiate between the highly magnetizable, volcanic-rich turbidites and the low magnetic organic and calcareous turbidites. Density and P-wave velocity appear mostly related to consolidation effects, but at a detailed scale shows variations related to the presence of thin, coarser-grained (silty) bases of some calcareous or organic-rich turbidite intervals.

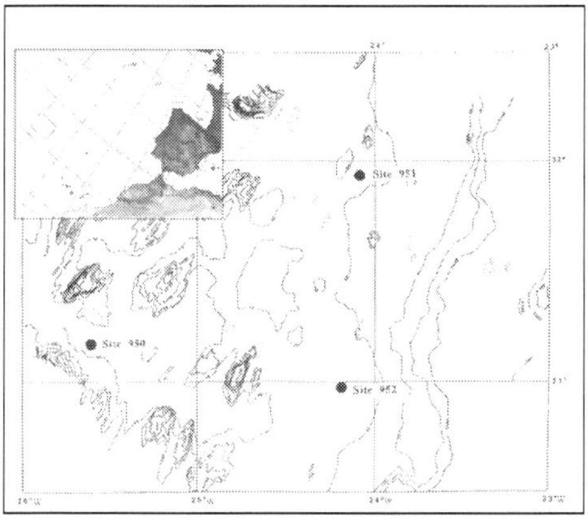

Fig. 14. Location map of ODP sites 950a, 951a and 952 in the Madeira Abyssal Plain.

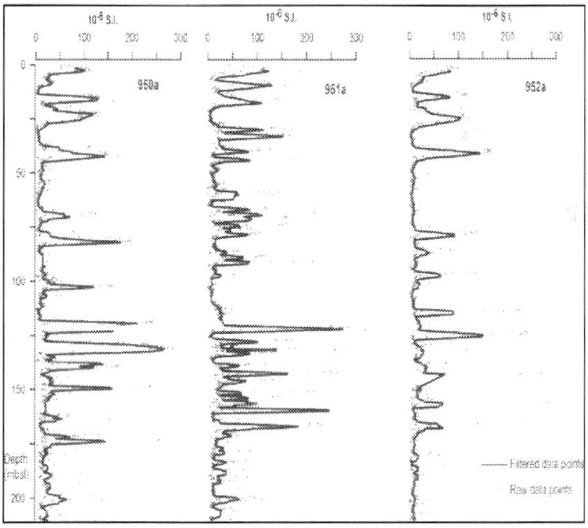

Fig. 15. Magnetic susceptibility profiles of the upper 200 mbsf for the three holes. From Casas et al. (2006).

Other properties as grain density, water content, porosity, undrained shear strength or calcium carbonate content show in general a range of variability associated to normally consolidated, fine-grained deep-sea sediments. Variations in grain density do not have a defined downcore trend, but they are mostly related to changes in composition (especially silica and carbonate) and primary bulk mineralogy. Grain densities are relatively higher in the volcanic turbidite intervals. The rest of index properties are mostly related to the decrease in porosity and increase in bulk density of the sediment due to compaction by overburden. Progressive consolidation due to overburden results in an expulsion of interstitial water and an increase of friction between particles on the sediment. The reduction of porosity and water content by progressive consolidation is the major factor controlling the increase in shear strength with depth (Schmincke et al., 1995). A plot of shear strength "versus" water content for samples from all holes (Fig. 16) shows the higher strength values for samples having water content around 40% and a sharp decrease in strength as water content increases to 60%. Nevertheless, changes in the rate of downcore increase/decrease of a given index properties may be related to compositional changes (carbonate and silica content).

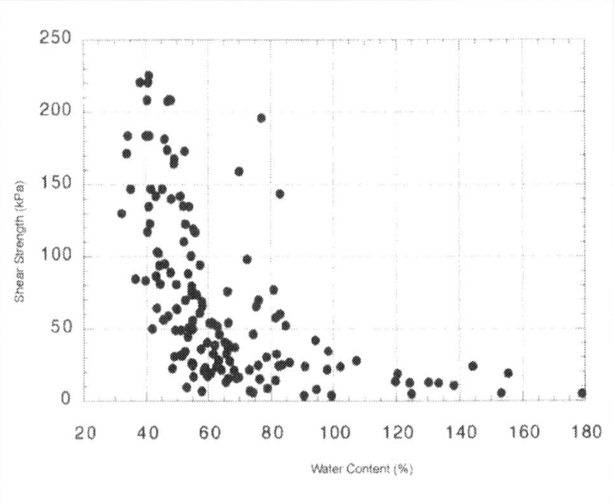

Fig. 16. Water content vs. shear strength for all samples studied. From Casas et al. (2006).

2.6 Sedimentary processes in sediments from mud volcanoes in the Anaximander Mountains (Eastern Mediterranean)

This case study focuses on the mud volcanoes Amsterdam, Kazan and Kula (Fig. 17) which are located in the Anaximander Mountains (SW Turkey continental margin) and are characterized by the presence of gas and gas hydrates (Woodside et al., 1998; Lykousis et al., 2004; Werne et al., 2004). A mud volcano is a positive relief constructed mainly of mud that typically emits a mixture of gas, water, and solid sediment derived from deep (Zitter et al., 2003). This sediment is usually composed by breccias comprising clasts of solid rock in a mud matrix (mud breccia). Mud volcanoes are dynamic and unstable sedimentary structures, accordingly sedimentary mass movements and gravitative flows may affect their flanks reworking the mud breccia.

Four cores with a maximum length of 131 cm were studied. Cores An05GC1, An07GC4 and An14GC1 are sited inside the crater of the Amsterdam, Kazan and Kula mud volcanoes respectively (2030, 1700 and 1636 meters water depth), and core An13GC1 at the outflowing masses that form the external flank of the Kula Volcano at 1636 meters water depth.

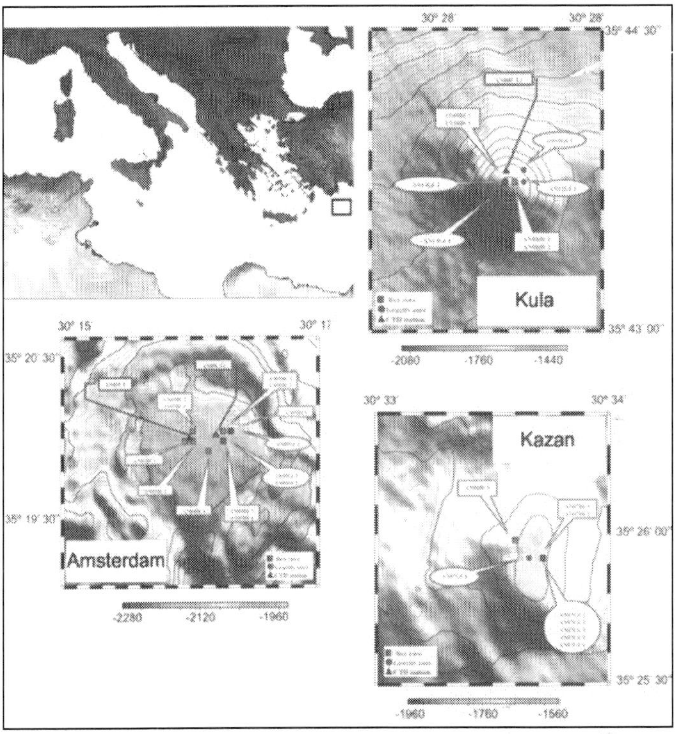

Fig. 17. Location of the Anaximander Mountains and mud volcanoes studied. Modified from Casas et al. (2006b).

The stratigraphy of cores An05GC1, An07GC4 and An14GC1 comprises the vertical stacking of mud breccia, whereas the core An013GC1 is defined by mud breccia that toward the top changes to hemipelagic mud (Fig. 18). Mud breccia is characterized by a high clay and silt content ranging between 67-56% and 19-30%, with a sand and gravel content of about 14%. Hemipelagic mud is characterized by high clay and silt content and a sand content lower than 3%.

Several correlations found between texture and density, magnetic susceptibility and P-wave velocity suggest that for the sediment cores located inside the mud volcanoes (An05GC1, An07GC4 and An14GC1) the physical properties are controlled by lithology and volcanic processes. This is supported for example by the correlations found between density and sand or silt (R= 0.47 and R=0.69 respectively) in core An07gc4 and between density and gravel or silt in core An14GC1(R=0.78, R=-0.75). The magnetic susceptibility of these cores seems to be controlled by the fine fraction. By contrast, the core located outside the Kula mud Volcano (An13GC1) displayed physical properties mostly related to consolidation effects and to the type of sediment at a detailed scale (hemipelagic mud vs. mud breccia), as occurs typically in deep sea fine-grained sediments. This is supported for example by the relationships found between density and core-depth (R=0.72). This suggests a restricted influence of volcanic processes outside the crater.

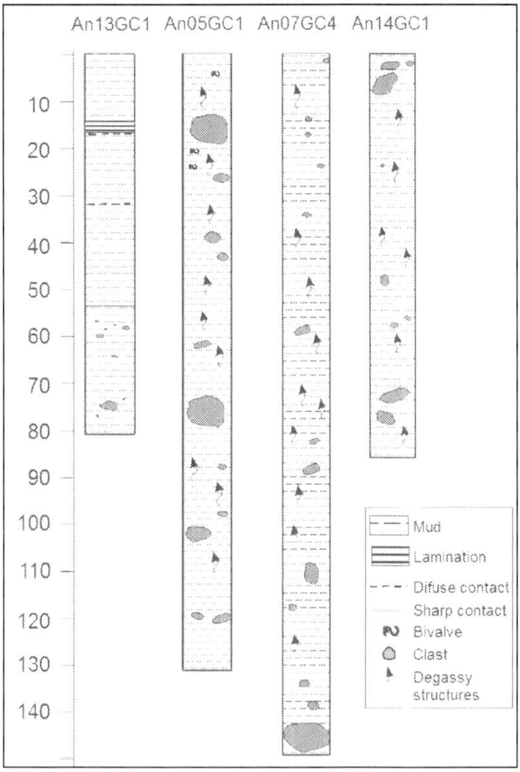

Fig. 18. Drawing of sediment logs showing the stratigraphy of the four cores studied.

Other variables measured as shear strength (Fig. 19) and its low or null correlation with, for example core-depth, could be reflect the effect of presence of porus gas-release structures resulting from depressurization after collection of samples containing gas. This could suggest recent fluid circulation, and therefore the possibility of current volcanic activity in the studied volcanoes. The depressurization may also be responsible for the relatively low strength values obtained. Mud breccia probably have greater absolute shear strength values because gas bubbles may affect sediment strength by decreasing grain-to-grain contacts (Briggs et al. 1996). However, the presence of gas hydrates could have the opposite effect.

2.7 Sediment stability on the Continental Slope and Basin of the Bransfield Basin (Antarctic Peninsula)

This case study is focused on the continental slope of the Antarctic Peninsula and adjacent deep sea areas, in the Bransfield Basin (Fig. 20). Type of sediments, sedimentary stratigraphy, and physical and geotechnical characterization of the sediments have been integrated. Four sediment gravity cores, located on the slope (at 1200 m water depth) and at the foot of the slope on the basin, at 1575 m water depth offshore the Antarctic Peninsula have been analysed (Fig. 20).

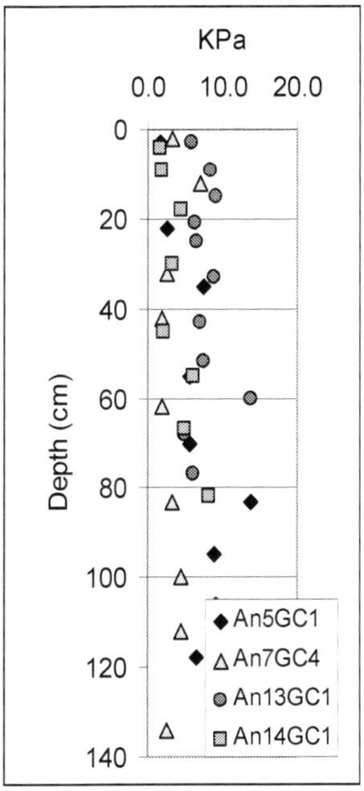

Fig. 19. Downcore variation of shear strength (kPa) for the Anaximander mud volcanoes sediments. From Casas et al. (2006b).

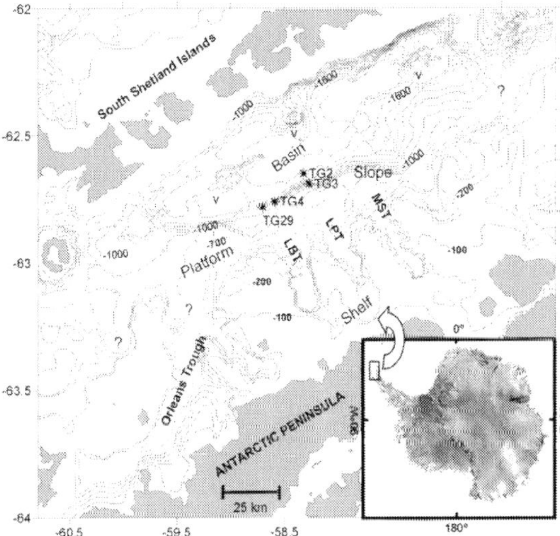

Fig. 20. Location of the study area and bathymetry of the Bransfield Basin. Black plots represent the position of studied sediment cores.

Based on the textural character and composition, the recovered sediments are defined as diamicton, unchannelised turbidite deposits (silty and clayey) and contourites which are associated to glacio-marine processes, gravitational flows and bottom currents, respectively. The stratigraphy indicates that sedimentation and related processes on the continental slope display lateral variations in short distances.

The physical properties, measured on the sediment cores (Fig. 21), appear to be controlled by textural differences and stratigraphy (sedimentary structures, vertical grain size trending). P-wave velocity decreases with increasing clay content (Fig. 22). Density and magnetic susceptibility values increase with the presence of gravel-size clasts or clusters of them. They also appear to be controlled by effects in two main macrofabric features: textural trend (vertical trending of grain size) and sedimentary structures (e.g. parallel and cross laminations). From a quantitative point of view, the statistical correlations found between textural and physical properties cannot explain some particular trends. This may imply that a secondary data set (e.g., microfabric, mineralogy, chemical activity, biological activity, or mechanical processes) is necessary to explain and understand variations in the physical properties.

Consolidation and shear strength properties are similar in all cores. Sediments in the Bransfield Basin are normally consolidated, except for slope core TG3, where lightly overconsolidation condition can be considered (Table 1). This overconsolidation results from the occurrence of mass wasting processes, a dominant process on the Bransfield slope (Ercilla et al., 1998; García et al., 2009). The loss of about 9 m of sediment overburden is suggested as the most probably cause. According to the stability under gravitational loading

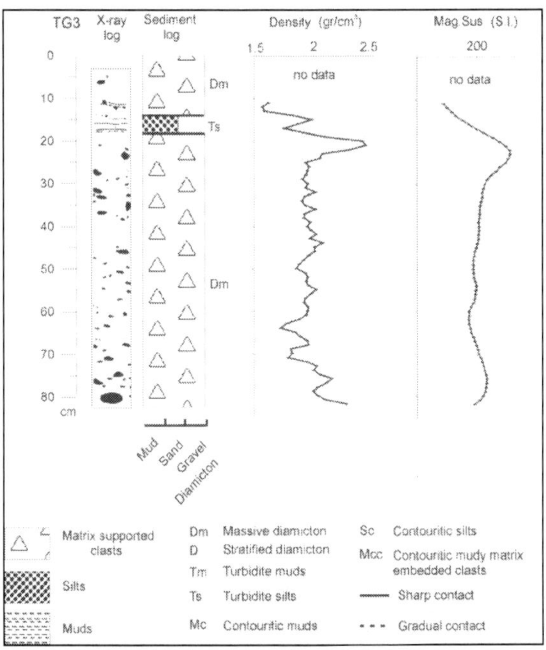

Fig. 21. X-ray log, sediment log, density and magnetic susceptibility records for the slope core TG3. From Casas et al. (2004).

concepts we can establish that the present slope with a maximum gradient of 20° is stable. This data is important because it could explain why the slope, as others in high latitudes are relatively steep with high gradients even when they have a sedimentary control. Nevertheless, different instability features have been seismically observed or deducted from the consolidation tests. To explain their presence we have to resort to external triggering since all sediments studied are stable by themselves.

Based on the geologic and oceanographic framework, volcanic activity, glacial loading-unloading, bottom currents and earthquakes are the triggers considered as potential mechanisms to induce sedimentary instability. Active mid-ocean ridge basalt volcanism characterizes the central Bransfield Basin (Gràcia et al., 1996). Volcanic emplacement may have caused earthquake shacking, triggering instability processes on the studied area. Ice sheet loading on a continental shelf may induce far-field pore-pressure effects (Mulder and Moran,1995), and the isostatic rebound related to the retreat of glacial loading could produce a crustal uplift with a probable influence on the instability processes (Anderson, 1999). Bottom-current-related processes have played a role in the most recent sedimentary history of the basin, favouring the formation of contourite drifts and associated moats (Fig. 23) at the foot of the continental slope (Ercilla et al. 1998). The formation of a moat at the foot of a slope as a consequence of an erosive current could become a trigger to initiate instability processes on the adjacent slope. Nevertheless the Bransfield Basin is considered to be a zone with moderate seismic activity with earthquake magnitudes varying between 4.8 and 6 (Ibáñez et al., 1997). Infrequent strong earthquakes with magnitudes greater than 6 can be tentatively considered as a major mechanism for instability features (Baraza et al., 1990). Likewise, the high activity of weaker earthquakes (Mb 2-4) registered in the area (Robertson et al., 2001) could be taking into account because they can reduce the sediment shear strength (Hampton et al., 1996).

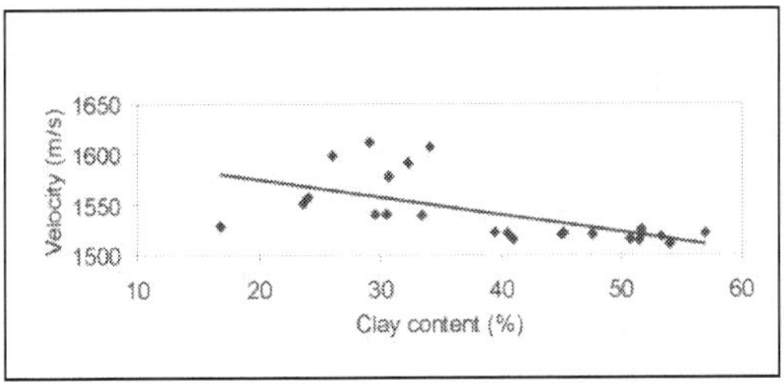

Fig. 22. Comparison of P-wave *vs* clay content. Vp = 1608-1.75 ∗ (% clay). From Casas et al. (2004).

Core	Depth (cm)	Density (gr/cm3)	OCR	σ'e
TG2	85	1.35	9.435	22.9
TG2	140	1.70	2.660	15.3
TG3	55	1.88	16.323	70.5
TG4	74	1.81	4.639	20.8
TG29	95	1.54	5.673	22.6

Table 1. Results of incremental consolidation tests. Legend: OCR (over consolidation ratio); σ'_e (excess maximum past stress).

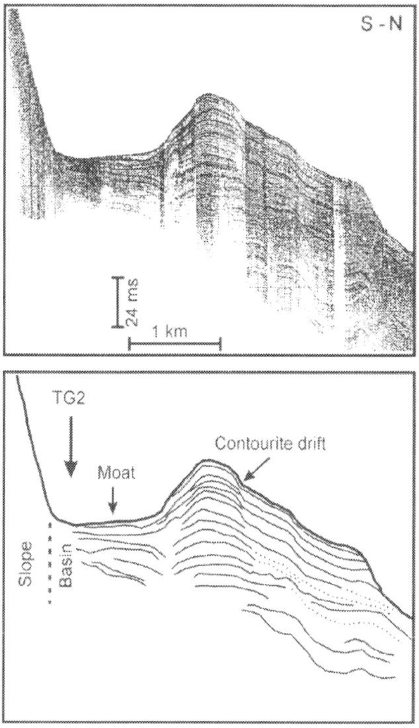

Fig. 23. Seismic profile illustrating the presence of a contourite drift and the associated moat at the foot of slope. Modified from Ercilla et al. (1998).

3. Contributions from multidisciplinary and multiscale analysis of mass-movement features and deposits

Each of the above case studies has been focused in one specific topic related to mass-movement processes and their resulting sedimentary products. They also represent multidisciplinary and multiscale approaches in different geological contexts. The integration of different study techniques is not an easy task because they have differences in relation to study scale, resolution and volume of data. In addition, commonly it must be considered the limited availability of the full range of data due to their economic cost or lack of technical and scientific development. Although this constraint, the overall integration of the main results obtained from the case studies give important inputs for the knowledge of submarine mass-movements:

Contribution 1. Characterization of a continental margin and historic register of submarine mass-movement features observed in a continental slope: The geologic characterization of a continental margin offers information about the potential key factors that may affect to sedimentary instability. The genesis of morpho-sedimentary and morpho-structural features or fluid dynamic in sediments can also explain the presence of mass-movements in an area surrounded by stable sediment. The study of continental margin also allows assessing the role played by mass-movements in the sediment transfer to deep sea areas as well as shaping the seafloor.

The complexity of failuring styles and evolution of sedimentary instabilities as well as the identification of triggering mechanisms require the joint analysis of all those geological parameters that characterize the area of study. In this sense, the above case studies establish that continental margin geological characterization provides information on those critical factors that can affect the stability of continental slope (bottom and sub-bottom), and allows defining essential key quantitative and qualitative parameters in the indirect study of the mass-movements. These parameters can be divided into five main groups, which comprise those critical factors to consider in any detailed analysis of seafloor and subbottom stratigraphy:

1. Physiographic parameters: gradients; width; water depth; type of boundaries between physiographic domains (sharp, gradual).
2. Morphologic parameters of mass-movements deposits (both plan view and dimensions): plan view and cross-sections shape measurements; water depth extension; curvature and/or irregularities of the downslope concavity of the scar and downslope convexity of the distal end of the slide; offset of the scar and lateral variations; seafloor relief.
3. Sedimentary parameters of mass-movements deposits: internal acoustic facies and their vertical and lateral variations; slide planes and their morphoacoustic characteristics in their both expression, plane or level; presence o lack of tensional depression; type of the distal end of the slide, frontal confined or unconfined.
4. Surrounding unfailed sedimentary systems: acoustic and seismic facies; sedimentary structure; growth pattern.
5. Deformational features of mass-movements deposits and surrounding unfailed sediments: structural, as folds and faults; fluid (water and gas)-dynamic features, and their lateral and vertical distribution.

On the other hand, the regional study of distribution, size and morphology of submarine sedimentary instabilities is also important to assess their effect on the slope sedimentary systems and their role as a mechanism to shape continental margins. Furthermore, they provide criteria for understanding the direction of travel, which is roughly perpendicular and sub-perpendicular to bathymetry. When quantifying the number of submarine mass-movements in terms of depth, it appears that they are more abundant on continental slope (Pratson and Laine, 1989, Booth et al., 1993). The case studies 2.2, 2.3, 2.5 and 2.7 also enable broadly to define two main areas on the continental slope: an evacuation or depletive area and a depositional or accumulative area. Depletive area is characterized by thinning of the sliding deposits, tensional structures, outflowing escape features, and formation of erosional surfaces such as canyons, gullies and scars. That is, upper slope are areas where erosive gravitational features predominate. The accumulative area is characterized by depositional features such as debrites, mass flow deposits, levees bordering channles, turbidite boulding up depositional lobes and the presence of compressional structures, and even outflowing escape features.

Contribution 2: Understanding the dynamics of failures: Regional studies of mass-movements indicate that similar sediments show differences in the potential stability and behavior. Distribution and variability of mass-movements and their related triggering should be studied individually in order to determine the local conditions of failure events. One way to explain differences is through the integration of the geotechnical models with the morphologic and sedimentary observations on seismic records.

Mass-movement is a common process in different margins, both in submarine canyons and open slope environments (e.g., case studies 2.2, 2.3, 2.6, 2.7). Seismic analysis offers indirect observations of the tectono-sedimentary framework where the mass-movement features occur and how we see them, being able to define slide plane, internal pattern, scale of failure, slide geometry, run-out distances, etc. Among the mass-movement deposits identified through the case studies, the Baraza Slide (Alboran Sea) and Torreblanca Slide (Ebro margin) are the best features for defining the sediment dynamics and their evolution.

On the basis of its well-defined morphology, geometry, internal pattern, and the vertical trend of the seismic facies and surfaces, the sediment dynamics and evolution of the Baraza Slide is defined by three stages: metastable, flowing, and sliding (Fig. 24). The flowing stage began when the metastable equilibrium was broken and occurred when the downslope-oriented shear stress exceeded the shear strength, resulting in the development of an instability plane within the stratified slope deposits. The sliding stage occurred when the Baraza Slide was reactivated with a different instability mechanism: a slide-type movement. This movement affected both the buried mass-flow deposits and the overlying sediments, which moved with a shear-dominated movement along the plane of the scar.

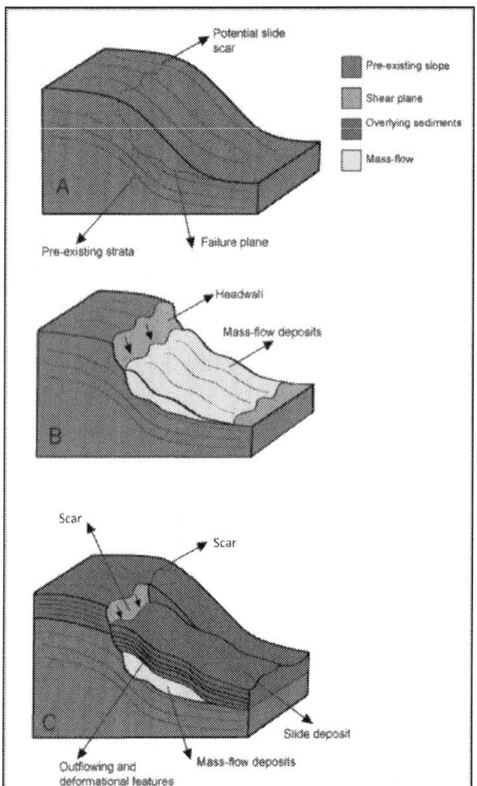

Fig. 24. Evolution of the Baraza Slide. Modified from Casas et al. (2011).

Regarding the dynamics of Torreblanca Slide its geometry and internal pattern allow defining its evolution in four stages: metaestable, triggering, sliding and stopping stages (Fig. 25). During the initial moment the slope was in metastable equilibrium mainly governed by downslope gravitational forces. Under this equilibrium some bedding planes acted as a zone of weakness. This is suggested because the subsequent shear plane is subparallel to older strata and is roughly parallel to the regional slope. The triggering stage occurred when the downslope-oriented shear stress exceeded the shear strength, resulting in the development of an instability bed within the stratified slope deposits. The initial combined rotational and horizontal displacement evolves to the sliding stage with a tensional regime which was responsible of the extensional features observed along the moving sediment package. The stopping stage began when the downslope-oriented forces decreased because the action of friction along the shear plane. The slowing of the motion produced a compressional regime which was responsible of the formation of compressional deformation on the toe of mass movement.

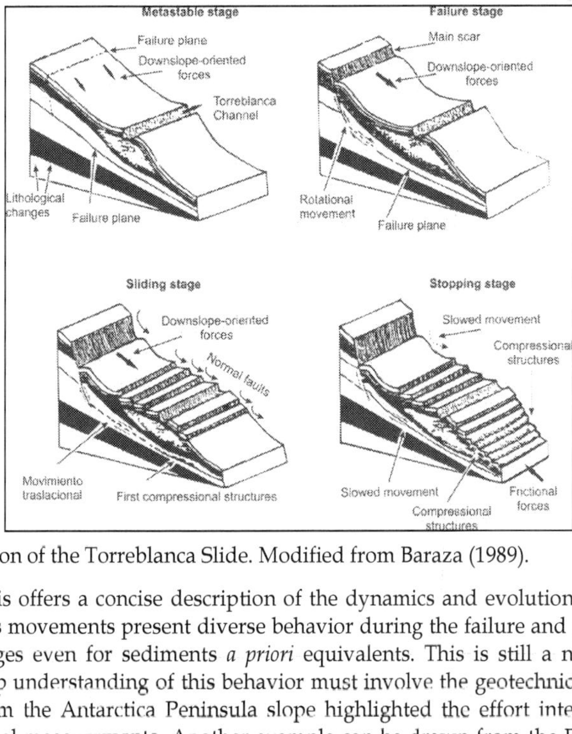

Fig. 25. Evolution of the Torreblanca Slide. Modified from Baraza (1989).

Seismic analysis offers a concise description of the dynamics and evolution of instabilities. Although mass movements present diverse behavior during the failure and post-failure (re-activation) stages even for sediments *a priori* equivalents. This is still a non-well known point. The deep understanding of this behavior must involve the geotechnical analysis. The case study from the Antarctica Peninsula slope highlighted the effort integrating seismic and geotechnical measurements. Another example can be drawn from the Ebro slope if we contrast our observations with a geotechnical model done by Baraza et al. (1990) based on surficial cores (<3 m long) on the central and southern sectors of that slope.

This model defines two different areas on the basis of physical and geotechnical properties: the upper slope (<500m) and the lower slope (>500m). On the upper slope, prodeltaic mud with high silt content is dominant. The average water content is 33% dry weight, slightly below the liquid limit, which is about 34%. The plasticity index is about 15% and the sediment is highly to moderately overconsolidated (OCR as high as 8).On the lower slope predominates hemipelagic. In addition, the water content is higher than in the upper slope (approaching 90%) and is above the liquid limit (ranging from 55 to 75%), the plasticity index is higher and the degree of overconsolidation is lower (OCR 2-3). Normalized strength parameter S (ratio of strength to consolidation stress) for normal consolidation are both lower, whereas the cyclic strength degradation factor (Ar) is higher than that of the upper slope. According to the geotechnical results, the upper slope deposits are slightly more stable under undrained static loading conditions while the lower slope is more stable under drained or very long term static conditions. Maximum slopes in Ebro area appear to be stable under static (gravitational) loading. Following this model nevertheless, localized instability might be produced between 200 and 700 m depth by a combination of oversteepening (5° and 10°) and infrequent, intense seismic loading.

The proposed geotechnical zonation could explain why most of debris flow deposits are concentrated in the lower slope, since there the sediment has a higher plasticity than the sediment in the upper slope (Casas et al., 2003c). Likewise, the geotechnical zonation could help to understand why the large scale slides have their scars on the upper slope; this is because between 200 and 700 m water depth the slope would the more susceptible area to failure triggered by seismic loading. But this model is certainly insufficient to explain the variability of settings, types, scales, and geometries of the mass movement features as well as depositional environments where they occur. This fact suggests that the distribution and variability of mass movement features and their probable triggering mechanisms should be studied individually from a geotechnical point of view in order to know local conditions of stability or failure. This proposed approach may be a good way to know why one region of seafloor remains intact whereas the neighbouring sector fails and why it fails in the way that it does. Likewise, in-situ geotechnical measurements (i.e. shear strength and pore pressures) could represent an important boost to better understand of submarine mass movements.

Contribution 3: The need of knowing the physical parameters: The physical properties are important for slope instability analysis because they offer basic information about sediment type and are necessary in the assessment of the potential stability of a slope area. Our studies suggest that sediment fabric is a key feature for the physical properties, and the mineralogical properties have a more relevant effect in the absolute value of those properties.

The continuous and high-resolution log of physical properties (basically bulk density, P-wave velocity and magnetic susceptibility) of marine sediments are important to understand the different sedimentary environments and the geological events that occur in them. Different studies have focused on the relationship between physical properties and textural parameters of marine sediments, since they depend largely on lithology, grain size and composition. The bulk density for example, is related to porosity and grain density but is also partially controlled by the grain size. P-wave velocity is controlled by the porosity, carbonate content or clay minerals. Magnetic susceptibility relates to the sediment composition and then changes in magnetic susceptibility can be important parameters to obtain information about sediment provenance, palaeoclimate, bottom-water flow conditions and regional stratigraphy. The physical properties are strongly influenced by diagenetic processes, such as the decrease of porosity by compaction, cementation or dissolution of carbonate. These properties also provide useful information related to geotechnical properties which are essential in estimating the stability of a particular area.

The physical properties obtained in our case studies have allowed us to study in detail the stratigraphy, vertical trends in cores and their lateral correlation. It has been useful also for refining the characteristics of the sediments described as well as into the stability assessment.

The knowledge or definition of one type of sediment exclusively from its physical properties is a complex exercise. The problem arises because in practice, the physical properties depend on many variables which in turn are interrelated in complex ways. We are therefore in front of a multi-variable problem, mathematically complex and do not have a single solution. But it is possible to state different statistically acceptable relationship between physical properties and sediment variables. These relationships are not constant, and depend largely on the characteristics of each type of sediment (geological environment). There are patterns in the records of density, magnetic susceptibility and P-wave velocity associated with the "style" of sedimentation. This can be observed on relatively similar records of cores recovered from the slope of the Antarctic Peninsula and the Anaximander mud volcanoes. In both cases the "style" of diamicton (glacial-marine origin) and the mud breccia (volcanic origin) is similar and contrasts with strictly marine sediments (e.g., turbidites). Obviously the diamicton and mud breccia have a completely different genesis, and their composition is not at all comparable. But they have in common to be sediment matrix containing clasts of various sizes and shapes. It thus appears that sediment fabric is of key importance in the physical properties and the mineralogical composition have a significant effect on the absolute value of these properties.

In our case studies where marine processes dominate, physical properties are basically controlled by the consolidation (compaction of the sedimentary column) with differences of minor order that roughly correspond to lithological changes. Besides that in cores with deposits associated to glacio-marine and mud-volcanism processes, the physical properties seem to be controlled by sediment characteristics; the effect of consolidation, if observed, is secondary.

Contribution 4: Forces controlling triggering of mass-movements: Destabilizing forces and processes that can trigger mass-movements depend on both geologic regional factors and specific processes of sedimentary environments. The characterization of revelant, pre-disposition and trigger factors could provide a good approach for defining and mapping the instability hazard.

The destabilizing forces and processes that originate the mass-movement deposits presented in the different case studies depend on both the regional framework and local factors. Both condition particular processes for each sedimentary environments. The characterization of revelant, pre-disposition and trigger factors could provide a good approach for defining and mapping the instability hazard (Leroueil et al., 1996). The revelant factors are evidences, for example, of previous mass movement, creep etc. The pre-disposition factors are those conditioning the stability of a slope. The trigger factors are the processes that effectively cause one mass movement (Table 2).

The spatial distribution, variability, types of mass movement deposits and their evolution in the Ebro and Alboran continental slope are influenced by independent factors. These factors include unequal contribution of sediment, the failure frequency, thickness of the affected sediment, slope gradients, and proximity to the epicenters of earthquakes. Relatively high slopes present in the Ebro (2.5° to 5.8°) and Alboran (3° to 3.5°) continental slopes and the rapid deposition directly onto the upper slope during periods of lowstand stages of sea level were probably the main conditioning factors destabilizing the slope. The high sedimentation rate is responsible of sub-consolidation sediments and their consequent decrease in shear strength. These elements are probably the main predisposing factors, however, failure could have been triggered by seismic shaking (or by tectonic-related activity), and also by bottom currents in the Alboran slope (Ercilla et al., 2011). The potential link between high slope gradients, under consolidation, earthquakes, and instability along continental margins has been discussed by numerous authors (e.g., Hampton et al. 1996).

Factors	Ebro	Alboran	Cádiz	Anaximander	Bransfield
Predisposition	Tectonics Sedimentation rates Slope gradients Lithostatic load	Tectonics Sedimentation rates Slope gradients Lithostatic load	Gas and gas hydrates in sediments Tectonics	Gas and gas hydrates in sediments	Seismic activity
Trigger	Lithostatic load ? Earthquakes Storms Internal waves	Lithostatic load? Earthquakes Storms Internal waves	Gas Tectonic movements		Volcanic activity Earthquakes
Revelant	Previous instabilities	Previous instabilities	Pockmarks Diapirs	Previous instabilities	Previous instabilities

Table 2. Main conditioning factors for slope instability identified in the different case studies presented.

In the continental slope sediments of the Gulf of Cadiz, the area affected by mass movement deposits occurs in a domain with sediments containing free gas and then it is the critical element in sediment stability of an area. This presence of gas has been evidenced by acoustic turbidity and also "bright spots" features in seismic records. The presence of pockmarks and diapirs are revelant factors, whereas the presence of gas hydrate could be a predisposing factor, although its limited presence mainly associated to diapirs, suggests that the dissociation of these hydrates as triggering factor is not relevant. In the area of the Anaximander Mountains, as happens in the Gulf of Cadiz, the distribution of gas hydrate in the sediment appears to be limited, although in this case it occurs within the mud volcanoes and the southern flank of the Volcano Amsterdam. This suggests that although their presence can be considered a pre-disposition factor, its dissociation would not be a relevant triggering.

In the continental slope of the Antarctic Peninsula, contrasting with the continental margin of the Ebro, lithostatic load (sub-consolidation) is not a pre-disposition factor. Based on the geotechnical analysis results, the maximum slope measured in that area (20 °) would be stable according to the concepts of stability under static load (lithostatic) and then instabilities observed, could not be explained by characteristics of the sediment because they are stable themselves. Based on the geological framework: volcanic activity, bottom currents, tidal currents, glacial loading-unloading and/or earthquakes can be considered as triggering factors.

4. Acknowledgements

The research was done in the framework of the Projects CONTOURIBER (Ref. CTM2008-06399-MAR), and MONTERA (CTM-14157-C02), of the Spanish Science and Innovation Ministry. The "Grupo de Márgenes Continentales" (GMC) also thank the Generalitat de

Catalunya for providing economic support in the framework of SGR- 2009 (Exp. 00707). Likewise, we thank the Spanish "Ministerio de Medio Ambiente y Medio Rural y Marino" for some of the bathymetric data presented and Seismic Micro-Technology, Inc. for supporting us with the Kingdom Suite Program.

5. References

Alonso, B. & Maldonado, A. (1990) Late Quaternary sedimentation patterns of the Ebro turbidite system (north-western Mediterranean): Two styles of deep-sea deposition. En: Nelson, C.H., Maldonado, A., (Eds.), The Ebro Margin. *Marine Geology* 95: 353-377.

Anderson, J.B. (1999). *Antarctic Marine Geology*. Cambridge: Cambridge University Press. 298 pp.

Arnau, P., Liquete, C. & Canals, M. (2004). River mouth plume events and their dispersal in the northwestern Mediterranean Sea. *Oceanography* 15: 22-30.

Baraza, J., 1989. Procesos de Edificación y Características Geotécnicas del Talud del Ebro. Tesis Doctoral. Universitat Politècnica de Catalunya.

Baraza, J., Lee, J., Kayen, R. & Hampton, M.A. (1990). Geotechnical characteristics and slope stability on the Ebro margin, Western Mediterranean. En C.H. Nelson, A. Maldonado (Eds.), The Ebro margin. *Marine Geology* 95: 379-393.

Baraza, J. & ODP Leg 157 Shipboard Scientific Party. 1996. Physical properties of sediments from Madeira Abyssal Plain: results from ODP Leg 157. Geogaceta 20 (2): 146-148.

Baraza J., Ercilla G. & Nelson C.H. (1999). Potential geologic hazards on the eastern Gulf of Cadiz slope (SW Spain). *Marine Geology* 155: 191-215

Bonnin, J., Olivet, J.L. & Auzende, J.M. (1975). Structure en nappe a l'ouest de Gibraltar. *Comptes Rendus de l'Académie des Sciences* 280 (5): 559–562.

Booth, J.S., O'Leary, D.W., Popenoe, P. & Danforth, W.W. (1993). U.S. Atlantic continental slope landslides: their distribution, general attributes, and implications. In: Submarine Landslides: Selected Studies in the U.S. Exclusive Economic Zone (EdsW.C.Schwab, L.J.Lee and D.C. Twichell), *US Geological Survey Bulletin*: 14–22.

Briggs, K.B. & M.D. Richardson (1996). Variability in situ shear strength of gassy muds. *Geo-Mar Letters* 16: 189-195.

Casas, D., Ercilla, G., Baraza, J., Alonso, B. & Maldonado, A. (2003a). Recent Mass Movement Processes on the Ebro Continental Slope (NW Mediterranean). *Marine and Petroleum Geology* 20: 445-457.

Casas, D., G. Ercilla & J. Baraza (2003b). Acoustic evidences of gas in the continental slope sediments of Gulf of Cadiz (E Atlantic). *Geo-Marine Letters* 23: 300-310.

Casas D., Ercilla G., Lee H., Alonso B., Maldonado A., & Baraza J. (2003c). Submarine Mass Movement on The Ebro Slope. *in* J. Locat and J. Mienert (Eds.), *Submarine Mass Movements and Their Consequences. Advances in Natural and Technological Hazards Research*. 19, 393-400. Netherlands: Kluwer academic publishers.

Casas, D., H. Lee, G. Ercilla, R. Kayen, F. Estrada, B. Alonso, J. Baraza & F. Chiocci. (2004). Physical and geotechnical properties and assessment of sediment stability on the continental slope and basin of the Bransfield Basin (Antarctica Peninsula). *Marine Georesources and Geotechnology* 22 (4): 253-278.

Casas D., Ercilla G., Lee H., Alonso B., Maldonado A., Baraza J. & Coakley, B. (2006). Physical properties and their relationship to texture and consolidation effects in

Pliocene-Quaternary sediments from Madeira Abyssal Plain. *Marine Georesources and Geotechnology* 24(4): 265-286.

Casas, D., Ercilla, G., Lykousis, V., Ioakim, Ch. & Perissoratis, C. (2006b). Physical porperties and their relationship to sedimentary processes and texture in sediments from mud volcanoes in the Anaximander Mountains (Eastern Mediterranean). *Scientia Marina* 70 (4): 643-659.

Casas, D., Ercilla, G., Yenes, M., Estrada, F., Alonso, B., García, M. & Somoza, L., (2011). The Baraza Slide: Model and dynamics. *Marine Geophysical Researches,* doi: 10.1007/s11001-011-9132-2.

Castellón, A., Fint, J. & Garcia Ladona & E. (1990). The Liguro-Provençal-Catalan current (NW Mediterranean) observed by Doppler profiling in the Balearic Sea. *Science on the March* 45: 269-276.

Dañobeitia, J.J., Alonso, B., and Maldonado, A. (1990). Geological framework of the Ebro continental margin and surrounding areas. *Marine Geology* 95: 265–287.

Dewey, J.F., Helman, M.L., Turco, E., Hutton, D.H.W. & Knott, S.D. (1989). Kinematics of the west Mediterranean: alpine tectonics. *in*: M Coward (ed.), Special Publication Geological Society of London, London, vol. 45, pp. 265–283.

Ercilla, G., Baraza, J., Alonso, B. & Canals, M. (1998). Recent geological processes in the Central Bransfield Basin (Western Antarctic Peninsula). *in*: Stoker, M. S., Evans, D., Cramps, A. (Eds.*), Geological Processes on Continental Margins: Sedimentation, Mass-Wasting and Stability*. Geol. Soc., Spec. Publ., London,129, pp. 205-216.

Ercilla, G., Estrada F., Yenes, M., Casas, D., Alonso, B., Farran, M. & Sagas team (2009). The Baraza landslide (Northern Alboran Sea); defining its dynamic. *Rendiconti Online della Società Geologica Italiana* 7: 163-165.

Ercilla, G., Juan, C., Estrada, F., Casas, D., Alonso, B., Palomino, D., Vázquez, J.T., Llave, E., Hernández-Molina, F.J., Medialdea, T., Gorini C. & CONTOURIBER and MONTEREA TEAMS (2011). A new model for recent sedimentation in the Alboran Sea (SW Mediterranean). IAS Se 28th IAS Meeting of Sedimentology. Abstract book, p. 16.

Field, E. & Gardener, V. (1990). Pliocene-Pleistocene growth of the Rio Ebro margin, northeast Spain: A prograding-slope model. *Geological Society American Bulletin* 102 (6): 721-733.

Frey Martinez, J., Cartwright J. & Hallw, B. (2005). 3D seismic interpretation of slump complexes:examples from the continental margin of Israel. *Basin Research* 17: 83–108.

Garcia, M., Ercilla, G. & Alonso B. (2009). Morphology and sedimentary systems in the Central Bransfield Basin, Antarctic Peninsula: sedimentary dynamics from shelf to basin. *Basin Research* 21: 295–314.

Gràcia, E., Canals, M., Farran, M., Prieto M.J. and Sorribas & J. GEBRA team (1996). Morphostructure and evolution of the Central and Eastern Bransfield Basins (NW Antarctic Peninsula). *Marine Geophysical Researches* 18: 429-448.

Hamilton, E.L. & Bachman R.T. (1982). Sound velocity and related properties of marine sediments. *Journal of the Acoustical Society of America* 72: 1891-1904.

Hampton, A.M., Lee & H., Locat, J. (1996). Submarine landslides. *Review of Geophysics* 34 (1): 33-59.

Hernández-Molina F.J., Llave E., Stow D.A.V., García M., Somoza L., Vázquez J.T., Lobo F.J., Maestro A., Díaz del Río V., León R., Medialdea T. & Gardner J. (2006). The contourite depositional system of the Gulf of Cádiz: A sedimentary model related to the bottom current activity of the Mediterranean outflow water and its interaction with the continental margin. *Deep Sea Research II* 53:1420-1463.

Hueneke H. & Mulder T. (2011). *Dep-Sea Sediments. Developments in Sedimentology*, 63. Series Editor, A.J. van Loon. Elsevier Science, pp. 750.

Ibáñez, J.M., Morales, J., Alguacil, G., Almendros, J., Ortiz, R., Pezzo, E., Posadas, A. & Luzón, F (1997). Regional seismic activity around Livingston and Deception islands from 1992 to 1996: shallow and intermediate-depth earthquakes. *Boletín de la Real Sociedad de Historia Natural* (Sección Geología) 93: 1-4.

Johson, G.R. & Olhoeft G.R. (1984). Density of Rocks and Minerals. *in*: Carmichael, R.S. (ed.), CRC Handbook of Physical Properties of Rocks (Vol. 3): Boca Raton, FL (CRC Press, Ins.), 1-38.

Judd, A. & Hovland, M. (2007). *Seabed Fluid Flow. The Impact on Geology, Biology and the Marine Environment.* Statoil, Norway, 475 pp.

Maldonado A., Somoza L. & Pallarés L. (1999). The Betic orogen and the Iberian-African boundary in the Gulf of Cadiz: geological evolution (central North Atlantic). *Marine Geology* 155: 9-43.

Masson, D., Harbitz, C.B., Wynn, R.B., Pedersen, G. & Lovholt, F. (2011). Submarine landlsides: processes, triggers and hazard prediction. *Philosophical Transactions of the Royal Society* A2006 364: 2009-2039.

Mienert, J. (1984). The importance of carbonate content in the acoustic stratigraphy of Panama Basin. *Marine Geology* 54: 237-247.

Mulder, T. & Moran, K. (1995). Relationship among submarine instabilities, sea level variations, and the presence of an ice sheet on the continental shelf: An example from the Verrill Canyon Area, Scotian Shelf. *Paleoceanography* 10: 137-154.

Nobes, D.C., H. Villinger, E.E. Davis & L.K. Law. (1986). Estimation of marine sediments bulk physical properties at depth from seafloor geophysical measurements. *Journal of Geophysical Research* 91: 14033-14043.

Nobes, D.C., Mienert, J. & Dirksen, G. J. (1991). Lithologic control of physical-property interrelations. In: Ciesielski, P.F., Kristoffersen, Y., et al ., Proc. ODP, Int. Repts., 114: College Stations, TX (Ocean Drilling Program): 657-669.

Nobes, D.C., Langseth, M.G., Kuramoto, S., Holler, P. & Hirata, N. (1992). Comparison and correlation of physical-property results from Japan Sea Basin and Rise Sites, legs 127 and 128. *In*: Tamaki, K., Suyehiro, K., Allan, J., McWilliams, M., et al., Proc. ODP, Int. Repts., 127/128, Pt. 2: College Station, TX (Ocean Drilling Program), 1275-1296.

Lastras, G., Canals, M., Hughes-Clarke, J.E., Moreno, A., DeBatist, M., Masson, D.G., & Cochonat, P. (2002). Seafloor imagery from the BIG'95 debris flow, western Mediterranean. *Geology* 30 (10): 871-874.

Leon, R. Somoza L., Giménez-Moreno, C. J., - Dabrio, C.J., Ercilla ., - Praeg, D., Díaz del Río, V., & Gómez Delgado, M. (2006)-. A predictive numerical model for potential mapping of the gas hydrate stability zone in the Gulf of Cadiz. *Marine and Petroleum Geology* 26: 1564-1579.

Leroueil, S., Vaunat, J., Picarelli, L., Locat, J., Lee, H., Faure, R., 1996. Geotechnical characterization of slope movements. In Proceedings of the 7th International Symposium on Landslides, Trondheim. 1: 53-74.

Lykousis, V., Alexandri, S., Woodside, J., Nomikou, P., Ioakim, Chr., Sakellariou, D., De Lange, G., Daehlman, A., Casas, D., Rousakis, G., Ballas, D. & Perissoratis, C. (2004). New evidence of extensive active mud volcanism in the Anaximander Mountains (Eastern Mediterranean): The ATHINA mud volcano. *Environmental Geology* 46(8): 1030-1037.

Pratson L.F. & Laine, E.P. (1989). The relative importance of gravity-induced versus current-controlled sedimentation during the Quaternary along the mideast U.S. outer continental margin revealed by 3.5 kHz echo character. *Marine Geology*, 89, 87-126.

Robertson, S.D., Wiens, D.A., Dorman, L.M. & Vera, E (2001). Seismicity and tectonics of the South Shetland Islands region from a combined land-sea seismograph deployment. Eos. Trans. AGU, 82 (47), Fall Meet. Suppl., Abstract S12B-0596.

Scmincke, H.U. Ewaver, P.P.E., Firth, J.V. et al. (1995). Proc. ODP, Int. Repts., 116: College Station TX (Ocean Drilling Program).

Suriñach, E. & Roca, E.(1982). Catálogo de terremotos de Cataluña, Pirineos y zonas adyanentes. *in*: La sismicidad en la zona comprendida entre 40°N-44°N y 3°W-5°E NE de la Península Ibérica. Publicaciones de la Cátedra de Península Ibérica, Universidad Complutense, Madrid, 190: 9–53.

Werne J.P., Haese, R.R., Zitter, T., , Aloisi, G., Bouloubassi, L., Heijs, S. Fiala-Medioni, A., Pancost, R.D., Damste, J.S.S., , De Lange, G., Forney, L.J., Gottschal, J.C., Foucher, J.P., Mascle J. & Woodside J. (2004). Life at cold seeps: a synthesis of biogeochemical and ecological data from Kazan mud volcano, eastern Mediterranean Sea. *Chemical Geology* 205 (3-4): 367-390.

Woodside, J.M., Ivanov, M.K., Limonov, A.F. & Shipboard Scientists of the Anaxiprobre expeditions (1998). Shallow gas and gas hydrates in the Anaximander Mountains region, eastern Mediterranean Sea. *in*: Henriet, J.-P. &Mienert, J. (eds.) Gas Hydrates: Relevance to World Margin Stability and Climate Change. Geological Society, London, Special Publications, 137: 177-193.

Zitter, T.A., Woodside, J.M. & Mascle, J. (2003). The Anaximander Mountains: a clue to the tectonics of southwest Anatolia. *Geological Journal* 38: 375-394.

Part 2

Geochemistry

7

Self-Organization of the Khibiny Alkaline Massif (Kola Peninsula, Russia)

Gregory Ivanyuk, Victor Yakovenchuk, Yakov Pakhomovsky, Natalya Konoplyova, Andrei Kalashnikov, Julia Mikhailova and Pavel Goryainov
Kola Science Centre of the Russian Academy of Science
Russia

1. Introduction

The world's largest Khibiny alkaline massif occupies the area of about 1327 km^2 in the extreme West of the Kola Peninsula, at the contact of rocks of the Imandra-Varzuga Proterozoic greenstone belt and the Archaean metamorphic complexes of the Kola-Norwegian megablock (Fig. 1). According to Pb-Pb, Rb-Sr and Sm-Nd dating (Arzamastsev et al., 2007), the age of the main rock types of the Khibiny massif is 380-360 million years.

About 70% of the massif area is occupied by nepheline syenites (foyaite) monotonous in composition which are, in most works, subdivided into two equal parts: foyaite proper (in the center) and "khibinite" (surrounding them), separated from each other by a zone rock complex of the Main Ring. Besides, practically every geological map of the Khibiny massif highlights the concentric zones of massive and trachytic khibinite and foyaite, along the edge and in the center, and on both sides of the Main Ring, respectively.

Within the Main Ring, foidolites (melteigite-ijolite-urtite), high-potassic poikilitic nepheline syenite (rischorrite) and less widespread malignite, as well as titanite-nepheline, titanite-apatite and apatite-nepheline rocks are of crucial importance. The same complex of rocks can be related to the so-called irregular-grained nepheline syenite ("lyavochorrite"), transitive to rischorrite in accordance with modal composition, texture-structural features and geological position (see Fig. 1). The rock complex of the Main Ring fills a conic fault in which the angle between the axis and generatrix varies between 50–70° close to the surface and 10–40° at the depth of more than 1 km. On the day surface, rocks of this complex occupy 30 % of the total area of the massif, the share of foidolites, rischorrite and lyavochorrite making up 10 vol. % each. Apatite-nepheline and titanite-apatite-nepheline rocks form fractal ore stockworks (Fig. 2) in the apical parts of the foidolite ring, being related to it by gradational transitions. The thickness of apatite-rich foidolites ranges from 200 m in the south-western part of the Main Ring up to the first meters in its north-eastern part.

Within the Main Ring and, especially, in the adjoining parts of nepheline syenites (on both sides of the Ring), there are a lot of xenoliths (from half a meter up to several kilometers across) of volcanogenic-sedimentary rocks metamorphosed to hornfels and fenitized (Korchak et al., 2011). Xenoliths, though occupying less than 1 % of the total day surface of the massif, are in constant association with the much wider spread fine-grained alkaline and nepheline syenites obviously representing the result of a more or less deep fenitization of volcanogenic-sedimentary rocks metamorphosed to hornfels.

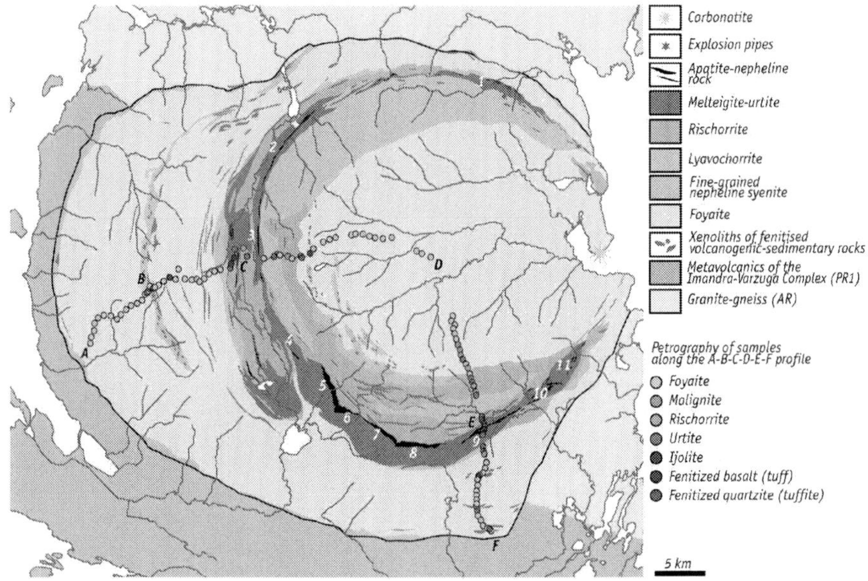

Fig. 1. Simplified geology map of the Khibiny massif (after Snyatkova et al., 1983). Apatite-nepheline deposits: 1 - Valepakhk, 2 - Partomchorr, 3 - Kuelporr, 4 - Snezhny Zirk, 5 - Kukisvumchorr; 6 - Yuksporr; 7 - Apatitovy Zirk; 8 - Plato Rasvumchorr; 9 - Koashva; 10 - Niorkpakhk; 11 - Oleniy Ruchei. *A–B–C–D–E–F* - profile with sampling points.

Fine-grained alkaline and nepheline syenites (3 % of the massif's total area) are concentrated within three ring zones: at the border of the massif, on the periphery of the Main Ring and within the Small Semiring (see Fig. 1). The latter, up to 500 m wide, located in the foyaite part of the massif external relatively the Main Ring, is composed of fine-grained alkaline and nepheline syenites (fenites?) with xenoliths of volcanogenic-sedimentary rocks and also by bodies of melteigite, urtite, and malignite. Fine-grained alkaline and nepheline syenites of the outer zone are bedded as separate, up to 200 m wide, lenses and strips; they usually have gradational contacts with foyaite and are often present as xenoliths within the latter.

Dyke rocks of the Khibiny massif are represented, for the most part, by hypabyssal analogues of its plutonic rocks: alkali-feldspar trachyte, phonolite and melanephelinite, mainly concentrated near the Main Ring, as well as by monchiquite and carbonatite composing veins and explosion pipes in its eastern part (Ivanyuk et al., 2009). Pegmatite and hydrothermal veins, which include an unusually great number of mineral species (about 300), are common throughout the massif, with their main concentration within rischorrite and foidolites of the Main Ring. In foyaite, there are ordinary clinopyroxene-nepheline-microcline veins, but, as the Main Ring is approached, their mineral composition becomes more and more varied - up to 80 minerals in a vein (Yakovenchuk et al., 2005).

Generally speaking, the geological events were mostly occurring within the Main Conic Fault and it is this structure that controls the bodies of apatite-nepheline rocks, brecciating

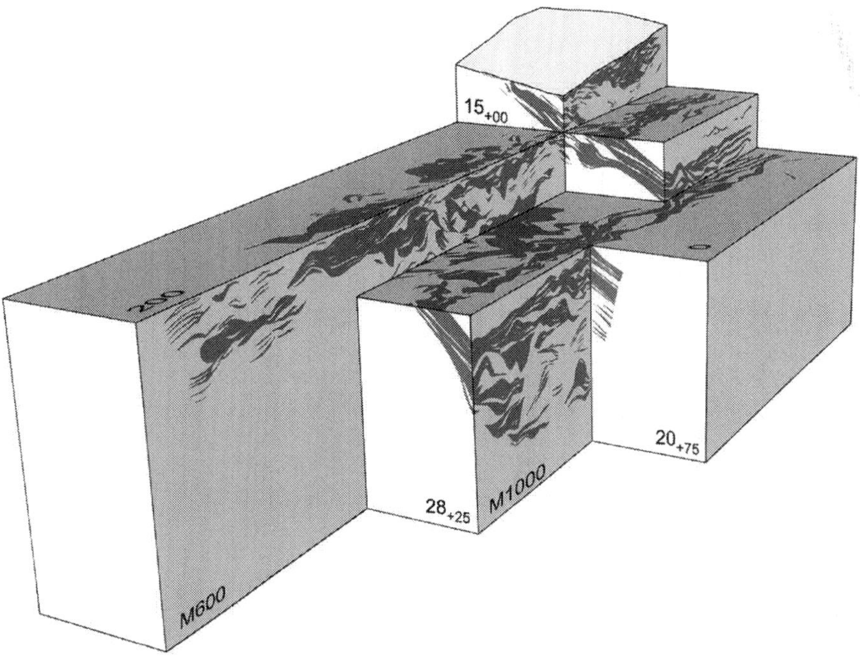

Fig. 2. Block-diagram of the Koashva deposit. Greenish-grey - ijolite-urtite, blue - apatite-nepheline rock (foidolites with $P_2O_5 \geq 4$ wt. %).

and folding zones, explosion pipes, dykes of phonolites, alkali-ultrabasic rocks and foidites; zones of albitization and aegirinization of nepheline syenites bearing eudialyte, astrophyllite, loparite, rinkite, and pyrochlore mineralization; pegmatite and hydrothermal veins, zones of contemporary mineral formation, and epicentres of earthquakes (Goryainov et al., 1998). Besides the Main Ring, the Small Semiring, and a number of some less pronounced conic tectonic structures, the massif also reveals later subvertical radial faults diverging from the extreme eastern point of the massif (the latter is marked, in addition, by stockworks of carbonatite veins) similar to cracks resulting from hammering on glass. Although no essential displacements were found within the radial faults, all zones of foliation, crushing, spreusteinization, and chemical decomposition of nepheline syenites and foidolites are related to them.

Similar to the majority of stretching fractures, the Main Ring fault has a fractal morphology of percolation clusters (${}^{3}D = 2.54$): fractal dimension of the foidolite cluster shown in Fig. 1 is ${}^{3}D \approx {}^{2}D + 1 = 2.5$, approximately equal to that of apatite-ore clusters of the Koashva deposit (see Fig. 2) ${}^{3}D \approx {}^{2}D + 1 = 2.7$ and fluorapatite clusters in apatite-nepheline rock ${}^{3}D \approx {}^{2}D + 1 = 2.6$ (Fig. 3) in an interval of scales from 0.001 up to 3 kms. Processes of fractal fault network formation kept recurring with accumulation of critical pressure in the rising massif - every time with a smaller power effect. Notably, the morphology and fractal dimension of the forming structures (pegmatites, hydrothermalites, and modern clusters of sodium carbonates) always corresponded to those of percolation clusters (Goryainov et al., 1998; Ivanyuk et al., 2009).

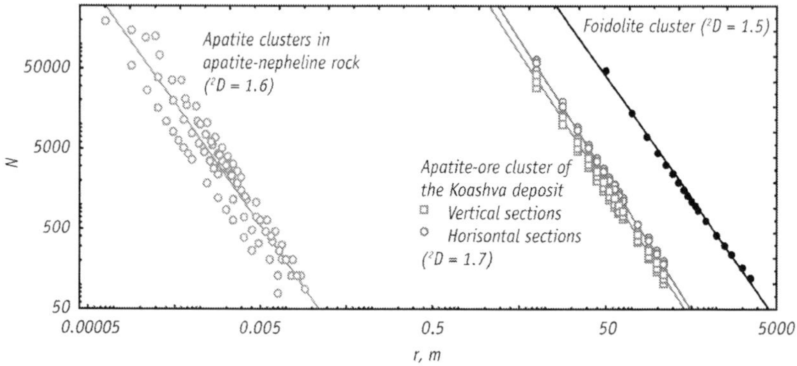

Fig. 3. The ln of number of boxes $N(r)$ with side length r needed to cover clusters of foidolite, apatite-nepheline rock and fluorapatite as a function of r.

Despite its rather simple structure, not only the genesis of the Khibiny massif but also the age ratio of the rock complexes discussed above, have generated heated debates till nowadays. A lot of genetic models have been proposed to explain the concentric-zoned structure of the massif (Fig. 4). The reason for such a variety of genetic models is ambiguity of the rock ratio and proximity of their absolute ages. In order to correlate the results of tectonic, geochemical and mineralogical researches within a uniform concept, we have carried out own systemic study of the Khibiny massif, including its mineral, petrographic and geochemical zonation along the profile from the Khibiny railway station (the *A* point in Fig. 1) to the massif center, Mt. Vantomnyutsk (point *D*) and further across the Koashva deposit (the *E* point) to the contact with host rocks at the foot of Mt. Kitchepahk (the *F* point). We also have re-interpreted the structures of the Khibiny apatite deposits on the basis of the data obtained during their mining. This work summarizes the results of research, which was partly published in a number of our works (Yakovenchuk et al., 2005, 2008, 2010a–d; Ivanyuk et al., 2009, 2010; Konopleva et al., 2008; Pakhomovsky et al., 2009).

2. Petrography

According to *QAPF* classification, the conventional boundary of nepheline syenites with foidolites passes at the line of $Fsp_{60}Ne_{40}$, and with alkaline syenites – at the line of $Fsp_{90}Ne_{10}$ (Fig. 5). According to the color index *M*, the nepheline syenites are subdivided into foyaite (0–30% of dark-coloured minerals), malignite (30–60%), and shonkinite (more than 60 % of dark-coloured minerals). For foidolites, a division into urtite (10–30% of dark-coloured minerals), ijolite (30–70%) and melteigites (70–90%) is generally accepted. The estimation of the volumetric proportion of minerals in samples selected by us along the *A–B–C–D–E–F* profile (see Fig. 1) was made by counting the areas occupied by these minerals on polished sample sections (about 20 × 20 cm). The minerals were diagnosed by using microprobe analysis and powder X-ray diffractometry.

Foyaite represents medium- to coarse-grained leucocratic greenish-grey rocks composed of tabular up to equant crystals of potassium feldspar (usually with albite perthites) whose interstices in the aggregate are filled with euhedral grains of nepheline and prismatic crystals of dark-coloured minerals, with domineering clinopyroxenes of aegirine-diopside series, and also alkaline and Na-Ca amphiboles (richterite-ferrorichterite, magnesiokatophorite-katophorite, magnesioarfvedsonite-arfvedsonite, etc.) among the latter. The rock-forming minerals are nepheline, microcline, orthoclase, albite, aegirine, augite, Na-Ca-and Na-amphiboles, aenigmatite, titanite, eudialyte, lamprophillite, and annite. The widest spread accessory minerals include magnetite, loparite-(Ce), pyrochlore, sodalite, rinkite, astrophyllite, barytolamprophyllite, ilmenite, lorenzenite, rosenbuschite, fluorapatite, fluorite, and pyrrhotite. Secondary minerals are represented mainly by zeolites (natrolite, gonnardite, analcime, etc.), illite, pyrite, anatase, and goethite.

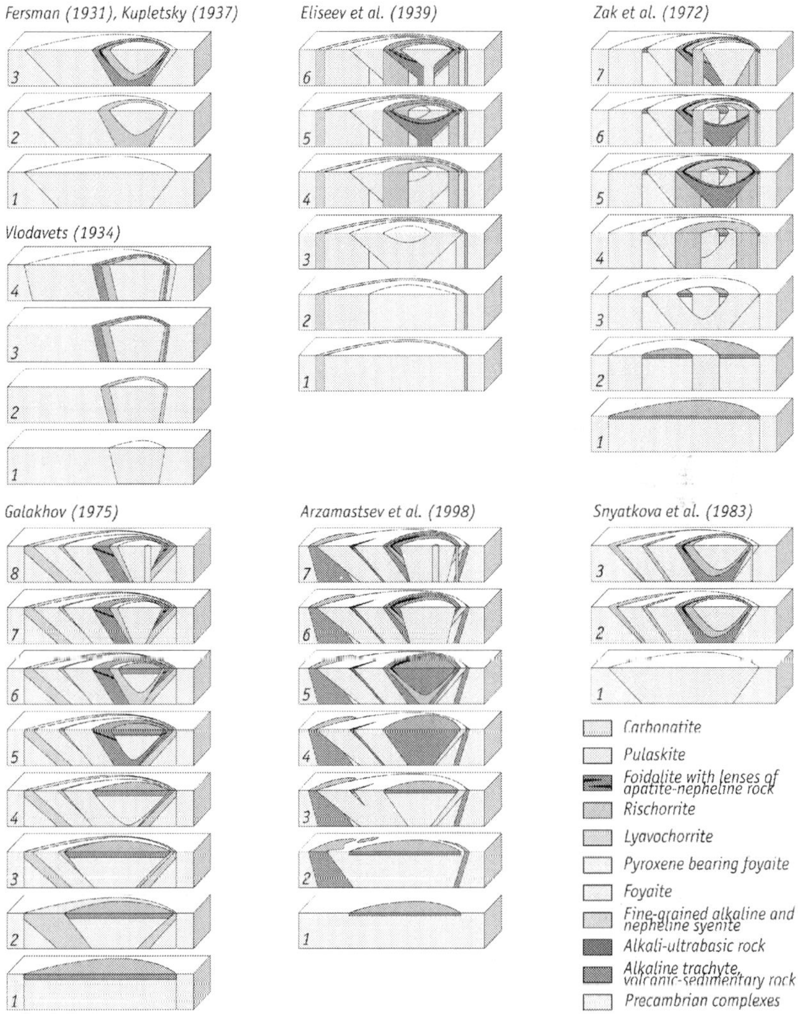

Fig. 4. Genetic models of the Khibiny massif.

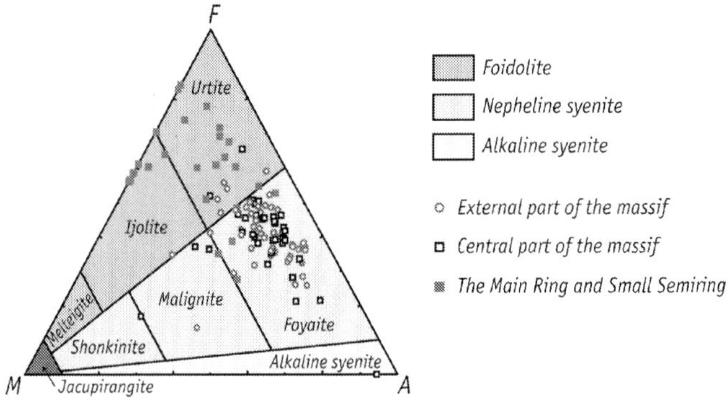

Fig. 5. Modal composition of alkaline rocks estimated using grain squares of feldspar (*A*), nepheline+kalsilite+sodalite (*F*), and dark coloured minerals (*M*) in polished hand-sized specimens sampled along the *A–B–C–D–E–F* profile.

The quantitative proportion of feldspar, nepheline and dark-coloured minerals in foyaite along the *A–B–C–D–E–F* profile varies over a rather wide range (see Fig. 5), generally corresponding to the composition of $Fsp_{44}Ne_{40}M_{16}$. The average mineral composition of foyaite in the part of the massif external relatively the Main Ring ("khibinite"), $Fsp_{44}Ne_{40}M_{16}$, is practically identical to that of the rocks in the central part of the massif, $Fsp_{43}Ne_{40}M_{17}$. It should be noted that, in accordance with classification *QAPF*, about 10 % of the samples corresponding to foyaite according to its position in the *A–B–C–D–E–F* profile, its structural-textural features, and the set and composition of minerals composing them, are referred to the foidolite field. Above all, these samples differ from the feldspar ijolite-urtite of the Main Ring, similar to them in modal composition, due to the tabular shape of microcline-perthite crystals and an utter absence of "web-footed" metacrysts of orthoclase with poikilitic inclusions of nepheline and kalsilite.

The content of major minerals in foyaite changes symmetrically relative to the Main Ring along the *A–B–C–D–E–F* profile (Fig. 6). The share of alkaline feldspar in the rock composition decreases towards the contacts with foidolites proportionally to the width of the latter in the profile: less intensively in the area at Mt. Marchenko Peak (the *C* point); more intensively in the area at Mt. Koashva (the *E* point). This change is compensated by increasing contents of nepheline in the *C* point and dark-coloured minerals in the *E* point. In the former case, this results in the formation of leucocratic foyaite transitive to urtite (or even urtite, still corresponding to foyaite in composition); in the latter case - in mesocratic nepheline syenite and malignite. The consecutive increasing of feldspar content (at the expense of nepheline) from the Main Ring towards the outer parts and centre of the massif results in the occurrence of alkaline syenites, described in (Ramsay & Hackman, 1894; Korobeynikov & Pavlov, 1990) under the name of umptekite *(A)* and pulaskite *(D)*.

On the whole, the distribution of accessory minerals in foyaite along the *A–B–C–D–E–F* profile appears as extremely non-uniform. However, practically all curves of the occurrences of certain mineral species (Fig. 7) proved to be symmetric relatively the center of the massif, this concerning both widespread minerals (biotite, sodalite, ilmenite, magnetite, etc.) and rare species (banalsite, rosenbuschite, strontioapatite, etc.). Titanite and fluorapatite are through minerals of foyaite, whereas lamprophillite, biotite, aenigmatite, astrophyllite, and lorenzenite mainly concentrate either in the central part of the massif (biotite, astrophyllite) or in its outer parts. The occurrence of the majority of accessories (barytolamprophyllite, rinkite, rosenbuschite, pyrochlore, ilmenite, magnetite, fluorite, etc.) decreases as the Main Ring is approached. Conversely, eudialyte and sodalite become here an integral attribute of the rocks.

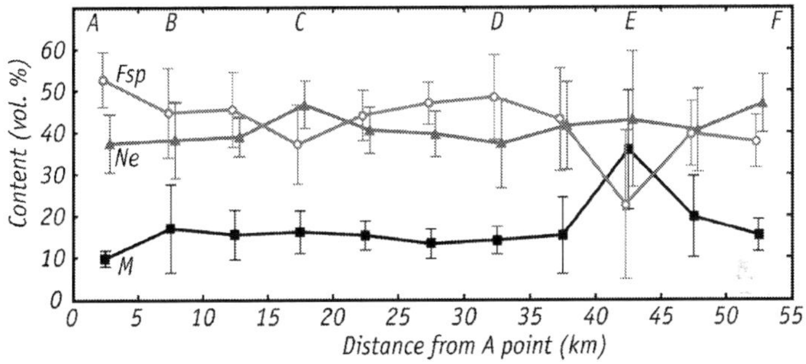

Fig. 6. Variation of alkaline feldspar (Fsp), nepheline (Ne) and dark coloured minerals (M) contents along the A–B–C–D–E–F profile (mean plus-minus standard deviation).

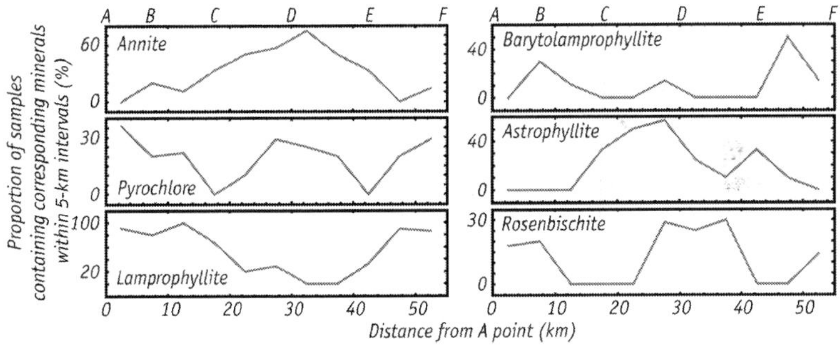

Fig. 7. Variation of accessory mineral contents along the A–B–C–D–E–F profile.

The texture of foyaite can be both massive and trachytic. Field observations of the majority of researchers have noted the trachytic variety of foyaite at contacts with foidolites of the Small Semiring and the Main Ring, and also to the western and southern contacts of the massif with host metamorphic rocks (Tihonenkov, 1963; Zak et al., 1972; Galakhov, 1975; etc.). At the same time, trachytic foyaite was often found (Snyatkova et al., 1983) to be alternating with massive foyaite. The definition of foyaite trachytoid structure, i.e. the definition of the degree of orientation of potassium feldspar tabular crystals in the rock, has been made by us in the polished sections of foyaite samples picked along the A–B–C–D–E–F profile. From the diagram showing the variation of standard deviation of the b axis orientation in potassium feldspar tabular crystals from their average direction, σ_{Fsp} (Fig. 8a), it follows that, contrary to the common belief, the degree of foyaite isotropism gradually increases moving from the outer and central parts of the massif towards the Main Ring, near which foyaite grades into entirely isotropic rischorrite.

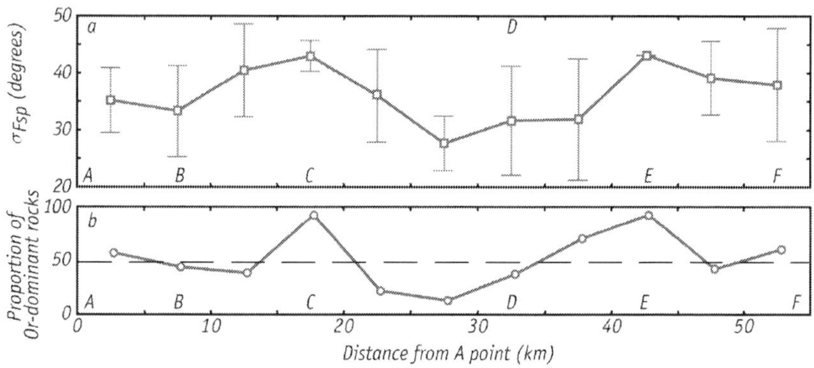

Fig. 8. Variation of standard deviation (a) of the *b* axis orientation in potassium feldspar tabular crystals from their average direction (σ_{Fsp}) and the presence of orthoclase-dominant rocks (b) along the *A–B–C–D–E–F* profile.

X-ray diffraction analysis of potassium feldspar has shown (Fig. 8*b*) that orthoclase-dominant foyaite occurs less frequently moving from the outer part towards the center of the massif and it is against this background that a sharp "orthoclase" maximum is observed near the Main Ring (Ivanyuk et al., 2010). Since orthoclase, compared to microcline, is a higher-temperature modification of potassium feldspar, its diminishing share towards the massif center allows to suggest gradual decreasing of the temperature of feldspar (re)crystallization in this direction while the foyaite intrusions were consolidating. Apparently, the transition of microcline foyaite to orthoclase foyaite, finally turning to rischorrite near the Main Ring, accompanying the textural isotropization of these rocks, occurred owing to the warming-up of fault parts of the massif by foidolite melt.

Data on the chemical composition of foyaite are presented in Table 1. As is peculiar to plutonic rocks, the empirical distributions of principal components in foyaite (Si, Al, Na, K and Fe oxides) correspond to the normal law. Logarithmically normal (exponential) distributions of MgO, CaO, SrO, P_2O_5, TiO_2, ZrO_2, F and Cl contents are due to enrichment of the foyaite parts contacting with the Main Ring with these components (such as fluorapatite, titanite, and other minerals).

Malignite, a melanocratic (M_{30-60}) variety of nepheline syenites, has a rather limited distribution in the Khibiny massif being mostly concentrated on the periphery of the rischorrite-foidolite ring. Contacts of malignite with ambient foyaite can be both gradational (more often) and sharp. The latter case is usually realized within various tectonic zones where malignite and foyaite often interchange. Macroscopically, malignite represents fine- or medium-grained composed of nepheline, orthoclase-perthite, aegirine, aenigmatite, and iron-dominant Na-Ca and alkaline amphiboles (ferroeckermannite, ferrorichterite, arfvedsonite, etc.). The average composition of malignite, observed by us along the *A–B–C–D–E–F* profile, makes $Fsp_{31}Ne_{28}M_{41}$ (see Fig. 5). Malignite is characterized by a gneissose structure caused by oriented arrangement of aegirine fibers, prismatic crystals of amphiboles and lamprophyllite, elongated segregations of aenigmatite and eudialyte. Eudialyte (up to 5 vol. %) and titanite (up to 7 vol. %) are typical constituents of malignite. Other characteristic accessories are fluorapatite, aenigmatite, lamprophillite, baryto-lamprophyllite, rinkite, ilmenite, and pyrrhotite.

	Distribution (r^2)	n	Content (wt. %)				
			Mode	Mean	Min	Max	σ
SiO_2	Normal (0.952)	143	53.75	54.74	47.48	63.67	2.15
TiO_2	Lognormal (1.000)	143	0.86	0.88	0.20	5.34	0.56
ZrO_2	Lognormal (0.955)	86	0.07	0.10	0.02	0.30	0.05
Al_2O_3	Normal (0.998)	143	21	20.95	13.61	28.6	1.91
Fe_2O_3	Lognormal (0.997)	143	1.5	2.26	0.24	7.57	1.10
FeO	Normal (0.992)	143	1.75	1.78	0.39	6.46	0.84
MnO	Lognormal (0.997)	139	0.14	0.19	0.08	0.71	0.10
MgO	Lognormal (0.989)	143	0.42	0.49	0.07	2.57	0.32
CaO	Lognormal (0.998)	143	1.12	1.18	0.45	6.53	0.73
SrO	Lognormal (0.933)	112	0.22	0.11	0.01	0.58	0.09
Na_2O	Normal (0.996)	143	9.50	9.47	5.78	13.15	1.13
K_2O	Normal (0.999)	143	5.50	5.83	1.14	9.91	0.8
P_2O_5	Exponential (0.999)	124	0.06	0.18	0.00	1.30	0.14
CO_2	Lognormal (0.993)	84	0.10	0.12	0.03	1.87	0.2
H_2O	Cauchy (0.991)	143	0.75	1.09	0.24	3.01	0.48
S_{tot}	Exponential (0.999)	64	0.01	0.04	0.01	0.26	0.05
Cl	Exponential (0.996)	123	0.01	0.04	0.00	0.33	0.05
F	Exponential (0.955)	120	0.13	0.15	0.00	0.64	0.09

Table 1. Chemical composition of foyaite (n - quantity of samples, σ - standard deviation).

Rischorrite, or poikilitic nepheline syenite, is a leucocratic massive medium- to coarse-grained rock easily identifiable during a field research owing to its characteristic poikilitic structure. This structure is due to the presence of large (up to 20 cm in diameter) equant metacrysts of orthoclase, overflown with poikilitic inclusions of nepheline, kalsilite and dark-coloured minerals, in fine-, medium-grained mass of euhedral grains of nepheline cemented by dark-coloured minerals (mainly aegirine and potassicarfvedsonite). The average composition of rischorrite taken by us along the *A–B–C–D–E–F* profile makes $Fsp_{36}Ne_{44}M_{20}$. It should be noted, however, that, similarly to the case with foyaite, the boundary between rischorrite and feldspar urtite was drawn absolutely formally, according to *QAPF* classification. Actually, they form a continuous series of rocks genetically related among themselves by processes of orthoclase poikiloblasts forming. The rock-forming minerals are nepheline, sodalite, orthoclase, aegirine, arfvedsonite, potassicarfvedsonite, potassic-ferroeckermannite, magnesioarfvedsonite, annite, titanite, aenigmatite, ilmenite, lamprophillite, astrophyllite, and fluorapatite. The accessories include eudialyte, lorenzenite, pectolite, rinkite, yuksporite, fersmanite, lomonosovite, murmanite, loparite-(Ce), ancylite-(Ce), magnetite, sphalerite, djerfisherite, and pyrrhotite. Also found are kalsilite, natrolite, goethite, wadeite, and catapleiite present as secondary minerals.

The data on the average chemical composition of rischorrite are given in Table 2. Empirical distributions of components in these rocks are essentially different from those in foyaite. So, normal distribution of FeO and logarithmically normal distribution of Fe_2O_3 in foyaite "change places" on transition to rischorrite. This inversion reflects a transition from essentially calcium-rich pyroxenes and amphiboles, dominating in foyaite, to aegirine and arfvedsonite peculiar to rischorrite. Normal distribution of Al_2O_3 and Na_2O concentrations in foyaite is replaced by *Q*-normal on transition to rischorrite, which suggests introducing of these components in protorischorrite in the course of their nephelinization on the contact with foidolites. The content of K_2O, CaO, P_2O_5, Cl, F, and H_2O in rischorrite is characterized by bimodal distributions in which the first maximum is quite comparable to the average content of the given components in foyaite. The second maximum is, obviously, related to introducing of considered components from foidolites during a fluid-metasomatic alteration of foyaite overlying them, and transformation of the latter into rischorrite.

	Distribution (r^2)	n	Content (wt. %)				
			Mode	Mean	Min	Max	σ
SiO_2	Cauchy (0.897)	80	51.5	51.48	47.02	56.93	2.33
TiO_2	Normal (0.997)	80	1.25	1.23	0.28	3.15	0.51
ZrO_2	Exponential (0.962)	13	0.05	0.12	0.02	0.60	0.18
Al_2O_3	Q-normal (0.984)	80	21.00	20.87	12.07	25.07	2.37
Fe_2O_3	Cauchy (0.965)	80	2.50	2.70	0.03	8.05	1.57
FeO	Lognormal (0.975)	80	1.96	2.50	0.90	10.66	1.94
MnO	Lognormal (0.995)	78	0.12	0.17	0.06	0.93	0.13
MgO	Normal (0.805)	80	0.62	0.68	0.07	2.37	0.41
CaO	Bimodal	80	0.75 and 1.75	1.40	0.19	3.81	0.73
SrO	Exponential (0.979)	61	0.06	0.15	0.01	0.93	0.13
Na_2O	Q-normal (0.952)	80	8.50	8.35	3.77	12.82	1.84
K_2O	Bimodal	80	<u>7.50</u> and 10.50	8.40	4.13	14.12	2.24
P_2O_5	Bimodal	70	<u>0.05</u> and 0.50	0.16	0.01	0.81	0.18
CO_2	Exponential (0.945)	55	0.05	0.10	0.02	1.01	0.20
H_2O	Bimodal	47	<u>0.70</u> and 1.10	1.00	0.16	2.48	0.50
S_{tot}	Exponential (0.995)	20	0.03	0.06	0.00	0.74	0.19
Cl	Bimodal	27	<u>0.03</u> and 0.14	0.04	0.01	0.15	0.04
F	Bimodal	66	<u>0.15</u> and 0.60	0.12	0.01	0.68	0.11

Table 2. Chemical composition of rischorrite (n - quantity of samples, σ - standard deviation. More intensive maximum of bimodal distribution is underlined).

It is remarkable that foidolites, having the lowest potassium contents in the Khibiny rocks, should cause a potassic metasomatism of ambient foyaite. It is known (Hayward et al., 2000) that silicon, superfluous relative to the stoichiometric values of 4 atoms of Si per molecule in nepheline, along with the inverse process, is compensated by removing or introducing of an equivalent quantity of potassium (sodium takes no part in this process). As is shown by the dependence of Na and K contents on the proportion of Si, Al and Fe^{3+} in nepheline from various rocks of the Khibiny massif (Fig. 9), potassium content in nepheline from foidolites of the Main Ring is extreme and the rest potassium must by accumulated in a residual fluid.

Autometasomatic alteration of foidolites (including apatite-nepheline rocks) by residual potassium-rich fluids during the intrusion cooling has resulted in substitution of nepheline for kalsilite, the formation of orthoclase, biotite, potassic amphiboles, wadeite and other K-minerals. Simultaneous steaming operation of overlying foyaite by such fluids is the reason of kalsilite-orthoclase poikiloblasts forming in these rocks: $NaAlSi_3O_8+K^+ \leftrightarrow KAlSi_3O_8+Na^+$ and $Na_3KAl_4Si_4O_{16}+3K^+ \leftrightarrow 4KAlSiO_4 + 3Na^+$. For this reason, the content of K_2O in rischorrite is in inverse proportion to the content of Na_2O, whereas foyaite exhibits no such correlation, and it is positive in foidolites. The releasing sodium takes part in the formation of albitites and aegirinites, spatially adjacent to rischorrite, as well as ultra-agpaitic veins abundant in rischorrite.

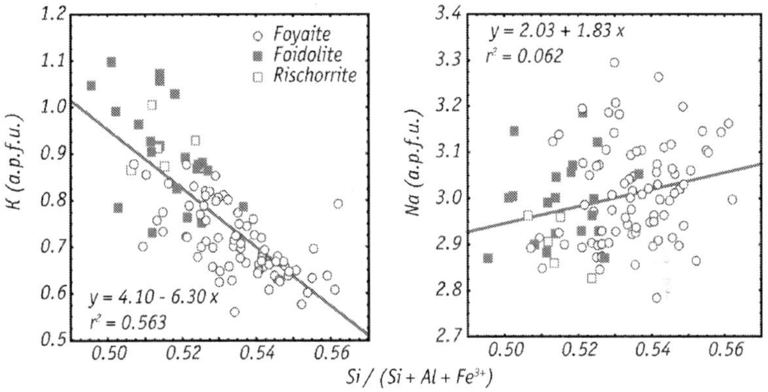

Fig. 9. Potassium (*a*) and sodium (*b*) dependence on Si–Al relation in nepheline.

Lyavochorrite (irregular-grained nepheline syenite) is a variety of foyaite which can be related to the rock complex of the Main Ring according to a number of features, including its geological position (see Fig. 1). Basically, these are medium- to coarse-grained greenish-grey leucocratic rocks composed of equant crystals of orthoclase-perthite (up to 3 cm in diameter), where interstices in the aggregate are filled with euhedral grains of nepheline, as well as aegirine-augite crystals and Na-Ca amphiboles (predominantly, richterite-ferrorichterite). Lyavochorrite is similar to rischorrite, primarily owing to the presence of poikilitic metacrysts of orthoclase. However, orthoclase poikiloblasts is not so widespread in lyavochorrite; its morphology is less pronounced; and it is often seen to coexist with tabular orthoclase and even with microcline-perthite. Lyavochorrite includes titanite, lamprophillite, eudialyte, rinkite, aenigmatite, götzenite, ilmenite, magnetite, loparite-(Ce), fluorapatite, and graphite as accessories; the secondary minerals are natrolite, cancrinite, sodalite, albite, goethite, wadeite, etc.

The mineral composition of lyavochorrite falls within the range of $Fsp_{43\text{-}77}Ne_{18\text{-}43}M_{4\text{-}23}$. A gradational transition from foyaite of the outer part relative to the Main Ring of the Khibiny massif (khibinite) through lyavochorrite to foyaite of the central part is observed (Tikhonenkov, 1963). Similar to foyaite, the lyavochorrites may contain parts with increased contents of nepheline which belong to foidolites (in accordance with *QAPF* classification). The majority of components of the lyavochorrite composition are characterized by bimodal distributions (Table 3), confirming the idea of hybridisms of these rocks (Tikhonenkov, 1963; Snyatkova et al., 1983). If foyaite has served as a basis for its formation, it is not surprising that one of the maxima in such distributions appears be close to the average content of this component in foyaite, and the second maximum is determined either by introducing or withdrawing of this component. Accordingly, lyavochorrite should be regarded as a variety of an "underdeveloped" rischorrite which, along with common rischorrite, was formed under the influence of foidolite intrusion upon the overlying foyaite, although only in places where the width of this intrusion was rather insignificant (see Fig. 1).

	Distribution (r^2)	n	Content (wt. %)				
			Mode	Mean	Min	Max	σ
SiO_2	Bimodal	18	54.00 and 55.75	53.66	51.2	56.2	1.43
TiO_2	Bimodal	12	0.98 and 1.13	1.13	0.91	1.26	0.11
ZrO_2		5	0.06	0.07	0.06	0.19	0.07
Al_2O_3	Bimodal	18	17.75 and 21.00	20.41	17.79	22.71	1.64
Fe_2O_3	Bimodal	18	1.50 and 5.50	3.18	1.74	8.32	1.76
FeO	Bimodal	18	1.90 and 2.70	1.88	1.29	3.53	0.51
MnO	Bimodal	12	0.10 and 0.14	0.13	0.08	0.18	0.03
MgO	Normal (0.840)	12	0.70	0.68	0.03	1.16	0.29
CaO	Q-normal (0.957)	12	1.45	1.34	1.04	1.48	0.15
SrO	Bimodal	12	0.01 and 0.06	0.07	0.01	0.08	0.02
Na_2O	Bimodal	18	8.75 and 10.25	9.87	7.70	10.8	0.92
K_2O	Bimodal	18	5.75 and 6.75	6.00	5.02	8.00	0.72
P_2O_5	Bimodal	13	0.15 and 0.35	0.19	0.1	0.57	0.13
CO_2		5	0.06	0.06	0.03	0.08	0.02
H_2O	Normal (0.778)	12	0.65	0.61	0.39	0.96	0.17
S_{tot}		7	0.04	0.03	0.00	0.18	0.06
Cl	Bimodal	9	0.08 and 0.30	0.07	0.01	0.31	0.10
F	Bimodal	8	0.05 and 0.15	0.07	0.04	0.19	0.05

Table 3. Chemical composition of lyavochorrite (n - quantity of samples, σ - standard deviation. More intensive maximum of bimodal distribution is underlined).

Melteigite-urtite of the Main Ring and Small Semiring. Unaltered urtite represents fine-, medium-grained greenish-grey rocks with either massive and gneissose texture. They are formed by euhedral crystals of a nepheline, in which interstices in the aggregate are filled with grains of clinopyroxenes of the diopside-aegirine-augite series, KNaCa-amphiboles (potassicrichterite, potassicferrorichterite, etc.), annite, titanite, magnetite, ilmenite, and eudialyte. Nepheline crystals quite often occur as poikilitic inclusions in large (up to 10 cm in diameter) metacrysts of aegirine-augite, potassicrichterite, and orthoclase. The content of orthoclase poikiloblasts in urtite varies continuously from zero up to the threshold value of urtite transition to rischorrite (see Fig. 5). Quite often one encounters here late albite forming fine-grained segregations and stringers in an intimate association with natrolite and aegirine. In ijolite and melteigite, nepheline becomes less widespread, its proportion to dark-coloured minerals remaining the same. These rocks are also characterized by a porphyritic shape owing to the presence of rather large short-prismatic crystals of nepheline in the fine-grained mass of nepheline and dark-coloured minerals. The characteristic accessory minerals of melteigite-urtite are forsterite, fluorapatite, aenigmatite, lamprophillite-barytolamprophyllite, astrophyllite, lorenzenite, pectolite, rinkite, loparite-(Ce), pyrochlore, fluorite, pyrrhotite, sphalerite, and chalcopyrite. The secondary minerals include kalsilite, natrolite, albite, orthoclase, and goethite all of which were formed after nepheline; aegirine, potassicrichterite, potassicarfvedsonite and anatase (after titanite and lamprophillite).

The content of main components of these rocks is characterized by symmetric distributions (normal or Cauchy); the minor ones have logarithmically normal or exponential distributions (Table 4). Distributions of Fe, Mg and Mn oxides are caused by an essential quantitative domination of urtite over their more melanocratic analogues. In the case of CaO, F, Cl and C, distributions are due to re-deposition of fluorapatite from melteigite-urtite into apatite-nepheline rocks.

	Distribution (r^2)	n	Content (wt. %)				
			Mode	Mean	Min	Max	σ
SiO_2	Cauchy (0.998)	268	42.5	43.13	10.8	60.47	4.61
TiO_2	Normal (0.976)	268	2.30	2.99	0.42	18.46	2.01
ZrO_2	Cauchy (0.985)	46	0.10	0.11	0.01	0.28	0.05
Al_2O_3	Normal (0.988)	261	19.5	18.85	1.81	29.89	5.46
Fe_2O_3	Exponential (0.941)	268	2.50	7.60	1.11	29.23	2.96
FeO	Exponential (1.000)	252	1.84	4.33	0.45	27.69	2.87
MnO	Lognormal (1.000)	232	0.10	0.21	0.04	1.56	0.18
MgO	Lognormal (0.994)	245	1.50	1.68	0.03	8.40	1.65
CaO	Lognormal (0.922)	246	5.50	5.67	0.31	19.74	3.46
SrO	Cauchy (1.000)	202	0.25	0.27	0.01	1.86	0.19
Na_2O	Normal (0.993)	267	10.50	10.15	0.64	15.21	2.37
K_2O	Normal (0.956)	269	5.00	4.70	0.25	11.21	1.67
P_2O_5	Exponential (0.998)	257	1.00	0.69	0.04	14.30	1.60
CO_2	Exponential (1.000)	185	0.10	0.22	0.00	2.15	0.17
H_2O	Normal (0.999)	235	0.66	0.80	0.00	3.29	0.45
S_{tot}	Exponential (0.998)	46	0.01	0.07	0.01	0.43	0.08
Cl	Exponential (0.999)	77	0.04	0.05	0.00	0.27	0.05
F	Exponential (0.956)	212	0.12	0.30	0.01	1.47	0.17

Table 4. Chemical composition of foidolites (n - quantity of samples, σ - standard deviation).

Apatite-nepheline rocks are foidolites more or less enriched with fluorapatite up to monomineral apatitolite. These are spotty greenish-grey rocks often forming a stockwork of later fluorapatite veinlets in melteigite-urtite. Major minerals of the considered rocks are the same as in ijolite-urtite: fluorapatite, nepheline, diopside–aegirine-augite, potassicrichterite, potassicferrorichterite, orthoclase, titanite, magnetite, and ilmenite. Accessory minerals include aegirine, potassicarfvedsonite, eudialyte, lamprophillite, annite, calcite, ancylite-(Ce), and burbankite. Anatase, astrophyllite, kalsilite, natrolite, illite, goethite, wadeite, gaidonnayite, and pyrite occur as secondary minerals.

Apatite-nepheline rocks (Fig. 10) are traditionally subdivided into a number of textured-substantial types differing, above all, by fluorapatite concentration; they are spotty, lens-shaped banded, block-shaped, reticulate, ingrained, and brecciated. All of these varieties of apatite-nepheline rocks exhibit a clear structural-textural anisotropy marked by an oriented arrangement of elongated crystals of fluorapatite and dark-coloured minerals; by fluorapatite veinlets and, accordingly, lens-shaped fragments of melteigite-urtite separated by these veinlets; by lens-shaped segregations of titanite, clinopyroxene, etc. On the whole, the structural-textural orientation of these rocks coincides with the contours of the Main Ring.

Among the specified ore types, the dominating varieties are reticulate and lens-shaped banded, differing only by the shape of ijolite-urtite blocks and the thickness of fluorapatite veinlets enveloping these blocks. The richest spotty ores are, as a matter of fact, the thickest (up to 1.5 м) layers of lens-shaped banded ores, while the poorest massive (ingrained) ores

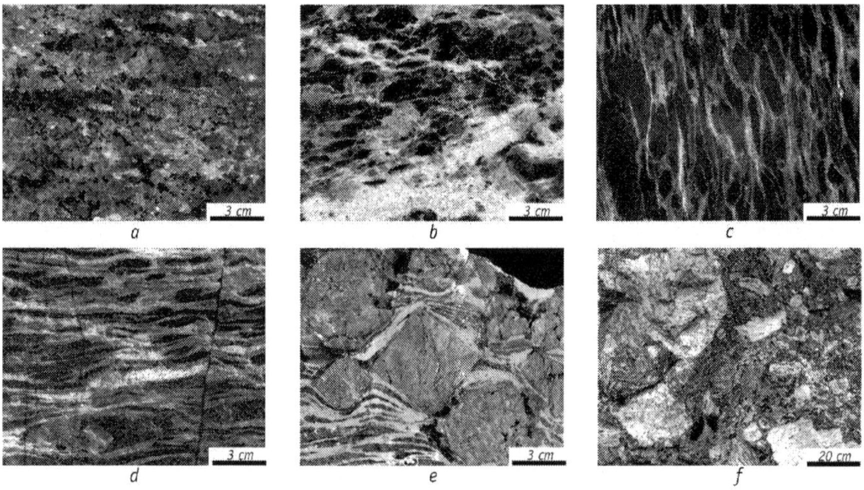

Fig. 10. Main textural types of apatite-nepheline rock: a - spotty, b - stockwork-like; c - reticulate; d - lens-shaped banded; e - block-shaped; f - brecciated.

are an implicitly pronounced variety of reticulate ores. Block-shaped ores are a result of recrystallization of reticulate or lens-shaped banded ores, in the course of which separate crystals of nepheline from ijolite-urtite lenses cluster in larger (up to 8 cm) metacrysts inside which relics of ijolite-urtite are often included. The geometry of apatite clusters on sections of these rock types is characterized by similar fractal dimension of $1.63 \leq {}^{2}D \leq 1.75$ (see Fig. 3). The value of their full fractal dimension determined by Mandelbrot's rule (Mandelbrot, 1983) is ${}^{3}D \approx 2.6$–2.7 and it is close to the dimension of apatite-ore and foidolite stockworks.

Brecciated ores, commonly found in all Khibiny deposits, occur as volumetric breccia organized similarly to "a blurred floe", in which fragments of apatite-nepheline rock of different sizes (from sub-centimeter up to several tens meters) and shapes are cemented by rather melanocratic foidolite (see Fig. 10f). The composition of the latter commonly corresponds to ijolite-urtite enriched with apatite and orthoclase. Sometimes, however, the cement is absent and differently oriented fragments appear to be compactly lap-fitted to each other. But, irrespective of the quantitative proportion of apatite-nepheline rock fragments and ijolite-urtite cement in breccias, the occurrence of fragments with increasing area decreases according to the power law with the parameter equal 0.6, i.e. the distribution of apatite-nepheline rock fragments in size is fractal with the dimension of $D \approx 1.2$.

Another textural variety of apatite-nepheline rocks atypical of magmatogene deposits is the plicated ores widespread in apatite-rich zones of Kukisvumchorr, Yuksporr, and Rasvumchorr deposits (Yakovenchuk et al., 2005; Ivanyuk et al., 2009). The fractal dimension of separate foidolite layers in plicated spotty-banded apatite-nepheline ores ranges from 1.0 up to 1.2 in a wide enough interval of scales (from several millimeters up to several meters), which makes the appearance of these rocks indistinguishable from those of iron quartzite and other metamorphic rocks.

The chemical composition of apatite-nepheline rocks varies in a wide range of values (Table 5) being limited, on the one hand, by the composition of original melteigite–urtite and, on the other hand - by fluorapatite content (up to 90 vol. %). The content of the main components of the rocks in question (SiO_2, Al_2O_3, Na_2O, K_2O, P_2O_5, Fe_2O_3, and F) is characterized by symmetric distributions (normal or Cauchy) reflecting a tendency of the mineral composition to an equilibrium fluorapatite+nepheline+aegirine association. As these three minerals compose 95 vol. % of these rocks, the contents of Si, Al, Na, K, on the one hand, and the contents of F, Ca, P, on the other hand, are in a negative linear dependence. If we extend the sampling by apatite-titanite rocks, one of the major components will become titanium, the content of its oxide being distributed according to the Cauchy law. The content of the majority of trace components is distributed according to the exponential law which is primarily caused by the essential predominance of urtite over more melanocratic constituents of the apatite-nepheline rock protolyte.

	Distribution (r^2)	n	Content (wt. %)				
			Mode	Mean	Min	Max	σ
SiO_2	Cauchy (0.980)	238	23.60	20.38	1.07	38.94	8.04
TiO_2	Exponential (0.999)	238	1.10	1.91	0.05	10.14	1.46
ZrO_2	Bimodal	10	<u>0.05</u> и 0.09	0.07	0.01	0.14	0.04
Al_2O_3	Cauchy (0.993)	238	12.50	11.08	0.47	27.28	5.17
Fe_2O_3	Cauchy (0.948)	238	2.20	2.67	0.08	16.22	2.20
FeO	Exponential (0.993)	225	1.30	2.20	0.00	15.41	1.46
MnO	Exponential (0.990)	139	0.07	0.12	0.02	0.46	0.08
MgO	Exponential (0.994)	224	0.50	0.66	0.02	3.06	0.47
CaO	Cauchy (0.957)	224	27.50	29.19	4.08	51.25	9.69
SrO	Bimodal	14	0.25 и <u>2.25</u>	2.11	0.02	4.43	1.14
Na_2O	Normal (0.995)	238	6.00	5.67	0.26	14.28	2.64
K_2O	Cauchy (0.960)	238	2.75	2.49	0.03	5.78	1.15
P_2O_5	Cauchy (0.956)	236	20.50	20.99	2.12	40.65	8.16
CO_2	Exponential (0.996)	126	0.12	0.22	0.00	1.27	0.21
H_2O	Lognormal (0.980)	139	0.49	0.56	0.11	1.96	0.37
S_{tot}	Exponential (0.872)	18	0.02	0.03	0.00	0.20	0.05
Cl	Exponential (0.990)	29	0.03	0.05	0.00	0.29	0.07
F	Normal (0.989)	139	1.90	1.88	0.23	3.34	0.77

Table 5. Chemical composition of apatite-nepheline rock (n – quantity of samples, σ – standard deviation. More intensive maximum of bimodal distribution is underlined).

The apatite-titanite and nepheline-titanite ores, generally occurring along the top contact of apatite-nepheline rocks, are formed as the result of enrichment of apatite-nepheline rocks and overlying melteigites-urtite with titanite {also with fluorapatite in melteigites-urtite}. Initially, titanite forms a fairly sparse network of veinlets in ijolite-urtite which, however, is quickly condensed, forming, already after some meters, apatite-titanite or nepheline-titanite rocks with relics of nepheline and oval segregations of recrystallized aegirine-augite. The characteristic accessories of these rocks are orthoclase, natrolite, magnetite, ilmenite, potassicferrorichterite, and calcite. The proportion between the contents of main components in apatite and nepheline-titanite ores is close to that in apatite-nepheline rocks. The TiO_2 content in these rocks is weakly ($r^2 < 0.5$) dependent on the quantity of any other component, linearly decreasing with increase in their total content (with the deduction of SiO_2 and CaO), because of a superimposed character of titanite mineralization.

Alkaline syenites in the Khibiny massif play an extremely subordinated role in comparison with nepheline syenites with which, as a rule, they are related by a gradational transition. As it has been noted, in extreme western (the *A* point in Fig. 1) and central (the *D* point) parts of the Khibiny massif is located foyaite with an increased content of alkaline feldspar at the expense of nepheline. In places, the proportion of these minerals exceeds the classification limit of Ne/(Fsp+Ne)=0.1 between alkaline and nepheline syenites (see Fig. 5). Alkaline syenites are white or light-grey fine-, medium-grained massive rocks made of an aggregate of tabular crystals of microcline- or orthoclase-perthite (75–99 vol. %) whose interstices are filled with grains of nepheline, aegirine-augite and magnesioarfvedsonite. As a rule, dark-coloured minerals form a uniform impregnation in the feldspar mass. Less frequently, they occur as separate clusters (up to 50 cm in diameter) imparting a taxite-like appearance to the rock. Accessory minerals of alkaline syenites are titanite, astrophyllite, phlogopite, sodalite, cancrinite, fluorapatite, fluorite, calcite, magnetite, and pyrrhotite.

Volcanogenic-sedimentary rocks metamorphosed to hornfels spatially related to rock complexes of the Small Semiring and Main Ring (see Fig. 1) occur in foyaite as lens-shaped xenoliths from half meter up to 3 kms long and up to 600 m wide, and are almost always separated from foyaite by a zone of fine-grained alkaline or nepheline syenites (fenites after hornfels). Macroscopically, hornfels are fine-grained rocks (the average grain size is 0.07 mm) with a characteristic conchoidal fracture the color of which varies from white, light-grey, pale-violet, blue, green and brown of different hues to dark-grey and black. The structure of these rocks changes from massive to high-contrast banded, lens-shaped banded, taxitic and porphyritic. The microtexture is typically hornfels, or poikiloblastic - in the presence of large metacrysts of amphiboles, corundum, nepheline and aenigmatite.

Hornfels are characterized by wide variations of mineral composition within an isolated xenolith, an individual sample, or even a separate thin section. In total, the considered rocks include more than 170 minerals, more than 30 of which are rock-forming (including crichtonite-loveringite, sekaninaite, topaz, fayalite and freudenbergite). However, despite the varied and rather exotic mineral composition of hornfels and fenites forming after them, recalculation of the chemical composition of these rocks to normative components using the CIPW method has revealed three groups of rocks. The first two of them correspond to more or less intensively fenitized quartzite and basalt, the third one corresponds to alkaline trachyte and phonolite (fenite.) In fact, the introduction of aluminium and alkaline metals in these rocks during their thermal metamorphism initially generates a variety of hornfels and fenites (Yakovenchuk et al., 2005; Korchak et al., 2011):

$$CaAl_2Si_2O_8 + CaFeSi_2O_6 + 2FeTiO_3 + 2SiO_2 + (9H_2O + K^+ + Na^+)_{aq} =$$

$$= 2(K_{0.5}Na_{0.5})AlSi_3O_8 + Fe_2SiO_4 + 2CaTiSiO_5 + FeSiO_3 + 9H_2;$$

$$27SiO_2 + 2KFe_3AlSi_3O_{10}(OH)_2 + (K^+ + 3Na^+ + 16AlO_2^- + 6H_2)_{aq} =$$

$$= 3Fe_2Al_4Si_5O_{18} + 6(K_{0.5}Na_{0.5})AlSi_3O_8 + 8H_2O, \text{ etc.}$$

3. Chemical zonation

The symmetric petrographic zonation of the Khibiny massif relatively the axial part of the Main Ring with bodies of apatite-nepheline rocks (see Fig. 1) predetermines the presence of a similar geochemical zonation. Such zonation, noted by many researchers (Fersman, 1931; Zak et al., 1972; Galakhov, 1975; etc.), is expressed in consecutive increasing of Ca, Sr, Ti, P and F contents from the outer and central parts of the massif to the axial zone of the Main Ring at the expense of Si, Al, Mg, Fe, Na and K. But our prime interest when stating the task was not this obvious zonation. We were interested in the hidden zonation of foyaite complex capable of throwing light on the Khibiny massif evolution.

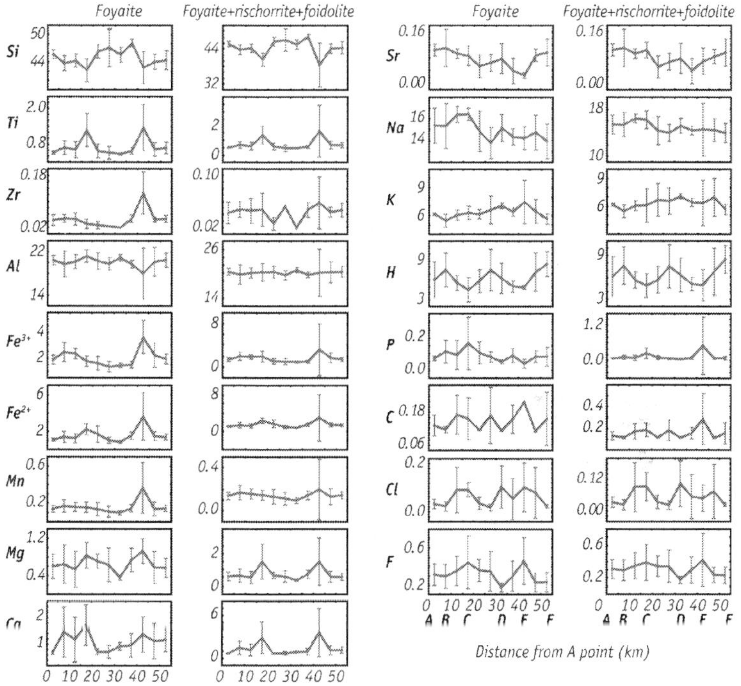

Fig. 11. Variation of chemical composition of alkaline rocks along the *A–B–C–D–E–F* profile.

Figure 11 represents a change in the chemical composition of rocks of the Khibiny massif, the data on which are 5-km interval-averaged along the *A–B–C–D–E–F* profile (see Fig. 1). Comparing these diagrams reveals that the Main Ring determines the composition of not only rischorrite but also foyaite, its influence becoming evident already at a distance of 5-10 kms away from the Main Ring! Consequently, of all the earlier proposed models of the Khibiny massif zonation origin, it is O. L. Snyatkova and colleagues' model (1983), in its turn developing the conceptions of A. E. Fersman (1931) about a relation of zonation with intruded foidolite melts along the ring fault (see Fig. 4), that represents the facts.

On the whole, in proximity of the Main Ring one observes substantial decreasing of the quantity of Si, Sr and H_2O in foyaite composition, compensated by increasing contents of all other elements, Ti, Zr, Ca, Mn, Fe, Mg, F, and Cl in the first place. At the same time, there are essential distinctions in foyaite composition adjacent to enriched in apatite-nepheline rock part of the Main Ring (Mt. Koashva, the *E* point) and its pure part (Mt. Marchenko Peak, the *C* point), mostly in regard to phosphorus and aluminium. Foyaite appears essentially enriched with Al, Na, P and Ca (i.e., accordingly, with nepheline and fluorapatite) in the area at Mt. Marchenko Peak, whereas foyaite from the Koashva deposit area (the *E* point) is obviously depleted in these elements due to increased contents of K, Fe, Mn, Zr and C (i.e., accordingly, biotite, K-amphiboles, kalsilite, orthoclase, and eudialyte). However, if the sampling data on composition first include rischorrite, and then ijolite-urtite, these distinctions practically disappear and the concentration profiles take a symmetric view relative to the center of the massif.

Otherwise, the rather monolithic "barren" areas of the Main Ring contain nepheline syenites essentially enriched with nepheline and fluorapatite as the result of steaming by foidolite-connected fluids. Foidolites in "ore" areas filled a great yawning fracture without causing such an intensive enrichment of host foyaite with nepheline and fluorapatite. But here emerge numerous processes of potassium metasomatism of ambient foyaite (up to formation of rischorrite), foidolites and even dykes of alkali-ultrabasic rocks. The formation of the Main Ring complex was completed when foidolites were fractured along the same "ring" filling this cluster with fluorapatite, which may partly be due to redistribution of the latter from ambient foidolites.

4. Typomorphic minerals

Studying the rocks of the Khibiny massif along the *A–B–C–D–E–F* profile, we have authentically identified more than 100 minerals, comprehensively described in (Yakovenchuk et al., 2005). In this work, prominence is given to typomorphic minerals: nepheline, feldspars, amphiboles, clinopyroxenes, and fluorapatite.

Nepheline is a rock-forming mineral of foidolites, nepheline and alkaline syenites, fenite and pegmatite-hydrothermal veins where its content varies from the first per cent up to 90 vol. % (see Fig. 5 and 6). Silica content in nepheline changes according temperature of rock forming (Yakovenchuk et al., 2010c): 4.3 a.p.f.u in hornfels and foyaite (about 900 °C), 4.2 a.p.f.u in foidolite (about 500 °C), 4.1 a.p.f.u in rischorrite and apatite-nepheline rock (about 400 °C), and 4.0 a.p.f.u in pegmatite-hydrothermal veins (about 200 °C). The chemical composition of nepheline in foyaite samples picked along the *A–B–C–D–E–F* profile (see Fig. 1) shows a regular change at moving from the outer and central parts of the massif towards the Main Ring (Fig. 12).

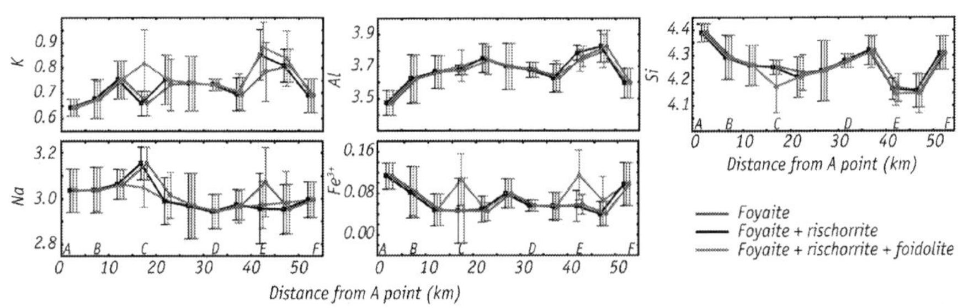

Fig. 12. Variation of nepheline composition along the *A–B–C–D–E–F* profile.

Thus, the change in nepheline composition from the outer and central parts of the massif towards the Main Ring can generally be expressed by the formula: $\square_B + (Si^{4+} + Fe^{3+})_T \leftrightarrow K^{+}_B + 2Al^{3+}_T$. Near the Main Ring, the substitution of Si for Al in nepheline from foyaite is accompanied by increasing of Na content in it (on the average, up to 3.2 *a.p.f.u.*), whereas in nepheline from foidolites the preferred substitution of Si for trivalent Fe is compensated by K and Ba: $\square_A + K_B + Si_T \leftrightarrow Na_A + \square_B + Al_T$ (in foyaite), and $Na_A + \square_B + Si_T \leftrightarrow \square_A + K_B + Fe^{3+}_T$ (in foidolites). As a result, the high-potassium and iron-rich nepheline of foidolites is essentially different in composition from the high-sodium and iron-poor nepheline of the adjacent foyaite and, partly, rischorrite. Besides, the initially potassium-rich ijolite-urtite melt seems to have caused not only the subsequent wide development of orthoclasizing and kalsilitizing processes within foidolites (including apatite-nepheline rocks), but the alteration of ambient nepheline syenites as well.

Microcline and *orthoclase* are major minerals of alkaline and nepheline syenites, feldspar-bearing ijolite-urtite, fenite, hornfels of volcanogenic-sedimentary rocks, phonolits, alkali-feldspar trachytes and pegmatite-hydrothermal veins. Orthoclase essentially dominates in rischorrite, foidolites, fine-grained nepheline syenites and hornfels (in hornfels, together with anorthoclase); microcline is more common in pegmatite veins, while foyaite contains both the modifications, although in different proportions (Ivanyuk et al., 2010). Orthoclase-bearing foyaite is generally concentrated within the 5-km outer zone of the massif, and also on the periphery of the Main Ring where its gradational transformation to lyavochorrite and rischorrite is observed. Changing of the potassium feldspar composition in samples of the Khibiny rocks picked along the *A–B–C–D–E–F* profile (see Fig. 1) is shown in Fig. 13.

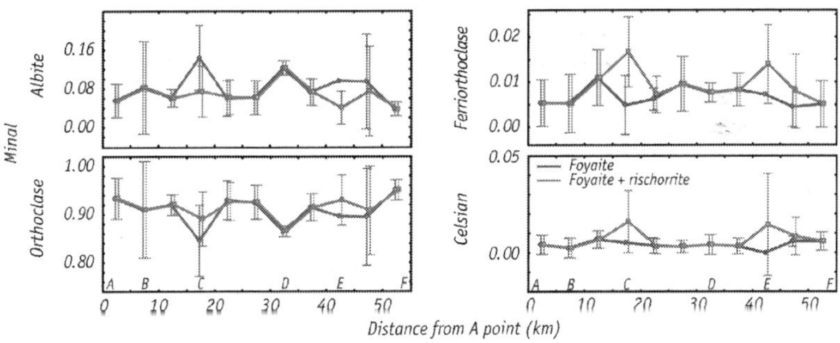

Fig. 13. Variation of K-feldspar composition along the *A–B–C–D–E–F* profile.

The "purest" orthoclase is concentrated in outer parts of the foyaite complex, whereas in the center of the massif and in the Main Ring area the feldspar is represented by a variety transitive to anorthoclase. Orthoclase of rischorrite, compared to that of foyaite, is essentially enriched with ferri-orthoclase and celsian constituents. The role of Ba and Fe increased at the final stages of orthoclase poikiloblasts formation owing to which the marginal zones of such poikiloblasts, both in rischorrite and in ijolite-urtite, are quite often represented by iron-rich hyalophane. Taking into account the data on foyaite complex textural zonation (see Fig. 8), it becomes obvious that this complex initially had monotonous zonation with an orthoclase-dominant outer zone and microcline-dominant central part, and that the orthoclase-bearing rocks near the Main Ring were formed due to warming-up and metasomatic alteration of ambient foyaite by fluidized foidolite melts filling the ring fault.

The *clinopyroxenes* of the Khibiny massif, predominating in the majority of rocks, are represented by diopside, hedenbergite, augite, aegirine-augite and aegirine (Yakovenchuk et al., 2005; Yakovenchuk et al., 2008). Diopside is a rock-forming mineral of alkali-ultrabasic rocks, alkali-feldspar trachytes, melteigite-urtite, metamorphosed to hornfels volcanogenic-sedimentary rocks of basalt composition and their host foyaite. Hedenbergite is observed in fenitized hornfels (after tuffite) where it together with aegirine forms parallel-columnar coronas around fayalite inclusions in albite. Aegirine-augite is the main mineral of all types of nepheline syenites (5–50 vol. %), foidolites (up to 90 vol. %), apatite-nepheline rocks, fenitized rocks of the massif frame and xenoliths of volcanogenic-sedimentary rocks in foyaite. In foyaite of the outer and central parts of the massif, it predominates among the other iron-magnesium-bearing silicates and quite often occupies a position subordinated in relation to aegirine, alkaline amphiboles, and annite in foyaite, lyavochorrite and rischorrite of the Main Ring zone. Aegirine is a ubiquitous primary and/or secondary mineral forming marginal zones around diopside-aegirine-augite crystals or separate needle-like crystals obviously formed later than the other clinopyroxenes (Yakovenchuk et al., 2005).

Diagrams of a change in clinopyroxene composition (Fig. 14) along the *A–B–C–D–E–F* profile (see Fig. 1) shows, above all, a different degree of rock differentiation in the Main Ring in its rich (the Koashva deposit, the *E* point) and poor ore (Mt. Marchenko Peak, the *C* point) parts. As the Main Ring is approached in the area at Mt. Marchenko Peak, the clinopyroxenes of foyaite feature increasing contents of Ca, Mg and Fe^{2+} at the expense of Na and Fe^{3+}, which proceeds on transition to rischorrite attaining the maximum in ijolite-urtite. Quite opposite is the situation in the area of the Koashva deposit, where the clinopyroxene of foyaite is represented by aegirine with traces of Ca, Mg and Fe^{2+}. However, the concentrations of these elements increase on transition to rischorrite and foidolites, with ijolite-urtite also containing clinopyroxenes of the diopside-hedenbergite series.

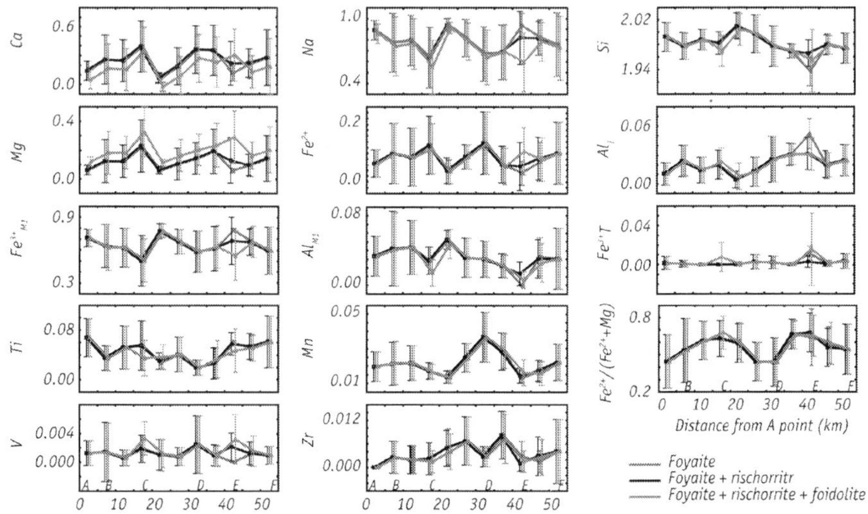

Fig. 14. Variation of clinopyroxene composition along the *A-B-C-D-E-F* profile.

Diagram of the change of the Mg and Fe^{2+} ratio in clinopyroxene composition demonstrates consecutive increasing of the hedenbergite constituent from the border of the massif to the Main Ring, where all rocks contain clinopyroxene with a maximal content of Fe^{2+}, followed by an abrupt decrease in Fe^{2+}concentration to the center of the massif. The Mn content in the clinopyroxene composition decreases from the border to the Main Ring, suddenly rising in the massif's central part. The local maxima of Ti and V contents in clinopyroxenes are confined to the rock complexes of the Main Ring, whereas the increased content of Zr shows the position of the albitization ring at the contact between rischorrite and foyaite in the central part of the massif. Moreover, clinopyroxene in all types of rocks found on the Main Ring area are deficient in silicon compensated by aluminium and/or iron.

Amphiboles of the Khibiny massif are rather variegated (Konopleva et al., 2008): the total number of members in this group, scattered within the massif according to its general zonation (Fig. 15), was found to be 25. Foyaite contains richterite, ferrorichterite, ferroeckermannite, arfvedsonite, magnesioarfvedsonite, katophorite, ferrikatophorite, magnesioferrikatophorite, ferrinyboite, and ferric-ferronyboite. Rischorrite is mostly characterized by the presence of potassicarfvedsonite, foidolites - of potassicrichterite. Dykes of alkali-ultrabasic rocks and alkali-feldspar trachytes contain pargasite, ferro-pargasite, hastingsite, magnesiohastingsite, and kaersutite; pegmatite-hydrothermal veins include potassicrichterite, potassicarfvedsonite, arfvedsonite, and magnesioarfvedsonite; in xenoliths of metamorphized volcanogenic-sedimentary rocks are present edenite, fluoredenite, magnesioferrikatophorite, arfvedsonite and ferric-ferronyboite.

Fluorapatite is a through accessory mineral of all the Khibiny massif rocks, becoming a rock-forming mineral in apatite-nepheline rocks. The content of fluorapatite is 0.2–1.0 vol. % in nepheline syenites, 1-7 vol. % in melteigite-urtite, achieving up to 98 vol. % in apatite-nepheline rocks. An examination of fluorapatite composition along the mentioned profile (Fig. 16) has shown that fluorapatite is released from Na, *REE*, and Si impurities in favour of Ca, Sr and P, as the foidolite ring is approached from the outer and central parts of the massif: $REE^{3+} + Si^{4+} \leftrightarrow (Ca, Sr)^{2+} + P^{5+}$ and $Na^{+} + REE^{3+} \leftrightarrow 2(Ca, Sr)^{2+}$. The behaviour of Ca

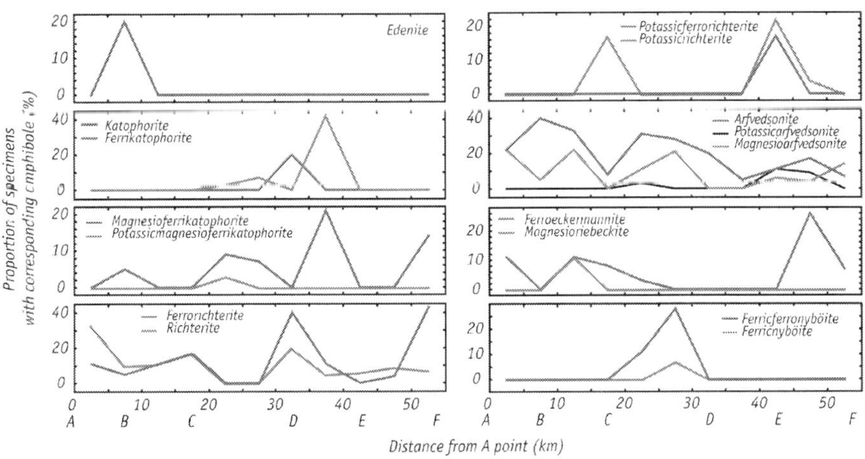

Fig. 15. Variation of amphibole composition along the *A–B–C–D–E–F* profile.

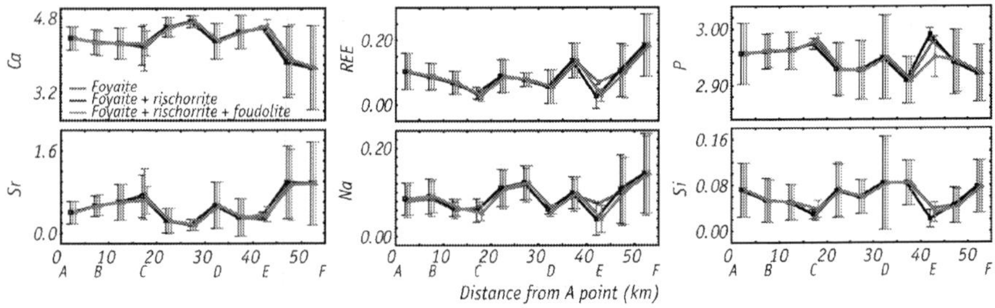

Fig. 16. Variation of fluorapatite composition along the *A–B–C–D–E–F* profile.

and Sr is different in the ore and barren parts of the Main Ring: the larger deposit the lower Sr content. It is important that similar behaviour of these elements was also observed within the apatite deposits: the higher the ore grade (high content of P_2O_5) in fluorapatite, the smaller the quantity of Sr in its composition.

The above cited data on the features of rock-forming and accessory minerals composition within the Khibiny massif indicate that the majority of "through" minerals change the composition as the Main Ring is approached. The extreme contents of some of the elements in mineral compositions, which are related to the Main Ring, are usually superimposed on original monotonous zonation of the foyaite complex, manifesting itself in gradational change of the contents of these elements from the border to the center of the massif. During the formation of substantial zonation of the Khibiny massif, there occurred both plain concentration of elements in the composition of suitable phases and their redistribution between coexisting minerals parallel with their self-cleaning from impurities. In the course of this process, the first to be formed are transitive metastable phases. At the next stage, numerous rare minerals are crystallized in situ, both in interstices of rock-forming minerals of the same rocks and in various types of pegmatite-hydrothermal veins. The zone of maximal differentiation in the mineral chemical composition is confined to the Main Ring, as is the zone of maximal differentiation of rocks.

The plot of quantity of rock-forming and accessory minerals in a rock along the *A–B–C–D–E–F* profile (Fig. 17) has an intensive minimum in the area of the Koashva deposit and a weak minimum in the area of the Marchenko deposit. These minimums correspond to the maximal quantity of mineral species known at these intervals. It means that the great mineral diversity of apatite deposits is related to pegmatites and zones of a later mineralization in both of which the impurities were moved during the ore zone formation. These impurities can be produced by accessory minerals destruction as well as by rock-forming minerals self-cleaning. The larger thickness of foidolite intrusion in the area of the Koashva deposit causes more long and intensive metasomatic and hydrothermal processes, longer chains of mineral transformations and, finally, larger mineral diversity.

Origin of the most of rare minerals by means of self-cleaning of rock-forming minerals causes good correlation between composition of host rocks, rock-forming minerals and mineral diversity (Fig. 18): the largest apatite deposit has the simplest mineral composition of ores, closest to ideal composition of rock-forming minerals, highest mineral diversity and longest list of firstly discovered minerals. Application of our rule to above described profile through the Khibiny massif helped us to discover 8 new minerals with interesting technological properties (see Fig. 17): cerite-(La) (Pakhomovsky et al., 2002), chivruaiite

(Men'shikov et al. 2006), ivanyukite-Na, ivanyukite-K and ivanyukite-Cu (Yakovenchuk et al. 2009), punkaruaivite (Yakovenchuk et al. 2010a); strontiofluorite and polezhaevaite-(Ce) (Yakovenchuk et al. 2010b,d).

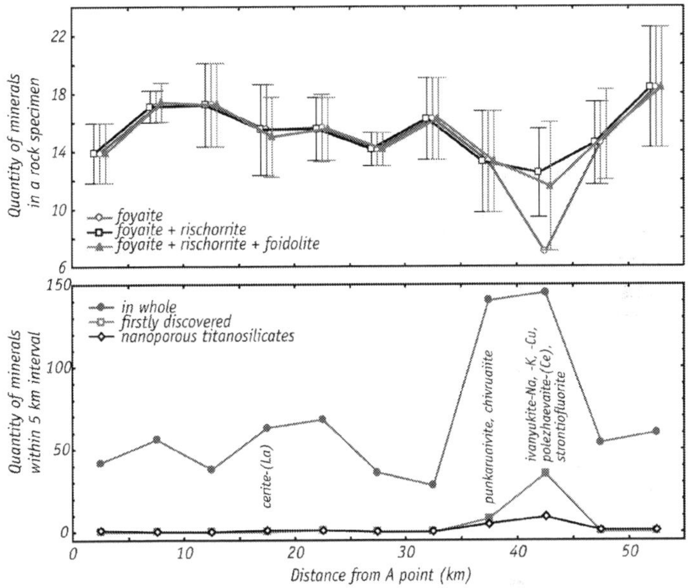

Fig. 17. Variation of quantity of mineral in alkaline rock along the *A-B-C-D-E-F* profile.

5. Conclusion

A thorough geological, petrographic, geochemical and mineralogical investigation of the world's largest Khibiny massif of nepheline syenites and foidolites has provided some essentially new information concerning this unique object and the genesis of its huge apatite deposits. The model of the Khibiny massif formation, in the light of the data obtained, can be seen as the following sequence of events: 1 - formation of a shallow-water mass of terrigene and volcanogenic-sedimentary rocks of the Lovozero suite (quartzite, sandstone, olivine basalt and their tufas); 2 - formation of foyaite massif with a monotonous zonation from the border to the center of the massif as a result of decreasing temperature of rock formation; 3 - formation of the Main and Small Conic faults in the consolidated day-surface part of the massif, owing to its expansion (dilatancy) near the day surface and filling of the faults by foidolite melts; 4 - consolidation and bursting of ijolite-urtite along the same ring, the position of which is determined by a stress field in the still extending Khibiny massif; extrusion to the fractures of residual fluid enriched with Ca, P, F, Cl, C, and H and development of fluorapatite stockworks; apatitization of ambient foyaite, kalsilite-orthoclase metasomatism (poikiloblasting process) and formation of the lyavochorrite-rischorrite series rock; 5 - development of zones of fractal plication and breccias, due to relieving of stresses still persisting along the Main Conic fault, and formation of fractal stockworks of pegmatite-hydrothermal veins within the day-surface parts of apatite-ore bodies; 6 - formation of a system of radial fractures, dykes of alkaline, alkali-ultrabasic rocks and carbonatites, explosion pipes and zones of a low-temperature hydrothermal alteration of the rocks concentrated near the day-surface part of the Main Ring; 7 - completion of formation of a characteristic fractal relief of the Khibiny Tundra due non-uniform uplifting of various parts of the massif accompanied by earthquakes and tremors; 8 - man-caused alterations due to excavation and moving of huge rock masses, accompanied by mountain bumps, earthquakes and intensive low-temperature mineral formation within the Main Ring.

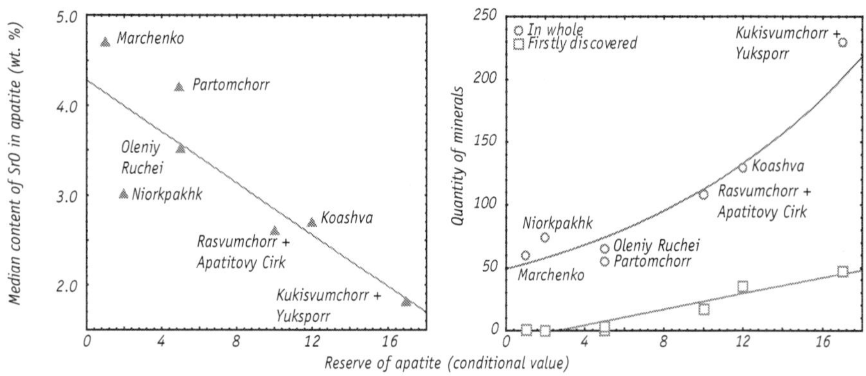

Fig. 18. Relation between size of apatite deposit, composition of apatite and quantity of minerals known in this deposit.

6. Acknowledgment

We are grateful to E.A.Selivanova for carrying out the X-ray phase analysis of minerals and N.I.Nikolaeva for the assistance in the preparation of the manuscript. The research was funded by "Apatit" Joint Stock Company and "Mineraly Laplandiay" Ltd.

7. References

Arzamastsev, A.A., Arzamastseva, L.V., Glaznev, V.N. & Raevsky, A.B. (1998). Deep structure and composition of the bottom horizons of the Khibiny and Lovozero complexes, Kola peninsula: petrological-geophysical model. *Petrology*, Vol. 6, pp. 478-496 (in Russian)

Arzamastsev, A.A., Arzamastseva, L.V., Travin, A.V., Belyatsky, B.V., Shamatrina, A.M., Antonov, A.V., Larionov, A.N., Rodionov, N.V. & Sergeev, S.A. (2007). Duration of Formation of Magmatic System of Polyphase Paleozoic Alkaline Complexes of the Central Kola: U-Pb, Rb-Sr, Ar-Ar Data. *Doklady Earth Sciences*, Vol. 413A, pp. 432-436.

Eliseev, N.A., Ozhinsky, I.S. & Volodin, E.N. (1937). Geology-petrographic studies of the Khibiny tundra), In: *Northern excursion. Kola Peninsula. The International Geological Congress. XVII session*, pp. 51-86, ONTI NKTP Publishing of the USSR, Moscow-Leningrad, Russia (in Russian).

Fersman, A.E. (1931). Geochemical arches of the Khibiny tundra. *Doklady Akademii Nauk. Series A,* No. 14, pp. 367–376 (in Russian).

Galakhov, A.V. (1975). *Petrology of the Khibiny alkaline massi,* Nauka, Leningrad (in Russian).

Goryainov, P.M., Ivanyuk, G.Yu. & Yakovenchuk, V.N. (1998). Tectonic percolation zones in the Khibiny massif: morphology, geochemistry, and genesis. *Izvestiya, Physics of the Solid Earth,* No. 10, pp. 822–827.

Hayward, S.A, Pryde, A.K.A., de Domba, l R.F., Carpenter, M.A. & Dove, M.T. (2000). Rigid Unit Modes in disordered nepheline: a study of a displacive incommensurate phase transition. *Physics and Chemistry of Minerals,* Vol. 27, pp. 285–290.

Ivanyuk, G.Yu., Goryainov, P.M., Pakhomovsky, Ya.A., Konoplyova, N.G., Yakovenchuk, V.N., Bazai, A.V. & Kalashnikov, A.O. Self-organization of ore-bearing complexes, Geokart-Geos, ISBN 978-5-89118-458-9, Moscow (in Russian).

Ivanyuk, G.Yu., Pakhmovsky, Ya.A., Konopleva, N.G., Kalashnikov, A.O., Korchak, Yu.A., Selivanova, E.A. & Yakovenchuk, V.N. (2010). Rock-Forming Feldspars of the Khibiny Alkaline Pluton, Kola Peninsula, Russia. *Geology of Ore Deposits,* Vol. 52, pp. 736–747.

Konopleva, N.G., Ivanyuk, G.Yu., Pakhomovsky,Ya.A., Yakovenchuk, V.N., Men'shikov Yu.P. & Korchak,Yu.A. (2008). Amphiboles of the Khibiny alkaline pluton, Kola Peninsula, Russia. *Geology of Ore Deposits,* Vol. 50, pp. 720–731.

Korchak, Yu.A., Men'shikov, Yu.P., Pakhomovskii, Ya.A., Yakovenchuk, V.N. & Ivanyuk, G.Yu. (2011). Trap Formation of the Kola Peninsula. *Petrology,* Vol. 19, pp. 87–101.

Korobeynikov, A.N. & Pavlov, V.P. (1990). Alkaline syenites of the eastern part of the Khibiny massif, In: *Alkaline magmatism of the North-East part of the Baltic shield),* pp. 4-19, Kola Science Centre of RAN Publishing, Apatity (in Russian).

Kupletsky, B.M. (1937). *Nepheline syenite formation of USSR* (Petrographiya SSSR. Series 2. No. 3), USSR Academy of Science Publishing, Leningrad.

Mandelbrot, B. (1983). *The fractal geometry of Nature,* W.H. Freeman, San Francisco.

Men'shikov, Yu.P., Krivovichev S.V., Pakhomovsky, Ya.A., Yakovenchuk, V.N., Ivanyuk, G.Yu., Mikhailova, J.A., Armbruster, T. & Selivanova, E.A. (2006). Chivruaiite, $Ca_4(Ti,Nb)_5[(Si_6O_{17})_2(OH,O)_5]\cdot 13\text{-}14H_2O$, a new mineral from hydrothermal veins of Khibiny and Lovozero alkaline massifs. *American Mineralogist,* Vol. 91, pp. 922–928.

Pakhomovsky, Ya.A., Men'shikov, Yu.P., Yakovenchuk, V.N., Ivanyuk, G.Yu., Krivovichev, S.V. & Burns, P. C. (2002). Cerite-(La), $(La,Ce,Ca)_9(Fe,Ca,Mg)(SiO_4)_3[SiO_3(OH)]_4(OH)_3$, a new mineral species from the Khibina alkaline massif: occurrence and crystal structure. *The Canadian Mineralogist,* Vol. 40, pp. 1177–1184.

Pakhomovsky, Ya.A., Yakovenchuk, V.N. & Ivanyuk, G.Yu. (2009). Kalsilite of the Khibiny and Lovozero Alkaline Plutons, Kola Peninsula. *Geology of Ore Deposits,* Vol. 51, pp. 822–826.

Ramsay, W. & Hackman, V. (1894). Das Nephelinsyenitgebiet auf der Halbinsel Kola. I. *Fennia.* B. 11, 1-225.

Snyatkova, O.L., Mikhnyak, N.K., Markitakhina, T.M., Prinyagin, N.I., Chapin, V.A., Zhelezova, N.N., Durakova, A.B., Evstaf'ev, A.S., Podurushin, V.F. & Kalinkin, M.M. (1983). *Report on the results of a geological study and geochemical exploration for rare metals and apatite on the scale 1:50000, carried out within the Khibiny massif and its surrounding area during 1979–1983).* Rosgeolfond, inv. no. 24440, Russia (in Russian).

Tikhonenkov, I.P. (1963). *Nepheline syenites and pegmatites of the North-East part of the Khibiny massif and the role of the post-magmatic phenomena in their formation,* AN SSSR Publishing, Moscow (in Russian).

Vlodavets, V.I. (1935) Pinuayvchorr-Yuksporr-Rasvumchorr. *Works of the Arctic Institute,* Vol. 23, pp. 5–55 (in Russian).

Yakovenchuk, V.N., Ivanyuk, G.Yu., Pakhomovsky, Ya.A. & Men'shikov, Yu.P. (Ed. F. Wall) (2005). *Khibiny,* Laplandia Minerals, ISBN 5-900395-48-0, Apatity.

Yakovenchuk, V.N., Ivanyuk, G.Yu., Pakhomovsky,Ya.A., Men'shikov, Yu.P., Konopleva, N.G. & Korchak,Yu.A. (2008). Pyroxenes of the Khibiny alkaline pluton, Kola Peninsula. *Geology of Ore Deposits,* Vol. 50, No. 8, pp. 732–745.

Yakovenchuk, V.N., Nikolaev, A.P., Selivanova, E.A., Pakhomovsky, Ya.A., Korchak, J.A., Spiridonova, D.V., Zalkind, O.A. & Krivovichev, S.V. (2009). Ivanyukite-Na-*T*, ivanyukite-Na-*C*, ivanyukite-K, and ivanyukite-Cu: New microporous titanosilicates from the Khibiny massif (Kola Peninsula, Russia) and crystal structure of ivanyukite-Na-*T*. *American Mineralogist,* Vol. 94, pp. 1450–1458

Yakovenchuk V.N., Ivanyuk G.Yu., Pakhomovsky Y.A., Selivanova E.A., Men'shikovYu.P., Korchak J.A., Krivovichev S.V., Spiridonova D.V. & Zalkind O.A. (2010a). Punkaruaivite, $LiTi_2[Si_4O_{11}(OH)](OH)_2 \cdot H_2O$, a new mineral species from hydrothermal assemblages, Khibiny and Lovozero alkaline massifs, Kola peninsula, Russia. *The Canadian Mineralogist,* Vol. 48, pp. 41–50.

Yakovenchuk V.N., Selivanova E.A., Ivanyuk G.Yu., Pakhomovsky Ya.A., Korchak J.A. & Nikolaev A.P. (2010b). Polezhaevaite-(Ce), $NaSrCeF_6$, a new mineral from the Khibiny massif (Kola Peninsula, Russia). *American Mineralogis,* Vol. 95, pp. 1080–1083.

Yakovenchuk, V.N., Ivanyuk, G.Yu., Konoplyova, N.G., Korchak, Yu.A. & Pakhomovsky, Ya.A. (2010c). Nepheline of the Khibiny alkaline massif (Kola Peninsula). *Proceedings of Russian Mineralogical Society,* No. 2, pp. 80–91 (In Russian).

Yakovenchuk, V.N., Selivanova, E.A., Ivanyuk, G.Yu., Pakhomovsky, Ya.A., Korchak, J.A. & Nikolaev, A.P. (2010d). Strontiofluorite, SrF_2, a new mineral species from the Khibiny massif, Kola peninsula, Russia. *The Canadian Mineralogist,* Vol. 48, pp. 1017–1022.

Zak S.I., Kamenev, E.A., Minakov, F.V., Armand, A.P., Mikheichev, A.S. & Petersil'e I.A. (1972). *Khibiny alkaline massif.* Nedra, Leningrad (in Russian).

Part 3

Seismology

8

Seismic Imaging of Microblocks and Weak Zones in the Crust Beneath the Southeastern Margin of the Tibetan Plateau

Haijiang Zhang[1], Steve Roecker[2], Clifford H. Thurber[3] and Weijun Wang[4]
[1]*Department of Earth, Atmospheric, and Planetary Sciences, Massachusetts Institute of Technology, Cambridge, MA*
[2]*Department of Earth and Environmental Sciences, Rensselaer Polytechnic Institute, Troy, New York*
[3]*Department of Geoscience, University of Wisconsin-Madison, Madison, WI*
[4]*Institute of Earthquake Science, China Earthquake Administration, Beijing,*
[1,2,3]*USA*
[4]*China*

1. Introduction

The southeast margin of the Tibetan Plateau lies between the heartland of the plateau to the west and the stable south China block to the east, spanning from western Sichuan to central Yunnan in southwest China. Based on low-gradient topographic slope and lack of large-scale young crustal shortening at the southeast plateau margin, Royden et al. (1997) and Clark and Royden (2000) proposed a channel-flow model in which a weak (low-viscosity) zone exists in the mid- to lower crust. Gravitational potential drives crustal materials from the Tibetan Plateau outward through the channel, creating a broad and topographically gentle margin and also accumulating stress near the strong crust of the Sichuan Basin. Using GPS data collected from the Crustal Motion Observation Network of China between 1998 and 2004, Shen et al. (2005) showed that the crust is fragmented into tectonic blocks of various sizes, separated by strike-slip and transtensional faults (Figure 1). They proposed a model for Tibetan Plateau deformation in which a mechanically weak lower crust experiences distributed deformation underlying a stronger, highly fragmented upper crust.

On May 12, 2008, a destructive Ms 8.0 earthquake occurred along the Longmen Shan Fault, located between the eastern margin of the Tibetan Plateau and the Sichuan Basin (Burchfiel et al., 2008). It ruptured mainly toward the northeast over a length of ~270 km along the northeast-trending fault, with coseismic slip mainly consisting of thrust- and right lateral strike-slip components (Wang et al., 2008b). No noticeable precursors were observed before the main shock, which was anticipated because GPS modeling showed very low right-slip (~1 mm/yr) and convergence (<~3 mm/yr) rates along the Longmen Shan boundary (Meade, 2007). A deep process involving channel flow is hypothesized to be responsible for the 2008 Wenchuan Ms 8.0 earthquake (Burchfiel, et al., 2008; Teng et al., 2008; Zhang et al., 2008). Other models than the channel flow model such as the block model were also proposed for causing this earthquake (e.g. Hubbard and Shaw, 2009).

Regional seismic tomography studies using body waves (Huang et al., 2002; Wang et al., 2003; Wang et al., 2007; Huang et al., 2009; Xu and Song, 2010) and surface waves (Yao et al., 2008, 2010; Huang et al., 2010; Li et al., 2010) found widespread low velocity zones in the mid- and lower crust, supporting the channel-flow model proposed by Clark and Royden (2000). Receiver function analysis on stations in southwest China also identified low velocity zones (LVZs) in the mid- and lower crust and high average Poisson's ratio in the crust (e.g. Xu et al., 2007; Wang et al., 2008a; Liu et al., 2009; Zhang et al., 2009c). In addition, magnetotelluric (MT) sounding detected low resistivity layers in the middle and lower crust (e.g. Sun et al., 2003; Zhao et al., 2008; Bai et al., 2010). These low velocity and low resistivity zones were interpreted to be caused by partial melt.

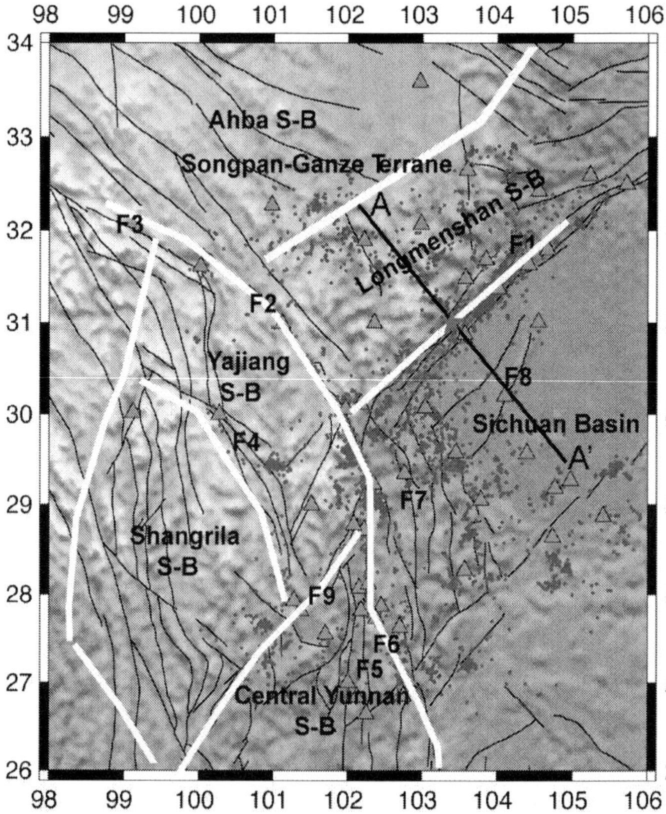

Fig. 1. Distribution of earthquakes (black dots) and stations (green triangles) for the study region. The black lines are mapped fault traces on surface. Red star indicates the 2008 Wenchuan Ms8.0 earthquake. White lines represent boundaries of deformation blocks from the surface GPS modeling (Shen et al., 2005). F1: Longmen Shan Fault; F2: Xianshuihe Fault; F3: Ganzi Fault; F4: Litang Fault; F5: Anninghe Fault; F6: Zemuhe Fault; F7: Daliangshan Fault; F8: Longquan Anticline; F9: Lijiang Fault.

In this article, we present the results of a joint inversion for Vp, Vs, and Vp/Vs models, applying a modified double-difference seismic tomography method to the catalog picks collected by the Seismological Bureau of Sichuan Province for the period 2001-2004. The joint interpretation of three models permits a more complete characterization of the mechanical properties and geological identity of crustal materials and therefore is helpful for better understanding the cause of the low velocity and low resistivity layers. Compared to the previous regional tomography studies in the Sichuan region, this is the first time that a Vp/Vs model is directly inverted from S and P arrival times instead of from dividing Vp by Vs. The three-dimensional (3D) shear-wave velocity model of Yao et al. (2008) indicated that the LVZs vary considerably in strength and depth range and faults may mark lateral boundaries of the LVZs. Our high-resolution 3D Vp, Vs, and Vp/Vs models are utilized to examine the spatial distribution of and interconnectivity between LVZs, which is important for understanding the tectonic block motions (Shen et al., 2005). For accurately calculating ray paths and travel times between events and stations in the case of strong velocity heterogeneity, a spherical-earth finite-difference (SEFD) travel time calculation method is developed and tested.

2. Spherical-Earth Finite-Difference (SEFD) travel time calculation

Since their introduction to seismology by Vidale (1988), finite difference solutions to the eikonal equation have enjoyed widespread application as a robust and efficient technique for computing travel times in heterogeneous media. To the extent that one can easily access the travel time tables produced by such techniques, they can be readily incorporated into earthquake location and tomographic imaging algorithms (e.g. Nelson and Vidale, 1990; Hole, 1992). With few exceptions (Fowler, 1994; Schneider, 1995), these finite difference algorithms solve the Cartesian form of the eikonal equation:

$$\left(\frac{dt}{dx}\right)^2 + \left(\frac{dt}{dy}\right)^2 + \left(\frac{dt}{dz}\right)^2 = s^2 , \tag{1}$$

where s is the local slowness. To the extent that there is no significant spatial regularity in the heterogeneity that we are attempting to parameterize, the bias that we introduce by a particular choice of grid system, Cartesian or otherwise, will not be significant or in the worst case will increase the level of model noise.

As a simple consequence of gravity and temperature, wavespeeds in the earth are primarily a function of depth; lateral variations in wavespeed often tend to be only a few percent. Over regional distances on the order of ~200 km or less, such depth variations should for most purposes be modeled adequately by a Cartesian grid. However, there is a potential for introducing a model induced signal into an inversion when at greater distances the radial variations in wavespeed do not correlate well with the Cartesian grid. One strategy for coping with sphericity is to employ earth flattening (e.g. Abers and Roecker, 1991) but the transformations for flattening are not appropriate for a laterally heterogeneous medium, and moreover there are issues with computing distance properly in the flattened frame (in particular they should always be computed along great circles). Another strategy is to simply put a round earth in a rectangular box, known as the sphere-in-a-box method (Flanagan et al., 2007), but this can artificially introduce anisotropy into the model because radial gradients are not represented the same way in all directions. Of course, such artifacts can be reduced by decreasing the grid spacing but resulting increase in the number of grid points could make the computations intractable.

As an alternative, one might consider solving the eikonal equation in a spherical coordinate system, so that radial gradients are parameterized equally throughout the model with a reduced number of grid points. The eikonal equation in spherical coordinates is:

$$\left(\frac{dt}{dr}\right)^2 + \left(\frac{1}{r}\frac{dt}{d\theta}\right)^2 + \left(\frac{1}{r\sin\theta}\frac{dt}{d\phi}\right)^2 = s^2, \tag{2}$$

where r is the radius from center of the earth, dr is positive away from the center, and $|dr| = h$; θ is the co-latitude (0° at north pole, 90° at equator), $d\theta$ is positive to the south, and $|d\theta| = \Theta$; ϕ is longitude, $d\phi$ is positive to the east, and $|d\phi| = \Phi$; and s is slowness.

To solve this system, we must be account for the differences in r, θ, and ϕ for each node in the mesh. For each node i we assign r_i, θ_i, ϕ_i, and also signs for directional purposes (Table 1). We derive expressions for each of the finite difference (FD) "stencils" used in the algorithm. For example, when applying Scheme A of Vidale (1990), we compute the time at one point given the times at 7 adjacent points in the 8-point cell.

Point	Position	r	θ	ϕ	r Sign (g)	θ Sign (n)	ϕ Sign (m)
0	Deep SE	r_1	θ_2	ϕ_2	-1	1	1
1	Deep SW	r_1	θ_2	ϕ_1	-1	1	-1
2	Deep NW	r_1	θ_1	ϕ_1	-1	-1	-1
3	Deep NE	r_1	θ_1	ϕ_2	-1	-1	1
4	Shallow SE	r_2	θ_2	ϕ_2	1	1	1
5	Shallow SW	r_2	θ_2	ϕ_1	1	1	-1
6	Shallow NW	r_2	θ_1	ϕ_1	1	-1	-1
7	Shallow NE	r_2	θ_1	ϕ_2	1	-1	1

Table 1. Convention on point numbering; the signs are the coefficients for the derivatives dt/dr, $dt/d\theta$, and $dt/d\phi$ as shown below.

Referring to Figure 2 and Table 1, the FD derivatives are:

$$\begin{aligned} dt/dr &= \left[(t_4 - t_0) + (t_5 - t_1) + (t_6 - t_2) + (t_7 - t_3)\right] / (4h) \\ 1/r\, dt/d\theta &= \left[(t_0 - t_3)/r_1 + (t_1 - t_2)/r_1 + (t_4 - t_7)/r_2 + (t_5 - t_6)/r_2\right] / (4\Theta) \\ 1/r\sin\theta dt/d\phi &= [(t_0 - t_1)/(r_1\sin\theta_2) + (t_3 - t_2)/(r_1\sin\theta_1) + (t_4 - t_5)/ \\ &\quad (r_2\sin\theta_2) + (t_7 - t_6)/(r_2\sin\theta_1)]/(4\Phi) \end{aligned} \tag{3}$$

From these equations, it can be shown that the eikonal equation for this stencil is

$$\begin{aligned} s^2 = &\left[\sum_{i=0}^{7} t_i^2 + 2\sum_{i=0}^{6} t_i g_i \sum_{j=i+1}^{7} t_j g_j\right] \Bigg/ 16h^2 + \left[\sum_{i=0}^{7} (t_i/r_i)^2 + 2\sum_{i=0}^{6} t_i n_i/r_i \sum_{j=i+1}^{7} t_j n_j/r_j\right] \Bigg/ 16\Theta^2 \\ &+ \left[\sum_{i=0}^{7} (t_i/r_i\sin\theta_i)^2 + 2\sum_{i=0}^{6} t_i m_i/r_i\sin\theta_i \sum_{j=i+1}^{7} t_j m_j/r_j\sin\theta_j\right] \Bigg/ 16\Phi^2 \end{aligned} \tag{4}$$

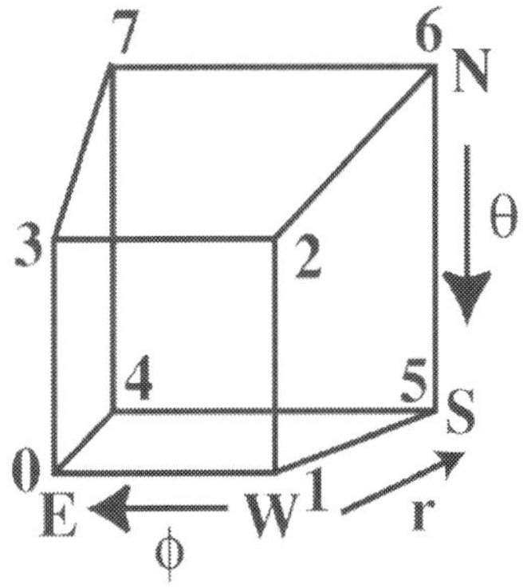

Fig. 2. Geometry of a basic cell for the spherical-earth FD calculation of travel times.

Given the values for t_0 through t_6, this expression can be rewritten in the form $at_7^2 + bt_7 + c = 0$ to solve for t_7, with coefficients a, b and c defined as follows:

$$a = \left(\frac{1}{h}\right)^2 + \left(\left(\frac{1}{\Theta}\right)^2 + \left(\frac{1}{\sin\theta_7\Phi}\right)^2\right)\Big/ r_7^2$$

$$b = 2\sum_{j=0}^{6} t_j \left(g_7 g_j \big/ h^2 + \left(n_7 n_j \big/ \Theta^2 + m_7 m_j \big/ \left(\sin\theta_7 \sin\theta_j \Phi^2\right)\right) \Big/ r_j r_7 \right)$$

$$c = \sum_{i=0}^{6}\left(t_i^2 \left(\frac{1}{h^2} + \left(\frac{1}{\Theta^2} + \left(\frac{1}{\sin\theta_i\Phi}\right)^2\right) \Big/ r_i^2 \right)\right) + 2\sum_{i=0}^{5} t_i \sum_{j=i+1}^{6} t_j \left(g_i g_j \big/ h^2 + \left(n_i n_j \big/ \Theta^2 + m_i m_j \big/ \sin\theta_i \sin\theta_j \varphi^2 \right) \Big/ r_i r_j \right) - 16s^2 \qquad (5)$$

Comparable equations, which are included in the Appendix, can be derived for the "edge" and "face" stencils of Vidale (1990).

One of the problems encountered with these finite difference techniques is that the travel times at the grid points in the immediate neighborhood of the starting point need to be assigned somehow. As long as the wavespeeds are not overly heterogeneous near the starting point, integration of slowness along a straight line path provides a reasonable estimate of travel time. This may not always be the case, however, and in any event as Vidale (1988) pointed out the finite difference approach does not work well when there is significant wavefront curvature over the size of the grid volume element. One efficacious way to solve both of these problems is to use a cascading approach by defining a fine grid in the vicinity of the starting point and a coarser grid outside that region. We have adopted this approach.

We tested the SEFD method by calculating travel times in an analytical velocity model $V=V_0(r_0/r)$, where V_0=4.0km/s, r_0 is the Earth's radius and r is the distance between the source and the Earth's surface. Figure 3a shows the analytical travel times for a source located at latitude 21.2° and longitude 121.75°. We discretized the model into a 3D grid with a grid interval of 0.1° in latitude and longitude and 10 km in depth. The source region is set up to be 3 grid nodes in which the travel times are calculated along a straight-line path. The differences in travel times compared to analytic times are shown in Figure 3b. The travel time error around the source is as much as 1.08 s. Outside the source region, the mean travel time error is 0.108 s, and is everywhere generally smaller than 0.3 s. Along the latitude

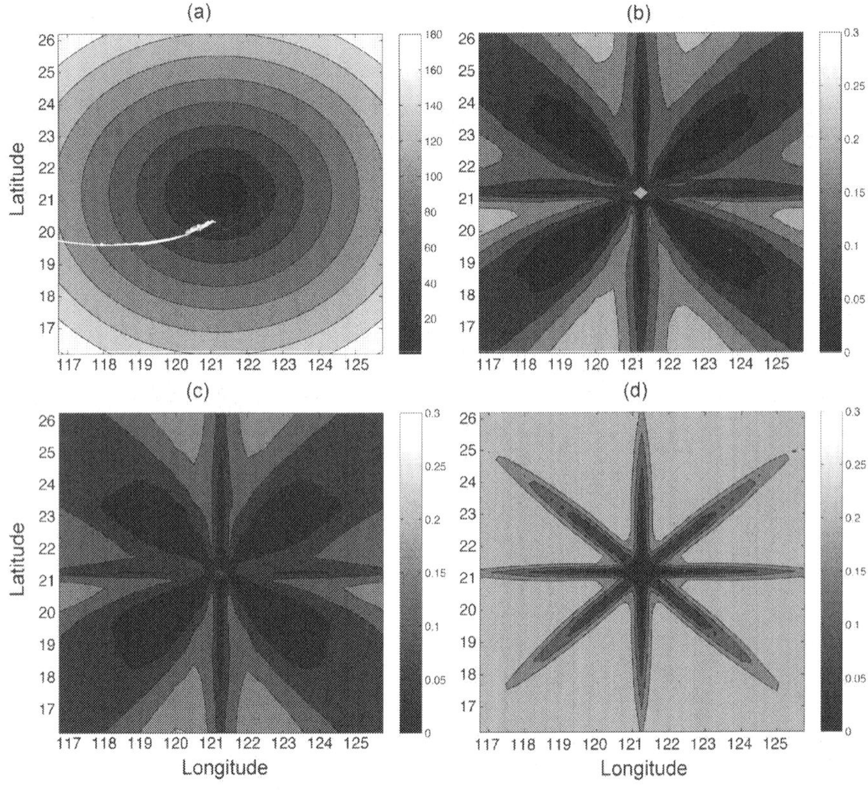

Fig. 3. (a) Analytic travel times from a source located at latitude 21.2° and longitude 121.75°. (b) Travel time errors for the SEFD method. The spherical grid intervals are 0.1° in latitude and longitude and 10 km in depth. (c) Travel time errors for the multi-grid SEFD method. The grid intervals are 0.01° in latitude and longitude and 1 km in depth around the source region. (d) Travel time errors from the FD travel time calculation method in Cartesian coordinates. The time unit is second.

and longitude directions and the directions between them, the travel time errors are relatively small due to the design of the stencils. To deal with the inaccuracy problem near the source region, we applied a cascading-grid strategy, in which a fine grid is used near the source region and a coarse one is used outside the source region. The grid interval inside the source region is 10 times smaller than that outside. The resulting travel time error near the source is much smaller than before, down to 0.17 s and the mean travel time error decreases to 0.087 s. The tests show that the cascading-grid strategy improves the travel time accuracy near the source region and can also decrease the travel time error away from the source region. We also calculated the travel times using the "sphere-in-a-box" method, in which the travel times are calculated on a 3D Cartesian grid with a uniform grid interval using the finite-difference eikonal solver of Podvin and Lecomte (1991). The velocity values on Cartesian grid nodes are linearly interpolated from 8 surrounding spherical grid nodes. The grid interval is set to be 5 km, about 2 times smaller than that used for the SEFD travel time calculation. The travel time errors from Cartesian grid FD method are plotted in Figure 3d. It can be seen that the travel time errors around the source region are small. This is because the FD scheme used in Podvin and Lecomte (1991) adopted an initialization procedure to accurately calculate the travel times around the source. Similar to our SEFD method, the travel time errors are small along latitude, longitude and their middle intersections. However, the travel time errors outside the source region are relatively large. The overall mean travel time error is 0.312 s, much greater than 0.108 s and 0.087 s for the two SEFD cases. This is mainly due to the inaccuracy in velocity values on Cartesian grid nodes when they are interpolated from the exact spherical grid nodes. Even when the Cartesian grid interval is finer, the travel time errors are still greater compared to the case using spherical grid.

3. Seismic tomography method

We employed a new version of the double-difference (DD) seismic tomography method that simultaneously solves for Vp, Vs, Vp/Vs and event locations using both absolute and differential P, S, and S-P times (Zhang, 2003; Zhang et al., 2009a, b). This new code, named tomoDDPS, avoids the pitfalls of inferring Vp/Vs from Vp and Vs models via division (Eberhart-Phillips, 1990). We briefly summarize the method as follows.

The P and S arrival times T_p and T_s from an earthquake i to a seismic station k are expressed using ray theory as path integrals

$$T_{pk}^{i} = \tau^{i} + \int_{i}^{k} u_p \, dl \tag{6}$$

$$T_{sk}^{i} = \tau^{i} + \int_{i}^{k} u_s \, dl \tag{7}$$

where τ^i is the origin time of event i, u_p and u_s are the P- and S-wave slowness fields and dl is an element of path length. The source coordinates (x_1, x_2, x_3), origin times, ray paths, and the slowness field are the unknowns. By assuming the ray paths of P and S waves are identical, which is true when Vp/Vs is constant, Vp/Vs can be determined from S-P arrival times $\mathrm{Ts} - \mathrm{Tp}$, as follows (Thurber, 1993),

$$\mathrm{Ts} - \mathrm{Tp} = \int_{path} \left(\frac{\mathrm{V_P}}{\mathrm{V_S}} - 1 \right) \frac{\mathrm{d}l}{\mathrm{V_P}} \,. \tag{8}$$

Note here because P and S waves from the same event share the same origin time, the unknown origin times are removed from this equation. In the simul2000 algorithm (Thurber and Eberhart-Phillips, 1999), Equations (6) and (8) are used to solve for Vp and Vp/Vs using P and S-P times and Vs is later calculated by dividing Vp by Vp/Vs. However, as noted by Wagner et al. (2005), the Vs model may be biased if calculated in this way because the anomaly in Vp may leak into Vs. In the new tomoDDPS algorithm, Vp, Vs, and Vp/Vs are determined simultaneously in a system using P, S, and S-P times based on Equations (6), (7) and (8) (Zhang, 2003; Zhang et al., 2009a, b). To meet the assumptions made for Equation (8), only S-P times from similar P and S ray paths are selected to solve for Vp/Vs.

Similar to the DD tomography code tomoDD, differential P and S times are also used in tomoDDPS to better constrain seismic event locations and Vp and Vs models (Zhang and Thurber, 2003). In addition, differential S-P times are also used to determine the Vp/Vs structure based on the differential time version of Equation (8), which can be directly constructed from differential P and S times. One advantage of using differential S-P times is to remove the effect of different ray paths of P and S waves outside the source region. Near the source region, P- and S-wave ray paths are generally close to each other. Smoothing weights are applied to P- and S-wave slowness perturbations and Vp/Vs perturbations for neighboring inversion grid nodes to stabilize the tomographic inverse problem. The complete tomographic system is represented as follows (Zhang et al., 2009a):

$$\begin{cases}
w_1 dr_k^i = w_1\left(\sum_{l=1}^{3}\frac{\partial T_k^i}{\partial x_l}\Delta x_l^i + \Delta\tau_i + \int_i^k \delta u ds\right) & \textit{Absolute S or P data} \\
w_2 dr_k^{i,j} = w_2\left(\sum_{l=1}^{3}\frac{\partial T_k^i}{\partial x_l}\Delta x_l^i - \sum_{l=1}^{3}\frac{\partial T_k^j}{\partial x_l}\Delta x_l^j + \Delta\tau_i - \Delta\tau_j + \int_i^k \delta u ds - \int_j^k \delta u ds\right) & \textit{Differential S or P data} \\
w_3 dr_{kSP}^i = w_3\left(\sum_{l=1}^{3}\left(\frac{\partial T_{kS}^i}{\partial x_l} - \frac{\partial T_{kP}^i}{\partial x_l}\right)\Delta x_l^i + \int_i^k \delta(V_p/V_s)\,{}^{ds}\!/_{V_p}\right) & \textit{Absolute } S-P \textit{ data} \\
w_4 dr_k^{i,j}{}_{SP} = w_4\left(\sum_{l=1}^{3}\left(\frac{\partial T_{kS}^i}{\partial x_l} - \frac{\partial T_{kP}^i}{\partial x_l}\right)\Delta x_l^i - \sum_{l=1}^{3}\left(\frac{\partial T_{kS}^j}{\partial x_l} - \frac{\partial T_{kP}^j}{\partial x_l}\right)\Delta x_l^j \right. & \textit{Differential } S-P \textit{ data} \\
\qquad \left. + \int_i^k \delta(V_p/V_s)\,{}^{ds}\!/_{V_p} - \int_j^k \delta(V_p/V_s)\,{}^{ds}\!/_{V_p}\right) & \\
w_5(\delta u_m - \delta u_n) = 0 & 1^{st} \textit{ order smoothing of slowness perturbation} \\
w_6\left(\delta\left(V_p/V_s\right)_m - \delta\left(V_p/V_s\right)_n\right) = 0 & 1^{st} \textit{ order smoothing of } Vp/Vs \textit{ perturbation}
\end{cases} \qquad (9)$$

where $dr_k^i = (T_k^i)^{obs} - (T_k^i)^{cal}$ is the absolute time residual, $dr_k^{ij} = (T_k^i - T_k^j)^{obs} - (T_k^i - T_k^j)^{cal}$ is the differential time residual, $\Delta\tau$ is the origin time perturbation, δu is the P or S slowness perturbation, $\delta(Vp/Vs)$ is the V_p/V_s perturbation, w_1 and w_2 are data weights for the absolute and differential P or S data, w_3 and w_4 are data weights for the absolute and differential S-P data, w_5 and w_6 are smoothing weights for slowness and V_p/V_s models, and m and n indicate neighboring inversion grid nodes. The complete system is solved using a damped least squares inversion method LSQR in which the weighted data residuals are minimized (Paige and Saunders, 1987).

4. Data and inversion details

For the Sichuan region, we collected ~38,600 P- and ~36,500 S-wave first arrival times from 4878 earthquakes observed on 55 stations for the period 2001 to 2004 (Figure 1). These arrival times are selected from the original catalog data based on the major trend of travel time curves (Figure 4). There are obvious 60-second clock shift errors and other reading errors in catalog picks. For each event included in the analysis, there are at least 6 P and 2 S observations, increasing the likelihood of reliable relocations. From the absolute P and S arrival times, we constructed ~269,000 P and ~261,000 S differential times. The average number of differential times (links) per event pair is 11 and the average hypocentral separation (based on catalog locations) for the linked event pairs is ~11 km.

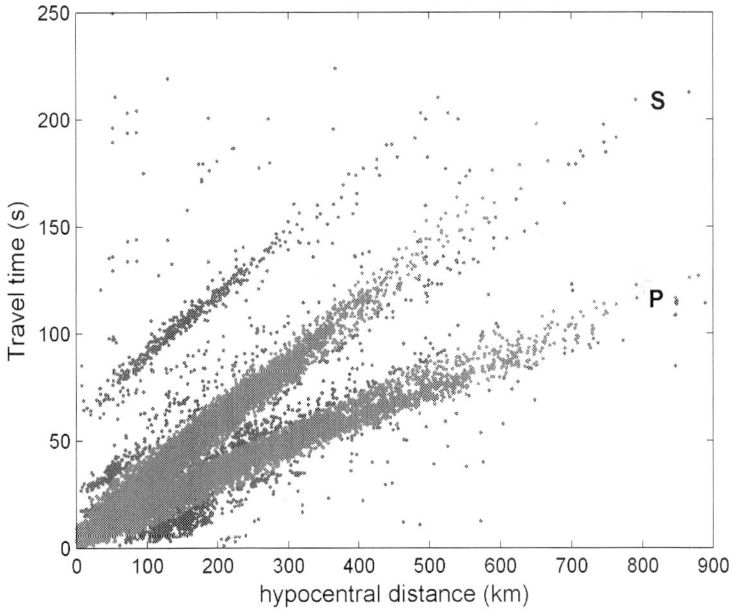

Fig. 4. P and S travel time curves for the original (blue) and selected (red) catalog data.

The inversion grid interval for the velocity model in latitude and longitude is 0.5°. In depth, the grid nodes were positioned at 0, 5, 10, 17.5, 25, 35, 45, 65, and 90 km. In the Sichuan region, the Moho depth varies from ~60 km in the Songpan-Ganze terrane to ~46 km in the Sichuan basin (Xu et al., 2007). Therefore our model mainly reflects the crustal structure of the southeastern Tibetan Plateau. We first derived a minimum one-dimensional (1D) velocity model for the region based on the regional 1D velocity model of Zhao et al. (1997)

(Figure 5). The travel times were calculated using the new SEFD method described above. Both damping and first-order smoothing were used to stabilize the inversion. A trade-off analysis between data variance and model variance was used to select optimum damping and smoothing parameters. The initial unweighted root-mean-square (RMS) travel time residual of 1.78 s was reduced to a final value of 0.48 s, a reduction of approximately 73%. We assess the model quality by a checkerboard resolution test. ±5% velocity anomalies were added to the final 3D Vp and Vs models with an anomaly size of one grid node (Figures 6 and 7). The velocity anomalies for Vp and Vs are made opposite in sign so that the Vp/Vs anomaly ranges from approximately -9% to 11% (Figure 8). A combination of constant noise for each station and random noise at a level comparable to the final inversion misfit is added to the absolute P and S times. The checkerboard resolution test showed that both Vp and Vs models are relatively well resolved for the depth range of 5 to 65 km except for the depth slice of 17.5 km. For the Vp/Vs model, it is also well resolved from a depth of 5 to 45 km except for the depth slice of 17.5 km. For the depth slice of 65 km, the Vp/Vs model has some resolution in the middle part of the model. All three models have poor resolution at depth 0 km.

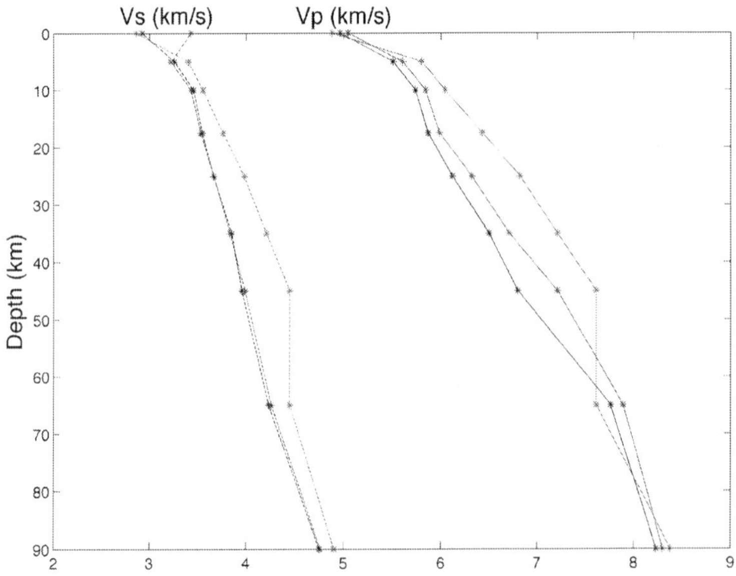

Fig. 5. Three different 1D Vp and Vs profiles for the Sichuan region. RRed: the 1D model of Zhao et al. (1997); Blue: the inverted 1D model; Black: the average 1D model from the 3D inverted model.

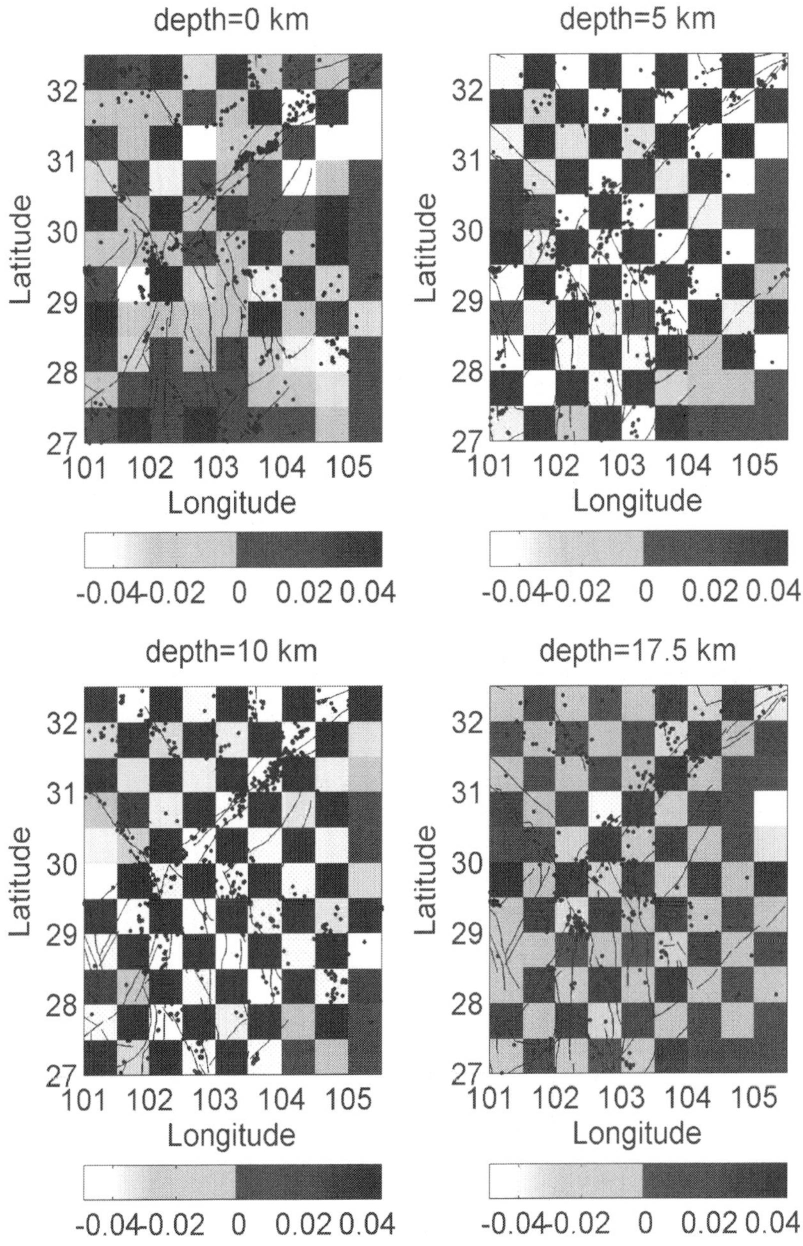

Fig. 6. Horizontal slices of the recovered Vp checkerboard model.

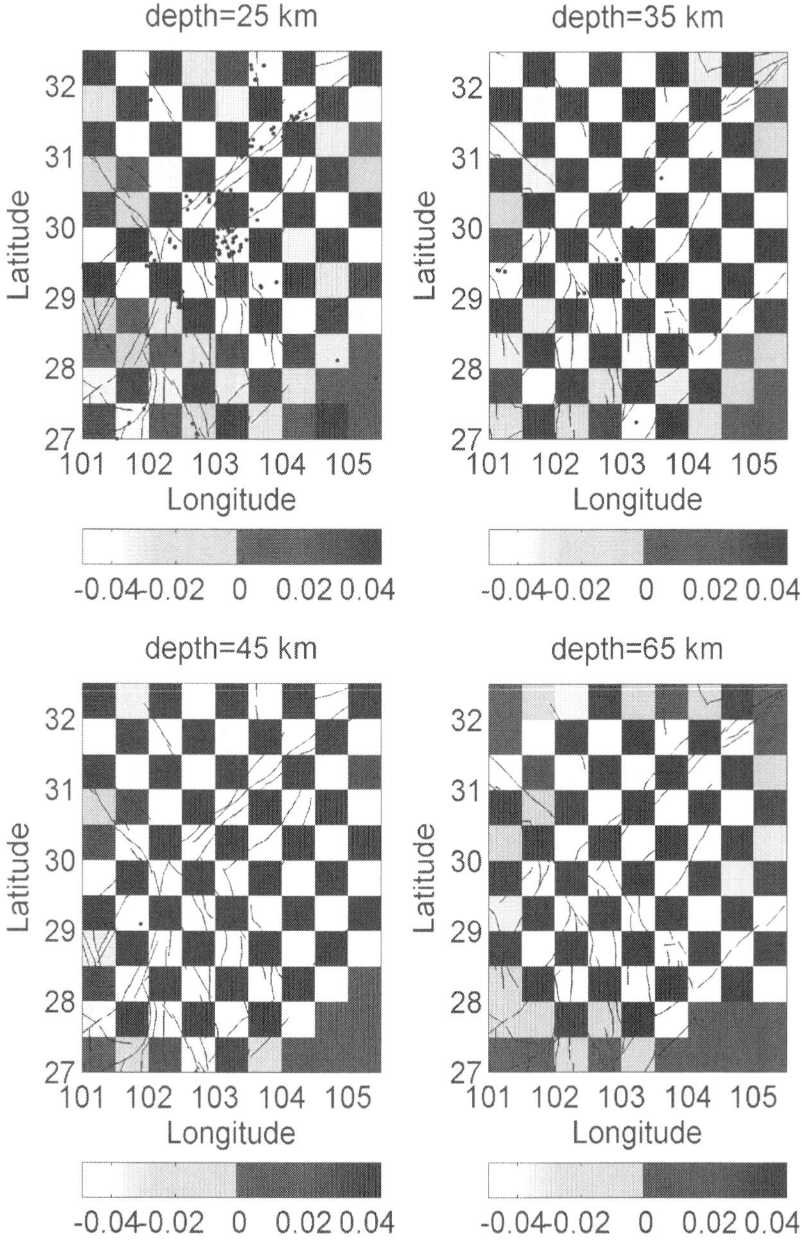

Fig. 6. (Continued)

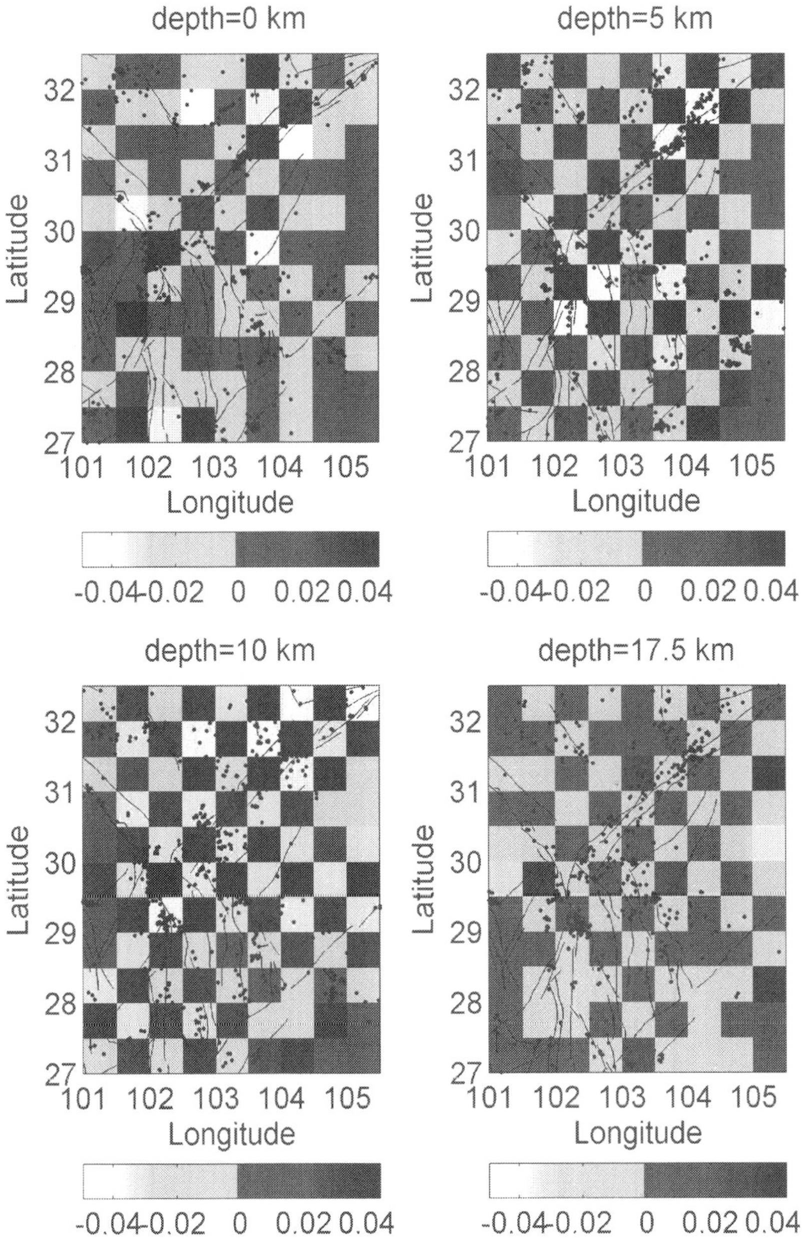

Fig. 7. Horizontal slices of the recovered Vs checkerboard model.

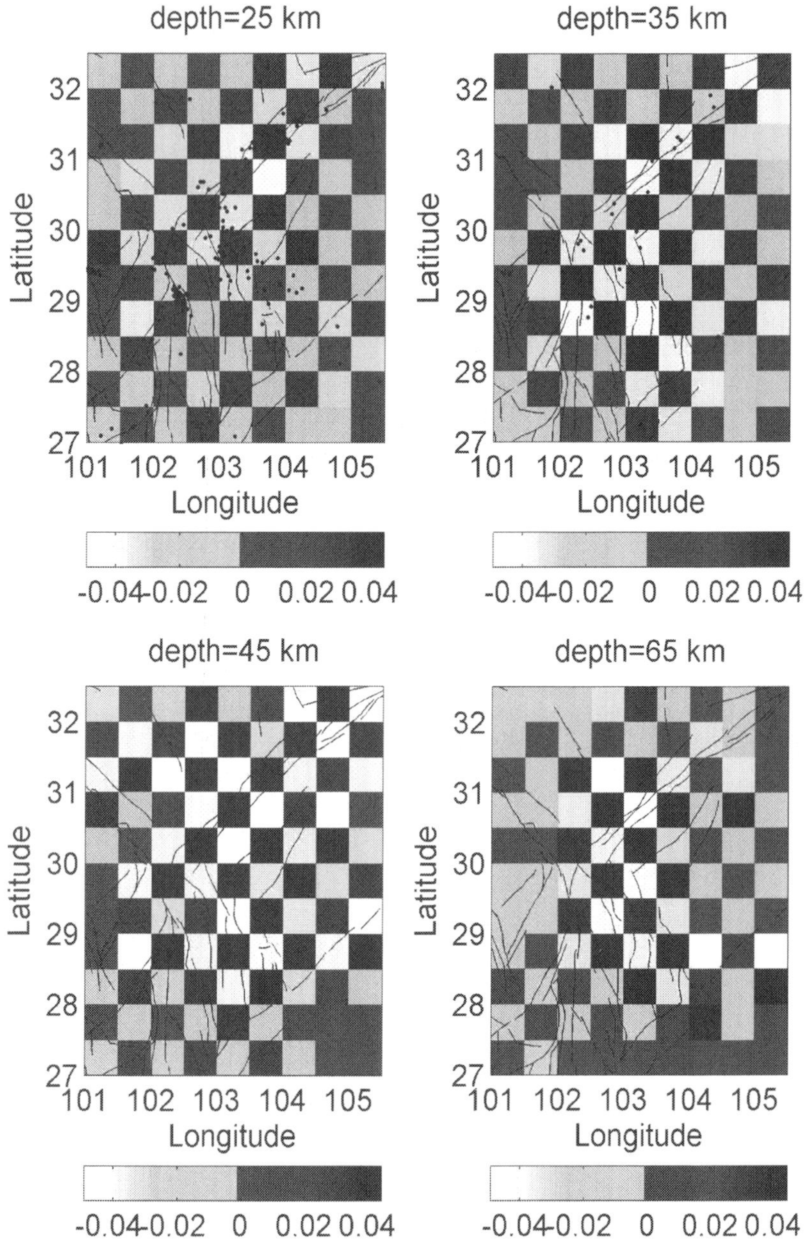

Fig. 7. (Continued)

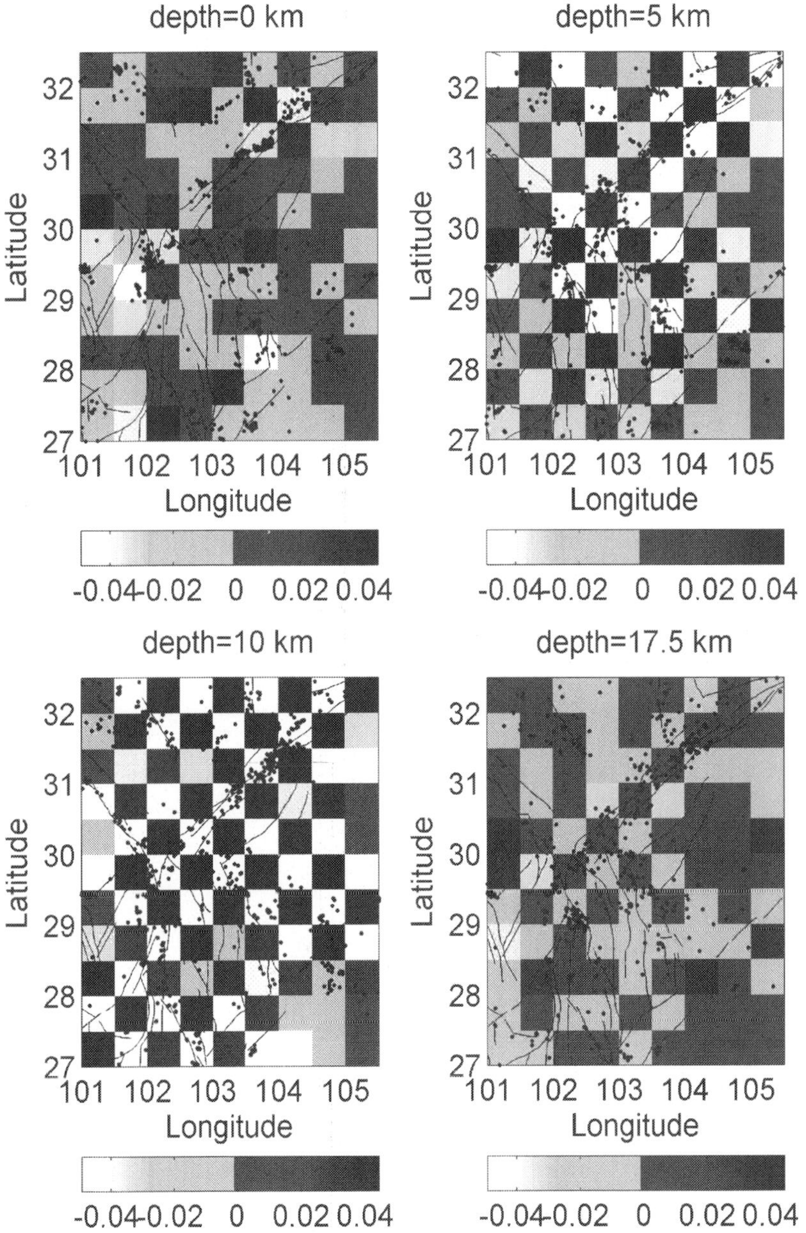

Fig. 8. Horizontal slices of the recovered Vp/Vs checkerboard model.

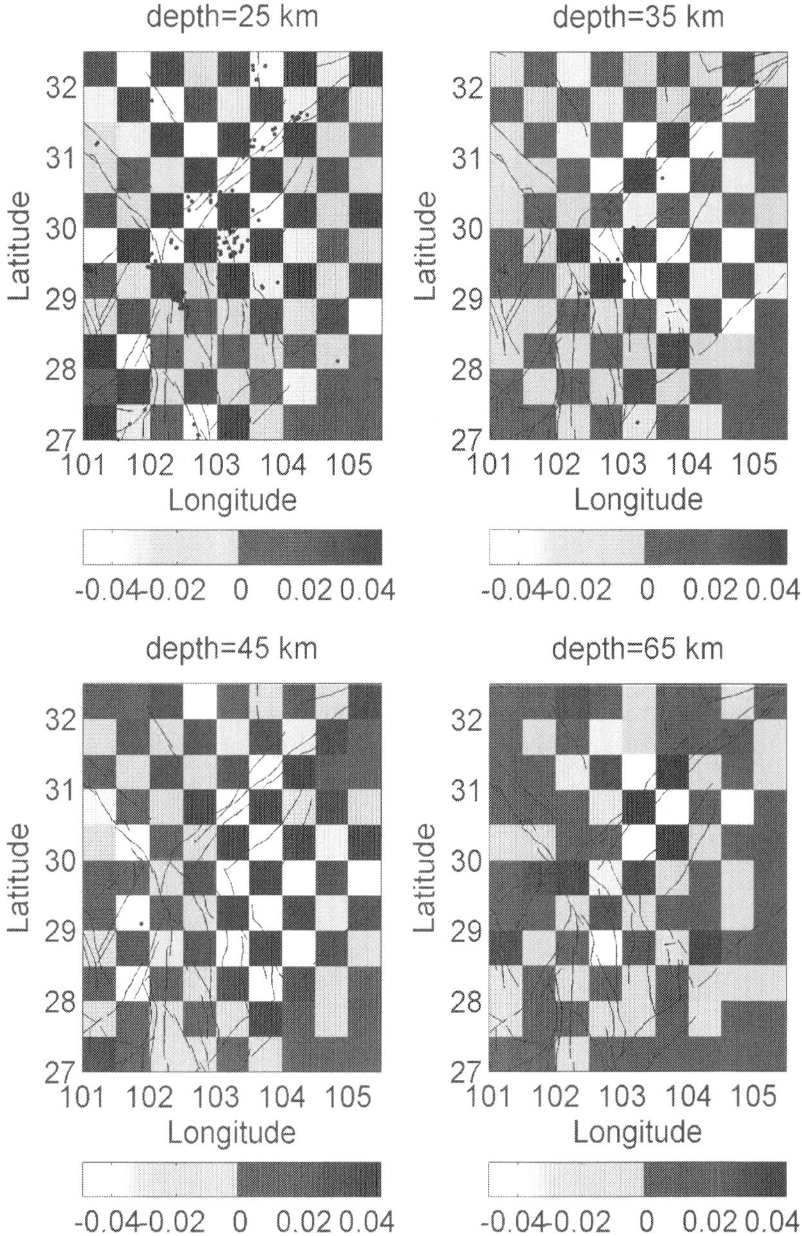

Fig. 8. (Continued)

5. Results and discussion

Figure 9 shows the horizontal slices of the Vp, Vs and Vp/Vs models at depths of 0, 5, 10, 17.5, 25, 35, 45 and 65 km. Figures 10, 11 and 12 show the cross sections of the Vp perturbations, Vs perturbations and Vp/Vs model at latitudes of 28°, 29°, 30°, 31° and 32°, respectively. Vp and Vs perturbations are with respect to the 1D Vp and Vs models averaged from the 3D models (Figure 5). Compared to previous tomography studies in the Sichuan region, to our knowledge this is the first time that the Vp/Vs model is directly determined from the S-P arrival times instead of being derived from the Vp and Vs models. Pei et al. (2010) used the same methodology of Zhang et al. (2009a, b) to solve for the Vp, Vs and Vp/Vs models around the Longmenshan Fault, but the depth extent of the model is only down to 30 km.

At shallow depths (0-5 km), our tomographic velocity models are consistent with the local geology. The Sichuan Basin is clearly imaged as a low velocity region with high Vp/Vs ratios (>1.9). The basin is mainly composed of Tertiary, Quaternary to Mesozoic sediments derived from uplift resulting from the collision. Outside the Sichuan Basin, velocities are generally higher and the Vp/Vs ratio is lower than 1.7, mainly corresponding to the Songpan-Ganze Terrane. One low velocity zone (low Vp and Vs and high Vp/Vs) in the Sichuan Basin is located between latitudes 31° to 32° and longitudes 104° to 105°, extending all the way from the surface down to the depth of a depth of 25 km. Although this low velocity zone is located around the model edge, the checkerboard resolution tests showed this area has good resolution. Previous tomography studies also identified this low velocity anomaly (Wang et al., 2009).

At deeper depths (10-25 km), strong velocity variations are present across the region. At 10 km depth, the Longmen Shan Fault (LMSF) separates a higher velocity region to the northwest from lower velocities to the southeast. At 17.5 km depth, the velocity contrast is still clear north of latitude ~30.5°, especially for Vs, indicating the LMSF may penetrate at least down to ~18 km. The Longquan anticline separates a relatively lower velocity region to the west and a higher velocity to the east at depths 10 and 17.5 km.

In the Songpan-Ganze Terrane, there are scattered low velocity regions bounded by high velocity bodies in the depth slices of 10 km and 17.5 km. Especially at 10 km, the velocity pattern resembles the deformation block model found by modeling the GPS data by Shen et al. (2005). The low velocity anomalies generally follow the derived sub-block boundaries. For example, the Songpan-Xihe deformation zone separating the Ahba block to the northwest and the Longmen Shan block to the southeast corresponds to a broad low velocity zone. The Yajiang block bounded by the Xianshuihe Fault, Lijiang Fault, and Litang Fault corresponds to a high velocity body, whereas these faults fall in low velocity zones. The Central Yunnan block bounded by the Anninghe Fault, Zemuhe Fault and Lijiang Fault is also associated with a high velocity body with relatively low velocities around it. The correspondence between low velocity zones and block boundaries indicates that the blocks themselves are strong and are surrounded by relatively weak zones, where deformation mainly occurs. At a depth of 17.5 km, there is a strong low Vp and Vs anomaly around latitude 30° and longitude of 102°, where the Longmen Shan Fault, the Xianshuihe Fault and the Anninghe Fault intersect. This low velocity body corresponds to a normal Vp/Vs value of ~1.7. Nakajima et al. (2001) also found similar low Vp, low Vs and average to low Vp/Vs patterns in the upper crust of the Japan Island, which they interpreted as being due to the presence of fluids.

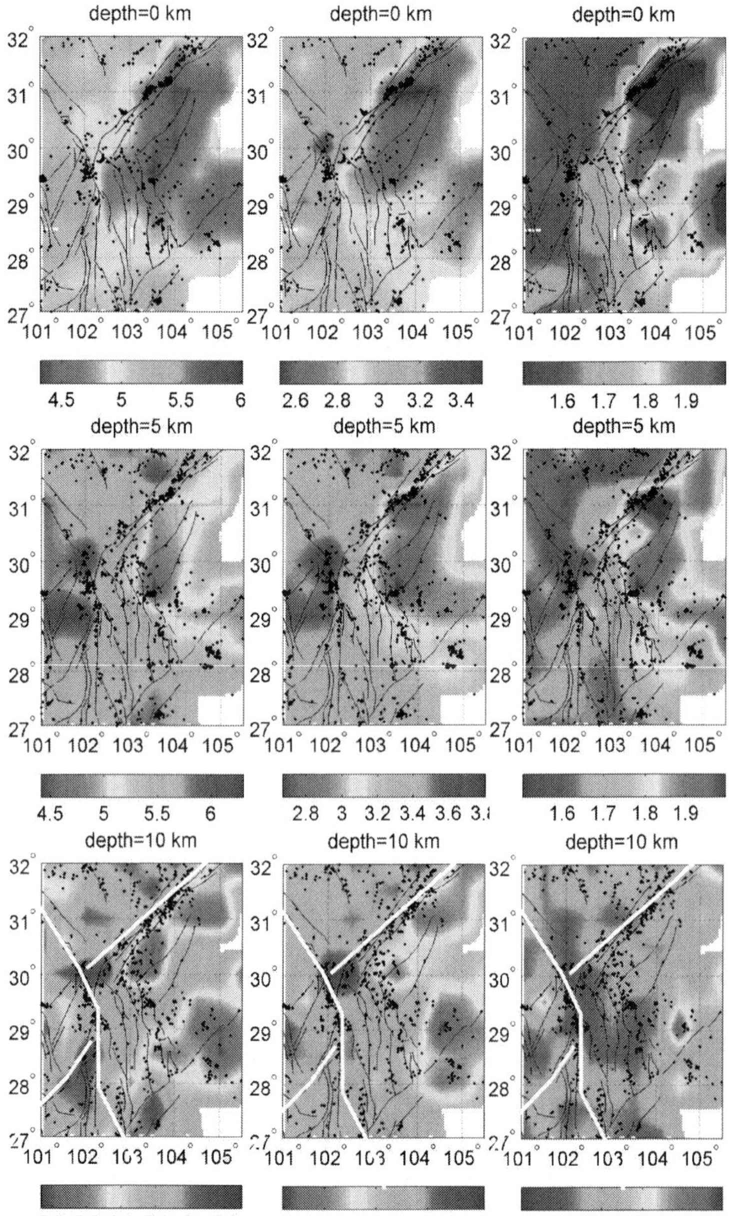

Fig. 9. Horizontal slices of the Vp, Vs and Vp/Vs models at depths 0 to 65 km.

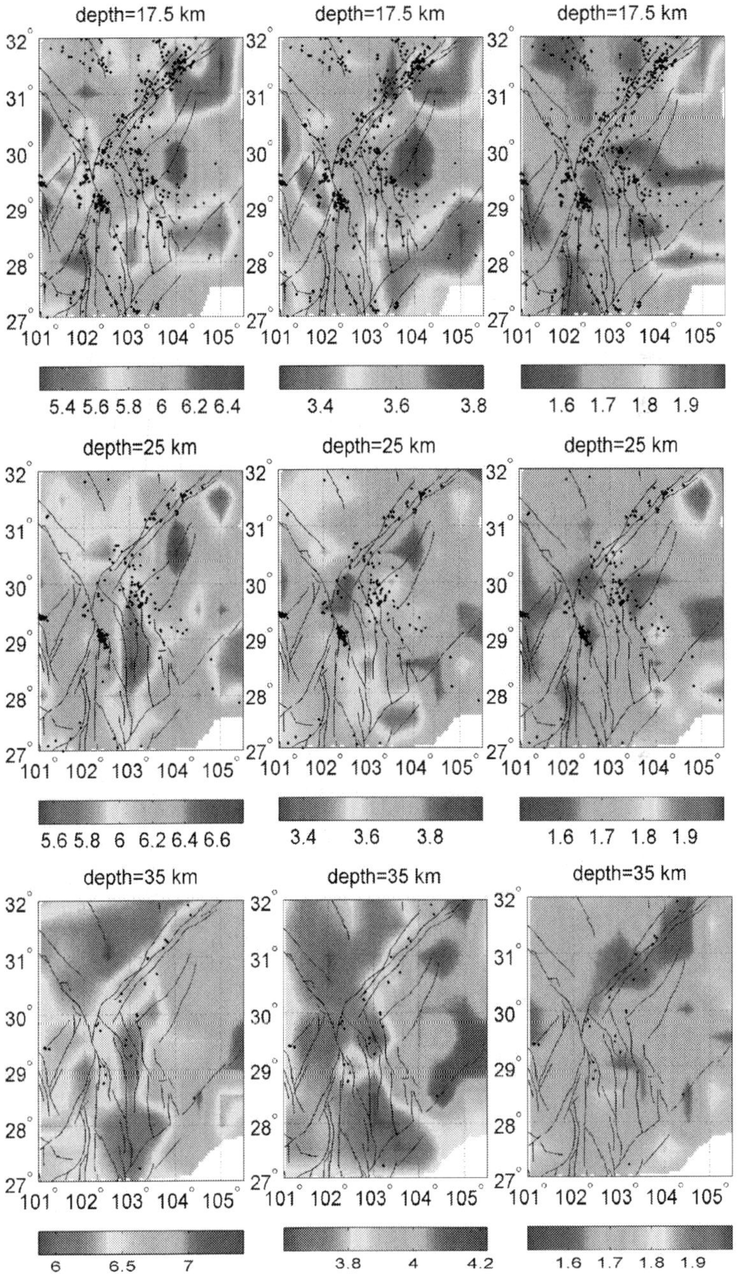

Fig. 9. (Continued)

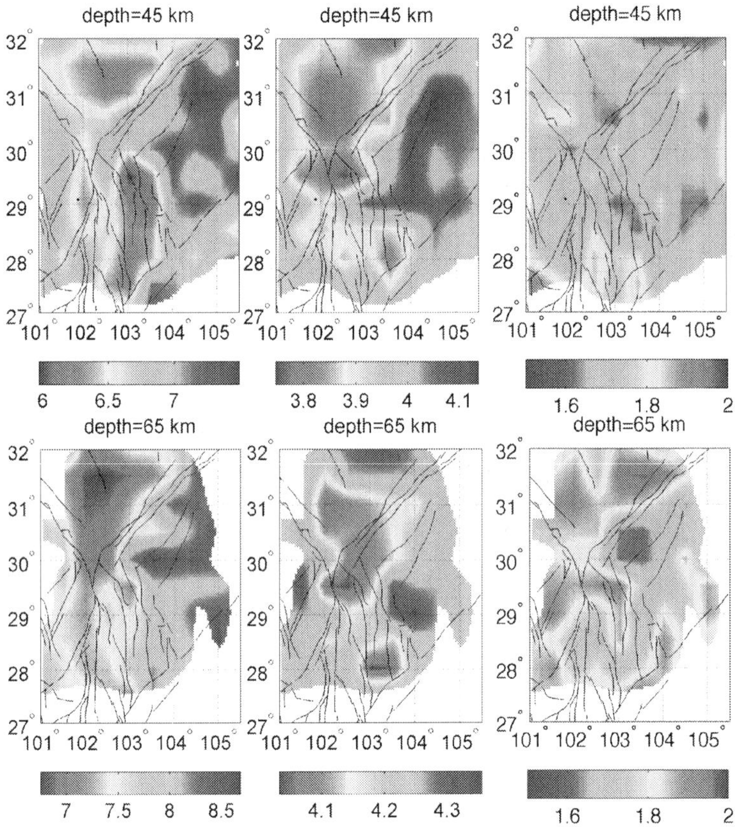

Fig. 9. (Continued)

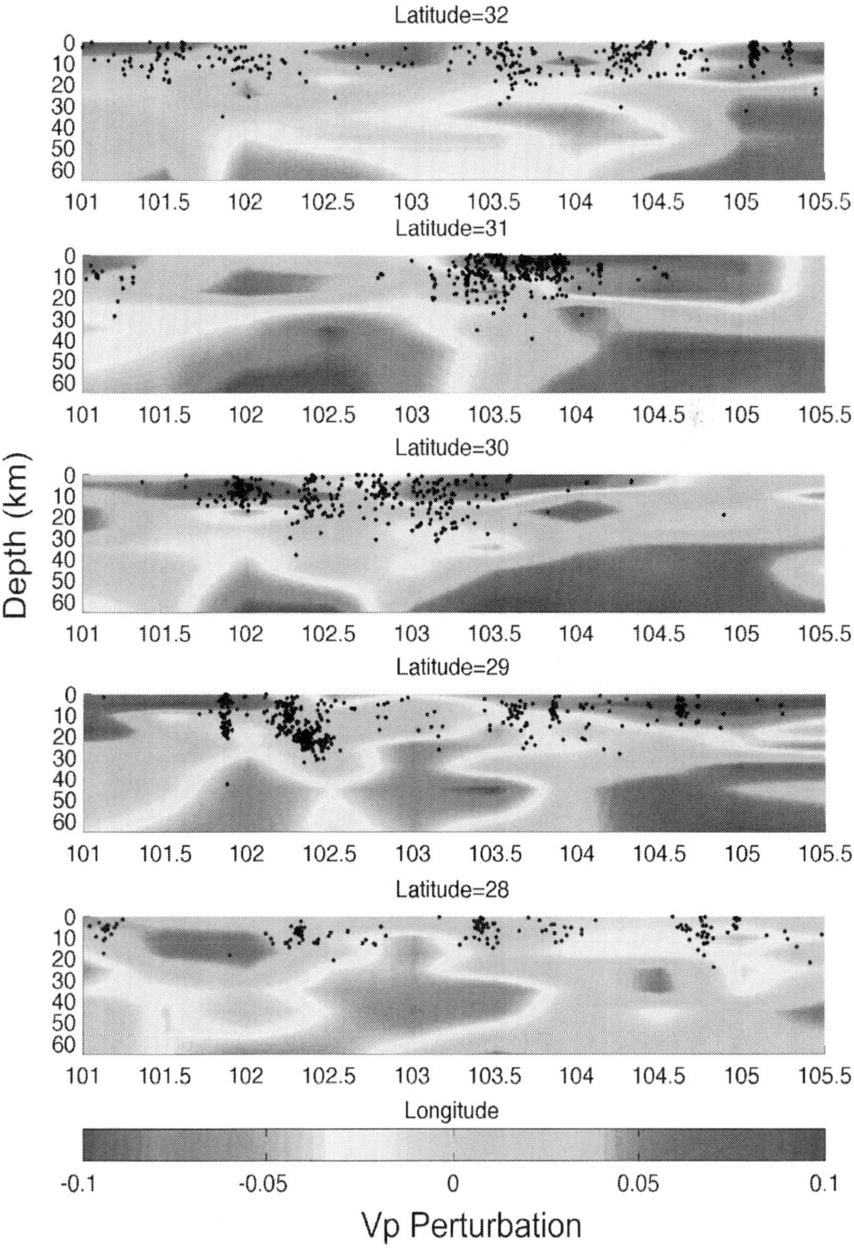

Fig. 10. Cross sections of the Vp perturbation model at latitudes 28, 29, 30, 31, and 32.

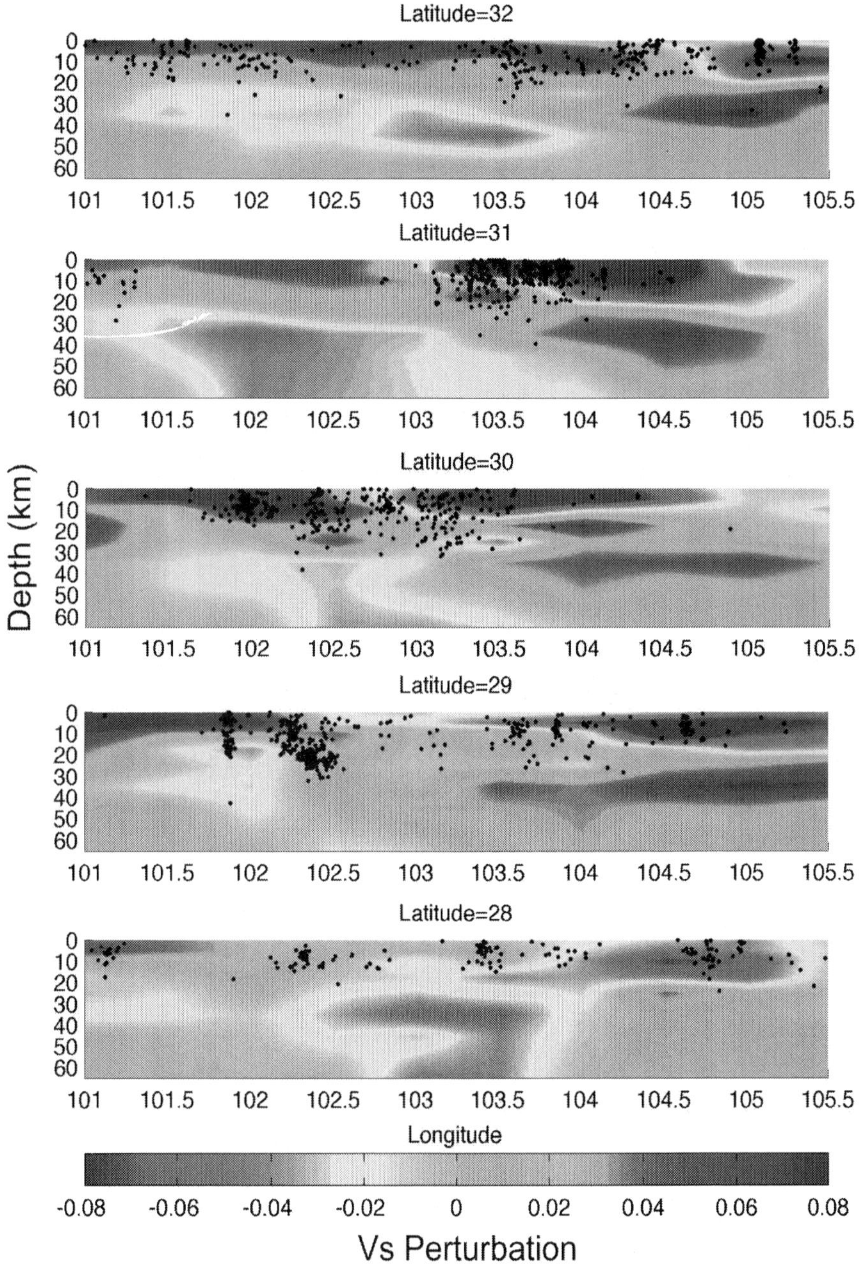

Fig. 11. Cross sections of the Vs perturbation model at latitudes 28, 29, 30, 31, and 32.

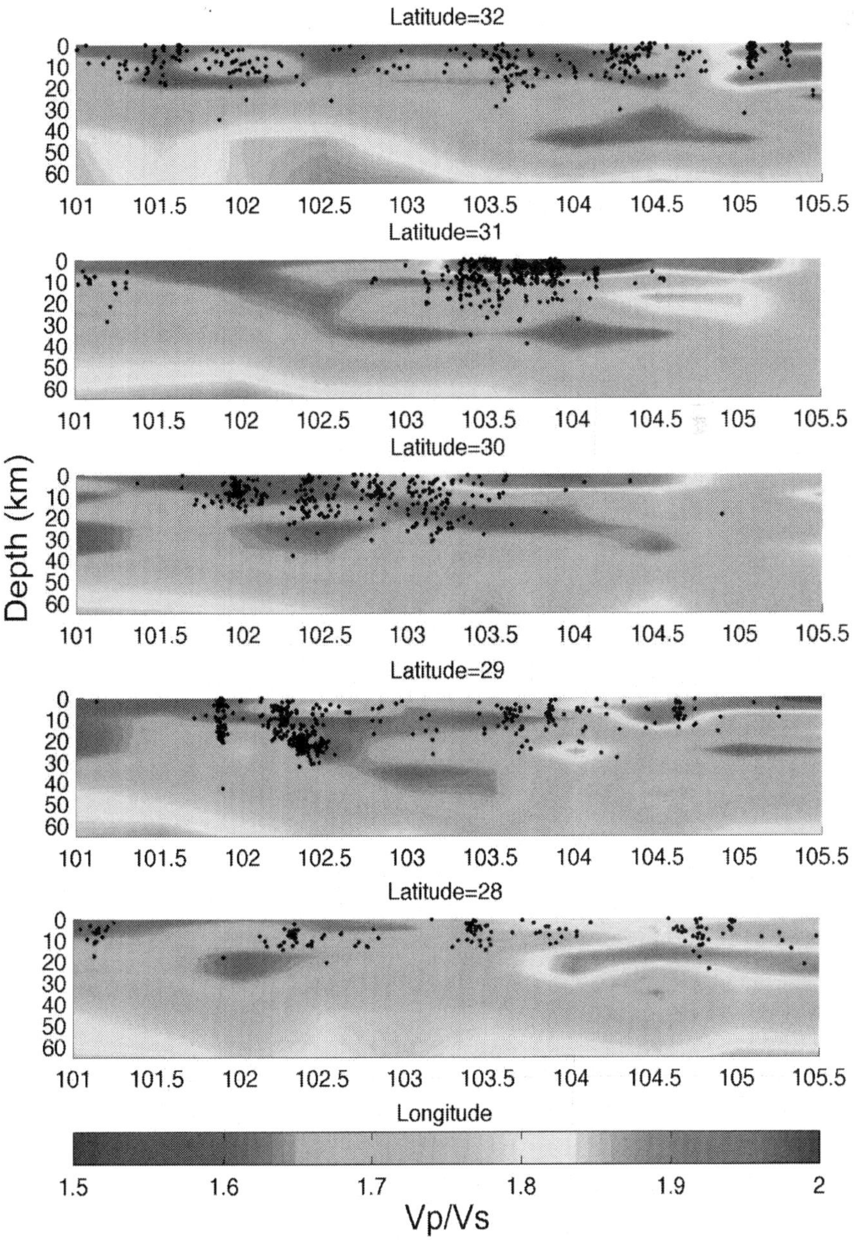

Fig. 12. Cross sections of the Vp/Vs model at latitudes 28, 29, 30, 31, and 32.

Starting from the depth slice of 25 km and down to 65 km, we see widespread low velocity zones outside the Sichuan Basin, which itself generally corresponds to a high velocity anomaly. These low velocity zones are not uniformly distributed but vary in amplitude and they are mostly connected to each other. Previous surface wave and body wave tomography studies also found crustal low velocity zones in this region and interpreted them as weak zones for possible channel flow (e.g. Yao et al., 2008, 2010; Wang et al., 2009). However, it is not clear how these low velocity zones are distributed. Are they bounded by the local structures such as faults, which may interrupt or deflect flow? From the Vp model, it is clear that the low velocity zones may be bounded by faults. For example, at depths of 25 km and 45 km, there is a low velocity zone clearly bounded by the Longmen Shan Fault and Xianshuihe Fault. This low velocity zone dips towards the south and is bounded by the Lijiang Fault and Zemuhe Fault. Another low velocity zone follows the Daliangshan Fault zone from a depth of 25 km to 65 km. The two low velocity zones seem not to be connected at a depth of 25 km and are connected at greater depths. Compared to the Vp model, the Vs model does not show such patterns as clearly. This could be due to larger data errors in the S arrival times and relatively poorer Vs model resolution.

From cross sections of Vp and Vs perturbations at different latitudes (Figures 10 and 11), we see low velocity anomalies below ~20 km depth underneath the Songpan-Ganze block. In comparison, the region beneath the Sichuan basin is shown as a high velocity body. For example, in the cross section of latitude 30°, there is an evident low velocity layer in both Vp and Vs around 20 km depth from longitude 101.5° to 102.5°. This low velocity layer was previously detected from a deep seismic sounding profile and by receiver function analysis (Wang et al., 2008) along latitude 30°. A magnetotelluric (MT) survey between longitudes 102° to 104° and slightly to the north of latitude 29° also showed a low resistivity layer around 20 km depth (Zhao et al., 2008). This low resistivity layer is associated with low Vp, low Vs and high Vp/Vs (Figure 12). Receiver function analysis at station MC09 (latitude 29° and longitude 102.8°) also showed a low Vs layer in the crust associated with a high Vp/Vs ratio (Xu et al., 2007).

Because of high regional surface heat flow values (Hu et al., 2000), the low Vp and Vs anomalies in the eastern Tibetan Plateau have been suggested by many researchers to be due to elevated temperatures or partial melt (e.g. Yao et al., 2008). However, the recent receiver function analysis by Robert et al. (2010) found low Vp/Vs (=1.69) beneath the Songpan-Ganze terrane, which is lower than the mean value for continental areas (Zandt and Ammon, 1995). This observation led them to dispute the existence of a thick and extensive zone of partial melt in the crust of the Songpan-Ganze terrane. In contrast, Wang et al. (2008) showed high Vp/Vs (or Poisson ratio) perturbations and Xu and Song (2010) showed high Poisson ratios in the middle and lower crust of the Songpan-Ganze terrane. Both of their models are obtained by directly dividing Vp by Vs. In comparison, our Vp/Vs model is obtained by directly inverting absolute and differential S-P times and thus is more reliable. Above ~35 km, most of the Songpan-Ganze terrane has a Vp/Vs value of ~1.6. Below ~35 km, the Vp/Vs value starts to increase and reaches up to ~1.85. By averaging from 0 to 60 km, the Vp/Vs value from our study is close to what Robert et al. (2008) found from the receiver function analysis. Although low velocity anomalies start from ~20 km depth in the crust of the eastern Tibetan Plateau, partial melt may not exist until 35 to 40 km depth where the Vp/Vs ratio is relatively high (Christensen, 1996). In the shallower part, the low velocity anomalies could be caused by the existence of aqueous fluids (Li et al., 2003).

The distribution of aqueous fluids in spheroidal pores can result in low Vp and Vs anomalies and an average to low Vp/Vs anomaly (Takei, 2002). In this study area, the earthquakes are mostly located above 30 km (Figure 13), supporting the argument that partial melt does not occur above this depth. Most of the earthquakes are located in the region with Vp>4.4 km/s and Vs>2.6 km.

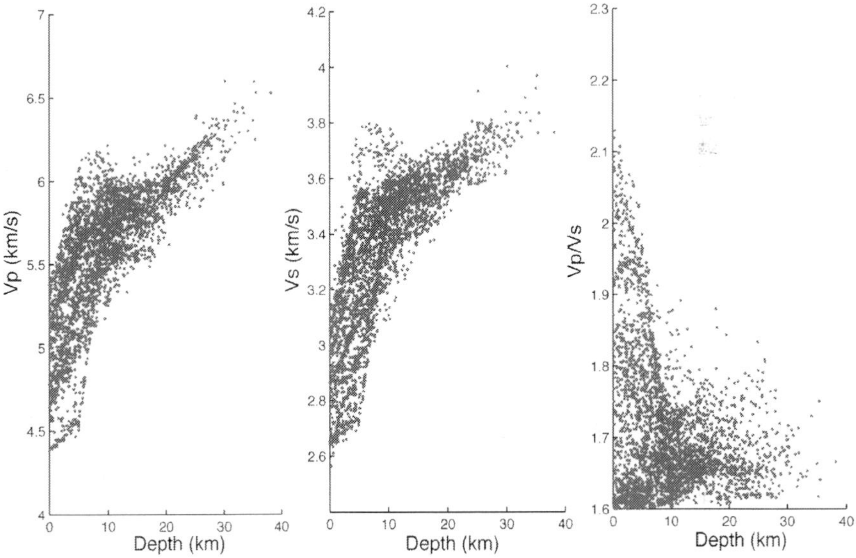

Fig. 13. Scatterplot of hypocenter depth versus Vp, Vs, and Vp/Vs at the hypocenter.

Laboratory experiments show that even a small percentage of melt dramatically reduces the viscosity of rock and it essentially loses its solid nature and behaves like fluid when the melt content reaches 20% to 55% (Kohlstedt and Zimmerman, 1996). A rock containing a fluid content greater than 5% is 10 times weaker than the surrounding material with the same composition (Rosenberg and Handy, 2005). An MT survey through latitude 30° showed a strong low resistivity anomaly below ~30 km depth, which may contain 5% to 20% of fluid content (Bai et al., 2010). The low velocity layer in the shallower part of the crust (such as depth 17.5 km) may play a role of decoupling the upper crust from the mid- and lower crust and the low velocity layer in the lower crust may decouple the crust from the upper mantle. If both low velocity layers exist, they can act as upper and lower sliding planes for the crustal materials to move eastward from the Tibetan Plateau (Teng et al., 2008). Jamieson et al. (2006) found that simply adding a weakened layer in the upper crust to a channel-flow model allowed the model to reproduce several geological observations. Our tomography results support the existence of low velocity zones in the shallower part of the crust.

The 2008 Wenchuan Ms 8.0 earthquake occurred at latitude 31° and longitude 103.4°, where there is a high velocity body seen in the 17.5 km and 25 km depth slices. This high velocity body may act as a local barrier to the channel flow so that it cannot flow to the east and north. As a result, the strain was continuously built up around the corner and the high velocity body acted as an asperity for the main shock. From a local-scale seismic tomography study around the Longmen Shan Fault using aftershocks of the 2008 Wenchuan earthquake, Pei et al. (2010) found two high velocity bodies around Wenchuan and Beichuan, associated with two large slip patches there. These two high velocity bodies act as asperities for the strain to accumulate and lead to large slip during earthquakes. These two high velocity bodies can also be identified in the depth slice of 17.5 km. There exists a relatively low velocity zone between two high velocity bodies along the Longmen Shan fault, where the aftershocks are relatively sparse (Pei et al., 2010).

6. Conclusions

New three-dimensional velocity models including Vp, Vs and Vp/Vs for the southeastern margin of the Tibetan Plateau covering most of Sichuan, China, provide new insights into the geodynamics of the region. The tectonic subblocks found by modeling the GPS data (Shen et al., 2005) are associated with high velocity (or strong) bodies, surrounded by low velocity (or weak) regions. Widespread low velocity zones are found below ~20 km depth in the crust with a complicated spatial distribution. At some depths, the low velocity zones are clearly bounded by faults. Aqueous fluids may exist in the mid-crust above ~35 km depth where the Vp/Vs values are low to average. Partial melt may only exist in the deeper part where the Vp/Vs values are high. The existence of aqueous fluids and/or partial melt can significantly reduce the strength of rock and allow the channel flow in the crust to occur beneath the southeaster Tibetan Plateau. The 2008 Wenchuan Ms 8.0 earthquake likely resulted from the strain accumulation around the high velocity region near the main shock when the channel flow was obstructed there.

7. Acknowledgements

The research presented here was supported by the Chinese government's executive program for exploring the deep interior beneath the Chinese continent (SinoProbe-02). The work was also supported by the Air Force Research Laboratory under contract number FA8718-05-C-0016, and by the Department of Energy under contract number DE-FC52-06NA27325 and under grant number DE-FG3608GO18190.

8. Appendix

In this appendix, we give details of "edge" and "face" stencils used in the spherical finite-difference travel time calculation method.

1. "Edge" Stencils

In this stencil (or Scheme B in Vidale (1990)), 3 previously known nodes and one new node are used to compute an unknown time. Typically this occurs on the edges of new faces. All possible stencils are shown in Figure A1.

Using the above stencils, all the derivatives are:

	dt/dr	$1/r\ dt/d\theta$	$1/r\sin\theta\ dt/d\phi$
T-N	$[(t_4 - t_1) + (t_5 - t_2)]/2h$	$[(t_4-t_5)/r_2 + (t_1-t_2)/r_1]/2\Theta$	$[(t_0-t_3)/r_1\sin\theta]/2\Phi$
T-S	$[(t_4 - t_1) + (t_5 - t_2)]/2h$	$[(t_5-t_4)/r_2 + (t_2-t_1)/r_1]/2\Theta$	$[(t_3-t_0)/r_1\sin\theta]/2\Phi$
T-E	$[(t_4 - t_1) + (t_5 - t_2)]/2h$	$[(t_0-t_3)/r_1]/2\Theta$	$[(t_5-t_4)/r_2 + (t_2-t_1)/r_1]/2\Phi\sin\theta$
T-W	$[(t_4 - t_1) + (t_5 - t_2)]/2h$	$[(t_3-t_0)/r_1]/2\Theta$	$[(t_4-t_5)/r_2 + (t_1-t_2)/r_1]/2\Phi\sin\theta$
B-N	$[(t_1 - t_4) + (t_2 - t_5)]/2h$	$[(t_4-t_5)/r_2 + (t_1-t_2)/r_1]/2\Theta$	$[(t_3-t_0)/r_1\sin\theta]/2\Phi$
B-S	$[(t_1 - t_4) + (t_2 - t_5)]/2h$	$[(t_5-t_4)/r_2 + (t_2-t_1)/r_1]/2\Theta$	$[(t_0-t_3)/r_1\sin\theta]/2\Phi$
B-E	$[(t_1 - t_4) + (t_2 - t_5)]/2h$	$[(t_3-t_0)/r_1]/2\Theta$	$[(t_5-t_4)/r_2 + (t_2-t_1)/r_1]/2\Phi\sin\theta$
B-W	$[(t_1 - t_4) + (t_2 - t_5)]/2h$	$[(t_0-t_3)/r_1]/2\Theta$	$[(t_4-t_5)/r_2 + (t_1-t_2)/r_1]/2\Phi\sin\theta$
W-T	$[(t_2 - t_1) + (t_5 - t_4)]/2h$	$[(t_0-t_3)/r_1]/2\Theta$	$[(t_2-t_5)/r_2 + (t_1-t_4)/r_1]/2\Phi\sin\theta$
W-B	$[(t_1 - t_2) + (t_4 - t_5)]/2h$	$[(t_3-t_0)/r_1]/2\Theta$	$[(t_2-t_5)/r_2 + (t_1-t_4)/r_1]/2\Phi\sin\theta$
W-N	$(t_0 - t_3)/2h$	$[(t_4-t_5) + (t_1-t_2)]/2r_1\Theta$	$[(t_1- t_4)/\sin\theta_1 + (t_2-t_5)/\sin\theta_2]/2r_1\Phi$
W-S	$(t_3 - t_0)/2h$	$[(t_5-t_4) + (t_2-t_1)]/2r_1\Theta$	$[(t_1-t_4)/\sin\theta_1 + (t_2-t_5)/\sin\theta_2]/2r_1\Phi$
E-T	$[(t_2 - t_1) + (t_5 - t_4)]/2h$	$[(t_3-t_0)/r_1]/2\Theta$	$[(t_5-t_2)/r_2 + (t_4-t_1)/r_1]/2\Phi\sin\theta$
E-B	$[(t_1 - t_2) + (t_4 - t_5)]/2h$	$[(t_0-t_3)/r_1]/2\Theta$	$[(t_5-t_2)/r_2 + (t_4-t_1)/r_1]/2\Phi\sin\theta$
E-N	$(t_3 - t_0)/2h$	$[(t_4-t_5) + (t_1-t_2)]/2\ r_1\Theta$	$[(t_4-t_1)\ /\sin\theta_1 + (t_5-t_2)/\sin\theta_2]/2r_1\Phi$
E-S	$(t_0 - t_3)/2h$	$[(t_5-t_4) + (t_2-t_1)]/2\ r_1\Theta$	$[(t_4-t_1)\ /\sin\theta_1 + (t_5-t_2)/\sin\theta_2]/2r_1\Phi$
N-T	$[(t_2-t_1) + (t_5-t_4)]/2h$	$[(t_2-t_5)\ /r_2 + (t_1-t_4)/r_1]/2\Theta$	$(t_3-t_0)/2r_1\sin\theta\Phi$
N-B	$[(t_1-t_2) + (t_4-t_5)]/2h$	$[(t_2-t_5)\ /r_2 + (t_1-t_4)/r_1]/2\Theta$	$(t_0-t_3)/2r_1\sin\theta\Phi$
N-W	$(t_3 - t_0)/2h$	$[(t_2-t_5) + (t_1-t_4)]/2r_1\Theta$	$[(t_1-t_2)/\sin\theta_1 + (t_4-t_5)/\sin\theta_2]/2r_1\Phi$
N-E	$(t_0 - t_3)/2h$	$[(t_2-t_5) + (t_1-t_4)]/2r_1\Theta$	$[(t_2-t_1)/\sin\theta_1 + (t_5-t_4)/\sin\theta_2]/2r_1\Phi$
S-T	$[(t_2-t_1) + (t_5-t_4)]/2h$	$[(t_5-t_2)\ /r_2 + (t_4-t_1)/r_1]/2\Theta$	$(t_0-t_3)/2r_1\sin\theta\Phi$
S-B	$[(t_1-t_2) + (t_4-t_5)]/2h$	$[(t_5-t_2)\ /r_2 + (t_4-t_1)/r_1]/2\Theta$	$(t_3-t_0)/2r_1\sin\theta\Phi$
S-W	$(t_0 - t_3)/2h$	$[(t_5-t_2) + (t_4-t_1)]/2r_1\Theta$	$[(t_1-t_2)/\sin\theta_1 + (t_4-t_5)/\sin\theta_2]/2r_1\Phi$
S-E	$(t_3 - t_0)/2h$	$[(t_5-t_2) + (t_4-t_1)]/2r_1\Theta$	$[(t_2-t_1)/\sin\theta_1 + (t_5-t_4)/\sin\theta_2]/2r_1\Phi$

Note that for each case:

T-N = +-(T-S) = +-(B-N) = +-(B-S)
T-E = +-(T-W) = +-(B-E) = +-(B-W)
W-T = +-(W-B)=+-(E-T)=+-(E-B)
W-N = +-(W-S) =+-(E-N) =+-(E-S)
N-T = +-(N-B)=+-(S-T)=+-(S-B)
N-W = +-(S-W) =+-(N-E) =+-(S-E)

So, because we square derivatives, we need only concern ourselves with expressions for 6 cases.

	dt/dr	$1/r\ dt/d\theta$	$1/r\sin\theta\ dt/d\phi$
T-N	$[(t_4 - t_1) + (t_5 - t_2)]/2h$	$[(t_4-t_5)/r_2 + (t_1-t_2)/r_1]/2\Theta$	$[(t_0-t_3)/r_1\sin\theta]/2\Phi$
T-W	$[(t_4 - t_1) + (t_5 - t_2)]/2h$	$[(t_3-t_0)/r_1]/2\Theta$	$[(t_4-t_5)/r_2 + (t_1-t_2)/r_1]/2\Phi\sin\theta$
W-T	$[(t_2 - t_1) + (t_5 - t_4)]/2h$	$[(t_0-t_3)/r_1]/2\Theta$	$[(t_2-t_5)/r_2 + (t_1-t_4)/r_1]/2\Phi\sin\theta$
W-N	$(t_0 - t_3)/2h$	$[(t_4-t_5) + (t_1-t_2)]/2r_1\Theta$	$[(t_1- t_4)/\sin\theta_1 + (t_2-t_5)/\sin\theta_2]/2r_1\Phi$
N-T	$[(t_2-t_1) + (t_5-t_4)]/2h$	$[(t_2-t_5)\ /r_2 + (t_1-t_4)/r_1]/2\Theta$	$(t_3-t_0)/2r_1\sin\theta\Phi$
N-W	$(t_3 - t_0)/2h$	$[(t_2-t_5) + (t_1-t_4)]/2r_1\Theta$	$[(t_1-t_2)/\sin\theta_1 + (t_4-t_5)/\sin\theta_2]/2r_1\Phi$

The general form of the equation is:

$4s^2 = [(t_4-t_1)/d_{14} + (t_5-t_2)/d_{25}]^2 + [(t_4-t_5)/d_{45} + (t_1-t_2)/d_{12}]^2 + [(t_0-t_3)/d_{01}]^2$

And we solve for t_5

$4s^2 = [(t_4-t_1)/d_{14}]^2 + [(t_1-t_2)/d_{12}]^2 + [(t_0-t_3)/d_{01}]^2$
$+ [(t_5^2 + t_2^2 - 2t_2t_5)/d_{25}^2 + 2(t_1t_2 - t_1t_5 - t_2t_4 + t_4t_5)/d_{14}d_{25}]$
$+ [(t_5^2 + t_4^2 - 2t_4t_5)/d_{45}^2 + 2(t_1t_4 - t_1t_5 - t_2t_4 + t_2t_5)/d_{12}d_{45}]$

Isolating t_5:

$4s^2 = [(t_5^2 - 2t_2t_5)/d_{25}^2 + 2(- t_1t_5 + t_4t_5)/d_{14}d_{25}] + [(t_5^2 - 2t_4t_5)/d_{45}^2 + 2(- t_1t_5 + t_2t_5)/d_{12}d_{45}]$
$+ [(t_4-t_1)/d_{14}]^2 + [(t_1-t_2)/d_{12}]^2 + [(t_0-t_3)/d_{01}]^2$
$+ [(t_2^2)/d_{25}^2 + 2(t_1t_2 - t_2t_4)/d_{14}d_{25}] + [(t_4^2)/d_{45}^2 + 2(t_1t_4 - t_2t_4)/d_{12}d_{45}]$

$4s^2 = t_5^2 [1/d_{25}^2 + 1/d_{45}^2] + 2t_5[(- t_2)/d_{25}^2 + (- t_1 + t_4)/d_{14}d_{25} + (- t_4)/d_{45}^2 + (- t_1 + t_2)/d_{12}d_{45}]$
$+ [(t_4-t_1)/d_{14}]^2 + [(t_2^2)/d_{25} + 2\ t_2\ (t_1 - t_4)/d_{14}d_{25}]$
$+ [(t_2-t_1)/d_{12}]^2 + [(t_4^2)/d_{45} + 2\ t_4\ (t_1 - t_2)/d_{12}d_{45}] + [(t_0-t_3)/d_{01}]^2$

$4s^2 = t_5^2 [1/d_{25}^2 + 1/d_{45}^2] + 2t_5[(- t_2)/d_{25}^2 + (t_4 - t_1)/d_{14}d_{25} + (- t_4)/d_{45}^2 + (t_2 - t_1)/d_{12}d_{45}]$
$+ [(t_4-t_1)/d_{14} - t_2/d_{25}]^2 + [(t_2-t_1)/d_{12} - t_4/d_{45}]^2 + [(t_0-t_3)/d_{01}]^2$

$4s^2 = t_5^2 [1/d_{25}^2 + 1/d_{45}^2] + 2t_5[(t_4 - t_1)/d_{14}d_{25} + (t_2 - t_1)/d_{12}d_{45} - t_2/d_{25}^2 - t_4/d_{45}^2]$
$+ [(t_4-t_1)/d_{14} - t_2/d_{25}]^2 + [(t_2-t_1)/d_{12} - t_4/d_{45}]^2 + [(t_0-t_3)/d_{01}]^2$

	d_{01}	d_{12}	d_{14}	d_{25}	d_{45}
TN	$r_2\sin\theta_2\Phi$	$r_2\Theta$	h	h	$r_1\Theta$
TW	$r_2\Theta$	$r_2\sin\theta_1\Phi$	h	h	$r_1\sin\theta_1\Phi$
NW	h	$r_1\sin\theta_2\Phi$	$r_1\Theta$	$r_1\Theta$	$r_1\sin\theta_1\Phi$
NT	$r_2\sin\theta_2\Phi$	h	$r_2\Theta$	$r_1\Theta$	h
WN	h	$r_2\Theta$	$r_1\sin\theta_2\Phi$	$r_1\sin\theta_1\Phi$	$r_2\Theta$
WT	$r_2\Theta$	h	$r_2\sin\theta_2\Phi$	$r_1\sin\theta_1\Phi$	h

In General:

$4s^2 = at_5^2 + b\ t_5 + c$

with

$a = 1/d_{25}^2 + 1/d_{45}^2$
$b = 2[(t_4 - t_1)/d_{14}d_{25} + (t_2 - t_1)/d_{12}d_{45} - t_2/d_{25}^2 - t_4/d_{45}^2]$
$b = 2[(t_4 - t_1)/d_{14} - t_2/d_{25}]/d_{25} + [(t_2 - t_1)/d_{12} - t_4/d_{45}]/d_{45}]$

When $d_{14} = d_{25}$ (TN, TW, NW)

$b = 2[(t_4 - t_1 - t_2)/d_{25}^2 + (t_2 - t_1)/d_{12}d_{45} - t_4/d_{45}^2]$

When $d_{12} = d_{45}$ (WN, WT, NT)

$b = 2[(t_2 - t_1 - t_4)/d_{45}^2 + (t_4 - t_1)/d_{14}d_{25} - t_2/d_{25}^2]$
$c = [(t_4-t_1)/d_{14} - t_2/d_{25}]^2 + [(t_2-t_1)/d_{12} - t_4/d_{45}]^2 + [(t_0-t_3)/d_{01}]^2 - 4s^2$

When $d_{14} = d_{25}$ (TN, TW, NW)

$c = [(t_4-t_1-t_2)/d_{25}]^2 + [(t_2-t_1)/d_{12} - t_4/d_{45}]^2 + [(t_0-t_3)/d_{01}]^2 - 4s^2$

When $d_{12} = d_{45}$ (WN, WT, NT)

$c = [(t_2\text{-}t_1\text{-}t_4)/d_{45}]^2 + [(t_4\text{-}t_1)/d_{14} - t_2/d_{25}]^2 + [(t_0\text{-}t_3)/d_{01}]^2 – 4s^2$

2. Face stencils (or Scheme C in Vidale (1990))

New face stencils solve for point 5 on a new face given 4 points on an existing face. The geometries are shown in Figure A2. Given the geometries shown, the derivatives are:

	dt/dr	1/r dt/dθ	1/rsinθ dt/dϕ
T	$(t_5\text{-}t_2)/h$	$(t_0\text{-}t_4)/2r_1\Theta$	$(t_3\text{-}t_1)/2r_2\sin\theta_1\Phi$
B	$(t_2\text{-}t_5)/h$	$(t_0\text{-}t_4)/2r_1\Theta$	$(t_1\text{-}t_3)/2r_2\sin\theta_1\Phi$
N	$(t_4\text{-}t_0)/2h$	$(t_2\text{-}t_5)/r_1\Theta$	$(t_1\text{-}t_3)/2r_2\sin\theta_2\Phi$
S	$(t_4\text{-}t_0)/2h$	$(t_5\text{-}t_2)/r_1\Theta$	$(t_3\text{-}t_1)/2r_1\sin\theta_2\Phi$
W	$(t_3\text{-}t_1)/2h$	$(t_0\text{-}t_4)/2r_1\Theta$	$(t_2\text{-}t_5)/r_1\sin\theta_1\Phi$
E	$(t_1\text{-}t_3)/2h$	$(t_0\text{-}t_4)/2r_1\Theta$	$(t_5\text{-}t_2)/r_1\sin\theta_1\Phi$

Note that T = +-B; N = +-S and W = +-E in each case, so it is enough to know (T, N, W).

The general form is:

$s^2 = [(t_5\text{-}t_2)/d_{25}]^2 + [(t_4\text{-}t_0)/2d_{02}]^2 + [(t_3\text{-}t_1)/2d_{12}]^2$

And we solve for t_5

$[(t_5\text{-}t_2)/d_{25}]^2 = s^2 - [(t_4\text{-}t_0)/2d_{02}]^2 – [(t_3\text{-}t_1)/2d_{12}]^2$

$t4_5 = t_2 + d_{25}(s^2 - [(t_4\text{-}t_0)/2d_{02}]^2 – [(t_3\text{-}t_1)/2d_{12}]^2)^{1/2}$

$t_5 = t_2 + d_{25}(s^2 - 0.25*((t_4\text{-}t_0)^2/d_{02}^2 + (t_1\text{-}t_3)^2/d_{12}^2))^{1/2}$

	d_{02}	d_{12}	d_{25}
T	$r_2\Theta$	$r_2\sin\theta_1\Phi$	h
N	h	$r_1\sin\theta_2\Phi$	$r_1\Theta$
W	$r_1\Theta$	h	$r_1\sin\theta_1\Phi$

Top Side

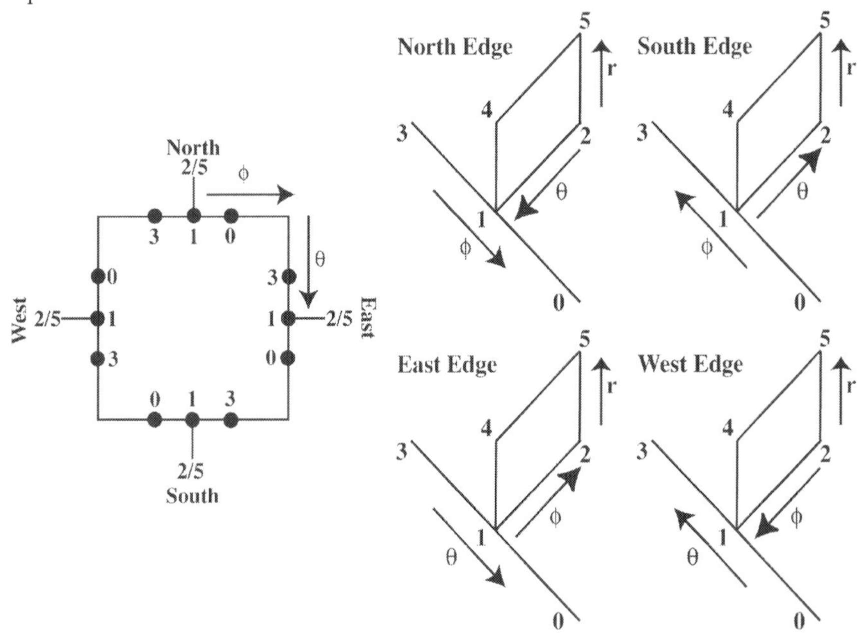

Bottom Side

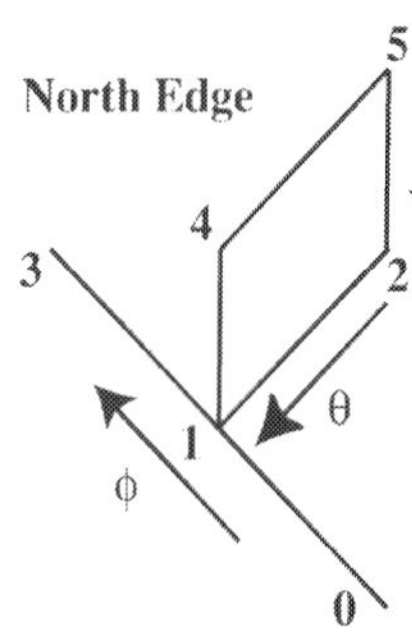

Top Side

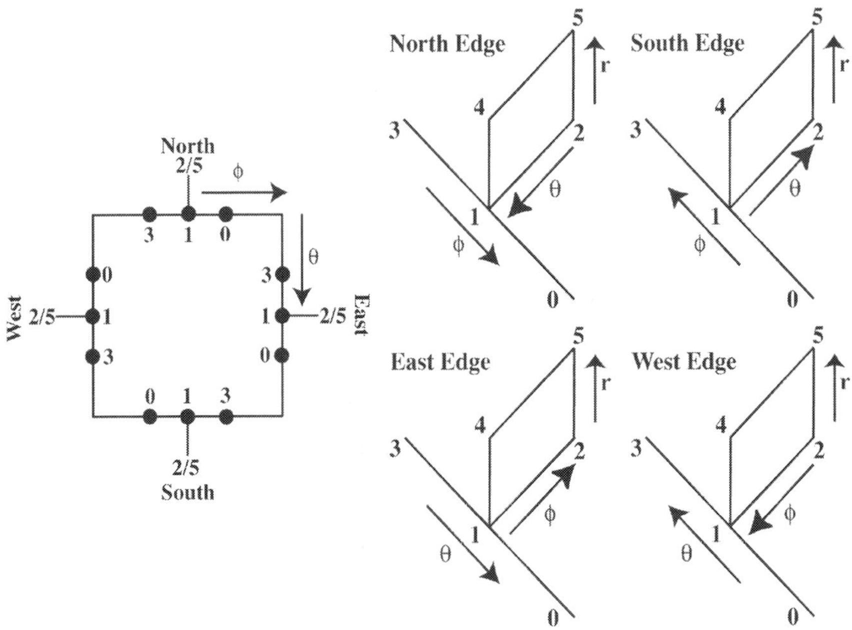
North
2/5
ϕ
θ
3 1 0
West
2/5
East
2/5
0 1 3
2/5
South
North Edge
South Edge
East Edge
West Edge
r

Bottom Side

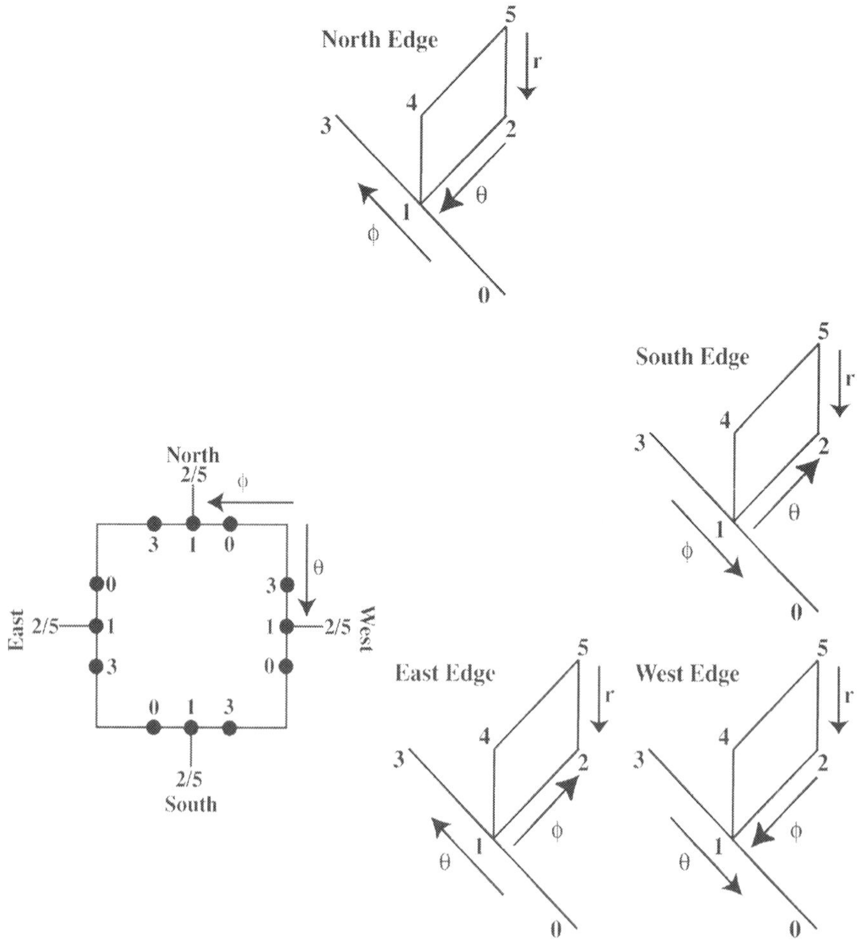
North Edge
South Edge
East Edge
West Edge
North
2/5
South
2/5
East
2/5
West
2/5
r
θ
ϕ
0
1
2
3
4
5

North Side

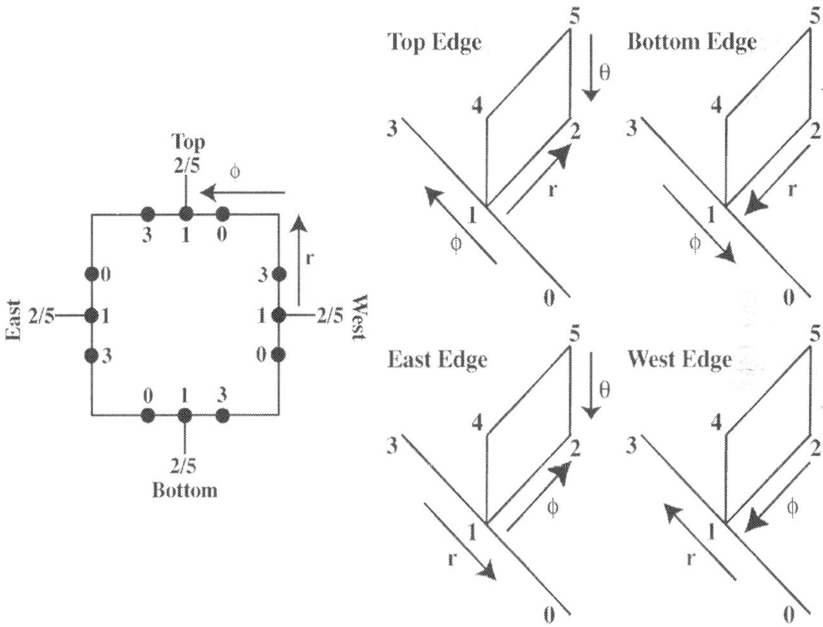
Top
2/5
ϕ
r
3 1 0
East
West
2/5
Bottom
Top Edge
Bottom Edge
East Edge
West Edge
θ
0
1
2
3
4
5

South Side

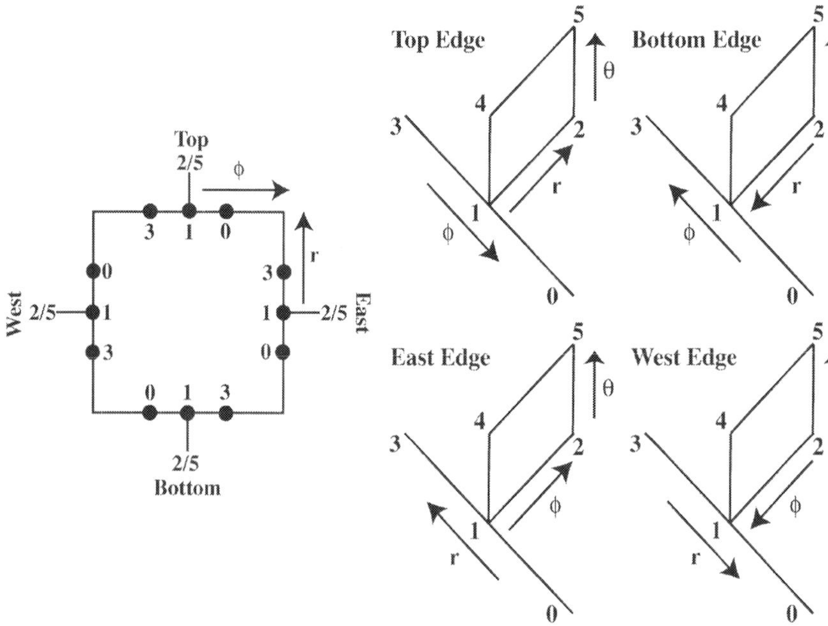
Top
2/5
ϕ
3 1 0
r
West
2/5
0
1
3
East
2/5
3
1
0
0 1 3
2/5
Bottom
Top Edge
Bottom Edge
East Edge
West Edge
5
4
3
2
1
0
θ
r
ϕ

West Side

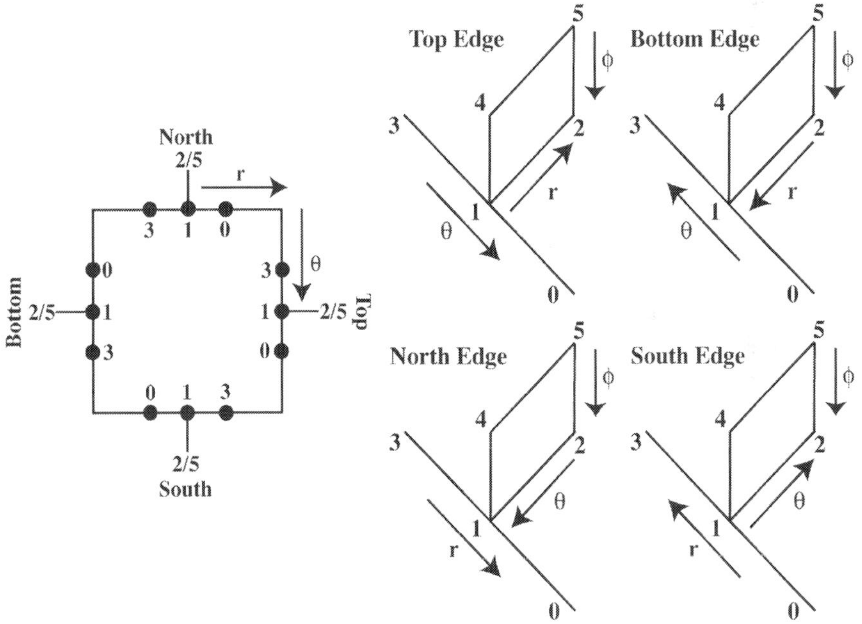
North
2/5
r
3 1 0
θ
Bottom
2/5
0
1
3
3
1
0
2/5
Top
0 1 3
2/5
South
Top Edge
Bottom Edge
North Edge
South Edge
5
4
3
2
1
0
ϕ
r
θ

East Side

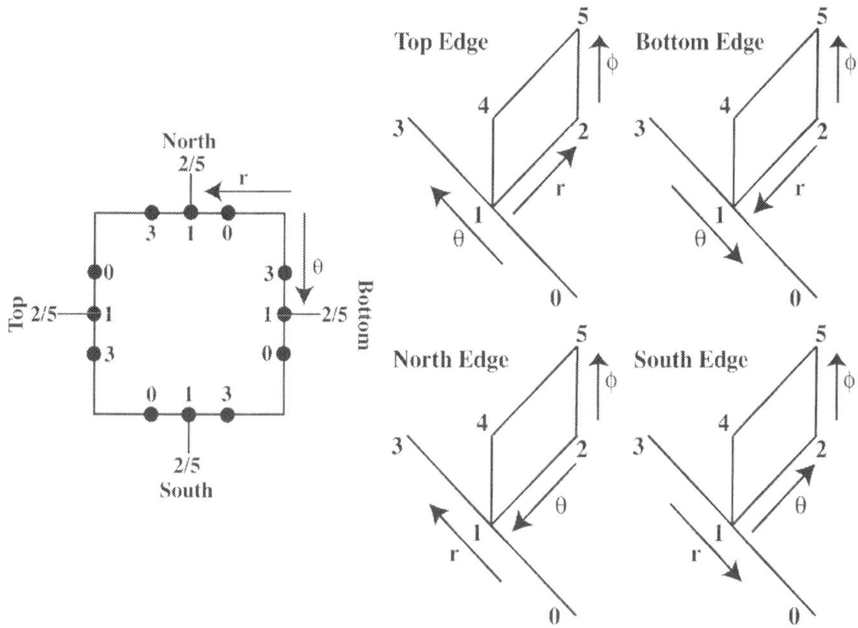
North
2/5
r
θ
Top
2/5
Bottom
2/5
South
Top Edge
Bottom Edge
North Edge
South Edge
ϕ

New Face Stencils

East Side

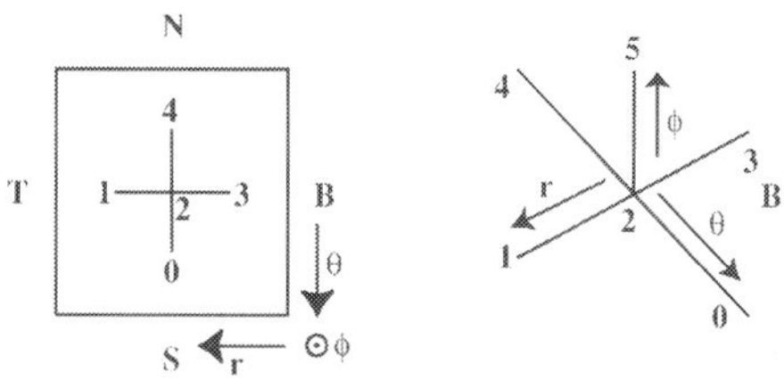

West Side

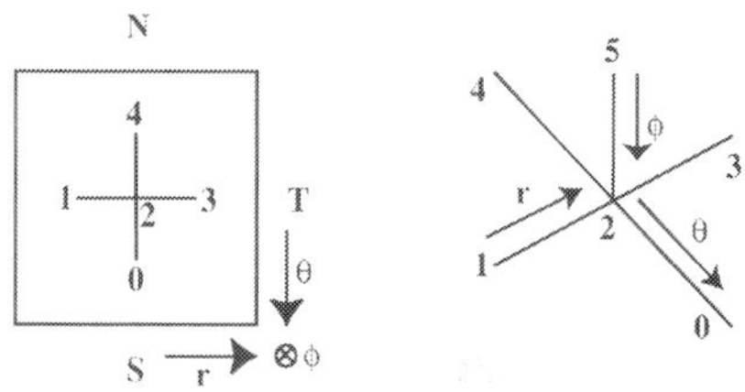

Top Side

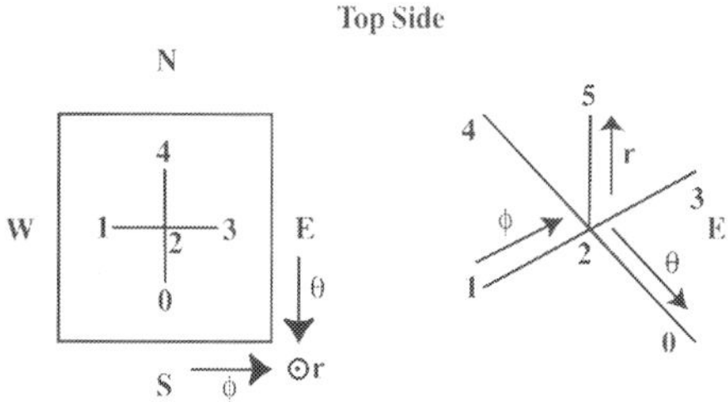

Bottom Side

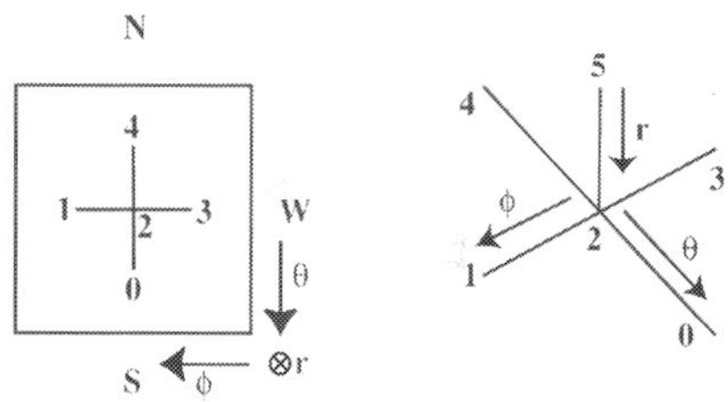

North Side

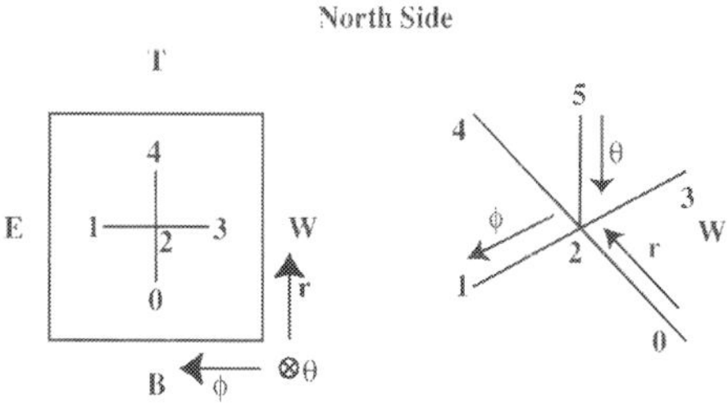

South Side

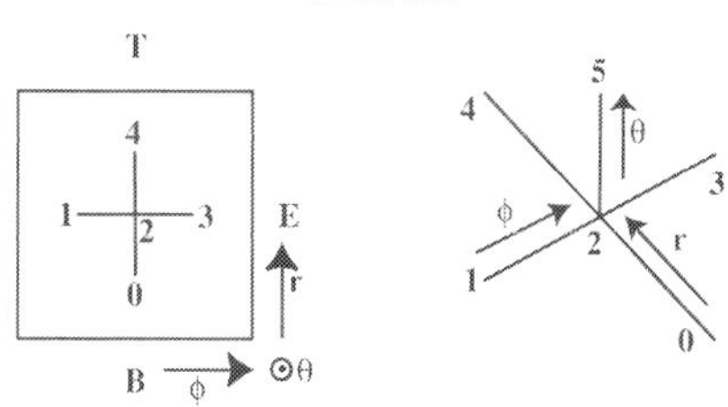

Two dimensional stencils

For each edge there are 8 stencils: 4 parallel to the face and 4 perpendicular to the face. We number the nodes (t_0, t_1, t_2, t_3), with t_0 being the unknown, t_1, t_2 being the adjacent points and t_3 being at the opposite corner.

```
t0-------t1
|        |
t2--------t3
```

The perpendicular stencils all have (0-2) as a common edge, and the parallel stencils all have point 0 in common and either (0-1) or (0-2) as a common edge.

It is easy to show that in each case the Eikonal equation can be written as:

$$4s^2 = [(t_0-t_2)/d_{02} + (t_1-t_3)/d_{13}]^2 + [(t_1-t_0)/d_{01} + (t_3-t_2)/d_{23}]^2$$

And so the solution consists of solving for t0 and identifying the proper distances (d01, etc) in each case. To solve for t0:

$$4s^2 = (t_0^2 + t_2^2 -2 t_2t_0)/ d_{02}^2 + [(t_1-t_3)/d_{13}]^2 + 2(t_0t_1 - t_0t_3 - t_1t_2 + t_2t_3)/d_{02}d_{13} +$$
$$(t_0^2 + t_1^2 -2 t_1t_0)/ d_{01}^2 + [(t_3-t_2)/d_{23}]^2 + 2(t_0t_2 - t_0t_3 - t_1t_2 + t_1t_3)/d_{01}d_{23}$$

Isolating t_0 :

$$4s^2 = (t_0^2 -2 t_2t_0)/ d_{02}^2 + 2(t_0t_1 - t_0t_3)/d_{02}d_{13} + (t_0^2 -2 t_1t_0)/ d_{01}^2 + 2(t_0t_2 - t_0t_3)/d_{01}d_{23} +$$
$$(t_2^2)/ d_{02}^2 + [(t_1-t_3)/d_{13}]^2 + 2(-t_1t_2 + t_2t_3)/d_{02}d_{13} +$$
$$(t_1^2)/ d_{01}^2 + [(t_3-t_2)/d_{23}]^2 + 2(- t_1t_2 + t_1t_3)/d_{01}d_{23}$$

$$4s^2 = (t_0^2)/ d_{02}^2 + (t_0^2)/ d_{01}^2 +$$
$$2t_0 [(-t_2)/ d_{02}^2 + (t_1 - t_3)/d_{02}d_{13} + (-t_1)/ d_{01}^2 + (t_2 - t_3)/d_{01}d_{23}]+$$
$$(t_2^2)/ d_{02}^2 + [(t_1-t_3)/d_{13}]^2 + 2 t_2(-t_1 + t_3)/d_{02}d_{13} +$$
$$(t_1^2)/ d_{01}^2 + [(t_3-t_2)/d_{23}]^2 + 2 t_1(- t_2 + t_3)/d_{01}d_{23}$$

$$4s^2 = t_0^2[1/ d_{02}^2 + 1/ d_{01}^2] - 2t_0 [t_2/ d_{02}^2 + (t_3 - t_1)/d_{02}d_{13} + (t_1)/ d_{01}^2 + (t_3 - t_2)/d_{01}d_{23}]+$$
$$(t_2^2)/ d_{02}^2 + [(t_1-t_3)/d_{13}]^2 + 2 t_2(t_3 - t_1)/d_{02}d_{13} +$$
$$(t_1^2)/ d_{01}^2 + [(t_3-t_2)/d_{23}]^2 + 2 t_1(t_3 - t_2)/d_{01}d_{23}$$

$$4s^2 = t_0^2[1/ d_{02}^2 + 1/ d_{01}^2] - 2t_0 [(t_2/d_{02} + (t_3 - t_1)/d_{13})/ d_{02} + (t_1/d_{01} + (t_3 - t_2)/d_{23})/ d_{01}] +$$
$$[t_2/d_{02} + (t_3-t_1)/d_{13}]^2 + [t_1/ d_{01} + (t_3-t_2)/d_{23}]^2$$
$$4s^2 = t_0^2[1/ d_{02}^2 + 1/ d_{01}^2] + 2t_0 [((t_1 - t_3)/d_{13} - t_2/d_{02})/ d_{02} + ((t_2 - t_3)/d_{23} - t_1/d_{01})/ d_{01}] +$$
$$[(t_1-t_3)/d_{13} - t_2/d_{02}]^2 + [(t_2-t_3)/d_{23} - t_1/ d_{01}]^2$$

Which is quadratic for t_0:

$$a = 1/ d_{02}^2 + 1/ d_{01}^2$$
$$b = 2*[((t_1 - t_3)/d_{13} - t_2/d_{02})/ d_{02} + ((t_2 - t_3)/d_{23} - t_1/d_{01})/ d_{01}]$$
$$c = [(t_1-t_3)/d_{13} - t_2/d_{02}]^2 + [(t_2-t_3)/d_{23} - t_1/ d_{01}]^2 - 4s^2$$

Perpendicular Stencils:

		d_{01}	d_{02}	d_{13}	d_{23}
T/B	NS	$r_1\Theta$	h	h	$r_2\Theta$
	EW	$r_1\sin\theta\Phi$	h	h	$r_2\sin\theta\Phi$
E/W	TB	h	$r\sin\theta_1\Phi$	$r\sin\theta_2\Phi$	h
	NS	$r_1\Theta$	$r\sin\theta_1\Phi$	$r\sin\theta_2\Phi$	$r_2\Theta$
N/S	TB	h	$r_1\Theta$	$r_2\Theta$	h
	EW	$r\sin\theta_1\Phi$	$r_1\Theta$	$r_2\Theta$	$r\sin\theta_2\Phi$

Parallel Stencils:

	d_{01}	d_{02}	d_{13}	d_{23}
T/B	$r\sin\theta\Phi$	$r_1\Theta$	$r_1\Theta$	$r\sin\theta_2\Phi$
E/W	h	$r_1\Theta$	$r_2\Theta$	h
N/S	$r\sin\theta_1\Phi$	h	h	$r_2\sin\theta\Phi$

Top and Bottom Faces. Stencils for the Top face shown below. Bottom differs only in sign of r and θ.

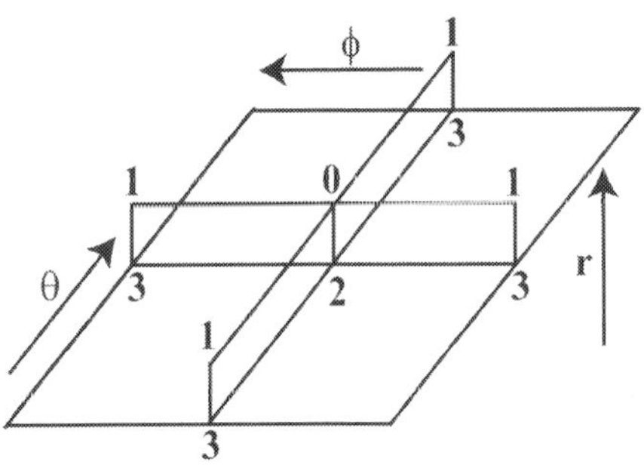

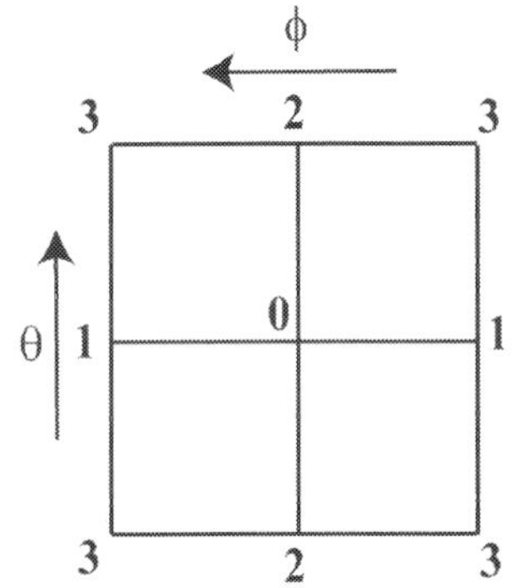

North and South Faces. Stencils for the North face shown below. South differs only in sign of ϕ and θ.

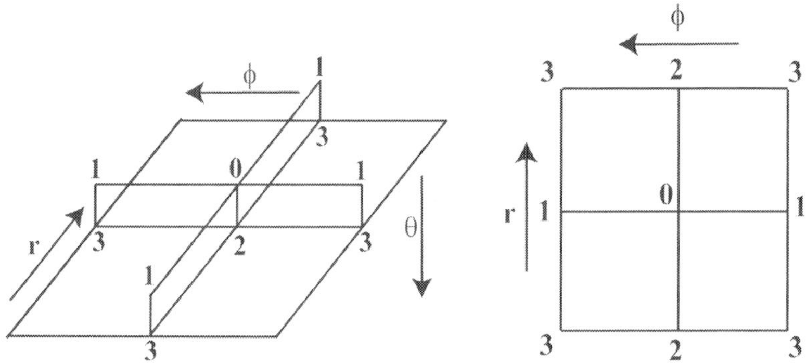

West and East Faces. Stencils for the West face shown below. East differs only in sign of ϕ and r.

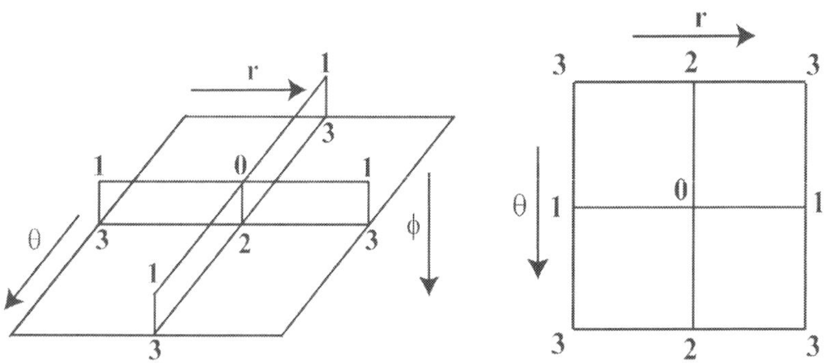

One Dimensional Stencils

Change in r: $t_0 = t_1 + hs$
Change in θ: $t_0 = t_1 + r\Theta s$
Change in ϕ: $t_0 = t_1 + r\sin\theta\Phi s$

Note that this needs to be along the edge of the volume element for energy that travels along the edge, in the same way as the Cartesian coordinate system.

There are three types, depending on which variables are along the boundary. In each case we have (t_0, t_1, t_2, t_3), with t_0 being the unknown, t_1, t_2 being the adjacent points and t_3 being at the opposite corner.

```
t0-------t1
|        |
t2--------t
```

9. References

Abers, G.A., and Roecker, S.W. (1991), Deep structure of an arc-continent collision: earthquake relocation and inversion for upper mantle P and S wave velocities beneath Papua New Guinea, J. geophys. Res., 96, 6379–6401.

Bai, D., et al. (2010), Crustal deformation of the eastern Tibetan plateau revealed by magnetotelluric imaging, Nat. Geosci., 3, 358–362, doi:10.1038/ngeo830.

Burchfiel, B.C., L.H. Royden, R.D. van der Hilst, B.H. Hager, Z. Chen, R.W. King, C. Li, J. Lu, H. Yao, E. Kirby (2008), A geological and geophysical context for the Wenchuan earthquake of 12 May 2008, Sichuan, People's Republic of China, GSA Today, 18, 4-11.

Christensen, N.I. (1996), Poisson's ration and crustal seismology, J. Geophys. Res. 101 (B2), 3139–3156.

Clark, M. K., and L. H. Royden (2000), Topographic ooze: Building the eastern margin of Tibet by lower crustal flow, Geology, 28, 703- 706.

Eberhart-Phillips, D. (1990), Three-dimensional P and S velocity structure in the Coalinga region, California, J. Geophys. Res., 95, 15,343–15,363, doi:10.1029/JB095iB10p15343.

Flanagan, M.P., S.C. Myers, and K.D. Koper (2007), Regional travel-time uncertainty and seismic location improvement using a three-dimensional a priori velocity model, Bull. Seismol. Soc. Am., 97, 804-825.

Fowler P.J. (1994), Finite-difference solutions of the 3D eikonal equation in spherical coordinates, 64th SEG meeting, Los Angeles, USA, Expanded Abstracts, 1394-1397.

Hole, J.A. (1992), Nonlinear high-resolution 3-dimensional seismic travel time tomography, J. Geophys. Res., 97, 6553-6562

Hu, S., He, L. & Wang, J., 2000. Heat flow in the continental area of China: a new data set, *Earth planet. Sci. Lett.*, 179, 407–419.

Huang, R., Z. Wang, S. Pei, and Y. Wang (2009), Crustal ductile flow and its contribution to tectonic stress in southwest China, Tectonophysics, 473, 476-489.

Huang, H., H. Yao, and R. D. van der Hilst (2010), Radial anisotropy in the crust of SE Tibet and SW China from ambient noise interferometry, Geophys. Res. Lett., 37, L21310, doi:10.1029/2010GL044981.

Huang, J., D. Zhao, and S. Zheng (2002), Lithospheric structure and its relationship to seismic and volcanic activity in southwest China, J. Geophys. Res., 107(B10), 2255, doi:10.1029/2000JB000137.

Hubbard, J. and J.H. Shaw (2009), Uplift of the Longmen Shan and Tibetan plateau, and the 2008 Wenchuan (M= 7.9) earthquake, Nature, 458, 194-197.

Jamieson, R.A., Beaumont, C., Nguyen, M.H., and Grujic, D. (2006). Provenance of the greater Himalayan Sequence and associated rocks: predictions of channel flow models. In: Law, R.D., Searle, M.P. & Godin, L. (eds) Channel Flow, Ductile Extrusion and Exhumation in Continental Collision Zones. Geological Society, London, Special Publications, 268, 165–182.

Li, H.Y., Su, W., Wang, C.-Y., Huang, Z.X., Lv., Z.Y. (2010), Ambient noise Love wave tomography in the eastern margin of the Tibetan plateau, Tectonophysics 491, 194–204.

Li, S. et al. (2003), Partial melt or aqueous fluid in the mid-crust of southern Tibet? Constraints from INDEPTH magnetotelluric data. Geophys. J. Int. 153, 289-304.

Liu, Q.Y., Y. Li, and J.H. Chen et al. (2009), Wenchuan Ms8.0 earthquake: preliminary study of the S-wave velocity structure of the crust and upper mantle, Chinese J. Geophys., 52(2), 309-319. (in Chinese)

Meade, B.J. (2007), Present-day kinematics at the India-Asia collision zone, Geology, 35, 81-84, doi:10.1130/G22942A.

Nakajima, J., T. Matsuzawa, A. Hasegawa, and D. Zhao (2001), Three-dimensional structure of Vp, Vs, and Vp/Vs beneath northeastern Japan: Implications for arc magmatism and fluids, J. Geophys. Res., 106, 21,843-21,857.

Nelson, K. D., et al. (1996), Partially molten middle crust beneath southern Tibet: Synthesis of project INDEPTH results, Science, 274, 1684–1688.

Nelson, G.D., and J.E. Vidale (1990), Earthquake locations by 3-D finite-difference travel-times, Bull. Seismol. Soc. Am., 80, 395-410

Paige, C. C., and M. A. Saunders (1982). LSQR: sparse linear equations and least squares problems, *ACM Trans. Math. Softw.* 8, no. 2, 195–209.

Pei, S., J. Su, H. Zhang, Y. Sun, M.N. Toksoz, Z. Wang, X. Gao, J.L. Zeng, and J. He (2010), Three-dimensional seismic velocity structure across the 2008 Wenchuan Ms 8.0 earthquake, Sichuan, China, Tectonophysics, 491, 211-217.

Podvin, P. and Lecomte, I. (1991), Finite difference computation of traveltimes in very contrasted velocity models: a massively parallel approach and its associated tools., *Geophys. J. Int.*, 105, 271-284.

Robert, A., J. Zhu, J. Vergne, R. Cattin, L.S Chan, G. Wittlinger, G. Herquel, J. de Sigoyer, M. Pubellier, and L.D. Zhu (2010), Crustal structures in the area of the 2008 Sichuan earthquake from seismologic and gravimetric data, Tectonophysics, 491, 205-210.

Rosenberg, C. L. and Handy, M. R (2005)m Experimental deformation of partially melted granite revisited: Implications for the continental crust. J. Metamorphic Geol. 555, 1-9.

Royden, L. H., B. C. Burchfiel, W. King, E. Wang, Z. Chen, F. Shen, and Y. Liu (1997), Surface deformation and lower crustal flow in Eastern Tibet, Science, 276, 788- 790.

Shen, Z.-K., J. Lu, M. Wang, and R. Burgmann (2005), Contemporary crustal deformation around the southeast borderland of the Tibetan Plateau, J. Geophys. Res., 110, B11049, doi: 10.1029.

Schneider, W. A. (1995), Robust and efficient upwind finite-difference traveltime calculations in three dimensions, Geophysics, 60, 1108-1117.

Sun, J., Jin, G.W., Bai, D.H. and Wang, L.F. (2003), Electrical structure of the crust and upper mantle and tectonics sense on the edge of the East Tibet, *Science in China* (Series D) (in Chinese), 33 (Suppl.) 173- 180.

Takei, Y. (2002), Effect of pore geometry on Vp/Vs: From equilibrium geometry to crack, J. Geophys. Res., 107 (B2), 2043.

Teng. J.W., D.H. Bai, H. Yang, et al. (2008), Deep processes and dynamic responses associated with the Wenchuan Ms8.0 earthquake of 2008, Chinese J. Geophys. (in Chinese), 51(5), 1385-1402.

Thurber, C. H. (1993), Local earthquake tomography: Velocities and Vp/Vs theory, in Seismic Tomography: Theory and Practise, edited by H. M. Iyer and K. Hirahara, pp. 563–583, Chapman and Hall, New York.

Thurber, C., and D. Eberhart-Phillips (1999), Local earthquake tomography with flexible gridding, Comput. Geosci., 25, 809–818, doi:10.1016/S0098-3004(99)00007-2.

Vidale, J. (1988), Finite-difference calculation of travel times, Bullet. Seismol. Soc. Am., 78, 2062-2076.

Vidale, J.E. (1990), Finite-difference calculation of traveltimes in three dimensions, Geophysics, 55, 521-526.

Wagner, L. S., S. Beck, and G. Zandt (2005), Upper mantle structure in the south central Chilean subduction zone (30 to 36 S), J. Geophys. Res., 110, B01308, doi:10.1029/2004JB003238.

Wang, C.-Y., W. Winston Chan, and W.D. Mooney (2003), Three-dimensional velocity structure of crust and upper mantle in southwestern China and its tectonic implications, J. Geophys. Res., 108(B9), 2442, doi:10.1029/2002JB001973.

Wang, C.-Y., W.-B. Han, J.-P. Wu, H. Lou, and W. W. C. Chan (2007), Crustal structure beneath the eastern margin of the Tibetan Plateau and its tectonic implications, J. Geophys. Res., 112, B07307.

Wang, C.-Y., H. Lou, Z. Lu, J. Wu, L. Chang, S. Dai, H. You, F. Tang, L. Zhu, and P. Silver (2008a), S-wave crustal and upper mantle's velocity structure in the eastern Tibetan Plateau - Deep environment of lower crustal flow, Sci. China Ser. D-Earth Sci., 51, 263-274.

Wang, W.M., L.F. Zhao, J. Li, and Z.X. Yao (2008b), Rupture process of the Ms 8.0 Wenchuan earthquake of Sichuan, China, Chinese J. Geophys. (in Chinese), 51(5), 1403-1410.

Wang, X.B., X.T. Zhu, N. Yu, et al. (2008), Magnetotelluric sounding new evidence to thrust belt structure in LongmenShan, submitted to Chinese J. Geophys. (in Chinese).

Wang, Z., Y. Fukao, and S. Pei (2009), Structural control of rupturing of the Mw 7.9 2008 Wenchuan earthquake, China, Earth Planet. Sci. Lett., 279, 131-138.

Wei, W., M. Unsworth, A. Jones, J. Booker, H. Tan, D. Nelson, L. Chen, S. Li, K. Solon, P. Bedrosian, S. Jin, M. Deng, J. Ledo, D. Kay, and B. Roberts (2001), Detection of widespread fluids in the Tibetan crust by magnetotelluric studies, Science, 292, 716-718.

Xu, L., S. Rondenay, R.D. van der Hilst (2007), Structure of the crust beneath the southeastern Tibetan Plateau from teleseismic receiver functions, Phys. Earth Planet. Int., 165, 176-193.

Xu, Z.J. and X. Song (2010), Joint inversion for crustal and Pn velocities and Moho depth in Eastern Margin of the Tibetan Plateau, Tectonophysics, 491, 185-193.

Yao, H., C. Beghein, and R.D. van der Hilst (2008), Surface wave array tomography in SE Tibet from ambient seismic noise and two-station analysis - II. Crustal and upper-mantle structure, Geophys. J. Int., 173, 205-219.

Yao, H., R. D. van der Hilst, and J.-P. Montagner (2010), Heterogeneity and anisotropy of the lithosphere of SE Tibet from surface wave array tomography, J. Geophys. Res., 115, B12307, doi:10.1029/2009JB007142.

Zhang, H., and C. H. Thurber (2003), Double-Difference Tomography: The Method and Its Application to the Hayward Fault, California, *Bull. Seism. Soc. Am*, 93, 1875-1889.

Zhang, H., S. Sarkar, M.N. Toksoz, H.S. Kuleli, and F. Al-Kindy (2009a), Passive seismic tomography using induced seismicity at a petroleum field in Oman, *Geophysics*, 74(6), WCB67, doi:10.1190/1.3253059.

Zhang, H., C.H. Thurber, and P. Bedrosian (2009b), Joint inversion for Vp, Vs, and Vp/Vs at SAFOD, Parkfield, California, *Geochemistry, Geophysics, Geosystems*, 10, Q11002, doi:10.1029/2009GC002709.

Zhang, P.Z., X.W. Xu, X.Z. Wen, et al. (2008), Slip rate and recurrence interval of the Longmen Shan active fault zone, and tectonic implications for the mechanism of the May 12 Wenchuan earthquake, 2008, Sichuan, China, Chinese J. Geophys. 51(4), 1066-1073. (in Chinese)

Zhang, Z., Y. Wang, Y. Chen, G. A. Houseman, X. Tian, E. Wang, and J. Teng (2009c), Crustal structure across Longmenshan fault belt from passive source seismic profiling, Geophys. Res. Lett., 36, L17310, doi:10.1029/2009GL039580.

Zhao, G., X. Chen, L. Wang, J. Wang, J. Tang, Z. Wan, J. Zhang, Y. Zhan, and Q. Xiao (2008), Evidence of crustal 'channel flow' in the eastern margin of Tibetan Plateau from MI measurements, Chinese Sciences Bulletin, 53, 1887-1893.

Zhao, Z. J. Fang, and S. Zhen (1997), Crustal velocity and relocated events in the Longmen Shan fault zone, Acta Seismol. Sin., 19, 615-612 (in Chinese)

Part 4

Petroleum Geology

9

Advances and Challenges of Reservoir Characterization: A Review of the Current State-of-the-Art

Ailin Jia, Dongbo He and Chengye Jia
Research Institute of Petroleum Exploration & Development, PetroChina
P. R. China

1. Introduction

Since the first technical paper which reported works of Jahns on two-dimensional description of reservoir heterogeneity using regression analysis on well testing data in 1966 (Jahns, 1966), reservoir characterization has attracted remarkable research efforts particularly in the past three decades (Doss et al., 2001; Wang, 2008). It is widely recognized now by oil industry that reservoir characteristics such as natural heterogeneity, spatial variability of permeability and porosity, porous media properties and spatial distribution of oil & water predominantly control the flow field, reservoir performance, development strategies and hence the economic returns of investments which are most concerned by oil companies.

Reservoir characterization is a combined technology associated with geostatistics, geophsics, petrophysics, geology and reservoir engineering and the main goals of reservoir characterization research are to aid field development and reservoir management teams in describing the reservoir in sufficient detail, developing 3D/4D data for reservoir development planning to obtain higher recoveries with fewer wells in better positions at minimum cost through optimization, increasing reserves, improving stimulation and completion practices and reducing to a minimum uncertainty in production forecasts (Haldorsen and Damsleth, 1993; Phillips, 1996; Johnston, 2004).

In recent years, technological innovations have lead to advances of reservoir characterization through methodology and instruments. Terrestrial scanning lidar and laser technology were applied to outcrop stratigraphic mapping with extremely accurate and efficient for digital outcrop modeling (Bellian et al., 2005; Buckley et al., 2009). Seismic imaging technologies either single-sensor seismic data processing or calibration of seismic amplitude and attributes showed advantages in porosity detection and reservoir modeling (Slatt and Mark, 2002; Refae et al., 2008). Surface microseismic monitoring was also used for hydraulic-fracture monitoring in reservoir characterization of development stage (Duncan and Eisner, 2010).

In this chapter, there are two main purposes of the author. One is to conduct a comprehensive review to the current state, challenges and developing trends of reservoir characterization technology. The second is trying to verify some phrases used in papers which are frequently used and affiliated to reservoir characterization and set up a practical procedure for people engaging or to be engaged in the field of reservoir characterization.

2. Terminologies

Varies of words or phases were currently used by researchers to decorate "reservoir characterization". In related papers or literatures, digital detailed, comprehensive, advanced, integrated and practical are most frequent words ahead of reservoir characterization (Phillips 1996, Castillo et al., 1998; Montgomery and Morea, 2001; Slatt and Mark, 2002; Johnston, 2004; Jackson 2005; Jia and Cheng, 2010). Although technical features may be implicit in them, misunderstanding or confusion would also be associated with these terms. In this section, discussions on the differences between these decorative words were proposed.

Phillips (1996) used the word "advanced" to decorate reservoir characterization. Three-dimensional deterministic and stochastic geologic models were developed integrated with pilot production and experiment. These works characterized heterogeneity of turbidite sands which do help to enhance sweep efficiency of steam injection for heavy oil recovery. Montgomery and Morea (2001) used the phase advanced reservoir characterization to summarize his work on three-dimensional earth modeling and flow simulation to evaluate CO_2 injection for enhanced oil recovery of Antelope shale, Buena Vista Hills field. Advanced reservoir characterization of the Antelope shale zone has involved a wide range of specific analyses aimed at delineating the detailed rock and production characteristics of this complex unit. Mineralogic and petrologic studies, normal and specialized log analysis, and core analyses were performed, in addition to fluid characterization, fracture analysis, cross-well seismology, 3-D modeling, and flow simulation. It can be informed obviously that works summarized by authors to be advanced reservoir characterization are more comprehensive than traditional in which geological models were integrated with laboratory experiments, reservoir simulation or pilot production.

Slatt and Mark (2002) purposed the challenges for independent operators with limited manpower and sources to select the right technique characterizing compartmentalized reservoir as reservoir performance is governed by complex features, which may be difficult to detect. Compartmentalized reservoirs may be resulted from primary stratigraphy or structure and are common to most types of sedimentary deposits. They can be discovered in fluvial deposits, eolian deposits, shoreface deposits, deltaic deposits and deep water (turbidite) deposits and practical methods to investigated flow units. Cores and logs are suggested to be the practical means for research on lithology, bed thickness and facies which affects porosity and permeability. Seismic analysis, borehole images and dipmeter reading used in combination can be effective for interwell petrophysical detection. Seismic imaging, core analysis and pressure tests can reveal complex fluvial and incised valley compartments. 3D modeling, sequence stratigraphy, outcrop analogs, interpretation of seismic data provide useful data on reservoir architecture, scale and connectivity of reservoir compartment. Reservoir compartments of sedimentary deposits were purposed and Practical reservoir characterization methods were summarized by selecting cost-effective technical means. Thus, practical reservoir characterization means a cost-effective manner to conduct reservoir characterization by selecting best available technologies to determine reservoir compartments.

Reservoir characterization with 4D and four component (4C) seismic which result in data with both lower and higher frequencies than traditional systems and has materially reduced cross-talk onto incorrect channels was depicted as comprehensive reservoir characterization by Fageraas et al. (2003). 4C ocean bottom seismic arrays has the advantage for its ability to acquire shear wave data directly which help address specific issues such as imaging through gas clouds and imaging low impedance reservoir. 4C/4D data can also provide a more accurate picture of dynamic reservoir processes because of its direct access to reservoir rock and fluid properties. Comparison between innovative 4C sensor technology and traditional systems reveals a significant improvement.

Castillo et al. (1998) purposed an integrated model for the optimization and iterative integration of geophysical, geological, petrophysical and reservoir engineering data. Reservoir interval architectures were determined through description, correlation and cartography of geological data. Genetic or stratigraphic units and reservoir compartments were identified by structural-stratigraphic interpretation of 3D seismic survey and well logs. Then, production was incorporated and integrated and reservoir model was generated. Dynamic simulation was conducted and potentially recoverable reserves were forecasted. In this work, an integrated geological model was developed with support of seismic 3D interpretation, advance sequence analysis, petrophysics and fluid data analysis and the author summarized it as integrated reservoir characterization.

Jia and Cheng (2010) summarized methods for detailed, digital and integrative reservoir characterization in the mid-late stages of oilfield development and specified the definition of detailed digital reservoir characterization. Compared with traditional approaches, detailed digital reservoir characterization is more closely related to "digital reservoir". It is characterized of visualization and integration with geological analysis, seismic interpretation, well logs, production and artificial intelligence. The author pointed out that detailed digital reservoir characterization aims to enhance oil recovery and remaining oil development at the middle or mature stage of field development and its core mission is to realize high precision reservoir prediction and quantitative assessment to reservoir architecture. It is characterized of quantitative, detailed, visible and integrated.

As words decorative to the phase "reservoir characterization", advanced, practical, comprehensive, integrated, digital and detailed have their unique technique contents or features. Advance reservoir characterization has more emphasis on integration of geological model with flow simulation, or pilot production and production forecast. Practical reservoir characterization aims at cost-effectiveness for independent operators to choose appropriate techniques conducting characterize reservoir compartments of different kinds of deposits. Comprehensive reservoir characterization has the features of 4C/4D technical innovation for improvement of reservoir imaging. Integrated reservoir characterization underlines the calibration and integration of different kinds of data, such as geophysical, geological, petrophysical and reservoir engineering data. Digital and detailed reservoir characterization has more particularity in visualization.

3. Advances of reservoir characterization technology

Reservoir characterization has attracted remarkable research efforts particularly over the past 20 years. Material improvements and technical innovations have lead to the advances of reservoir characterization technology. These improvements or innovations which improve reservoir characterization technology can be summarized into two aspects. One is that advanced instruments and technologies applied in data gathering, processing and monitoring have improved quality and reliability of test data. The second is that the development of related technology such as computer science and information science has realized comprehensive integration of reservoir characterization with outcrop analogues, seismic, geology and well logs and made the subsurface reservoir model, output of reservoir characterization study into digital 3D visible model. In this section, some technological advances were selected and introduced by the author according his knowledge degree.

3.1 Reservoir characterization and monitoring with 4D seismic

Seismic time-lapse reservoir monitoring (4D) was used to access the petrophysical properties and performance of reservoir at stage of field development in recent years (Fageraas et al., 2003; Kovacic and Poggiagliomi, 2003). Reservoir characterization and monitoring with 4D seismic data involves both comparison and analysis of repeated seismic surveys shot over the same location during the production life of a field. Compared with borehole based measurements (logs, pressure, temperature, etc), property distributions obtained from seismic attributes have more advantages in accuracy and spatial resolution.

Basic methodologies, dealing with some of the most critical issues and phases encountered in a seismic time-lapse project, were developed by Kovacic and Poggiagliomi: (1) feasibility studies to assess the suitability of time-lapse seismic technology to monitor the performance of a specific reservoir. This includes fully integrating laboratory measurements performed on core samples, well data analysis (editing and normalization of well logs), rock mechanical modeling, rock physics data and petroacoustic processing and Its purpose is to assess whether acoustic changes, related to time-lapse fluid movements within the production intervals, are of sufficient magnitude to be detected seismically.; (2) accurate homogenization of legacy 3D seismic surveys by means of wavelet equalization. In this phase, two 3D surveys were processed and wavelet-equalized by extracting from all the traces in trace-sets, located at the same well position, with the same well-log derived reflectivity sequence.; (3) calibration of absolute acoustic impedance volumes to reservoir pressure and petrophysical properties (porosity, saturation, clay content, etc.) to derive the spatial pattern of fluid movements.

Through reservoir characterization with 4D seismic, partially drained areas can be recognized by analysis to pattern of acoustic impedance and quantified by petroacoustic calibration. Thus, remaining reserves can be evaluated which is one of the main purposes of reservoir characterization study.

3.2 Outcrop study with application of digital data capture technology

Outcrop studies have long been employed as a mechanism of studying analogues and understanding petroleum fields (Collinson, 1970; Glennie, 1970; Breed and Grow, 1979). Depositional architectures are well preserved in outcrops and outcrop analogues offer the opportunity to enhance the understanding of subsurface reservoir architecture, geometry, and facies distributions (Pringle et al., 2006). Compared to studies of modern sedimentary events or laboratory-based experiments and process-based modeling, outcrops are more geologically comparable to the subsurface reservoir architecture and capture large enough scale of heterogeneity.

Outcrop studies have developed from qualitative to quantitative. Traditional outcrop studies were focused on collecting outcrop data, such as sand width, thickness, to populate inter-well areas by stochastic or object-based methods (Dreyer et al., 1993; Bryant and Flint, 1993; Chapin et al., 1994; Clark and Pickering, 1996; Reynolds, 1999; Floris and Peersmann, 2002). However, traditional outcrop studies can hardly provide useful data especially when it needs to be integrated into reservoir engineering database or be visualized in 3D. Accurate

Digital data collection method	Typical accuracy	Typical application	Advantages	Disadvantages	Typical cost
Aerial digital photogrammetry	~5-25m	Mapping large scale stratigraphy & generate digital model framework	Fast, usually third party acquisition (minutes); large areas covered & fast remote mapping (days)	Slow time processing (days); relatively low resolution & poor on near vertical outcrop faces	High if survey has to be commission ed. Cheap if existing photos are used
Ground-based digital photogrammetry	~0.1-0.5m	Detailed study of complex outcrop faces	Fast acquisition (minutes); less detailed fieldwork needed	Medium time processing (days) & interpretation	Relatively cheap
Calibrated photo logs	~0.2m	Rapid collection of facies thickness and relative surface from cliff sections	Fast acquisition (minutes), fast processing (hours) & rapid model creation	Can suffer from photograph distortion, no high resolution logging	Very cheap
Hand-held GPS	~1-5m	Sample point location & regional mapping	Instant locational fix	Significant 'Z' positional error (up to 30m)	Very cheap
RTK dGPS	Better than 10mm	Attribute collection, surveying outcrops & accurate base stations	Instant point collection allows 'walking out' of key surfaces, medium time processing (typically a day)	Not possible on near vertical cliff-faces	Expensive
Reflectorless Total Station	~3mm at 200m range	Attribute collection, surveying outcrops, good for vertical faces	Instant point collection, data capture on near-vertical cliff faces	Slow to acquire, dGPS data needed to convert to UTM co-ordinates	Moderately expensive
Ground-based LIDAR (laser scanner)	~5mm at 200m range	Very rapid collection of outcrop surface topography	Relatively rapid acquisition (minutes)	Significant post processing (days)	Expensive
Bore-hole data	~1mm (from core)	Drilled behind outcrop to extend horizons into 3D	Very high resolution data, comparable to outcrop information & reservoir logs	Very slow acquisition (weeks), processing and interpretation (weeks)	Very expensive
Near-surface geophysics (GPR in this case)	~0.1-0.5m	Acquired behind outcrop to extend correlated horizons into 3D	Allow 3D information behind outcrop to be acquired	Slow acquisition (days), only works in specific site conditions	Moderately expensive

Table 1. Summary of outcrop analogue data collection methods (From McCaffery et al., 2005; Pringle et al., 2006)

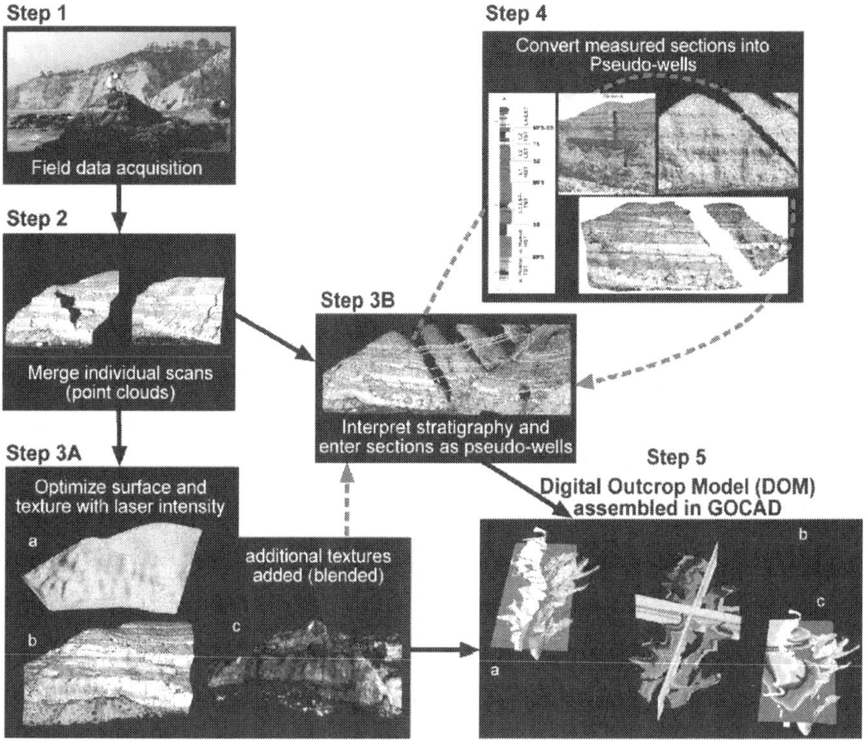

Fig. 1. Workflow diagram for digital outcrop study. Black arrows indicate flow direction and red arrows indicate feedback. (From Bellian et al., 2005).

and quantitative outcrop analogue supported by digital data capture technique has been developed in recent years and it realized 3D reconstruction to aid or modify subsurface reservoir model.

Varies of digital data capture techniques are listed in table 1 and application of advance techniques allow rapid acquisition of more accurate and denser digital datasets from outcrop. In table 1, accuracy, application condition, advantages and disadvantages for specific method have been summarized by McCaffrey et al. (2005) Currently, ground-based LIDAR (Light Detection and Ranging) or laser scanning is the preferred technology (Pringle et al, 2006).

Lidar scanner uses laser light to measure distance with extreme precision, whereas radar scanner uses radio waves. They use "time of travel" to measure the distance. Resolution of radar scanner is about 8mm at range of 350m and it is almost 5mm at range of 200m for lidar scanner. Data extracted from digital outcrop model can be incorporated into reservoir model. Details for scanning lidar and radar technology with its application in digital outcrop study can be referred in paper purposed by Bellian et al. (2005) and Buckley et al. (2009).

3.3 Reservoir characterization using downhole/surface microseismic monitoring

Microseismic imagining has been applied for downhole monitoring especially to image fracture network deformation of hydraulic fracture operations (Bailey, 1973; Duncan and Eisner, 2010; Maxwell et al., 2010). To plot the estimates of the event hypocenter locations on an event-by-event basis over time is currently the common practice for reporting the result of microseismic monitoring. As shown in Figure 2, the hypocenter locations for a multiwell frac for five horizontal wells in Marcellus Shale play are estimated.

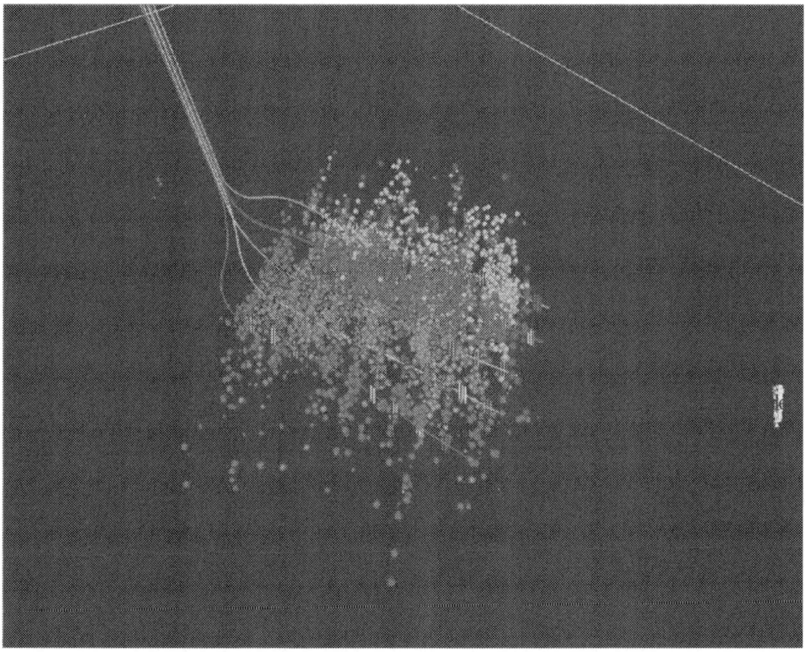

Fig. 2. Perspective view of the microseismic monitoring results from treating five wells completed in the Marcellus Shale in Pennsylvania. The dots represent the estimated event hypocenters. The colors of the dots match the color of the treated well to which they correspond. (From Schisselé and Meunier, 2009; Duncan and Eisner, 2010).

Maxwell et al. (2010) listed three general classes of techniques for locating microseismic events: (1) hodogram techniques based upon the particle motion of direct arrivals, which is the simplest method and using only one three-component (3C) sensor (Albright and Hanold, 1976), (2) triangulation schemes based upon arrival times of direct waves by combinations of P- and/or S-waves at multiple stations (Gibowicz and Kijko, 1994), and (3) semblance methods based upon stacking waves without arrival-time picking. All three classes of location techniques can be used in conjunction with surface or downhole sensors (Duncan and Eisner, 2010).

However, many researchers have developed other approaches on passive seismic emission tomography such as long-time-interval stacking similar to semblance (Kuznetsov et al., 2006; Kochnev et al., 2007) and picking the maximum amplitude of the P-wave migration as

the imaging condition (Chambers et al., 2008, 2009a, 2009b; Robein et al., 2009). A method to improve resolution for amplitude picking through recognizing the vertical distribution of false hypocenter estimates for has been purposed by Duncan et al. (2008).
Surface monitoring technique of hydraulic fracture stimulation also has been developed (Abbott et al., 2007; Kochnev et al., 2007; Barker, 2009; Hall and Kilpatrick, 2009; Keller et al., 2009; Robein et al., 2009). Usually linear groups of vertical phones are laid out along the spokes of a wheel centered on the wellhead of the treatment well. Details of data gathering, processing and migration about the technique have been investigated by Duncan and Eisner (2010).

4. Practical procedures for reservoir characterization technology

Reservoir characterization is a comprehensive technology. The main goal of reservoir characterization is for high precision reservoir predictions and quantitative depictions of reservoir architecture and properties to aid field development and reservoir management. It involves geology, geophysics, petrophysics, and reservoir engineering. Depositional background and sedimentary environment of research area should be investigated through geological study firstly. Then the structural model and reservoir architecture is provided through outcrop analogues and 3D seismic survey. Reservoir petrophysical properties can be determined by well logs or laboratory analysis. Finally, 3D reservoir model can be obtained with integration of reservoir framework and properties. Practical reservoir characterization for compartmentalized reservoirs still presents a challenge to technical staffs. In this section, practical workflow to carry out reservoir characterization research has been conducted and it is supposed to be helpful to technical staffs or researchers. This procedure is comprised of eight steps.

4.1 Depositional background and sedimentary environment analysis

As geology is a first order control on reservoir architecture and petrophysical properties, thorough study on depositional background and sedimentary environment is the priority. Depositional mechanism and genesis study are the tools to investigate reservoir features on the macroscopic scale or for regional deposition researches. Regional structure characteristics and flooding surface fluctuation reveal much about the subsurface reservoir architecture. e.g. for large constructive fluvial-deltaic depositional system with a gentle slope and relatively low flood surface at the geological age, study reveals that the fluvial system still remains high-energy and extends forward into the lake and the mouth bar is not developed. As result of the depositional environment, sand is thin in thickness and laminated sand and shale occurs. Depositional background and flooding surface is the control factors to reservoir scale and sand distribution patterns, which provide a basis for sand distribution prediction. Study of the regional depositional background and sedimentary environment can provide basic and qualitative knowledge to reservoir architecture and is the foremost step for reservoir characterization.

4.2 Isochronous stratigraphic framework and structure modeling

Stratigraphic division, correlation and structural interpretation are the basis for study of reservoir architecture. Appropriate stratigraphic framework can accurately characterize reservoir architecture and improve the precision and reliability of reservoir prediction. At present, the most effective method is to establish an isochronous stratigraphic framework and conduct fine structural interpretation by means of inter-well seismic correlation, which are guided by high resolution sequence stratigraphy, and combining with core, well logs, seismic imaging and dynamic production data.

High resolution sequence stratigraphy theory and its application and structure interpretation are the key techniques. Application of high resolution sequence stratigraphy theory is effective in reservoir prediction and evaluation during the mature stage of field development. Using data from core, well logs and seismic data, high resolution sequence stratigraphy framework can be established through base-level cycle identification, isochronous correlation and internal structure analysis.

Structure interpretation includes micro-structure interpretation and low-order fault interpretation. Using high precision survey and processed seismic data, techniques integrated with 3D visualization, coherence analysis and seismic horizontal slice can effectively improve the precision and identify micro-amplitude structure and low-order faults.

4.3 Establishment of reservoir sedimentary models

Genetic units, sand spatial distribution and superimposed pattern can be identified through microfacies subdivision. Microfacies analysis is the key technique in this step and study on sedimentary facies is essential throughout stages of oilfield exploration and development. Especially in the recent years, advances in outcrop analogue, dense well correlation, well logs and seismic survey resulted in innovation in sedimentary facies study.

Microfacies subdivision gives sands the meaning of genesis. Therefore, contact relationship of different facies, sand superimposition patterns and spatial distribution pattern for different sedimentary system can be established through outcrop analogue and study of modern sedimentary events. Then reliable reservoir sedimentary model can be established which is used to provide guidance to reservoir prediction, precisely characterize sand spatial distribution pattern and investigate more subtle reservoir heterogeneity.

Taking a block in Daqing Oilfield, China as an example, detailed microfacies study shows that the major oil zone develops in the delta plain environment which can be further subdivided into main distributary channel, branch distributary channel, abandoned channel, overbank sediment (natural levees and crevasse splays) and inter-distributary bay (Figure 3). Furthermore, the distributary channel sands are the major reservoir and the main target for detailed microfacies study. The distributary channel sand is characterized of sheet-like or isolated patterns. In order to thoroughly depict the sand distribution pattern, the abandoned channel can identified firstly according to some special features, such as curve shape of well logs, sand thickness variation and microfacies assemblages. Thus sheet-like sands developing in different channels can be identified and classified clearly. Point bar can be further identified which are prepared for correlation of the lateral accretion sand and thorough investigation of sand architecture.

4.4 Establishment of prototype model and geological database

The purpose to establish a prototype model and geological database is aimed to get parameters, such as sand geometry and scale (e.g. length, width, thickness and proportion) of genetic units and then provide quantitative references for inter-well sand prediction. Generally, seismic data is important for inter-well prediction, but it has limitation for low

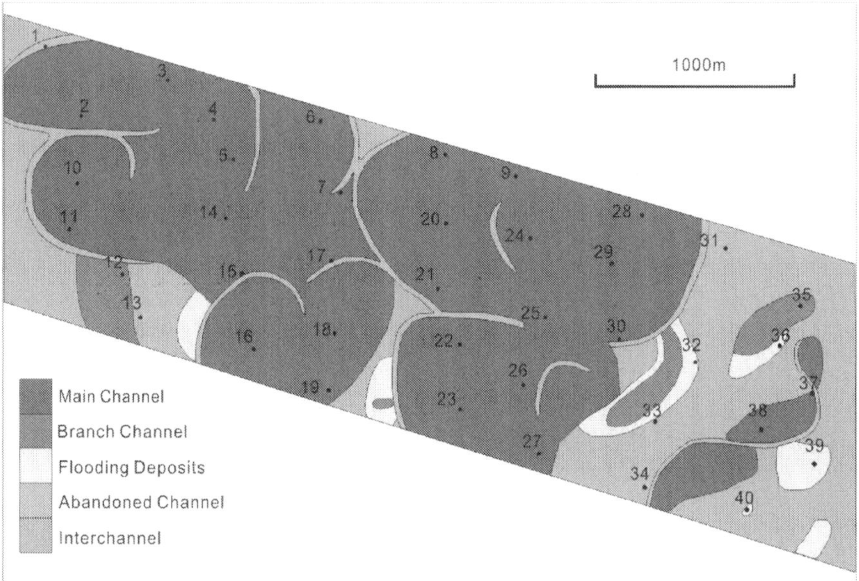

Fig. 3. Microfacies distribution of P I31 sub-layer in a block of Daqing oilfield, China

resolution (usually more than 5m in thickness). Predictions for thin sand and subdivision for thick sand are still difficult for seismic survey. The key technique solving this problem is to establish a reservoir prototype model and geological database. The so-called prototype model refers to the detailed model for outcrop analogue(s) which is geologically comparable to the system being studied, dense well pattern block in mature fields or the modern sedimentary event analogues.

Establishment of reservoir prototype model and geological database is an important part for reservoir characterization. Quantitative characterization for sand can provide quantitative parameters for inter-well reservoir prediction and establishment of 3D reservoir model.

The prototype model and geological database refers to parameters which quantitatively characterize the spatial distribution, boundary patterns and petrophysical properties of genetic units, and some qualitative sedimentary patterns. The database mainly includes lithologic/lithofacies knowledge, depositional environment and microfacies knowledge, sand geometry and diagenesis knowledge (Table 2) for a specific outcrop analogue. The primary prototype model should be study of the Gypsy profile sponsored by BP, which greatly enriches the geologic database of fluvial deposits (Doyle and Sweet, 1995). Similar study was also carried out in China on outcrop analogues in Datong and Luanping respectively. Qualitative description, quantitative measurement, sample analysis and well logs were conducted. Interior architecture of Luanping fan delta deposits and Datong braided channel deposits are thoroughly investigated. Therefore, prototype models and geological database for fan delta and braided channel deposits are established. Furthermore, in combination with stochastic modeling technique, methods for sand distribution prediction of this two kinds of deposits are developed (Jia and He, 2003).

Methods for establishment of prototype model and geological databases include detailed outcrop analogue, dense well correlation, modern sedimentary event analogue and laboratory based experiments. Combined application of two or more methods is the best way to establish the model. But as the subsurface reservoirs are usually lack of outcrop analogous and studies of modern sedimentary events has limitation for characterization for changes in accommodation through time, commonly available data are core and well logs in oil and gas industry. Although prototype model and geological database base on study of dense well pattern development block using core and well logs data is of less precision compared with that based on outcrop analogues or modern sedimentary event analogues, it is much more practical especially for study in oil and gas industry.

Genesis unit	Shape	Size		Barrier bed	Depositional structure	Grain size	Sorting
		Width/m	Thickness/m				
Plain mudflow		1-3	10.0-30.0	Horizontal argillaceous siltstone	Massive bedding Large trough cross bedding		
Braided channel		300-500	1.5-8.0	Horizontal mudstone Horizontal siltstone	Massive bedding Parallel bedding Tabular bedding		
Near-bank channel		200-1 000	3.0-16.0	Horizontal mudstone Wavy argillaceous siltstone	Parallel bedding Cross bedding (trough, tabular)		
Far-bank channel		100-500	0.5-5.0	Horizontal mudstone	Parallel bedding Cross bedding		
Natural dam		50-200	0.5-1.5	Wavy argillaceous siltstone	Parallel bedding Cross bedding		
Mouth bar		30-100	2.0-4.0	Wavy argillaceous siltstone	Parallel bedding Wavy bedding		
Distal bar		20-7	1.0-3.0	Wavy argillaceous siltstone	Small cross bedding		
Sand sheet		300-2 000	0.5-2.0	Horizontal shale	Massive bedding Parallel bedding Wavy bedding		
Collapse sediments		50-200	2.0-6.0	Horizontal shale Inclined mudstone	Massive bedding		
Flooding sediments		100-300	1.0-1.5	Horizontal mudstone	Wavy bedding		

Table 2. Expert knowledge and geologic database for Luanping fan delta deposits. Genetic units are qualitatively and quantitatively described.

4.5 Reservoir prediction with application of high resolution seismic

Seismic is always an important method for reservoir prediction. For reservoir characterization, seismic data provides not only fine structure interpretation but also inter-well information. Based on the previous four steps which focus on sedimentary studies and in combination of seismic data, inter-well reservoir distribution prediction should be more reliable.

Currently, seismic resolution is still low and improvement of seismic resolution is the precondition for application of seismic data into reservoir characterization at the development stage. There are abundant data such as core, well logs and production data at the development stage which are advantages for development seismic. Therefore, seismic data processing and interpretation should incorporate existing data and geological knowledge as much as possible to improve its precision.

Vertical Seismic Profiling (VSP) technique is an effective method used to distinguish small structure near wellbore, identify lithology and obtain fracture parameters. Combined with the seismic data interpretation, this technique can provide near wellbore petrophysical parameters for reservoir characterization. Time lapse seismic (4D) technique is specialized in dynamic monitoring through seismic responses changing with time. This technique plays an important role in remaining reserves identification in mature field (Mezghani et al., 2004). Inter-well seismic technique focus on reservoir lateral heterogeneity and provide inter-well peterophysical parameters for reservoir modeling (Justice et al., 2000). Applications of Inter-well seismic technique in the Daqing and Shengli oilfields shows that layers with 2~3m in thickness can be identified, sometime even to 1m, and small faults have a clear response (Figure 4).

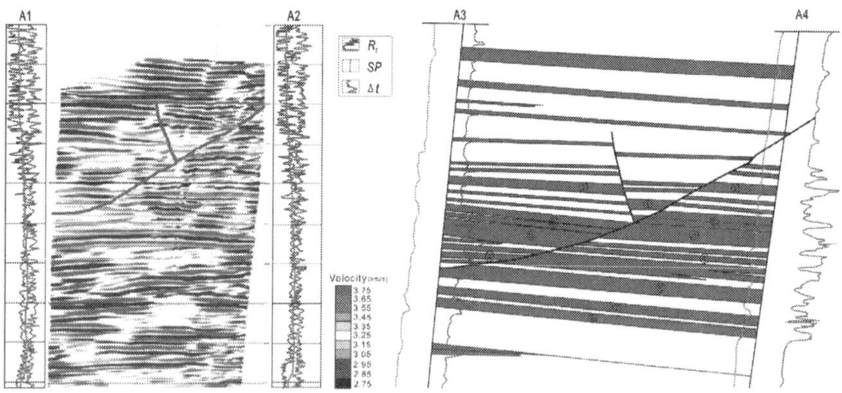

Fig. 4. Application of inter-well seismic survey for reservoir prediction in some oilfield, China. The picture on the left shows the result of ineterwell seismic imaging and the right one is about interwell seismic interpretation. It illustrates that interwell reservoir heterogeneity can be investigated and small scale faults can be identified clearly.

4.6 Establishment of comprehensive reservoir architecture model

Main tasks of reservoir characterization at different stage of field development vary a lot. At early stage of field development or at exploration stage, the priority for reservoir characterization should be investigation for reservoir architecture and characterization of reservoir heterogeneity. Whereas, at mature stage of field development, the key target for reservoir characterization should be focused on the reservoir compartments which are separated or laminated. Especially for water injection, separated or laminated reservoir compartments are usually not swept areas and hold most of the remaining reserves. Establishment of reservoir architecture model can further depict the interior features of genetic units and investigate separated or laminated reservoir compartment. Figure 5 shows a cross-section or correlation for a point bar reservoir architecture in a block of Daqing Oilfield, China.

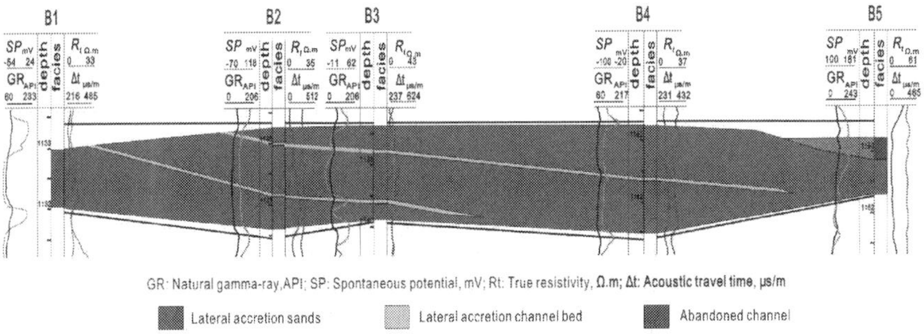

Fig. 5. Cross-section profile for reservoir architecture of single channel sand in a block of Daqing Oilfield, China. Sand lateral accretion can be identified and subdivision of the single point bar into multiple lateral accretionary sands was initialized. It further reveals that genetic units inside reservoir architecture are separated or laminated.

In this step, reservoir architectural elements should be analyzed thoroughly. Miall et al. (1985) studied the subdivision of bounding surfaces inside reservoir architecture and proposed a full set of methods and theories about hierarchical order to the bounding surfaces. Hierarchical boundary analysis is from point view of system theory and studies the hierarchy and structure of the deposition system, highlighting isochronous and intermittent characteristics of the sedimentary process. Thus the hierarchical boundaries are compatible with the stratigraphy sequences and can be subdivided infinitely but always fitting the depositional genesis analysis. On this basis, reservoir architecture elements are widely studied for thoroughly investigation to the reservoir interior in China and aboard.

Research on fluvial deposits is highly developed. In recent years, researchers from China have carried out a lot of research in this area. Jia and Cai (1992) and Zhang (1992) introduced their ideas and methods of reservoir architecture from different aspects. Although the terminology is different, their works have much similarity in conceptual system with reservoir architecture. Hierarchies of reservoir architecture were studied. Subdivision and identification methods for architecture elements were developed. Following researchers such as Yin et al. (2001; 2002), He et al. (2005a; 2005b) and Liao (2006) studied the delta depositional sands from point view of reservoir architecture. These works show that researches on reservoir architecture provides a sedimentary basis and a new approach to subdivision of architecture elements.

4.7 Evaluation of reservoir petrophysical properties and fluid distribution

The previous six steps aim to establish reservoir framework model. However petrophysical properties and fluid distribution models should be integrated with reservoir framework model to obtain a comprehensive reservoir model. Methods of reservoir petrophysical properties characterization include quantitative description, influencing factor analysis, spatial distribution analysis, inter-well prediction and variation pattern analysis. Research in fluid distribution rules is defined as qualitative or quantitative descriptions for fluid types, properties and saturations.

Currently, reservoir petrophysical properties are derived from laboratory-based core analysis and well logs. Reservoir fluid can be identified through seismic survey or sampling.

Their spatial distribution predictions are reached through seismic interpretation or extrapolation of well data. It can be found that above works are based on static data without consideration of reservoir dynamic characteristics and suitable for oilfield at exploration stage or early stage of development. But for oilfield at the middle or mature stage of development, studies on reservoir petrophysical properties and fluid distribution are more complicated, especially for oilfield through long-term water flooding. Reservoir petrophysical properties change a lot with the advance of field development. Oil zone permeability, rock cementation, production rate, oil viscosity are control factors to petrophysical property changes. High permeable channel (large pores) gradually develops in highly flooded zone, causing ineffective water circulation and seriously affecting the displacement efficiency. Remaining reserves are still abundant in non-swept areas. Whereas water blockage will take place in low permeable zone which causes additional permeability reduction. Thus, reservoir characterization at the middle or mature stage of field development should take the changes of peterophysical properties into consideration, especially distribution of high permeable channel.

4.8 Establishment of comprehensive and integrated reservoir geological model

A subsurface reservoir model is a computer based representation of petrophysical parameters such as porosity, permeability, fluid saturation, etc (Pringle et al., 2006). Through previous steps, various parameters needed have been well prepared. In this step, 3D subsurface reservoir model can be finally established with calibration of existing data and the key factor is the optimization of modeling method and integration of geologist's expert knowledge. In the past two decades, different stochastic methods (pixel-based methods e.g. Gaussian-related stochastic models, object-based modeling methods e.g. Boolean models, and process-based methods) and deterministic methods have been developed for reservoir geological modeling (Hu, 2003).

5. Challenges and development trend of reservoir characterization technology

With everlasting high heterogeneity and diversity of reservoir rocks, characterization methods for reservoir framework, pore structure and petrophysical flow units are still great challenges. Today, there is still no substitute for high quality seismic interpretation which may reveal essential geological features and important heterogeneities. Seismic acquisition, processing and interpretation techniques have been refined steadily over the years. Methods e.g. Vertical Seismic Profiling (VSP), inter-well seismic, time lapse seismic (4D) technologies etc. have been calibrated with seismic processing and interpretation. But seismic technique is still short of resolution and can not effectively diagnose small-scale structures or architecture elements. Just as Haldorsen and Damsleth (1993) remarked: The industry wants 1-m resolution at typical North Sea reservoir depths before the turn of the century! Achieve a dramatic seismic resolution improvement should still be goal of the oil and gas industry. In other words, there are still challenges in seismic survey and methods to improve resolution will be one of development trends.

Crosswell (or interwell) techniques have been developed in these years. Crosswell data were calibrated into reservoir models. Crosswell seismic imaging to provide images of reflectivity, velocity and formation properties by amplitude-versus-angle (AVA) inversion has been developed (Yu et al., 2008) and interwell seismic imaging was also developed for hydraulic fracture monitoring (Bailey, 1973; Duncan and Eisner, 2010; Maxwell et al., 2010). Whereas, interwell petrophysical properties of reservoir geological model are still determined by extrapolation of adjacent well data or derived directly from seismic interpretation. But because of low resolution, petrophysical properties and hydrocarbon distribution derived from seismic data usually can not fulfill the need of development planning. Crosswell petrophysical properties and hydrocarbon distribution are still great challenges.

As sustained recovery of oil or gas from porous media, reservoir petrophysical properties, pressure and fluid distribution are changing with time. With initialization and development of 4D seismic technology, methods to update petrophysical properties and fluid distribution of geological model have been highlighted. Whereas fluid distribution and petrophysical properties changes a lot and becomes more complicated through field development especially for oilfield through long-term water flooding. High permeable channel (large pores) gradually develops in highly flooded zone, causing ineffective water circulation and seriously affecting the displacement efficiency. Whereas water blockage will take place in low permeable zone which causes additional permeability reduction. Remaining reserves are still abundant in non-swept areas. History matching and 4D seismic are current methods to update reservoir geological model with advances of field development. Neither history matching nor 4D seismic methods can characterize the changes of reservoir petrophysical properties. Effective way to characterize petrophysical properties changes with field development should be challenge and one of emerging trends.

In summary, challenges of reservoir characterization remain in three aspects: resolution to identify small scale reservoir compartments, e.g. fracture, small-scale structure and thin layers, which aim to establish a detailed reservoir framework; crosswell petrophysical properties and hydrocarbon prediction, which are basis for spatial distribution of reservoir properties; characteristics of physical property dynamic changes with advances of field development, which are very important to reservoir performance. Thus, in the author's point of view, resolution improvement for seismic survey, effective methods to appropriately characterize crosswell petrophysical properties and hydrocarbon distribution and effective way to characterize dynamic changes of petrophysical properties should be three main development trends for reservoir characterization technology.

6. Conclusions

In this paper, discussions were carried out from four aspects and progresses of this work can be summarized as follows: (1) the words, e.g. advanced, practical, comprehensive, integrated, digital and detailed which are used to decorate the phase "reservoir characterization" , were discussed. Technique contents and features for each word (phase) were verified. (2) Three techniques have been selected and their advances have been reviewed. 4D seismic, digital data capture technology applied in outcrop study and microseismic monitoring can improved reservoir characterization from different aspects. Reservoir characterization with 4D seismic can characterize the dynamic of reservoir performance and remaining reserves can be evaluated. Digital data capture technology can be calibrated with outcrop study and it has advantages for obtaining more accurate and denser digital datasets from outcrop. Microseismic monitoring can effectively image the downhole fracture network deformation of hydraulic fracturing and capture dynamic changes of reservoir. (3) Practical procedures for reservoir characterization have been initiated. It is consist of eight steps. The technique feature and the main task were clearly defined. It is a hoped to be helpful to technical staffs or researchers. (4) Challenges and development trends were reviewed. Resolution improvement for seismic survey, effective methods to appropriately characterize crosswell petrophysical properties and hydrocarbon distribution and effective way to characterize dynamic changes of petrophysical properties should be three main development trends for reservoir characterization technology.

Finally, as reservoir characterization is a comprehensive and highly integrated technology, it is important to state that the author has limited knowledge and if there is any mistake or something improper please feel free to point it out.

7. Acknowledgements

This work is supported by research funds from National Grand Research Project (No. 2011ZX05015) and Science & Technology Research Project of PetroChina (No. 2011B-15). The authors are extremely grateful to PetroChina Co. Ltd., for permission to publish this work.

8. References

Abbott, D., C. Neale, and J. D. Lakings, et al., 2007, Hydraulic fracture diagnostics in theWilliams Fork Formation, Piceance basin, Colorado using surface microseismic monitoring technology: Proceedings of the Rocky Mountain Oil & Gas Technology Symposium, Society of Petroleum Engineers, Paper 108142.

Albright, J. N., and R. J. Hanold, 1976, Seismic mapping of hydraulic fractures made in basement rocks: Proceedings of the Energy Research and Development Administration (ERDA) Symposium on Enhanced Oil and GasRecovery, v. 2, Paper C-8.

Bailey, J. R., 1973, Mapping of earth fractures induced by hydrofracturing: U. S. Patent 3 739 871.

Barker,W., 2009, Increased production through microseismic monitoring of hydraulic fracturing over amultiwell program: Proceedings of the Annual Technical Conference and Exhibition, Society of Petroleum Engineers, Paper 124877.

Bellian, J.A., C. Kerans, and D.C.Jennette, 2005, Digital outcrop models: application of terrestrial scanning lidar technology in stratigraphic modeling: Journal of Sedimentary Research, v. 75, no. 2, p. 166-176.

Breed, C.T., and T. Grow, 1979, Morphology and distribution of dunes in sand seas using LANDSAT (ERTS) Imagery: In McKee, E.D. (Ed), A study of global sand seas. United States Geological Survey (USGS) Professional Paper, 253-303.

Bryant, I.D., and S.S. Flint, 1993, Quantitative clastic reservoir geological modelling: problems and perspectives In: Flint, S. & Bryant, I.D. (Eds) The geological modelling of hydrocarbon reservoirs and outcrop analogues. International Association of Sedimentologists Special Publication, 15, 3-20.

Buckley, S.J. J.A. Howell, and H.D. Enge et al., 2008, Terrestrial laser scanning in geology: data acquisition, processing and accuracy considerations: Journal of the Geological Society, v.165, no. 3, p. 625-638. DOI: 10.1144/0016-76492007-100.

Castillo, J.J., M. Sanchez, and M.Rodriguez, et al., 1998, Integrated reservoir characterization and volumetric analysis of the analysis of the Arecuna field (3D seismic area) Orinoco belt: presentation at the 7th UNITAR Conference on Heavy Crude and Tar Sands.

Chambers, K., S. Brandsberg-Dahl, and J.M. Kendall, et al., 2008, Testing the ability of surface arrays to locate microseismicity: 78th Annual International Meeting, SEG, ExpandedAbstracts, 1436–1440.

Chambers, K., O. Barkved, and J.-M.Kendall, 2009a, Imaging microseismic activity using the LoFS permanent sensor surface array: 79th Annual International Meeting, SEG, Expanded Abstracts, 1612–1616.

Chambers, K., J.-M., Kendall, and S. Brandsberg-Dahl, et al., 2009b, The detectability of microseismic events using surface arrays: Workshop on Passive Seismic, EAGE, A25.

Chapin, M.A., S. Davies, and J.L. Gibson, et al., 1994, Reservoir architecture of turbidite sheet sandstones in laterally extensive outcrops, Ross Formation, Western Ireland: In GCSSEPM 15th Annual Research Conference Proceedings, Submarine Fans and Turbidite Systems., 53-68.

Clark, J.D., and K.T. Pickering, 1996, Submarine channels, processes and architecture: Vallis Press.

Collinson, J.D., 1970, Deep channels, massive beds and turbidite current genesis in the central Pennine Basin: Proceedings of the Yorkshire Geological Society. 37, 495-520.

Doss, S.K., S., Ezzedine and R.J. Gelinas et al., 2001, Forward-inverse adaptive techniques for reservoir characterization and simulation: Theory and applications: Proceedings of the 2001 SPE Annual Technical Conference and Exhibition, Society of Petroleum Engineering, Paper 71600.

Doyle, J. D., and M. L. Sweet, 1995, Three-dimensional distribution of lithofacies, bounding surfaces, porosity, and permeability in fluvial sandstone; Gypsy Sandstone of northern Oklahoma: AAPG Bulletin, v. 79, p. 70-96.

Dreyer, T., L. Falt, and T. H. Knarud, et al., 1993, Sedimentary Architecture of Field Analogues for Reservoir Information (SAFARI): a case study of the fluvial Escanilla Formation, Spanish Pyrenees. In: S.S. Flint, and I.D. Bryant (Eds), The Geological Modelling of Hydrocarbon Reservoirs and Outcrop Analogues International Association of Sedimentologists Special Publication. London, 15, 57-79.

Duncan, P.M., and L. Eisner, 2010, Reservoir characterization using surface microseismic monitoring: Geophysics, v.75, no. 5, 139-146.

Fageraas, B., P. Behn and J. Tessman et al., 2003, Comprehensive reservoir characterization "From inside and out" using high-fidelity coupling techniques and point receivers: Proceeding of the 2003 Offshore Technology Conference. OTC Paper 15073.

Floris, F.J.T. and M.R.H.E. Peersmann, 2002, Integrated scenario and probabilistic analysis for asset decision support. Petroleum Geoscience, 8, 1-6.

Gibowicz, S. J., and A. Kijko, 1994, An introduction to mining seismology: Academic Press Inc.

Glennie, K.W., 1970, Desert sedimentary environments: Elsevier Co., Amsterdam.

Haldorsen, H.H. and E. Damsleth, 1993, Challenges in reservoir characterization: Geohorizons, v. 77, no. 4, p. 541-551.

Hall, M., and J. Kilpatrick, 2009, Surface microseismic monitoring of slick water and nitrogen fracture stimulations, Arkoma basin, Oklahoma: 79th Annual International Meeting, SEG, Expanded Abstracts, 1562–1565.

He W., Wu S., Tang Y., et al., 2005a, The architecture analysis of the underground point bar—taking Gudao Oilfield as an example. Journal of Mineralogy and Petrology, 25(2): 81-86.

He W., Wu S., Tang Y., et al., 2005b, Detailed architecture analysis of debouch bar in Shengtuo oilfield, Jiyang Depression. Petroleum Exploration and Development, 32(5): 42-46.

Hu L.Y., 2003, History matching of object-based stochastic reservoir models: Proceeding of the SPE 13th Middle East Oil Show & Conference, Society of Petroleum Engineering, Paper 81503.

Jackson, M.D., S. Yoshida, A.H. Muggeridge et al., 2005, Three-dimensional reservoir characterization and flow simulation of heterolithic tidal sandstones; AAPG Bulletin, v. 89, no. 4, p. 507-528.

Jahns, H.O., 1966, A rapid method for obtaining a two-dimensional reservoir description from well pressure response data: SPE Journal, 6(4), 315-327.

Jia Z, and Cai Z., 1992, Research methods of architecture: Geological Science and Technology Information, 11(4): 63-68.

Jia A., He D., He W., et al., 2003, Application of outcrop geological knowledge database to prediction of inter-well reservoir in oilfield. Acta Petrolei Sinica, 24(6): 51-53.

Jia A., and Cheng L., 2010, The technique of digital detailed reservoir characterization: Petroleum Exploration and Development, v. 37, no. 6, p. 709-715.

Johnston, D., 2004, Reservoir characterization improves stimulation, completion practices: Oil & Gas Journal, 102(4), p. 60-63.

Justice J H, Woerpel J C, Watts G P, et al., 2000, Interwell seismic data for reservoir characterization: Proceeding of SPE Permian Basin Oil and Gas Recovery Conference, Society of Petroleum Engineering, Paper 59695.

Keller, W. R., B. J. Hulsey, and P. M. Duncan, 2009, Correlation of surface microseismic event distribution to water production and faults mapped on 3D seismic data: A West Texas case study: 79th Annual International Meeting, SEG, Expanded Abstracts, 1524–1526.

Kochnev, I. V., V. S. Polyakov, and I. Murtayev, et al., 2007, Imaging hydraulic fracture zones from surface passive microseismic data: First Break, 25, 77–80.

Kovacic, L., and E. Poggiagliomi, 2003, Integrated time-lapse reservoir monitoring and characterization of Cervia field: a case study: Petroleum Geoscience, v. 9. p. 43-52.

Kuznetsov, O. L., L. N. Chirkina, and G. A. Belova, et al., 2006, Seismic location of emission centers-A new technology for monitoring the production of hydrocarbons: 68th Conference & Technical Exhibition, EAGE, Extended Abstracts, B018 (in Russian).

Liao G., 2006, Analysis on sedimentology system and reservoir architecture of delta-shallow lake facies of Fanzhuang in Jinhu depression. Journal of Southwest Petroleum Institute, 28(2): 4-7.

Maxwell, S.C., J. Rutledge, and R. Jones, et al., 2010, Petroleum reservoir characterization using downhole mircroseismic monitoring., Geophysics, v. 75, no. 5, p. 129-137.

McCaffrey, K.J.W., R.R. Jobnes, R.E. and Holdsworth, et al., 2005, Unlocking the spatial dimension: digital technologies and the future of geoscience fieldwork: Journal of the GeologicalSociety of London 162, 1-12.

Mezghani M., A. Fornel, and V. Langlais, et al. 2004, History matching and quantitative use of 4D seismic data for an improved reservoir characterization: Proceedings of SPE Annual Technical Conference and Exhibition, Society of Petroleum Engineering, Paper 90420.

Miall A D., 1985, Architectural-element analysis: A new method of facies analysis applied to fluvial deposits: Earth Science Review, 22(2): 261-308.

Montgomery, S.L., and M.F. Morea, 2001, Antelope shale (Monterey Formation), Buena Vista Hills field: Advanced reservoir characterization to evaluate CO_2 injection for enhanced oil recovery: AAPG Bulletin, v. 85, no.4, p. 561-585.

Phillips, C., 1996, Enhanced thermal recovery and reservoir characterization: AAPG Pacific Section-Old Oil Fields and New Life: A Visit to the Giants of the Los Angeles Basin, p. 65-82.

Pringle, J.K., J.A. Howell and D. Hodgetts et al., 2006, Virtual outcrop models of petroleum reservoir analogues: a review of the current state-of-the-art: First break, v. 24, p. 33-42.

Refae, A.T., S. Khalil, and B. Vincent et al., 2008, Increasing bandwidth for reservoir characterization with single-sensor seismic data: a case study from Libya's challenging Lehib field. Petroleum Africa, 41-44.

Reynolds, A.D., 1999, Dimensions of Paralic Sandstone Bodies. AAPG Bulletin, v. 83, no. 2, p. 211-229.

Robein, C. E., F. Drapeau, and D. L. Maurel, et al., 2009, Multi-network microseismic monitoring of fracturing jobs—Neuquen TGR application: 71st Conference & Technical Exhibition, EAGE, Extended Abstracts, X009.

Schisselé, E., and J. Meunier, 2009, Detection ofmicro-seismic events using a surface receiver network:Workshop on Passive Seismic, EAGE, A12.

Slatt, R., and S. Mark, 2002, Practical reservoir characterization for the independent operator, computer technology for comprehensive reservoir characterization: Petroleum Technology Transfer Council, 512-514.

Wang J., 2008, Greeting the great development of theories and applications of reservoir modeling: Earth Science Frontiers, 15(1): 16-25.

Yin T., Zhang C., Tang J., et al., 2001, Analysis on reservoir hierarchical structure in Machang Oilfield: Journal of Jianghan Petroleum Institute, 23(4): 19-21.

Yin T., Zhang C., Fan Z., et al., 2002, Establishment of the prediction models of reservoir architectural elements: Journal of Xi'an Petroleum Institute(Natural Science Edition), 17(3): 7-10, 14.

Yu G., B. Marion, and B. Bryans et al., 2008, Crosswell seismic imaging for deep gas reservoir characterization: Geophysics, v. 73, no. 6, p.117-126, DOI: 10.1190/1.2980417

Zhang C., 1992, Hierarchy analysis in reservoir researches: Oil & Gas Geology, 13(3): 344-350.

10

Mechanisms and Effective Prevention of Damage for Formations with Low-Porosity and Low-Permeability

Jienian Yan and Jiaojiao Geng
China University of Petroleum, Beijing
China

1. Introduction

With the progress of the petroleum industry and the increased demands of the oil and gas resources, the focus of petroleum exploration and development has been shifted from the formations with high permeability to the ones with low-porosity and low-permeability all over the world. As a newly emerging force to increase the oil and gas reserves and energy supplies, the development of the formations with low-porosity and low-permeability has provided significant potential for the steady development of the petroleum industry. Since the 1990s', the reserves of this type of reservoir have occupied a large percentage of the total reserves, which has already reached up to 60% to 70% according to statistical data, with an accumulative total of 4.349 billion ton, distributing extensively in 21 blocks of the major oilfields in China.

Due to the already inherently low permeability, and to the fact that the reservoirs with low-porosity and low-permeability are commonly characterized by the tiny size of the pore throats, high content of clay minerals, high capillary pressure and high flow-resistance, this type of reservoir is susceptible to formation damage that is difficult to remove and easy to result in the loss of the industrial capacity, compared with the conventional reservoirs. Therefore, the mechanisms and effective prevention of damage for target formations with low-porosity and low-permeability have attracted high attention in petroleum engineering and have important significance for the increase in reserves and production. So far, the conventional techniques and methods which are only applicable to higher permeability formations have been still adopted for the formation damage control of this type of reservoir, and there is no systematic and deep-going research on this subject.

This chapter provides a summary of recent work which has been conducted in the analysis of formation damage associated with the formations with low-porosity and low-permeability. Although little solid invasion has been observed in formations with low-permeability, the invasion of filtrates of drilling fluids often induces the damage of various kinds of sensitivities easily, especially the water-sensitivity damage caused by clay swelling. Also, water blocking is one of the most common mechanisms of damage for formations with low-porosity and low-permeability. This phenomenon has been observed as a particularly severe problem in reservoirs with ultra-low permeability or where the original water saturation is lower than the irreducible water saturation formed by the invasion fluids.

Numerous studies have shown that the damage induced by water-blocking may decrease the effective permeability to oil by 70% to 90%.

Generally, the permeability is usually too low to be carried out the core flooding tests easily for the formations with low-permeability. Sometimes it is difficult for the engineers to retrieve typical core samples and conduct detailed evaluation tests. Hence, it is significant to establish the mathematical models for predicting the mechanisms of formation damage for this type of reservoir. The new mathematical models were developed to predict the five kinds of sensitivities of formations and the damage induced by water blocking quantitatively in this study on the basis of numerous experimental results, applying the theory of petrophysics and interfacial chemistry for reservoirs and the intelligent method of ANN (Artificial Neural Network). Extensive applications have shown that more than 85% of the predicting results are in agreement with the experimental results.

From the aspect of drilling fluids, the effective techniques of prevention for the formations with low-porosity and low-permeability have been proposed as follows: (1) The inhibition of drilling fluids should be enhanced in order to reduce the formation sensitivity; (2) Effecient surfactants should be optimized so as to reduce the oil/water interfacial tension, and minimize formation damage induced by water blocking; (3) Adopt the technique of ideal packing compounding with film-forming to form quickly ultra-low permeable mud cake so as to minimize the filtrate of drilling fluids; (4) All kinds of treating agents used in drilling fluids should be compatible with formation rocks and fluid.

Specific successful application of the drilling fluid systems suitable for protecting the reservoirs with low-porosity and low-permeability used in a certain oilfield is presented in this paper, which can improve the returned permeability of core samples to more than 85% and enhance productivity considerably.

2. Damage mechanisms of formations with low-porosity and low-permeability

2.1 Solid invasion and fines migration

Since the pore size is approximately in a direct proportion to the permeability, the pores of which the size is smaller than 1μm, occupy as much as 35%-90% in formations with low-porosity and low-permeability; while there are nearly no solid particles with the size smaller than 2μm in the current drilling fluid systems. Therefore, the majority solid particles (bentonite, barite, bridging agents, drilling cuttings, etc.) in drilling fluids are too large to invade into the formations. Even if there is a little solid phase invasion, the perforation will be eliminate the corresponding damage (Bennion, D., 2000). On the other hand, the mobile fines are capsuled by the still water, namely, the wetting phase in the formations with low-porosity and low-permeability, which isolates the particles from the flowing oil and gas. In addition, there are only relatively few mobile fines present in pores and throats since this type of formation often buried deeply and accompanied by the compaction and diagenesis. Hence, the solid phase invasion and fines migration are usually not the main mechanisms of damage in this type of formation.

2.2 Formation sensitivity

Although little solid invasion has been observed in formations with low-permeability, the invasion of filtrates of drilling fluids often induces the damage of various kinds of sensitivities easily, especially the water-sensitivity damage caused by clay swelling. Usually, there is high content of active clay minerals, such as smectite and mixed layer of illite/smectite that can react with the filtrate of water-based fluids in the formations with low-porosity and low-permeability. Expandable clays can swell 600 to 1000 times their original volume after exposure to incompatible fluids, expanding to plug pores and pore throats, reducing permeability of formations drastically and leading to formation damage (Brian, D., 2004, Erwom, M.D., 2003, Bennion, D., 2002). The fines generated during swelling can also migrate and plug at pore throats, reducing permeability further. Hence, it is important to enhance the inhibition of drilling fluid and make sure that the drilling fluids are compatible with formation rock and fluid so as to minimize the damage of formation sensitivity.

2.3 Water blocking

Water blocking, which has been discussed extensively in the literature from both a theoretical and field production perspective, is one of the most common mechanisms of damage for formations with low-porosity and low-permeability, and the damage induced by water-blocking may decrease the effective permeability to oil by 70% to 90% (Shu Y., 2009, Geng J.J., 2010, Zhang H.X., 2010). The water blocking occurs due to the capillary effect of micro-pores for this type of formation when the filtrate of drilling fluids invades formation, as shown in **Fig.1**. The capillary force can be expressed by the Laplace's equation:

$$\Delta P = 2\sigma\left(\frac{1}{R_1} - \frac{1}{R_2}\right) \tag{1}$$

Where, the ΔP is the differential pressure of the surface; σ is the oil/water interfacial tension; and R_1, R_2 is the radius of the two surfaces, respectively.

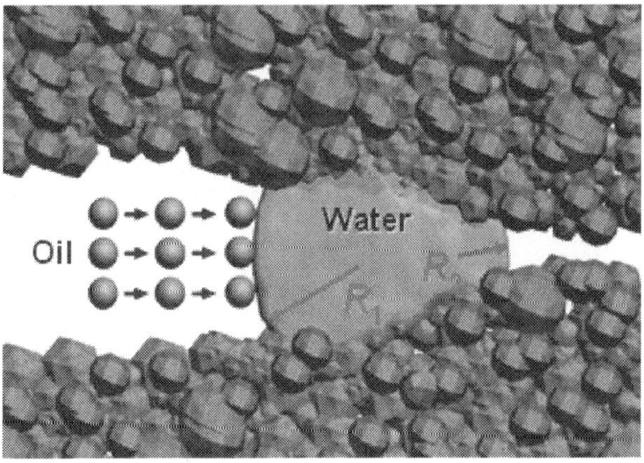

Fig. 1. Schematic diagram for water blocking effect

In the formations with low-porosity and low-permeability, originally, water usually adsorbs on the surface of water-wet rocks or occupies at the corner of micro-pores, while the oil and gas are in the middle areas and afford the flow passage. As the invasion of well fluids, the water content in the pore structure increases sharply due to the capillary effect of micro-pores. Then the water blocking occurs and the number of the flow passage has been reduced. This phenomenon has been observed as a particularly severe problem in reservoirs with ultra-low permeability ($1\times10^{-3}\mu m^2\sim0.1\times10^{-3}\mu m^2$) or where the initial water saturation (S_{wi}) is lower than the irreducible water saturation (S_{wirr}) formed by the invasion fluids. As shown in **Fig. 2**, when the S_{wi} is lower than the S_{wirr} which is obtained from a conventional water-gas drainage capillary pressure test, the water blocking occurs inevitably. The larger the difference between the S_{wi} and S_{wirr} is, the greater the decline of permeability is, and the more severity the formation damage is.

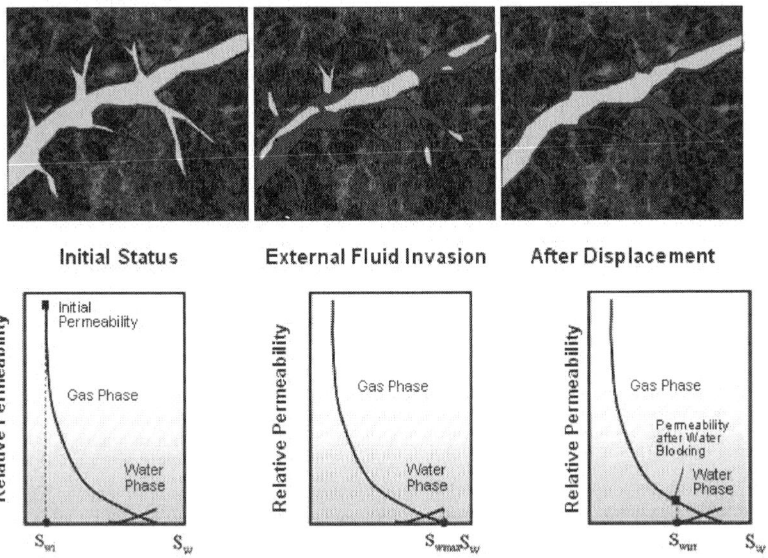

Fig. 2. Changes of relative permeability of water-gas

K_g $\times10^{-3}\mu m^2$	Damage degree of water blocking (R_s)				
	S_{wi} <10%	S_{wi} 10-20%	S_{wi} 20-30%	S_{wi} 30-50%	S_{wi} >50%
$K_g<0.1$	Severe	Severe	Moderate	Moderate	Weak
$0.1<K_g<1$	Severe	Moderate	Weak	Weak	Weaker
$1<K_g<10$	Severe	Moderate	Weak	Weaker	No damage
$10<K_g<100$	Moderate	Weak	Weaker	No damage	No damage
$100<K_g<500$	Weak	Weak	No damage	No damage	No damage
$K_g>500$	Weaker	No damage	No damage	No damage	No damage

Table 1. Correlation between water blocking effect and the K_g/S_{wi}
Note: Severe- R_s >90 %; Moderate-50 %< R_s ≤90%; Weak-20 %< R_s ≤50 %; Weaker-10 %< R_s ≤20 %; No damage- R_s ≤10 %.

The severity of water blocking is highly influenced by: (1) initial fluid saturations in the reservoir, (2) rock wettability, (3) pore system geometry, (4) fluid type, composition and interfacial tension, (6) invasion depth of fluid into formation. It can be seen from the statistics results shown in **Table 1** that the lower the permeability to gas (K_g) and the S_{wi} are, the more serious the water blocking phenomenon is; while the permeability is higher than $100\times10^{-3}\mu m^2$, there will be no water blocking. Since the inherent properties can not be changed, to alleviate water blocking effect, the filtration loss of drilling fluids should be controlled as low as possible. Meanwhile, adding some surfactants or alcohols in drilling fluids is quite helpful to minimize the oil/water interfacial tension, and then reduce the capillary resistance and prevent damage induced by water blocking.

3. Prediction of formation sensitivities

So far, the conventional techniques and methods which are only applicable to higher permeability formations have been still adopted for the evaluation of formation damage degree of this type of reservoir. However, the permeability is usually too low to be carried out the core flooding tests easily for the formations with low-permeability. And sometimes it is difficult for the engineers to retrieve typical core samples and conduct detailed evaluation tests. Hence, it is significant to establish mathematical models for predicting the mechanisms of formation damage for this type of reservoir.

3.1 Comparison and selection of the predicting method

The five sensitivities of formations include the sensitivity to flow rate, to water, to salinity, to alkalinity and to acidity. These sensitivities are influenced by various factors which interact on each other complexly. From the aspect of lithology and physical property analysis, a considerable effort has been made to explain and evaluate the damaging extent of formation sensitivities. However, it is very difficult to establish a specific structural model between formation sensitivities and influencing factors according to the principle of chemical balance. So far, a number of mathematical models have been established to predict formation sensitivities, such as Multigroup Discriminant Analysis (MDA), Multiple Regression Analysis (MRA) and Fuzz Mathematical (FM).

The MDA method is to divide the known core samples into several groups and build the discriminant function of the formation sensitivity degree and the parameters of core samples for each group. Then by putting the parameters of the unknown samples into the corresponding function, the formation sensitivity degree will be judged. Although the predicting result is objective, the MDA only generates qualitative results, which is wide of the truth and very difficult to use in practice.

The MRA method is assumed firstly that there is a certain function between the formation sensitivity degree and the parameters of the core sample. Then the undetermined coefficient and the regression equation are obtained by regressing the known samples. It is very convenient to evaluate the formation sensitivity degree by substituting the parameters of the unknown sample into the regression equation. However, the MRA depends on human factors, and there will be a big error if the regression equation selected is not proper.

The FM method adopts the inverse problem of synthetic judgment in fuzzy mathematics to solve the fuzzy relationship between the influencing factors and the formation damage degree which is only suitable for the simple case. Due to the various influencing factors and the complex reaction in this situation, the FM exhibits low efficiency.

We can see that methods discussed above are not accurate enough to predict the formation sensitivity. The Artificial Neural Networks (ANN) method has been increasingly used for prediction of complexes non-linear systems with good precision (Kalam, M.Z., 1996;

Nikravesh, M., 1996; Zuluaga, E., 2000). The ability of ANN systems is to spontaneously learn from examples, reason over inexact and fuzzy data, and provide adequate responses to new information not previously seen. It consists of a large number of simple interconnected artificial neurons. As shown in **Fig.3**, the artificial neuron (i) takes information from other neurons ($x_1, x_2, x_3 ... x_n$), performs very simple operations on this data, and passes results (y_i) on to other artificial neurons. Every artificial neuron meets the following equations:

$$s_i = \sum_{j=1}^{n} w_j x_j - \theta_i \tag{2}$$

$$u_i = g(s_i) \tag{3}$$

$$y_i = f(u_i) \tag{4}$$

Where, equation (2) is the accumulating potential value of the artificial neuron (i) after synapse; θ_i is the processing elements threshold; and w_i is the interlayer connection weights. Equation (4) is the relational expression between input and output values, in which u_i is the state of the artificial neuron (i). Neural networks operate by virtue of many artificial neuron data in this manner.

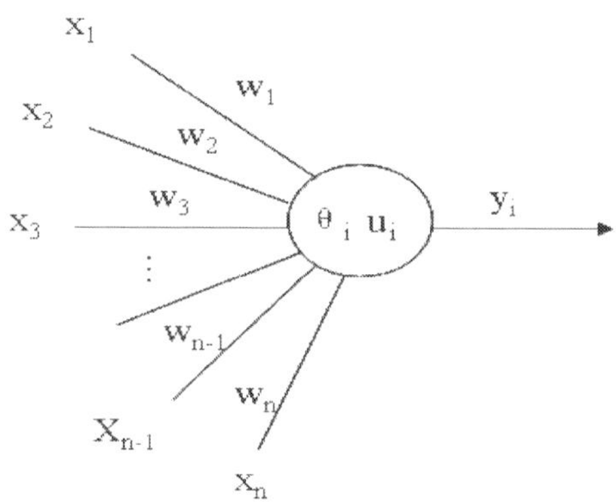

Fig. 3. Sketch Diagram of Artificial Neuron

The specific operation has two steps. One is the training process and another is the testing process. In the process of training, the ANN model has to be trained to recognize the relationships between the input and the desired output values by adjusting the connection weights between the different neurons. This process continues until weights converge to the desired error level or the output reaches an acceptable level. In the testing process, the developed ANN model is tested with several sets of experimental values, which are not used in the training of the model, to judge its performance. The developed model can memorize the correct output once input data is given. Compared with other methods, the ANN method has many advantages and has been introduced to predict the formation sensitivity.

3.2 Development of the predicting model

Determinate the influencing factors of formation sensitivities. As we all know, the formation sensitivity depends on many factors, such as the rock structure, the composition of the formation, the formation fluid, etc. Based on the theoretical analysis and experimental study, the influencing factors have been determined. Take the water-sensitivity damage as an example, the mainly influencing factors are shale content, smectite content, illite content, illite/smectite content, illite/bentonite ratio, porosity, permeability, cemented type, mineral size, and salinity.

Collect and process the data. The data in ANN model can be divided into qualitative data and quantitative data. In order to meet the requirements of the model, the qualitative data must be quantified firstly, and then all the data must be normalized. The examples are presented.

(1) The particle sorting

Generally, the sorting coefficient of rock is expressed by Fowke Watder's Standard Deviation, the normalization is as followed:

$$X_i = \begin{cases} 0.0690\delta_i < 0.35 \\ \delta_i / (4.0 + 3.5) 0.35 \le \delta_i \le 4.0 \\ 0.94326\delta_i > 4.0 \end{cases}$$

where X_i is the normalized value; δ_i is the Fowke Warder Standard Deviation.

(2) The normalization of the contents of smectite and illite/ smectite

$$X_i = 1.0 - \exp(-0.6701 S_i)$$
$$S_i = M_i + 0.6 I_i$$

where M_i is the content of smectite, %; I_i is the content of illite, %; X_i is the normalized value.

(3) The evaluation criteria of water sensitivity

According to the standard of core flooding experiment SY/Y5358-2002, the criteria are shown in **Table 2**.

Degree of Water Sensitivity	Water Sensitivity Index (I_w)
No sensitivity	$I_w \le 0.05$
Weak	$0.05 < I_w \le 0.30$
Intermediate-weak	$0.30 < I_w \le 0.50$
Intermediate-strong	$0.50 < I_w \le 0.70$
Strong	$0.70 < I_w \le 0.90$
Very strong	$I_w > 0.90$

Table 2. Evaluation Criteria of Water Sensitivity

Select and modify the algorithm. The standard BP (Back Propagation) model is one of the ANN models that consists of a three-layer neurons, an input layer, an output layer and a hidden layer, and it is usually used in formation sensitivity prediction. Although the BP algorithm is valid in the standard BP model, there are some problems in adjusting the interlayer connection weights, such as the low learning speed, likely to converge to a local minimum point, etc. Hence, the modified BP algorithm has been developed by the use of appending momentum factors, adjusting automatically learning factors and batch processing of connection weights adjustment, which proved to be more effective. Finally, the model for predicting the formation sensitivities has been established by using the modified BP algorithm in this study.

3.3 Application of the predicting model

Based on the ANN model established above, the practical software has been developed using the Visual Basic for Windows 6.0. It is very convenient for use in the field only by inputting the related parameters of formations and obtaining the indices of the formation sensitivities automatically, which can be instead of the complicated and time-consuming laboratory tests. The accuracy of the predictions can be significantly enhanced with model training using more precise reservoir data in the field applications. Neural networks have immense potential in predicting sensitivities and thereby assessing formation damage in reservoirs. The software has been tested using the production data. As shown in **Table 3,** more than 85% of the predicting results are in agreement with the experimental results.

Core No.	Shale %	Quartz %	Smectite %	I/S %	I/S Ratio %	Particle Size AVG %	Porosity %	Kg $10^{-3}\mu m^2$	Salinity g/L	Cementation Type	Predicting Index	Test Index
2	5.21	25	0.10	2.86	25	0.132	14.5	1.46	10.57	Contact	0.316	0.315
3	4.74	45	0.40	0.00	0	0.168	15.4	14.2	3.95	Porosity	0.425	0.457
6	7.72	40	0.12	5.56	20	0.150	24.5	7.06	5.00	Contact	0.502	0.500
7	2.94	40	0.45	0.71	30	0.175	16.6	41.6	6.72	Contact-porosity	0.539	0.531
9	5.00	40	0.00	0.75	40	0.500	8.37	2.80	7.26	Porosity	0.326	0.323
11	9.45	30	1.40	6.80	30	0.120	21.1	37.1	5.00	Contact-porosity	0.717	0.743
34	5.08	46	0.72	2.03	25	0.149	14.7	23.1	3.95	Contact-porosity	0.611	0.588
56	3.78	40	0.80	1.51	25	0.179	15.6	43.4	3.95	Contact-porosity	0.637	0.635

Table 3. Partial Results of Water Sensitivity Prediction

4. Evaluation and prediction of water blocking

4.1 Evaluation of water blocking

Water blocking has been recognized as one of the major causes of formation damage for the sandstone reservoirs with low-permeability, which may result in tremendous productivity reduction. Because of the complexity and variability of the pore structures, it is difficult to calculate the capillary force by using the Laplace's equation, and then judge the extent of damage induced by water blocking. Hence the experimental study on the formation damage induced by water blocking has been conducted in the laboratory.

In order to remove the interference, a certain amount of core samples with no or minor formation sensitivity, taken from the typical formations with low-porosity and low-permeability, have been chosen for the evaluation tests. The testing procedure was as follows: Firstly, the permeability to nitrogen gas was measured by using dry core, and the cores were vacuum saturated by standard brine, then the cores was displaced with kerosene to leaving irreducible water remaining in the core, and then core permeability was measured by displacing with kerosene in the forward direction after it was contaminated by brine in the reverse direction. Finally the rate of permeability damage was calculated. The intelligent core flooding apparatus was used for displacement tests, while the the oil/water interfacial tension was measured by using XZD-3 entire measuring range tension apparatus. It can be seen from the test results shown in **Table 4** that the extent of damage induced by water blocking is not only related to permeability, but also related to the initial water saturation, porosity, oil/water interfacial tension, etc. As a general rule, the lower the permeability or the higher the interface tension is, the more serious water blocking will happen. This rule is summarized by assuming that the other influencing factors kept constant when the influence of an individual factor was considered. However, there will be a dimly defined relationship between water blocking and each influencing factor if a variety of factors are changing at the same time. In order to achieve the quantitative prediction of water blocking in formations with low-porosity and low-permeability, it is essential to develop the mathematical model of the damage extent induced by water blocking and the influencing factors, by using the experimental data.

K_g $10^{-3}\mu m^2$	φ %	S_{wi} %	σ mN/m	K_o, $10^{-3}\mu m^2$	K_{oa}, $10^{-3}\mu m^2$	R_s %
11.96	16.8	49.29	2.081	8.64	4.32	50.00
6.78	9.34	14.75	2.081	4.90	2.53	48.32
19.86	9.16	65.84	0.868	15.51	13.01	16.12
26.81	14.8	40.02	0.652	19.37	17.43	9.99
9.22	22.2	51.79	0.469	5.43	4.38	19.34
4.67	16.0	38.95	0.360	3.85	3.27	15.14
10.66	20.5	55.26	0.350	7.63	6.84	10.35

Table 4. Predicted Results of Water Blocking
Note: K_o, K_{oa} are the permeabilities to oil before and after water blocking damage, respectively. R_s is the permeability loss rate caused by water blocking damage, which can be calculated by $(K_o-K_{oa})/K_o$.

4.2 Development of the grey neutral network model

Water blocking may be regarded as a grey process containing known and unknown factors. Theory and practice have proven that the Grey GM (0, n) prediction model is much better than the traditional MRA models for a complicated process. The GM (0, n) model uses the accumulated data instead of the original data to establish the prediction model, which can weaken the randomness or eliminate the errors of the original data to some extent. However, the static Grey GM (0, n) model is still a linear model that can not obtain accurate and satisfying result aiming at the prediction of water blocking. The problem can be solved by introducing the BP neural network with highly nonlinear and extrapolation into the Grey GM (0, n) via referring to formation sensitivity prediction mentioned above. A grey neural network model used to predicting the formation damage induced by water blocking for formations with low-permeability has been established (Zhang, Z.H., 2001, Gruber, N.G., 1996).

Based on the test results, we assumed that the water blocking mainly depends on permeability to gas K_g, porosity φ, initial water saturation S_{wi} and oil /water interface tension σ. Firstly, the grey predicting model Grey GM (0, 5) was established to gain the accumulated data (see **Appendix A**), and then the gained data was input into the modified

BP neural network model and thus the purpose of prediction was achieved. The predicted results using the grey neural network model are listed in **Table 5**. It can be seen that the grey neutral network model used to predict water blocking is reliable and has satisfactory accuracy and practicability. The evaluation criteria of water blocking damage are shown in **Table 6**.

K_g $10^{-3}\mu m^2$	R_s %	Predicting R_s %	Experimental results	Predicted results
11.96	50.00	49.6	Intermediate	Intermediate
6.78	48.32	53.2	Intermediate	Intermediate
19.86	16.12	14.1	Weak	Weak
26.81	9.99	7.63	Weak	Weak
9.22	19.34	22.8	Weak	Intermediate-weak
4.67	15.14	19.7	Weak	Weak
10.66	10.35	12.5	Weak	Weak

Table 5. Predicted Results of Water Blocking
Note: Experimental results and Predicted results is the degree of water blocking damage.

R_s	≤ 20%	20%-40%	40%-60%	60%-80%	> 80%
Degree of damage	Weak	Intermediate-weak	Intermediate	Intermediate-strong	Strong

Table 6. Evaluation Criteria of Water blocking Damage

5. Protecting drilling fluid technique for the formations with low-porosity and low-permeability

From the aspect of drilling fluids, the effective techniques of prevention for the formations with low-porosity and low-permeability have been proposed as follows: (1) The inhibition of drilling fluid should be enhanced in order to reduce the formation sensitivity; (2) Effecient surfactants should be optimized so as to reduce oil/water interfacial tension, and minimize formation damage induced by water blocking; (3) Adopt the technique of ideal packing compounding with film-forming to form quickly ultra-low permeable mud cake so as to minimize the filtrate of drilling fluids; (4) All kinds of treating agents used in drilling fluids should be compatible with formation rocks and fluid.

Take the typical formations with low-porosity and low-permeability located in Jilin Oilfield as an example. The effective porosity of the target formations is less than 13.3% and the average permeability is less than $18.39\times10^{-3}\mu m^2$. According to the geological and formation characters in this area, the potassium chloride (KCl) /polymer drilling fluid was selected. Based on the results of extensive tests, the basic formulation of drilling fluids was optimized as follows: 4% bentonite + 4.5%KCl +0.3%KPAM + 1.5%NPAN + 1% anti-complex salt filtration control agent +2%SMP-1 + 1%sulfonated-asphalt (Formulation 1#).

5.1 Optimization of surfactants

Adding proper surfactants in drilling fluids is able to minimize the oil/water interfacial tension, and then prevent or removal the water blocking timely. Five commonly used surfactants have been evaluated by measuring the contact angle in laboratory. From the **Table 7** and **Fig 4** we can see that the contact angles of OP-10 and ABSN are relatively small, which mean that the OP-10 and ABSN are strong in reducing the surface tension. However, OP-10 is water-soluble poorly. Hence the ABSN was recommended as the best water-blocking prevention agent.

Solution	**Distilled water**	**0.4%T-80**	**0.4%ABS**	**0.4%Span-80**	**0.4%OP-10**	**0.4%ABSN**
θ	62.30	60.99	44.08	29.71	13.32	18.08

Table 7. Contact angles for aqueous solutions of different kinds of surfactants (Ambient temperature, 20℃)

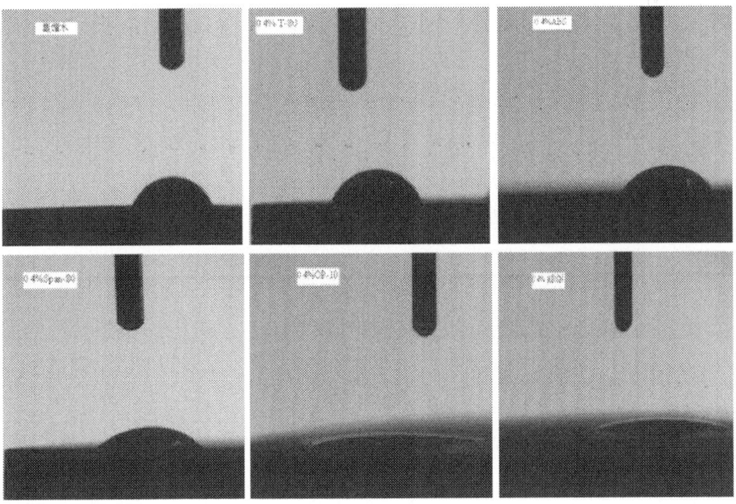

Fig. 4. The contact angles for aqueous solutions of different kinds of surfactants on water-wet rocks

It is shown from **Fig.5** that ABSN exhibits excellent effects on reducing the interfacial tension and surface tension of filtrates with a very low concentration of its aqueous solution. The oil/water interfacial tension can be reduced to 0.65 mN/m when the concentration of ABSN is 0.2%. Considering the lost amount of adsorption on cuttings, the recommended concentration of ABSN added in drilling fluids is 0.4%.

5.2 Optimization of Temporary Bridging Agents (TBA)

The TBA in drilling fluids can quickly form thin and tough mud cake to prevent small particles and filtrates from invading formations. In this study, a mixed TBA has been selected according to the principle of Ideal Packing Theory (IPT) (Zhang, J.B., 2004). Adopting designing software developed by our research group, the optimized TBA (a combination of calcium carbonate with different particle size) was determined as below: 45μm (30%), 23μm (50%) and 13μm (20%).

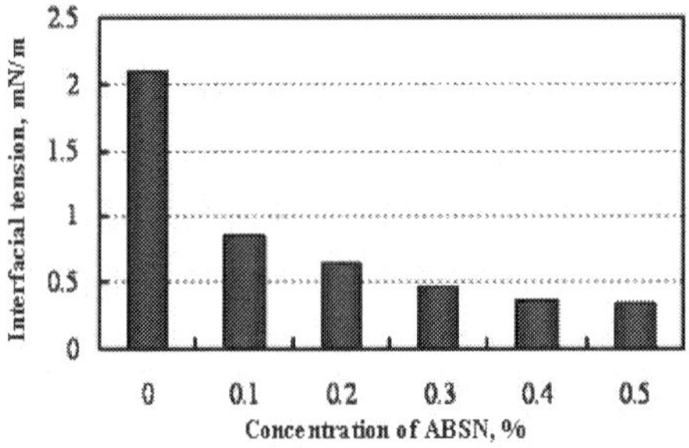

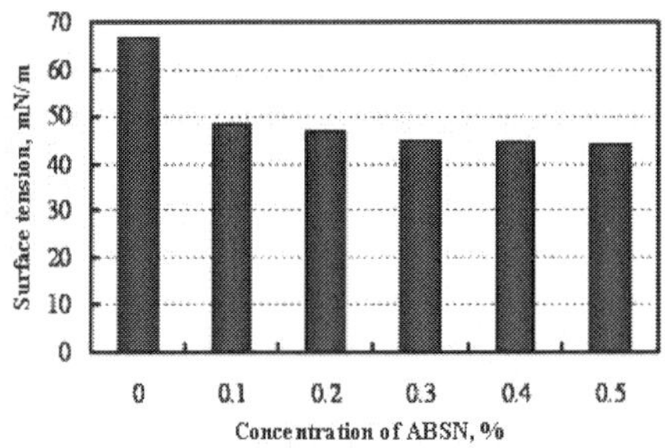

Fig. 5. Effect of ABSN concentration on o/w interfacial tension and surface tension of mud filtrates

The properties of drilling fluids after adding different amount of optimized TBA were measured as shown in **Table 8**. It is found that the addition of TBA has little impact on rheological parameters, but is helpful to reduce the filtration rate. Meanwhile, the spurt loss will decrease drastically, indicating that the presence of TBA in drilling fluids is beneficial to forming mud cake quickly. The relative amount of optimized TBA in drilling fluids is recommended as 2%~3% that is different from the formations with high-permeability for which the relative amount of TBA should be 4%~5%.

Fluid properties		ρ g/cm^3	*AV* mPa·s	*PV* mPa·s	*YP* Pa	Gel Pa/Pa	API FL ml
1#	Room temperature	1.14	33	22	12	1.5/3.0	7.2
	120℃/16h	1.14	31	20	10	1.5/3.5	8.0
1#+2%TBA	Room temperature	1.15	41	28	18	2.5/4.5	6.4
	120℃/16h	1.15	38	32	15	4.0/5.0	7.0
1#+3%TBA	Room temperature	1.16	50	38	20	3.5/6.5	5.1
	120℃/16h	1.15	45	33	17	4.0/6.0	4.8
1#+4%TBA	Room temperature	1.16	64	48	24	5.5/9.0	4.8
	120℃/16h	1.16	58	44	21	4.5/8.0	4.9
1#+5%TBA	Room temperature	1.18	72	52	27	5.0/9.5	4.5
	120℃/16h	1.17	67	53	24	6.5/10	4.8

Table 8. Effect of the addition of TBA on properties of drilling fluids

5.3 Optimization of film-forming agents

Two kinds of film-forming drilling fluid techniques have been developed in recent years (Tan C.P., 2002, Sun J.S., 2003, Yuan C., 2004, Sun J.S., 2005, Bai X.D., 2006). One is the semi-permeable membrane, and its typical agent is JYW-1. By using the peculiar polymer agent, it can concentrate into micelle and thus form semi-permeable membrane at the rock surface so as to seal a wide range of the pore throat. The other is the isolating membrane, and its typical agent is CMJ-2. The polymer adsorbs, concentrates and forms into no permeable membrane at the rock surface, which is called isolating membrane that any solid and liquid phase can not transport.

Both kinds of film-forming techniques have their advantages and disadvantages. It takes 40 minutes to form the isolating membrane, while the filtrate of drilling fluid may invade into formations and induce damage during this period. In addition, it is difficult to form into the isolating membrane when the size of the pore throat is larger than 114μm. The sealing capacity of the semi-permeable membrane is poor compared with the isolating membrane. However, the semi-permeable membrane can be formed quickly which usually takes only 1 to 3 minutes, and it can seal the large size of pore throat or the fracture (as large as 3mm) effectively.

Hence, given the characteristics of the two film-forming techniques, we have developed the double film-forming technique by compounding the two kinds of film-forming agents, which is synergistic. The mechanism of action is that the semi-permeable membrane is formed quickly and firstly, entering the large sizes of pore throat, and then the isolating membrane with no permeable is formed, supported by the wellbore wall and the semi-permeable membrane. As for the formations with low-porosity and low-permeability, the proper concentrations of film-forming agents are 1.5%CMJ-2 + 1%JYW-1, based on the extensive test results.

5.4 The synergistic effect of the ideal packing and the film-forming techniques

The TBA can block the large sizes of pore throat and fractures, while the double film-forming agents can form a sealing layer with zero permeability at the rock surface quickly. Compounding the techniques of Ideal Packing and Film-forming makes full use of the advantages of the two aspects and overcomes their own disadvantages, which can form a sealing layer with high pressure bearing capability at the rock surface, realizing the low-damage or no-damage drilling. The schematic diagram is shown in **Fig.6**.

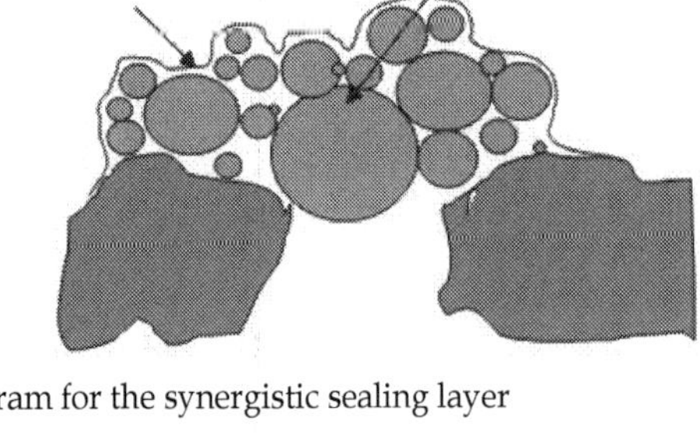

Fig. 6. Schematic diagram for the synergistic sealing layer

Two core samples with similar permeability, A and B, were contaminated by distilled water + 1.5%CMJ-2+ 1%JYW-1 (Formulation 2#) and distilled water + 3%IP-TBA + 1.5%CMJ-2+ 1%JYW-1 (Formulation 3#), respectively. And then the Pressure Bearing Capacity Measurement Apparatus was used for the tests. As we can see from **Fig.7**, the pressure bearing capability of the core sample contaminated by 2# solution was improved to 3.5 MPa, while the one contaminated by 3# solution was improved to 7.48MPa. The experimental results proved that the synergistic effect of the ideal packing and the film-forming techniques is helpful to formation protection and wellbore stability, which can prevent or reduce effectively the filtrate of fluid, and decrease the transportation of pore pressure.

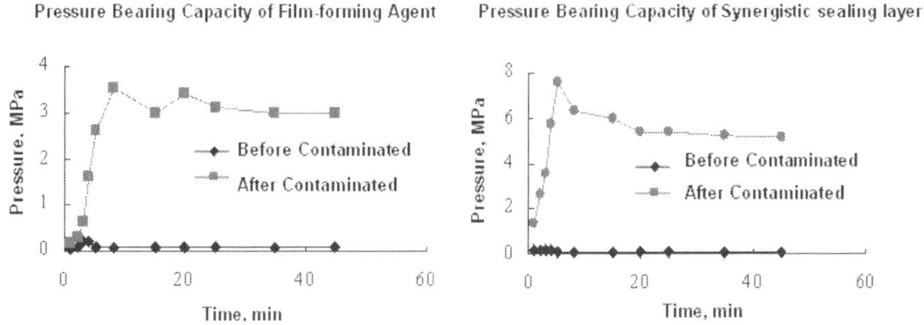

Fig. 7. The evaluation results of the pressure bearing capacity

Added 3% TBA and 1.5%CMJ-2 + 1%JYW-1 into the 1#, respectively. And then the dynamic filtrate of the two drilling fluids was tested so as to evaluate the synergistic sealing effect of the ideal packing and the film-forming techniques. The result was shown in **Fig.8**. We can see that the filtrate of drilling fluid adding the optimized film-forming agents and the TBA, was ultra-low and the initial filtrate was almost zero, which showed excellent formation protection, compared with the one adding the TBA only.

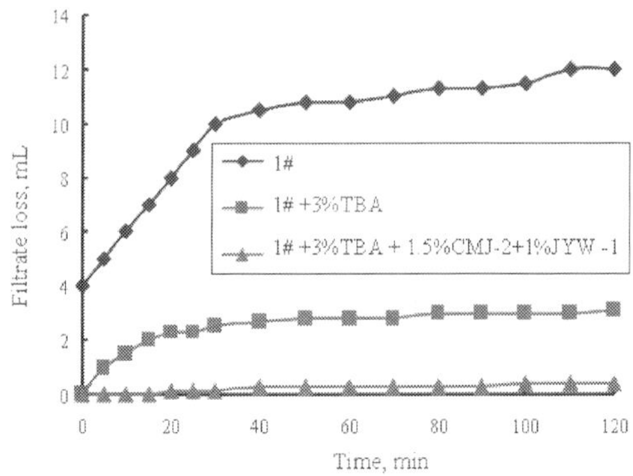

Fig. 8. Dynamic loss of drilling fluids with IP-TBA and film forming agents

6. Properties of newly developed drilling fluid

Based on the optimization results, a novel low-damage drilling fluid has been developed. It is composed of polymers, water blocking preventing surfactants, TBA, film-forming agents and other additives. The typical formulation is as follow: 4% bentonite + 4.5%KCl +0.3%KPAM + 1.5%NPAN + 1% anti-complex salt filtration control agent +2%SMP-1 + 1%sulfonated-asphalt + 0.4%ABSN + 3% TBA + 1.5%CMJ-2 +1%JYW-1 (Formulation 4#).
The conventional properties of the novel drilling fluid were evaluated and the results are shown in **Table 9.** It can be seen that the novel drilling fluid has good rheological properties and the API filtration rates before or after rolling at 120°C for 16h are less than 5mL, which can meet the requirements of drilling operations.

Conditions	*AV* mPa s	*PV* mPa s	*YP* Pa	Gel Pa/ Pa	pH	API FL mL
Room temperature	28	24	7.0	2.5/4.0	8.0	4.0
120℃/16h	31	27	8.5	2/4.5	8.0	3.6

Table 9. Properties of the newly developed drilling fluid

The dynamic core flooding tests were conducted using the newly developed drilling fluid, compared with Formulation 1#. The core samples were taken from a target formation with low-permeability in the Qian-231 well. The LH-2 HTHP dynamic core flooding apparatus manufactured by Lu Hai Co., Ltd was used for tests. The tests were performed at temperature of 90°C, differential pressure of 3.5MPa and shear rate of $150s^{-1}$. It is shown from **Table 10** that the returned permeability of the core contaminated with the newly developed drilling fluid is 88.1%, compared with the returned permeability of 77.4% contaminated by the former drilling fluid, indicating an exellent effectiveness of protecting the formations with low-porosity and low-permeability.

Core No.	Drilling fluid System	φ, %	K_g, $10^{-3}\mu m^2$	K_o, $10^{-3}\mu m^2$	K_{oa}, $10^{-3}\mu m^2$	Returned permeability, %
1	1#	10.45	7.93	2.75	2.13	77.4
2	4#	10.08	6.56	2.15	1.90	88.1

Table 10. Results of the returned permeability for cores

7. Conclusions

The main mechanisms of damage for the formations with low-porosity and low-permeability are usually water-sensitivity and water blocking. The extent of water blocking damage is related to various influencing factors such as permeability, initial water saturation, porosity, and oil/water interfacial tension.
The Artificial neural network (ANN) model and software have been developed successfully to predict sensitivities of formations. Extensive applications show that more than 85% of predicted results are in agreement with the measured results. The grey neutral network model takes the advantage of grey static models and the non-linear of neural network, and can be used to predict water blocking with more precise results.
Some surfactants or alcohols are helpful to minimize surface tension of filtrates and oil/water interfacial tension, and reduce the capillary resistance and prevent water blocking

damage. Compounding the techniques of Ideal Packing and Film-forming makes full use of the advantages of the two aspects and overcomes their own disadvantages, which can form a sealing layer with high pressure bearing capability at the rock surface, realizing the low-damage or no-damage drilling. The newly developed drilling fluid is suitable for protecting the typical reservoirs with low-permeability and has excellent performance. The returned permeability of the contaminated with this drilling fluid is higher than 88%.

8. Appendix A-establishment of GM (0 N)

8.1 Assumption

X_1 – Damage ratio caused by water-blocking, %; X_2 —Gas permeability, $10^{-3}\mu m^2$; X_3— Initial saturation, %; X_4— Oil / water interfacial tension , mN/m; X_5— Porosity,%. Establish the GM (0, 5) related to X_1 as following equation:

$$X_1^{(1)}(i) = b_1 X_2^{(1)}(i) + b_2 X_3^{(1)}(i) + b_3 X_4^{(1)}(i) + b_4 X_5^{(1)}(i) + a$$

The parameter list is yet to be determined:

$$\hat{b} = [b_1, b_2, b_3, b_4, a]$$

Data matrix is formed as follows:

$$B = \begin{bmatrix} X_2^{(1)}(2), & X_3^{(1)}(2), & X_4^{(1)}(2), & 1 \\ X_2^{(1)}(3), & X_3^{(1)}(3), & X_4^{(1)}(3), & 1 \\ \cdots & \cdots & \cdots & \\ X_2^{(1)}(n), & X_3^{(1)}(n), & X_4^{(1)}(n), & 1 \end{bmatrix}.$$

$$Y_N = \left[X_1^{(1)}(2), X_1^{(1)}(3), \cdots, X_1^{(1)}(n-1), X_1^{(1)}(n)\right]^T$$ (*n*, the number of samples)

8.2 The calculative process

The original data is listed below:

Original data	$k = 1$	$k = 2$	$k = 3$	$k = 4$	$k = 5$	$k = 6$	$k = 7$
$X_1^{(0)}(k)$	50	48.32	16.12	9.99	19.34	15.14	10.35
$X_2^{(0)}(k)$	11.96	6.78	19.86	26.81	9.22	4.67	10.66
$X_3^{(0)}(k)$	49.29	14.75	65.84	40.02	51.79	38.95	55.26
$X_4^{(0)}(k)$	2.081	2.081	0.868	0.652	0.469	0.36	0.35
$X_5^{(0)}(k)$	16.8	9.34	9.16	14.8	22.2	16	20.5

$X_i^{(1)}(k)$ $(i = 1,2,3,4 \quad k = 1,2,\cdots,6,7)$ is the one time accumulating value of $X_i^{(0)}(k)$. According to

$$X_i^{(1)}(k) = \sum_{j=1}^{k} X_i^{(o)}(j)$$

Original data	$k = 1$	$k = 2$	$k = 3$	$k = 4$	$k = 5$	$k = 6$	$k = 7$
$X_1^{(0)}(k)$	50 •	98.32	114.44	124.43	143.77	158.91	169.26
$X_2^{(0)}(k)$	11.96	18.74	38.6	65.41	74.63	79.3	89.96
$X_3^{(0)}(k)$	49.29	64.04	129.88	169.9	221.69	260.64	315.9
$X_4^{(0)}(k)$	2.081	4.162	5.03	5.682	6.151	6.511	6.861
$X_5^{(0)}(k)$	16.8	26.14	35.3	50.1	72.3	88.3	108.8

Then the data matrix is gained as follows:

$$Y_N = \left[X_1^{(1)}(2), X_1^{(1)}(3), \cdots, X_1^{(1)}(6), X_1^{(1)}(7)\right]^T = [98.32, 114.44, 124.43, 143.77, 158.91, 169.26]$$

$$B = \begin{bmatrix} 18.74 & 64.04 & 4.162 & 26.14 & 1 \\ 38.60 & 129.88 & 5.030 & 35.3 & 1 \\ 65.41 & 169.90 & 5.682 & 50.1 & 1 \\ 74.63 & 221.69 & 6.151 & 72.3 & 1 \\ 79.30 & 2\,60.64 & 6.511 & 88.3 & 1 \\ 89.96 & 3\,15.90 & 6.861 & 108.8 & 1 \end{bmatrix}$$

The relevant parameters are calculated:

$$\hat{b} = \left[B^T B\right]^{-1} B^T Y_N = [-0.7959, -0.2137, 45.3895, 0.7132, -80.6178]^T$$

Then b_1= -0.7959, b_2=-0.2137, b_3=45.3895, b_4=0.7132, a=-80.6178
The final equation G (0, 5) is gained as follows:

$$\hat{X}_1^{(1)}(k) = -0.7959X_2^{(1)}(k) - 0.2137X_3^{(1)}(k) + 45.3895X_4^{(1)}(k) + 0.7132X_5^{(1)}(k) - 80.6178$$

$$\hat{X}_1^{(0)}(k) = \hat{X}_1^{(1)}(k) - \hat{X}_1^{(1)}(k-1) \qquad (k \geq 2)$$

The original data is invalid after the accumulation above. Then the value of number k+1 must be calculated on the basic of the independent variables k, generated by the GM (0, 5).

9. Acknowledgements

We gratefully acknowledge the financial support for the research project provided by the China National Natural Science Foundation (Project No.50974129).

10. References

Bai X.D., Pu X.L., Evolution of membrane forming technology of water-based mud [J]. Atural Gas Industry, 2006,26(08):75-77.

Bennion, D. B., "An Overview of Formation Damage Mechanisms Causing a Reduction in the Productivity and Injectivity of Oil and Gas Producing Formations", Journal of Canadian Petroleum Technology, Volume 41, No.11 Nov. 10-15 2002

Bennion, D. B., Thomas, F.B., and Ma,T.,"Formation Damage Processed Reducing Productivity of Low Permeability Gas Reservoirs, " SPE 60325, SPE Rocky Mountain Regional Low Permeability Reservoirs Symposium, Denver, Mar.12-15,2000

Brian, D., William, D.W., "Maximizing Economic Return by Minimizing or Preventing Aqueous Phase Trapping During Completion and Stimulation Operations" SPE 90170, presented at the SPE Annual Technical Conference and Exhibition held in Houston,Texas, Sep. 26-29,2004.

Erwom, M D., Riersom, C.R., and Bennion, D.B., "Brine Inhibition Damage in the Colville River Field, Alasla, " SPE 84320, SPE Annual Technical Conference, Denver, Oct. 5-8,2003

Geng J.J., Yan J.N., et al. Mechanisms and prevention of damage for formations with low-porosity and low-permeability [R]. SPE130961, 2010.

Gruber, N.G., "Water Block Effects in Low Permeability Gas Reservoirs", presented at the 47th Annual Technical Meeting of The Petroleum Society in Calgary, Alberta, June 10-12,1996

Kalam, M.Z., Al-Alami, S.M., and Al-Mukheini, M., "Assesment of Formation Damage Using Artificial Neural Networks", paper SPE 31100 presented at the 1996 Formation Damage Symposium, Lafayetter,Feb.14-15

Nikravesh, M., Kovscek, A.R., Johnston, R.M., and Patzek, T.W., "Prediction of Formation Damage During Fluid Injection into Fractured, Low Permeability Reservoirs via Neural Networks", paper SPE 31103 presented at the 1996 Formation Damage Symposium, Lafayetter,Feb.14-15

Shu Y., Yan J.N., Xiong C.M., et al. Mechanism and preventive techniques of fluid-block damage for tight sandstone condensate gas reservoirs with low-permeabilit[J]. Petroleum Exploration and Development, 2009,36(5): 628-634.

Sun J.S., Wang S.G., Zhang Y., et al. Study on membrane generating technology of water-based drilling fluid[J] Driling Fluid and Completion Fluid, 2003, 20(06):6-10.

Sun J.S., Tang J.P.,Zhang B.,,Wang S.G., et al. Study on ultra-low permeability drilling/completion fluid[J]. Driling Fluid and Completion Fluid, 2005, (01):1-4.

Tan C P, Mody F K, Tare V A. Novel high membrane efficiency water-based drilling fluids for alleviating problems in troublesome shale formations[R]. SPE77192, 2002.

Tan C P, Mody F K, Tare V A. Development and laboratory verification of high membrane efficiency water-based drilling fluids with oil-based drilling fluid-like performance in shale stabilization[R].SPE78159, 2002.

Yuan C., Sun J.S., Wang P.Q., et al. Development of CMJ-1—a high temperature film-forming fluid loss additive and the properties[J]. Petroleum Drilling Techniques, 2004, 32(2):30-32.

Zuluaga,E., "Prediction of Permeability Reduction by External Particle Invasion Using Artificial Neural Networks and Fuzzy Models",presented at the Petroleum Society's Canadian International Petroleum Conference 2000, Calgary, Alberta, June 4-8,2000

Zhang, Z.H., Yan,J.N., Wu, Y.M., The Model of gray neural network for predicting water blocking damage of low permeability sandstone reservoirs[J]. Drilling & Production Technology, 2001, 24 (1): 38-40

Zhang, J.B., Yan J.N., New theory and method for optimizing the particle size distribution of bridging agents in drilling fluids[J]. Acta Petrolei Sinica, 2004, 25 (6):88-91

Zhang H.X, Yan J.N., Shu Yong, Xue Yuzhi. High-performance polyalcohol drilling fluid applied to protection of ultra-deep reservoir with low permeability[J]. Acta Petrolei Sinica, 2010,31(1): 129-133

Part 5

Hydrology

11

Application of Hagedoorn's Plus-Minus Method to Hydrology Study

Jiandang Ge
ION Geophysical Corp., Houston
USA

1. Introduction

The Memphis aquifer has been the major source of water for the City of Memphis municipal, industrial, and commercial uses for the past 100 years, and is considered to be among the highest quality water reservoirs in the nation. Above the Memphis aquifer are the confining unit (aquitard) of the Memphis aquifer and the surficial aquifer (Figure 1). The surficial aquifer is exposed to the surface and is prone to pollution due to industrial and human activities. The potential for contamination of the Memphis aquifer is exacerbated in areas where the aquitard is missing or thin. Recent studies indicated that the drinking aquifer might be at risk for contamination due to aquitard breaches existing in the confining unit of the Memphis aquifer. Aquitard breaches in the Memphis area have been identified through the correlation of stratigraphic picks from borehole data (Parks and Mirecki, 1992). The lack of uniform data coverage has restricted the study of breaches in Shelby County to areas proximal to the well fields. Although accurate, direct and reliable, the study does not provide crucial information about aquitard breaches, such as their extent, orientation, origination, and matrix characterization. Indirect methods (e.g. shallow seismic methods) can provide critical information that can help identify the possible causes responsible for the formation of the breaches (Ge et al., 2010, Part II). In this paper, the Hagedoorn's (1959) plus-minus method was applied to the seismic refraction data acquired in a walkaway test to map the top of the confining unit and identify possible aquitard breaches.

2. Hagedoorn's plus-minus method

The Hagedoorn's (1959) plus-minus method provides a simple and fast tool to interpret refraction data and calculate the geometry and velocity of the first refractor. The procedure is remarkably straightforward: the arrival times of the refracted waves from two reciprocal shots are simply added to find the depth to the refractor at all geophone stations and subtracted to find the velocity of the wave propagating through the refractor (Overmeeren, 2001). The Hagedoorn method has been shown to be a cost-effective and efficient means of mapping the shallow subsurface velocity structure (Overmeeren, 2001). Overmeeren (2001) used Hagedoorn's plus-minus method in a regional groundwater study and found that this method not only can provide a detailed section, but also produces additional information to reduce ambiguity in the interpretation of other geophysical data (e.g., vertical electrical soundings). In Hagedoorn's (1959) classic paper, he utilized wave front reconstruction, usually by graphical means, to demonstrate the principle of the method. The derivation started from a model of one horizontal layer with velocity V_1 over a half space with velocity V_2 ($>V_1$). Two shots, A and B, (Figure 2A) are so far enough from the receiver spread that the first arrivals at each receiver are all from refracted waves (not direct waves). The red

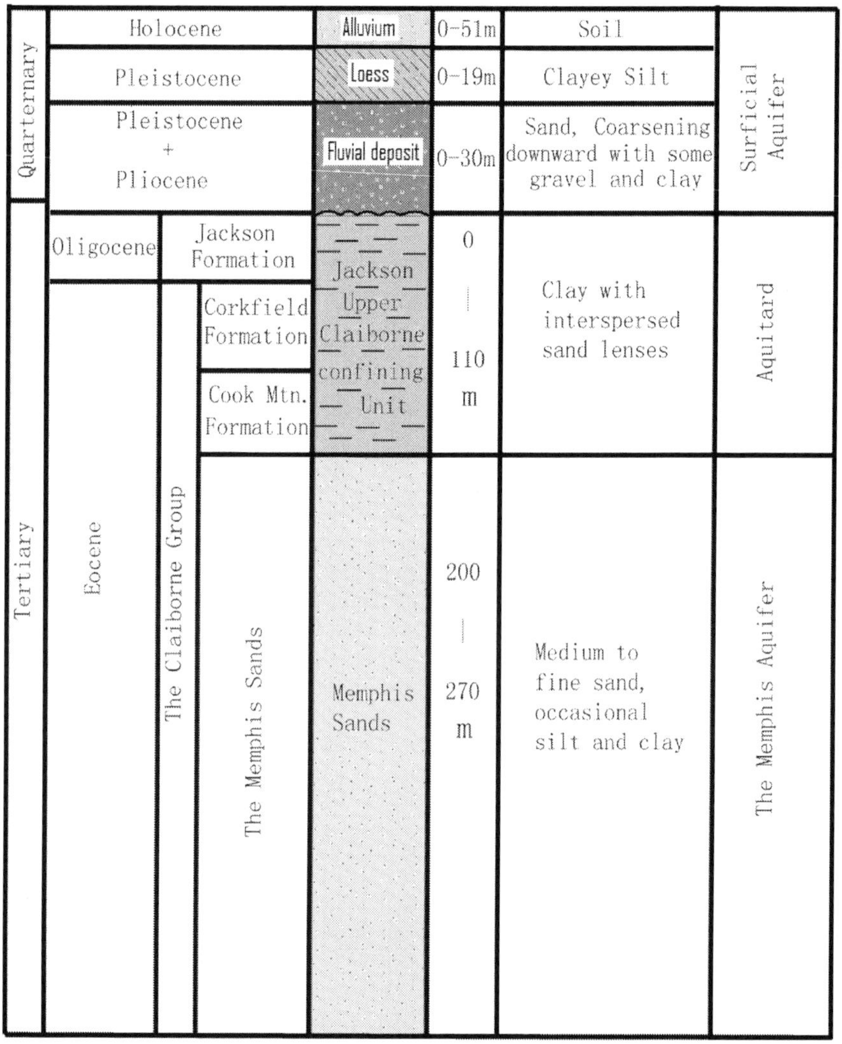

Fig. 1. Geology stratigraphy, lithology, and hydrologic significances in the Memphis area (modified from Parks and Mirecki, 1992).

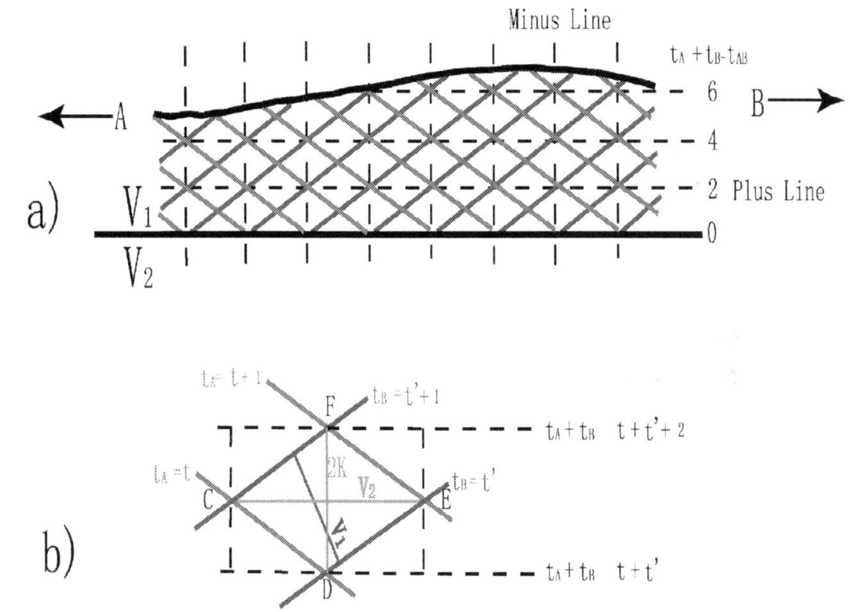

Fig. 2. Schematic wave fronts used in the Hagedoorn's plus-minus method (Hagedoorn, 1959). A) One layer over halfspace model with wave fronts drawn from two reciprocal shots; B) wave fronts composing the diamond-shaped region showing the constant plus value along the plus line. See text for details.

lines in the figure represent wave fronts generated by source A and propagating to the right; the blue lines represent wave fronts generated from source B and propagating to the left. The time intervals between all the neighboring wave fronts from each source are all the same, and regarded as unit time 1. Since the refractor is horizontal, the intersecting wave fronts drawn form diamond-shaped figures (Figure 2B). For the wave fronts propagating from source A to the right (CD and EF), the traveltimes are t and t+1, respectively; for the wave fronts propagating from source B to the left (DE and CF), the traveltimes are t′and t′+1, respectively. Note that vertex C is the intersection of wave fronts CD and CF; the summation of traveltimes of the two wave fronts at this intersection is t+t′+1. For intersection E, the summation is also t+t′+1. Hence, for the horizontal vertices (intersections) of the diamonds, the summation of travel times from the two wave fronts is constant. This is true for all the diamonds and results in what Hagedoorn called the "plus" lines, drawn as horizontal dashed lines in Figures 2A and B. The plus value is calculated by adding the two travel times at each intersection and subtracting t_{AB}, the travel time from source A to source B. The resulting values equal 0 on the refractor, 2 on the horizontal line through the first set of intersections vertically above those defining the refractor, 4 on the next line up, and so on. Note that any of the "plus lines" can be used to plot the refractor shape (structure). The plus values can be calculated on the surface. At each receiver station, t_A and t_B are the first arrival times picked from the two reciprocal shot records. The travel time from source A to source B, t_{AB}, may not be recorded in a typical refraction survey, but since t_{AB} is a constant for the two reciprocal shot records, it does not affect the shape of the refractor if t_{AB} is not included in the calculation. As shown in diamond CDEF, the distance between the two wave fronts CF and DE is v_1 (because the time difference between the two neighboring wave fronts is unit time) and similarly the length of CE is v_2. Let $2k$ be the length of DF (the distance between the horizontal, dashed 'plus' lines), then $2k$ can be calculated from equation 1 (Hagedoorn, 1959),

$$2k = \frac{v_1}{\sqrt{1-(v_1^2 / v_2^2)}} \tag{1}$$

Since the length of DF corresponds to a difference of one time unit for both wave fronts and the plus value difference between D and F is 2, consequently, the product of k and its "Plus" value is the actual height of a point above the boundary (Hagedoorn, 1959). Consequently, the product of k and the difference of two plus values at two points gives the actual distance between the two points.

Similarly, the difference between the travel times between shot A and B at an intersection is called the "minus" value. The minus value is constant along vertical lines passing through the intersections of wave fronts. In Figure 2, the minus lines are shown as vertical dashed lines spaced at a distance interval equal to the value of the velocity below the boundary, V_2 (because the time intervals between the neighboring wave fronts is the unit time) and their minus values differ by two time units. Hagedoorn (1959) also demonstrated this method for more specific cases (e.g., a refractor with a change in velocity and curved refractors).

3. Application

Hagedoorn's method was applied to the data collected in the study area to model the arrivals from the first refractor. The arrival from the first refractor observed on shot gathers has an apparent velocity of ~1458 m/s (Figure 3), which corresponds to the velocity of the confining unit in this area (Liu et al., 1997). According to Liu et al., (1997), the P-wave velocity (Figure 4) increases abruptly across this layer, giving rise to the first refracted energy observed in Figure 3A. In the Memphis area, the Pliocene strata directly overlies the confining unit (Eocene and Oligocene) and Miocene deposits are missing, indicating that after the deposition of the Jackson formation (the upper stratigraphic element of the confining unit), the area may have undergone significant erosion within the fluvial depositional system (Van Arsdale and TenBrink, 2000), and that erosional features (e.g. paleochannels) might be preserved at the top of the confining unit.

Based on the ~21 m crossover distance observed on shot gathers, 6 reciprocal shots were selected to perform the calculation. Each pair of reciprocal shots was located at both sides of the spread and at the same distance from the center of the spread. First arrival times were manually picked on unprocessed shot gathers for each shot pair and plotted in Figure 5. No data were recorded from one reciprocal shot location to the other (i.e. t_{AB}), and the summation of the reciprocal first arrival times was used to plot the geometry of the first refractor. Figure 6 shows the shape of the refractor calculated from the 6-shot pairs. Note that although the first arrival picks are very scattered (Figure 5), the shape of the refractor obtained from different shot pairs was very consistent, corroborating the robustness of this method. However, since t_{AB} is not available, the absolute depth cannot be calculated. The interpretation of the geometry so obtained (Figure 6) can only be regarded as the general trend or shape of the refractor. The small scale oscillations on Figure 6 should not be interpreted as the detailed structure of the refractor. A comparison between the first arrival picks and the geometry of the refractor (Figure 5 and Figure 6) suggests that these scattered

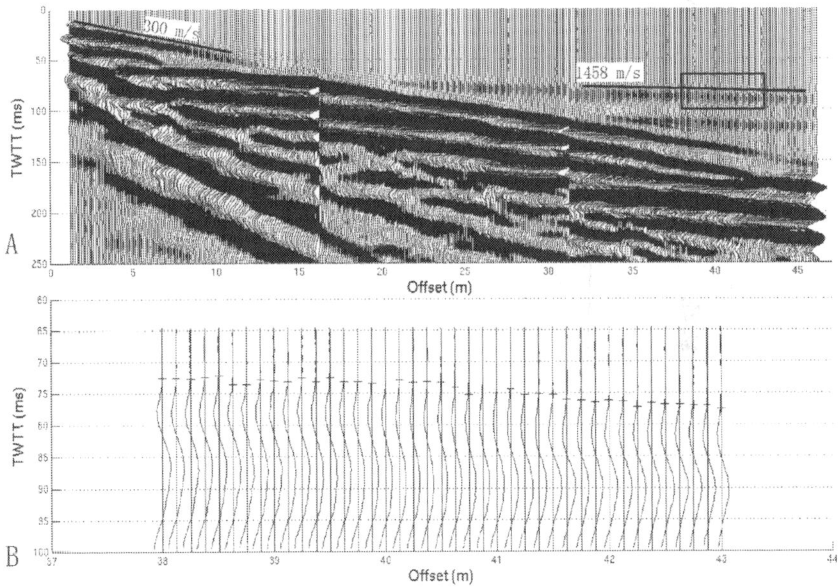

Fig. 3. A: composite shot gather, showing the refracted arrival with an apparent velocity of 1458 m/s, interpreted as the top of the Upper Claiborne clay layer (aquifer); B: close up of the rectangular area in A showing the data quality and the first break picks.

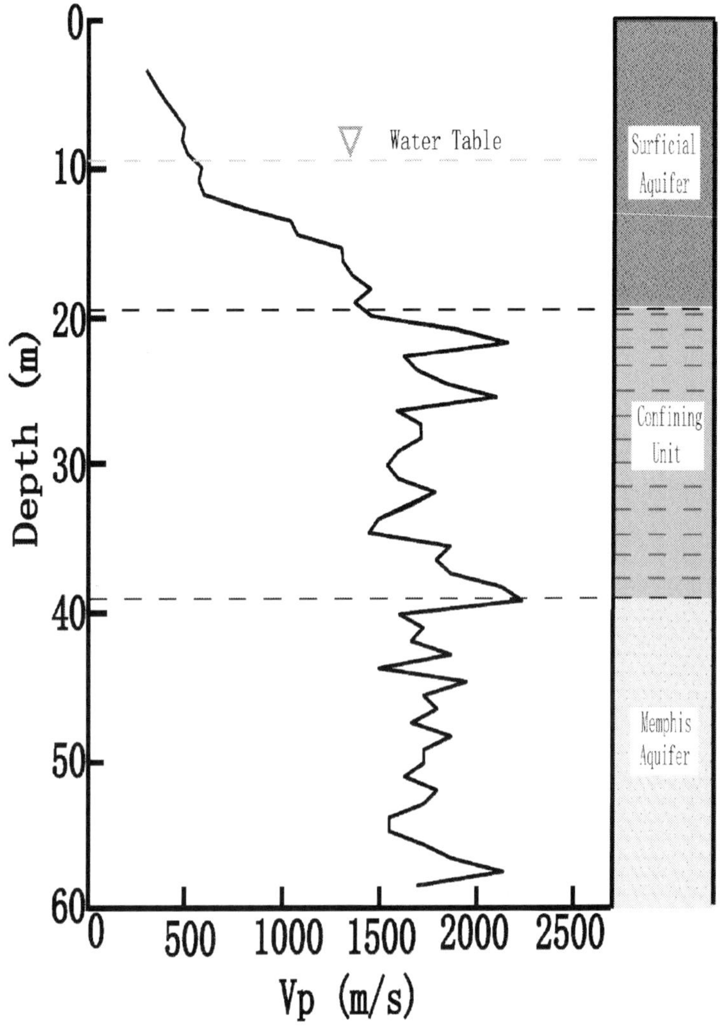

Fig. 4. VSP P-wave velocity profile and hydrological units for a borehole in Shelby County (Liu et al., 1997).

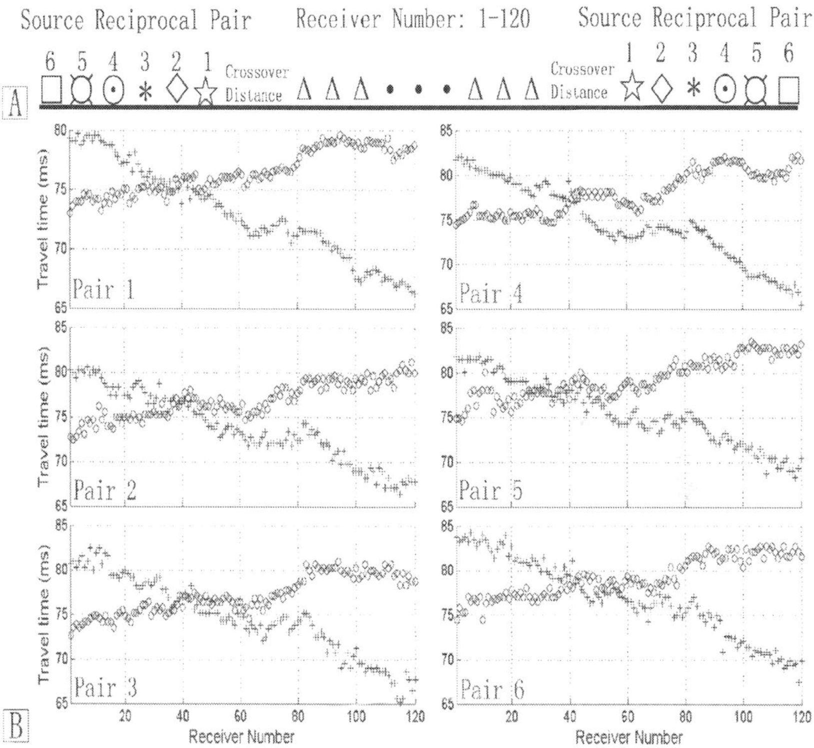

Fig. 5. A: geometry of the 6 reciprocal shot pairs selected for the analysis; B: first arrival picks for the first refracted arrival for the 6 reciprocal shots.

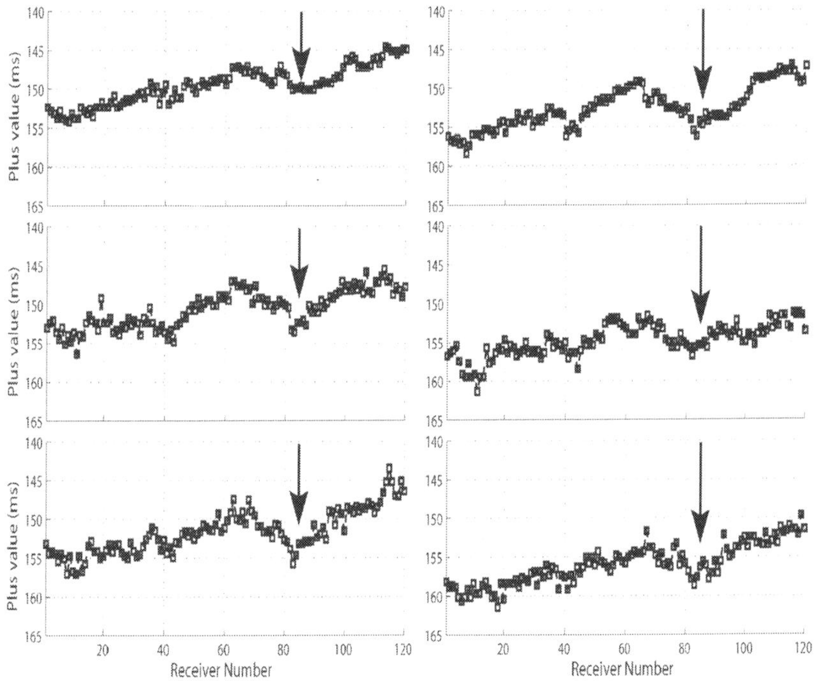

Fig. 6. Geometry of the first refractor resulting from the plus-minus method applied to the 6 reciprocal shots in Figure 5. Arrows show location of the depression.

saw-shape details are likely due to the scattered character of the first arrival picks (see Figure 5). In the field, the 120-receiver spread formed a straight line with a receiver spacing of 0.125 m (Ge et al., 2010, Part I). The clear hyperbolic moveout of the reflection events in Figure 3A indicates that the geophones were planted in the proper positions (all the receivers formed a straight line with a receiver spacing of 0.125 m) in the field. The scattered first break picks from the refracted wave field thus are not due to any inaccurate positioning of geophones but probably resulted from the low S/N ratio (the weaker amplitude of the refracted waves and the relatively higher amplitudes of background noise) and the heterogeneity of the surficial layer. The general trend of the geometry of the first refractor observed in all the reciprocal shots, shows a depression of about 9 ms around receiver number 85 (Figure 6). The velocity of the first layer can be estimated by measuring the slope of the direct wave in Figure 3A, which gives the first layer velocity (v_1) of around 300 m/s. The velocity of the refractor, v_2, is about 1458 m/s (Figure 3A). Using equation 1, k can be calculated to be 153.3, which results in a depth of the depression of about 1.4 m (153.3 * 0.009). The width of the depression visible on the reciprocal shot pairs is ~6 m. The apparent velocity of the first layer was calculated using the offset gather of 1.25 m because the information derived from the plus-minus method is relative to the refractor right below the receiver spread, not elsewhere. The velocity of the surficial layer below the receiver spread was therefore used to estimate the depth of the refractor. The observation that the depression is visible across all of the pairs and occurs at the same place suggests that the result obtained from different shot pairs is reliable and that this method is robust in imaging the geometry of the first refractor. If the pick error is 2 samples (corresponding to 1 ms), the maximum error of the plus value will be 2 ms. By using the same procedure used to calculate the depth of the depression, the corresponding uncertainty is calculated to be about 0.3 m.

Based on the geometry of the first refractor, which corresponds in this area to the top of the confining unit, and considering the fluvial depositional environment that characterized the study area in the Pliocene, the observed V-shaped depression is interpreted as a paleochannel resulting from river erosion likely associated with the Wolf river fluvial system, a branch of Mississippi river and a major river system in the study area.

4. Conclusions

Hagedoorn's plus-minus method was applied to the dataset to map the first refractor, represented by the top of the confining unit. Although the first arrival picks from different pairs of shots are very scattered, the calculated geometry of the top of the aquitard is consistent among the reciprocal shots. This suggests that this method is robust in mapping the structure of the first refractor. The geometry of the mapped first refractor reveals the presence of a depression that is interpreted as a paleochannel, consistently with the fluvial depositional environment and the presence of extensive erosional events that postdate the sedimentation of the Jackson formation.

This study shows that Hagedoorn's plus-minus method can provide a simple and fast tool to interpret refraction data and calculate the geometry and velocity of the first refractor. It has proved to be a cost-effective and efficient geophysical method in hydrology and ground water studies.

5. References

Ge, J.; Magnani, M.; Waldron, B. A., 2010, Imaging a shallow aquitard with seismic reflection data in Memphis, Tennessee, USA. Part II: data analysis, interpretation and traveltime tomography, Near Surface Geophysics, Vol 8, 341-351

Ge, J.; Magnani, M.; Waldron, B. A., 2010, Imaging a shallow aquitard with seismic reflection data in Memphis, Tennessee, USA. Part I: source comparison, walk-away tests and the plus-minus method, Near Surface Geophysics, Vol 8, 331-340.

Hagedoorn, J.G., 1959. The plus-minus method of interpreting seismic refraction section sections: Geophysical Prospecting, 2, 85-127.

Liu Hsi-Ping, Hu, Y., Dorman, J., T. Chang, and Chui, Jer-Ming, 1997, Upper Mississippi embayment shallow seismic velocities measured in situ: Engineering Geology, 46, 313-330.

Overmeeren, V. R. A., 2001, Hagedoorn's plus-minus method: the beauty of simplicity: Geophysical Prospecting, 49, 687-696.

Parks, S. W. and Mirecki, E.J., 1992, Hydrogeology, ground-water quality, and potential for water-supply contamination near the Selby County Landfill in Memphis, Tennessee: U.S.G.S, Water-Resources Investigations Report 91-4173.

Van Arsdale, R. B. and R. K. TenBrink (2000), Late Cretaceous and Cenozoic geology of the New Madrid seismic zone: Bull. Seism. Soc. Am., 90, 345-356.

12

Responses of River Deltas to Sea-Level and Supply Forcing: Autostratigraphic View

T. Muto[1], A.L. Petter[2], R.J. Steel[3], J.B. Swenson[4], A. Tomer[1] and G. Parker[5]

[1]*Nagasaki University*

[2]*St. Anthony Falls Laboratory, University of Minnesota, Minneapolis*

[3]*Department of Geological Sciences, University of Texas at Austin*

[4]*Department of Geological Sciences, University of Minnesota, Duluth*

[5]*Departments of Civil & Environmental Engineering and Geology, University of Illinois, Urbana*

[1]*Japan*

[2,3,4,5]*USA*

1. Introduction

A long-standing geological notion that dates back to Huttonian theory of the late 18th century (Schlager, 1993) suggests that (1) there can exist a balanced state between the effect of relative sea level rise and the effect of sediment supply to the depositional system, for example evidenced by coastal aggradation and a vertical shoreline trajectory, and that (2) regression and transgression reflect imbalances between the two primary drivers: i.e. regression when sediment supply dominates sea level rise, and transgression when sea level rise dominates sediment supply (Fig. 1A). More specifically, it has been taken as axiomatic that given steady external forcing by constant sediment supply (rate Q_S) and constant relative sea level rise (rate R_{slr}), a river delta grows to achieve an equilibrium configuration, produces a particular sediment-stacking pattern and maintains a constant rate of shoreline migration in a particular direction (Weller, 1960; Van Andel & Curray, 1960; Sloss, 1962; Curray, 1964; Swift, 1968; Swift et al., 1971; Curtis, 1970; Vail et al., 1977; Mitchum et al. 1977; Brown & Fisher, 1977; Posamentier et al., 1988; Galloway, 1989; Swift & Thorne, 1991; Shanley & McCabe, 1994; Stanley & Warne, 1994; Myers & Milton, 1996; Neal & Abreu, 2009). We refer to this mode of stratigraphic response as *equilibrium response*, by which steady external forcing results in steady stratigraphic pattern of deposition.

Autostratigraphy, a fairly new arrival in the field of geology, suggests that this presumed mode of stratigraphic response does not hold true in general, but instead that (1) even with steady forcing, river deltas generally fail to sustain a constant and uniform stratigraphic pattern of deposition (Fig. 1B), and (2) unsteady forcing can result in uniform stratigraphic configuration. Exploring such *nonequilibrium responses* (see below) is essential if we are to elucidate the complex stratigraphy that river deltas produce at different time scales. Introducing principles of autostratigraphy and related basic notions, the present chapter outlines these recent discoveries and gives a synthetic understanding of the origin of regression and transgression and of aggradation and degradation in deltaic settings.

2. Deterministic autogenesis

Autogenesis has conventionally been associated with responses that are local (a small part of the system), stochastic and cyclic, such as typically illustrated with river avulsion or delta-lobe switching. There is also another type of autogenesis that is global (i.e. the entire system), deterministic and non-cyclic, as has been noticed recently (Fig. 2). A primary aim of autostratigraphy is to explore the latter and their stratigraphic responses, thereafter to identify allogenic stratigraphic products and responsible unsteady dynamic external forcing. Although stratigraphic records are generally composed of both autogenic and allogenic products, conventional stratigraphy has been apt to ignore the importance of autogenesis and thus to overrate allogenic processes.

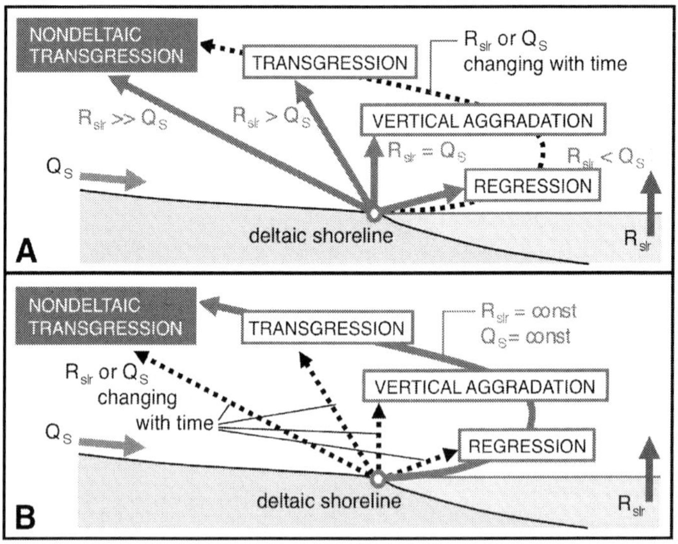

Fig. 1. Two different views of the origin of regression and transgression during relative sea level rise. (A) Conventional geology of river deltas, inherently based on the recognition of equilibrium response, suggests that there exists a balanced state between the effect of relative sea level rise (rate R_{slr}) and the effect of sediment supply (rate Q_S), and that given magnitudes of the two factors, the shoreline migrates at a constant rate in a particular direction. (B) A new view, provided by autostratigraphy, claims that regression, vertical aggradation and deltaic transgression all reflect transient states of a river delta that given enough time must become a nondeltaic transgressive system. Such a shoreline trajectory curve as shown in the diagram has conventionally been attributed to temporal change in R_{slr} or Q_S rather than nonequilibrium response. According to the new viewpoint advocated here, the constant linear shoreline trajectories shown in (A) must be due to unsteady dynamic forcing.

The concept of *deterministic autogenesis* (Muto & Steel, 2002a; formally defined by Paola et al., 2009) has given rise to innovative thinking in regard to the geology of river deltas. This new concept applied to, for example, typical regressive-transgressive/flooding successions such as those shown in Fig. 3A, illustrates that such successions can form solely as autogenic

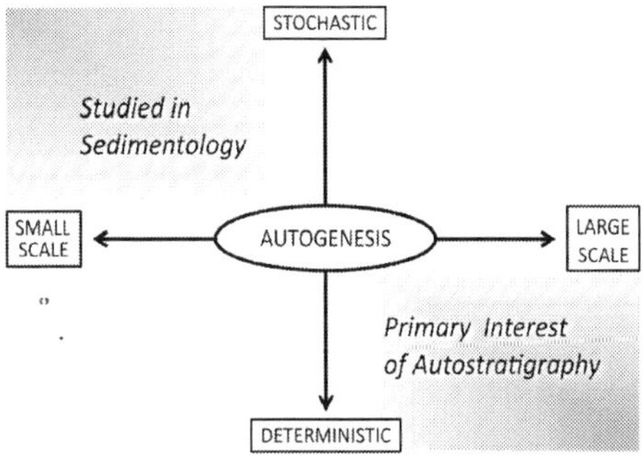

Fig. 2. A conceptual division of the entire field of autogenesis in terms of whether it is on large-scale or small-scale and whether stochastic or deterministic. The primary interest of autostratigraphy is to explore large-scale and deterministic autogenesis, whereas small-scale and stochastic autogenesis has been well studied in conventional sedimentology.

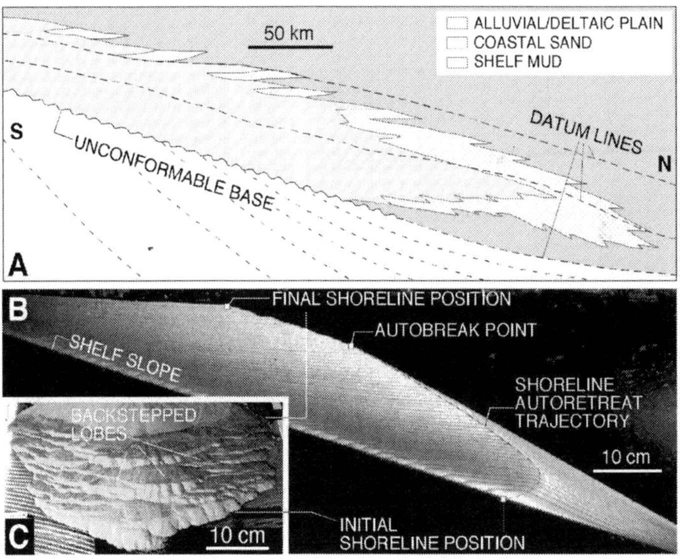

Fig. 3. (A) Schematic N-S cross-section through the Middle Jurassic Brent Delta, northern North Sea, showing an overall regressive-transgressive succession associated with back-stepping delta lobes. Simplified from Graue et al. (1987). (B) Longitudinal profile of a delta that was built during an experimental run conducted with constant rates of sediment supply and sea-level rise. See Muto (2001) for details of the experiments. Note that the stratigraphic architecture of the Brent Delta is similar to a significant degree to that of the experimental delta. (C) Repeated stochastic autogenesis (lobe switching) interacts with longer term deterministic autogenesis to form the details of the shoreline migration pattern reflected in the "shazzam" facies boundaries of (A). From Muto & Steel (2001).

responses to steady rise of relative sea level (e.g. constant subsidence without eustatic fluctuation) in conjunction with steady sediment supply, a result that is reproducible in flume/tank experiments (Fig. 3B; Muto, 2001). Consideration of the various forms of such deterministic autogenic behavior that may be manifested in nature leads us to believe that many existing stratigraphic studies on river deltas require thorough re-examination in light of this new perspective. In the following sections we argue the impact of deterministic autogenesis on the interpretation of stratigraphy, and as a result, strongly suggest that the science of river deltas currently stands at a crossroads for further major advances.

3. External forcing and stratigraphic responses

Conventional understanding of river deltas inherently relies on the assumption that equilibrium response holds true in general, and consequently is apt to favor the interpretation that any large-scale facies break or change in the stratigraphic pattern within a deltaic succession reflects unsteady external forcing such as temporal changes in R_{slr} or Q_S (*allogenic general response*). However, equilibrium response is not the only response to steady forcing, nor even necessarily the expected response. Theoretically, there are two other modes of stratigraphic response in such a cause-and-effect relationship. These are *autogenic nonequilibrium response* (unsteady stratigraphic configuration caused by steady forcing) and *allogenic nonequilibrium response* (steady stratigraphic configuration maintained by unsteady forcing) (Fig. 4). Nonequilibrium responses essentially arise from downstream transformation of the sediment-supply signal from constant to variable due to systematic deposition and erosion along the path of transport. Unfortunately, stratigraphic interpretation of equilibrium response can often be flawed due to a failure to appropriately consider nonequilibrium responses.

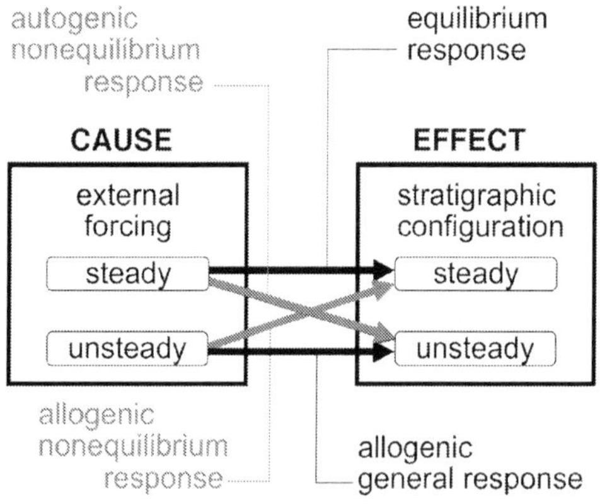

Fig. 4. Stratigraphic response of a depositional system to external forcing. From the viewpoint of a cause-and-effect relationship, we can imagine four different modes of stratigraphic response: *equilibrium response* (steady stratigraphic configuration by steady forcing), *autogenic nonequilibrium response* (unsteady stratigraphic configuration by steady forcing), *allogenic nonequilibrium response* (steady stratigraphic configuration by unsteady forcing), and *allogenic general response* (unsteady stratigraphic configuration by unsteady forcing). The importance of nonequilibrium responses has only recently become widely recognized in the geological community.

New insight into the problem has been obtained via experimental research on stratigraphy in general (Paola et al., 2009), as well as numerical (Muto & Steel, 1992; Milton & Bertram, 1995; Ritchie et al., 1999; Swenson et al., 2000; Parker et al., 2008a) and experimental research (Paola, 2001; Muto, 2001; Muto & Steel, 2001, 2004; Kim et al., 2009) that specifically addresses the question of how river deltas react to steady and unsteady forcing. This work, when combined with site-specific field application (Muto & Steel, 2002a; Parker et al., 2008b), provides a view of the problem that differs rather markedly from the conventional model in three ways. First, although equilibrium response is possible, it is restricted to very specific conditions which are rare in natural systems (Muto & Swenson, 2006). Second, steady sea-level forcing is much more likely to generate autogenic nonequilibrium response than equilibrium response (Swenson & Muto, 2007). Third, deltaic systems can have different stratigraphic responses to the same external forcing depending on geomorphic conditions (Petter & Muto, 2008).

A long-standing geological notion suggests that steady external forcing results in a steady stratigraphic pattern of deposition (equilibrium response; Fig. 4), and that this mode of stratigraphic response is true in general. Prior to the recognition of nonequilibrium responses, any large-scale unsteady stratigraphic features were attributed to unsteady external forcing (i.e. allogenic general response). Autostratigraphy suggests that unsteady stratigraphic configuration can be caused by steady forcing (autogenic nonequilibrium response) and steady stratigraphic configuration can be maintained by unsteady forcing (allogenic nonequilibrium responses).

4. Regression and transgression as nonequilibrium responses

The nonequilibrium view of river deltas, along with the idea of deterministic autogenesis, has led to the following understanding of regression and transgression, two of the basic building blocks of stratigraphy. With constant R_{slr} (>0) and constant Q_S (>0), it is inevitable that a river delta initially experiencing regressive growth must eventually turn around into a transgressive mode, which is referred to as *autoretreat* (Muto & Steel, 1992; Swenson et al., 2000). After the onset of shoreline transgression, the subaqueous slope of the delta (foreset) may continue to accrete for some time. As sea level rise continues, however, the delta inevitably meets a critical event (*autobreak*; Fig. 3B) in which sediment supply to the delta front eventually drops to zero, the delta foreset is abandoned, and the shoreline undergoes rapid transgression by drowning (Muto, 2001; Parker et al., 2008a). After this time, the depositional system is no longer deltaic because sediment is not delivered beyond the shoreline, but has instead become an estuary (*sensu* Darlymple, 1992). In fact, the stratigraphic record is full of flooding events similar to those that arise in response to autobreak (e.g. Fig. 3A). These flooding events, usually defining parasequences in sequence stratigraphy, need not be due to eustatic fluctuation but rather may arise naturally either from deterministic autogenesis in response to steady subsidence (R_{slr} = const) or stochastic autogenesis (e.g. channel avulsion), or a combination of the two (Muto & Steel, 2001; Fig. 3C).

The primary causes for this nonequilibrium response to steady sea level rise are (1) progressive expansion of the river delta both basinwards and laterally due to continuing sediment supply, but with increasing tendency for the sediment to deposit landward of the shoreline, and (2) continuing rise in relative sea level. Suppose that R_{slr} and Q_S are kept constant with time. Cumulative sea level elevation increases with time, whereas the aggradation rate of the delta averaged over the entire surface area progressively decreases as t^{-n}, where t denotes time and n can vary between 2 and 3. Because of this behavior, a prograding delta is intrinsically unable to sustain a constant response to steady sea level rise. For the depositional system to maintain its original progradation as a delta, it would be necessary for Q_S or R_{slr} to change in a specific manner with time. This, by definition, would lead to an allogenic nonequilibrium response. Any river delta subjected to steady sea level rise cannot therefore sustain a particular depositional style indefinitely, but will inevitably experience a nonequilibrium response. If sea level rise continues for a sufficiently long duration, the river shoreline may become nondeltaic, for example the upstream end of an estuary or drowned valley. The magnitude of R_{slr} relative to Q_S does play an important role, as these parameters can be used to characterize intrinsic length and time scales of the river delta such that the nonequilibrium response is delayed or hastened. Under conditions of the same constant sea level rise, a small depositional system fed with low Q_S will experience transgression and become an estuary in a shorter time, whereas a larger system fed with high Q_S will maintain a regressive delta behavior for a longer period of time before transgression and drowning (Parker et al., 2008b). Since both of the afore-mentioned systems can be coeval, there is thus little basis for correlating a particular deltaic stratigraphic pattern to a particular segment of a sinusoidal curve of sea level change. For example, the Sabine and Trinity Rivers became nondeltaic (estuarine) during the Postglacial sea-level rise, while less than 100 km to the west, the Colorado and Brazos Rivers deposited a succession of backstepping delta lobes during the same period (Anderson et al., 1996).

The autogenic nonequilibrium response of a delta displays variation depending upon the initial downstream length of their feeder alluvial river(s) (as measured from e.g. a bedrock-alluvial transition point). There exists a critical magnitude of alluvial length (L_{crt}) for which, given R_{slr}, Q_S is precisely as large as required to maintain aggradation over the entire length of the existing alluvial reach of the river (Tomer et al., 2011). In case a pre-existing alluvial length exceeds L_{crt}, the shoreline abruptly migrates landward at the onset of sea level rise as an estuary rather than a delta. This is because under such conditions Q_S is no longer sufficient to cover the entire length of the existing alluvial river, and thus no river sediment reaches the shoreline (i.e., substantially the same as autobreak). Such nondeltaic transgression is expected to proceed very rapidly initially, but subsequently decelerate as the alluvial length approaches L_{crt}. Even though R_{slr} and Q_S are held constant, the shoreline inevitably follows a concave-upward trajectory as a manifestation of the nonequilibrium response. Fig. 5 shows shoreline trajectories estimated with the autoretreat-autobreak model (Muto, 2001) for five natural rivers (Fly, Mekong, Mississippi, Brahmaputra and Ebro) during Postglacial sea level rise, on the assumption that (1) prior to sea level rise, they had extended to the present shelf edge or thereabouts and built deltas there, (2) Q_S was constant but different for each river, (3) R_{slr} was significantly decreased, or became zero, around 8–6 kaBP, and (4) the shoreline passed through reference points that are specified based on separate evidence from published literature. The simulation suggests that every one of the five alluvial systems became nondeltaic and transgressive as soon as the sea level began to rise. This is because each of the Glacial lowstand river systems built such an alluvial reach with length that far exceeded L_{crt}, prior to sea level rise. In the case of the Fly River during Postglacial sea level rise, for example, L_{crt} is estimated to have been 24 km. Nevertheless the river had extended over 900 km to the shelf edge before sea level started to rise. Evidence for intense transgression associated with this "overextension" can be found in the modern Fly system, which appears to be in a recovery process of deltaic sedimentation starting from when sea level rise decelerated (Parker et al., 2008b). Each modeled system possesses a unique shoreline trajectory despite similar relative sea-level histories.

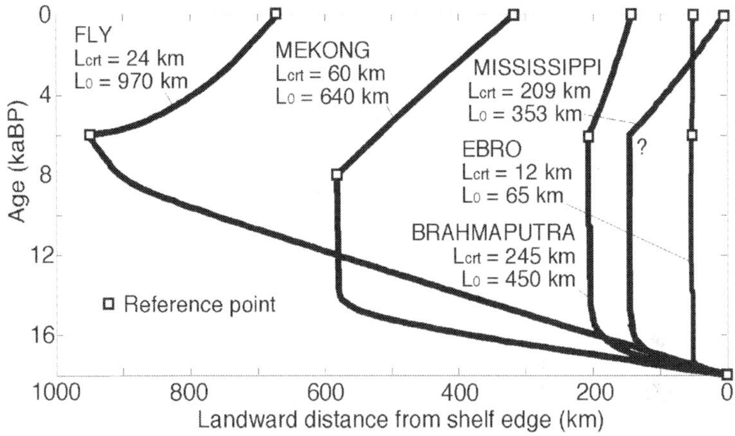

Fig. 5. Shoreline trajectories estimated via numerical simulation based on the autoretreat-autobreak geometrical model of Muto (2001). Relevant assumptions are as follows: (1) prior to sea level rise, the rivers had extended to present shelf edge positions or thereabouts and built deltas there, (2) Q_S was constant but unique to each river, (3) R_{slr} was also unique to each river, but significantly decelerated around 8-6 kaBP in each case, according to the characteristics of that case, and (4) the shoreline migrated via reference points that can be specified with separate evidence. The reference points and/or related data were adopted from information in Parker et al. (2008b) for the Fly River; Coleman et al. (1998), Harmar & Clifford (2007) for the Mississippi River; Tamura et al. (2009), Liu et al. (2009), Xue et al. (2010) for the Mekong River; Somoza et al. (1998), Rovira & Ibanez (2007) for the Ebro River; and Goodbred & Kuehl (2000), Goodbed et al. (2003), Mikhailov and Dotsenko (2006), Liu et al. (2009) for the Ganges-Brahmaputra River system. Note that the L_{crt} and L_0 values are not related to the scale at the distance from shelf edge.

5. Aggradation, degradation and grade during falling sea level

Aggradation and degradation of river deltas with falling sea level is another fundamental issue comparable to the question of regression and transgression with rising sea level. It is well documented that both aggradation and degradation of river deltas can take place during sea-level fall (Schumm, 1993; Blum & Törnqvist, 2000; Van Heijst & Postma, 2001; Browne & Naish, 2003; Strong & Paola, 2008). However, the rationale for this apparent complexity of behavior remains partially obscure. Recent physical experiments (Muto & Steel, 2004; Swenson & Muto, 2007; Petter & Muto, 2008) suggest that nonequilibrium response can account for some of this behavior in a straightforward way.

An understanding of the stratigraphic response of river deltas to falling sea level requires a clarification of the concept of *grade*, the state of a river at which neither net deposition nor net erosion take place in spite of continuing sediment supply. Grade therefore precisely defines the critical condition discriminating between aggradational and degradational river systems. This concept, originally advocated by G. K. Gilbert in the late 19th century, is often presented as the consequence of long-term equilibrium response of a river system subject to stationary sea level. Common beliefs based on equilibrium response (Thorne & Swift, 1991; Holbrook et al., 2006) are that (1) alluvial rivers in deltaic settings aggrade in response to sea-level rise and degrade in response to sea-level fall, (2) as long as sea level remains stationary, the rivers eventually become graded, and thus (3) grade represents the equilibrium configuration of an alluvial river under conditions of stationary sea level.

Such conceptual models of graded rivers downplay the fate of the sediment bypassed through the "graded" reach, and in particular, how its sequestration in the deltaic environment affects the dynamics of the attached river system. However, if sea level remains stationary, rivers continue to aggrade in response to delta progradation, and consequently never attain grade. Model experiments to examine the dynamics of the downstream and upstream boundaries of alluvial rivers building deltas have shown that alluvial grade is physically possible only under rather specific conditions pertaining solely to sea-level fall (Jordan & Flemings, 1991; Nummedal et al., 1993; Leeder & Stewart, 1996; Muto & Swenson, 2005). Alluvial grade arises in two distinct ways depending on geomorphic conditions and characteristics of sea-level fall, for which alluvial slope S_a and basin slope S_b are particularly influential (Figs. 6, 7). Where $S_a < S_b$, alluvial grade is attained and sustained by allogenic nonequilibrium response through a particular style of decelerating sea-level fall (Muto & Swenson, 2005). If sea level instead falls at a constant rate in this geomorphic setting, the river aggrades at an early stage but later degrades by autogenic nonequilibrium response (Swenson & Muto, 2007). Where $S_a = S_b$, alluvial grade is attained by equilibrium response at any constant rate of sea level fall (Muto & Swenson, 2006). Where $S_a > S_b$, grade is never attained, and the alluvial system simply continues to aggrade during sea level fall and the alluvial river finally detaches from the receding shoreline, so that the depositional system becomes nondeltaic via autogenic nonequilibrium response (Petter & Muto, 2008). Thus, aggradational river deltas tend to undergo autogenic nonequilibrium response to constant sea-level fall whereby they eventually become nondeltaic ($S_a > S_b$) or degradational ($S_a < S_b$). Thus, rivers building deltas, in general, cannot maintain a particular growth style for prolonged periods of time, during either sea-level rise or fall, and the manner in which sediment is distributed across a basin depends heavily upon the geomorphic conditions of the alluvial river and basin (Figs. 6, 7).

6. Timescales

Most present-day large deltas have existed through the past 50–60 Ma. However, during this time they have continually evolved and changed at much shorter time scales (i.e. the autogenic focus of the present argument). Whereas stochastic autogenic responses in river-delta systems (not discussed in this work) are commonplace at very short timescales (0.1-1 ka), deterministic autogenic responses require a longer time and minimum basin length. Deterministic autogenic responses involving cross-shelf regression and wholesale retreat of deltaic complexes have been shown to operate at shelf-transit time scales of 50-200 ka (Burgess & Hovius, 1998; Muto & Steel, 2002b; Carvajal & Steel, 2006; Steel et al., 2008). Quaternary eustatic sea-level curves show 10–20 m amplitude changes at ka-cyclicity over interglacial-to-glacial eustatic fall intervals of 10–100 ka duration (e.g. Stages 4–2; Lambeck et al., 2002), though Holocene sea-level rise was relatively steady over a period as long as 15 ka. Past greenhouse climate conditions would likely have yielded more prolonged periods of sea-level stability. Short allogenic cycles do not give river deltas sufficient time to adjust to the changes in boundary conditions, and therefore cannot be expected to significantly change the autogenic response of a system. Likewise, low-magnitude allogenic forcing may not alter the boundary conditions sufficiently to deter autogenic response. The following questions therefore remain outstanding: 1) do allogenic boundary conditions remain stable for long enough periods to allow deterministic autogenic responses to run their course; and 2) what are the threshold amplitudes and frequencies of perturbations in allogenic boundary conditions of river deltas that are sufficient to interrupt these responses? These questions have implications for paleo-environmental interpretation of the stratigraphic record as well as for predicting the long-term fate of modern river deltas under the effect of climatic change and human impact (Ericson et al., 2006; Kim et al., 2009; Syvitski et al., 2009).

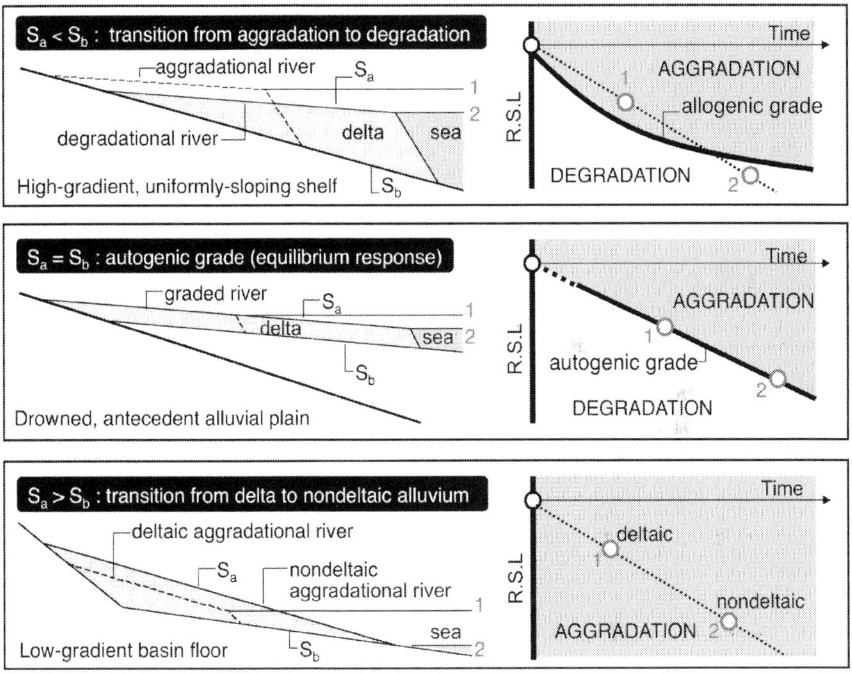

Fig. 6. Autostratigraphic view of alluvial aggradation and degradation during sea level fall. Alluvial grade is physically possible but can be attained and sustained only during sea level fall unless the river delta has a fixed downstream boundary. Patterns of sea level fall that allow the attainment of grade depend on geomorphic conditions of the deltaic system (alluvial slope S_a and basin slope S_b, particularly). Where $S_a < S_b$, alluvial grade can be attained and sustained only with sea level fall of a particular decelerative pattern. If sea level drops at a constant rate in this geomorphic condition, the feeder alluvial system aggrades in the early stage, but with enough time inevitably becomes degradational. Where $S_a = S_b$, a river delta steadily progrades and sustains grade autogenically only during constant sea level fall. Where $S_a > S_b$, the feeder alluvial system never attains grade, but instead continues to aggrade, so evolving from a deltaic system to a nondeltaic system as long as sea level continues to fall.

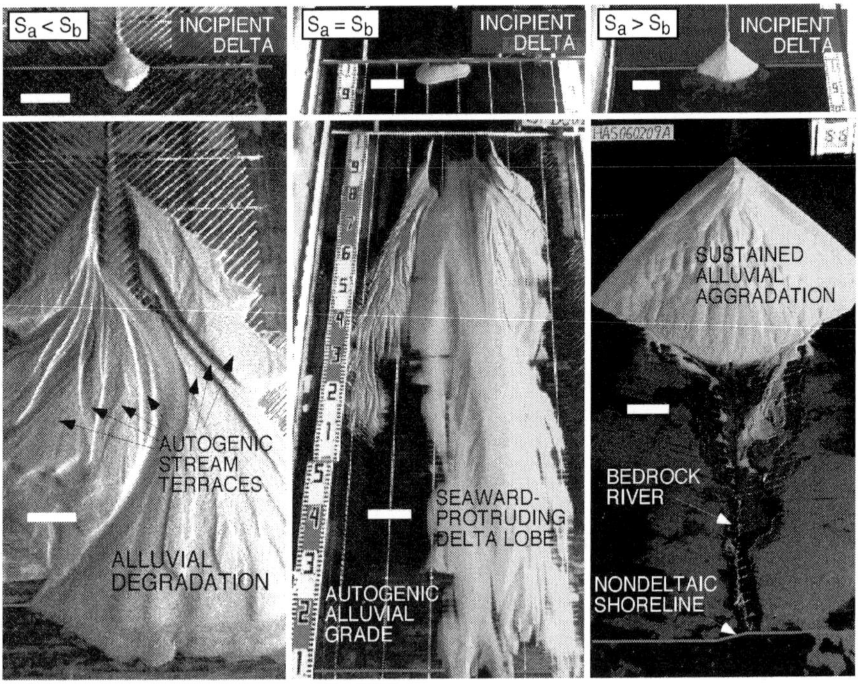

Fig. 7. Experimental illustration of the three types of autogenic nonequilibrium response to constant sea level fall depending upon geomorphic conditions. The upper and lower images represent states 1 and 2, respectively, of the same geomorphic conditions in Fig. 6. Note that different river/basement slope relationships can give rise to different patterns of response of the systems to constant sea-level fall. Photos adopted from Muto & Steel (2004) and Petter & Muto (2008).

7. How can autogenic and allogenic response be distinguished in the stratigraphic record?

Long-term stratigraphy encapsulates the composite signal of both autogenic and allogenic responses. However, the intrinsic nature of autogenic responses makes them a constant and predictable signal in the stratigraphic record, and therefore, allogenic responses should be interpreted only after the autogenic framework has been established. How should this be done?

The dependence of nonequilibrium response upon the shape of the basin and depositional surface is such that it can be easily simulated using geometric models (Paola, 2000; Muto et al., 2007; Petter et al., 2011). This requires input concerning basement and fluviodeltaic gradients, as well as rates of sea-level change and sediment supply. Comparison of the modeled shoreline trajectory with observed trajectories allows the identification of deviations from autogenic nonequilibrium response (Muto & Steel, 2002a; Petter et al., 2010; Wolinsky et al., 2011). Trajectories resulting from this response are recognized as smooth, concave-up or concave-down curves indicative of decelerating progradation as the system expands through time. At each point of deviation from the predicted trajectory, a general allogenic response is interpreted, and the geometric model can be reset to new boundary conditions at this point. Successive breaks in boundary conditions due to paleo-environmental changes are thus reconstructed by repeated application of this procedure. The parameters required for modeling are readily interpreted from regional geologic datasets (Petter et al., 2011).

8. River deltas and their stratigraphy

River deltas constitute the single most important agent delivering clastic sediment from land to sea. They drive the sedimentary growth of continental margins and fill basins of various types. The recognition of nonequilibrium responses in the development of coastal stratigraphy has given rise to a new framework of genetic stratigraphy, autostratigraphy (Muto et al., 2007), that encompasses both equilibrium and nonequilibrium responses, and takes full account of both steady and unsteady external forcing. Autostratigraphic responses in river deltas tend to prevent any prolonged continuity of particular growth styles, whether during rising or falling sea level forcing. It has also become increasingly clear that nonequilibrium response plays a key role in the variety of observed shoreline stacking patterns (Kim et al., 2006). Consequently, changes in R_{slr} or Q_S condition need not be interpreted based on the presence of certain stacking patterns, and said stacking patterns do not necessarily predict subsequent stacking patterns (e.g. Neal & Abreu, 2009) since they are not manifestations of allogenic response. The stratigraphic record of river deltas therefore reflects the extent to which nonequilibrium behavior proceeds between periodic changes in boundary conditions caused by external forcing (i.e. tectonic, climatic, or eustatic events). Stratigraphic interpretation of coastal plain and shallow-marine strata should be conducted with an acute awareness of the intrinsic intermittent character of river-delta growth style.

9. Conclusion

Recent developments in experimental stratigraphy and geomorphology have cast doubt on a long-standing principal theorem in geology, i.e. that given steady external forcing by constant sediment supply and constant relative sea level change, a river delta grows to achieve an equilibrium configuration and produces a particular sediment-stacking pattern. A new, alternative view that is provided by autostratigraphy tells that (1) even with steady forcing, river deltas generally fail to sustain a constant and uniform stratigraphic pattern of deposition due to their inherent deterministic autogenesis, and (2) unsteady forcing can result in uniform stratigraphic configuration. Exploring such nonequilibrium response is essential if we are to elucidate the complex stratigraphy that river deltas produce at different time scales. This ongoing change in how we view river deltas and their stratal products brings a whole new understanding of the origin of regression and transgression and of aggradation and degradation in deltaic settings.

10. Acknowledgment

The present work was financially supported in part by a 2008-2010 Japanese Grant-in-Aid for Scientific Research (B2034140) and RioMAR Industry Consortium at University of Texas in Austin.

11. References

Anderson, J.B.; Abdulah, K., Sarzalejo, S., Siringan, F. & Thomas, M.A. (1996). Late Quaternary sedimentation and high-resolution sequence stratigraphy of the east Texas shelf, In: *Geology of Siliciclastic Shelf Seas: Geological Society of London Special Publication* 117 (January 1996), M. De Batist & P. Jacobs, (Eds.), 95-124, ISBN 9781897799710

Blum, M.D. & Törnqvist, T. (2000). Fluvial response to climate and sea-level change: review and look forward. *Sedimentology*, Vol.47, Supplement 1 (February 2000), pp. 2-48

Brown, L.F., Jr. & Fisher, W. (1977). Seismic-stratigraphic interpretation of depositional systems: Examples from Brazilian rift and pull-apart basins. In: *Seismic stratigraphy: applications to hydrocarbon exploration: American Association of Petroleum Geologists Memoir* 26 (June 1977), C.E. Payton, (Ed.), 213-248, ISBN 978-0891813026

Browne, G.H. & Naish, T.R. (2003). Facies development and sequence architecture of a late Quaternary fluvial-marine transition, Canterbury Plains and shelf, New Zealand: implications for forced regressive deposits. *Sedimentary Geology*, Vol.158, No.1-2 (May 2003), pp. 57-86, ISSN 00370738

Burgess, P.M. & Hovius, N. (1998). Rates of delta progradation during highstands: consequences for timing of deposition in deep-marine systems. *Journal of Geological Society London*, Vol.155, No.2 (April 1998), pp. 217-222, ISSN 0016-7649

Carvajal, C. & Steel, R.J. (2006). Thick turbidites successions from supply-dominated shelves during sea-level highstand. *Geology*, Vol.34, No.8 (August 2006), pp. 665-668, ISSN 0091-7613

Coleman, J.M.; Roberts, H.H. & Stone, G.W. (1998). Mississippi River delta: an overview. *Journal of Coastal Research*, Vol.14, No.3 (Summer 1998), pp. 698-716, ISSN 0749-0208

Curray, J.R. (1964). Transgressions and regressions. In: *Papers in marine geology*, R.L. Miller, (Ed.), 175-203, Macmillan, New York

Curtis, D.M. (1970). Miocene deltaic sedimentation, Louisiana Gulf Coast. In: *Deltaic sedimentation, modern and ancient: Society of Economic Paleontologists and Mineralogists Special Publication*, J.P. Morgan, (Ed.), 293-308

Dalrymple, R.W.; Zaitlin, B.A. & Boyd, R. (1992). Estuarine facies models: conceptual basis and stratigraphic implications. *Journal of Sedimentary Petrology*, Vol.62, No.6 (November 1992), pp. 1130-1146, ISSN 1527-1404

Ericson, J.P.; Vorosmarty, C.J., Dingman, S.L., Ward, L.G. & Meybeck, M. (2006). Effective sea-level rise and deltas: Causes of change and human dimension implications. *Global and Planetary Change*, Vol.50, No.1-2 (February 2006), pp. 63-82, ISSN 092118181

Galloway, W.E. (1989). Genetic stratigraphic sequences in basin analysis—part I: architectures and genesis of flooding-surface bounded depositional units. *American Association of Petroleum Geologists Bulletin*, Vol.73, No.2, pp. 125-142, ISSN 0149-1423

Goodbred, S.L., Jr. & Kuehl, S.A. (2000). The significance of large sediment supply, active tectonism, and eustasy on margin sequence development: Late Quaternary stratigraphy and evolution of the Ganges-Brahmaputra delta. *Sedimentary Geology*, Vol.133 (June 2000), pp. 227-248, ISSN 0037-0738

Goodbred, S.L., Jr.; Kuehl, S.A., Steckler, M.S. & Sarker, M.H. (2003). Controls on facies distribution and stratigraphic preservation in the Ganges-Brahmaputra delta sequence. *Sedimentary Geology*, Vol.155 (February 2003), pp. 301-316, ISSN 0037-0738

Graue, E.; Helland-Hansen, W., Johnsen, J., Lomo, L., Nottvedt, A., Ronning, K., Ryseth, A. & Steel, R. (1987). Advance and retreat of Brent Delta system, Norwegian North Sea. In: *Petroleum Geology of North West Europe*, J. Brooks & K. Glennie, (Eds.), 915-937, Graham & Trotman, London

Harmar, O.P. & Clifford, N.J. (2007). Geomorphological explanation of the long profile of the Lower Mississippi River. *Geomorphology*, Vol.84, No.3-4 (February 2007), pp. 222-240, ISSN 0169-555X

Holbrook, J.; Scott, R.W. & Oboh-Ikuenobe, F.E. (2006). Base-level buffers and buttresses: A model for upstream versus downstream control on fluvial geometry and architecture within sequences. *Journal of Sedimentary Research*, Vol.76, No.1 (January 2006), pp. 162-174

Jordan, T.E. & Flemings, P.B. (1991). Large-scale stratigraphic architecture, eustatic variation, and unsteady tectonism: A theoretical evaluation. *Journal of Geophysical Research*, Vol.96, pp. 6681-6699.

Kim, W.; Dai, A., Muto, T. & Parker, G. (2009). Delta progradation driven by an advancing sediment source: Coupled theory and experiment describing the evolution of elongated deltas. *Water Resources Research*, Vol.45, W06428, doi:10.1029/2008WR007382

Kim, W.; Mohrig, D., Twilley, R., Paola, C. & Parker, G. (2009). Is it feasible to build new land in the Mississippi River Delta? *EOS Transaction*, Vol.90, No.42 (October 2009), pp. 373-374

Kim, W.; Paola, C., Swenson, J.B. & Voller, V.R. (2006). Shoreline response to autogenic processes of sediment storage and release in the fluvial system. *Journal of Geophysical Research*, Vol.111, F04013, doi:10.1029/2006JF000470

Lambeck, K.; Yokoyama, Y. & Purcell, T. (2002). Into and out of the Last Glacial Maximum: sea-level change during Oxygen Isotope Stages 3 and 2. *Quaternary Science Review*, Vol.21, No.1-3 (January 2002), pp. 343-360

Leeder, M.R. & Stewart, M.D. (1996). Fluvial incision and sequence stratigraphy: Alluvial responses to relative sea-level fall and their detection in the geological record. *Geological Society Special Publication* 103, pp. 25-39

Liu, J.P.; Xue, Z., Ross, K., Wang, H.J., Yang, Z.S., Li, A.C. & Gao, S. (2009). Fate of sediments delivered to the sea by Asian large rivers: Long-distance transport and formation of remote alongshore clinohems. *The Sedimentary Record*, Vol.7, No.4, pp. 4-9

Mikhailov, V.N. & Dotsenko, M.A., (2006), Peculiarities of the hydrological regime of the Ganges and Brahmaputra river mouth area. *Water Resources*, Vol.33, No.4 (July 2006), pp. 353-373, ISSN 0097-8078

Milton, N.J. & Bertram, G.T. (1995). Topset play types and their controls. *American Association of Petroleum Geologists Memoir* 64, pp. 1-9

Mitchum, R.M., Jr.; Vail, P.R. & Sangree, J.B. (1977). Seismic stratigraphy and global changes of sea level, Part 6: Stratigraphic interpretation of seismic reflection patterns in depositional sequences. In: *Seismic stratigraphy: applications to hydrocarbon exploration: American Association of Petroleum Geologists Memoir* 26 (June 1977), C.E. Payton, (Ed.), 117-133, ISBN 978-0891813026

Muto, T. (2001). Shoreline autoretreat substantiated in flume experiments. *Journal of Sedimentary Research,* Vol.71, No.2 (March 2001), pp. 246–254

Muto, T. & Steel, R.J. (1992). Retreat of the front in a prograding delta. *Geology,* Vol.20, No.11 (November 1991), pp. 967–970, ISSN 0091-7613

Muto, T., & Steel, R.J. (2001). Autostepping during the transgressive growth of deltas: Results from flume experiments. *Geology,* Vol.29, No.9 (September 2001), pp. 771–774, ISSN 0091-7613

Muto, T. & Steel, R.J. (2002a). Role of autoretreat and A/S changes in the understanding of deltaic shoreline trajectory: a semi-quantitative approach. *Basin Research,* Vol.14, No.3 (September 2002), pp. 303–318, ISSN 1365-2117

Muto, T. & Steel, R.J. (2002b). In defense of shelf-edge delta development during falling and lowstand of relative sea level. *Journal of Geology,* Vol.110, No.4 (July 2002), pp. 421–436, ISSN 0022-1376

Muto, T. & Steel, R.J. (2004). Autogenic response of fluvial deltas to steady sea-level fall: Implications from flume-tank experiments. *Geology,* Vol.32, No.5 (May 2004), pp. 401–404, ISSN 0091-7613

Muto, T. & Swenson, J.B. (2005). Large-scale fluvial grade as a non-equilibrium state in linked depositional systems: Theory and experiment. *Journal of Geophysical Research,* Vol.110 (F), F03002 (July 2005), DOI: 10.1029/2005JF000284

Muto, T. & Swenson, J.B. (2006). Autogenic attainment of large-scale alluvial grade with steady sea level fall: An analog tank/flume experiment. *Geology,* Vol.34, No.3 (March 2006), pp. 161–164, ISSN 0091-7613

Muto, T.; Steel, R.J. & Swenson, J.B. (2007). Autostratigraphy: a framework norm for genetic stratigraphy. *Journal of Sedimentary Research,* Vol.77, No.1 (January 2007), pp. 2–12, ISSN 1527-1404

Myers, K.J. & Milton, N.J. (1996). Concepts and Principles, In: *Sequence Stratigraphy,* D. Emery & K.J. Myers, (Eds.), (October 1996) 11-41, Blackwell, ISBN 0-632-03706-7, Oxford

Neal, J. & Abreu, V. (2009). Sequence stratigraphy hierarchy and the accommodation succession method. *Geology,* Vol.37, No.9 (September 2009), pp. 779–782, ISSN 0091-7613

Nummedal, D.; Riley, G.W. & Templet, P.L. (1993). High-resolution sequence architecture: A chronostratigraphic model based on equilibrium profile studies. In: *Sequence Stratigraphy and Facies Associations, International Association of Sedimentologists Special Publication* 18 (December 1993), H.W. Posamentier, C.P. Summerhayes, B.U. Haq & G.P. Allen, (Eds.), pp. 55- 68, ISBN 9780632035489

Paola, C. (2000). Quantitative models of sedimentary basin filling. *Sedimentology,* Vol.47, Supplement 1 (February 2000), pp. 121–178

Paola, C. (2001). Experimental stratigraphy. *GSA Today,* Vol.11, No.7 (July 2001), pp. 4–9.

Paola, C.; Straub, K., Mohrig, D. & Reinhardt, L. (2009). The "unreasonable effectiveness" of stratigraphic and geomorphic experiments. *Earth Science Review,* Vol.97, No.1-4 (December 2009), p.1-43, doi:10.1016/j earscirev.2009.05.003.

Parker, G.; Muto, T., Akamatsu, Y., Dietrich, W. & Lauer, J.W. (2008a) Unraveling the conundrum of river response to rising sea level from laboratory to field. Part I. Laboratory experiments. *Sedimentology,* Vol.55, No.6 (December 2008), pp. 1643–1655

Parker, G.; Muto, T., Akamatsu, Y., Dietrich, W. & Lauer, J.W. (2008b). Unraveling the conundrum of river response to rising sea level from laboratory to field. Part II. The Fly-Strickland River system, Papua New Guinea. *Sedimentology*, Vol.55, No.6 (December 2008), pp. 1657-1686

Petter, A.L. & Muto, T. (2008). Sustained alluvial aggradation and autogenic detachment from shoreline in response to steady fall of relative sea level. *Journal of Sedimentary Research*, Vol.78, No.2 (February 2008), pp. 98-111

Petter, A.; Kim, W., Muto, T. & Steel, R. (2011). Comment on "Clinoform quantification for assessing the effects of external forcing on continental margin development". *Basin Research*, Vol.23, No.1 (February 2011), pp. 118-121, ISSN 1365-2117

Posamentier, H.W.; Jervey, M.T. & Vail, P.R. (1988). Eustatic controls on clastic deposition I—Conceptual framework. In: Sea-level changes: An integrated approach: SEPM Special Publicatiôn 42 (April 1989), C.K. Wilgus, B.S. Hastings, C.G.St.C. Kendall, H.W. Posamentier, C.A. Ross & J.C. Van Wagoner, (Eds.), 125-154, ISBN 0918985749

Ritchie, B.D.; Hardy, S. & Gawthorpe, R.L. (1999). Three-dimensional numerical modeling of coarse grained clastic deposition in sedimentary basins. *Journal of Geophysical Research*, Vol.108, pp. 17759-17780

Rovira, A. & Ibanez, C. (2007). Sediment management options for the Lower Ebro River and its delta. *Journal of Soils and Sediments*, Vol.7, No.5 (November 2007), pp. 285-295, ISSN 1439-0108

Schlager, W. (1993). Accommodation and supply—a dual control on stratigraphic sequences. *Sedimentary Geology*, Vol.86, No.1-2 (July 1993), pp. 111-136, ISSN 0037-0738

Schumm, S.A. (1993). River response to base level change: implications for sequence stratigraphy. *Journal of Geology*, Vol.101, No.2 (March 1993), pp. 279-294, ISSN 0022-1376

Shanley, K.W. & McCabe, P.J. (1994). Perspectives on the sequence stratigraphy of continental strata. *American Association of Petroleum Geologists Bulletin*, Vol.78, No.4 (April 1994), pp. 544-568

Sloss, L.L. (1962). Stratigraphic models in exploration. *Journal of Sedimentary Petrology*, Vol.32, No.3 (September 1962), pp. 415-422

Somoza, L.; Barbolas, A., Arasa, A., Maestro, A., Rees, J.G. & Hernandez-Molina, F.J. (1998). Architectural stacking patterns of the Ebro delta controlled by Holocene high-frequency eustatic fluctuations, delta-lobe switching and subsidence processes. *Sedimentary Geology*, Vol.117, No.1-2 (April 1998), pp. 11-32, ISSN 0037-0738

Stanley, D.J. & Warne, A.G. (1994). Worldwide initiation of Holocene marine deltas by deceleration of sea-level rise. *Science*, Vol.265, No.5169 (July 1994), pp. 284-231

Steel, R.J; Carvajal, C, Petter A.L. & Uroza, C. (2008). Shelf and shelf-margin growth in scenarios of rising and falling sea level. *SEPM Special Publication* 90, pp. 47-71

Strong, N. & Paola, C. (2008). Valleys that never were: time surfaces versus stratigraphic surfaces. *Journal of Sedimentary Research*, Vol.78, No.8 (August 2008), pp. 579-593

Swenson, J.B. & Muto, T. (2007). Response of coastal-plain rivers to falling relative sea level: allogenic controls on the aggradational phase. *Sedimentology*, Vol.54, No.1 (February 2007), pp. 207-221, ISSN 0037-0746

Swenson, J.B.; Voller, V.R., Paola, C., Parker, G. & Marr, J.G. (2000). Fluvio-deltaic sedimentation: A generalized Stefan problem. *European Journal of Applied Mathematics*, Vol.11, issue No.5 (November 2000), pp. 433–452, ISSN 1110-757X

Swift, D.J.P. (1968). Coastal erosion and transgressive stratigraphy. *Journal of Geology*, Vol.76, No.4 (July 1968), pp. 444–456, ISSN 0022-1376

Swift, D.J.P. & Thorne, J.A. (1991). Sedimentation on continental margins, I: A general model for shelf sedimentation. In: *Shelf Sand and Sandstone Bodies: Geometry, Facies and Sequence Stratigraphy, International Association of Sedimentologists Special Publication* 14 (December 1991), D.J.P. Swift, G.F. Oertel, R.W. Tillman & J.A. Thorne, (Eds.), 3-31, ISBN 9780632032372

Swift, D.J.P.; Stanley, D.J. & Curray, J.R. (1971). Relict sediments on continental shelves: a reconsideration. *Journal of Geology*, Vol.79, No.3 (May 1971), pp. 322-346, ISSN 0022-1376

Syvitski, J.P.M.; Kettner, A.J., Overeem, I., Hutton, E.W.H., Hannon, M.T., Brakenridge, G.R., Day. J., Vörösmarty, C., Saito, Y., Giosan, L. & Nicholls, R.J. (2009). Sinking deltas due to human activities. *Nature Geoscience*, Vol.2, doi:10.1038/NGE629 (September 2009)

Tamura, T.; Saito, Y., Sieng, S., Ben, B., Kong, M., Sim, I., Choup, S. & Akiba, F. (2009). Initiation of the Mekong River delta at 8 ka: evidence from the sedimentary succession in the Cambodian lowland. *Quaternary Science Reviews*, Vol.28, No.3-4 (February 2009), pp. 327–344.

Thorne, J.A. & Swift, D.J.P. (1991). Sedimentation on continental margins, II: application of the regime concept. In: *Shelf Sand and Sandstone Bodies: Geometry, Facies and Sequence Stratigraphy, International Association of Sedimentologists Special Publication* 14 (December 1991), D.J.P. Swift, G.F. Oertel, R.W. Tillman & J.A. Thorne, (Eds.), 33–58, ISBN 9780632032372

Tomer, A.; Muto, T. & Kim, W. (2011), Autogenic hiatus in fluviodeltaic successions: geometrical modeling and physical experiments. *Journal of Sedimentary Research*, Vol.81, No.3 (March 2011), pp. 207-217, ISBN 978-0891813026

Van Andel, T.H. & Curray, J.R. (1960). Regional aspects of modern sedimentation in northern Gulf of Mexico and similar basins, and paleogeographic significance. In: *Recent sediments, northwest Gulf of Mexico: American Association of Petroleum Geologists*, F.P. Shepard, F.B. Phleger & T.H. Van Andel, (Eds.), 345–364

Van Heijst, M.W.I.M. & Postma, G. (2001). Fluvial response to sea-level changes: a quantitative analogue, experimental approach. *Basin Research*, Vol.13, No.3 (September 2001), pp. 269–292, ISSN 1365-2117

Weller, J.M. (1960). *Stratigraphic principles and practice*, Harper & Row, New York.

Wolinsky, M.A.; Swenson, J.B., Litchfield, N. & McNinch, J.E. (2010). Coastal progradation and sediment partitioning in the Holocene Waipaoa Sedimentary System, New Zealand. *Marine Geology*, Vol.270, No.1-4 (April 2010), pp. 94–107, ISSN 0025-3227

Xue, Z.; Liu, J.P., DeMaster, D., Nguyen, L.V., Oanh & Ta, T.K.O. (2010). Late Holocene Evolution of the Mekong Subaqueous Delta, Southern Vietnam. *Marine Geology*, Vol.269, No.1-2 (February 2010), pp. 46–60, ISSN 0025-3227

Part 6

Hydrogeology

13

Numerical Geodynamic Modeling of Continental Convergent Margins

Zhonghai Li[1,2,3], Zhiqin Xu[2] and Taras Gerya[3]
[1]FAST Laboratory, CNRS/University of Paris 6 and 11
[2]State Key Lab of Continental Tectonics and Dynamics, Institute of Geology, Chinese Academy of Geological Sciences
[3]Institute of Geophysics, ETH-Zurich
[1]France
[2]China
[3]Switzerland

1. Introduction

The continental convergence (subduction/collision) normally follows the oceanic subduction under the convergent forces of lateral ridge push and/or oceanic slab pull (Turcotte and Schubert, 2002). During these scenarios, a large amount of positively buoyant materials enter the trench causing slow down of the convergence that, eventually, may stop. However, before collision ceases, convergence between the plates can continue actively for tens of millions of years after ocean closure as it is testified by the 50 *Ma* active collisions in the Western Alps and Himalayas (e.g. Yin, 2006).

A remarkable event during the early continental collision is the formation and exhumation of high-pressure to ultra-high-pressure (HP-UHP) metamorphic rocks, which is one of the most provocative findings in the Earth sciences during the past three decades. Occurrences of UHP terranes around the world have been increasingly recognized with more than 20 UHP terranes documented (e.g. Liou et al., 2004), which have repeatedly invigorated the concepts of deep subduction (>100 *km*) and subsequent exhumation of crustal materials (e.g. Chopin, 2003). It has been suggested that the HP-UHP metamorphism can be considered as a "hallmark" for the modern plate tectonics regime characterized by colder subduction and started from a Neoproterozoic time (e.g. Brown, 2006, 2007).

The understanding of the dynamics of continental convergent margins implies several different but strictly correlated processes, such as continental deep subduction, HP-UHP metamorphism, exhumation, continental collision and mountain building. Besides the systematic geological/geophysical studies of the continental convergent zones, numerical modeling becomes a key and efficient tool (e.g. Burov et al., 2001; Yamato et al., 2007; Gerya et al., 2008; Warren et al., 2008a,b; Li and Gerya, 2009; Beaumont et al., 2009; Li et al., 2011). The tectonic styles of continental subduction can be either one-sided (overriding plate does not subduct) or two-sided (both plates subduct together) (Tao and O'Connell, 1992; Pope and Willett, 1998; Faccenda et al., 2008; Warren et al., 2008a), as well as several other possibilities, e.g. thickening, slab break-off, slab drips etc (e.g. Toussaint et al., 2004a,b). Models of HP-UHP rocks exhumation can be summarized into the following groups: (1) syn-collisional exhumation of a coherent and buoyant crustal slab, with formation of a weak zone at the entrance of the subduction channel (Chemenda et al., 1995, 1996; Toussaint et al., 2004b; Li and Gerya, 2009); (2) episodic ductile extrusion of HP-UHP rocks from the subduction channel to the surface or crustal depths (Beaumont et al., 2001; Warren et al., 2008a); (3) continuous material circulation in the rheologically weak subduction channel stabilized at the plate interface, with materials exhumed from different depths (Burov et al., 2001; Stöckhert and Gerya, 2005; Yamato et al., 2007; Warren et al., 2008a).

In this chapter, the processes and dynamics of continental subduction/collision and HP-UHP rocks exhumation are investigated by the method of large-scale numerical geodynamic modeling. First the numerical method is described, which is followed by the numerical model setup and systematic thermo-mechanical numerical experiments. The discussion section covers a broad range of topics related to the continental subduction and exhumation. Finally a concluding part is presented.

2. Numerical modeling method

2.1 Governing equations and numerical implementation

The momentum, continuity and heat conservation equations for a 2D creeping flow including thermal and chemical buoyant forces are solved:

(i) Stokes equation

$$\frac{\partial \sigma'_{xx}}{\partial x} + \frac{\partial \sigma'_{xz}}{\partial z} = \frac{\partial P}{\partial x} \tag{1}$$

$$\frac{\partial \sigma'_{zx}}{\partial x} + \frac{\partial \sigma'_{zz}}{\partial z} = \frac{\partial P}{\partial z} - g\,\rho(C,M,P,T)$$

where the density ρ depends on composition (C), melt fraction (M), pressure (P) and temperature (T); g is the acceleration due to gravity.

(ii) Conservation of mass is approximated by the incompressible continuity equation

$$\frac{\partial v_x}{\partial x} + \frac{\partial v_z}{\partial z} = 0 \tag{2}$$

(iii) Heat conservation equations

$$\rho\, C_p\,(\frac{DT}{Dt}) = -\frac{\partial q_x}{\partial x} - \frac{\partial q_z}{\partial z} + H_r + H_a + H_s + H_L \tag{3}$$

$$q_x = -k(C,P,T)\,\frac{\partial T}{\partial x},\quad q_z = -k(C,P,T)\,\frac{\partial T}{\partial z}$$

$$H_a = T\,\alpha\,\frac{\partial P}{\partial t},\quad H_s = \sigma'_{xx}\,\dot{\varepsilon}_{xx} + \sigma'_{zz}\,\dot{\varepsilon}_{zz} + 2\,\sigma'_{xz}\,\dot{\varepsilon}_{xz}$$

where D/Dt is the substantive time derivative. x and z denote the horizontal and vertical directions, respectively. The deviatoric stress tensor is defined by σ'_{xx}, σ'_{xz}, σ'_{zz}, whilst the strain rate tensor is defined by $\dot{\varepsilon}_{xx}$, $\dot{\varepsilon}_{xz}$, $\dot{\varepsilon}_{zz}$. q_x and q_z are heat flux components. ρ is the density. $k(C,P,T)$ is the thermal conductivity as a function of composition (C), pressure (P) and temperature (T). C_p is the isobaric heat capacity. H_r, H_a, H_s, H_L are radioactive, adiabatic, shear and latent heat production, respectively (see Table 1 for details of these parameters).

To solve the above equations, the I2VIS code is used (Gerya and Yuen, 2003a). It is a two-dimensional finite difference code with marker-in-cell technique which allows for non-diffusive numerical simulation of multi-phase flow in a rectangular fully staggered Eulerian grid. I2VIS accounts for visco-plastic deformation and several geological processes that are described below. All abbreviations and units used in this chapter are listed in Table 1.

Symbol	Meaning	Unit
A_D	Material constant (viscous rheology)	$MPa^{-n}\,s^{-1}$
C	Cohesion (plastic rheology)	MPa
C_p	Isobaric heat capacity	$J\,kg^{-1}\,K^{-1}$
E	Activation energy	$kJ\,mol^{-1}$
g	Gravitational acceleration	$m\,s^{-2}$
G	Plastic potential	Pa
H_a, H_r, H_s, H_L	Heat production (adiabatic, radioactive, viscous, latent)	$W\,m^{-3}$
k	Thermal conductivity	$W\,m^{-1}\,K^{-1}$
M	Volume fraction of melt	$Dimensionless$
n	Stress exponent	$Dimensionless$
P	Dynamic pressure	Pa
P_{fluid}	Pore fluid pressure	Pa
P_{lith}	Lithostatic pressure	Pa
q_x, q_z	Horizontal and vertical heat fluxes	$W\,m^{-2}$
Q_L	Latent heat of melting	$kJ\,kg^{-1}$
t	Time	s
T	Temperature	K
$T_{liquidus}$	Liquidus temperature of the crust	K
$T_{solidus}$	Solidus temperature of the crust	K
v_e, v_s	Erosion and sedimentation rate	$m\,s^{-1}$
v_x, v_z	Horizontal and vertical components of velocity	$m\,s^{-1}$
V	Activation volume	$J\,MPa^{-1}\,mol^{-1}$
x, z	Horizontal and vertical coordinates	m
a	Thermal expansion coefficient	K^{-1}
β	Compressibility coefficient	Pa^{-1}
γ	Strain rate	s^{-1}
$\dot{\varepsilon}_{ij}$	Components of the strain rate tensor	s^{-1}
$\dot{\varepsilon}_{II}$	Second invariant of the strain rate tensor	s^{-2}
η	Viscosity	$Pa\,s$
κ	Thermal diffusivity	$m^2\,s^{-1}$
λ	Pore fluid pressure coefficient: $\lambda = P_{fluid}/P$	$Dimensionless$
μ	Shear modulus	Pa
ρ	Density	$kg\,m^{-3}$
σ'_{ij}	Components of the viscous deviatoric stress tensor	Pa
σ_{II}	Second invariant of the stress tensor	Pa
σ_{yield}	Yield stress	Pa
τ	Shear stress	Pa
ψ	Internal frictional angle	$Dimensionless$
χ	Plastic multiplier	s^{-1}

Table 1. Abbreviations and units of the variables used in this chapter.

2.2 Boundary conditions

For the 2D numerical models presented in this chapter, the velocity boundary conditions are free slip at all boundaries except the lower one, which is permeable (Burg and Gerya, 2005;

Li et al., 2010). Infinity-like external free slip conditions along the lower boundary imply free slip to be satisfied at 1000 km below the bottom of the model. As for the usual free slip condition, external free slip allows global conservation of mass in the computational domain and is implemented by using the following limitation for velocity components at the lower boundary: $\partial v_x/\partial z = 0, \partial v_z/\partial z = -v_z/\Delta z_{external}$, where $\Delta z_{external}$ is the vertical distance from the lower boundary to the external boundary where free slip ($\partial v_x/\partial z = 0, v_z = 0$) is satisfied. The thermal boundary conditions have a fixed value (0 °C) for the upper boundary and zero horizontal heat flux across the vertical boundaries. For the lower thermal boundary, an infinity-like external constant temperature condition is imposed, which allows both temperatures and vertical heat fluxes to vary along the permeable box lower boundary, implying constant temperature condition to be satisfied at the external boundary. This condition is implemented by using the limitation $\partial T/\partial z = (T_{external} - T)/\Delta z_{external}$ where $T_{external}$ is the temperature at the external boundary and $\Delta z_{external}$ is the vertical distance from the lower boundary to the external boundary (Burg and Gerya, 2005; Li et al., 2010).

2.3 Rheological model

A viscoplastic rheology is assigned for the model in which the rheological behaviour depends on the minimum differential stress attained between the ductile and brittle fields. Ductile viscosity dependent on strain rate, pressure and temperature is defined in terms of deformation invariants as:

$$\eta_{ductile} = (\dot{\varepsilon}_{II})^{\frac{1-n}{n}} F (A_D)^{-\frac{1}{n}} \exp(\frac{E + PV}{nRT}) \tag{4}$$

where $\dot{\varepsilon}_{II} = 0.5\, \dot{\varepsilon}_{ij}\, \dot{\varepsilon}_{ij}$ is the second invariant of the strain rate tensor. A_D, E, V and n are experimentally determined flow law parameters (Table 2). F is a dimensionless coefficient depending on the type of experiments on which the flow law is based. For example: $F = [2^{(1-n)/n}]/[3^{(1+n)/2n}]$ for triaxial compression and $F = 2^{(1-2n)/n}$ for simple shear.
The ductile rheology is combined with a brittle/plastic rheology to yield an effective visco-plastic rheology. For this purpose the Mohr-Coulomb yield criterion (e.g. Ranalli, 1995) is implemented as follows:

$$\begin{aligned} \sigma_{yield} &= C + P\sin(\varphi_{eff}) \\ \sin(\varphi_{eff}) &= \sin(\varphi)\,(1-\lambda) \\ \eta_{plastic} &= \frac{\sigma_{yield}}{2\,\dot{\varepsilon}_{II}} \end{aligned} \tag{5}$$

where σ_{yield} is the yield stress. $\dot{\varepsilon}_{II}$ is the second invariant of the strain rate tensor. P is the dynamic pressure. C is the cohesion. φ is the internal frictional angle. λ is the pore fluid coefficient that controls the brittle strength of fluid-containing porous or fractured media (Brace and Kohlstedt, 1980). φ_{eff} can be illustrated as the effective internal frictional angle that integrates the effects of internal frictional angle (φ) and pore fluid coefficient (λ). λ is the pore fluid coefficient that controls the brittle strength of fluid-containing porous or fractured media.
The effective viscosity of molten rocks ($M \geq 0.1$) was calculated using the formula (Pinkerton and Stevenson, 1992; Bittner and Schmeling, 1995):

$$\eta = \eta_0 \exp[2.5 + (1-M)(\frac{1-M}{M})^{0.48}] \tag{6}$$

where η_0 is an empirical parameter depending on rock composition, being $\eta_0 = 10^{13}\ Pa\,s$ (i.e. $1 \times 10^{14} \leq \eta \leq 2 \times 10^{15}\ Pa\,s$ for $0.1 \leq M \leq 1$) for molten mafic rocks and $\eta_0 = 5 \times 10^{14}\ Pa\,s$ (i.e. $6 \times 10^{15} \leq \eta \leq 8 \times 10^{16}\ Pa\,s$ for $0.1 \leq M \leq 1$) for molten felsic rocks. Successfully tested for a broad range of suspensions with various bubble or crystal conventions, this formula takes into account, other than concentration, particle shape and size distribution.

2.4 Partial melting model

The numerical code accounts for partial melting of the various lithologies by using experimentally obtained *P-T* dependent wet solidus and dry liquidus curves (Gerya and Yuen, 2003b). As a first approximation, volumetric melt fraction M is assumed to increase linearly with temperature accordingly to the following relations (Burg and Gerya, 2005):

$$M = 0,\ when\ T \leq T_{solidus}$$

$$M = \frac{T - T_{solidus}}{T_{liquidus} - T_{solidus}},\ when\ T_{solidus} < T < T_{liquidus} \tag{7}$$

$$M = 1,\ when\ T \geq T_{liquidus}$$

where $T_{solidus}$ and $T_{liquidus}$ are the wet solidus and dry liquidus temperature of the given lithology, respectively (Table 3). Consequently, the effective density, ρ_{eff}, of partially molten rocks varies with the amount of melt fraction and *P-T* conditions according to the relations:

$$\rho_{eff} = \rho_{solid} - M(\rho_{solid} - \rho_{molten}) \tag{8}$$

where ρ_{solid} and ρ_{molten} are the densities of the solid and molten rock, respectively, which vary with pressure and temperature according to the relation:

$$\rho_{P,T} = \rho_0[1 - \alpha(T - T_0)][1 + \beta(P - P_0)] \tag{9}$$

where ρ_0 is the standard density at $P_0 = 0.1\ MPa$ and $T_0 = 298\ K$; α and β are the thermal expansion and compressibility coefficients, respectively (Tables 1 and 3).
The effects of latent heat H_L (e.g. Stüwe, 1995) are accounted for by an increased effective heat capacity (C_{Peff}) and thermal expansion (α_{eff}) of the partially molten rocks ($0 < M < 1$), calculated as

$$C_{Peff} = C_P + Q_L(\frac{\partial M}{\partial T})_P \tag{10}$$

$$\alpha_{eff} = \alpha + \rho\frac{Q_L}{T}(\frac{\partial M}{\partial P})_T \tag{11}$$

where C_P and α are the heat capacity and the thermal expansion of the solid crust, respectively, and Q_L is the latent heat of melting of the crust (Table 1).

2.5 Topographic model

The spontaneous deformation of the upper surface of the lithosphere, i.e. topography, is calculated dynamically as an internal free surface by using a low viscosity (e.g. $10^{18}\ Pa\,s$), initially 8-12 *km* thick layer (thickness of this layer changes dynamically during experiments) above the upper crust. The composition is either "air" ($1\ kg/m^3$, above water level) or "water" ($1000\ kg/m^3$, below water level). The interface between this weak layer and the underlying crust is treated as an internal erosion/sedimentation surface which evolves according to the

Eulerian transport equation solved in Eulerian coordinates at each time step (Gerya and Yuen, 2003b):

$$\frac{\partial z_{es}}{\partial t} = v_z - v_x \frac{\partial z_{es}}{\partial x} - v_s + v_e \tag{12}$$

where z_{es} is the vertical position of the surface as a function of the horizontal distance v_x. v_z and v_x are the vertical and horizontal components of the material velocity vector at the surface. v_s and v_e are the sedimentation and erosion rates, respectively, which correspond to the relation: $v_s = 0, v_e = v_{e0}$, when $z_{es} <$ erosion level; $v_s = v_{s0}, v_e = 0$, when $z_{es} >$ erosion level; where v_{e0} and v_{s0} are the imposed constant large scale erosion and sedimentation rates, respectively. The code allows for marker transmutation that simulates erosion (rock markers are transformed to weak layer markers) and sedimentation (weak layer markers are transformed to sediments).

3. Numerical model design

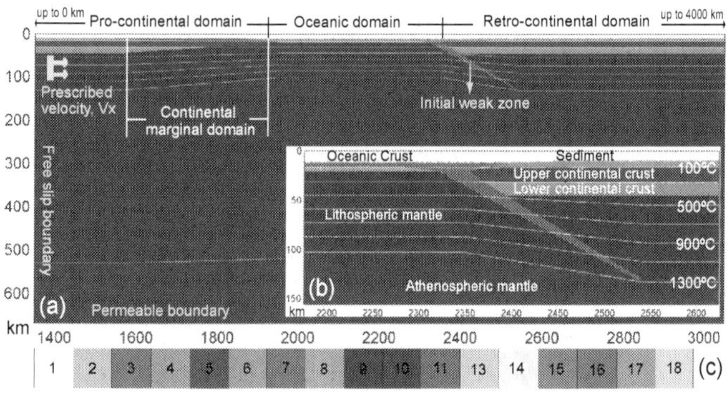

Fig. 1. Initial model configuration and boundary conditions. (a) Enlargement (1700 × 670 km) of the numerical box (4000 × 670 km). Boundary conditions are indicated in yellow. (b) The zoomed domain of the subduction zone. White lines are isotherms measured in °C. (c) The colorgrid for different rock types, with: 1-air; 2-water; 3,4-sediment; 5-upper continental crust; 6-lower continental crust; 7-upper oceanic crust; 8-lower oceanic crust; 9-lithospheric mantle; 10-athenospheric mantle; 11-weak zone mantle; 13,14-partially molten sediment (3,4); 15,16-partially molten continental crust (5,6); 17,18-partially molten oceanic crust (7,8). The partially molten crustal rocks (13, 14, 15, 16, 17, 18) are not shown in this figure, which will appear during the evolution of the model. In our numerical models, the medium scale layering usually shares the same physical properties, with different colors used only for visualizing plate deformation. Detailed properties of different rock types are shown in Tables 2 and 3.

Large scale models (4000 × 670 km, Fig. 1) are designed for the study of dynamic processes from oceanic subduction to continental collision associated with HP-UHP rocks formation and exhumation. The non-uniform 699 × 134 rectangular grid is designed with a resolution varying from 2 × 2 km in the studied collision zone to 30 × 30 km far away from it. The lithological structure of the model is represented by a dense grid of 7 million active Lagrangian markers used for advecting various material properties and temperature (Gerya et al., 2008; Li et al., 2010). The subducting plate is pushed rightward by prescribing a constant

ID symbol	Flow Laws	E	V	n	A_D	$\eta_0^{(a)}$
A^*	Air/water	0	0	1.0	1.0×10^{-12}	1×10^{18}
B^*	Strong wet quartzite	154	0	2.3	3.2×10^{-6}	1.97×10^{19}
C^*	Plagioclase An_{75}	238	0	3.2	3.3×10^{-4}	4.80×10^{22}
D^*	Dry olivine	532	8	3.5	2.5×10^{4}	3.98×10^{16}
E^*	Wet olivine	470	8	4.0	2.0×10^{3}	5.01×10^{20}
$F^{*(b)}$	Molten felsic	0	0	1.0	2.0×10^{-9}	5×10^{14}
$G^{*(b)}$	Molten mafic	0	0	1.0	1.0×10^{-7}	1×10^{13}

Table 2. Viscous flow laws used in the numerical experiments. (a) η_0 is the reference viscosity, which is calculated as $\eta_0 = (1/A_D) \times 10^{6n}$. Other parameters are illustrated in Table 1. (b) The molten felsic (F^*) is used for partial molten sediment and continental crust. The molten mafic (G^*) is used for partial molten oceanic crust. The rheological data in this table are from Ranalli (1995).

Material$^{(a)}$	ρ_0	C_p	$k^{(b)}$	$T_{solidus}^{(c)}$	$T_{liquidus}^{(d)}$	Q_L	H_r	Viscous$^{(e)}$	Plastic$^{(f)}$
Air	1	100	20	-	-	-	0	A^*	0
Water	1000	3300	20	-	-	-	0	A^*	0
M_1-Solid (3,4,5)	2700	1000	k_1	T_{S1}	T_{L1}	300	2	B^*	0.15
M_1-Molten (13,14,15)	2500	1500	k_1	T_{S1}	T_{L1}	300	2	F^*	0.06
M_2-Solid (6)	3000	1000	k_1	T_{S1}	T_{L1}	300	0.5	C^*	0.15
M_2-Molten (16)	2500	1500	k_1	T_{S1}	T_{L1}	300	0.5	F^*	0.06
M_3-Solid (7,8)	3000	1000	k_2	T_{S2}	T_{L2}	380	0.25	C^*	0.15
M_3-Molten (17,18)	2900	1500	k_2	T_{S2}	T_{L2}	380	0.25	G^*	0.06
M_4-Solid (9,10)	3300	1000	k_3	-	-	-	0.022	D^*	0.6
M_4-Molten (11)	3200	1500	k_3	-	-	-	0.022	E^*	0.06
Reference$^{(g)}$	1,2	-	3	5	5	1,2	1	4	-

Table 3. Material properties used in the numerical experiments. (a) M_1 is used for sediment and continental upper crust. M_2 is used for continental lower crust. M_3 is used for oceanic crust. M_4 is used for lithospheric and athenospheric mantle. Numbers in the brackets are corresponding to material colors in Figure 1. (b)
$k_1 = [0.64 + 807/(T_K + 77)] \exp(0.00004 P_{MPa})$,
$k_2 = [1.18 + 474/(T_K + 77)] \exp(0.00004 P_{MPa})$,
$k_3 = [0.73 + 1293/(T_K + 77)] \exp(0.00004 P_{MPa})$. ($c$)
$T_{S1} = \{889 + 17900/(P + 54) + 20200/(P + 54)^2 \; at \; P < 1200\,MPa\} \; or \; \{831 + 0.06P \; at \; P > 1200\,MPa\}$, $T_{S2} = \{973 - 70400/(P + 354) + 778 \times 10^5/(P + 354)^2 \; at \; P < 1600\,MPa\} \; or \; \{935 + 0.0035P + 0.0000062P^2 \; at \; P > 1600\,MPa\}$. ($d$) $T_{L1} = 1262 + 0.09P$, $T_{L2} = 1423 + 0.105P$. (e) Parameters of viscous flow laws are shown in Table 2. (f) This column shows the values of $\sin(\phi_{eff})$, which is the effective internal frictional angle implemented for plastic rheology. The plastic cohesion is zero in all the experiments. (g) 1=(Turcotte and Schubert, 1982); 2=(Bittner and Schmeling, 1995); 3=(Clauser and Huenges, 1995); 4=(Ranalli, 1995); 5=(Schmidt and Poli, 1998). In this table, meanings of all the variables are shown in Table 1. Thermal expansion coefficient $\alpha = 3 \times 10^{-5}\,K^{-1}$ and Compressibility coefficient $\beta = 1 \times 10^{-5}\,MPa^{-1}$ are used for all the rocks.

convergence velocity (V_x) in a small internal domain that remains fixed with respect to the Eulerian coordinate (Fig. 1).

In the numerical models, the driving mechanism of subduction is a combination of plate push (prescribed rightward convergence velocity) and increasing slab pull (temperature-induced density contrast between the subducted lithosphere and surrounding mantle). This type of boundary condition is commonly used in numerical models of subduction and collision (e.g. Toussaint et al., 2004b; Burg and Gerya, 2005; Currie et al., 2007; Yamato et al., 2007; Warren et al., 2008b) and assumes that in the globally confined three-dimensional system of plates, local external forcing coming either from different slabs or from different sections of the same laterally non-uniform slab can be significant.

Following previous numerical studies performed with similar geodynamic settings (e.g. Warren et al., 2008a; Li and Gerya, 2009) we design numerical models consisting of three major domains (from left to right, Fig. 1): (1) a pro-continental domain, (2) an oceanic domain, and (3) a retro-continental domain. The subducting pro-continent comprises a marginal unit and an interior unit. In the continental domain, the initial material field is set up by a 35 km thick continental crust composed of sediment (6 km thick), upper crust (14 km thick) and lower crust (15 km thick), overlying the lithospheric mantle (85 km thick) and subjacent mantle (540 km thick). The oceanic domain comprises an 8 km thick oceanic crust overlying the lithospheric mantle (82 km thick) and subjacent mantle (570 km thick). The material properties of all layers (Fig. 1) are listed in Table 3. The initial thermal structure of the lithosphere (white lines in Fig. 1) is laterally uniform with $0\,^{\circ}C$ at the surface and $1300\,^{\circ}C$ at the bottom of the lithospheric mantle (both continental and oceanic). The initial temperature gradient in the asthenospheric mantle is around $0.5\,^{\circ}C/km$.

4. Model result

4.1 Reference model

The reference model is designed with prescribed convergence velocity (V_x) of 5 cm/y. All the other configurations and parameters are shown in Figure 1 and Table 3.

At the initial stages, the relatively strong oceanic plate subducts along the weak zone to the mantle (Fig. 2a).The continental margin subducts to >100 km depth, following the high-angle oceanic subduction channel (Fig. 2b). The significant characters are the detachment of subducting upper/middle crust at the entrance zone of the subduction channel with a series of thrust faults formed (Fig. 2b-d). A small amount of crustal rocks located in the lower part of the channel are detached from the plate at asthenospheric depths, indenting into the mantle wedge and forming a compositionally buoyant plume (Fig. 2c). Such sub-lithospheric plumes are discussed in detail in Currie et al. (2007) and Li and Gerya (2009). In addition, a partially molten plume forms in the deeply subducted oceanic plate and moves up vertically until it collapses at the bottom of the overriding lithospheric mantle (Fig. 2c,d). As subduction continues, another partially molten plume forms in the deeply subducted continental plate. It also moves up vertically until it collapses at the bottom of the overriding lithospheric mantle (Fig. 2e,f). The characteristics and 2D and 3D dynamics of this kind of plume are studied in detail in Gerya and Yuen (2003b) and Zhu et al. (2009).

As continental subduction continues, partially molten rocks accumulated in the subduction channel extrude upward to the crustal depths (Fig. 2d,e). Then these UHP rocks exhume buoyantly to the surface forming a dome structure (Fig. 2f). The exhumed UHP rocks are mainly located near the suture zone with a fold-thrust belt formed in the foreland extending for about 300-400 km (Fig. 2f). P-T paths (Fig. 2, inset) show that peak P-T conditions

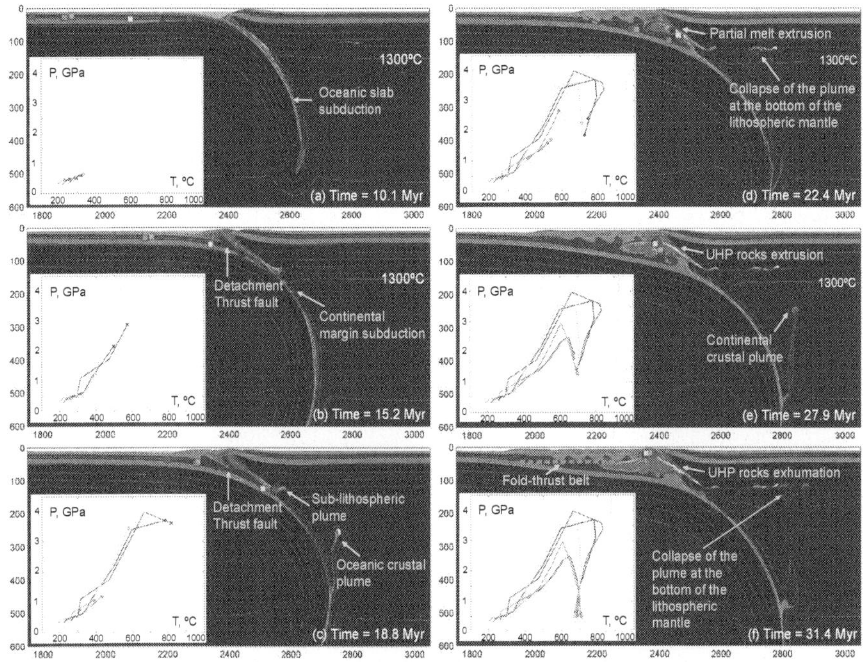

Fig. 2. Enlarged domain evolution (1300 × 600 km) of the reference model. Colors of rock types are as in Figure 1. Time (Myr) of shortening is given in the figures. White numbered lines are isotherms in °C. Small colored squares indicate positions of representative markers (rock units) for which P-T paths are shown (inset). Colors of these squares are used for discrimination of marker points plotted in P-T diagrams and do not correspond to the colors of rock types.

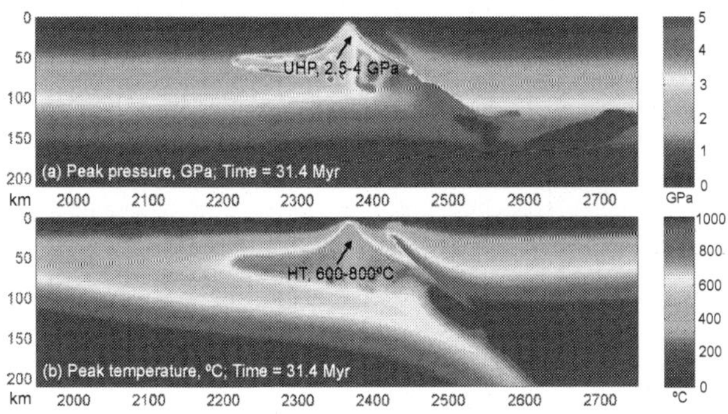

Fig. 3. Peak metamorphic conditions of the reference model. (a) Peak pressure condition in GPa; (b) Peak temperature condition in °C.

of the exhumed rocks are 2.5-4 GPa and 600-800 °C, respectively (also see Figure 3 for the peak pressure and temperature conditions of the collision zone). This indicates the UHP metamorphic rocks are formed and exhumed from a depth >100 km.

4.2 Models with variable convergence velocity

The reference model is further investigated with lower convergence velocity (2.5 cm/y) and higher convergence velocity (10 cm/y). All the other parameters are the same as in Tables 2 and 3.

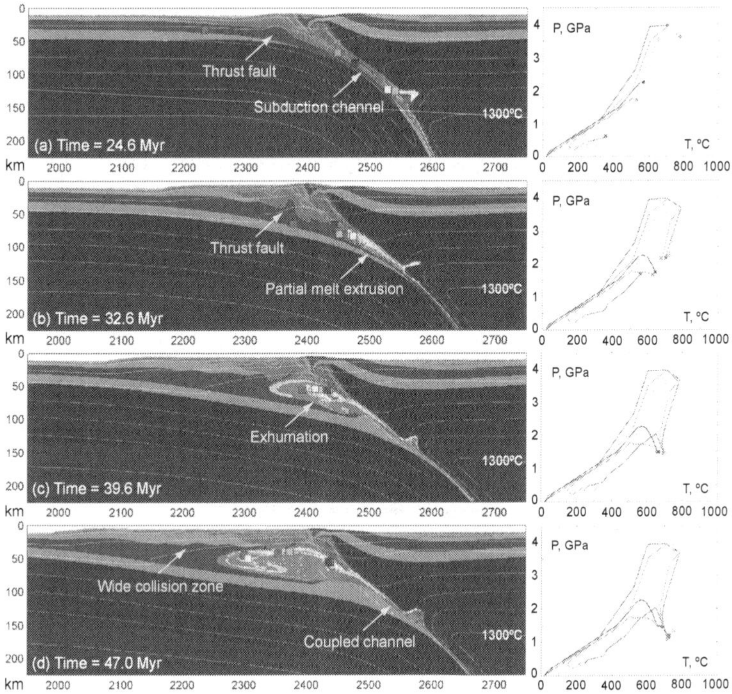

Fig. 4. Enlarged domain evolution (800 × 225 km) of the model with lower convergence velocity $V_x = 2.5\,cm/y$. Colors of rock types are as in Figure 1. Time (Myr) of shortening is given in the figures. White numbered lines are isotherms in °C. Small colored squares indicate positions of representative markers (rock units) for which P-T paths are shown (right). Colors of these squares are used for discrimination of marker points plotted in P-T diagrams and do not correspond to the colors of rock types.

In the slow convergence regime, the continental margin also subducts to >100 km depth along the high-angle oceanic subduction channel to the bottom of the lithospheric mantle (Fig. 4a). The subducting upper/middle crust at the entrance zone of the subduction channel detaches with thrust faults formed (Fig. 4a-d). With continued continental subduction, partially molten rocks accumulated in the subduction channel extrude upward to the crustal depth (Fig. 4c,d). P-T paths (Fig. 4) show that peak P-T conditions of the exhumed rocks are 2-4 GPa and 600-800 °C.

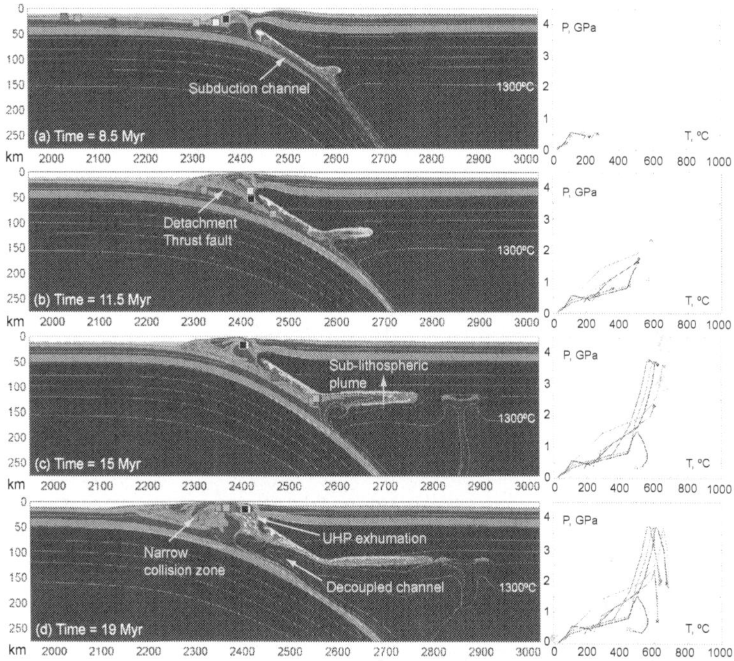

Fig. 5. Enlarged domain evolution (1075 $\times$ 275 km) of the model with higher convergence velocity $V_x = 10\,cm/y$. Colors of rock types are as in Figure 1. Time (Myr) of shortening is given in the figures. White numbered lines are isotherms in °C. Small colored squares indicate positions of representative markers (rock units) for which P-T paths are shown (right). Colors of these squares are used for discrimination of marker points plotted in P-T diagrams and do not correspond to the colors of rock types.

In the fast convergence regime, the continental domain continues subducting along the high-angle oceanic subduction channel to the bottom of the lithospheric mantle (Fig. 5a). Then the crustal rocks in the lower part of the channel detach from the plate at asthenospheric depths, intrude into the mantle wedge, and form a horizontal compositionally buoyant plume (Fig. 5a-c). In addition, a partially molten plume forms in the deeply subducted plate and moves up vertically until it collapses at the bottom of the overriding lithospheric mantle (Fig. 5c,d), which is similar to behavior of the reference model. After the convergence ceases (1500 km shorting, 15 Myr), the subducted continental crustal rocks in the sub-lithospheric plume extrude upward to the surface forming a dome structure (Fig. 5d). P-T paths (Fig. 5) indicate that peak P-T conditions of the exhumed rocks are 3-4 GPa and 600-800 °C, respectively.
This parameter sensitivity studies indicate that the slower convergence produces very small sub-lithospheric plume (Fig. 4a), coupled subduction channel and wide collision zone (Fig. 4d). In contrast, the faster convergence results in very large sub-lithospheric plume (Fig. 5a), decoupled subduction channel and narrow collision zone (Fig. 5d). Both of the models can obtain UHP rocks exhumation. However, the convergence velocity changes the amount of crustal rocks subducted to and exhumed from UHP depth (c.f. Figs 4 and 5).

4.3 Models with variable thermal structure of the oceanic lithosphere

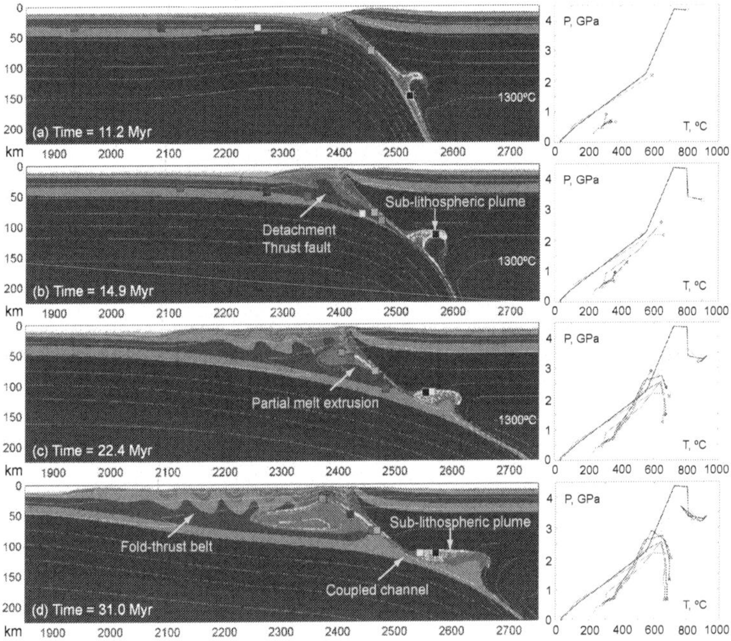

Fig. 6. Enlarged domain evolution (900 × 225 km) of the model with higher temperature of the oceanic lithosphere (hot model). Colors of rock types are as in Figure 1. Time (Myr) of shortening is given in the figures. White numbered lines are isotherms in °C. Small colored squares indicate positions of representative markers (rock units) for which *P-T* paths are shown (right). Colors of these squares are used for discrimination of marker points plotted in *P-T* diagrams and do not correspond to the colors of rock types.

The lithospheric thermal structure plays an important role on subduction/collision processes (e.g. Toussaint et al., 2004a, b). Therefore we investigate the sensitivity of oceanic thermal gradient for the reference model. In the hot model, initial thermal structure of the oceanic lithosphere is linearly interpolated with 0 °C at the surface ($\leq 10\,km$ depth) and 1300 °C at 70 km depth (compared to 1300 °C at 100 km depth in the reference model). In contrast, the initial thermal structure of oceanic lithosphere in the cold model is linearly interpolated with 0 °C at the surface ($\leq 10\,km$ depth) and 1300 °C at 130 km depth.

In the hot model, the continental margin subducts following the oceanic subduction channel to the bottom of the lithospheric mantle (Fig. 6a). Then the crustal rocks in the lower part of the channel detach from the plate at asthenospheric depths, intrude into the mantle wedge, and form a horizontal compositionally buoyant plume (Fig. 6b). The subducting upper/middle crust at the entrance zone of the subduction channel detaches with thrust faults formed (Fig. 6b,c). With continued continental subduction, partially molten rocks in the middle channel extrude upward to the crustal depth (Fig. 6c,d). The subduction channel is highly coupled. As a result, the partially molten rocks in the sub-lithospheric plume stay at the bottom of the overriding lithosphere (without exhumation).

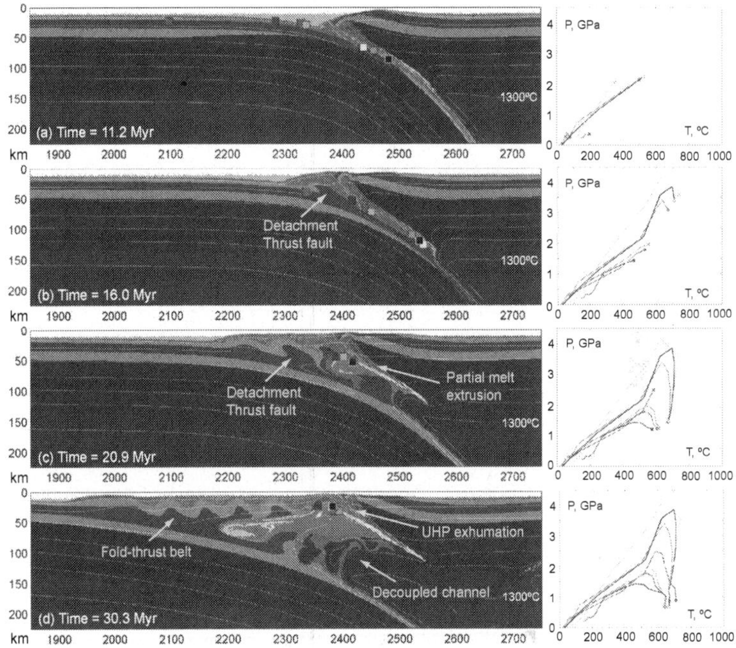

Fig. 7. Enlarged domain evolution (900 × 225 *km*) of the model with lower temperature of the oceanic lithosphere (cold model). Colors of rock types are as in Figure 1. Time (*Myr*) of shortening is given in the figures. White numbered lines are isotherms in °C. Small colored squares indicate positions of representative markers (rock units) for which *P-T* paths are shown (right). Colors of these squares are used for discrimination of marker points plotted in *P-T* diagrams and do not correspond to the colors of rock types.

In the cold model, the continental margin subducts following the oceanic plate to the bottom of the lithospheric mantle (Fig. 7a). In this case, there is no sub-lithospheric plume formed. Consequently, the subduction channel is thicker and thicker. The subducting upper/middle crust at the entrance zone of the subduction channel detaches with thrust faults formed (Fig. 7b,c). The subducted continental crustal rocks extrude upward to the surface forming a dome structure (Fig. 7c,d). The subduction channel is highly decoupled.
The shape and characteristics of the subduction channel in the hot model (Fig. 6) is similar to that in the slow convergence model (Fig. 4). It indicates that both the higher temperature and the slower convergence can increase the rheological coupling at plate interface. As a result, coupled subduction channel is produced in these two models. In contrast, decoupled channels are formed in the colder model (Fig. 7) as well as in the faster convergence model (Fig. 5).

5. Discussion

5.1 Flow modes in the subduction channel

To a first approximation, viscous channel flow can be analysed using lubrication theory (e.g. England and Holland, 1979; Cloos, 1982; Cloos and Shreve, 1988a, 1988b; Mancktelow, 1995;

Raimbourg et al., 2007; Warren et al., 2008b; Beaumont et al., 2009). Under the lubrication approximations, channel flow velocity is

$$u(x,y) = -\frac{1}{2\eta}\frac{\partial P}{\partial x}(yh - y^2) + U(1 - \frac{y}{h}) \quad (13)$$

where η is the assumed uniform viscosity in the subduction channel, $\partial P/\partial x$ is the effective down-channel pressure gradient, with x measured in the down-dip direction, y is the position in the channel measured normal to the base, h is channel thickness and U is the subduction velocity of the underlying lithosphere (Fig. 8a). The overlying lithosphere is assumed to be stationary.

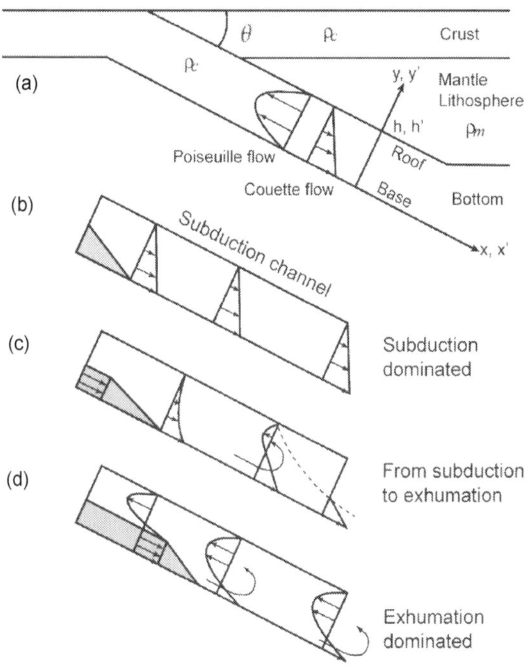

Fig. 8. Schematic diagram showing subduction/exhumation channel flow behavior in terms of dominating Couette (subduction) and Poiseuille (exhumation) flows (after Warren et al., 2008b; Beaumont et al., 2009). (a) General nomenclature. (b-d) Flow types identified in the models. (b) Couette flow (subduction) dominates. All flow is directed downward. (c) Buoyant materials stagnate at bottom of channel with the Poiseuille flow effect increasing. It is characterized as the transition from subduction-dominated to exhumation-dominated channel (d) Poiseuille flow (exhumation) dominates. Buoyancy-driven exhumation starts at channel bottom and propagates upward.

When nondimensional variables $u' = u/U$, $h' = h/H$, $x' = x/h$ and $y' = y/h$ are used, Equation 13 reduces to

$$u' = -\frac{E h'^2 (y' - y'^2)}{2} + (1 - y') \quad (14)$$

where

$$E = \frac{H^2(\partial P/\partial x)}{\eta_{eff} U} \quad (15)$$

$$H = (\frac{2\,\eta_{eff} U}{\partial P/\partial x})^{\frac{1}{2}} \quad (16)$$

where η_{eff}, the effective viscosity and $\partial P/\partial x$ are averaged quantities measured at the channel-scale used to estimate H. This parameter H is the characteristic channel thickness for $E \sim 1$, the balance point between downward and return flows.

Numerical models of the subduction channels can be conveniently interpreted in terms of the characteristic exhumation number E (Raimbourg et al., 2007; Warren et al., 2008b; Beaumont et al., 2009). The corresponding flow modes associated with burial and exhumation of UHP rocks are shown in Figure 8. The first order dynamics can be approximated by the above-mentioned lubrication theory for creeping flows, and characterized in terms of the competition between down-channel Couette flow ($U(1 - y/h)$ in Eq. 13), caused by the drag of the subducting lithosphere and the opposing up-channel Poiseuille flow ($(1/2\eta)(\partial P/\partial x)(yh - y^2)$ in Eq. 13), driven by the buoyancy of low-density subducted crustal material. This competition can be expressed through the exhumation number E, which is a force ratio derived from the non-dimensional channel flow equation (Eq. 14). The actual values that determine E (Eqs 15, 16) will depend on the particular problem and its evolving solution. For a channel with deformable walls and no tectonic over/under-pressure, $\partial P/\partial x = (\rho_m - \rho_c)\,g\sin(\theta)$ where θ is channel dip (Fig. 8a).

Along with the characteristic E, defined at the scale of the subduction channel, space-time variations in the channel flow can be interpreted in terms of the local exhumation number $E(x,t)$ and corresponding flow modes (Warren et al., 2008b; Beaumont et al., 2009). During continental subduction, $E(x,t)$ evolves from <1 during subduction (c.f. Fig. 8b and Fig. 2a), to ~1 during detachment and stagnation in the subduction channel (c.f. Fig. 8c and Fig. 2b,c), to >1 at the onset of and during exhumation (c.f. Fig. 8d and Fig. 2d-e). Buoyancy is a necessary, but not sufficient, condition for UHP exhumation. Among other controlling factors (Fig. 8), decreasing viscosity (η_{eff}) is typically most important for driving E beyond the exhumation threshold. In general, $E(x,t)$ should be regarded as a measure of local exhumation potential, even where the local threshold value is exceeded (E>1), efficient exhumation may be impeded by constrictions (small h) or high viscosities (large η_{eff}) further up the channel.

5.2 Coupled and decoupled subduction channel

The numerical results show that the coupled subduction channel favors lower convergence velocity (Fig. 4) and hotter oceanic lithosphere (Fig. 6). It is characterized by continuous accretion of the weak upper continental crust resulting in the development of a thick and broad crustal wedge. In contrast, the higher convergence velocity (Fig. 5) and colder oceanic lithosphere (Fig. 7) result in decoupling of the convergent plates. Transition from coupled to decoupled regime occurs always at the early stages of continental collision indicating that insertion of rheologically weak crustal material in the subduction channel is critical for the subsequent evolution of the collision zone (Faccenda et al., 2009). The numerical models confirm that HP-UHP complexes can be formed in both coupled and decoupled channels in the wide range of convergence scenarios (Figs 2-7).

As discussed in Faccenda et al. (2009), coupled collision zones (which can be either retreating or advancing) are characterized by a thick crustal wedge and compressive stresses (i.e. Himalaya and Western Alps), while decoupled end-members (which are always retreating) are defined by a thin crustal wedge and bi-modal distribution of stresses (i.e. compressional in the foreland and extensional in the inner part of the orogen, Northern Apennines).

5.3 Thrust fault formation and exhumation of (U)HP units

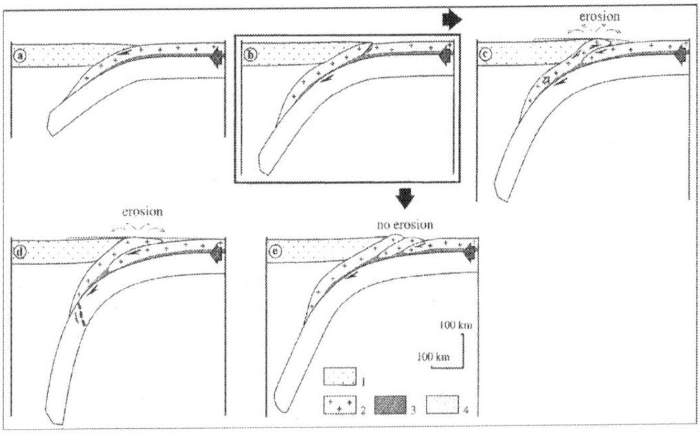

Fig. 9. Highly-compressional regime of continental subduction (low pull force) after Chemenda et al. (1995, 2000). (a), (b) and (e) are successive stages of continental subduction in experiments without erosion. (a)-(d) show continental subduction in experiments with erosion. 1, overriding plate; 2, upper crust; 3, lower crust, 4, eroded material (sediments).

One of the most important characteristics of the numerical models presented in this study is the formation of the thrust faults (rheological weak zones), which is followed by exhumation processes (e.g. Fig. 2). Similar detachment phenomenon is also documented in analogue models of continental subduction (Fig. 9; Chemenda et al., 1995, 1996, 2000).
The behavior of the subducted continental crust depends on two competing effects: upward buoyancy and downward subduction drag as discussed in Section 5.1. Subduction drag within the crust and mantle drives the subduction of buoyant crustal materials into larger depths (Fig 2a). At the same time the buoyancy forces and also the deviatoric stresses increase in the subduction channel. When the materials are no longer strong enough to sustain the accumulated buoyancy and deviatoric stresses, the subducted continental crust will yield with forming the rheological weak zone (thrust fault) (Fig. 2b,c) followed by the detachment and exhumation of the buoyant crustal materials (Fig. 2d-f) and release of the accumulated buoyancy and deviatoric stresses.

5.4 Upper crustal structure of the HP-UHP terrane

The upper-crustal settings of many UHP terranes share a number of structural characteristics (Fig. 10a; Beaumont et al., 2009): (1) a dome structure cored by the UHP nappe, (2) domes flanked by low-grade accretionary wedge and/or upper crustal sedimentary rocks, (3) overlying and underlying medium- to high-pressure nappes, (4) suture zone ophiolites and (5) foreland-directed thrust faults and the syn-exhumation normal faults. Our numerical models reproduce the general characteristic upper crustal structures (Fig. 10b), especially the dome structure of the HP-UHP cores, the flanked and overlaid low-grade accretionary wedge,

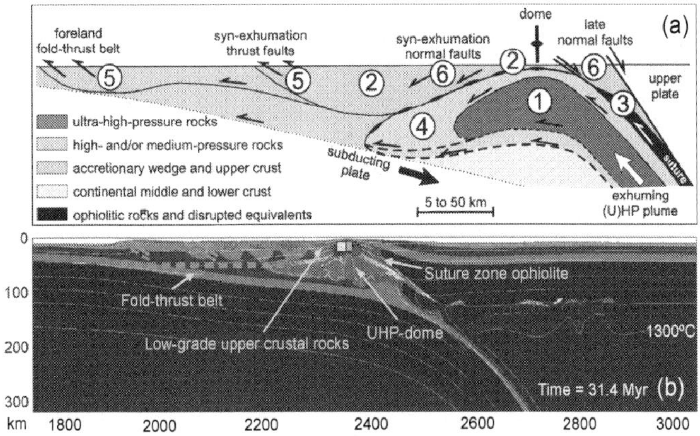

Fig. 10. (a) General characteristics of UHP complexes and surrounding upper crust (Beaumont et al., 2009). 1, structural dome cored by UHP nappe; 2, overlying lower grade rocks; 3, suture zone ophiolites; 4, medium- to high-pressure nappes; 5, foreland-directed thrust faults; 6, syn-exhumation normal faults. Variable scale reflects size range of UHP complexes. (b) Structure of UHP units and surrounding upper crust in the reference numerical model (Fig. 2f).

the foreland-directed thrust faults and the fold-thrust belt. The ophiolites are distributed near the suture zone in the reference model (Fig. 10b).

5.5 Influence of tectonic overpressure on the P-T paths of (U)HP rocks

The principle of lithostatic pressure is habitually used in metamorphic geology to calculate burial/exhumation depth from pressure given by geobarometry. However pressure deviation from lithostatic, i.e. tectonic overpressure/underpressure due to deviatoric stress and deformation, is an intrinsic property of flow and fracture in all materials, including rocks under geological conditions (e.g. Rutland, 1965; Brace et al., 1970; Mancktelow, 1993, 1995, 2008; Petrini and Podladchikov, 2000). Therefore, one important question is whether the principle of lithostatic pressure is applicable in subduction/collision zones where crystallization and exhumation of HP-UHP rocks take place. Some authors have argued that rocks under geological conditions are too weak to support significant overpressure (e.g. Brace et al., 1970; Ernst, 1971; Burov et al., 2001; Renner et al., 2001; Green, 2005). Yet, Stüwe and Sandiford (1994) suggested that petrologically derived P-T paths may not record depth changes only but stress changes. In addition, lithospheric-scale numerical models reveal regions where pressure may be hundreds of MPa or even several GPa higher or lower than lithostatic values (Mancktelow, 1993, 1995; Petrini and Podladchikov, 2000; Burg and Gerya, 2005). The analytical solutions show that the tectonic overpressure can be as high as 60% of the lithostatic value in the brittle regime. In contrast, the ductile overpressure is normally <10% of the lithostatic value (Li et al. 2010).

Li et al. (2010) conducted systematic numerical simulations of continental subduction/collision zones with variable brittle and ductile rheologies of the crust and mantle. In the numerical model (Figs 11, 12), the uppermost lithospheric mantle that can be considered as the wall of the subduction channel shows the largest tectonic overpressure ($\geq$1 GPa and $\geq$50% of the lithostatic pressure). However, these overpressured zones rarely

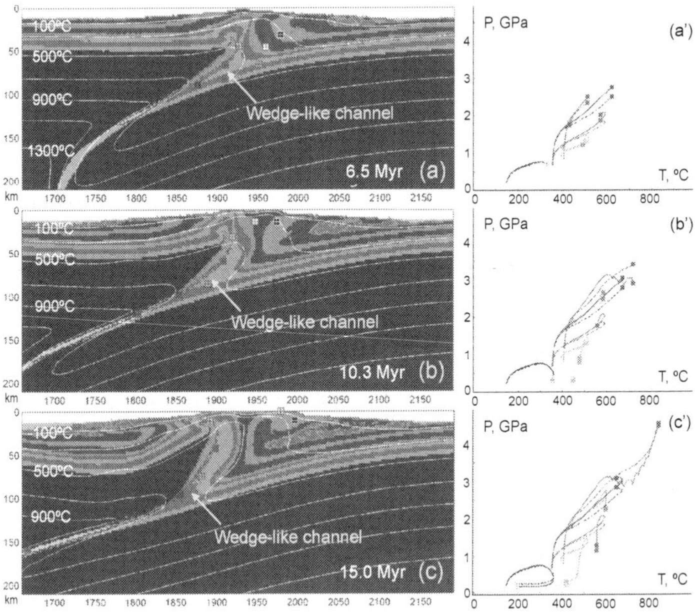

Fig. 11. Evolution of the wedge-like subduction channel within enlarged 530 × 210 km domain of the original 4000 × 670 km model. Time (Myr) of shortening is given in the figures. White numbered lines are isotherms in °C. Small coloured squares (with '+' in them) show positions of representative markers (rock units) for which *P-T* paths are shown (right). Colours of these squares are used for discrimination of marker points plotted in the *P-T* diagrams and do not correspond to the colours of rock types. The solid curves show *P-T* paths with dynamic pressure, while the dashed curves show that with lithostatic pressure.

or never participate in the exhumation processes. Hence, they do not influence the *P-T* conditions of geologically distributed HP-UHP rocks in nature. The main overpressure region that may influence the *P-T* paths of HP-UHP rocks is located in the bottom corner of the wedge-like confined channel (Fig. 11) with the characteristic magnitude of pressure deviation on the order of ~0.3 GPa and 10-20% from the lithostatic values (Fig. 12). The degree of confinement of the subduction channel is the key factor controlling this magnitude. The models show that the overpressure is small (~10% lithostatic) and should not affect in a crucial way the metamorphic mineral equilibria of the exhumed UHP rocks. The challenge would be to identify the geological record to actually measure precisely such minor deviations. An important unresolved issue concerns the question of how the tectonic overpressure could be potentially recorded by mineral equilibria in natural rocks. Thermodynamic tools used for thermobarometry of natural rocks are mainly based on experimental data obtained under conditions of an isotropic stress state (i.e. in the absence of significant deviatoric stresses) and may not be directly applicable for recording dynamic pressure in strongly stressed rocks. Indeed, not all overpressured rocks should be strongly internally stressed. For example, weak (e.g. reacting, fluid rich) rock inclusions in strong overpressured stressed lithosphere will also have strong overpressure, but the stress state will

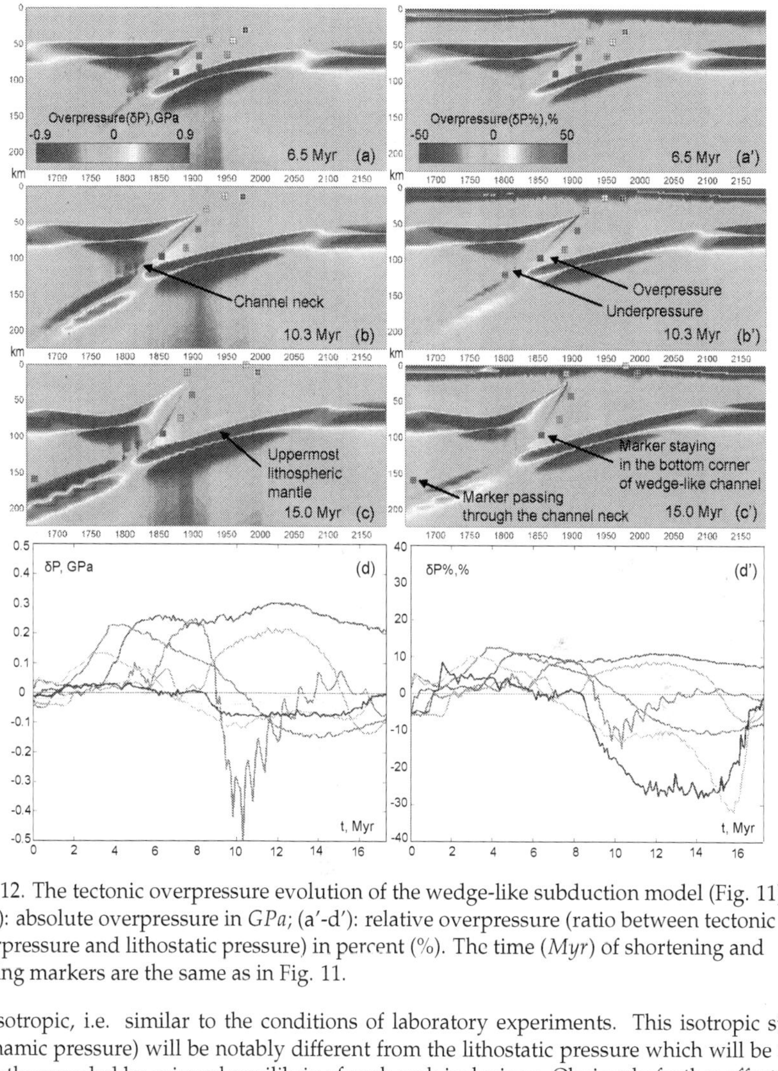

Fig. 12. The tectonic overpressure evolution of the wedge-like subduction model (Fig. 11). (a-d): absolute overpressure in *GPa*; (a'-d'): relative overpressure (ratio between tectonic overpressure and lithostatic pressure) in percent (%). The time (*Myr*) of shortening and tracing markers are the same as in Fig. 11.

be isotropic, i.e. similar to the conditions of laboratory experiments. This isotropic stress (dynamic pressure) will be notably different from the lithostatic pressure which will be then directly recorded by mineral equilibria of such rock inclusions. Obviously further efforts are needed to experimentally study mineral equilibria in stressed rocks.

The tectonic overpressure is also investigated for several different subduction/collision and exhumation scenarios (Fig. 13). In these numerical models, the overpressures are not significant in the mature subduction channel and/or the inner collision belt, which suggests that the overpressure that can possibly affect the HP-UHP rocks is mainly related to the wedge-like confined subduction channel (Figs 11, 12).

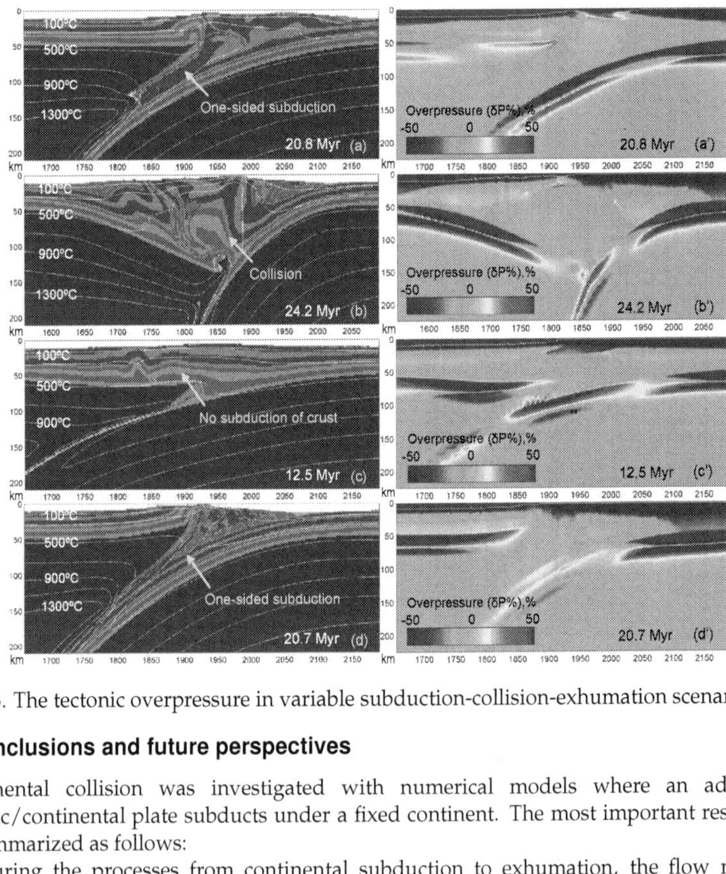

Fig. 13. The tectonic overpressure in variable subduction-collision-exhumation scenarios.

6. Conclusions and future perspectives

Continental collision was investigated with numerical models where an advancing oceanic/continental plate subducts under a fixed continent. The most important results can be summarized as follows:

(1) During the processes from continental subduction to exhumation, the flow mode in the subduction channel changes from Couette-dominated to Poiseuille-dominated flows. Numerical models of the subduction channels can be conveniently interpreted in terms of the characteristic exhumation number.

(2) The coupled subduction channel is formed in the models with slower convergence or hotter oceanic lithosphere. Whereas faster convergence or colder oceanic lithosphere result in decoupling of the converging plates.

(3) The continental margin can subduct following the oceanic plate to >100 *km* depth, and then exhume to the surface forming a HP-UHP dome. The upper crustal structures of the collision zone in the numerical models are consistent with both the analogue model and the natural UHP zones. The exhumation of UHP rocks occurs in a large variety of numerical parameters.

(4) The main tectonic overpressure region that may influence the *P-T* paths of HP-UHP rocks is located in the bottom corner of the wedge-like confined channel with the characteristic magnitude of pressure deviation on the order of $\sim 0.3\,GPa$ and 10-20% from the lithostatic values. The degree of confinement of the subduction channel is the key factor controlling this magnitude. The tectonic overpressure should not affect in a crucial way the metamorphic

mineral equilibria of the exhumed UHP rocks. The challenge would be to identify the geological record to actually measure precisely such minor deviations.

The numerical modeling method is demonstrated to be a great tool to study the geodynamic processes in the continental convergent margins. Most of the existed numerical models in the relevant topics are based on the two-dimensional regimes with the along-strike variations ignored. However, the tectonic processes remain inherently three-dimensional. So it is quite significant to conduct 3D numerical modeling to investigate the dynamics of continental collision with applicable to the natural tectonic settings (e.g. Alpine and Himalayan collision belts).

7. Acknowledgments

The research leading to this work receives funding from Crystal2Plate, a FP7-funded Marie Curie Action under grant agreement number PITN-GA-2008-215353, to ZHL. N. Ribe is thanked for fruitful discussion.

8. References

Beaumont, C.; Jamieson, R.A., Nguyen, M.H. & Lee, B. (2001). Himalayan tectonics explained by extrusion of a low-viscosity crustal channel coupled to focused surface denudation. *Nature*, Vol. 414, pp. 738-742

Beaumont, C.; Jamieson, R.A., Butler, J.P. & Warren, C.J. (2009). Crustal structure: A key constraint on the mechanism of ultra-high-pressure rock exhumation. *Earth and Planetary Science letters*, Vol. 287, pp. 116-129

Bittner, D. & Schmeling, H. (1995). Numerical modeling of melting processes and induced diapirism in the lower crust. *Geophysical Journal International*, Vol. 123, pp. 59-70

Brace, W.F.; Ernst, W.G. & Kallberg, R.W. (1970). An experimental study of tectonic overpressure in Franciscan rocks. *Geological Society of America Bulletin*, Vol. 81, pp. 1325-1338

Brace, W.F. & Kohlstedt, D.T. (1980). Limits on lithospheric stress imposed by laboratory experiments. *Journal of Geophysical Research*, Vol. 85, pp. 6248-6252

Brown, M. (2006). Duality of thermal regimes is the distinctive characteristic of plate tectonics since the Neoarchean. *Geology*, Vol. 34, No. 11, pp. 961-964

Brown, M. (2007). Metamorphic conditions in orogenic belts: A record of secular change. *International Geological Review*, Vol. 49, pp. 193-234

Burg, J.-P. & Gerya, T.V. (2005). The role of viscous heating in Barrovian metamorphism of collisional orogens: thermomechanical models and application to the Lepontine Dome in the Central Alps. *Journal of Metamorphic Geology*, Vol. 23, pp. 75-95

Burov, E.; Jolivet, L., Le Pourhiet, L. & Poliakov, A. (2001). A thermomechanical model of exhumation of high pressure (HP) and ultra-high pressure (UHP) metamorphic rocks in Alpine-type collision belts. *Tectonophysics*, Vol. 342, pp. 113-136

Chemenda, A.I.; Mattauer, M., Malavieille, J. & Bokun, A. (1995). A mechanism for syn-collisional rock exhumation and associated normal faulting: Results from physical modeling. *Earth and Planetary Science letters*, Vol. 132, pp. 225-232

Chemenda, A.I.; Mattauer, M. & Bokun, A. (1996). Continental subduction and a mechanism for exhumation of high-pressure metamorphic rocks: new modelling and field data from Oman. *Earth and Planetary Science letters*, Vol. 143, pp. 178-182

Chemenda, A.I.; Burg, J.-P. & Mattauer, M. (2000). Evolutionary model of the Himalaya-Tibet system: geopoem based on new modelling, geological and geophysical data. *Earth and Planetary Science letters*, Vol. 174, pp. 397-409

Chopin, C. (2003). Ultrahigh-pressure metamorphism: tracing continental crust into mantle. *Earth and Planetary Science letters*, Vol. 212, pp. 1-14

Clauser, C. & Huenges, E. (1995). Thermal conductivity of rocks and minerals, In: *Rock physics and phase relations*, Ahrens, T.J. (Ed.), pp. 105-126, American Geophysical Union, Reference Shelf 3, Washington D.C.

Cloos, M. (1982). Flow melanges: Numerical modeling and geologic constraints on their origin in the Franciscan subduction complex, California. *Geological Society of America Bulletin*, Vol. 93, pp. 330-344

Cloos, M. & Shreve, R.L. (1988a). Subduction-channel model of prism accretion, melange formation, sediment subduction, and subduction erosion at convergent plate margins: Part 1, Background and description. *Pure Applied Geophysics*, Vol. 128, pp. 455-500

Cloos, M. & Shreve, R.L. (1988b). Subduction-channel model of prism accretion, melange formation, sediment subduction, and subduction erosion at convergent plate margins: Part II, Implications and discussion. *Pure Applied Geophysics*, Vol. 128, pp. 501-545

England, P.C. & Holland, T.J.B. (1979). Archimedes and the Tauern eclogites: The role of buoyancy in the preservation of exotic eclogite blocks. *Earth and Planetary Science letters*, Vol. 44, pp. 287-294

Ernst, W.G. (1971). Domineral parageneses reflect unusually high-pressure conditions of Franciscan metamorphism? *American Journal of Sciences*, Vol. 270, pp. 81-108

Faccenda, M.; Gerya, T.V. & Chakraborty, S. (2008). Styles of post-subduction collisional orogeny: Influence of convergence velocity, crustal rheology and radiogenic heat production. *Lithos*, Vol. 103, pp. 257-287

Faccenda, M.; Minelli, G. & Gerya, T. (2009). Coupled and decoupled regimes of continental collision: Numerical modeling. *Earth and Planetary Science letters*, Vol. 278, pp. 337-349

Gerya, T.V. & Yuen, D.A. (2003a). Characteristics-based marker-in-cell method with conservative finite-differences schemes for modeling geological flows with strongly variable transport properties. *Physics of the Earth and Planetary Interior*, Vol. 140, No. 293-318

Gerya, T.V. & Yuen, D.A. (2003b). Rayleigh-Taylor instabilities from hydration and melting propel cold plumes at subduction zones. *Earth and Planetary Science letters*, Vol. 212, pp. 47-62

Gerya, T.V.; Perchuk, L.L. & Burg, J.-P. (2008). Transient hot channels: Perpetrating and regurgitating ultrahigh-pressure, high-temperature crust-mantle associations in collision belts. *Lithos*, Vol. 103, pp. 236-256

Green, H.W. (2005). Psychology of a changing paradigm: 40+ years of high-pressure metamorphism. *International Geological Review*, Vol. 47, pp. 439-456

Li, Z. & Gerya, T.V. (2009). Polyphase formation and exhumation of HP-UHP rocks in continental subduction zone: Numerical modelling and application to the Sulu UHP terrane in eastern China. *Journal of Geophysical Research*, Vol. 114, No. B09406, doi:10.1029/2008JB005935

Li, Z.H.; Gerya, T.V. & Burg, J.-P. (2010). Influence of tectonic overpressure on P-T paths of HP-UHP rocks in continental collision zones: thermomechanical modelling. *Journal of Metamorphic Geology*, Vol. 28, pp. 227-247

Li, Z.H.; Xu, Z.Q. & Gerya, T.V. (2011). Flat versus steep subduction: contrasting modes for the formation and exhumation of high- to ultrahigh-pressure rocks in continental collision zones. *Earth and Planetary Science letters*, Vol. 301, pp. 65-77

Mancktelow, N.S. (1993). Tectonic overpressure in competent mafic layers and the development of isolated eclogites. *Journal of Metamorphic Geology*, Vol. 11, pp. 801-812

Mancktelow, N.S. (1995). Nonlithostatic pressure during sediment subduction and the development and exhumation of high-pressure metamorphic rocks. *Journal of Geophysical Research*, Vol. 100, pp. 571-583

Mancktelow, N.S. (2008). Tectonic pressure: Theoretical concepts and modelled examples. *Lithos*, Vol. 103, pp. 149-177

Petrini, K. & Podladchikov, Y. (2000). Lithospheric pressure-depth relationship in compressive regions of thickened crust. *Journal of Metamorphic Geology*, Vol. 18, pp. 67-77

Pinkerton, H. & Stevenson, R.J. (1992). Methods of determining the rheological properties of magmas at subliquidus temperatures. *Journal of Volcanology and Geothermal Research*, Vol. 53, pp. 47-66

Pope, D. & Willett, C. (1998). Thermal-mechanical model for crustal thickening in the central Andes driven by ablative subduction. *Geology*, Vol. 26, No. 6, pp. 511-514

Raimbourg, H.; Jolivet, L. & Leroy, Y. (2007). Consequences of progressive eclogitization on crustal exhumation, a mechanical study. *Geophysical Journal International*, Vol. 168, pp. 379-401

Ranalli, G. (1995). *Rheology of the Earth, Deformation and flow processes in geophysics and geodynamics, 2nd edition*, Chapman & Hall, London

Renner, J.; Stöckhert, B., Zerbian, A., Roller, K. & Rummel, F. (2001). An experimental study into the rheology of synthetic polycrystalline coesite aggregates. *Journal of Geophysical Research*, Vol. 106, pp. 19411-19429

Rutland, R.W.R. (1965). Tectonic overpressures, In: *Controls of Metamorphism*, Pitcher, W.S. & Flinn, G.W. (Eds), pp. 119-139, Oliver and Boyd, Edinburgth

Schmidt, M.W. & Poli, S. (1998). Experimentally based water budgets for dehydrating slabs and consequences for arc magma generation. *Earth and Planetary Science letters*, Vol. 163, pp. 361-379

Stöckhert, B. & Gerya, T.V. (2005). Pre-collisional high pressure metamorphism and nappe tectonics at active continental margins: a numerical simulation. *Terra Nova*, Vol. 17, pp. 102-110

Stüwe, K. & Sandiford, M. (1994). Contribution of deviatoric stresses to metamorphic P-T paths: an example appropriate to low-P, high-T metamorphism. *Journal of Metamorphic Geology*, Vol. 12, pp. 445-454

Stüwe, K. (1995). Thermal buffering effects at the solidus, Implications for the equilibration of partially melted metamorphic rocks. *Tectonophysics*, Vol. 248, pp. 39-51

Tao, W. & O'Connell, R. (1992). Ablative subduction: a two-sided alternative to the conventional subduction model. *Journal of Geophysical Research*, Vol. 97, No. B6, pp. 8877-8904

Toussaint, G.; Burov, E. & Avouac, J.-P. (2004a). Tectonic evolution of a continental collision zone: A thermomechanical numerical model. *Tectonics*, Vol. 23, No. TC6003, doi:10.1029/2003TC001604

Toussaint, G.; Burov, E. & Jolivet, L. (2004b). Continental plate collision: Unstable vs. stable slab dynamics. *Geology*, Vol. 32, pp. 33-36

Turcotte, D.L. & Schubert, G. (1982). *Geodynamics*, John Wiley, 450, New York

Turcotte, D.L. & Schubert, G. (2002). *Geodynamics, Second Edition*, Cambridge University Press, United Kingdom

Warren, C.J.; Beaumont, C. & Jamieson, R.A. (2008a). Modelling tectonic styles and ultra-high pressure (UHP) rock exhumation during the transition from oceanic subduction to continental collision. *Earth and Planetary Science letters*, Vol. 267, pp. 129-145

Warren, C.J.; Beaumont, C. & Jamieson, R.A. (2008b). Formation and exhumation of ultra-high pressure rocks during continental collision: Role of detachment in the subduction channel. *Geochemistry, Geophysics, Geosystems*, Vol. 9, No. Q04019, doi:10.1029/2007GC001839

Yamato, P.; Agard, P., Burov, E., Le Pourhiet, L., Jolivet, L. & Tiberi, C. (2007). Burial and exhumation in a subduction wedge: Mutual constraints from thermomechanical modeling and natural P-T-t data (Schistes Lustres, western Alps). *Journal of Geophysical Research*, Vol. 112, No. B07410, doi:10.1029/2006JB004441

Yin, A. (2006). Cenozoic tectonic evolution of the Himalayan orogen as constrained by along-strike variation of structural geometry, exhumation history, and foreland sedimentation. *Earth-Science Review*, Vol. 76, pp. 1-131

Zhu, G.; Gerya, T.V., Yuen, D.A., Honda, S., Yoshida, T. & Connolly, J.A.D. (2009). 3-D Dynamics of hydrous thermalchemical plumes in oceanic subduction zones. *Geochemistry, Geophysics, Geosystems*, Vol. 10, No. Q11006, doi:10.1029/2009GC002625

14

The Thermogeographic Model in Paleogeography: Application of an Abiotic Model to a Plate Tectonic World

Lee-Ann C. Hayek[1] and Walter H. Adey[2]
[1]*National Museum of Natural History, Statistics and Mathematics,*
[2]*National Museum of Natural History, Department of Botany,*
Smithsonian Institution, Washington D.C.
USA

1. Introduction

1.1 Marine biogeography and paleoecology

Recent coastal marine biogeography has a long and venerable history. (Ekman, 1953; Briggs, 1974, 1995; Vermejj, 1978). However, it has lacked a substantial theory and has been the subject of widespread criticism (Longhurst, 1998; Rosenzweig, 1995). The most recent research, driven by the need for coastal conservation, has added satellite remote sensing and constructions based on present day sea surface temperature (Spalding et al, 2007). Based on largely subjectively-created assemblages of organisms and sea surface temperatures (SST), these studies are increasingly sophisticated creating hundreds of ecoregions where previously biogeography researchers had created dozens. However, mostly they fail to consider that macro-organism distribution, the focus of virtually all biogeographical analyses, has not developed solely based on SST characteristics that have lasted for only a few millennia at most. The evolution of stable assemblages of macro-organisms requires geological time, since it is based in the evolution of its component organisms.

In paleoecological analysis, understanding the present has always been regarded as a key to the past and modern studies normally make that connection (Brenchley and Harper, 1998). However, paleoecological analyses by definition, treats change with geological time and is based in the evolution of organisms, typically over millenia to millions of years. When the analog is coastal biocoenoses, the modern analog studies are mostly fixed Intime. Coastal Thermogeographic Regions as defined by the Adey and Steneck (2001) five-dimensional abiotic model (TM). The principal variables were mean minimum and maximum surface temperature, with coastal area over time (present and 18K) appearing as contours. Isolation by oceans and continents, i.e., northern and southern and Atlantic and Pacific/Indian Oceans, were introduced by separating the main diagram into quadrants, and then stretching some overlapping coasts. The strength of each Region is represented by the number of contours of coastal area that is constant over Pleistocene time.

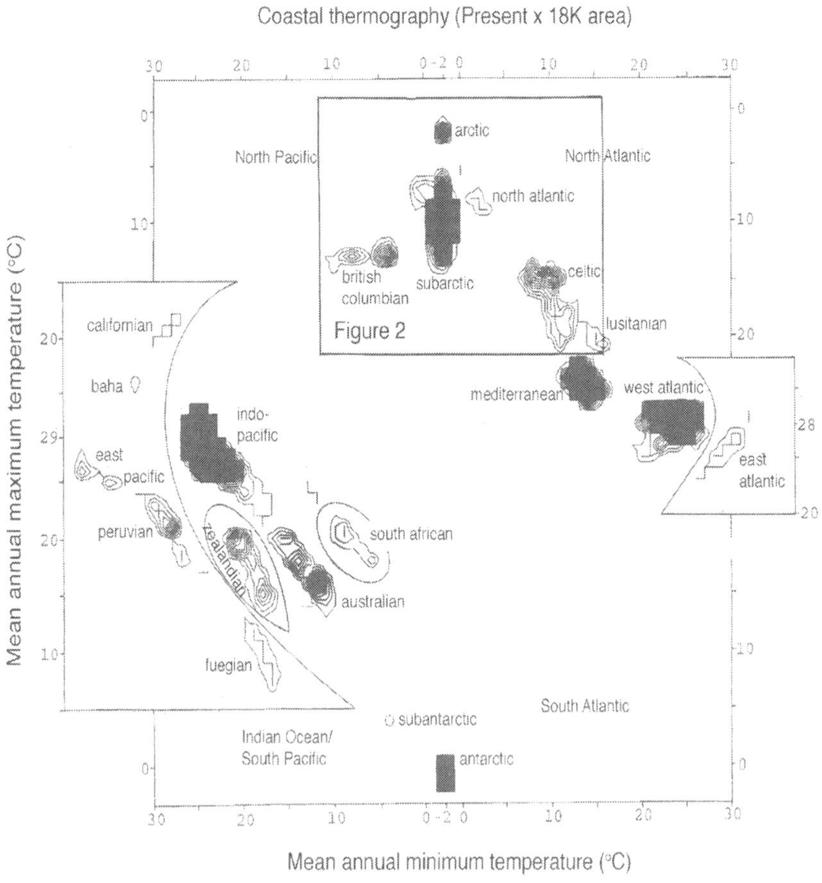

Fig. 1. Coastal Thermogeographic Regions

1.2 Thermogeography over time

Adey and Steneck (2001) developed a time-integrated thermogeographic model to demonstrate why benthic marine algal assemblages of coastal rocky marine, sublittoral zones, develop biogeographic patterns in their distribution and abundance. The TM is a predictive, abiotic model in which the maximum and minimum sea surface temperatures (sea climate) are tabulated and plotted for each nautical mile of rocky coastline for both the present (1955) and the Last Glacial Maximum (LGM) (CLIMAP, 1976; see Adey and Steneck, 2001). These two alternate states (glacial and interglacial) characterize the principal climatic character of the global marine realm since late Pleistocene time (0.7–1.8 Ma); it is during this time that most living species have evolved (Briggs, 1995). Some sea climates (specific thermal regimes) have a large amount of coast; others none. Some sea climates with large coastal areas remain more or less constant through Pleistocene time, while other sea climates are absent or ephemeral (lacking major coastlines over Pleistocene time). When the sea climate/coastal area contour diagrams for LGM (glacial) and Present (interglacial) times are mathematically integrated, a contour diagram results (Fig. 1.) which is the Thermogeographic Model. In paleoecological and ecological study, the identification of unique biofacies boundaries and the structure of the assemblages on either side of those boundaries is critical to identifying environmental changes with time. With the application of SHE we show an expansion of the results of the TM and provide a framework for examining the distinct biofacies based upon their species composition.

The resulting thermogeographic model (TM; Fig. 1) defines 20 regions that correspond with the cores of 24 traditionally recognized biogeographic regions determined by published distributions of organisms. The four remaining classical regions were weak and disputed or lacked significant rocky shore. In the colder North Atlantic, the primary traditionally defined regions (e.g., Briggs, 1974), the Western Atlantic Boreal and the Eastern Atlantic Boreal are represented by equivalent regions in the Thermogeographic Model (TM) (Subarctic-west; Boreal or Celtic-east). However, the boundaries of those regions in the western Atlantic are markedly different between the classical organism-defined approach (Cape Hatteras to the Strait of Belle Isle; Briggs 1974), and the physical/time model (Newfoundland and northern Gulf of St. Lawrence to central Labrador, TM, Adey and Steneck, 2001). (Fig. 2).

Thermogeographic regions (TRs), although clearly correlated with most classical biogeographic regions, have many shapes in area/sea climate space, from elliptical to two- or three-lobed. In a few cases, two relatively strong TRs are conjoined by a narrow isthmus. Most striking, however, is that 10–20% of the world's coastlines do not belong to any TR. These transitional zones result from shifting Pleistocene climates and their contained coastlines. An important finding of the TM is that the rocky western North Atlantic Coast from Long Island Sound through the Gulf of Maine, and the southern Gulf of St. Lawrence comprises one of these transitional zones (Adey and Steneck, 2001, see their Fig. 7).

Adey and Steneck (2001) also proposed that biogeographic patterns should be determined by quantitatively analyzing community assemblages. They tested the efficacy of thermogeographic regions, as determined by the abiotic TM, with abundance-determined patterns in the distribution of crustose coralline red algae (Rhodophyta/Corallinales) in the colder part of the northern hemisphere. The corallines have special relevance to paleoecology in that they have a fossil history dating back to the Mesozoic and perhaps earlier (Johansen, 1981). By mid-Tertiary they are rock-formers and important index fossils (Bassi and Nebelsick, 2010). Adey and Hayek (2011) also validated the TM by using a large data set for fleshy seaweed assemblages in the same regions in the NW Atlantic. Their analysis used several test statistics and models for validation, and provided a pictorial summary diagram based in the Bray Curtis similarity matrix (Fig. 3).

Herein we use the biomass sample data from Adey and Hayek (2011) to test the biofacies determined by the TM. We analyze this data and compare the individual depth biofacies to the TM results. We use an approach called SHE (Hayek and Buzas, 1997; 2010) Analysis that was developed specifically for biofacies identification Buzas and Hayek, 1998) uses all aspects of the taxon data as advocated by Adey and Steneck (2001). Using this method in combination with a new conditional biofacies boundary indicator of taxon importance termed CoBBI our results show strong support for the TM at all depths in this biogeographic region of the North Atlantic.

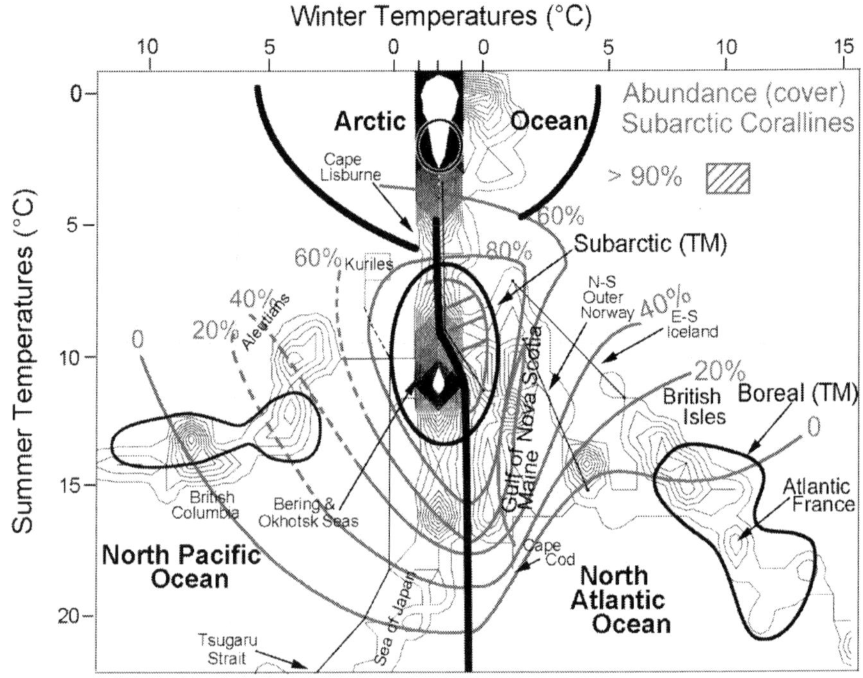

Fig. 2. The distribution of Subarctic crustose coralline algal cover (blue contours) on a background of present-day coastal area distribution as a function of summer-winter temperatures as determined in the Adey and Steneck (2001) TM. The background diagram shows the temperate to Arctic coastal area/temperature distribution with the resultant thermogeographic regions (ellipses) superimposed. The location of "core" Subarctic Coast (northern Gulf of St. Lawrence; northeastern NF and Labrador) is shown as a light black line on the left side of the North Atlantic Subarctic ellipse. The line extends nearly to the Arctic circle, as it includes northern Labrador. The Gulf of Maine and Nova Scotia are shown as red lines with their intersection representing the Bay of Fundy.

Based in repeated successful tests of the validity of the TM, and our rapidly increasing knowledge of past shoreline configuration, grounded in plate tectonics, and our knowledge of past SST, centered on oceanic sediment microfossils, it is now possible to extend the biogeography of the TM into the past, provide an over-arching theory for paleobiogeography and link the very patchy information of marine paleoecology.

2. Methods

2.1 Sampling

With the intention of comparing seaweed populations in the Subarctic Region (NLQ) (as determined by the Thermogeographic Model) with those in the Subarctic/Boreal transition coasts of Nova Scotia (SNS) and the Gulf of Maine (GOM),, sets of sampling stations ranging from exposed shores through those of intermediate exposure to highly protected sites were occupied every 50–100 km along each of those coasts (Fig. 4). Adey and Hayek (2011) provide locations and description of those stations.

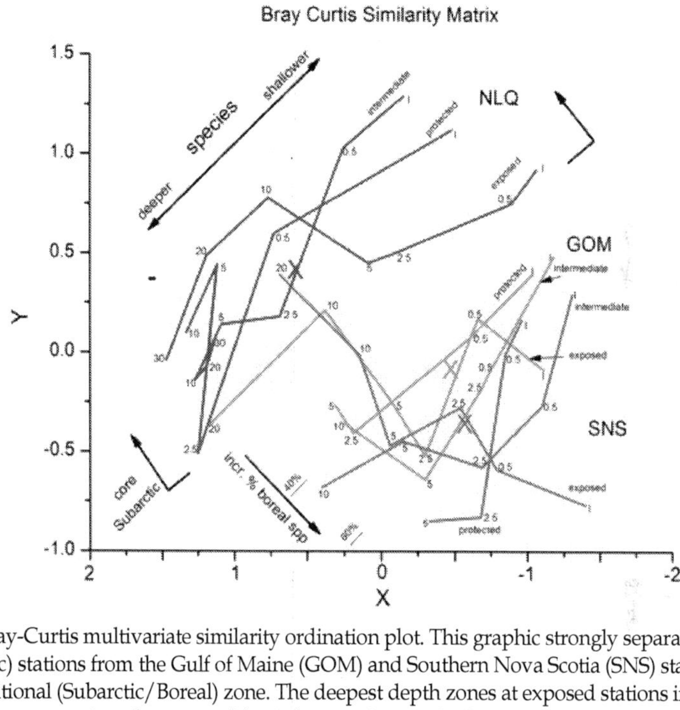

Fig. 3. Bray-Curtis multivariate similarity ordination plot. This graphic strongly separates NLQ (Subarctic) stations from the Gulf of Maine (GOM) and Southern Nova Scotia (SNS) stations in the transitional (Subarctic/Boreal) zone. The deepest depth zones at exposed stations in GOM and SNS crossover into the range of the Subarctic; this results from the dominance of Subarctic species related to a strong thermocline, and colder temperatures at depth.

At most stations, one or two one-meter-square PVC quadrats were dropped by SCUBA at each of six depth zones (0.5, 2.5, 5, 10, 20, 30 m) depending mostly on substrate availability (see Adey and Hayek, 2011; Fig. 18) or occasionally the apparent photic limit (especially for GOM stations). In some cases, three or more replicate quadrats were taken at some depth-zones when time was available.

The infralittoral zone (i.e., area between low water spring tides and low water neap tides) was also occupied at most stations, but a 1/10th m^2 quadrat was used because of the frequent narrowness of the zone. Quadrat location at each station-depth zone was random, as typically the PVC square meter was dropped when the bottom came in view to the descending diver. After the quadrat was set on the bottom, a diver removed all macroscopic fleshy algae with a dive knife and placed them in a small mesh plastic dive bag.

Within each region, as discussed above, stations were selected for their wave exposure characteristics. As we shall demonstrate, wave exposure was a critical factor in determining macroalgal assemblages, and regional comparisons need to be based on similar exposure characteristics. All exposed stations occupied in this study occur on the open coast where they are subject to sea and swell at very large fetch and wide angle (23°-204°).

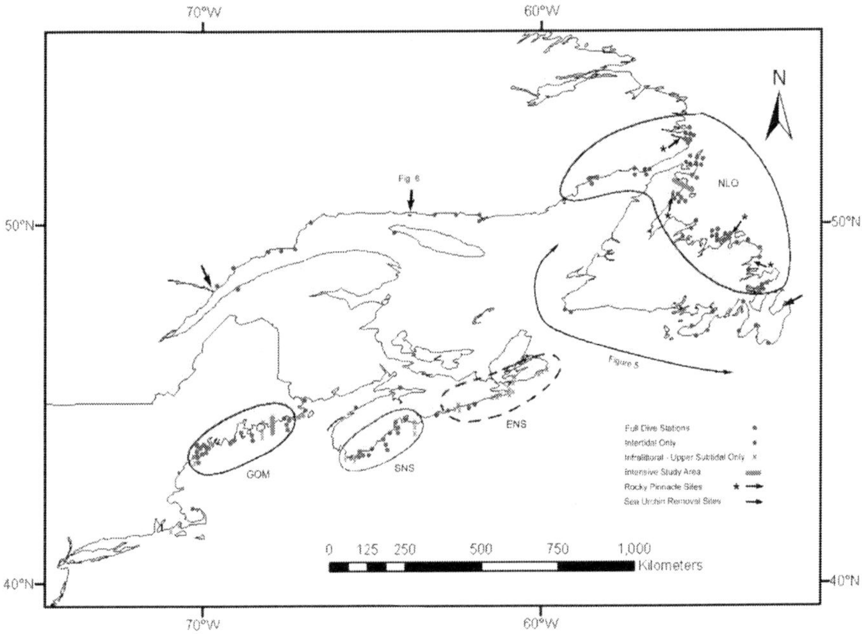

Fig. 4. Stations occupied in the Adey and Hayek (2005) project in the northwestern North Atlantic. The solid red dots are the intertidal stations occupied in the initial survey. The X's are infralittoral/0.5 m stations occupied primarily to compare Atlantic Nova Scotia, sea-ice affected shores with non sea ice shores. The solid blue dots are full dive stations and the two long dashes represent areas of numerous local stations and bottom mapping; the dash on the east side of the northern peninsula of NF represents two separate areas: Lunaire Road to the north and Canada Bay/Englee to the south

For the intermediate stations, open water exposure is more limited, a mean of 11-25° as compared to 103-117° for the exposed shores; however, such stations are set back from the outer shore 1.0, 2.9, and 5.9 km respectively for NLQ, GOM, and SNS. In summary, open ocean exposure angle drops from a mean of 109° for exposed stations to 17° for intermediate stations to 2° for protected stations, while the exposure distance changes respectively from 0 to 3 km to 7 km. There is no significant difference between NLQ, GOM, and SNS regions for exposed stations (Adey and Hayek, 2011).

2.2 Data

The data set consists of numerical abundances on 70 algal taxa collected at 7 depths in exposed stations, 5 depths in intermediate stations and 5 depths at protected stations. We chose to examine only 5 depths for each station type (exposed, intermediate and protected), so that comparisons could be made across taxa. Figure 4 shows the location of stations within depth zones.

A total of 68 samples were taken in the infralittoral, 20 samples were taken in the protected, 24 intermediate and 24 exposed. There were 71 samples taken at 0.5m, 21 protected area samples, 25 intermediate and 25 exposed area samples. Sixty nine samples were taken at a depth of 2.5m, 16 protected, 27 intermediate and 26 exposed. At a depth of 5m, 75 samples were taken, with 20 from protected sites, 26 from intermediate and 29 from exposed.

2.3 SHE Analysis for biofacies identification (SHEBI)

SHE analysis is the information - theoretic approach for biofacies identification and biodiversity analysis (Hayek and Buzas, 1997; 2010), composed of two variants. The first (SHEBI) variant or usage of SHE is for biofacies identification. After biofacies identification is complete, then within each biofacies so identified the second aspect termed SHECSI can be used to identify the community structure and ecological or evolutionary health status of the contained assemblage.

SHE is a methodology derived upon a conditional probabilistic basis with theorems from statistical entropy theory. No statistical testing of biofacies formation is required since each designated biofacies is a mathematical, closed, dynamic, faunal system and therefore replicable. For SHE, samples are accumulated (Hayek and Buzas, 1997) at separate depths over stations along the gradient from northernmost Newfoundland, through the Gulf of Maine to the southernmost zones of Nova Scotia (program available in Past ver 1.78 (2001) or from the author). At each step in this accumulation the values for additional new taxa, S, along with an information measure we denote as H, and an evenness measure E, which uses taxon proportions, are each calculated. The boundary of each biofacies is determined from the changing values of estimated evenness of the biomass proportions as we accumulate over samples (Buzas and Hayek, 1998; Osterman et al. 2002; Hayek et al, 2007; Wilson, 2008). The entropy of the system is examined as the expected value of this information measure (Hayek and Buzas, 2010) and a comprehensive snapshot of the depth-related fauna is obtained. As evenness decreases, dominance increases, randomness over the taxon space increases and uncertainty and entropy must decrease within any unified system (Hayek et al., 2007; Hayek and Buzas, 2010). When we observe an increase on a plot of log standardized N versus E, a new biofacies boundary has been reached.

2.4 Conditional Biofacies Boundary Index (CoBBI)

At any biofacies boundary, there is a total amount of change in species abundances, recognizable intuitively or quantitatively. This change is distinct from the subtle gradational taxon range overlap that is present throughout any ecological or paleoecological data set but never to our knowledge separately evaluated. After the boundary for each biofacies is determined, we developed a new index we call CoBBI (conditional biofacies boundary index). This index provides an examination of the taxon composition and abundance pattern only at the boundary. Calculations are based upon the total taxon assemblage accumulated just prior to and just after the biofacies boundary. In this way a total taxon change between biofacies is obtained. The change in abundance of each taxon across the boundary is used relative to the total change at the boundary. This index does not, as is usual, use the 100% change in the total assemblage over the entire set of samples and all its biofacies. This boundary-specific change is conditioned on only the chosen biofacies boundary, or specific faunal break, and thereby gives a picture of the entire assemblage within the biofacies of interest. With CoBBI we examine the prevalence of each species in the assemblages on either side of a biofacies boundary to obtain a total assemblage change between the biofacies

3. Results

The SHE methodology for biofacies recognition was applied for each of the 5 depths at each of the set of stations that were designated Protected, Intermediate and Exposed. We accumulated through the longitudinal gradient from Newfoundland through the Gulf of Maine to Nova Scotia. It is of considerable importance in palaoecology that this is not a straight longitudinal line: the climate change shifts eastwards due to oceanic versus continental effects. In general, biofacies boundaries were determined that separated each of these regions from the other. In some cases, in particular for the infralittoral, the Newfoundland samples did not always form a continuum within a single biofacies. After determination of biofacies and boundaries between these regions, we applied CoBBI to obtain a total percentage change or difference between each pair of regions geographically and then to characterize the taxon assemblage within each biofacies and the change at the boundary. Results are given below and summarized in Tables 1 and 2. The results for each biofacies boundary and total change are presented in Table 1. The taxa contributing to each of the boundaries are listed in Table 2A and B with their percentage contribution to the total amount of change at the boundary.

Depth Site	Protected	Intermediate	Exposed
Infralittoral	1. 139%	199%	199%
	2. 75%*1	75%*2	188%
0.5m	1. 199%	200%	182%
	2. 115%	157%	188%
2.5m	1. 196%	200%	100%
	2. 191%	198%	156%
5m	1. 195%	200%	200%
	2. 191%	176%	146%

*1 There was not a biofacies boundary in the infralittoral zone between the Gulf of Maine and Nova Scotia. There was a change in the magnitude of the slope to indicate an assemblage change of less magnitude than that of a biofacies break. The biofacies boundary determined by SHE was between Newfoundland plus the first sample from Gulf of Maine and the remainder of Gulf of Maine. 168% difference was observed for this break.

*2 There was not a boundary between Gulf of Maine and Nova Scotia samples. There was a boundary determined by SHE between the first 3 and the last 3 samples from Gulf of Maine. Then between Gulf of Maine and Nova Scotia there was a change in slope.

Table 1. Change in the taxon assemblage composition by depth and site between biofacies determined by SHE. 1. Newfoundland and Gulf of Maine biofacies ; 2. Gulf of Maine and Nova Scotia biofacies

3.1 Newfoundland and Gulf of Maine

3.1.1 Biofacies depth changes

Although especially for the infralittoral and 0.5m depths, the entirety of the samples from Newfoundland did not always form a single biofacies, the actual boundaries determined by SHEBI between the end of the sample stations in Newfoundland and the beginning in Gulf of Maine were always clear cut. There was a total change in the taxon assemblage of 139% between those biofacies in protected sites and 199% for intermediate and exposed sites (Table 1). Table 2A shows the most prominent contributors to this overall compositional change. In each of the infralittoral areas, *Chondrus crispus*, *Chordaria flagelliformis* and

Devaleraea ramentacea were among those most important contributors to this overall total biofacies change but provided distinctly different contributions. While *C. flagelliformis* was the most dominant in protected infralittoral, this species dropped to second then third most dominant as sites increased in exposure. Likewise C *crispus* was only 11% of the total of the 139% change over protected sites while it increased to 45% then 47% from intermediate to exposed. D *ramentacea* remained about 7% in protected and intermediate infralittoral but increased substantially to 27% dominance at the exposed infralittoral boundary between Newfoundland and Gulf of Maine.

A. Newfoundland to Gulf of Maine

1. Infralittoral

Protected	**Intermediate**	**Exposed**
C flagelliformis 28% (NLQ)	*C crispus* 45% (GOM)	*C crispus* 47% (GOM)
C purpureum 18% (GOM)	*C flagelliformis 33%* (NLQ)	*C flagelliformis 18%* (GOM)
C crispus 11% (GOM)	*D ramentacea* 7% (NLQ)	*D ramentacea* 27% (NLQ)
N harveyi 11% (GOM)	*Ceramium spp* 5% (GOM)	
A arcta 7% (NLQ)	*A arcta* 5% (NLQ)	
D ramentacea 7% (NLQ)		
F distichus 8% (NLQ)		

2. 0.5m

Protected	**Intermediate**	**Exposed**
C crispus 29% (GOM)	*C crispus* 30% (GOM)	*A esculenta* 31% (NLQ)
F distichus 14% (NLQ)	*A esculenta* 18% (NLQ)	*S longicrurus* 27% (GOM)
D foeniculaceus (NLQ)	*C flagelliformis* 15% (NLQ)	*S latissima* 13%(GOM)
D viridis 10% (NLQ)	*C officinalis* 17%(GOM)	*L digitata* 10% (GOM)
A clathratum 9%(NLQ)	*S dermatodea* 9% (NLQ)	*C flagelliformis* 8% (NLQ)
A esculenta 9% (NLQ)		*D ramentacea* 5% (NLQ)
P purpurea 6%(GOM)		*A arcta* 5% (NLQ)
C flagelliformis 5% (NLQ)		

3. 2.5m

Protected	**Intermediate**	**Exposed**
A clathratum 61% (NLQ)	*A clathratum* 33% (NLQ)	*S latissima* 41% (GOM)
D viridis 36% (NLQ)	*C crispus* 27% (GOM)	*A esculenta* 18% (NLQ)
	D viridis 17% (NLQ)	*L digitata* 10% (GOM)
	L digitata 13% (GOM)	*S longicrurus* 7% (GOM)
	C rubrum 8% (GOM)	*A clathratum* 5% (GOM)
	D viridis 5% (NLQ)	*D foeniculaceus* 5% (NLQ)

4. 5m

Protected	**Intermediate**	**Exposed**
D viridis 27% (NLQ)	*D viridis* 30% (NLQ)	*S latissima* 49% (GOM)
S latissima 27% (GOM)	*S longicruris* 43% (GOM)	*L digitata* 26% (GOM)
A clathratum 23% (NLQ)	*A clathratum* 20% (NLQ)	*A esculenta* 24% (NLQ)
S longicruris 21% (GOM)		

B. Gulf of Maine to Nova Scotia

5. Infralittoral

Protected	Intermediate	Exposed
C crispus 40% (SNS)	*C crispus* 34%	*C crispus* 25% (SNS)
F vesiculosus 23% (SNS)	*C officinalis* 44% (SNS)	*C officinalis* 19% (SNS)
F distichus 14% (SNS)	*A arcta* 8% (GOM)	*D ramentacea* 11% (GOM)
S latissima 9% (SNS)		
N harveyi 11% (GOM)		
A esculenta 10% (GOM)		
F distichus 7% (SNS)		
N multifidum 6% (SNS)		

6. 0.5m

Protected	Intermediate	Exposed
C officinalis 21% (SNS)	*C crispus* 24% (SNS)	*A esculenta* 41% (GOM)
S latissima 25% (SNS)	*C officinalis* 44% (SNS)	*C officinalis* 28% (SNS)
C crispus 20% (GOM)	*C purpureum* 10% (GOM)	*L digitata* 11% (SNS)
D foeniculaceus 8% (SNS)	*S latissima* 15% (SNS)	*F serratus* 5% (SNS)
D viridus 5%(GOM)	*A sp. 5%* (SNS)	

7. 2.5m

Protected	Intermediate	Exposed
S dermatodea 28% (SNS)	*F serratus* 26% (SNS)	*S longicruris* 24% (GOM)
L digitata 20% (SNS)	*C officinalis* 23% (SNS)	*L digtata* 20% (SNS)
S latissima 14% (GOM)	*S latissima* 18% (SNS)	*S latissima* 18% (GOM)
C crispus 9% (SNS)	*S longicrusis* 14% (GOM)	*A esculenta* 12% (GOM)
F serratus 6% (SNS)	*L digitata* 9% (SNS)	*F serratus* 7% (SNS)
C fragile 5% (SNS)		
C crispus 5% (SNS)		

8. 5m

Protected	Intermediate	Exposed
A clathratum 40% (GOM)	*L digitata* 21% (SNS)	*S latissima* 33% (GOM)
L digitata 8% (GOM)	*F serratus* 18% (SNS)	*L digitata* 34% (SNS)
P pseudoceranoides 5% (SNS)	*S latissima* 15% (SNS)	*C rubrum* 5% (SNS)
S latissima 5% (GOM)	*S longicruris* 14% (GOM)	*F serratus* 5% (SNS)
N harveyi 13% (GOM)		
P pseudoceranoides 8% (SNS)		

Table 2. Taxon contributors to the biofacies change in assemblage composition of A. Newfoundland to Gulf of Maine; B. Gulf of Maine and Nova Scotia. Relative percentages are only those that are over 5% of the total.

At a depth of 0.5m Table 1 shows that CoBBI found well over 150% change at each of the biofacies for protected, intermediate and exposed sites. Here as Table 2A shows *C crispus* was the most dominant contributor to the protected and intermediate biofacies boundaries but was not relevant for exposed. *C flagelliformis,* though found at each type of station set at

this depth was a minor contributor (5%) in protected and exposed (8%) but second most dominant (18%) after *C crispus* in intermediate. In the exposed areas at 0.5m depth, *Alaria esculenta* (31%) and *Saccharina longicurus* (27%) were the two most important contributors.
Samples taken at 2.5m formed distinct protected (196% total change), intermediate (200%) and exposed (100%) biofacies between Newfoundland and Gulf of Maine. *Agarum clathratum* was identified by CoBBI as the most important contributor to these total changes in protected (61%) and intermediate (33%) areas while it was a minor player at only 5% in exposed. *Desmarestia viridis* was 36% in protected, 17% in intermediate and only 5% in exposed. *C crispus* was only a contributor (27%) in intermediate and not relevant at this depth for other exposures (Table 2A).
Finally at 5m, the highest depth observed for this study, *D viridis* was the most important contributor for both protected (27%) and intermediate (30%), while not relevant in exposed areas (Table 2A). *Saccharina latissima* (27%) in protected, *S longicruris* (43%) intermediate and *L digitata* (26%) in exposed were the second highest contributors. *A clathratum* was the third most important contributor to the biofacies total changes in protected (23%) and intermediate (20%) yet not relevant in exposed at this depth.

3.2 Gulf of Maine and Nova Scotia

3.2.1 Biofacies depth changes

The boundaries between Gulf of Maine and Nova Scotia were in general clear cut except for infralittoral (Table 1). In cases in which SHE Analysis did not identify a boundary at the last station in Gulf of Maine, the Gulf of Maine plus one or two of the first Nova Scotia samples showed a change in slope rather than a distinct boundary. A change in the regression slope of lnN versus lnE indicates a community disruption of a lesser magnitude than the total change that results in a biofacies boundary. However, this was resolved when we re-ran SHEBI without replicates and used only the separate sampled stations; these totals are noted in Table 1. Table 2B shows that in the infralittoral *C crispus* was again prominent at all exposure level sites comprising 40%, 34% and 25% respectively of each of the total changes identified by CoBBI from protected to exposed. *Corallina officinalis,* though not relevant in protected sites, was the most important contributor for intermediate (44%) and second most (19%) in exposed. Neither *C flagelliformis* nor *D ramentacea* were contributors to the infralittoral biofacies boundary changes in these areas, except for exposed when *D ramentacea* was 11%.
Each of the total boundary changes was well over 100% for the 0.5m depth. *Saccharina latissima* (25%), *C. officinalis* (44%) and *Alaria esculenta*(41%) provided the most important contributions to the total changes for protected, intermediate and exposed respectively. *C officinalis* was second at both protected (21%) and exposed (28%). *C. crispus* was 20% in protected and 24% in intermediate but had no involvement for exposed sites (Table 2B).
At the 2.5m depth, sampling between the biofacies of Gulf of Maine and Nova Scotia there was less dominance at each level of exposure. At protected sites, *Saccorhiza dermatodea* (28%) and *L digitata* (20%) were the two taxa providing the most input into the total biofacies change. The top two contributors were *Fucus serratus* (26%) and *Corallina officinalis* (23%) for intermediate with *A longicruris* (24%) and *L digitata* (20%) for exposed (Table 2B).
For the 5m depth, protected sites only, *A clathratum* at 40% of the total change of 195% contributed importantly. For the intermediate sites, *L digitata* was 21% of the total and *F serratus* was 18%, with *S latissima* (15%) and *S longicruris* (14%) and *Neosiphonia harveyi* (13%) of equivalent importance to the total biofacies change of 176%. For exposed 5m sites *S*

(Venetian area northeast Italy) *Paleogeography Palaeocliminatology and Palaeoecology* 295: 258-280.

Brenchley, P. & Harper, D, 1998. *Palaeoecology: Ecosystems, environments and evolution.* Chapman and Hall. London. 402pp.

Briggs, J. 1974. *Marine Zoogeography.* McGraw Hill. New York. 475 pp.

Briggs, J. 1995. *Global Biogeography.* Elsevier. Amsterdam. 452 pp.

Buzas, M.A., & Hayek, L.C. 1998, SHE Analysis for Biofacies Identification, *Journal of Foraminiferal Research*, v. 28, no. 3, p. 233-239.

CLIMAP Project Members. 1976. The surface of the ice-age earth. *Science* 191: 113–7.

Gressley, A. 1838. Observations geologiques sur le Jura Soleurois. *Schweizer Gesellgesamten Naturwiss.* Neue Demkschr. 2: 1-112.

Hammer, O., Harper, D.A.T., & Ryan, P.D. 2001. PAST: Palaeontological Statisticis software package for education and data analysis. *Palaeontologia Electronica.* 4(1):9. Version 1.78, 2008.

Hargreaves, J., Paul, A., Ohgaito,R., Abe-Ouchi, A., & Annan, J. 2011. Are paleoclimate model ensembles consistent with the MARGO data synthesis? Clim. Past? Discuss. 7775-2011.

Hayek, L. C. & Buzas, M.A., 1997a, *Surveying Natural Populations*, Columbia Univ. Press, NY, 563 p.

Hayek, L.C. & M.A. Buzas. 1997b. SHE Analysis: An integrated approach to the analysis of forest biodiversity. Pages 311-322 In F. Dallmeier and J.A. Comisky, eds. *Forest Biodiversity Research, Monitoring and Modeling: Conceptual Background and Old World Case Studies.* Washington D.C. Unesco, Paris and The Parthenon Publishing Group.

Hayek, L.C. & M.A. Buzas, 2010 *Surveying Natural Populations: Quantitative tools for biodiversity assessment,* Columbia Univ. Press, NY, 590p.

Hayek, L.C., Buzas, M.A. & Osterman, L. 2007. Community structure of foraminiferal communities within temporal biozones in the Western Arctic Ocean, Journal of Foraminiferal Research, v. 37, p. 33-40.

Johansen, H.W. 1981. Coralline Algae, a First Synthesis. CRC Press. Boca Raton. 239 pp.

Kominz, M. 2001. Sea Level Variations Over Geological Time. Academic Press. doi 10.1006.2001.0255.

Longhurst, A. 1998. *Ecological Geography of the Sea.* Academic Press. New York. 398 pp.

Mana, D., 2005, A test application of the SHE method as a biostratigraphical parameter, *Geo. Alp.*, v. 2, p. 99-106.

MARGO Project Members. 2009. Constraints on the magnitude and patterns of ocean cooling at the last glacial maximum. *Nature GeoScience* 2: 127-132.

Osterman, L., Buzas, M. A., & Hayek, L. C., 2002, SHE analysis for biozonation in western Arctic sampling. *Palaios*, v. 17, p. 297-303.

Rosenzweig, M. 1995. *Species Diversity in Space and Time.* Cambridge University Press. Cambridge. 436 pp.

Vermeij. C. 1978. *Biogeography and Adaptation.* Harvard Univ. Press. Cambridge, Mass. 332 pp.

Wilson, B., 2008, Using SHEBI (SHE Analysis for Biozone Identification): To Proceed from the Top Down or the Bottom Up? A Discussion Using Two Miocene Foraminiferal Successions from Trinidad, West Indies: *Palaios,* v. 23, p. 636-644.

Part 7

Minerology

15

Mesothermal Lode Gold Deposit Central Belt Peninsular Malaysia

Kamar Shah Ariffin
School of Materials and Mineral Resources Engineering,
Universiti Sains Malaysia
Malaysia

1. Introduction

Malaysia had already established itself as one of the important gold producer long before the development of the great gold-fields such as in South Africa, Australia and USSR (Chu & Singh, 1986; Becher, 1983; Santokh Singh, 1977). Prior to the Portuguese conquest of Malacca in 1511, the country was known as the "Aurea Chesonese" or "Golden Peninsular". Malaysia has a long history of widespread small-scale gold mining throughout the country, especially in the Central Belt of Peninsular Malaysia and highly potential region for gold mining industry.

The Central "Gold" Belt is a 20km wide, a major N-S trend of gold mining districts that shows the important role of hydrothermal fluids in the development of gold in Peninsular Malaysia, especially in the North Pahang and Kelantan area (Ariffin & Hewson, 2007; Yeap, 1993; Lee et al. 1986; Proctor, 1972; Richardson, 1939).

The majority of the gold production apparently came from the states of Pahang and Kelantan within the Central Belt (Fig. 1). A study of literatures covering the geology of the Central Belt goldfield shows the important role of hydrothermal fluids in the formation of gold deposits (Yeap, 1993; Lee et al. , 1982 , 1986; Alexander, 1949; Proctor, 1972; Richardson, 1939, 1950; Scrivenor, 1931, 1928, 1911). In Kelantan which is located in the north, gold mineralization typically associated with hydrothermal quartz vein system, skarn and volcanogenic massive sulphides (Teoh, et al., 1987; Chu & Singh, 1986; Chu, 1983). The regional geochemical survey for gold, carried out by Mineral and Geosciences Department of Malaysia over the Central Belt in North Pahang and Kelantan, has defined a 20-km-wide, north-south-trending gold mineralization in the Raub-Kuala Medang-Lipis-Merapoh area in Pahang, including Ulu Sokor-Sungai Sok-Katok Batu-Pulai in Kelantan (Figs 1 and 2). Gold mineralization in the Central Gold Belt is generally categorized as a low mesothermal lode gold deposit due to its tectonic and geological setting.

2. Gold mining history and prospects in Central Belt

During the British reign between 1880s and 1940s, major gold production generally came from the state of Pahang, Kelantan and Negeri Sembilan within Central Gold Belt .During this booming period Raub, Selinsing, Kechau-Tui, Katok Batu, Penjom and Batu Bersawah goldfields (Fig. 2) were the important underground lode gold mines. Between 1889 and 1960 some 30 tonnes of gold was mined from underground working from the historic Raub Australian Gold Mine (RAGM) and some 1100kg (over 1 million oz) of gold was extracted mainly from underground works at Bukit Koman, Raub goldfield (Richardson, 1939).

As quoted from Free Press Mercantile news (F.M.S Gold Mining, 1903 in The Straits Times, 17 June 1904, Page 10) indicated that some 1,421 ounces of gold was extracted from 7,000 tons of treated tailing at Selinsing goldfield. Treatment of the tailings using heap leach extraction produced 6,624 oz of gold between 2003 and 2005. Whilst, the small Kechau-Tui goldfield where the gold is recovered from a shaft sunk to the depth of 30m which cutting the lode at this depth produced merely 48.5 oz from 1440 tons treated ore, whilst, Batu Bersawah gold mine, which is located at the southern part of the Central Belt, contributed some 180kg of gold between 1890 and 1910 through operation by the Batu Bersawah Gold Mining Company ore (The Straits Times, 5 May 1905 (page 6).

In the last 15 years of active evaluation of gold mineralization and development activities at the former Raub Gold mine-Tersang-Tenggelan-Chenua belt have witnessed a few new modern and bigger scale; open pit gold mined have been developed. In Raub, Selinsing, Kerchau-Tui, Pulai, Rubber hill, Buffalo Reef, and Tersang are among the old alluvial mining goldfields which are actively being revisited for the existent of low grade bulk-mineable gold deposits. Most of these newly discovered goldfields are located in the heart of Central Gold Belt.

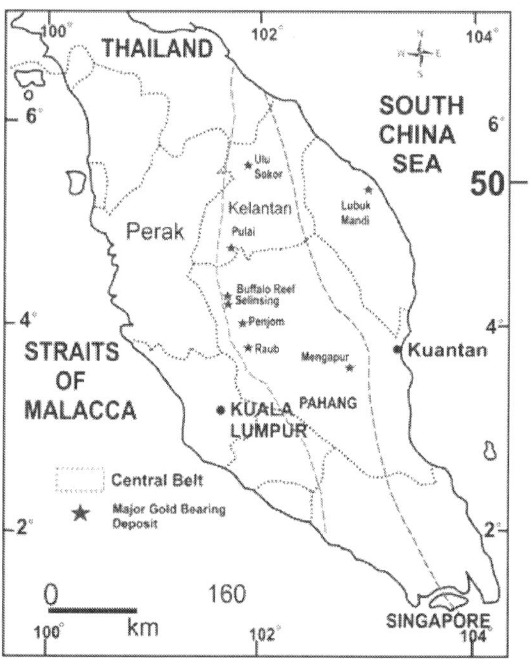

Fig. 1. Central mineral gold belt of Peninsular Malaysia with major gold bearing deposits.

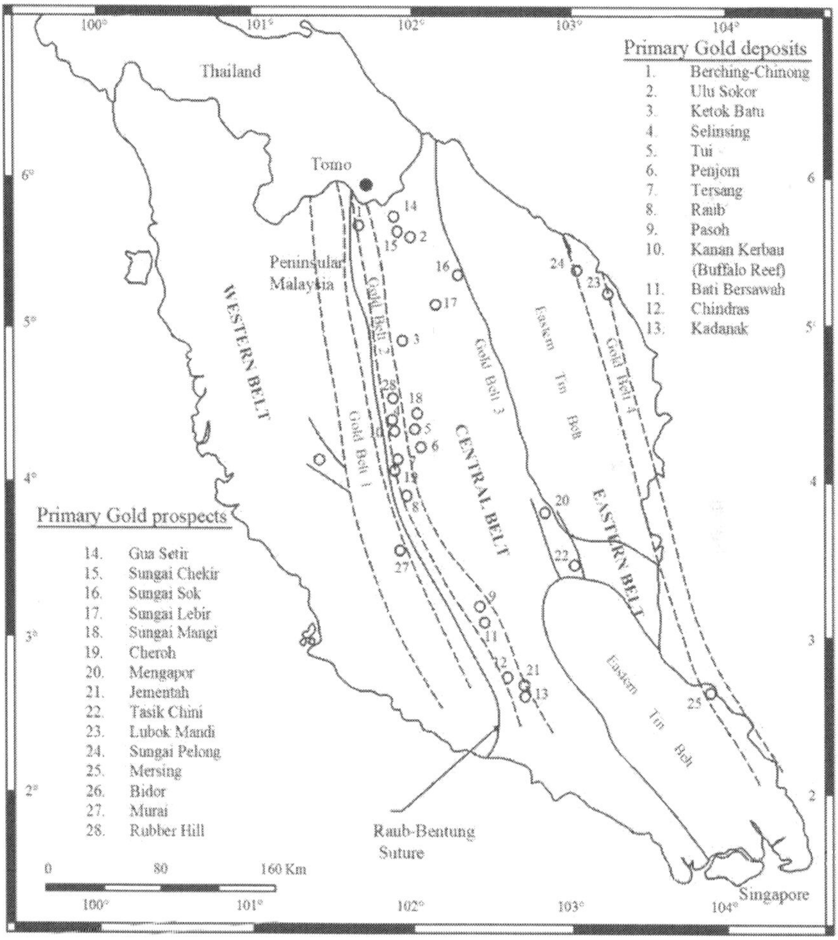

Fig. 2. Peninsular Malaysia showing mineral belts and primary gold occurrences (after Ariffin & Hewson, 2007 and Yeap, 1993)

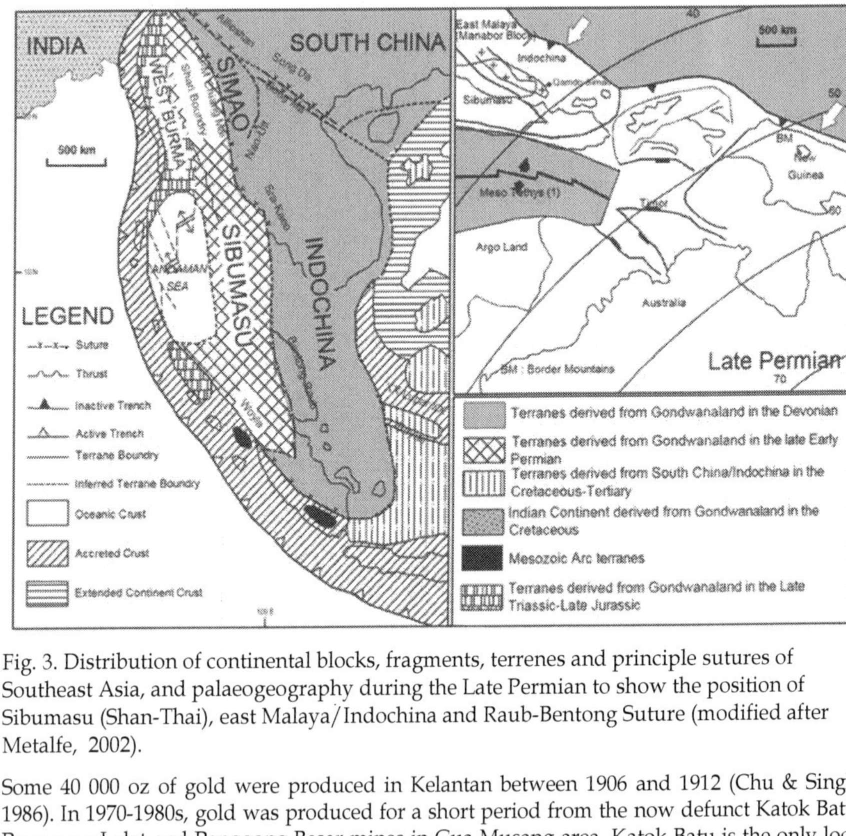

Fig. 3. Distribution of continental blocks, fragments, terrenes and principle sutures of Southeast Asia, and palaeogeography during the Late Permian to show the position of Sibumasu (Shan-Thai), east Malaya/Indochina and Raub-Bentong Suture (modified after Metalfe, 2002).

Some 40 000 oz of gold were produced in Kelantan between 1906 and 1912 (Chu & Singh, 1986). In 1970-1980s, gold was produced for a short period from the now defunct Katok Batu, Panggong Lalat and Panggong Besar mines in Gua Musang area. Katok Batu is the only lode mine in Kelantan that produced 102 tahils or 4530gm (1 tahil = 37.8gm) of gold in 1934. In late 1903, Pahang and Negeri Sembilan produced 12 400 and 2 664 ounces of gold, respectively with the total amount of 15 070 ounces. Widespread alluvial gold occurrences have been long recognized in Pahang and Kelantan where there is a total in excess of one million ounces of gold has been recovered after this period. Over hundred prospecting permit/mining leases were issued for gold exploration in Pahang especially within Kuala Lipis-Raub districts to about 39 companies, mainly for alluvial gold mining mainly before 1990s.

Renewed interest in intensive exploration and mining for gold within the Central Belt has come into being since 1985 after the collapse of tin price. Its attraction lies in the good possibilities of finding the existence of a sizeable tonnage of low grade gold deposits, amenable to exploration by low cost, modern techniques of bulk mining (open-pit) with heap leaching and CIP/CIL (Carbon in Pulp/CIL Carbon in Leach) treatments. Many pre-world War II abandoned small scale alluvial and alluvial gold mining spots which were worked intermittently and have been targeted for re-evaluation since 1990.

The Penjom Gold Mine is the first, largest and the modern open pit gold mine that uses modern extraction methods and processing in Malaysia since its operation in 1996 (3.99 million tonnes, grading 3.78 g/t Au (484100 ounces of gold)) (Flindell, 2003).

Table 1 gives the Malaysia's production of raw gold between 2006 and 2009, indicated over 90% of gold production is originated from Pahang (Penjom gold mine since 1996, Avocet Mining PLC), and recently from Selinsing gold mine (Monument Mining Limited), and Raub Gold Mine (Peninsular Gold Limited). Between this period, Pahang contributed over 12 million grams of gold, especially from Penjom and Selinsing open pit gold mine.

State	2006		2007		2008		2009	
	Grams	Mines	Grams	Mines	Grams	Mines	Grams	Mines
Pahang	3,415,446	6	2,832,489	6	2,449,961	4	2,771,770	6
Kelantan	78,189	3	80,151	5	37,159	3	49,844	4
Terengganu	-	-	-	-	2,876	3	32,553	2
Johor	3,606	1	-	-	-	-	-	-
Total	**3,497,241**	**10**	**2,912,640**	**11**	**2,489,996**	**10**	**2,794,167**	**12**

Table 1. Malaysia's Production of Raw Gold 2006-2009
Source: Department of Mineral and Geoscience Malaysia, Mineral Year Book 2009

3. Tectonic setting

Peninsular Malaysia is a part of the east Eurasian Plate and tectonically located to the north of currently active subduction arc zones of the Sunda arc. Gold discovery distinctly in this region is always associated with Tertiary volcanic and hydrothermal activities, and appears to be very broadly related to tectonic boundaries in this region (Fig. 3).

The Malay Peninsular may be divided into two tectono-stratigraphic terranes which form part of the Sunda shelf, namely the East Malay (Eurasian plate-Indochina) and Sibumasu (Shan-Thai) terranes respectively. The Eurasian Terrane (Manabor block) has been interpreted as Permo-Triassic island arc system which never been far separated from Shan-Thai block. Stratigraphic, palaeontological and palaeomagnetic evidences suggest a possible origin of these terranes by the rifting of the north-east margin of the ancient Gondwanaland landmass in the Late Permian to Latey Triassic that responsible for the formation of the Central Belt and the Raub-Bentong Suture (Cocks et al., 2005; Metcalfe, 2002, 2000, 1988; Campi et al., 2002; Spiller and Metcalfe, 1995; Schwartz et al., 1995; Mitchell, 1977; Yeap, 1993; Tan, 1996, 1984; Tjia, 1989,1987; Khoo and Tan, 1983; Kobayashi and Toriyama, 1970).

Fig. 4 shows the conceptual cross-section that illustrating the formation of Central Belt to the east of Bentung Suture line as accretionary complex (Metcalf, 2000). Thus a thin and irregular strip of continental lithosphere and island arc sequence developed in front of it. These detachments later collided with the accreting Asian landmass and fused along the Raub-Bentong Suture. Peninsular Malaysia to the east of the suture belongs to Cathaysia.

A collision structure overprint has generated major N-S or NW-SW trending left slip fault and dilational Riedal and subsidiary shears and numerous splays associated with these fault (Hewson and Crips, 1992; Tjia and Zaitun, 1985). The Raub-Bentong Suture is a deep rooted 13km wide tectonic zone that runs generally in the N-S direction from Tomo in Thai Peninsular (Fig. 2) along the east side margin of the Main Range to the Malacca-Johore border (Cocks et al., 2005; Metcalfe, 2002, 2000, 1992, 1988; Yeap, 1993; Tan, 1996; Tjia, 1989).

This N-S zone is located some 20km to the west of Penjom gold deposit and next to the Selinsing gold mine. This zone is characterized with the presence of schist, cherts with small serpentine bodies, argillite, olistostrome and mélange (Metcalfe, 2000; Tjia, 1987). It is also a zone of parallel steeply dipping N-S faults with several periods of reactivation. The "Gold belt" lies in the East Malay/Indochina Block (Fig. 4), subdivided into Eastern Belt and Central Belt.

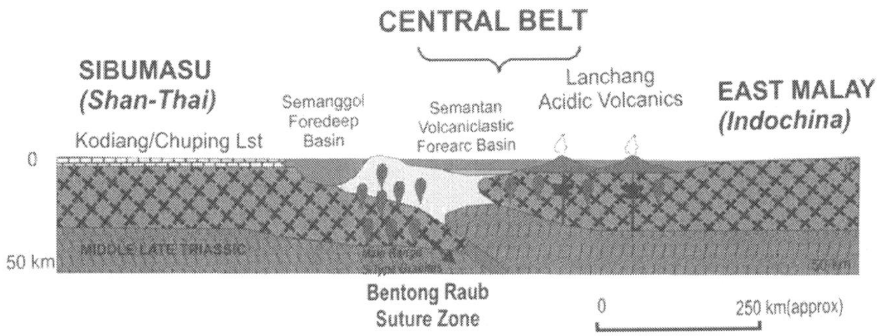

Fig. 4. Conceptual cross-section during Middle-Late Triassic illustrating the formation of Central Belt to the east of Bentung-Raub Suture line as accretionary complex modified after Metcalfe, 2000).

4. Regional geology

The Central Belt consists mainly of Permo-Triassic, a low-grade metasediments, deep to shallow marine clastic sediments and limestone with abundant intermediate to acid volcanics and volcaniclastics, deposited in paleo-arc basin (Metcalfe, 2002; Leman, 1994; Richardson, 1939; Gobbet and Hutchison, 1973; Proctor, 1972).

Acid and intermediate intrusive rocks were emplaced east and parallel to Raub-Bentong Suture (Figs. 4 and 5). Batholiths in the Eastern Belt are smaller than those of Sibumasu, but are, in comparison, compositionally expanded. The Jurassic-Late Cretaceous batholiths, dominantly monzogranitic suite are of I-type affinity and carry both precious metal and base metal mineralizations.

Magmatism in the Central Belt is markedly less common and consists of an alkali series ranging from gabbro- diorite (157 Ma) monzonite (163 Ma) to quartz syenite (127 Ma), and a later calc-alkali series of granodiorites and granites (Yong et al., 2004; Mohd Rozi and Syed Sheikh Almashoor, 2000; Khoo and Tan, 1983; Jaafar Ahmad, 1979; Bignell and Snelling 1977; Hutchison, 1977). The Central Belt granitoids (slab break-off), which lie critically close to the Raub-Bentong Suture line have very high large ion lithophile (LIL) elements, that is, Ba and Sr, nearly to 1000 times rock/mantles and classified as mantle plume type magmatism (Azman et al., 2006; Mustafa Kamal and Azman, 2003). Fig 5 shows the general geology of the Central Belt along the major N-S trend goldfields in Kelantan and North Pahang. Seting granite, Stong igneous complex and Mahang Granite are the major granite intrusives in Kelantan. The Benom Plutonic Complex (Early Jurassic) which comprises Bukit Lima, Bukit Tujuh and Damar granite are the major shoshonitic granitoid in central Pahang (Fig 6), characterized by high K_2O content.

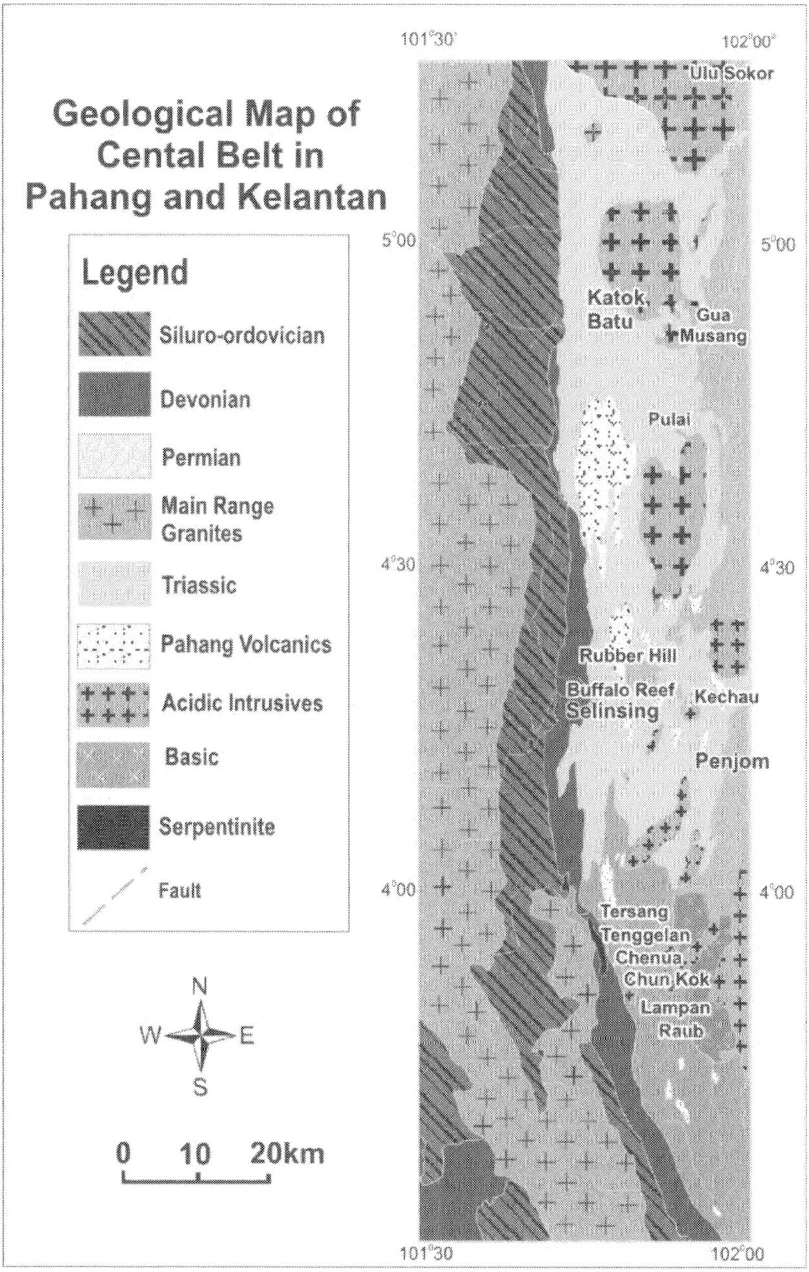

Fig. 5. Geological map of Central Belt in northern Pahang and South Kelantan

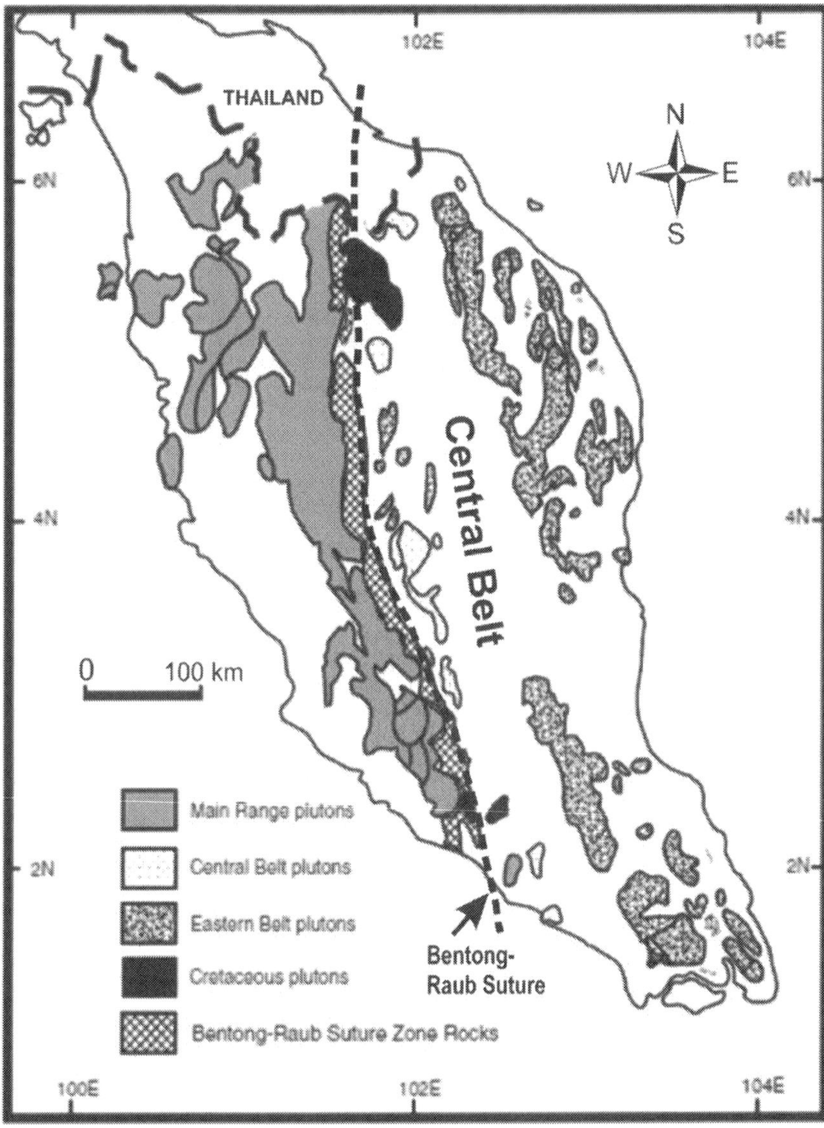

Fig. 6. Granitoids of Peninsular Malaysia shown in relation to the Bentong-Raub suture zone (modified after Metcalfe et al., 2000; Cobbing et al., 1992)

5. Gold mineralization in Central Belt

Mineralization in Central Belt is dominated by gold. Most of the gold is mined from quartz lode and stokwork deposits that are much associated with the accretionary prism along the terrain boundary known as the Raub-Bentong Suture. Mineralization took place within a low grade Permo-Triassic island arc system composed of meta-(sedimentary) and volcanic rocks accompanied by extensive deformation (brittle-ductile and shearing zone), metamorphism, and magmatic events that created the favorable environment for source and trap for the gold mineralization.

Most of the gold mineralization took place within a low-grade meta-sedimentary-volcanic terrain formed during the collision of the Sibumasu block underneath the East Malaya (Indochina) block through the Permian to late Triassic. Fig. 7 shows the typical representation of crustal environments of orogenic gold deposits in term of depth of formation and structural setting within attracted terrain that illustrated the resemblance to Central Belt gold formation (Groves et al., 1998).

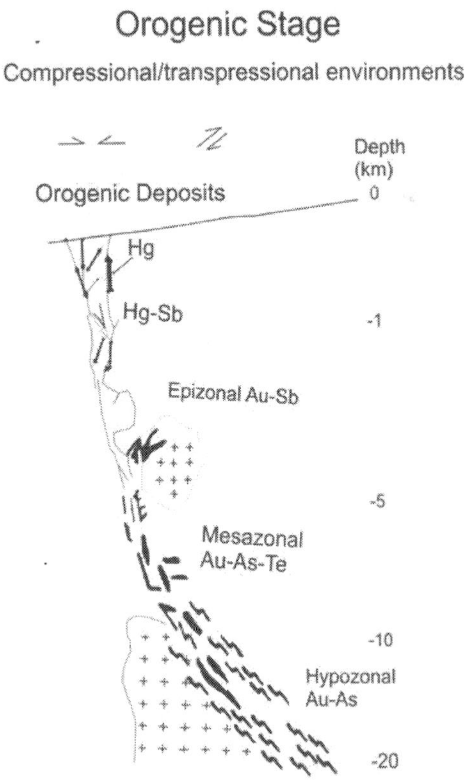

Fig. 7. Schematic representation of crustal environments of orogenic gold deposits in term of depth of formation and structural setting within accreted Terrance (modified after Groves et al., 1998)

Several mines being worked extensively were alluvial deposits developed on vein stockworks in altered, brecciated and sheared intrusive or adjacent country rocks (Scrivenor, 1911). Attraction lies on the good possibilities of the existence of sizeable tonnage of low-grade gold deposit. The old Raub gold mine lies within the western side of the Central Belt, whereas the Mengapur copper-gold porphyry skarn prospect on the north-eastern side (Figs. 1 and 2). Both deposits carry significant gold mineralization. Therefore, this Central Belt is well-known as "The Gold Belt". As signified in Fig. 2, major primary gold mineralization patterns within Central Belt can be grouped into two types: type I (gold belt 2) and type II (gold belt 3), respectively.

The type I deposits consist of significantly large quartz reefs/lodes and parallel swarms of vein, traversing metasediments and granite. This type I mineralization belt is also identified as the gold geochemical zone (Lee et al., 1986, 1982). The mineralization is confined within brittle - ductile shear or brecciated zones. This gold belt is located immediately to the east of the Main Range granite and Raub - Bentong line (Yeap, 1993). Two major goldfields within the type I belt are the Buffalo reef (Kanan Kerbau) and further south, the Selinsing gold mine and the Tersang alluvial goldfields. Enhanced level and occurrences of stibnite and scheelite are common characteristics of the Buffalo reef, Selinsing and Raub goldfields, whereas ilmenite and cassiterite occurrence is considerable at the Tersang goldfield (Kamar Shah & Khairun Azizi, 1995; Pereira, 1993; Pereira et al., 1993).

However, elevated As and Sb are considered a common trend of these goldfields. Type II, which is located immediately to the east of the type I deposits, exhibits a broader variety of gold mineralization, bounded to gold disseminated within a stockwork of quartz veins affiliated with intrusive bodies and volcanogenic exhalative sulphides within a shear zone system. Dilated quartz veins and Au-Ag-bearing skarn carry significant amount of sulfides (Sinjeng, 1993). The type II belt is also designated as the silver zone (Lee et al., 1986). Ulu Sokor goldfield which is located in Gold belt 3 was geochemically delineated as Gold-base metals mineralization zone (Teoh, 1987; Goh et al., 2006). Table 2 presents common elemental or geochemical composition of common epizonal (Au-Sb) and Mesazonal (Au-As-Te) types of orogenic gold formation in the Central Gold Belt from Rubber hill prospects, Buffalo Reef, Tersang goldfield and Raub gold mine (Bukit Koman).

Sample no	Unit	K001	K002	K003	K004	K005	K006	K007	K008	K009	K010	K011	K012
Location Description		Rubber Hill - Trenching samples				Bufflao Reef	Selinsing Quartz vein	Buffalo Reef	Buffloo Reef	L S Chong Pulai, Tersang	Bkt. Koman Quartz	Buffalo Reef Breccia	Rubber Hill
Element		(Red-Brown)	(Yellow-Brown)	(Yellow-Brown)	(Red-Brown)	(Brown-Yelow)	(Redish-Yellow)	(Dark Black)	(Grey)	(Greyish-Brown)	Shale (Grey)	(Brown-Yelow)	(Gold Bearing)
Gold *(Au)*	ppm	0.01	0.11	0.01	0.07	145	0.07	4.44	0.11	0.03	1.18	2.5	363
Silver *(Ag)*	ppm	---	---	---	---	6	---	5.5	2	---	4	2	57
Antimony *(Sb)*	ppm	---	10	---	---	---	30	---	---	---	---	---	6
Antimony *(Sb)*	%(w/w)	---	---	---	---	---	---	10	0.13	---	---	—	---
Lead *(Pb)*	ppm	---	27	---	---	---	---	—	---	10	---	---	150
Zinc *(Zn)*	ppm	---	180	---	---	---	---	---	—	2	---	---	29
Copper *(Cu)*	ppm	---	64	---	---	---	---	—	---	14	---	---	16
Arsenic *(As)*	ppm	—	570	—	—	—	—	---	—	---	—	---	425

Note : Soil and quartz vein samples from severals deposit/prospects

Table 2. Geochemical composition of selected grab and trenching samples represent gold deposits and prospects from Northern Pahang.

Gold belt 4 (Lubok Mandi-Mersing Belt) is located in the eastern part of Peninsular Malaysia and it is juxtaposed with the Eastern Tin Belt. The Lubok Mandi gold deposit is an 8-km gold–quartz lode hosted in weakly metamorphosed and folded slate, phyllite and metaarenites. In the Mersing area the primary gold mineralization was observed as several discontinuous, approximately 350° striking, gold-quartz veins cutting strongly folded meta-argillites and arenites (Yeap, 1993).

6. Major goldfields, mines and prospect in Central Belt

Pahang and Kelantan were the first promulgated to be attractive and favorable economically for attracting foreign investment for systematic gold exploration in the states by the granting of prospecting licenses of designated concession blocks and consequently mining leases for systematic and modern techniques of gold ventures. The total production of gold bullion from Pahang for the period 1889 to 1985 is reported to be at least 36 tonnes. At present, 88% of the total production in Peninsular Malaysia is from Pahang state.
Penjom, Raub, Selising and Buffalo reef are the currently active gold mines in Pahang. At present, the Penjom Gold Mine is the highest production of gold in Malaysia with concentration about 6.0g/ton in Central Belt. The Penjom gold deposit lies within the western margin of Central Belt (Gold belt 3). Other less important gold bearing deposits are Mengapur, Tersang, Rubber Hill (Pahang) and Batu Bersawah in Negeri Sembilan. Ulu Sokor and Pulai are the only active gold mining and exploration projects in Kelantan, and mostly restricted to alluvial deposit.

6.1 Penjom gold mine

The geology of the Penjom gold deposit is dominated with widespread occurrences of marine clastic sediments, intermediate to acid volcaniclastics, and subordinate rhyolitic lava sequences. Its is belongs to so-called Padang Tengku Formation of the , a Raub-group rock assemblage and the Pahang Volcanic series. This volcaniclastic and sedimentary association is intruded by a few shallow dipping sheets of tonalite unit as narrow sills and minor dykes of quartz porphyryies and running almost parallel to the main mineralized shear zone. Tonalite is a major igneous intrusion complex within the area. The Raub-Bentong Suture has accommodated considerable strike-slip movement (Fig. 8). Structural analysis has indicated a regular geometrical pattern of repeated district scale fault trends (Kelau fault) which can be observed within most the goldfields in the Central Belt (Tjia & Zaitun, 1985).
Mineralization at the Penjom gold deposit is structurally controlled and erratic laterally and vertically. The Penjom thrust is the dominant feature controlling the distribution of ore at Penjom and generally strikes NE (35°) and dips to the southeast (30°-40°). Considerable shear stresses along the Penjom thrust have remobilized much of the carbon within the shale sequence to form a graphitic "alteration" zone. This, together with sheared and milled rock (fault gouge materials), makes the Penjom thrust an impermeable zone (Ariffin & Hewson, 2007; Flindell, 2005; Mustaffa Kamal, et al., 2003; Sonny et al., 2001; Kidd, 1998; Kamar Shah, 1995; Kamar Shah et al., 1995; Hewson & Crips, 1992). Major gold mineralization took place within the footwall of this thrust (Figs. 9 & 10).
Both the veining and massive ores can be subdivided on the basis of their mineral constitution into (1) gold-galena-tetrahedrite-tellurides (especially altaite) ore, (2) gold-arsenopyrite-pyrite ore, and (3) pyrite. At Penjom, the ore systems display permeability controlled or governed by lithology, structure and breccias and changes in wall rock alteration (quartz, carbonate, sericite, chlorite, fuchsite and clay). Gold mineralization was believed to form at the homogenization temperature higher than 270°C of hydrothermal fluid which is typical for mesothermal vein deposits (Ariffin & Hewson, 2007; Wan Fuad and Heru Sigit, 2003, 2001; Kamar Shah, 1995; Herrington, 1992).

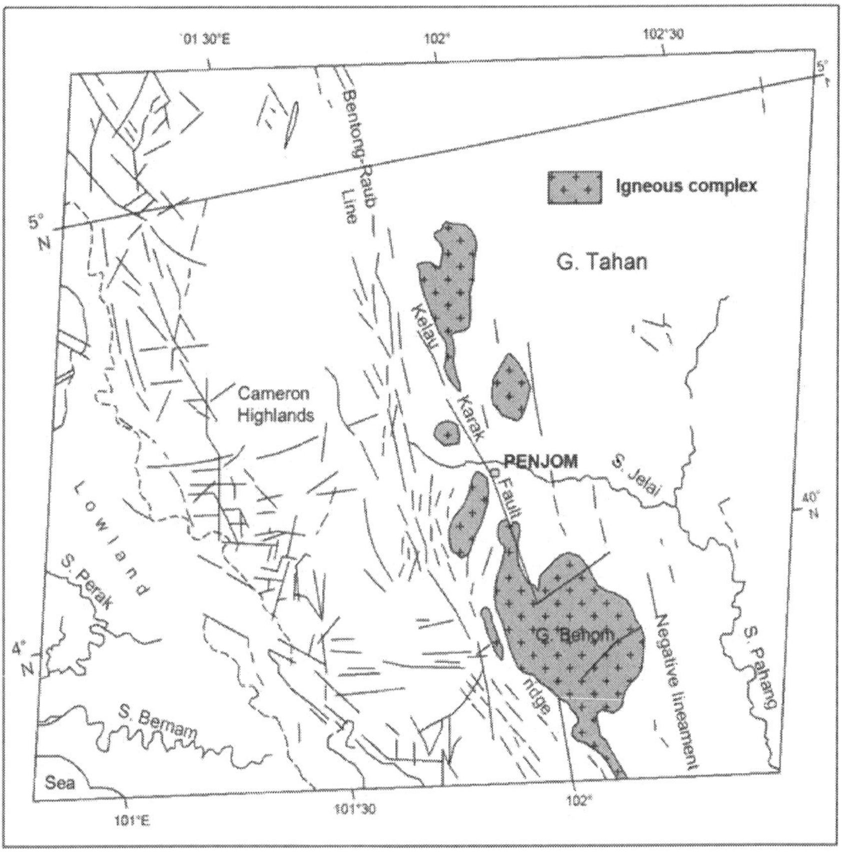

Fig. 8. Regular geometrical pattern of repeated district scale fault trends and numerous splays running along the Central Gold Belt with major granitoid emplacement (after Tjia and Zaitun, 1985).

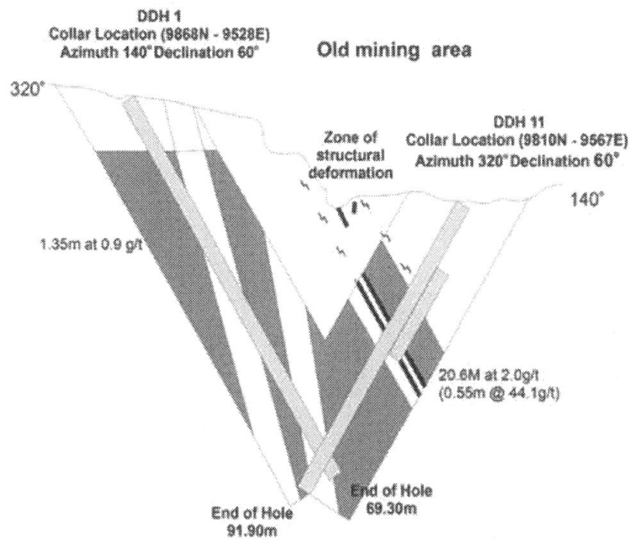

Fig. 9. An example of exploratory drillholes (DDH 1 and DDH 11) that cut the significant mineralization section within the Penjom trust (shear zone) (Kamar Shah, 2007)

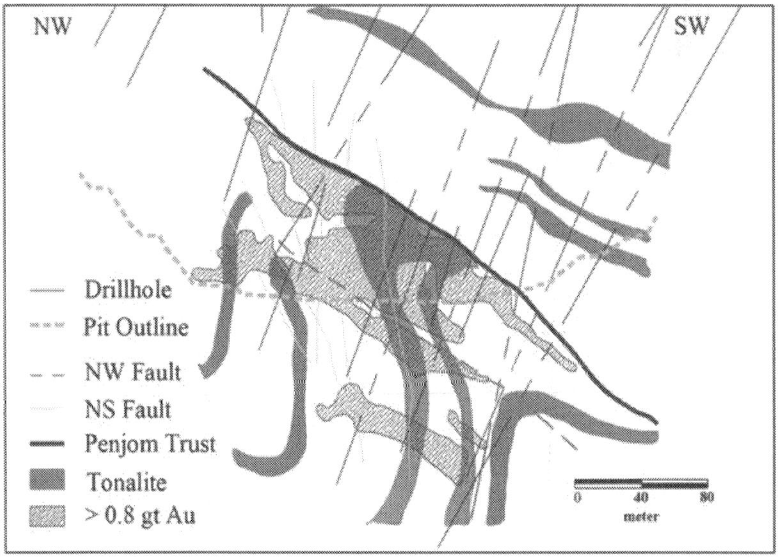

Fig. 10. Cross-section across the Penjom ore body running through the centre of the main deposit (Ariffin & Hewson, 2007; Flindell, 2003).

Multi-elemental distribution patterns with respect to the depth, litho-geochemistry and structural features of the Penjom gold deposit such as from DDH-3 and DDH-11 (Fig. 9) show that most of the gold-rich samples are proportionally elevated in arsenic. Ag, As, Te, Sb, and Bi except Hg as shown in DDH-3 (Table 3 and Fig. 11) and DDH-11(Fig. 11 bottom) are most elevated in segments associated with sulphide-gold mineralization. Gold has a marked affinity for Te and Bi and less for Sb. Two of the analyzed mineralized samples hosted within tuff of DDH-11, which are characterized by fault gouge materials have shown compelling occurrence of As (80 000 ppm), Au (18 - 47 ppm), Ag (4-8 ppm) and Te (8.5 ppm). Based on the elemental relationships, the Penjom gold deposit can be classified as an Au-Cu-Ag-Pb deposit.

Site No	Field description	BH No	Top Depth	Bottom Depth	Au	Ni	Co	Ag	Mo	Cu	Pb	Zn	Fe	Mn	As	Sn	W	Hg	Sb	Bi	Ba	Cr	Se	Te
1127	tuff	3	43.00	43.10	0.001	14	8	0.10	3.4	27	3	63	3.4	290	5.0	5	2.0	0.02	1	6	101	24		
1128	agglom, siderite alteration	3	45.00	45.30	0.001	11	13	0.20	3.4	24	5	54	3.8	1560	10.0	5	2.0	0.02	1	4	173	24	0.1	0.5
1129	agglom, siderite alteration	3	52.00	52.35	0.001	15	26	0.20	3.2	31	4	134	8.1	5480	15.0	5	2.0	0.02	2	5	352	24	0.1	0.7
1130	agglom, siderite alteration	3	54.40	54.50	0.001	15	21	0.30	2.8	10	5	121	7.9	5040	15.0	5	2.0	0.02	1	6	261	25		
1131	tuff, foliated	3	58.85	59.25	0.001	16	14	0.05	2.0	10	2	58	3.5	362	25.0	5	8.0	0.02	1	4	109	19		
1132	agglom, rubble with q.vein	3	67.00	67.50	57.463	22	31	34.00	14.2	300	185	136	13.0	1910	10.0	5	16.0	0.08	2	25	149	49	0.5	2.9
1121	tuff	3	91.65	92.20	0.001	2	4	0.05	2.9	5	11	51	1.7	584	3.0	5	2.0	0.06	1	7	561	33	0.1	0.3
1122	silic/cherty zone within tuff	3	103.00	103.40	0.008	6	2	0.10	4.8	20	14	21	0.9	361	15.0	5	2.0	0.02	1	5	197	83	0.1	0.1
1123	tuff with 1% euhedral pyrite	3	105.60	105.85	0.007	6	5	0.05	3.9	17	9	59	1.5	1580	10.0	5	2.0	0.02	5	4	668	35		
1124	calc, tuff, banded	3	106.60	106.80	0.048	6	6	0.20	2.9	9	13	40	1.9	1390	25.0	5	24.0	0.02	7	4	358	22	0.1	0.7
1125	tuff, silica	3	108.00	108.15	1.378	7	3	1.00	3.4	64	36	45	1.5	650	20.0	5	2.0	0.02	6	7	182	50	0.2	0.3
1126	tuff, silica	3	109.80	110.10	0.003	2	2	0.30	3.7	11	6	36	1.5	652	5.0	5	2.0	0.02	1	4	97	21	0.1	0.2

Table 3. Results of multi-element analyses data from DDH 3 of Penjom gold deposit from the early exploration stage (1990-1992; after Kamar Shah, 1995)

Ores from the Penjom deposit can be broadly divided into four groups, namely vein, dissemination, massive and fragmental. Sulphide minerals mainly arsenopyrite and pyrite are dominant constituents embedded in quartz-carbonate veins. There are widespread occurrences of pyrite, arsenopyrite, sphalerite, galena, chalcopyrite, molybdenite, tetrahedrite and tellurides (Fig. 12). The gangue minerals are commonly associated with gold mineralization include quartz, feldspars, carbonates (calcite, ankerite, dolomite, siderite and minrecordite), epidote, manganite, graphite and muscovite (sericite), talc, chlorites, fuchsite, goethite, limonite, fluorite, carbonaceous matter, pyrolusite and kaolinite. They basically coincide with the formation of polymetallic gold-silver ore that is transitional to higher crustal level carbonate-base metal class.

6.2 Selinsing goldfield

This is an active goldfield located in NW Pahang, Peninsular Malaysia. Mining at Selinsing commenced prior to 1888 and has operated intermittently through to 1966 (Johnston, 1998). Underground and open cut mining, together with tailings treatment, has produced an estimated 85,000 ounces of gold during this period.

Lithology of the area consists of low-grade metamorphosed sedimentary and volcanic rocks of Gua Musang Formation of Late Permo-Triasic age (Figures 5 & 13). Wall rock alteration in Selinsing Gold mine shows a direct relation with hydrothermal solution, structures, formation of quartz veins and gold mineralization.

The Selinsing deposit occurs along the north striking Raub Bentong Suture. The deposit is hosted by a series of auriferous quartz veins and stockworks of quartz veinlets in a package of sheared calcareous epiclastic sediments. The gold mineralisation occurs in quartz veins that cut the host rocks and wall rocks with intensive alteration that are related to the N-S and NE-SW lateral faults and shear zones (Mohd Basril et al., 2009). Gold mineralization at Selinsing is associated with high grade quartz veining and accompanied by strong sericitization and silicification within a major shear zone. Formations of quartz veins are mainly related to the right lateral faults. Minerals often associated with gold mineralization are pyrite, arsenopyrite, chalcopyrite, tetrahedrite and sphalerite (Wan Fuad et al., 2008; Ariffin & Hewson, 2007; Pereira, 1993; Pereira, et al., 1993).

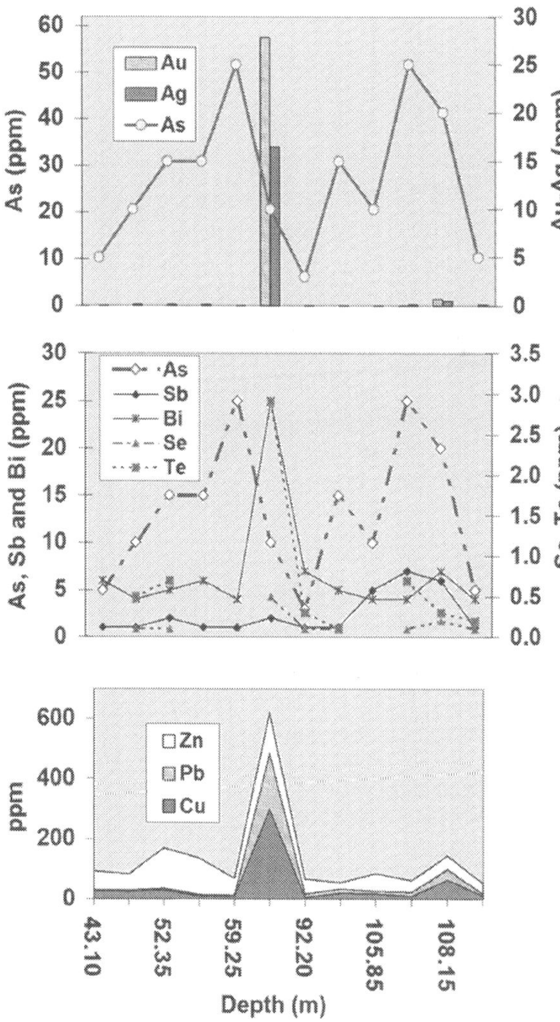

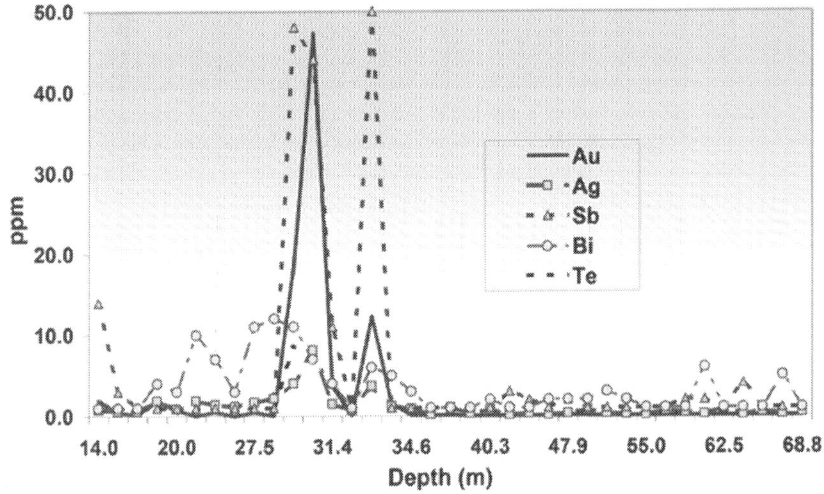

Fig. 11. Multi-elemental distribution patterns against the depth of the Penjom gold mineralization rock formation from DDH-3 and DDH-11 (after Ariffin & Hewson, 2007 and Kamar Shah, 2007).

Brecciation and cataclasite within shear zone are very prominent and show sensible relationship with mineralization. The host rocks of the deposit consist of a series of argillaceous and arenaceous of likely epiclastic origin, which have undergone low temperature regional metamorphism. This metamorphism has had little effect on their original mineralogy or texture. The ore samples show epigenetic gold and sulphide mineralisation in quartz veins with no direct relationship to the sediments hosting the quartz veins.

6.3 Buffalo Reef

The Buffalo Reef prospect lies close to the eastern flank of the Raub-Bentong Suture and main range granitic intrusions which also share the similar structural styles with Selinsing gold mine that located just to the north (Fig. 13). Gold mineralization is hosted by widespread occurrence of low regional grade metamorphism (greenschist to locally amphobolite facies) of marine clastic sediment, consist chiefly of pale to dark grey phyllitic carbonaceous and calcareous shale, lesser amounts of tuffaceous rocks, limestone and fine-grained schistose sandstone of Permian and subordinate of Devonian conglomerate. The conglomerate unit is belonging to Bentong group, whilst the Permian sedimentary rock formation is to the Raub Groups, respectively.

Gold mineralization is mainly confined to the marine clastic rock sequence, which is generally striking in N-NW direction and dipping towards east between 65° and 70° (Kamar Shah et al., 1995; Pereira, 1993). Some irregular and fractured quartz-carbonate veining also occurs throughout. Quartz veins are found mostly parallel to the bedding. The most significant feature with respect to gold mineralization is N-S aligned shear zone in the Raub Group. Significant gold mineralization often confined to the N-S trending sheared zone composed of metamorphosed, brecciated and hydrothermally altered calcareous graphitic shale with minor interbedded fine-grained sandstone and tuffaceous rocks.

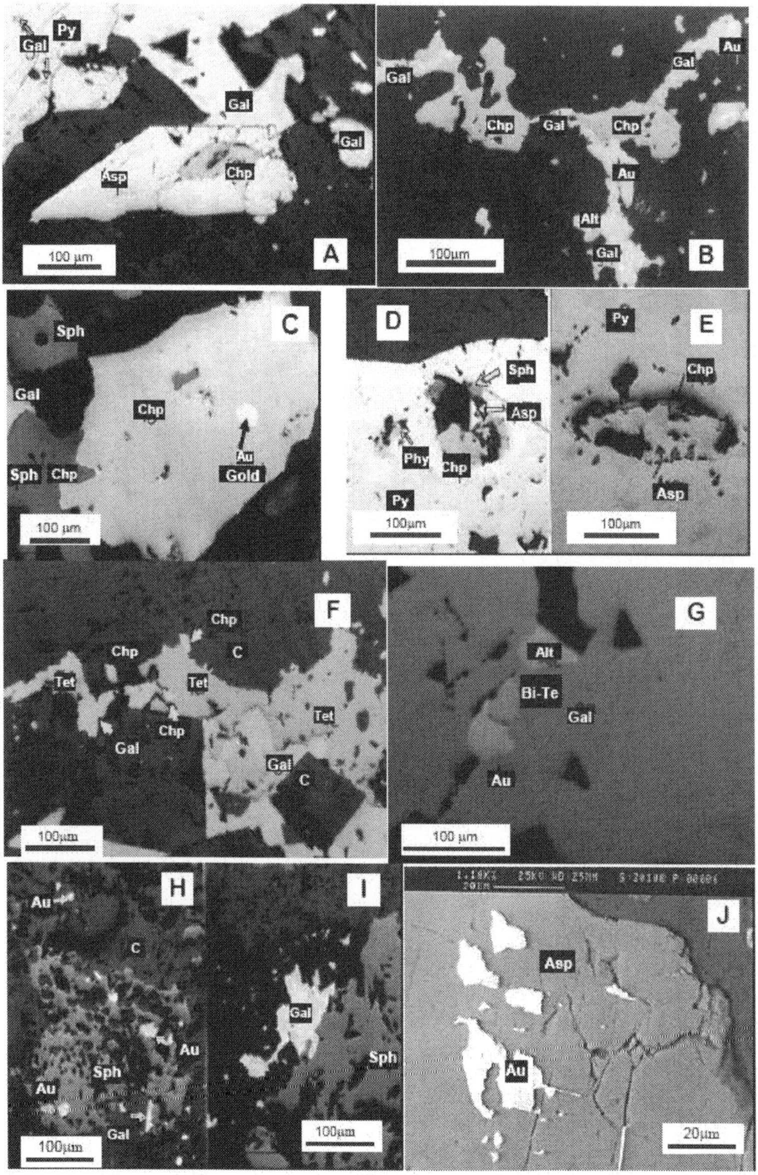

(a) Chalcopyrite (Chp) is seen occupying a pit in subhedral arsenopyrite (Asp). Galena (Gal) is seen occupying tiny pits in pyrite (Py) and replacing both sulfides,

(b) Carbonate matrix of quartz - carbonate vein shows fractures infi lled by chalcopyrite (Chp), while altaite (Alt) and galena (Gal) are intergrown, replacing chalcopyrite (Chp) with the occurrences of tiny electrum,

(c) Chalcopyrite (Chp) is being replaced by sphalerite (Sph), which in turn is replaced by galena (Gal). An inclusion of submicroscopic gold (15 μm) embedded in chalcopyrite is also visible,
(d,e) Arsenopyrite (Asp), chalcopyrite (Chp), sphalerite (Sph) and pyrrhotite (Phy) occupying fractures and pits in pyrite,
(f) Jagged tetrahedrite (Tet), which is being replaced by galena (Gal), and chalcopyrite (Chp) occupied the interstices in the calcite (C) of quartz - carbonate vein,
(g) Electrum (Au) is seen associated with light brownish grey bismuth - telluride BiTe(Pb)- tetradymite and altaite(Alt) enclosed in massive galena (Gal),
(h) Sphalerite intergrown with galena and gold blebs locally within the quartz - carbonate vein (C-calcite), (i) irregular shaped sphalerite (Sph) intergrown with (Gal), and
(j) Back-scattered electron image shows gold (Au) infilling the interstitial spaces and fractures of arsenopyrite (Asp).

Fig. 12. Photomicrographs of ore minerals in the Penjom deposit (after Ariffin & Hewson, 2007).

From north to the south of the prospects, in the north section, the gold mineralization are characterized by structurally bounded of steeply dipping quartz lode within complex dilated and silicified zone with dimension of 100m wide and 360m long. Two main NW Redial shear structures that composed of 40m wide x 300m long quartz lode and another narrower 500m long, east dipping extension lode in the central section. In the south, shear-parallel lode structure with 550m long and 70m wide are the main mineralization features of the area that is open to the south (Snowden, 2008; Pereira, 1993).

6.4 Raub gold deposit

The Raub deposit lies along the same Raub-Bentong suture 50 km south of Selinsing and Buffalo Reef goldfields. Raub Gold mine (RAGM) which is located within Bukit Koman vicinity comprises mainly interbedded sedimentary and metasedimentry rock strata flanked by the Kajang's granite porphyry of Triassic age which is exposed 5.5 km to the west of the mine. The sedimentary formation consist mainly of interbedded carbonaceous, silicified or calcareous types of grey to black shales, limestone, marble, and tuff which have experienced low grade metamophism and belong to Raub Group of carbonaceous age (Gunn et al, 1993; Richardson, 1939). The shale is almost pyritic throughout and generally striking northwards and mainly steeply dipping eastwards. The rock often isoclinally folded, with compression, tension and oblique faults. Hard and occasionally jointed quartzite is the major metasedimentary rock that occupies an N-S orientation hill.

The Raub mine has been the site of extensive historic gold mining, as well as limited modern operations and currently hosts a proven reserve of 202,000 ounces in 8.6 million tonnes of tailings (Snowden, 2008; Howe, 2004). Between 1889 and 1961 approximately 400kg which accounted for 85% of gold annual output in Pahang was from Raub, until recently some 32 tonnes (1Moz) at grade 4.2g/t of gold have been mined. The Raub gold deposit is hosted in a 6 km long vertical mesothermal quartz-carbonate veins system. Recent investigation, In addition a further 218,000 ounces of gold has been identified to date in an area known as the East Lode oxides, comprising 136,000 ounces in the measured and indicated categories and 82,000 ounces inferred (per the JORC standard). The target is delineated to contain over 1 million ounces of gold resources. Lampan is another adjacent prospect located to the NW of Raub gold mine.

Gold mineralization is mainly discovered occupying the two 300m apart of N-S trending fault zones. Gold ores mostly extracted from steeply-inclined faulted and folded zone and

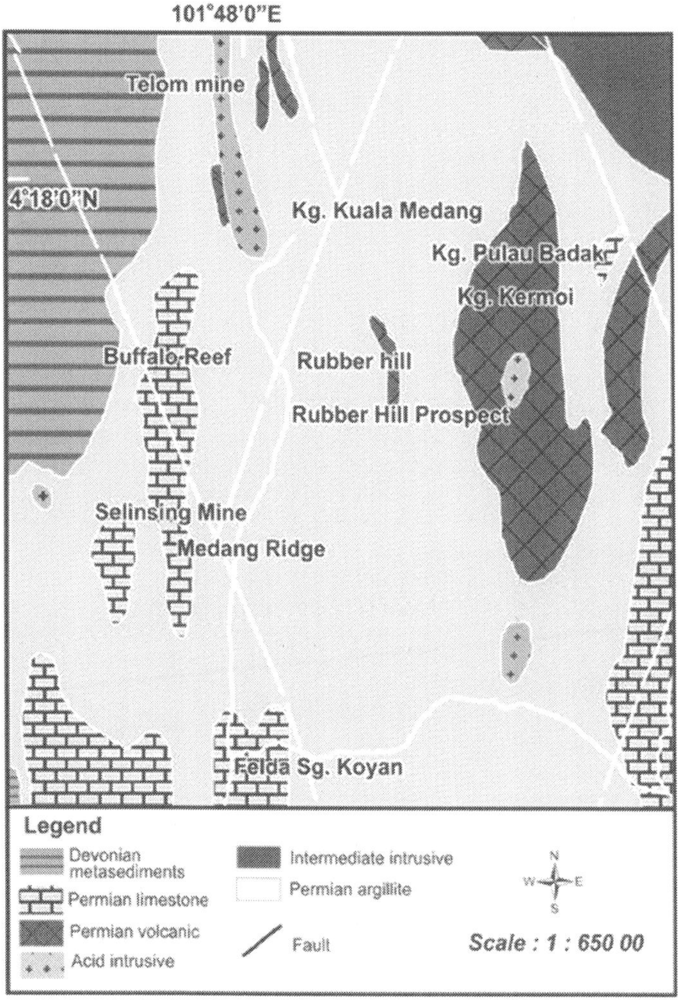

Fig. 13. Geological map of Selinsing gold mine and some prospects in the State of Pahang

lodes that occupying the central part of folded structure and proximity. Most of the underground working, alluvial and open cast mining efforts in the area are centred at the easternmost of the two fault zones along an area over 5km N-S to maximum 335m depth. Complexes quartz-carbonate veining stockworks has resulted intensive gold mineralization within cross-cutting, fissures and brecciated and silicified textures with high tenor in narrow zones of 2m wide. Sporadically from high-grade ore-shoots that featuring complex discontinuity and branching vein system. Gold fines is more than 981.16 with average Au (96-99%), Ag (1.44-1.91), As (0.04-0.12%), Hg (0.14-0.22) and Te (< 0.07%) (Henny et al., 1995).

6.5 Tersang Zone

Tersang is located about 20km north of Raub and along the same regional mineralized strike. The N-S trending Tersang gold deposit consists of a large and extensive outcroping quartz veins stockworks cutting within the 2km elongated of hydrothermally altered pyritised aplite-rhyolitic dyke as well as the metasediments. Hydrothermal alteration is shown by silicification, disseminated pyrite and arsenopyrite in the felsites with eminent sericitisation. The thickness of quartz veins varying from 1-20 cm and can up to 1 m in width at 80°/60° to the south. It is currently hosts an inferred gold resource of 528,000 tailings (Snowden, 2008; Howe, 2004). Assay results of stream sediment have return as high as 410 ppb Au and Sb anomaly. Other nearby prospects within the Tersang zone is Tenggelan, Chenua and Chun Kok in the south.

6.6 Kechau-Tui

Gold mineralization at old Kechau Tui (Ajmal mine), Kuala Lipis, Pahang is characterized by the quartz veins transecting the Permian limestone Bedrocks of Gua Musang Formation (Wan Fuad, 2008; Cheang, 1988). The veins are generally steeply dipping at about 70° to the west along N-S and NE-SW trending faulted and sheared zones, occurred as isolated or multiple parallel of a few cm to 20-30cm wide vein system.

The primary gold mineralization in the area seems to be related to the igneous intrusion of the area emplaced during Triassic-Jurassic period. To the north is a small pluton, named Bukit Tujuh granite and to the north-east is the Bukit Damar pluton. Gold mineralization in the Mine is significantly different from that of Penjom and Bukit Mandi (Wan Fuad & Heru Sigit, 2008, 2002; Gunn, 1993) as it contains less arsenopyrite and pyrite. The gold bearing veins are sulphide-poor. There are two type of mineralized veins in this deposit, viz i) sulphide rich, gold poor quatz veins and ii) sulphide poor gold bearing quartz veins. The main gold mineralization occurs in N-S trending sulphide-poor veins. Mineralogically, the later, consist mainly of late tetrahedrite, galena and traces of chalcopyrite and sphalerite. Gold is found as tiny free gold and isolated specks within these milky white quartz veins, away from main galena mass, and also as fine gold specks in the tetrahedrite. The gold was deposited earlier stage, and later, together with tetrahedrite. Wall-rock alteration is hardly visible. At the vein quartz-limestone contact, there is a narrow transition zone where the white colour is gradually changed to grey colour of limestone, and often near the contact it consist mainly of dolomite, followed by quartz and calcite.

6.7 Mengapur deposit

The Mengapur Copper mine in Maran, Pahang is a typical Cu-Fe of gold-bearing distal skarn deposit located in the Central Mineralization Belt. The study shows the ore deposit at Mengapur is a contact-metasomatic type associated with Triassic granodiorite. It is confined mainly to the extensive contact-metasomatic skarn aureole formed within the Permian calcareous metasediments and volcanic surrounding the Botak granite intrusion. The skarn rocks comprise a spectrum of garnet and pyroxene-rich types with the gold mineralization preferentially concentrated in the pyroxene-rich types varieties. Quartz-veins stockwork are common within the skarn and hosts for vein-type mineralization. The dominant metallic mineral assemblages of the skarn deposit are pyrrhotite, magnetite, chalcopyrite and arsenopyrite. In the veins the assemblage is more varied and includes pyrite, chalcopyrite. pyrrohtite, chalcocite, covellite, digenite, galena, sphalerite, molybdenite, bismuth, arsenopyrite, stibnite, boulangerite, schellite and gold. Gold occurs as fine to medium grains, infilling fractures and fissures in pyrite and arsenopyrite, and also intergrown or enclosed by galena. Minute gold grains 5 to 10 micron infilled the interstices of intermediate quartz associated with fine carbonaceous stylolitic streaks and fragments of phyllite (Sinjeng, 1993). Formed at the homogeneous temperature of 169.2 to 313.7°C (NaCl wt% 2.4-8.0%) at shallow depth (Goh et al., 2003).

7. Potential of gold in Kelantan

Detailed accounts on gold mineralization in Kelantan was discussed by Goh et al. (2006); Chu & Singh (1986), Chu (1983 & 1980). In Kelantan, basically, most of the gold mines are working on placer deposits; they contribute approximately 10% of the annual gold production of Malaysia. Gold mineralization in Kelantan is mainly distributed in the central part of the state, bounded by Stong Igneous Complex and Seting Granite on the west, Kemahang granite in the north and Boundry Range granite in the east. Gold was mined from early times in the Pulai Districts, Galas, Pergau, Lebir and Kelantan River. Significant gold mineralizations mainly occur in sedimentary-metasedimentary rock of Perm-Triassic age. Gold mineralization typically associated with hydrothermal quartz vein system, skarn and volcanogenic massive sulphides. The main factors contributing succession of gold mineralization are source rocks, heating chamber as well as depositional structure.

The principle source rocks are Permian-Triassic volcanic rocks that are associated with sedimentary rocks (Fig. 14). The heating chamber that induced the hydrothermal fluids is the granitoid bodies that intruded under the volcanic-sedimentary rock, whilst structures which allow the infiltration and deposition of gold are sheared and faulted zones originating from depth.

Based on the type of ore deposits, geochemical data, and geological setting, the study area can be divided in to five mineralization zones (Goh et al., 2006; Teoh et al., 1987). Namely, hydrothermal vein gold mineralization zone, gold-base metal associated with volcanic exhalative zone, gold-silver-mercury zone (hydrothermal veins). Available data suggests that in Kelantan primary gold mineralization is associated with Ag-Au quartz veins, Massive sulphide bodies, pyritiferous and carbonaceous metasediments, skarn-type mineralization, and sulphide-bearing volcanic and volcaniclastic rocks. The most significant Ag-Au massive pyritic Pb-Zn sulphide bodies remaining are those at Ulu Sokor. These oxidised bodies show supergene gold enrichment, with the oxidised zones displaying a higher gold tenor than primary sulphides. Much of the gold in the primary sulphides is locked in pyrite (Batcher, 1994; Chu & Singh, 1986).

The intrusive rocks show some sign of gold mineralization. Gold mineralization is also occurs in shear zones in granite (Schroeder & Cameron, 1996) and not really significant (Batchelor, 1994; Chu 1983; Chu et al., 1980). Quartz veins are well-developed along these shear zones and cut through the sheared granitoid. These types of deposits can be seen in Katok Batu Mine, Pulai and Batu Melintang (Goh et al., 2006).

So far most of gold in Pulai were from placer type. The Pulai fluviatile gold placer deposit stretches along 17 km of the upper reaches of Sungai Galas. The valley alluvium ranges up to 1200 m wide and averages 6.2 m in thickness. Malaysia Mining Corporation had proved-up sizeable reserves following drilling and bulk testing during 1979-1983 (Batchelor, 1994). Six types of hydrothermal quartz veins can be recognized in the state of Kelantan, namely:

a. low sulphide quartz veins
b. high sulphide quartz veins
c. quartz veins in sheared granite zones
d. quartz vein at the boundary of sedimentary rocks
e. structurally controlled quartz veins in volcanics
f. sedimentary rocks
g. metamorphic segregation quartz veins, as shown in Fig. 15.

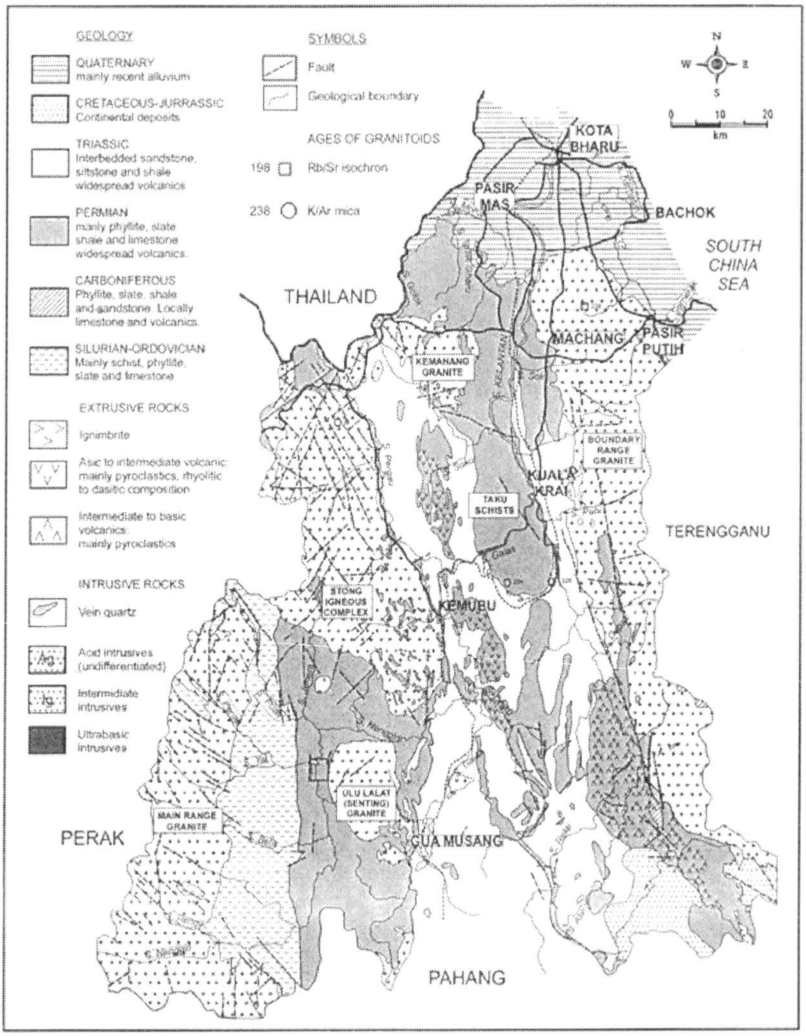

Fig. 14. General Geological map of Kelantan (after Teoh et. al., 1987; Goh et al., 2006)

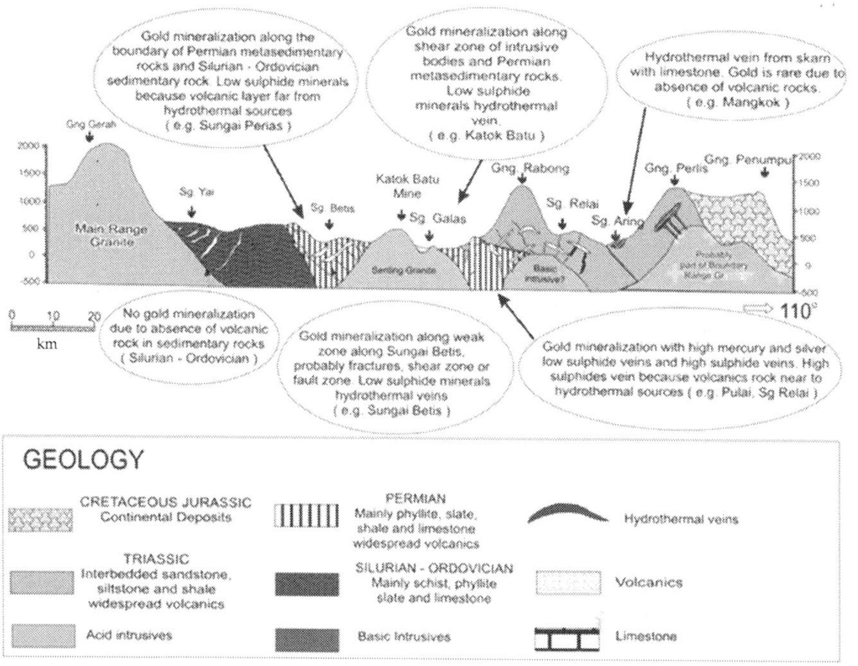

Fig. 15. Gold mineralization model across Main Range granite on the west to Sungai Relai, in the south-eastern corner of Kelantan (after Goh et al., 2006)

8. Potential of gold mineralization in Eastern Belt

The first significant discovery of primary gold of phanerozoic mesothermal lode gold mineralization type in the east coast of Peninsular Malaysia, within the Eastern Tin-gold belt is from Lubuk Mandi, near Rusila in Terangganu. As stated by Yeap (2000, 1993), most of the potential gold occurrences and deposits are located within Gold belt 3 and 4. Both Lubuk Mandi and Sungai Pelong (Bukit Panji) golfields are Gold Belt 4 (Fig. 2). Geology of the Eastern Tin-gold belt are predominantly consists of Carbon to Permian metasedimentary rock sequence (interbedded phyllite, and minor meta-arenite) with subordinate of volcanic which later intruded by the Late Permian to early Triassic granitoids. Interbcdded siltstone and meta-sandstone are the younger, Jurassic-Cretaceous age, rock formation overlying a few areas.

A few new gold prospects have been recognized for follow-up study in the northern part of Terangganu during exploration programs carried-out by the Department of Mineral and Geosciences between 1991 and 2005. Following this programme, Sungai Tapah, Sungai Setiu, Sungai Pelong and Sungai Pelagat were identified as potential targets for integrated follow-up and detailed appraisal (Mohamad Sari, et al., 2005). Preliminary appraisal of geochemical data indicated several drainage basins within this area are gold and tin enrichment, where illegal gold panning activities was perceived. In Sungai Tapah 2m wide quartz vein, striking N-S within N-S shear zone, cutting the graphtic phyllite and slate bedrock of the area were observed with assay result of 3.6ppm.

8.1 Lubuk Mandi deposit

The mineralization occurs in the structurally control, deformed and brecciated carbonaceous metasidement and main quartz veins (345°/80°) within of N-S trending brittle deformation zone of 5 to 10m wide. The quartz veins belonging to more than one generation are widespread throughout the sedimentary sequence, folded, discontinuous, strained, and complexes in nature, always concordant to the fabric of host rock and often extending less than 1m.

The country rocks generally comprise low-grade, chloritic alteration metasediments of Carbonaceous age, mainly grey to black laminated phyllite and shale unit with subordinate siltstone and sandstone of Sungai Perlis Beds, and generally dipping steeply at about 80-85° to the east. Post-date mineralization, up to 4m wide dolerite dyke is the only igneous intrusive encountered, undeformed and free of quartz-vein· (Gunn et al., 2000). Mineralization took place at 196.2°C homogenization temperature (salinity 4.2wt.) at a depth of 156m at 16kbar pressure (Wan Fuad & Heru Sigit, 2003).

Free gold (5 to 400μm) was observed to occur close to and within streaks and clasts of graphitics shale and phyllite, incorporated into the veins and where stylolitic texture is developed (Gunn et al., 1993). Gold was deposited after the introduction of the most abundant pyrite and arsenopyrite and almost contemporaneously with sphalerite and galena including traces amount of schellite, casseterite, hematite and other secondary iron. Elevated gold abundance in the veins is accompanied by enhancement of As and Pb, and to a lesser extent, of Ag, Zn, Bi, W, Mo, Cu, Te and Se. Based on the elemental relationships, the Lubuk Mandi gold deposit can be classified as a Au-Cu deposit.

8.2 Mersing

Mersing is located on the eastern coast of Peninsular Malaysia (eastern gold tin belt), in Johor. Gold has been reported in two distinct settings in the area, quartz veins in the Permo-Carboniferous shale sequences and placer type overlying Jurassic conglomerates that contain Ag < 2%. The results show mesothermal sediment-hosted quartz-vein is a major type of gold mineralization indicated by the presence of pyrite, arsenopyrite and galena which is the common type in many other parts of Malaysia. Other minor components are ultramafic/mafic rock type and red-bed type unconformity (palaeo-placer) related mineralization. EPMA analysis on these alluvial gold grains indicated the samples often contain, maximum 0.18% Hg, 0.12% As, 0.08% Te and occasionally Se in Tiemannite (Styles et al., 1994).

9. Conclusion

Gold mineralization in the Central Gold Belt is generally categorized as a low mesothermal lode gold deposit due to its tectonic and geological setting. Most of the gold mineralization took place within a low-grade meta-sedimentary-volcanic terrain formed during the collision of the Sibumasu block underneath the East Malaya (Indochina) block through the Permian to late Triassic. Gold mineralization in Central Belt is much associated with the accretionary prism along the North-south trending terrain boundary known as the Raub-

Bentong Suture. A collision structure overprint has generated major N-S or NW-SW trending left slip fault; and dilational Riedel and subsidiary shears and numerous splays associated with these faults. These structures have of great consequence in hosting many mesothermal quartz lodes within the Central Belt.

Mineralization took place within a low grade Permo-Triassic island arc system composed of meta-(sedimentary- green facies and occasionally upgraded to amphibolite facies) and volcanic rocks accompanied by extensive deformation (brittle-ductile and shearing zone), metamorphism, and magmatic events that created the favorable environment for source and trap for the gold mineralization. In many occasion, the most significant feature with respect to gold mineralization in Central Belt is a north-south aligned shear zone in carbonaceous metasedimentary rock sequences. Significant gold mineralization often confined to the north-south trending sheared zone composed of metamorphosed, brecciated and hydrothermally altered calcareous graphitic shale with minor interbedded fine-grained sandstone and tuffaceous rocks. The gold mineralization in Central Belt is regarded as high grade quartz-carbonate-gold type related to a phyllic propylitic alteration of granite intrusion complex that intrudes weakly metamorphosed green schist facies sedimentary strata. In many deposits, ore systems display permeability controlled or governed by lithology, structure and breccias and changes in wall-rock alteration.

Arsenopyrite, pyrite, galena and sphalerite are the common sulphides. Signs of mineralization such as extensive wall-rock alteration that gave rise to carbonate and alkali metasomatism are evident such as conspicuous occurrence of sericitization, fuchsite, potassic albitization, chloritization as well as sulphidation. In Kelantan and Mengapur, gold mineralization typically associated with hydrothermal quartz vein system including as skarn and volcanogenic massive sulphides. Most of the gold-rich samples are proportionally elevated in arsenic. Ag, As, Te, Sb, and Bi except Hg are evident in segments associated with sulphide-gold mineralization. Gold has a marked affinity for Te and Bi and less for Sb except at Buffalo Reef. Other signatures conspicuously associated with gold included Ba, Mo, Co, Ni, W and Se. Based on elemental analyses epizonal (Au-Sb) and mesazonal (Au-As-Te) types of orogenic gold (Carline type) formation is favorable in Central Gold Belt (Groves et al., 1998).

In general, these deposits often characterized by a potassic radiometric anomaly (Nor'aini Surip, et al., 2003) and often accommodated within tightly folded sedimentary rock, associated with tonalitic intrusion and various scales of thrust faults, and appeared running parallel to the Raub Bentong Suture and nearby, district scale, splays. Mineralization often hosted within structurally control sheared thrust and veins within fold axes, bedded thrust associated mineralization type, stockworks in felsite, and finally steep, cross-cutting fractures. Mineralization of gold was believed to be formed at the homogenization temperature higher than 270-280°C hydrothermal fluid which is typical in epizonal and mesothermal lode and quartz veining within the metamorphogenic deformational terrain.

10. References

Alexander, J.B. (1949). *Progress report on geological work in southwest Pahang and in part of northwest Selangor.* Malaya Geo. Survey, Annual Report, Govt. Press, Kuala Lumpur, pp. 19-24.

Ariffin, K. S. & Hewson, N. J. (2007). *Gold-Related Sulfide Mineralization and Ore Genesis of the Penjom Gold Deposit, Pahang, Malaysia,* Resource Geology, 57(2), pp. 149–169.

Azman, A. G., Ramesh, V., Yong, B.T. & Khoo, T.T. (2006). *Geochemistry and petrology of syenite, monzonite and gabbro from the Central Belt of Peninsular Malaysia.* Geological Society of Malaysia Bulletin, 49, 25-30.

Barton, P.B., Jr. and Bethke, P.M. (1987). *Chalcopyrite disease in sphalerite: Pathology and epidemiology.* American Mineralogist., 72, pp. 451-467.

Batchelor, D.A.F. (1994). *Geological characteristics of the Pulai alluvial gold deposit, South Kelantan,* Malaysia Journal of Southeast Asian Earth Sciences, 10, 1-2, p. 101-108.

Becher, H.M. (1983). *The gold - quartz deposits in Pahang (Malay Peninsular),* Quaternary Journal Geology Soc., London, 49, pp. 84-88.

Bignell, J.D. and Snelling, J.B. (1977). *The Geochronology of Malayan granite,* Overseas Geology and Mineral Res. Institute of Geological Science, London, 47, pp. 358-359.

Bogie, I. (2002*). Petrology Report on 15 Samples from the Penjom Gold Mine, Malaysia,* Internal Report for Specific Resources Sdn. Bhd. (unpublished report).

Boyle, R.W. (1979). *The Geochemistry of Gold and its deposits,* Canada Geological Survey Bulletin, 280, 584pp.

Campi, M.J., Shi, J.R. & Leman, M.S. (2002). *The Leptodus Shales of central Peninsular Malaysia: Distribution, age and palaeobiogeographical affinities,* Journal of Asian Earth Sciences, 20, pp. 703-717.

Cheang, K.K. (1988*). Gold-bearing quartz veins from Ajmal Mine, Kuala Lipis Pahang, Malaysia,* Warta Geologi, 14, pp. 30.

Cherry, M.E. (1983). The association of gold and felsic intrusion - examples from the Abitibi Belt, In: *Geology of Gold in Ontario* - Ontario Geological Survey Miscellaneous Paper, National Academy Press Washington.

Chu L.H., Muntanion, H. Aziz bin Sidek, Chand, F. and Troup, A. (1983). *Regional Geochemistry of South Kelantan,* Geological Survey Malaysia Geochemical Report 1.

Chu, L. H. (1980). *Preliminary report on geochemistry, Anomaly 3503, Kelantan,* Geo. Survey Malaysia Report, (unpublished).

Chu, L.H. (1983). *Exploration in the Sok Base Metal Prospect, Kelantan,* Geological Survey Malaysia Report (unpublished).

Chu, L.H. & Singh, D.S. (1986). The Nature and Potential of Gold Mineralization in Kelantan, Peninsular Malaysia. In : *GEOSEA V. Proc.,* Geological Society of Malaysia, Bulletin, 19, pp. 431-440.

Cobbing E.J., Pitfield P.E.J., Darbyshire D.P.F., Mallick D.I.J. (1992). *The granites of the South-East Asian tin belt. British Geological Survey Overseas* Memoir 10, London, 369pp.

Cocks, L.R.M, Fortey, R.A. and Lee, C.P. (2005). *A review of Lower and Middle Palaeozoic biostratigraphy in west Peninsular Malaysia and southern Thailand in its context within the Sibumasu Terrane,* Journal of Asian Earth Sc., 24, pp. 703-717.

Corbett, G. (2002). *Epithermal Gold for Explorationist,* AIG Journal, Paper 2002-01, February 2002, 26pp.

Fillis, P. (2000). *Penjom Resource Model,* Internal Report for Specific Resources Sdn. Bhd (Unpublished).

Flindell, P. (2003). *Avocet Mining - Exploration and Development Across Central and Southeast Asia,* Australia Institute of Geoscientists (AIG), Mineral Exploration Discussion Group (SMEDG). 10 Oct., Sydney, 8pp.

Gobbett, D.J. & Hutchison, C.S. (1973). *Geology of the Malay Peninsular,* West Malaysia and Singapore. Wiley Interscience, New York, 438pp.

Goh, S. H., Teh G.H. & Wan Fuad Wan Hassan (2006). *Gold Mineralization and Zonation In The State of Kelantan Abstract: Gold mineralization in Kelantan.* Geological Society of Malaysia Bulletin 52, June 2006, pp.129-135

Goh, S.H., Teh, G.H. & Mohd Rozi Umor. (2003). *Fluid Inclusion studies of Bukit Botak Skarn deposit, Mengapur, Pahang,* Geological Society of Malaysia, Bull. 46, pp. 231-238.

Groves, D.I., Goldfarb, R.J., Gebre-Mariam, M., Hageman, S.G. & Robert, F. (1998). *Orogenic gold deposits: A proposed classification in the context of their crustal distribution and relationship to other gold deposit,* Ore Geology Reviews, 13, pp. 7-27.

Gunn, A.G., Sinjeng, P.P. & Zulkipli Che Kasim (2000). Geochemical and mineralogical studies of primary gold mineralization at Lubuk Mandi, Rusila, Terangganu, Malaysia, In (Ed,) Khoo K.K., *Technical papers,* Mineral and Geosciences Department, Volume 1. Malaysia, pp. 1-25.

Gunn A.G, Halim Hamzah, Ponar Sinjeng, P. (1993). Geochemical and mineralogical studies at the Kim Chuan gold mine, Raub, Pahang, Malaysia. Geo. Survey Malaysia- BGS *Gold Sub-Programme Report 93/3.* Ipoh. pp. 46.

Herrington, J.R. (1992). *Report on the Eight Rock Sample for Evocet Ventures Inc.,* Fluid Inclusion Group, Natural Historical Museum, London (unpublished report).

Hewson, J.N. & Crips, D.A. (1992). *Final Report Phase One Exploration Block 7, Pahang, Malaysia,* Luckfrost Limited, London (unpublished internal report).

Howe, (2004). *Valuation of the Raub Mining Lease in Pahang State, Malaysia.* Unpublished report prepared for Raub Australian Gold Mining Sdn Bhd by ACA Howe Int. Limited. August, 2004, pp. 15.

Hutchison, C.S. (1971). *The Benta Magmatite Complex: Petrology of two important localities,* Geo.logical Society of Malaysia Bulletin, 4, pp. 49-70.

Hutchison, C.S. (1977). *Granite emplacement and their tectonic subdivision of Malay Peninsular,* Geological Society of Malaysia Bulletin, 9, pp. 187-207.

Jaafar Ahmad (1979). *The Petrology of the Gunung Benom Igneous Complex,* Geo. Survey Malaysia, Special paper, 2, 141p.

Johnston, G.G. (1998). *Selinsing gold project, Pahang Darul Makmur, Malaysia: Ore resource statement.* Unpublished Report, Syarikat Selinsing Mining Sdn. Bhd.

Kamar Shah, A. (2007). *Low grade, Mesothermal Gold Deposit Characteristics of the Penjom Gold Deposit, Kuala Lipis, Pahang,* Warta Geologi, 33(3), pp. 102-110.

Kamar Shah, A. (1995). *Geology and Mineralogy of the Penjom Gold Mineralization Prospect,* Universiti Sains Malaysia (unpublished M.Sc.), 176pp.

Kamar Shah, A. and Khairun Azizi, M. A. (1995). An overview of the mineralization and mineralogical characteristic of the goldfields from Central Belt of Peninsular Malaysia, *Proceeding of the International Conference on Geology, Geotechnology and Mineral Res. of Indochina,* Khon Kheau, Thailand, pp. 188-199.

Khoo, T.T. and Tan, B.K. (1983). Geological evolution of Peninsular Malaysia, *Proceeding of the Workshop on Stratigraphic Correlation of Thailand and Malaysia,* 1, pp. 253-290.

Kidd, R. (1998). *Exploration, Geology, Mineralization and Resource Estimate of the Penjom Gold Mine* (unpublished internal report for Specific Resources).

Kobayashi, T. & Toriyama, R. (1970). *Geology and Palaeontology of South East Asia.* Univ. Tokyo Press, Japan, pp. 1-25.

Kwong Y.T. and Crocket, J.H. (1978). *Back ground and anomalous gold in rocks of an Archean greenstone assemblage, Kagaki Lake Area, N.W. Ontario,* Economic Geol., 73, pp. 50-63.

Lee, A.K., Gonzales, R.A., Tyebally, F.H., Chand, F. & Troup, A. (1982). *Regional geochemistry of North Pahang, Geochemical Report 2*, Geo. Survey Malaysia, , 86pp.

Lee, A.K., Khong, Y. and Hock, O.W. (1986). *Gold Mineralisation and Prospects in North Pahang*, Geological Survey Malaysia, Geochemistry Report, 4, 50 pp.

Leman, M.S. (1994). *The significance of Upper Permian brachiopods from Merapoh area, northwest Pahang*, Geological Society of Malaysia Bulletin, 35, pp. 113-121.

Metcalfe, I. (1988). Origin and assembly of Southeast Asian continental terranes, Gondwana and Tethys. *in* Audley-Charles, M.G. and Hallam, A. (Ed.), *Geological Society of London Special Publication*, 37, pp. 101-118.

Metcalfe, I. (1992). *Lower Triassic (Smithian) Conodon from northwest Pahang, Peninsular Malaysia*, Journal of Micropaleontology, 11, pp. 13-19.

Metcalfe, I. (2000). *The Bentong-Raub Suture Zone*, Journal of Asian Earth Sciences, 18,pp. 691-712.

Metcalfe, I. (2002). *Permian tectonic framework and palaeogeography of SE Asia*, Journal of Asian Earth Sc., 20, pp. 551-566.

Minerals and Geosciences Department Malaysia, (2009). *Malaysian Minerals Yearbook 2009*, Issue 19, 125pp.

Mitchell, A.H.G. (1977). *Tectonic setting for emplacement of southeast Asian tin granite*, Geo. Soc. Malays, 9, pp. 123-140.

Mohamad Sari H., Yusoff Ismail, Mohamad Fauzi, A. (2005). *Potentail of Gold Mineralization in Northern Terengganu, Malaysia*, Mineral and Geoscience Dept, Malaysia, Report No. JMG.TGG (SGR) 02/2005, 66pp.

Mohd Basril Iswadi B., Ibrahim Abdullah & Wan Fuad W. H. (2009). *The relationships of structure and deformation to the gold mineralization in Selinsing Gold Mine, Pahang*, Geological Society of Malaysia, Bull. 55, pp. 33 - 37.

Mohd Rozi Umor & Syed Sheikh Almashoor (2000). Jujukan Usia Batuan di Dalam Kompleks Benta, Pahang Berdasarkan Cirian Lapangan dan Penentuan Usia Batuan Secara K/Ar Keseluruhan Batuan, *Proceeding Annual Geological Conference*, Geological Society of Malaysia, Penang, pp. 87-95.

Mustaffa Kamal, S. and Azman, A. G. (2003). *'Mantle Plume' type magmatism in the Central Belt Peninsular Malaysia and its tectonic implications*, Geological Society of Malaysia Bulletin, 46, pp. 365-371.

Nancy, Te (1972). *Carboniferous fossils from "Raub Group" near Chuak Village, Western Pahang, West Malaysia*, Geological Society of Malaysia Bulletin, 34, pp. 5-9.

Noraini Surip, Ab. Halim Hamzah, Mohammad Roston Zakaria, Azlikamil Napiah, & Jasmi Ab.Talib (2003). *Mapping of Gold In Densely Vegetated Area Using Remote Sensing and GIS Technique in Pahang, Malaysia*, 15pp.

Nuraiteng T.A. (1993). The occurrence of upper Permian foraminifera in NW Pahang. *Proceeding of the 24th Annual Geological Conference, Technical Paper*, 3, 25-33.

Pereira, J.J. (1993). *Geology, Mining and Tailing Characteristic of the Selinsing Gold mine, Pahang*, Warta Geologi, Newsletter of the Geology Society of Malaysia, Mac-April, 19(2), pp. 35-41.

Pereira, J.J. Yeap, E.B. and Ng, T.F. (1993). *Application of soil geochemistry to the detection of Sb-Au mineralization in the Buffalo Reef Area*, Geological Society of Malaysia Bulletin, 33, pp. 1-10.

Proctor, W.D. (1972). *Geology and Mineral Resources Benta Area, Pahang,* Geo. Survey of Malaysia, Map Bulletin, 4, 25pp.

Richardson, J.A. (1939). *The Geology and Mineral Resources of the Neighbourhood of Raub, Pahang with an Account of the Geology of the Raub Australian Gold Mine,* Geological Survey Malaysia, Kuala Lumpur, 3, 166pp.

Richardson, J.A. (1950). *The Geology and Mineral Resources of the Neighborhood of Cegar Perah and Merapoh,* Pahang, Geological Survey Malaysia, Memoir 4, 162pp.

Santokh Singh, D. (1977). Gold Mineralization in Peninsular Malaysia, National seminar on the mining industry, Ministry of Primary Industries, Malaysia, 8pp.

Schroeder, T.G. and Cameron, R. (1996). Alkalic Intrusion-associated Au-Ag. *in* Lefebure, D.V. and Hoy, T. (Eds.) *Selected British Columbia Mineral Deposits Profiles, Vol. 2 - Metallic Deposits,* British Columbia Ministry of Employment and Investment, Open File 1996-13, pp. 49-51.

Schwartz, M.O., Rajah, S.S., Askury, A.K., Putthapiban, P. and Djaswadi, S. (1995). *The Southeast Asian Tin Belt,* Earth Science Reviews, 38, pp. 95-293.

Scrivenor, J.B. (1911). *The Geology and Mining Industries of Ulu Pahang,* Gov. Press, Malaya.

Scrivenor, J.B. (1928). *The Geology of the Malaysia Ore Deposit,* MacMillan & Co., London, 216pp.

Scrivenor, J.B. (1931). *The Geology of Malaya,* MacMillan, London, 250pp.

Sinjeng, P.P. (1993). Mineralogy of gold bearing rocks from the Mengapur Ore deposit. *Proceeding of the 24th Annual Geological Conference, Technical paper 3,* pp. 25-33.

Snowden (2011). *Monument Mining Ltd: Buffalo Reef Gold Deposit, Malaysia,* N143-101 Technical Report, Project No. 2015. Mineral Resource Estimate May 2011, 80pp.

Sonny, L.T.C., Sharafuddin, M., Sulaiman, M., Teh, G.H. and Abdul Aziz, J.H. (2001). *Geology, structure, mineralization and geochemistry of the Penjom gold deposit, Penjom, Pahang,* Geological Society of Malaysia Bulletin, 44, pp. 61-63.

Spiller, F.C.P. and Metcalfe, I. (1995). *Late Palaeozoic radiolarians from the Bentong-Raub suture zone and Semanggol Formation, Peninsular Malaysia - initial finding,* Journal of Southeast Asian Earth Sciences, 11, pp. 217-224.

Tan, B.K. (1984). *The tectonic framework and evaluation of the Central Belt and its margin, Peninsular Malaysia,* Geological Society of Malaysia, Bulletin, 17, pp. 307-322.

Tan, B.K. (1996). *"Suture Zone" in Peninsular Malaysia and Thailand: Implications for paleotectonic reconstruction of Southeast Asia,* Journal of Southeast Asian Earth Science, 13, pp. 243-249.

Teoh, L.H., Lee, A.K. and Foo, K.Y. (1987). *Gold Mineralization and Prospects in Kelantan,* Minerals and Geoscience Department Malaysia (unpublished report).

Tjia, H.D. (1987). *Olistostrome in the Bentong area, Pahang,* Warta Geologi, 13, pp. 105-111.

Tjia, H.D. (1989). The Bentong Suture, In: Situmorang, B. (Ed.), *Proceedings of the 6th Regional Conference on Mineral and Hydrocarbon Resources SE Asia,* pp. 73-85.

Tjia, H.D. and Zaitun H. (1985). *Regional Structures of Peninsular Malaysia,* Sains Malaysiana, 14, pp. 95-107.

Wan Fuad W. H. (2003). *Relationship between gold and associated elements at gold mines in Penjom, Pahang and Lubuk Mandi, Terengganu,* Sains Malaysiana 32, pp. 1-12

Wan Fuad, W.H. (2008). *The mineralogy of gold mineralization of the Ajmal Mine, Kechau Tui, Pahang Darul Makmur,* Geological Society Malaysia, 54, pp. 59-62.

Wan Fuad, W.H. & Heru Sigit, P. (2001). Perubahan batuan dinding berkaitan dengan permineralan emas di Penjom Gold Mine, Pahang, Malaysia, *Proceedings, Geo. Society of Malaysia, Annual Geo. Conf*, Pangkor, Pangkor, Malaysia, pp. 13-17(*in Malay language)*.

Wan Fuad, W.H. & Heru Sigit, P. (2003). *Analisis bendalir terkepung pada terlerang kuarza yang mengandungi emas di kawasan lombong Penjom, Kuala Lipis, Pahang dan Lubok Mandi*, Geo. Society of Malaysia Bulletin, 46, pp. 359-363 (*in Malay language)*.

Yeap, E.B. (1993). *Tin and gold mineralization in Peninsular Malaysia and their relationships to the tectonic development*, Journal of Southeast Asian Earth Sciences, 8, pp. 329-348.

Yong, B.T., Azman, A.G., Khoo, T.T. & Shafari, M. (2004). *Benom Complex: Evidence of magmatic origin*, Geological Society of Malaysia Bulletin, 47, pp. 55-60.

Part 8

Mining

16

Computer Aided Ore Body Modelling and Mine Valuation

Kaan Erarslan
Dumlupinar University, Mining Engineering Department, Kutahya Turkey

1. Introduction

Mine valuation can be defined as the process of determining the worth of a specific mineral deposit and capability of making a return by a prospective investment (SME, 2005). Although the definition is very brief and compact, actually, it has a very wide content. Determination of worth of an underground asset requires a plenty of works to explain physical, structural and economical properties of it (Kennedy, 1990).

Underground assets are invisible bodies whose shapes, quality compositions and quantities are unknown. Geological explorations and investigations aim at determining all these unknowns (Sinclair and Blackwell, 2004). At the beginning of process, topographical and lithological data are gathered and a database is generated. Depth, thickness and grade changes, overburden structure, ore volume, shape and extensions, footwall and hanging wall properties are determined by various mathematical approaches using this database. All numerical estimations and visual supports help bringing out ore body model (Singer and Menzie, 2010).

The most concrete data to define shape, location, quality and quantity of an ore body is drill hole cores. GPS data is mostly used to draw topographical maps and surfaces. Additionally, underground maps such as thickness and grade contours are drawn as well. When topographical coordinates are combined with stratigraphical information, a three dimensional data set is handled. Eventually, after following several mathematical techniques, three-dimensional model of ore body can be obtained (Hustrulid and Kuchta, 2006). Beside physical ore model, quality composition should also be known. This is crucial because further engineering activities have an economical aspect. Mine design and production schedule is fairly related to both physical structure and quality composition of ore (Hartman, 1992).

Surveying data include three-dimensional components x, y, z (easting, northing, altitude/elevation) which enable surface modelling. Drill hole data including depth and layer information contribute to explain how geological structure is in the third dimension (Torries, 1998). Drill holes also carry the information of ore grade or calorific value. Geological interpretation of stratigraphical layers provides three-dimensional ore body model (Nieuwland, 2003).

Major instruments for computer aided mine valuation are;

- drill hole logs,
- contour maps,
- cross-sections,

- surfaces (topography, thickness, grade, etc.),
- surface sections,
- solid models (three dimensional models), volume and reserve estimation.

Some of outputs of mine valuation are visual and some of them are numerical. Visual outputs help researchers see drill hole sections, how ore body is, how it extents, how its shape looks like, how ore body and overburden relation is, how quality and thickness of deposit changes thru axes. Numerical outcomes are generally, area and volume reports, drill hole lengths, survey coordinate sets, composite calculations, economical assessments, etc (Torries, 1998).

In the last several decades, many approaches have been developed to clear up geological modelling problems (Agoston, 2005). The purpose is to estimate unknown values regarding limited data in hand. The methods such as geostatistics and neural networks have complicated mathematical and statistical bases and are utilised to model topography, ore body, and grade distribution (McKillup and Dyar, 2010). Iterative structure of computations makes computer use necessary for many of recent methods. There are various specialised and expert software to support geological and mining engineers. Ore body modelling with computer aid is faster and more reliable. New information addition is simple and updating is much quicker. Different scenarios can be studied and better decisions can be made. General work plan and flowchart of computer aided mine valuation is shown in Fig. 1.

Once ore deposit is visually and numerically modelled, next step is mine design and production scheduling. Engineering economics and optimisation concepts are considered at this stage. Optimisation and simulation methods such as graph theory, dynamic programming, linear and goal programming, mixed integer programming, moving cones, genetic algorithm and network analyses are main techniques that can be counted. Scope of optimisation and simulation is generally open pit limits and production (Erarslan and Celebi, 2001).

Purpose of geological database generation is to describe physical and quality properties of ore bodies and preparation economical of database. Limited data should be optimally utilised and unknowns should be cleared. Last decades have shown great improvements in ore body visualisation and figuring. Buried underground body with so many unknowns can be visualised with animations and virtual reality applications by the help of computer software. Ore body is represented as a three dimensional entity laying under topography.

Modelling approaches utilise topographical information and drill hole database. Frequently and representatively recorded GPS values and aerial views help to visualise photo realistic views of topography. Next step is to define ore shape in space. Drill holes values are the most crucial and critical data set for this purpose. Problem at this stage is to determine what happens between sample points. Answer of the question requires two-dimensional and three-dimensional representations of ore body. This may look easy for regular structures. However, in most cases the picture is just reverse. Geological interpretation is highly needed. Computer systems help interpreters at this point. Computer programs, in general trend, provide two alternatives to interpreters for 3D determinations; geological parallel cross-sections and block models.

In this chapter, the basic ideas and some mathematical approaches of computer aided mine valuation are given. How GPS and drill hole data are used to figure out an entire ore body model in visual and numerical aspects are explained.

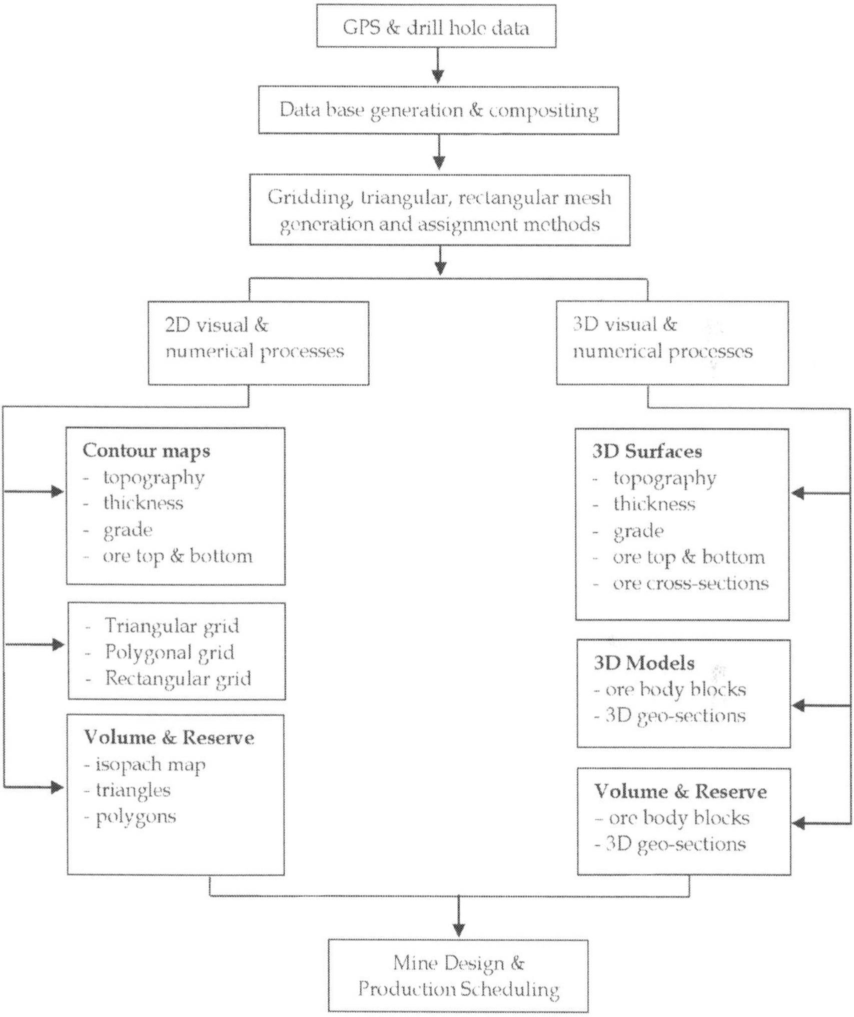

Fig. 1. Mine valuation work flow.

2. Database processes

Database is the base of every further study. Health of projects entirely depends on health of samples. GPS records are very crucial to model topography. However, the most concrete data that can be taken from field is drill hole cores. Mine valuation process starts with database building. It is composed of spatial coordinates of drill holes, geological formations that they intersect, depths of formations and their assay values. Surface determinations are, almost entirely related to GPS data.

2.1 Database structure

Input material for mine valuation systems is mainly topographical surveying data and drill hole cores. It is possible to categorise this database as; i- collar data, ii- survey data, iii- stratigraphical/lithological data, iv- assay values/grade analyses.
Collar data keeps (x,y,z) coordinates of drill ring. Survey database may include also physical coordinates of drill holes, depth information, azimuth and bearing angles. Stratigraphy database contains the information of geological formations through each hole. Assay database has the quality values of ore formations. Drill hole log/stamp is like an identity card of them (Fig. 2).

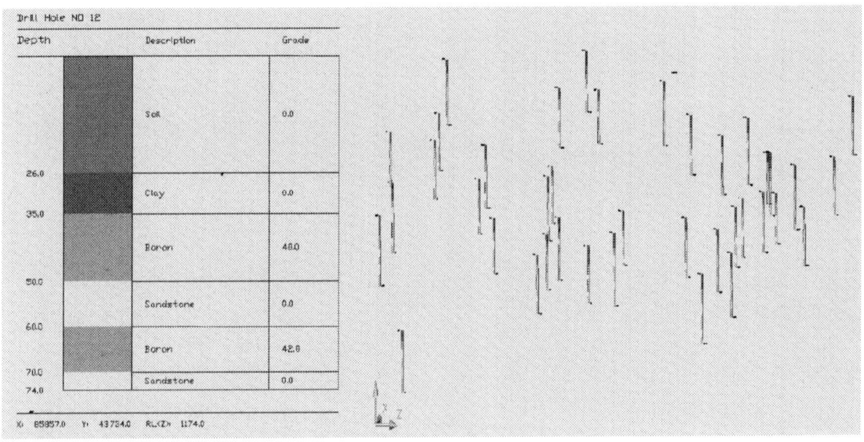

Fig. 2. A single drill hole log and drill holes in three dimensions (Erarslan, 2007).

Surveying data may also include three-dimensional components x, y, z (easting, northing, altitude/elevation) which enable surface modelling. However, GPS data that have been taken with frequent intervals enables photo realistic models. If data can take representatively, topography model looks like the field itself. Drill hole data including depth and layer information contribute to explain how its shape is in the third dimension. Drill holes also carry the information of ore grade or calorific value. Geological interpretation of stratigraphical layers provides three-dimensional ore body model.

2.2 Drill hole compositing

Drill holes cut intersect downwards successive layers. Mechanical properties of waste layers and quality values of ore layers are determined by applying several tests on hole cores. Generally, mineral formations are not monolithic and single piece bodies. Inter-burden layers may intersect mass or mineralisation may occur with waste layers in alternating forms. In other words, valuable mineral layers may exist in different thickness and quality amounts. During numerical calculations, a single thickness and grade value may be needed. Some classical reserve estimation techniques such as triangulation and polygon methods need composite values. In that case, what is the net thickness of valuable part and its quality? Here, compositing computation is applied to have total thickness and a unique grade (quality) value for each drill hole. Regarding a cut-off grade, ore thicknesses are summed up to give ore thickness. On the other hand, grade value is the thickness-weighted average (Hustrulid and Kuchta, 2006).

$$t_{total} = \sum_{i=1}^{n} t_i \tag{1}$$

$$g_{comp} = \frac{\sum_{i=1}^{n} g_i t_i}{\sum_{i=1}^{n} t_i} \tag{2}$$

where, t_{total} is total ore thickness, g_{comp} is composited grade, t_i is thickness and g_i is grade of ith core piece within n core pieces. After obtaining a single thickness and grade a value for each drill holes, volumes and reserve amounts in superimposed triangles or polygons can be calculated.

Another type of compositing is called as bench or level compositing. Ore field under investigation is divided into parallel horizontal levels and parts of drill holes corresponding to those levels are composited instead of compositing entire hole (Fig. 3).

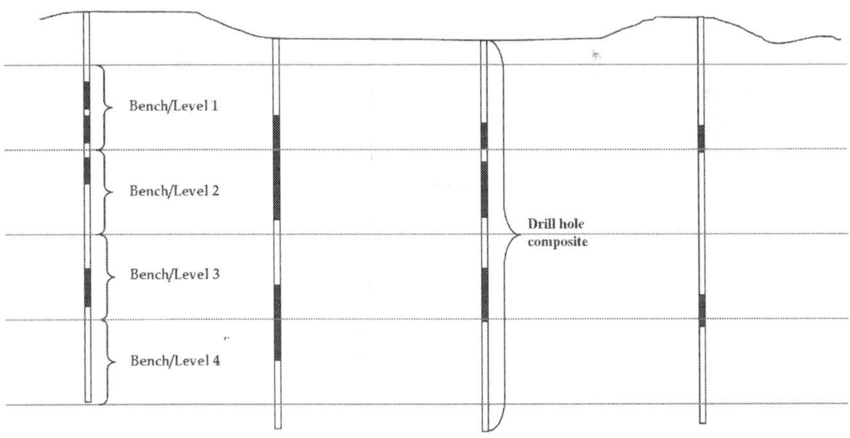

Fig. 3. Bench/Level Compositing

Level elevations may be same with bench elevations of prospective open pit. Such a compositing process gives ability and base to further block modeling and bench/level reserve estimation. Bench reserves help for production planning in further stages.

3. Data extension on a network

Drill hole data looks like spotting points in bird's eye view. Problem is to know how structure changes between these sample points. Sample points, where several parameters such as thickness and grade are already known, should be used to estimate these parameters at points where no sampling is available. This process can be named as data extension. Data can be extended thru two-dimensional planes or three-dimensional space. During extension process, square or rectangular grids, wireframes are imposed onto area. Triangulation is another type of artificial net, applied on field. Then, extension methods such as inverse distance square, geostatistics and artificial neural networks are applied.

3.1 Gridding

Although there are several methods to figure underground treasures such as aerial and seismic surveys, drill holes are still the most reliable and decisive data givers. In plan view, they look as irregularly distributed point data.

Grid is a regular network formed by triangle, square or rectangles generally. It is the result of a regular mesh need. Irregularly distributed sample points get a regular form if node points are assigned parameter values such as thickness and grade. In other words, thickness and grade values are estimated at node points, which results in a regular data structure (Fig. 4). Plenty of node points are generated by superimposing a wireframe grid onto a field (Knudsen, 1990, Parker, 1990).

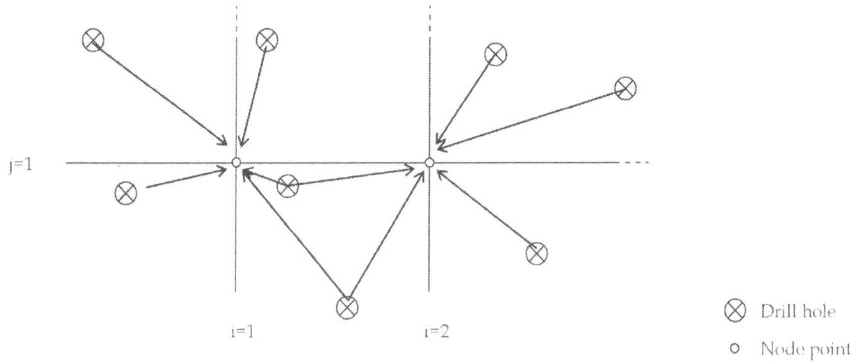

Fig. 4. Assignment to node points.

Irregularly distributed sample values (drill holes) are extended/distributed to field by several approaches;

a. Classical methods (triangle, polygon)
b. Inverse distance methods
c. Geostatistical methods
d. Artificial intelligence (neural networks)

These methods are generally employed to assign node values by using drill holes. Nodes are aligned through triangular or rectangular grids/networks (Fig. 5, Fig. 6).

Calculation of coordinates of grid nodes is a simple mathematical process. Let number of nodes in x direction is n and in y direction is m. Total number of nodes on grid is mxn. If each point is represented as $p(i,j \mid i=1,2,\ldots,n \; j=1,2,\ldots,m)$, then $p(i,j)$'s are function of x_i and y_i coordinates;

$$p(i,j) = f(x_{i,}, y_j) \tag{3}$$

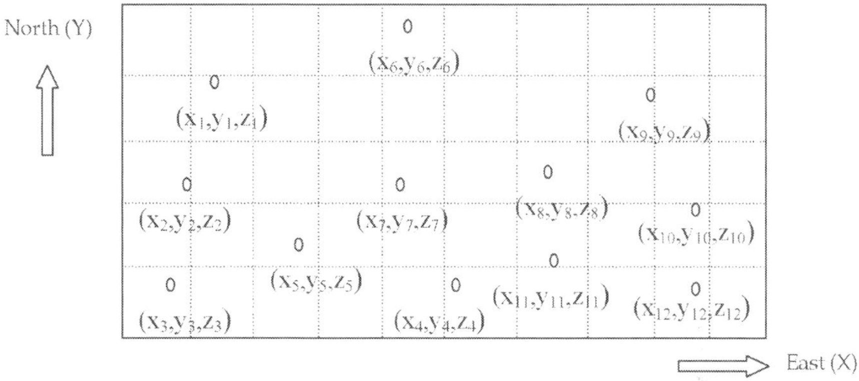

Fig. 5. Grid superimposed onto irregularly distributed drill holes.

Fig. 6. Polygon and triangle generation.

where, x_i is east coordinate of ith column and y_j is north coordinate of jth row. Then x_i and y_j can be calculated as;

$$x_i = s_x + i \cdot \Delta x \tag{4}$$

$$y_j = s_y + j \cdot \Delta y \tag{5}$$

where, Δx and Δy are spacing thru x and y lines and s_x and s_y are origin coordinates for east and north respectively. After computing x and y coordinates of all nodes, next step is calculating the third coordinate which may be thickness, grade, elevation or anything else. Third coordinate/parameter comes from surrounding sample points by carrying/extending drill hole parameters to these node points. For this purpose, there are several methods such as inverse distance square, geostatistics and neural networks.

Once a grid system is built, parametric values like thickness and grade are assigned. Then by means of rotation around an axis, third dimension can be sensed (Fig. 7).

3.2 Data extension in 2D

Data extension is a result of need to interpret how ore body behaves between present samples. Major two-dimensional extension applications of drill holes are triangular and polygonal area generation and of course, contour maps. The purpose is to carry pointwise data (drill hole) values into areas by certain mathematical methods and acceptations.

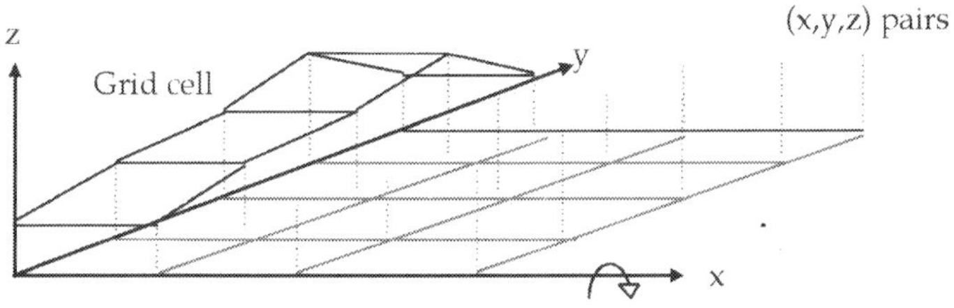

Fig. 7. Rotated grid system referring origin (s_x, s_y).

Additionally, in order to get a regular data structure, a grid is superimposed onto area. Each node point is assigned parametric values after computations. By this way, each grid node may act as a sample point.

3.2.1 Triangles and triangulation

Triangles are generated by joining drill holes. Area enclosed between three adjacent drill holes is calculated. Next step is to calculate volume (V - m^3) by multiplying triangle area (A - m^2) with average thickness (t - m) of three drill hole composited thicknesses.

$$V=A \cdot t_{ave} \tag{6}$$

$$t_{ave} = \frac{t_1 + t_2 + t_3}{3} \tag{7}$$

It is also possible to find triangular reserve (R - ton) if tonnage factor (f - ton/m^3) and thickness weighted average grade (g - %) of three composited grades is known.

$$R = V \cdot f \cdot g_{ave} \tag{8}$$

$$g_{ave} = \frac{g_1 \cdot t_1 + g_2 \cdot t_2 + g_3 \cdot t_3}{t_1 + t_2 + t_3} \tag{9}$$

Triangle method is a classical and rough estimation one with certain acceptations and could be employed only for horizontally and regularly bedded fields. Sedimentary type fields such as coal may be applied. Each triangle corner is a drill hole and composite thickness and grade values are used for computations (Fig. 8).

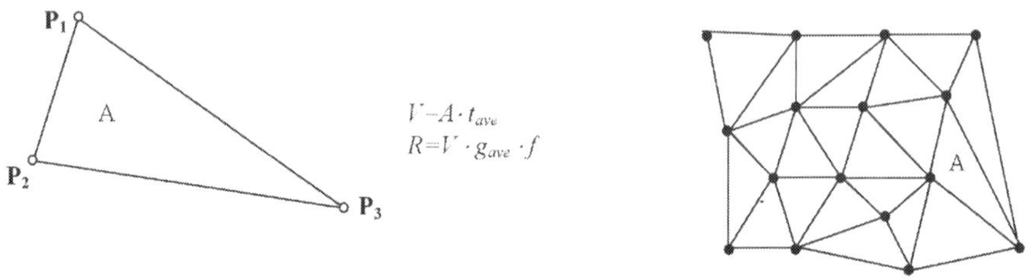

Fig. 8. Triangular grid by drill holes and local reserve.

Triangulation method can also be applied to produce sub-triangles and find coordinates of sub-triangles' corners. DeLaunay, Voronoi and Thiessen approaches can be employed during triangulation (Şen, 2009). New triangular network is assigned parametric values by several methods such as inverse distance square, geostatistics and neural network. It can be employed not only for reserve estimation but also for surface drawing.

During node assignments, a question may arise; which sample points should be included in calculations? There should be a limitation for drill holes and answer is *radius of influence*. Samples within radius of influence area are used in calculation of weighted averages. Radius of influence may be result of trial-error attempts or may be geostatistical variogram ***range***, which will be explained later.

3.2.2 Polygonal gridding

Polygon method is based on linear influence area concept. They are geometrically defined by the perpendicular bisectors of the lines between all points (Sen, 2009). Influence distances between drill holes are at their mid-points (Fig. 9).

Therefore, joint points of lines perpendicular to mid-points form a polygon and polygonal area can be calculated. Thickness and grade values within the area are assumed same with that drill hole's values. Volume of polygon is multiplication of area and thickness and reserve is product of volume, specific gravity and grade.

4. Assignment methods

Grid or mesh generation is followed by assignment stage. Nodes in 2D and blocks in 3D are assigned parametric values. This structure is the base for 2D and 3D models. 2D and 3D ore body modelling can be accomplished by several approaches. Joining cross-sections taken thru ore body and block modelling techniques are mostly used methods. The models do not only give the shape of ore body but also provide volume and reserve amount. Either triangular or rectangular mesh generation requires an assignment method. Here, commonly used mathematical approaches are;

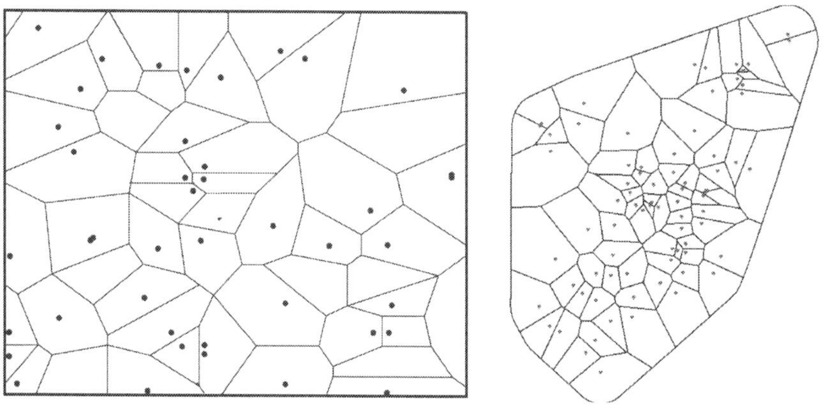

Fig. 9. Polygonalisation (Sen, 2009) and Voronoi polygons (BGIS, 2011).

i. Inverse distance,
ii. Geostatistics,
iii. Artificial Neural Networks.

4.1 Inverse distance method

Inverse distance method is actually a weighted average method. However, weights are calculated inversely. As distance of a sample point (drill hole) to an assignment point (node) gets more distant, its contribution on the average result gets less. In other words, closer sample point means more effect on assigned value. The basic idea is providing effect of a drill hole to point on which assignment is to be performed, is inversely related with the distance between them. This is realised by taking weighted average of parametric values by distances as shown below:

$$z(i,j)=\frac{\sum_{k=1}^{K}\frac{z_k}{(\Delta_{(i,j)}^{k})^{\alpha}}}{\sum_{k=1}^{K}\frac{1}{(\Delta_{(i,j)}^{k})^{\alpha}}} \tag{10}$$

where,

$z(i,j)$ = assigned value (i.e. grade, thickness) at node point on i^{th} column and j^{th} row.
z_k = parameter value carried by k^{th} drill hole (sample).
$\Delta^k_{(i,j}$ = distance between k^{th} drill hole and (i,j) node.
α = power of inverse distance process (α=2 in general and method is called as inverse distance square).

Distance between drill hole and node points can be calculated simply by;

$$\Delta_{(i,j)}^{k}=\sqrt{(\delta x(i,j)-D_x^n)^2-(\delta y(i,j)-D_y^n)^2} \tag{11}$$

where,

$\delta x(i,j)$ = x co-ordinate of node (i,j).
$D^n{}_x$ = x co-ordinate of n'th drill hole.
$\delta y(i,j)$ = y co-ordinate of node (i,j).
$D^n{}_y$ = y co-ordinate of n'th drill hole.

This method has an acceptation that geological structure has a linear behaviour. In case of sudden thickness or grade changes, inverse distance method may fail in determining a healthy result. On the other hand, its application is not only limited to 2D calculations but it can also be employed in 3D computations.

4.2 Geostatistical methods

Basic concept of geostatistics is regional variability of parameters (Matheron, 1971, 1963; Krige, 1984). Not only deterministic and descriptive manner but also probability and statistics take place in calculations with geostatistics (Mallet, 2002). If there is mathematically explainable structure of an ore body within a limited region, it can be possible to state that behaviour by equations and use them for estimation (Sarma, 2009). When variability cannot be formulated by mathematical models then it is decided that sample values show random behaviour and probabilistic methods of statistics can be applied there.

Geostatistics do not only consider distances between sample points and assignment points but also their position and direction with each other (Webster and Oliver, 2007). Additionally, during estimation, not only variances between sample points and assignment points are taken into account but also variances of samples within themselves are regarded (David, 1977, Davis, 1973). This means, geostatistics regards not only distance relation between sample points and node point but also variation between them in relation with distance and direction (David, 1988). Main two stages of geostatistical process are variogram modelling and kriging.

4.2.1 Variogram modelling

Main tool of geostatistics is variogram model (Diggle and Ribeiro, 2007). It can be defined as the graph representing the relation between distance and parametric variance in a certain direction (Deutsch and Journel, 1998). In estimation stage, kriging interpolation, geostatistics utilises co-variances between sample points and point/area/volume on which assignment is to be performed, and co-variances between sample points affecting that point for extending a value (Kitanidis, 2003).

Initially, a variogram model representing the relation between distance and variance is defined. Variance is;

where,

$$\gamma(h) = \frac{\sum_{n=1}^{N} (z - z')^2}{2n} \qquad (12)$$

$\gamma(h)$ = variance of sample pairs located h distance (lag) apart.
z = value at sample z.
z' = value at sample, h distance (lag) apart from z.
n = total number of pairs located h distance (lag) apart.

For each different h distance (lag), a variance $\gamma(h)$ is calculated (Fig. 10). According to distribution of h-$\gamma(h)$ pairs, a mathematical model is tried to fit. General trend is to use one of predefined variogram models such as, Spherical, Exponential, Gaussian, Linear, etc., (Davis, 1973; Krige, 1978).

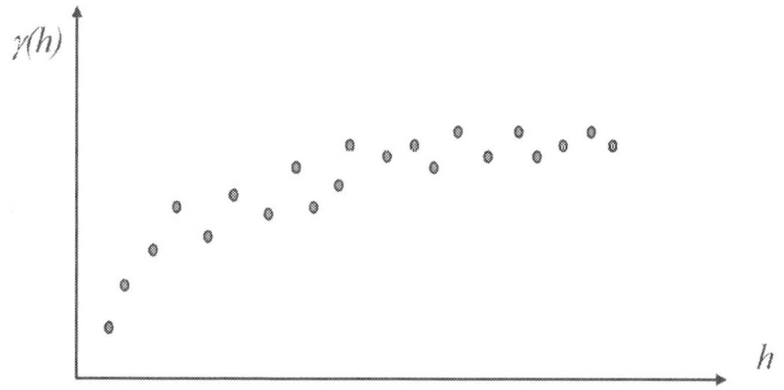

Fig. 10. Variance values marked for h distanced (lag) sample pairs.

For this purpose, sample pairs are found for a certain trigonometric direction θ° and lag distances *h, 2h, 3h,..., ∂h*. As it may not be possible to find pairs having direction and distance conditions absolutely, a tolerance distance $\pm\Delta h$ and direction (angle) $\pm\Delta\theta°$ is considered. Parameters, such as thickness and grade carried by sample points, are used in the formula and variance value is calculated. After marking variances on h-γ graph, ***experimental variogram*** is drawn (Leuangthong and Deutsch, 2004). Then, variogram is a graph explaining variance versus distance relation.

Experimental variogram points show a trend and needs interpretation for deciding which model best fits. Spotted points may have a mathematical determination in equation form, which is also called ***variogram model***. It is similar to regression modelling. (Fig. 11).

In the graph C_o is called as ***nugget effect***, ***C*** is ***sill*** and ***a*** is ***range***. Nugget is the variance value at zero distance. Normally it has to be zero however due to sampling problems, tolerances in pairing and structural inconsistencies, nugget gets over zero. Best graph and its mathematical equation (model) for *h-γ(h)* pairs are decided by visual interpretation regarding several predefined models such as spherical model (Matheron), exponential model, De-Wisjian, model, linear model.

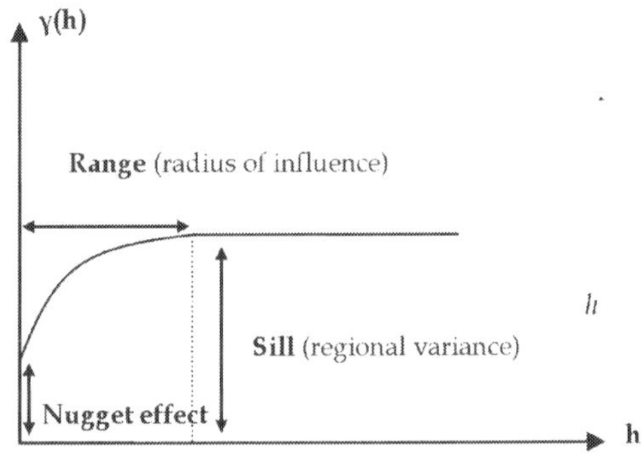

Fig. 11. Variogram model fitted for spotted points showing distance-variance; *h-γ(h)* relation.

Models are classed as models with ***sill*** and ***without sill.*** Sill is defined as the platform where variance gets parallel to *h* axis. Distance where graph reaches at sill level is called as ***range.*** Range may also be considered as *radius of influence*. Models are given below;
Spherical Model

$$\gamma(h) = \begin{cases} c\left(\dfrac{3h}{2a} - \dfrac{h^3}{a^3}\right) + c_0 & h \le a \\ c + c_0 & h > a \end{cases} \tag{13}$$

Linear Model

$$\gamma(h) = a \cdot h \tag{14}$$

Exponential Model

$$\gamma(h) = c\,(1 - e^{-h/a}) \tag{15}$$

where,

c = Sill value (constant value where variances get parallel to lag axis).

c_0 = Nugget effect (variance where distance is zero).

a = Range (distance where variance reaches at sill value)

Another concept in geostatistics is direction and variability of variograms according to directions. If variogram models developed for different directions can be accepted tolerably identical to each other, then ore body is called ***isotropic*** and ***anisotropic*** vice versa. Same variogram model can be used during kriging estimations for all directions if existence of an isotropy is decided.

4.2.2 Kriging

In order to extend sample data and use it for the prediction of unknown values, kriging is applied. Kriging is an extension and estimation method where parameters are assigned to assignment points by means of surrounding sample points. Aim is to calculate a weight a_n for each sample point.

Aim of developing a variogram model is to state variance as a function of distance and determination of mathematical equation that will be used to compute covariance matrix terms that will be used in kriging stage. Kriging process, developed by Krige (1966), is an estimation method regarding covariances between assignment point-sample points and within sample points mutually as well. In equation and matrix form after derivations, respectively;

$$z^*(i,j) = \sum_{k=1}^{n} w_k \cdot z(x_k) \tag{16}$$

$$\sum_{k=1}^{n} w_k = 1 \tag{17}$$

where,

$z^*(i,j)$ = assigned value to (i,j) node point regarding surrounding n sample points (drill holes)

$z(x_k)$ = parametric value of x_k sample point

w_k = weight of sample point k.

Simple representation of matrices is below:

$$[\sigma_{xx}]\cdot[w_x]=[\sigma_{vx}] \tag{18}$$

where,

$[\sigma_{xx}]$ = matrix including variances between samples points mutually

$[\sigma_{vx}]$ = matrix including variances between assignment point and sample points.

$[w_x]$ = weight matrix

Weight matrix is computed and weights are used to estimate unknown parameter at assignment point. σ variance values in matrices are calculated by putting distance amounts between sample points mutually (Δ_{xx}) and sample-to-assignment points (Δ_{vx}) into h term in vaiogram model.

Geostatistics has a wide application field in geological modelling. Main problem of the method is visual evaluation and interpretation of experimental variogram during model

development, which may be controversial and relative. Number of samples should be an acceptable amount; may be twelve and mostly over thirty (Sen, 2007).
One of the advantages of geostatistics is error estimation ability;

$$e^2 = (Z^* - Z)^2 \quad (19)$$

where, e^2 is estimation error (variance), Z^* is estimated value, Z is actual value (Sinclair and Blackwell, 2004).

4.3 Artificial neural network method

One of most recent methods used to extend sample data is artificial neural networks (Rabuñal and Dorado, 2006; Haupt, et.al., 2008). Initially, the neural system is trained to learn the structure of samples. Each sample is assigned a random weight and total error is recorded. Iteration by iteration, error is distributed to approximate actual sample values. An acceptable error level is succeeded and weights of sample points are calculated (Fig 12).

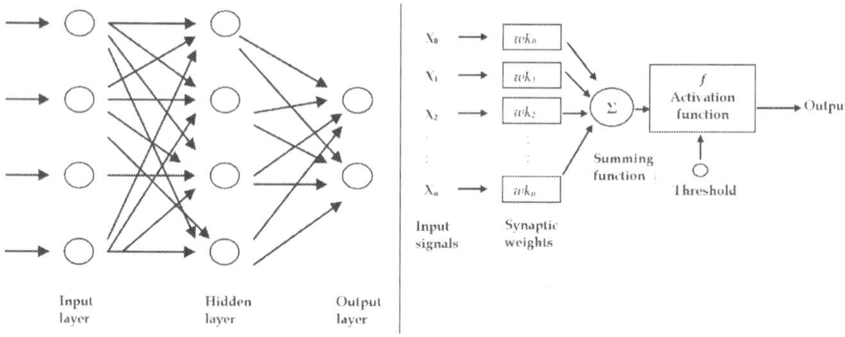

Fig. 12. Typical artificial neural network structures.

Then accordingly, a value is assigned to any desired location through area or space. In formulated form;

$$y = w_1 x_1 + w_2 x_2 + \ldots\ldots + w_n x_n = \sum_{j=1}^{n} w_j x_j \quad (20)$$

where, y is value to be assigned, w_j is the weight assigned to j'th sample and x_j is value of j'th sample. As explained earlier, calculation of w_j's requires an iterative process, which is called as *training*. A random weight assignment is followed by measurement of error between present input and desired output. An error above acceptable limit causes propagation of error through weights. During "squashing" the limit, an activation function, called as sigmoidal-function is employed;

$$f(x) = \frac{1}{1+e^{-x}} \quad (21)$$

New weights are tested if they fulfil condition of error limit. When the system fits error level, weights can be used for assignment (Fausett, 1994).

Total error gets less and less at each iteration and process continues as expected error level is reached. In order to achieve minimum error, hundred thousands, even millions of iterations may be needed. At that stage, it is decided that system has learned database structure and ready for further interpretations and estimations. Any more, w_j coefficients are available for assignments (Freeman and Skapura, 1991; Dowla and Rogers, 1995).

5. Two-dimensional visual and numerical processes

After generation of triangular and rectangular nets, they can be used for different purposes such as volume and reserve estimation, surface representation and forming a contour database. Another very basic visual instrument to show behaviour of ore through area is contour map. Sample data is extended to area as isolines.

5.1 Contour maps

Contour maps are generally based on triangulated or gridded networks, which are superimposed onto mine area to represent topography in computerised environment. Node points on network wire are assigned several values such as topographical elevation, composited thickness and grade, ore seam upper or bottom surface elevations, etc. (Watson, 1992). Main idea and purpose is to estimate several parametric values at node points using sample values obtained by drill holes. Here, inverse distance square, geostatistics and artificial neural networks are applied to assign node values. Thereafter topographical, thickness and grade contour maps can be drawn (Fig. 8). Bézier curves, B-Splines and Cubic Splines are primarily applied mathematical techniques (Mortenson, 1999; Comnino, 2006; Foley et.al., 1990; Vince, 2005; Vince 2010). Points on grid nodes with (x,y) coordinates get third dimension coordinate as well after assigning a parametric value as z coordinate such as thickness, grade, etc. Then, point data set is ready for contour (curve) fitting.

Bézier curves employ *Bernstein polynomials* (Vince, 2010). Its general equation is given below:

$$B_i^n(t) = \binom{n}{i} t^i (1-t)^{n-1} \tag{22}$$

$$\binom{n}{i} = \frac{n!}{(n-i)!i!} \tag{23}$$

where $\left(\frac{n}{i}\right)$ is shorthand for the number of selections of i different items from n distinguishable items when the order of selection is ignored and the coordinates of any point on the circumference in terms of some parameter t. Bézier curves may also be in *quadratic* and *cubic Bernstein polynomials* form (Vince, 2010).

B-splines also use polynomials to form a curve segment. However, B-splines use a series of control points determining the curve's local geometry. This feature provides and ensures that only a small portion of the curve is changed with movement of a control point (Vince, 2010). Splines can be classed as *uniform* and *non-uniform* and also *rational* and *non-rational*.

Cubic splines enable continuity between segments, which puts them one-step away from simple quadratics. Cubic splines can also be considered as *piecewise polynomials* (Fig. 13).

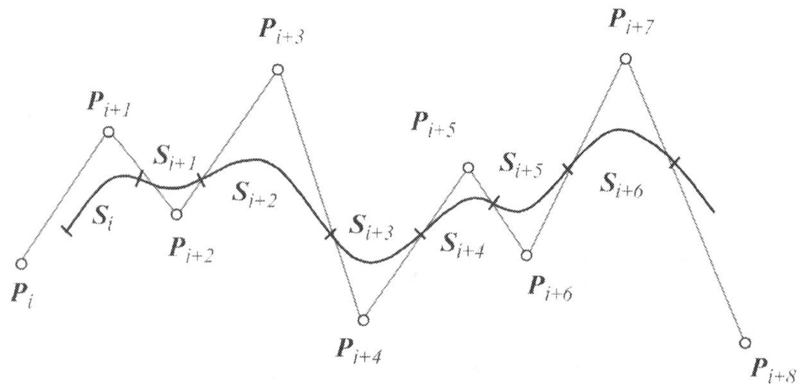

Fig. 13. Construction of a uniform non-rational B-Spline curve (Vince, 2010).

Here, curve segment S_i is under influence of points P_i, P_{i+1}, P_{i+2}, P_{i+3}, and curve segment S_{i+1} is related to points P_{i+1}, P_{i+2}, P_{i+3}, P_{i+4}. There exist $(m+1)$ control points and $(m-2)$ curve segments. Hence, a particular segment $S_i(t)$ of a B-spline curve is defined by

$$\mathbf{S}_i(t) = \sum_{r=0}^{3} \mathbf{P}_{i+r} B_r(t) \quad \text{for} \quad [0 \le t \le 1] \tag{24}$$

where,

$$\begin{aligned} B_0(t) &= \frac{-t^3 + 3t^2 - 3t + 1}{6} = \frac{(1-t)^3}{6} \\ B_1(t) &= \frac{3t^3 - 6t^2 + 4}{6} \\ B_2(t) &= \frac{-3t^3 + 3t^2 + 3t + 1}{6} \\ B_3(t) &= \frac{t^3}{6}. \end{aligned} \tag{25}$$

These are the basic functions of cubic splines (Rogers and Adams, 1990). Finally, topography, ore seam upper and bottom contours, thickness (isopach), grade (isograde), etc., can be drawn (Fig. 14).

Curve fitting methods are modified for also surface fitting in three-dimensional environments. Eventual aim of computer systems is to display ore body in 3D space.

6. Three dimensional operations

3D ore body modelling can be accomplished by several approaches. Joining cross-sections taken thru ore body and block modelling techniques are mostly used methods. The models do not only give the shape of ore body but also provide volume and reserve amount.

Drill hole cores explain researchers definite coordinates of underground layers. Topographical information is combined with depths to yield three-dimensional coordinates.

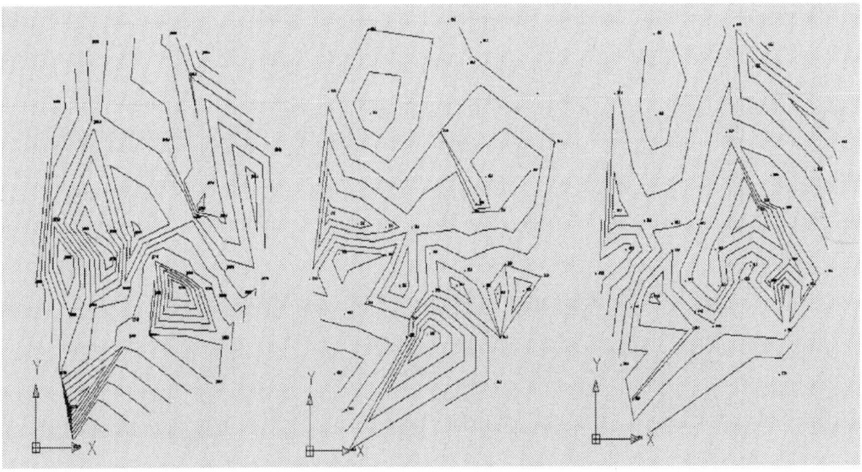

Fig. 14. Contour maps; topography, thickness and grade maps, respectively (Erarslan, 2007).

Major instruments for 3D computer aided mine valuation are;

- surfaces
- cross-sections,
- solid models (three-dimensional models).

6.1 Surfaces

Surfaces are important visual outputs of computer aided mine valuation systems. Surface can be prepared for any parameter assigned to grid system. (x,y) coordinate pairs of each node can be calculated easily. Fig. 15 shows how third dimension is sensed after rotation of 2D grid around y-axis.

Third coordinate may be topographical elevation, thickness, grade or anything else. Surface will be named according to its third parameter such as topographical surface, thickness surface, grade surface, etc. 3D visualisation of surfaces helps researcher imagine parametric changes.

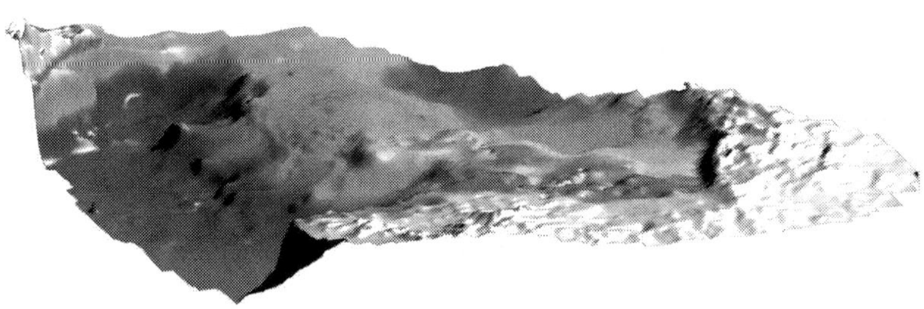

Fig. 15. Three-dimensional surface model (Pirsa, 2011).

Surface fitting is applied also on triangular or rectangular grid data. Assignment of third coordinate to (x,y) pairs of grid nodes is realised by already mentioned methods. After obtaining (x,y,z) coordinate set, *bicubic, planar surface patch, quadratic Bézier surface patch* and *cubic Bézier surface patch, B-Spline surface* approaches can be utilised (Hill, 1990).
Surface can be in wireframe (fishnet) or rendered form. During rendering action, several materials such as soil, several rock types, etc., and, also sky views for background can be imposed (Fig. 16). By this way, photo realistic appearances can be obtained. Some virtual reality program supports such as *OpenGl* and *GlView* allow users walk or fly over developed model (RealTech, 2011; GlView, 2011).

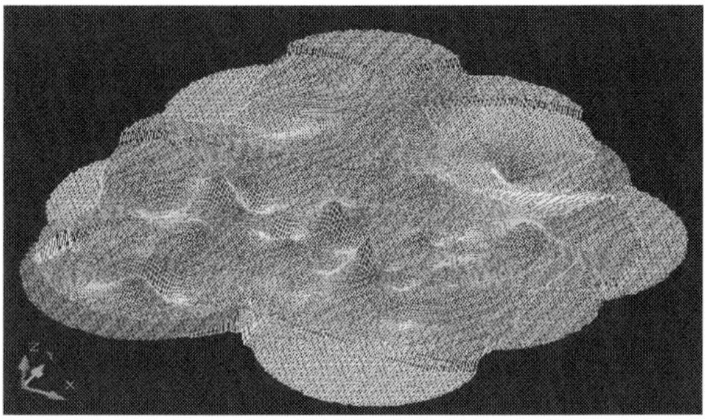

i)

ii)

Fig. 16. 3D surface model; i- wireframe appearance (Erarslan, 2003), ii- rendered with texture (Golden, 2011).

6.2 Geological parallel cross-sections

Drill hole sections aligned thru or near a cross-section line are used to draw geological section view. Softwares show users hole lithologies and let them make interpretation on them. Closed polygons regarding stratigraphy determine ore body cross-section thru that section line (Figure 17).

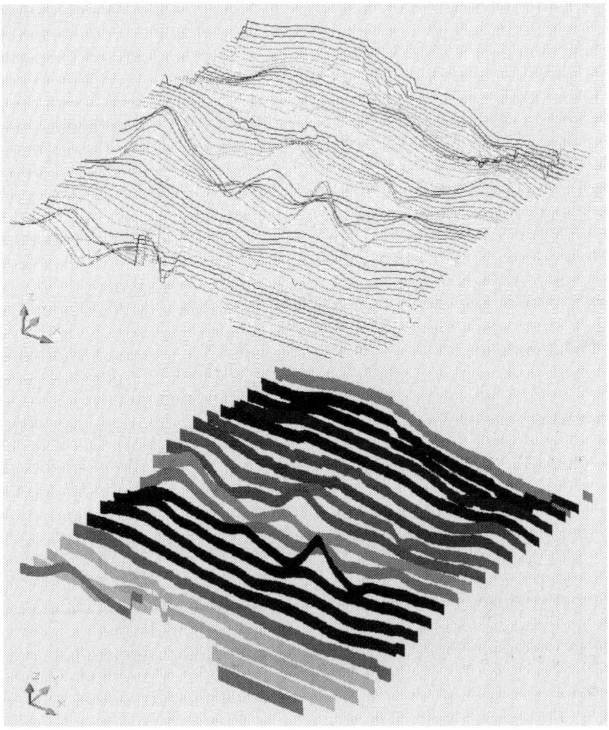

Fig. 17. Cross-section interpretation thru a section line (Erarslan, 2007).

6.3 Three-dimensional ore body models

Three dimensional ore body cross-sections do not only give an idea about structure of deposit but also prepare a base for three-dimensional ore model. Successive and parallel cross-sections are interpreted along ore body. Software can later on combine parallel cross-sections to form a 3D ore body model. Some software is also capable of combining cross-section, which are not parallel, and maybe intersecting each other. This type of sections are visualised on so-called fence diagrams (Fig. 18). Parallel cross-sections of ore body are fundamental tools for 3D modelling. They can be horizontal or vertical (Fig. 18).

Computer programs help researchers display sections and join them to build a 3D appearance. Each closed polygon enables calculation of section area and by means of average areas method, volume of ore body can be calculated. Gauss-Green area formulation can be employed to estimate sectional area:

$$A = \frac{\sum_{i=1}^{n-1} (y_{i+1} - y_{i-1}) x_i}{2} \tag{26}$$

where, A is polygonal area, x_i and y_i terms are x and y coordinates of polygon nodes.

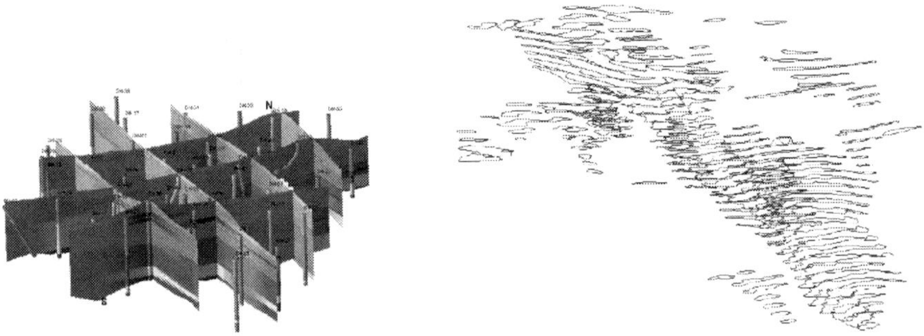

Fig. 18. Fence diagram (Rockworks, 2011) and wireframes through Zambujal Ore body (Lundin, 2007).

6.4 Block models

Another very frequently used ore body modelling method is block models. Block models could be thought as three-dimensional forms of 2D gridding. Field is divided into blocks, physical properties and quality composition are represented by this geometric form (Fig. 19). Centre and corner coordinates of each block can be calculated as (x,y,z) data sets. Reference/origin point is at top or bottom. In some cases, prospective open pit benches define height between levels (h). Regarding block width, length and height and referring to origin, coordinate computations are performed. Determination of physical position of each block is followed by thickness and grade assignments. Block assignments are performed by several mathematical approaches and *geological block model* is generated. Similar to calculations carried out on grid nodes, process is repeated for each elevation level. Geostatistics and neural networks are most advanced estimation models. Inverse distance square method can also be applied for more regular geological structures such as sedimentary beddings. According to assigned grade, blocks are coloured in software systems to enable observing quality tableau better.

6.5 Volume and reserve estimation

Volume and reserve of deposits is crucial subject of mine valuation as well as the visual aspect. There are various methods for numerical estimations. However, recent methods are computer dependent. Several approaches for ore volume and reserve estimations are given below (Hartman, 1992):

i. The area enclosed by ore limits are multiplied by the average thickness. The volume is also multiplied with average tonnage factor to give inventory. Percent grade gives how much of this inventory is ore; that is reserve.
ii. Triangular net is used to calculate total reserve. Each triangular area is found and average thickness is multiplied by area to calculate volume of ore in triangle. Product of volume and weighted averages of tonnage factor and grade give triangular reserve. Summation of reserve of triangles yields total reserve.
iii. Field is divided into grids. Node points are assigned thickness, grade, etc. by inverse distance square, geostatistics and neural network methods. By this way, corner vertices of each node cell are assigned a value. Multiplication of cell area that is enclosed by grid

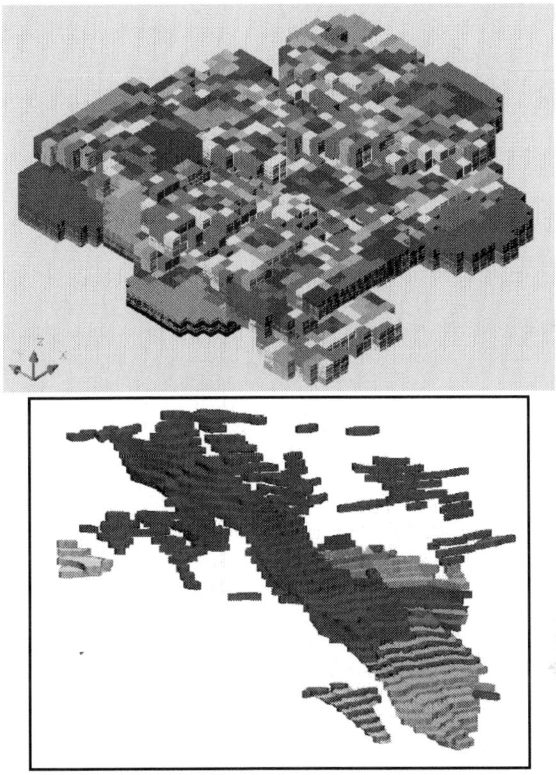

Fig. 19. 3D Ore body block model and Solid Model of Zambujal Ore body (Lundin, 2007).

nodes, thickness of ore in that cell and cell grade yields cell reserve. Total reserve is the total of all grid cells.

iv. Areas of parallel geological sections are calculated by Gauss-Green formula. Volumes between successive sections are estimated by average areas method.

v. Ore body block model gives also numerical results as well as visual. Each block reserve can be calculated. Total rcscrvc is the reserve total of all blocks.

Block volume is simply multiplication of length, width and height of it. However, after thickness assignment to a block, thickness of ore should be considered rather than its geometrical height. Tonnage factor and ore grade give block reserve. Total reserve is the summation of all block reserves (Taylor, 1994, 1993).

$$\textit{Block reserve (t)} = \textit{block volume } (m^3) \cdot \textit{tonnage factor } (t/m^3) \cdot \textit{grade } (\%) \qquad (27)$$

Volume inside the open pit and ore volume bench by bench can later be estimated (Taylor, 1992). After determining open pit limits, border of each bench can be thought as a polygon and by means of Gauss-Green formula pit volume is computed. Stripping volume can also be calculated to find stripping ratio (Taylor, 1991).

An important factor in reserve estimation is *cut-off grade*, which can be described as the grade where excavated material is classed as ore or waste (Taylor, 1986, 1985). It may be considered as breakeven point as well (Taylor, 1972). So many researches and models have been developed to determine accurate cut-off grade. However, this economically crucial subject is special and unique to each field and it should be studied particularly (Wellmer et.al., 2008).

6.5.1 Isopach maps for volume calculation

Beside triangle and polygon methods, a special technique using isopach contours is employed for volume computation. Contour maps can be thought as two-dimensional extension and interpretation of drill holes. Isolines are drawn for several parameters provided by holes. Thickness and grade values are very crucial to determine ore body. It is possible to see where thickness increases and decreases or where ore is rich in grade and less valuable. Isopach maps, which are isolines for thickness, can also be used for volume calculation; volume under isopach maps are ore volume itself (Fig. 20).

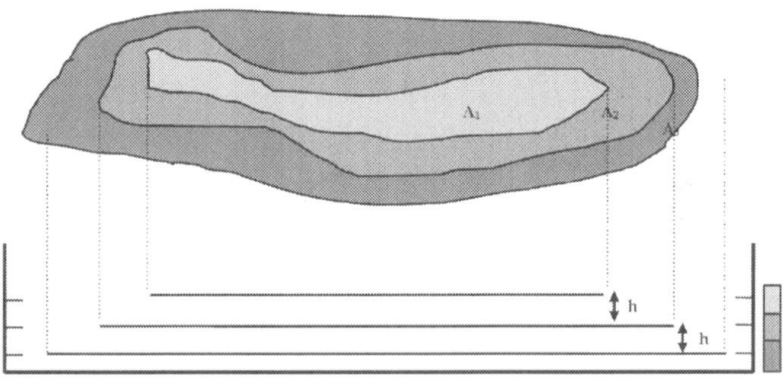

Fig. 20. Cross-section of isopach maps and volume calculation by using average areas rule.

Areas, at each thickness level are estimated and volume between each successive area pairs is calculated by using average areas rule:

$$V_j = \frac{A_j + A_{j+1}}{2} \cdot h \qquad (28)$$

where, V_j is volume at level j, A_j and A_{j+1} are successive areas of thickness level j and h is height (depth difference) between thickness isolines.

6.6 Mine design, production planning and mineral economics

Final stage of mine valuation is to decide if an ore deposit is worth making investment. Geological structure provides a database for economical assessment. Regarding physical and geological outputs of computerised systems, mine can be designed and production planning can be accomplished (Fig. 21).

Possible investment, annual costs and incomes are considered to estimate and foresee how that underground asset can be extracted optimally. Present worth, future worth, annual worth and rate of return of the net cash flows are calculated and reported. During calculations, straight line, double declining balance and sum of the years' digits methods are used for depreciation (Steiner, 1992). Taxes, salvage values, royalty costs, etc. are all taken into account.

Fig. 21. Computer aided mine design (Beck, 2011).

Optimisation is another wide application branch of mine valuation. Geological block models are used to generate *economical block models* by using unit costs and income (Erarslan, 2001). As volume of a block, thickness and grade of ore at each particular block is known, then it becomes possible to convert this information to economical aspect. Multiplication of volume, tonnage factor and grade give block reserve. Unit production cost and expected income are considered and an economical value is assigned to each block. Economical block models have visual and numerical results. 3D appearances of them give an idea where ore body is rich and how quality changes. Mine design and optimisation applications such as optimum pit limit and optimum production planning are comprehensive numerical assessment methods (Erarslan, 2001).

7. Computer software for ore body modelling and mine valuation

Generally, size of the database may be too bulky to manage studies with hand effort. Hence, numerical algorithms and mathematical approaches necessitate computer applications to overcome huge computational time and processes. Today, many software and computer aided systems serve for geological modelling and mine valuation in this sense. The accuracy and speed of computers enable evaluation of various scenarios within reasonably short times.

Commercial softwares in general have robust database management capability. After building a healthy database structure, computer programs are ready for ore body modelling. Many mathematical models and approaches in literature take place in software packages to determine shape, location and quality composition of the entire body. Visual appearance of geological body is supported by numerical data such as ore reserve amount and quality composition, which are vital parameters for mine design and scheduling.

There are several integrated commercial packages for this purpose (MineSight, 2011; Gemcom, 2011; GDM, 2011; Techbase, 2011; Datamine 2011; Lynx Mining, 2011; RockWorks, 2011). They can successfully handle real cases, which may be fairly complex structures. Their processing and graphical capabilities and utilities have improved year and year (Fig. 22).

Computer systems make engineering designs, project preparation and management easier. Raw data such as GPS and drill hole cores are converted to contour maps, three-dimensional solid structures, volume and reserve reports, open and underground mine plans, economical assessments and production schedules. Various scenarios are examined quickly. Eventually, underground asset with so many unknowns becomes visual. Its properties thru space are clear. Million dollars are invested regarding this tableau. Thus, computer systems are vital partners of geological and mining engineers.

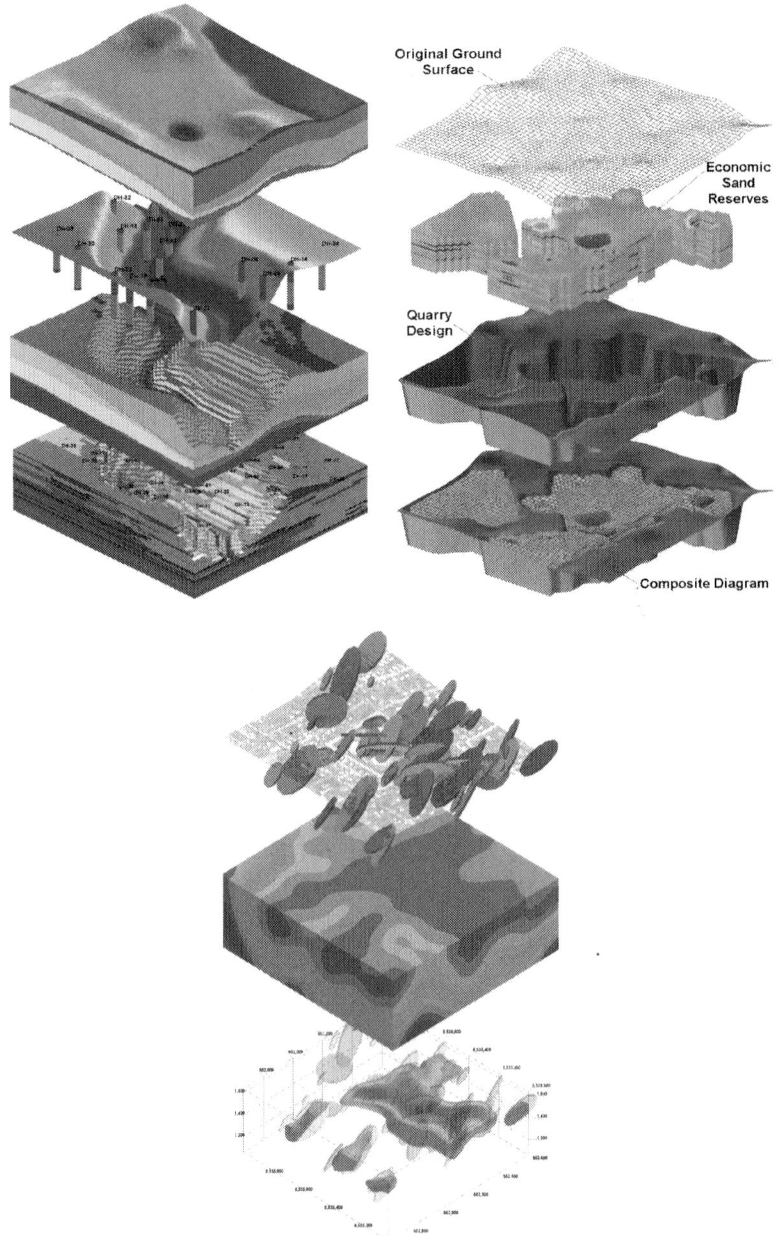

Fig. 22. Computer software functions for ore body modelling (RockWorks, 2011).

8. Conclusion

As mining is an industry requiring millions of dollars for investment and further operations, mine valuation is a crucial stage, which provides basic information for future stages. Before deciding on a mining investment, the preliminary process includes exploration, data gathering and valuation, determining geological structure, ore body modelling, mine design and planning.

In the last several decades, a number of mathematical and computational approaches have been developed to give the most accurate information related to ore bodies under investigation. In parallel, many computer programs have been developed to accomplish these complicated processes. This means that all investments are dependent on physical and economical characteristics of ore deposit. Similarly, profitability of the investment is strictly related to mine design and planning (Hartman, 1992). Eventually, data evaluation and ore body modelling is a very critical and basic process and mining operations are based on its results (Kennedy, 1990).

Drill hole database, geometrical and numerical analyses, contour maps, surfaces, sections, three dimensional ore body block models, volume and reserve calculations, economical assessment, classical valuation methods, triangulation, polygons, gridding, geostatistical approaches, neural network method, etc. are headings and instruments of mine valuation.

Computer aid has become inevitable and vital for contemporary ore body modelling and mine valuation applications. Long and complicated process starting with data base constitution and ending with mine design and production planning requires software support due to iterative mathematical structure of computations. Drill hole composites, triangulation, polygonal and grid nets, contouring, cross-sections, three-dimensional ore models, various volume and reserve estimations, geological and economical block models, mine design and production scheduling, optimisation and simulation works are highly computer dependent. At each stage of the processes, many developed mathematical approaches are utilised. One who deals with computer aided ore body modelling is faced with many and many studies and methods. However, geostatistics, neural networks, inverse distance methods are very basic subjects to be comprehended. Scientists work to improve present methods and develop new ones as well. Additionally, many commercial software are under a continuous development for a better service

9. References

Agoston, M.K. (2005). *Computer Graphics and Geometric Modeling*, Springer, ISBN 1852338172, USA.

Beck (2011). Beck Engineering Pty Ltd, Australia, http://www.beckengineering.com.au

BGIS (2011). Boston Geographical Information System, http://www.bostongis.com/

Comninos, P. (2006). *Mathematical and Computer Programming Techniques for Computer Graphics*, Springer, ISBN-10: 1-85233-902-0, USA.

Datamine (2011). Datamine Studio v. 2, CAE Mining, Montreal, Canada, http://www.cae.com

Davis, J.C. (1973). *Statistics and Data Analysis in Geology*, John Wiley and Sons, Inc., USA.

David, M. (1988). *Handbook of Applied Advanced Geostatistical Ore Reserve Estimation*, Elsevier, Amsterdam.

Deutsch C.V. & Journel A.G. (1998). *GSLIB Geostatistical Software Library and User's Guide*, 2nd Ed., Oxford University Press, ISBN 0-19-510015-8, New York.

Diggle P.J. & Ribeiro Jr., P.J. (2007). *Model-Based Geostatistics*, Springer, ISBN-10: 0-387-32907-2, New York.

Dowla F.U. & Rogers L.L. (1995). *Solving Problems in Environmental Engineering and Geosciences with Artificial Neural Networks,* The Massachusetts Institute of Technology, ISBN 0-262-04148-0, USA.Erarslan, K. (2007). Geology and Mining System (GMS) for Computer Aided-Mine Valuation, *Computer Applications in Engineering Education, Vol:15, No:1, 78-8.*

Erarslan, K. (2003). The Geology and Mining System (JMS) and Valuation of a Copper Field, *The Journal of Chamber of Mining Engineers of Turkey*, 42, (4), 3-14.

Erarslan K. & Çelebi, N. (2001), A Simulative Model for Optimum Open Pit Design, *Canadian Inst. Min. and Metall. CIM Bulletin*, 94, No. 1055, October, 59-68, Canada.

Fausett, L. (1994). *Fundamentals of Neural Networks*, Prentice Hall, ISBN 0133341860, USA.

Foley, J.D., van Dam, A., Feiner, S.K., Hughes, J.F. (1990). *Computer Graphics Principles and Practice*, 2nd Ed., Addison-Wesley, ISBN 0-201-12110-7, USA.

Freeman J.A. & Skapura, D.M. (1991). *Neural Networks Algorithms, Applications, and Programming Techniques,* Addison-Wesley Publishing Company, ISBN 0-201-51376-5, USA.

GDM (2003). BRGM International Division, Orléans Cedex, France. www.brgm.fr

GemCom (2011). GEMCOM, Inc., Canada. www.gemcomsoftware.com

GlView (2011). CAE Systems GMBH, Witten, Germany, http://www.glview.de

Golden Software (2011). Golden Software Inc., Colorado, http://www.goldensoftware.com/products/surfer/surfer-3Dsurface.shtml

Haupt S.E.; Antonello Pasini A. & Marzban C., Eds. (2008). *Artificial Intelligence Methods in the Environmental Sciences*, Springer, ISBN 978-1-4020-9117-9, USA.

Hartman, H.L. (1992). *Mining Engineering Handbook*, 2nd Edition, SME, AIME, Littleton.

Hill, Jr, F.S. (1990). *Computer Graphics*, McMillan Pub. Co., ISBN 0-02-354860-6, New York.

Hustrulid W. & Kuchta M. (2006). *Open Pit Mine Planning and Design*, Taylor & Francis, ISBN 9780415407410, USA.

Kennedy, B.A. (1990). *Surface Mining*, 2dn Edn, SME, AIME, Littleton.

Kitanidis, P. K. (2003). *Introduction to Geostatistics: Applications to Hydrogeology,* Cambridge University Press, ISBN 0 521 58312 8, UK.

Krige, D. G. (1984). Geostatistics and the definition of uncertainty; *Trans. Inst. Min. Metall.* Sect. A, v. 93, pp. A41–A47.

Krige, D. G. (1978). Lognormal–deWijsian geostatistics for ore evaluation; *South African Inst. Min. Metall.*, Monograph Series, Johannesburg, South Africa, 50 pp.

Leuangthong, O. and Deutsch, C.V. (2004). *Quantitative Geology and Geostatistics - Geostatistics Banff 2004*, Springer, ISBN-10 1-4020-3515-2, Netherlands.

Lynx Mining (2011). Lynx Geo Systems, S.A. Pty. Ltd., South Africa, http://www.lynxgeo.com/

Mallet J.L. (2002). *Geomodeling,* Oxford University Press, ISBN 0-19-514460-0, USA.

Matheron, G. (1971) The theory of regionalized variables and its applications; *Les cahiers du Centre de Morphologie Mathematique, Fontainebleau*, No. 5, 211 pp.

Matheron, G. (1963). Principles of geostatistics; *Economic Geololgy*, v. 58, pp. 1246–1266.

McKillup, S. & Dyar, E.D. (2010). *Geostatistics Explained- An Introductory Guide for Earth Scientists*, Cambridge University Press, ISBN-13 978-0-521-76322-6, New York.
MineSight (2011). MinTech, Inc., Tucson, Arizona, USA. www.mintec.com; www.minesight.com
Mortenson, M.E. (1999). *Mathematics for computer graphics applications*, 2nd Ed., Industrial Pres Inc., ISBN 0-8311-3111-X, USA.
Nieuwland D. A., Ed. (2003). *New Insights into Structural Interpretation and Modelling*, The Geological Society of London, ISBN 1-86239-133-5, London.
Pirsa (2011). Pirsa Minerals, Adelaide, Australia, http://www.pir.sa.gov.au/minerals/geology/3d_geological_models
Rabuñal, J.R. & Dorado J. (2006). *Artificial Neural Networks in Real Life Applications*, Idea Group Publishing, ISBN 1-59140-902-0, USA
RealTech (2011). RealTech-Virtual Reality Co., USA. http://www.realtech-vr.com
RockWorks (2011). RockWare, Inc., Golden, Colorado, USA, http://www.rockware.com/product/gallery.php?id=165
Rogers, D.F. & Adams, J.A. (1990). Mathematical Elements for Computer Graphics, McGraw Hill, Intl. Ed.2, ISBN 0-07-053529-9, USA.
Sarma D.D. (2009). *Geostatistics with Applications In Earth Sciences*, 2nd Ed., Springer, ISBN 978-1-4020-9 379-1, India.
Sen, Z. (2009). *Spatial Modeling Principles in Earth Sciences*, Springer, ISBN 978-1-4020-9671-6, New York.
Sinclair A.J. & Blackwell G.H. (2004). *Applied Mineral Inventory Estimation*, Cambrigr University Press, ISBN 0-521-79103-0, UK.
Singer D.A. & Menzie W.D. (2010). *Quantitative Mineral Resource Assessments an Integrated Approach*, Oxfrod University Press, ISBN 978-0-19-539959-2, New York.
SME (2005). *Exploring Opportunities: Careers in the Mineral Industry*, Society of Mining Engineers, Career Guidance Booklet, Littleton.
Steiner, H.M. (1992). *Engineering Economic Principles*, McGraw Hill International Editions, Singapore, 559p.
Taylor, H. K. (1994). Ore reserves, mining and profit; *Can. Inst. Min. Metall. Bull.*, v. 87, no. 983, pp. 38–46.
Taylor,H. K. (1993). Reserve inventory practice-reply; *Can. Inst. Min. Metall. Bull.*, v. 86, no. 968, pp. 146–147.
Taylor, H. K. (1992). The guide to the evaluation of gold deposits; integrating deposit evaluation and reserve inventory practices - discussion; *Can Inst. Min. Metall. Bull.*, v. 85, no. 964, pp. 76–77.
Taylor,H. K. (1991). Ore reserves - the mining aspects; *Trans. Inst. Min. Metall.*, v. 100, pp. 146–158.
Taylor, H. K. (1986). Ore Reserve Estimation, Methods, Models and Reality, *Can. Inst. Min. Metall.*, Montreal, May 10–11, pp. 32–46.
Taylor, H. K. (1985). Cutoff grades - some further reflections; *Trans. Inst. Min. Metall.*, v. 94, sect. A, pp. A204–A216.
Taylor, H. K. (1972). General background theory of cutoff grades; *Trans. Inst. Min. Metall.*, v. 81, sect. A, pp. A160–A179.
Techbase (2003). Techbase International, Lakewood, Colorado, USA. www.techbase.com

Torries T.F. (1998). *Evaluating Mineral Projects*, Society for Mining, Metallurgy, and Exploration, Inc., ISBN 0-87335-159-2, USA.

TUHH (2011). E-Learning Platform for IFM, Tech. Univ. Of Hamburg-Harburg, http://daad.wb.tu-harburg.de/?id=279

Vince, J. (2010). *Mathematics for Computer Graphics*, Springer, 3rd Ed., ISBN: 978-1-84996-022-9, New York.

Vince, J. (2005). *Geometry for Computer Graphics Formulae, Examples and Proofs*, Springer, ISBN 1-85233-834-2, USA.

Watson, D.F. (1982). Acord: Automatic Countouring of Raw Data, *Computers and Geosciences*, 8, (1), 97-101.

Wellmer F.W.; Dalheimer M. & Wagner M. (2008). *Economic Evaluations in Exploration*, Springer, ISBN 978-3-540-73557-1, New York.

Webster, R. & Oliver, M.A. (2007). *Geostatistics for Environmental Scientists*, 2nd Ed., John Wiley & Sons Inc., ISBN-13: 978-0-470-02858-2, England.

Part 9

Volcanology

17

Mud Volcano and Its Evolution

Bambang P. Istadi[1,*], Handoko T. Wibowo[2], Edy Sunardi[3], Soffian Hadi[4] and Nurrochmat Sawolo[1]
[1]*Energi Mega Persada*
[2]*Independent geologist*
[3]*Universitas Padjajaran*
[4]*Sidoarjo Mudflow Mitigation Agency*
Indonesia

1. Introduction

The term mud volcano refers to topographical expressions of naturally occurring volcano-shaped cone formations created by geologically excreted liquefied sediments and clay-sized fragments, liquids and gases. Ejected materials often are a mud slurry of fine solids suspended in liquids which may include water and hydrocarbon fluids. The bulk of released gases are methane, with some carbon dioxide and nitrogen. Mud volcanoes may be formed by a pressurized mud diapir which breaches the Earth's surface or ocean bottom. Flowing temperatures at the ocean bottom may be as low as freezing point and are associated with the formation of hydrocarbon hydrate deposits. Flowing temperatures can also be hot if associated with volcanic gases and heat escaping from deep magma which can turn groundwater into a hot acidic mixture that chemically changes rock into mud and clay-sized fragments. These mud volcanoes are built by a mixture of hot water and fine sediment that either pours gently from a vent in the ground like a fluid lava flow; or is violently ejected into the air as a lava fountain of escaping mud, volcanic gas, stream and boiling water.

Mud volcanoes are most abundant in areas with rapid sedimentation rates, active compressional tectonics, and the generation of hydrocarbons at depth. Typically they are also found in tectonic subduction zones, accretionary wedges, passive margins within deltaic systems and in active hydrothermal areas, collisional tectonic areas, convergent orogenic belts and active fault systems, fault-related folds, and anticline axes. These structures act as preferential pathways for deep formation fluids to reach the surface. (see Pitt and Hutchinson, 1982, Higgins and Saunders, 1974; Guliyiev and Feizullayev, 1998; Milkov, 2000; Dimitrov, 2002; Kopf, 2002, Mazzini, 2009).

The existence of mud volcanoes are controlled by tectonic activity where fluid escapes from areas undergoing complex crustal deformation as a result of transpressional and transtensional tectonics. Collisional plate interactions create abnormal pressure condition and consequently overpressured buildup of deep sedimentary sediment which in turn result in formation of diapirs. Over pressured zones typically are under-compacted sedimentary layers which have lower density than the overlying rock units, and hence have an ability to

* Corresponding Author

flow. They are the product of rapid deposition where the connate water is trapped, unable to escape as the surrounding rock compacts under the lithostatic pressure caused by overlying sedimentary layers. In thick, rapidly deposited shale dominant sedimentary sequence, the low and reduced porosity and permeability due to compaction inhibit the expulsion of water out of the shale. As burial continues, fluid pressure increases in response to the increasing weight of the overburden. This Non-equilibrium compaction is believed to be the dominant mechanism in formation of overpressured sediments. Over pressure however can also result from maturing organic rich predominantly clay sediments which are generating methane and other heavier gases that are still trapped within the sedimentary sequences. The above geological elements that result in diapirsm and mud volcano is often known as "elisional" basin mainly characterized by rapid deposition of thick young sediments, presence of abnormally high formation pressure or overpressures fluids, under-compacted sediments, petroleum generation, compressional setting, high seismicity and occurrence of faults (see Milkov, 2000, and Kholodov, 1983).

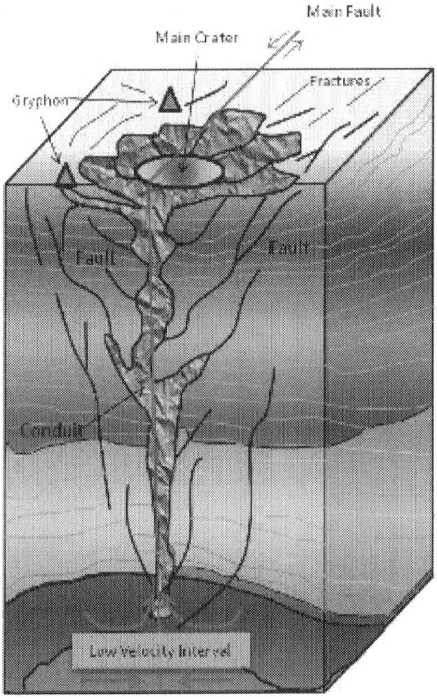

Fig. 1. Basic structure and anatomy of a conical mud volcano. The mud volcano is formed by the escaping natural gas that rises to the surface when it finds a conduit (strike slip fault) and carries mud which has a lower density (and typically found as low velocity interval) than the surrounding sedimentary succession. Fluid, gas, and surface water are ejected in a cone shape like a mountain and forms craters, mud pools (salses) and cones (gryphons). Tectonic movement is very influential, as well as rapidly deposited sediments and burial of organic rich sediments. Strike-slip faults in active tectonic regions are the most ideal place for the formation of mud volcanoes.

Overpressure buildup mechanisms contribute to the brecciation of the deep sedimentary units include for example the dewatering of thick clay-rich sedimentary units, and geochemical reactions in sedimentary units with high temperature gradients are the mechanism for the eruption (Mazzini, 2009). He further suggest that when the subsurface overpressure reaches a threshold depth where the overburden weight is exceeded, fracturing and breaching of the uppermost units occur, sometimes facilitated by external factors such as earthquakes. The upward movement of the mud to surface is due to buoyancy and differential pressure.

Geological structures like faults and anticlines where mud volcanoes are commonly found are easily perturbed by earthquakes as they represent weak regions for the seismic wave's propagation. This mechanism is well described by Miller et al. (2004) where earthquakes initiating local fluid movements cause fractures that propagate to the surface manifesting with a time delay from the main earthquake. Miller et al. (2004) propose a link between earthquakes, aftershocks, crust/mantle degassing and earthquake-triggered large-scale fluid flow where trapped, high-pressure fluids are released through propagation of coseismic events in the damaged zones created by the mainshock. The resulting disturbance of the gravitational instability triggers the beginning of flow, while the pressure drops and the lower cohesion media is easily fluidized and ultimately vacuumed to the surface through piercement structures which provide the conduits for high pressure mud/fluid and gas release.

The geometry of mud volcanoes is variable. They can be up to a few kilometers in diameter and several hundred meters in height. The main morphological elements of a mud volcano are the crater(s), hummocky periphery mud flows, irregularly shaped terrains, gryphons, and mud lakes or salses. A classification of mud volcanic edifices morphology was proposed by Kholodov, 2002 (in Akhmanov and Mazzini, 2007), these are: (1) "classic" conic volcanic edifice with main crater and mud flow stratification reflecting periodical eruption; (2) Stiff mud neck protrusion, typically due to its high viscosity and hence able to form steep hills; (3) swamp-like area; contrary to no (2), due to its low viscosity the mud spreads over a large area; (4) "collapsed synclinal" depression; and (5) crater muddy lake, is the most abundant type in various mud volcanic areas. It is often that mud volcano morphology shows a combination of the common types described above depending on the viscosity of the mud and the stage of its development.

Mud volcanoes show different cyclic phases of activity, including catastrophic events and periods of relative quiescence characterized by moderate activity. It appears that each eruptive mud volcano has its own period of catastrophic activity, and this period is variable from one volcano to another. The frequency of the eruptions seems essentially controlled by local pressure regime within the sedimentary sequences, while the eruptive mechanism and evolution seem strongly dependent on the state of consolidation and gas content of the fine-grained sediments. This is shown in the compilation of historical data onshore Trinidad as described by Deville and Guerlais (2009).

Approximately 1,100 mud volcanoes have been identified on land, in shallow as well as deep waters. It has been estimated that well over 10,000 may exist on continental slopes and abyssal plains. The largest known structures are 10 km in diameter and reach 700 m in height. Occurrences of mud volcanoes on the seafloor have been documented more frequently since the intensive use of side scan sonar began in the late 1960's. Mud volcanoes have been found in many parts of the world, and have been documented in Rumania, Italy, Iran, Iraq, New Zealand, India, the Myanmar, Malaysia, Gulf of Mexico, Trinidad,

Venezuela, Colombia and the USSR. The largest number of mud volcanoes is found in the Azerbaijan trend which continues into the Southern Caspian area. In the Indonesian region, mud volcanoes are found on the Islands of Sumatera, Nias, Pagai, Sipora, East Java, East Kalimantan '(Borneo), Rote, Barbar, Aru, Timor, Tanimbar, Yamdena and Papua. They are found in high rate subsidence basins such as Madura-East Java Basin, Kutai Basin, in high seismicity areas such as islands in the Banda Sea and in tectonically complex areas such as Timor and Papua (Sukarna, 2007).

In Papua, Indonesia, mud volcanoes are found along a zone of disruption 400 km long and nearly 100 km wide, occupying hilly terrain with low relief scarred by landslip. They are aligned along structural trends of up to 50 km long and 25 km wide. Individual mud volcanoes range from 3 m to 2.5 km in diameter and reach a maximum height of about 110 m. The ejected mud consist of mud stone containing various shapes and sizes of clasts of older rock assumed as exotic block, which is believed as part of mélange diapirsm (Sukarna, 2007).

Information on mud volcanoes can be used to study the subsurface condition and used as pathfinders of the conditions indicative of subsurface hydrocarbon accumulations in unexplored areas. Gas geochemical data from mud volcanoes can be examined for possible presence of source rocks and their maturity levels. Mud volcanoes are often related to active petroleum systems, especially if the released gas shows a deep thermogenic character. A global data-set of more than 140 onshore mud volcanoes from 12 countries shows that in 76% of cases the gas is thermogenic, with 20% mixed and only 4% purely microbial (Etiope et al., 2009). The thermogenic nature of most of mud volcanoes is related to the relatively high thermal maturity of gas-generating organic-rich rocks. On the other hand, mud volcanoes which release large amounts of CO_2, such as those related to magmatic activity, may not indicate the presence of significant hydrocarbon reservoirs (e.g., Milkov, 2005). Many large onshore hydrocarbon fields were discovered after drilling around mud volcanoes in Europe, the Caspian Basin, Asia and the Caribbean (see Etiope et al, 2009, Link, 1952; Guliyev and Feyzullayev, 1997). Gas origin, composition and secondary post-genetic processes such as secondary methanogenesis which follows anaerobic biodegradation of petroleum or heavy hydrocarbons however, are fundamental factors for determining depth and quality of the related petroleum system, especially in frontier or unexplored areas (see Etiope et al., 2009).

Apart from providing information and evidence of hydrocarbon potential and a working petroleum system, mud volcanoes also provide useful data about the sedimentary section which can be determined by examination of ejected rock fragments incorporated in mud volcano sediments (breccia).

Mud volcanoes, depending on their size and activity, can pose ecological hazards and disaster to the environment as well as to the population of the surrounding area. Mud volcanoes typically ejected breccia and/or mud flows and/or flame in temporal association with earthquakes. Active mud volcanoes can vent a large amount of carbon dioxide and flammable methane, and may influence global climate. Large eruptions are known to have occurred in the Black Sea and in areas around the Caspian Sea where gas exploded in a flame several hundred meters high that burns vegetation within the vicinity of the mud volcano. Mud volcanoes may also pose a geohazard for drilling and platform constructions due to the potentially violent release of large amounts of hydrocarbons and mud breccia. When the viscosity of the mud breccias is low, it may flood large area and inundate villages, homes, roads, rice fields, and factories and displace people from their homes.

2. Evolution of LUSI Mud volcano

Kusumadinata, 1980 described mud volcanoes as any extrusion on the earth's surface in the form of clay or mud in which morphology forms a cone in which there is a lake and coupled with the discharge of water and is driven by strong gas flow. Often the release of gas is followed by an explosion and burns, thus the extrusion appearance greatly resembles a magmatic volcano. Apart from this description, Sangiran dome (e.g. Watanabe and Kadar, 1985) and 1936 Dutch maps (see Duyfjes, 1936), very little has been published in journals and scientific papers on mud volcanoes in Western Indonesia, unlike the ones found in Eastern Indonesia such as in Timor (e.g. Barber et al., 1986). In fact, mud volcanism in Indonesia, particularly in East Java is poorly understood. This lack of understanding changed when the LUSI mud volcano (Lumpur "mud"- Sidoarjo), was born as it offered a unique opportunity to study the dynamic development of a mud volcano from its birth on 29th May 2006. In contrast, studies on mud volcanism are typically conducted during the dormant periods between eruptions of already existing mud volcanoes.

The subsequent eruption of LUSI mud volcano has been closely observed and analyzed by the geological community (see Mazzini et al., 2007, 2009; Sunardi et al, 2007; Kadar et al., 2007; Sudarman and Hendrasto, 2007; Kumai and Yamamoto, 2007; Hutasoit, L, 2007; Sumintadireja et al., 2007; Deguchi et al, 2007; Abidin et al. 2007, 2008; Satyana, 2007, 2008; Satyana and Asnidar, 2008; Fukushima, 2009; Mori and Kano, 2009; Hochstein and Sudarman, 2010; Sutaningsih, 2010 etc). Its birth occurred just 2 days after a devastating Yogyakarta earthquake in May 2006. The trigger of LUSI mud volcano is controversial, and has been the center of debate among geoscientists and drilling engineers as it is located near the Banjarpanji-1 oil and gas exploration well drilling that was being drilled. (i.e. Mazzini et al., 2007; Davies et al., 2007, 2008; Manga, 2007; Manga et al., 2009; Tingay et al., 2008, 2010; Sawolo et al., 2008, 2009, 2010; Istadi et al., 2008, 2009). Mazzini proposed that LUSI was caused by fracturing following the May 27th earthquake and accompanied depressurization of > 100 °C pore fluids from > 1700 m depth. This resulted in the formation of a quasi-hydrothermal system with a geyser-like surface expression and with an activity influenced by the regional seismicity. Davies, on the other hand, suggested that an underground blowout in the Banjarpanji-1 well breached to the surface, thus creating a conduit for the high pressured fluid to escape and created a mud volcano.

The LUSI mud volcano eruption has continued for over five years, and potentially will continue for many years to come, impacting an ever larger area. The mud eruptions occurred in at least five separate locations forming a NNE-SSW lineament about 200 m away from the Banjarpanji-1 (BJP-1) exploration well in Sidoarjo, approximately 30 km south of Surabaya, East Java, Indonesia (Fig. 2 & 8). The approximately 700 m lineament is contiguous with the Watukosek fault zone (Istadi et al., 2009).

Hot mud erupted at 5,000 m^3 a day at the beginning, increased to 50,000 m^3/day in the initial months, then escalated to 125,000 m^3/day and reached a high rate of 156,000 m^3/day by October 2006. The high flow rates coincide with a series of earthquakes. In less than one year after its birth, LUSI displaced some 24,000 people, inundated toll road, several villages, housing estates, paddy fields and farm land, factories, schools, mosques, shops, offices, destroyed a gas pipeline, killing 13 workers and has covered about 700 Ha of land. The extent of the damage caused by LUSI is substantial and has caused subsidence of up to 5.53 cm/day in the area next to the main eruption. Simulation shows that it will affect an area of 3 km radius, disrupt the subsurface condition and subsidence of up to 60 meters. In a highly populated area of Sidoarjo, this could potentially disrupt more than 10,000 families; a major issue to the people, infrastructures and environment. The eruption rate however, has decreased to less than 10,000 m^3/day at the time of writing in July 2011; perhaps LUSI is now entering a new phase, from an eruptive one to a mature and quiescence phase.

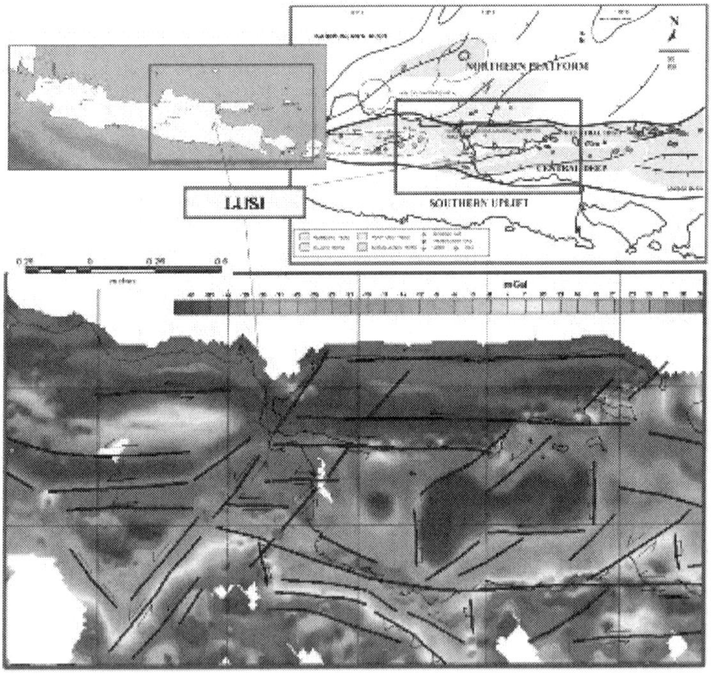

Fig. 2. LUSI is located in East Java, about 30 km South of Surabaya (top left). Regional tectonic framework of East Java (top right) shows major NE-SW and E-W major fault trends. Bouguer Gravity map (processed with 2.65 g/cc density) of East Java (bottom) showing East Java Basin's depositional centers (blue) which are controlled by the major faults in the area.

2.1 Stratigraphic setting of the LUSI Mud volcano

The LUSI Mud Volcano is located about 10 km northeast of Penanggungan Mountain, in Reno Kenongo village, Porong District, Sidoarjo Regency, East Java. Its location is in the Southern part of the hydrocarbon prolific East Java inverted back-arc Basin which was formed during the Oligocene- Early Miocene (Sribudiyani et al, 2003), on the Eastern tip of the Kendeng Zone (De Genevraye and Samuel,1972). The Geology of the area is characterized by the rapid deposition of thick organic rich sediment as part of the Brantas delta, influenced by the extensional tectonic regime (Willumsen and Schiller, 1994, Schiller et al, 1994). Due to the rapid deposition, shales in the area are undercompacted and overpressured (Mazzini et al., 2007). The geological condition is similar to other areas where mud volcanoes are found such as the Caspian and the Black Sea (Planke et al, 2004, Mazzini et al., 2007; Tingay et al., 2008)

Java Island, located at the southern part of the Sundaland, was formed by rock assemblages associated with an active margin of plate convergence. The island has recorded plate convergence between the Australian plate and the Sundaland continental fragment since Late Cretaceous. Therefore, the island is made up of complex of plutonic-volcanic arcs, accretionary prisms, subduction zones, and related sedimentary rocks (Satyana and Armandita, 2004). The structural history is divided into two phases: a Middle Eocene to Oligocene extensional phase, and a Neogene compressional or inversion phase. Grabens and half-graben structures were developed during the extensional phase, which was followed in the Neogene by compressional deformation with some wrenching. The most recent sedimentation in the East Java Basin occurred during the Late Pliocene to Holocene (3.6–0 Ma), during which time the southern part of the basin (Kendeng depression zone) was affected by north verging thrusts and uplift. The depression developed as a response to the isostatic compensation of the uplift of the southern Oligo-Miocene volcanic arcs. The uplift was accompanied by an influx of volcaniclastic rocks from the southern volcanic arc provenance and were deposited into the depression and causing the depression to subside.

Other provenance of the Kendeng Depression is the northern uplifted area that filled the basin with shallow-marine carbonates and marine muds from the Oligocene to the Holocene. Very thick sediments from the two provenances were deposited rapidly into the Kendeng Depression mostly as turbiditic deposits. The East Java geosyncline has thick Tertiary sediments of more than 6000 m (Koesoemadinata, 1980) with an estimated sedimentation rate of 2480 m/ma in the vicinity of LUSI (Kadar et al.,1997). The high sedimentation rates followed by rapid subsidence caused non-equilibrium compaction, and along with the maturation of organic materials resulted in the overpressured sediments within the Kendeng zone (see Willumsen and Schiller, 1994; Schiller et al., 1994). The overpressured sediments were later compressed, become mud diapirs and pierced the overlying sediments in many parts of East Java as mud volcanoes.

Outcrops of sedimentary rocks are very rare as they are covered by alluvial sediements and weathering. Therefore fresh rock outcrops at the rock mine in Karanggandang Village, 28 km to the northwest of the LUSI (Kadar et al, 2007) is important to complete the stratigraphic column of LUSI and the Banjarpanji-1 well.

The stratigraphy at LUSI (figure 3) consists of (1) alluvial sediments, (2) Pleistocene alternating sandstone and shale of the Pucangan Formation (to about 500 m depth), (3) Pleistocene clay of the Pucangan Formation (to about 1000 m depth), (4) Pleistocene bluish gray clay of the Upper Kalibeng Formation (to 1871 m depth), and (5) Late Pliocene volcaniclastic sand of at least 962 m thickness. The stratigraphy below the Late Pliocene sand is not well known; however, when the Banjarpanji-1 well reached 2834 m depth, cuttings from the bottom of the well did not contain limestone fragments indicating that drilling of Banjarpanji-1 well had not reached the carbonate reservoir target. Davies et al. (2007) suggested that these porous rocks were Kujung Formation limestone and that this formation is the source of the fluids erupting at LUSI. Seismic correlations from the Porong-1 well, 6.5 km to the northeast of LUSI, indicate that the rocks underlying the Late Pliocene volcaniclastic sands are carbonates which contain coralline red algae fossils, corals and foraminifera fragments. The strontium isotope (Sr) of the carbonates shows the absolute age of 16-18 ma Early Miocene, and therefore the carbonates are correlated with the Tuban Formation outcrops found extensively in the western part of East Java basin which show age range from 15.2 ma to 20.8 ma based on analysis of strontium isotopes (Sharaf et al., 2005).

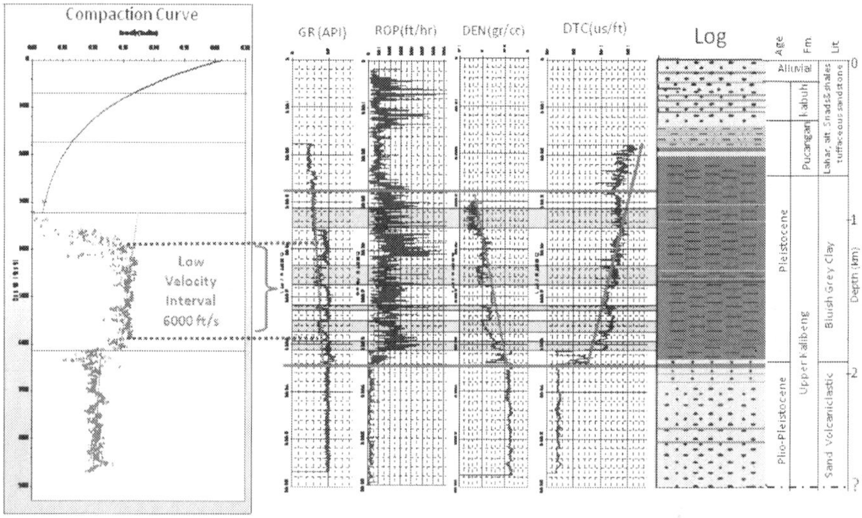

Fig. 3. Stratigraphy column in BJP #1 well nearby LUSI MV. Density, GR, ROP and DT suggest the presence of overpressured zones. These are probably **highly plastic,** undercompacted shale, controlled by rapid sedimentation that trapped water and resulted in an overpressured condition.

The seismic cross-section of this trajectory (figure 4) has a path through three wells, and track from SW to ENE. The wells are Banjarpanji-1 (BJP-1), Tanggulangin-1 (TGA-1) and Porong-1 (PRG-1). At the bottom of the well of Banjarpanji-1 well is a mass of shale that appears mounded and has a large structural dimension. Faulting in the form of positive flower structure between the wrench faults that continuously cut the low velocity intervals, is interpreted as a mud diapir (a depth> 9292 ft or 2834 m).

The structural feature that resembles a flower structure suggests the presence of a wrench fault, thus horizontally sliding components and oblique movements are suggested. The presence of overpressured zones probably consists of highly plastic, undercompacted shale. On seismic it is correlated with a Low velocity zone, and it is characterized by a chaotic discontinues pattern.

The collapse structure adjacent to the well Porong-1, located approximately 7 km from LUSI forms a depression around the crater. This structure likely represents an extinct mud volcano that, once it terminated its activity, gradually collapsed around a vertical feeder channel. The multiple contemporaneous faults indicated in the seismic section is due to gravitational slumps and intrusive structure suggesting piercement from the upward moving over-pressured sediments of the underlying shale diapir are evident (see Istadi, 2009). By the same mechanism, the multiple faults at LUSI may have been reactivated and served as conduit for the mud eruptions along the fault planes.

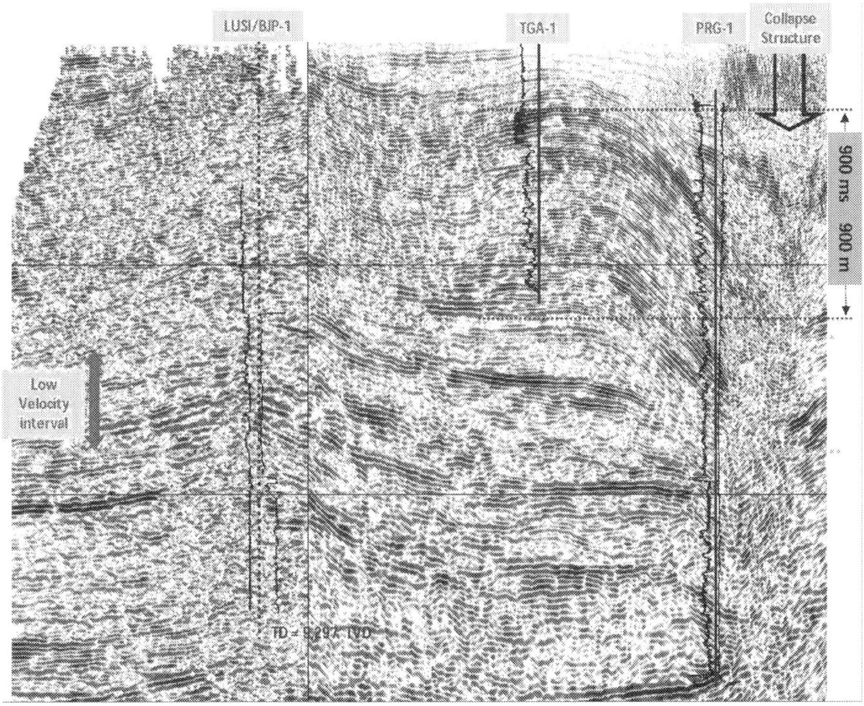

Fig. 4. Seismic section of LUSI - Banjarpanji-1 - Tanggulangin-1 - Porong-1 - Porong collapse structure. The Porong collapse structure located approximately 7 km from LUSI is a paleo mud volcano where subsidence is evident and the multiple faults present likely served as conduits for the mudflow. Similarly, the multiple faults near the BJP-1 well (200 m from LUSI) may have been reactivated and served as conduit for the mud eruptions and escaping gas, hence the appearance of gas bubbles along fault lines.

2.2 Controversy on the trigger

The trigger of the LUSI mud volcano is controversial because of its location and time of birth. The insinuating factors include:

1. The surface location of LUSI is approximately 200 meters away from an exploration well, the Banjarpanji well (Davies et al., 2007, 2008 and 2009). This suggests a possible connection between the drilling of the well and LUSI.
2. A major 6.4 Richter scale magnitude Yogyakarta earthquake rocked Java Island two days earlier (Mazzini et al., 2007 and 2009). This may have reactivated local fault and allowed an overpressured mud to escape to the surface.
3. The exploration well suffered two major drilling problems after the earthquake; two mud losses following the earthquake and a kick during the drill string removal (Sawolo et al., 2009 and 2010). This suggests a possible link between the earthquake and the change in the subsurface condition of the region.

4. A number of mud volcanoes in the vicinity of Watukosek fault were reported to suddenly become active at the time of the earthquake (Mazzini et al., 2009)
5. Three mud eruptions that appeared to be in line with the direction of Watukosek Fault (Sawolo et al., 2009)
6. The limited amount of well data available in the public domain especially on the events prior and following the initial eruptions as well as discrepancies over the interpretation of drilling data from the Banjarpanji-1 well (Tingay, M., 2010).

The process of LUSI mud volcano creation is generally believed to be a natural process of an over pressured shale eruption through seismically reactivated faults as conduits (Mazzini et al., 2009). The trigger of the eruption, however, is still a hotly debated issue within the engineering community. Was the creation triggered by a reactivation of nearby Watukosek fault or an underground blowout in Banjarpanji well? A well blowout generally means an uncontrolled flow of formation fluid into a wellbore that travelled up the well to the wellhead. An underground blowout, on the other hand, is when the flow of reservoir fluid does not exit at the wellhead but flows to a low pressured formation in the well and eventually breaches the surface.

Banjarpanji-1 is an exploration well that was drilled in March 2006 in a densely populated part of East Java, in the Sidoarjo region. In May 2006, the well was drilling toward its target, the Kujung limestone. There was little subsurface drilling problems during the drilling process as the well was drilled using a Synthetic Oil Based mud that eliminated much of the problems in the thick overpressured shale section. The schematic of the well is shown in figure 5.

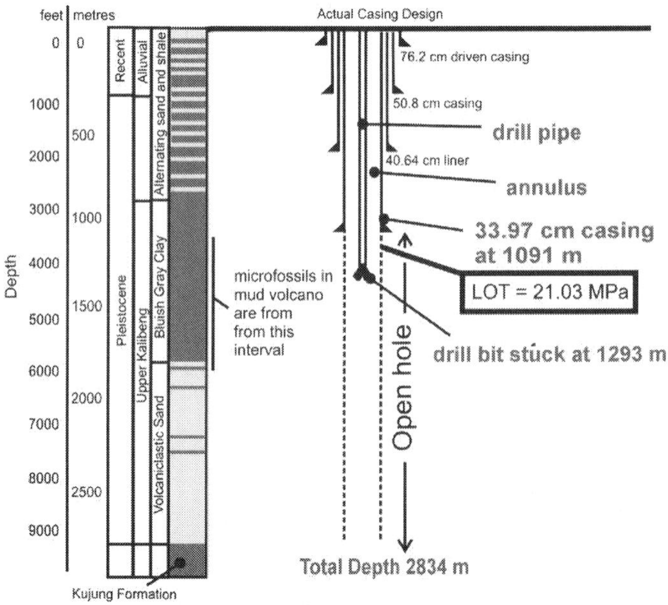

Fig. 5. Summary of the stratigraphy drilled by the Banjarpanji-1 well and the casing setting depths (Davies et al.,2008).

The problem free drilling condition changed in May 27th 2006 when a 6.4 magnitude earthquake strike near Yogyakarta about 250 km away. Within ten minutes of the earthquake, a 20 bbls of mud loss was observed in the well. Another mud loss occurred following two major aftershocks of 4.8 and 4.6 magnitude where a 130 bbls of drilling mud was lost to the open hole. Drilling mud loss is a serious problem in drilling as the mud hydrostatic pressure is the only mechanism to balance the reservoir pore pressure. Such major mud losses typically occur when drilling mud flows into a newly drilled cavity or a fractured zone or newly formed fractures.

2.3 The underground blowout hypothesis

An underground blowout is not an uncommon problem during well drilling, especially in exploration wells where the geology and reservoir pore pressure is uncertain. If the effective mud weight used to drill the well falls below the pore pressure, a kick (an influx of formation fluid into the well bore) can happen. This kick displaces the heavy drilling mud which reduces the hydrostatic pressure of the mud column and causes an even larger influx of formation fluid. This kick must be killed correctly and a proper hydrostatic pressure restored in order to retain the control of pressure in the well. If the kick is not controlled properly and the well is subjected to a pressure above its critical fracture pressure, then the weakest formation, generally immediately below the casing shoe, may fracture. When there is sufficient force behind the kick, this fracture can propagate upward into shallower formations, or even breach to the surface. This is the mechanism of an underground blowout during drilling of a well.

The possibility that the mud eruption was caused by an underground blowout in the Banjarpanji-1 well was initially suggested by the media immediately after the eruption. There was very little official information of its cause and very little well data was available in the public domain at the time. A similar charge was later iterated in the Davies et al., 2007 and 2008 papers with a pressure analysis that suggested a fracture at the casing shoe. Davies proposed that the well was subjected to a pressure above its critical fracture pressure that caused its weakest formation to fracture. The hydro-fracturing of the formation was propagated and finally it breached to the surface and caused LUSI.

In his papers, Davies proposed a number of operational sequences during the drilling of the Banjarpanji-1 well that caused a kick that leads to the underground blowout. Originally he proposed that the over pressured Kujung limestone was drilled, and this provided the high pressure fluid that caused a kick in the well (Davies et al., 2007). In his later paper (Davies et al., 2008), Davies refined his proposition to the kick was caused by a pressure under-balance in the well since 46 stands of drill pipe were removed without replacing the lost volume due to the removal of the drill pipes. And in his latest discussion paper (Davies et al., 2009) he proposes that the kick was due to a 'swabbing' effect. Swabbing means a negative pressure due to an upward movement of drill string.

Davies propositions of Underground Blowout hypothesis is rejected as pressure inside the well is too low to break the formation at the casing shoe of the Banjarpanji-1 well (Nawangsidi, D., 2007). The assumption used by Davies to develop his hypothesis is incorrect as the well is not full of mud but contains water as well as gas; this in effect lower the specific gravity and the hydrostatic pressure in the well. With a lower specific gravity and the published annulus pressure cited by Davies, it is impossible to fracture the casing shoe.

Davies's postulations on what happened, are not supported by well data and the actual condition of Banjarpanji-1 well (Sawolo et al., 2009). In his paper, Sawolo showed that (i) the

Kujung carbonate was not penetrated by the well, as the calcimetry data during drilling remained low - suggesting that no carbonate formation is drilled, and the well experienced a loss of mud (the opposite of a kick). (ii) The charge that mud was not pumped to compensate for the pulling of drill string out is incorrect. Automatic data recorder showed that mud was pumped appropriately to compensate for the pulling of the drill string. (iii) Analysis on speed when pulling out of the hole and the condition of the wellbore preclude the possibility of swabbing. And finally, (iv) pressure analysis based on the automatic data recorder (Real Time Data-RTD) showed that the pressure exerted on the casing shoe is too low to be able to fracture the formation. It is therefore concluded that the well was intact, no underground blowout occurred in the well and the Banjarpanji well could not have been the trigger of LUSI.

The geologists and drillers in charge of drilling the well have reviewed all the data and cannot support the underground blowout theory. They charge that it is an over simplification to the actual condition (Sawolo et al., 2010). Their approach in analyzing the trigger of LUSI was broader than that of Davies. Instead of relying on limited data and assumptions to fill the missing gap, they collected all available drilling data, logs and cross checked it with other reports to come up with an enriched credible set of data. The various data were assembled to form a mosaic that clearly showed that the well remained intact and an underground blowout in the Banjarpanji-1 well did not occur.

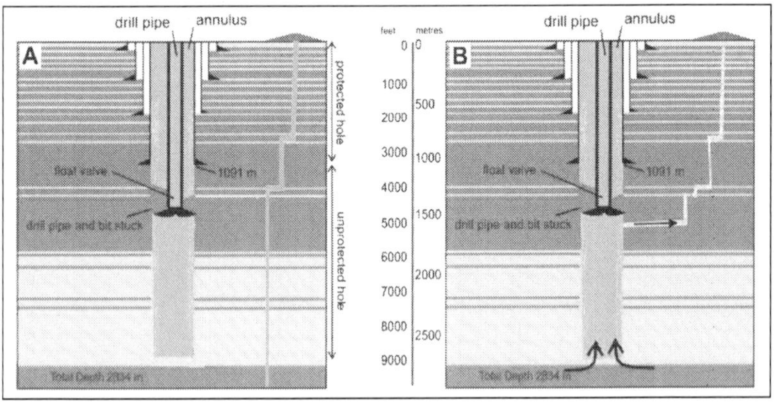

Fig. 6. The controversy on LUSI's trigger; Sawolo's view (A) - where the well is intact, the mud flow did not pass through the well. Davies (B), contends that the mud passed through the well, fracturing the casing shoe and triggered the LUSI mud volcano (Sawolo et al., 2010).

The geologists believe that LUSI is more likely to be related to seismic activity. The loss of drilling mud that coincided with the time of the earthquake suggests that the earthquake may have altered the subsurface condition of the region.

The two opposing views of the LUSI trigger as a seismic related event (Sawolo et al) and an underground blowout (Davies et al.,) are shown in figure 6. The scenario proposed by Sawolo is shown in A, where the well is intact and the mud flow did not pass through the well. Davies view is shown in B, where the mud flow passed through the well and fractured the casing shoe.

The root cause of the difference of opinion between Davies and Sawolo is believed to be access to well information; especially the quality and quantity of data used in the pressure analysis. Analysis based on limited data can result in a misleading conclusion. It should be understood that the well information on the public domain at the initial stages of the eruption was limited as majority of data was kept confidential by the oil and gas company. It was not until 2009 that Sawolo et al. published for the first time a wide range of well data and observations previously not available in the public domain; this includes important data on formation strength such as the Leak Off Test, Bottom Hole Pressure, minute by minute well pressures and fluid density in the well. The intent of opening the data to the scientific community was that future research on LUSI could be based on actual and credible data in order to minimize assumptions.

The engineering debate on the trigger of LUSI continued until 2010, when Tingay, 2010 published a paper outlining the pros and cons of the competing theories. The paper presented the first balanced overview of the LUSI mud volcano by identifying critical uncertainties of the plumbing system, events prior and following the initial eruptions as well as discrepancies over interpretation of petroleum engineering data from the Banjarpanji-1 well. These uncertainties caused much of the trigger controversy. It is obvious that more studies are needed before LUSI's actual trigger can be identified.

2.4 The fault reactivation hypothesis

Mud volcanoes are spatially associated with both major and secondary faults within the regional stress field. Surface and subsurface geology shows that faults exist in the region that crosses LUSI areas with the NE-SW and NW-SE direction trends. Faulting in the NE-SW direction is known as the Watukosek fault, an oblique strike-slip fault, whereas the NW-SE trend dextral strike-slip fault pattern is the Siring fault. These faults are partly buried by alluvial sediments.

Eruption of mud volcanoes along the fault line is one of several possible subsurface hydrological responses to earthquakes. Manga, however, dismissed the possibility that LUSI was triggered by the Yogyakarta earthquake that occurred two days earlier because of its distance and magnitude (Manga, M. 2007). An empirical plot from past earthquakes in the world shows a distinct relation between the earthquake's magnitudes to the distances from their epicentres (figure 9).

When the data-point from the Yogyakarta 27th May 2006 earthquake was entered, it shows that the Yogyakarta earthquake was unlikely to have caused the LUSI mud volcano. It is interesting to note that two past earthquakes that were larger and closer to the area failed to cause a mud volcano in the region.

Mellors has a similar view on the relationship between large earthquakes and triggering effect on mud volcanoes (Mellors et al., 2006). This is because large earthquakes produce strong static and dynamic stress changes near its centre that could trigger mud volcanoes. His research is based mainly on a 191 years record of mud volcano eruption and large earthquake in the Caspian Sea. It suggests that the triggering effect is strongest where the shock at the mud volcano has the intensity of approximately Mercalli 6 and above and at a distance of less than 100 km. However, he also found that even when the intensities exceed the apparent threshold, only a fraction of active volcanoes erupt. This indicates that other factors also play an important role.

However, there are too many uncertainties in predicting the triggering effect of earthquakes on mud volcanoes (Mori and Kano, 2009). Mori pointed out that until recently, seismologists would not believe that triggering due to earthquakes over hundreds of kilometres is

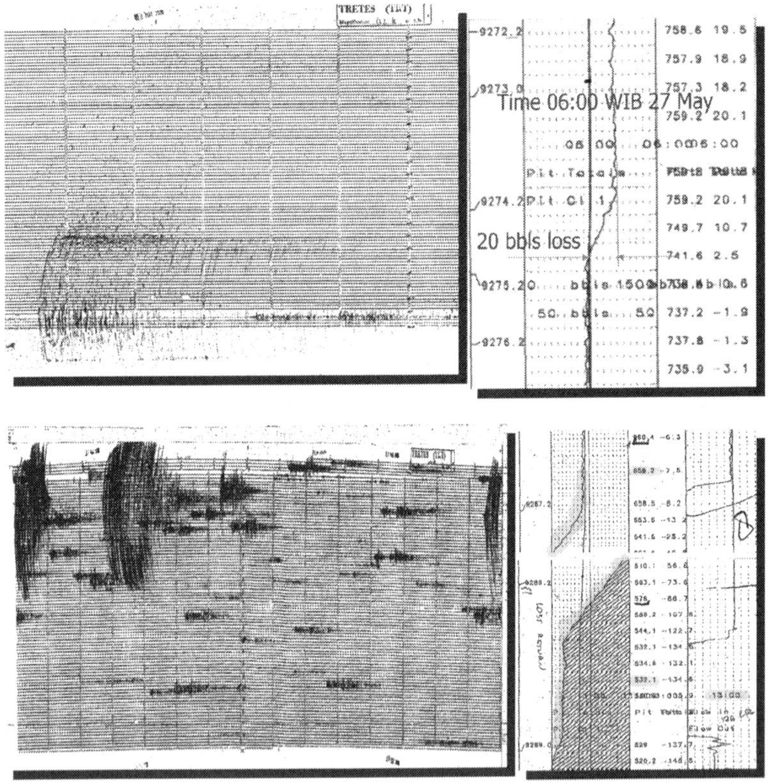

Fig. 7. Losses of mud after the Yogyakarta earthquake. The top left picture shows the seismograph reading at Tretes BMG station about 15 km away and the right picture shows the 20 bbls loss seven minutes after the main earthquake. The bottom left showed the aftershocks and the right shows the 130 bbls complete loss of circulation from the wellbore that happened two hours after two aftershocks (Sawolo et al., 2009).

possible. This view has changed especially in the hydrothermal areas. When Mori entered the Yogyakarta earthquake and LUSI data in Fisher's Dynamic Stress vs. Frequency plot (figure 10), it is located within the value that has triggered mud volcanoes in other regions.
Mazzini, one of the very few earth-scientists who conducts field work and data measurement right after the mud eruption, also disagrees with Manga's conclusion (Mazzini et al., 2007). He cited that eruptions can be affected by earthquakes several thousands of kilometres away and that a delay of few days between the time of the earthquakes and the eruption is not uncommon.
His field work showed that a regional fault, the strike-slip Watukosek fault, crosses The LUSI area. A number of extinct mud volcanoes are found aligned with this fault from Java to Madura island. In addition, seismic profiles acquired prior to the eruption shows evidence of a vertical piercement structure with upwards dipping strata around the LUSI conduit zone. He argued that this could be interpreted as evidence for a long history of

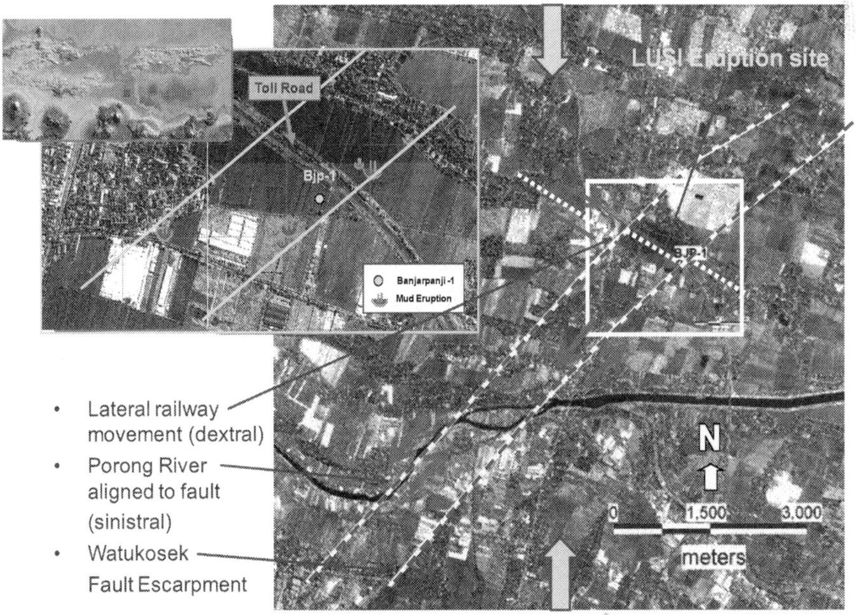

Fig. 8. Watukosek fault, consisting of 2 parallel faults where the Porong River is aligned along the fault line, while the Watukosek fault escarpment represents the up thrown fault block. LUSI eruption sites are along the Watukosek fault line. The Watukosek fault, striking from the Arjuno volcanic complex, crosses the LUSI mud volcano and extends towards the northeast of Java island.

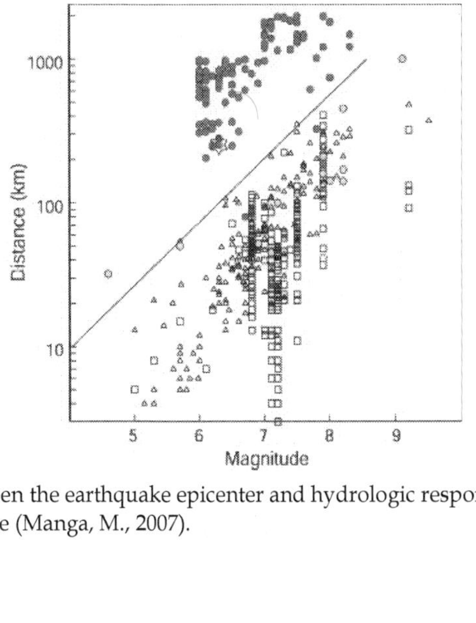

Fig. 9. Distance between the earthquake epicenter and hydrologic response as a function of earthquake magnitude (Manga, M., 2007).

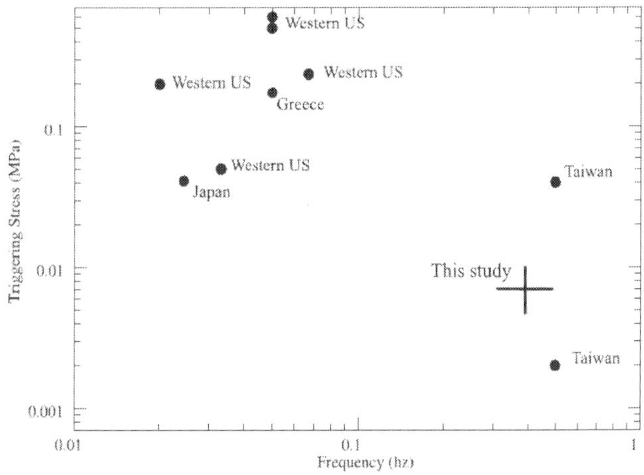

Fig. 10. Values for dynamic stress and frequency of seismic waves that have triggered small seismic events, compiled by Fisher et al. (2008). The cross shows the estimate for the Yogyakarta earthquake at LUSI. Source: Mori and Kano, 2009

active vertical movements of mud underneath LUSI, possibly with former eruptions or as a disturbed signal due to the fault that crosses this area. He suggested that the Yogyakarta earthquake ultimately triggered the eruption through the already overpressured subsurface piercement structure. This is supported by a partial loss of well fluid recorded in the Banjarpanji well nearby 10 minutes after the earthquake, and a major loss of well fluid after two major aftershocks (see previous chapter on The Underground Blowout Hypothesis - figure 7). These mud losses, he argued, could be the result of movements along the fault that was reactivated, lost its sealing capacity and become the passageways for overpressured subsurface fluid to escape. These fluids ultimately reached the surface at several locations aligned NE-SW in the Watukosek fault zone direction.

Davies disagreed with Mazzini's conclusion that the Yogyakarta earthquake reactivated the Watukosek fault and triggered LUSI mud volcano (Davies et al., 2007). He argued that the earthquake was too small and too distant to trigger an eruption when in the recent past, two bigger and closer earthquakes failed to trigger an eruption. He considered the static and dynamic stresses caused by the magnitude 6.4 earthquake too small to trigger LUSI.

Mazzini backed his hypothesis by presenting further field data that support his hypothesis that a strike-slip faulting was the trigger mechanism that released overpressure fluids through already present piercement structures (Mazzini et al., 2009). He presented several observations on the fault reactivation evidence, among others:

- Residents close to the Gunung Anyar, Pulungan, and the Kalang Anyar mud volcanoes, located along the Watukosek fault almost 40 km NE of LUSI (Fig. 1), reported increased venting activity of the mud volcanoes after the Yogyakarta seismic event. Simultaneously, boiling mud suddenly started to erupt in Sidoarjo, later forming the LUSI mud volcano.
- A 1200 m long alignment of several erupting craters formed during the early stages of the LUSI eruption. The direction of these aligned craters coincides with the Watukosek

fault. The craters were formed during May-early June 2006, but were later covered by the main LUSI mud flows.

- Large fractures several tens of centimetres wide and hundreds of meters long were observed in the proximity of the BJP-1 exploration well with identical NE-SW orientation. However no fluids were observed rising through these fractures, which suggests a shear movement rather than a deformation from focussed fluid flow.
  The intersection of the fault with the nearby railway clearly indicates lateral movement. The observed lateral movement recorded at the railway during the first four months was 40- 50 cm. The lateral movement recorded at the neighbouring GPS stations during the same time interval reveals at total displacement of 22 cm (2 cm in July, 10 cm in August, 10 cm in September) (see figure 11). This later displacement was possibly related to the gradual collapse of the LUSI structure. In any case, the difference between these two records shows that an initial 15-20 cm of displacement that must have occurred during the early stages (i.e. end of May-June) related to the Watukosek fault shearing. Since 27th May earthquake, the rails have had to be repaired four times. Two of these repairs were done within the first three months after the earthquake to remove the bending due to the continuous shearing.

Fig. 11. Shear stress have damaged nearby infrastructures such as the dextral movements of a railway, bursting of a gas pipeline and numerous breakages of water pipelines at the same location further supports displacements along faults. (A)The railway bent to the west of main vent on September 2006. Offsets that occurred approximately 40 cm with orientation direction NW - SE. (B) At the same location, the railway was bent again in October 2009, with an offset of approximately 45 cm. The bending of the railway line is due to fault reactivation that often has differential movements which created shear stress.

- A water pipeline experienced significant bending and ruptures at the intersection with the fault (Fig. 5A-B). Since the May 2006 earthquake occurred, the pipeline has been repaired sixteen times. Note that neither the rails nor the water pipeline had kink problems before the earthquake.

He also found seismic sections taken in the 1980s that showed a dome-shaped piercement structure; the most spectacular is the collapse structure in the nearby Porong 1 well (Istadi et al., 2009) (see figure 4). This structure is likely to represent an extinct mud volcano that gradually collapsed around its own vertical feeder channel.
Mazzini further showed shear-induced fluidization mechanism through experiment that a relatively small displacement resembling a fault movement can turn a pressurized sand box model from once sealing layers, to become non-sealing. He demonstrated that the critical fluid pressure required to induce sediment deformation and fluidization is dramatically reduced when strike-slip faulting is active. (see Mazzini et al., 2009).

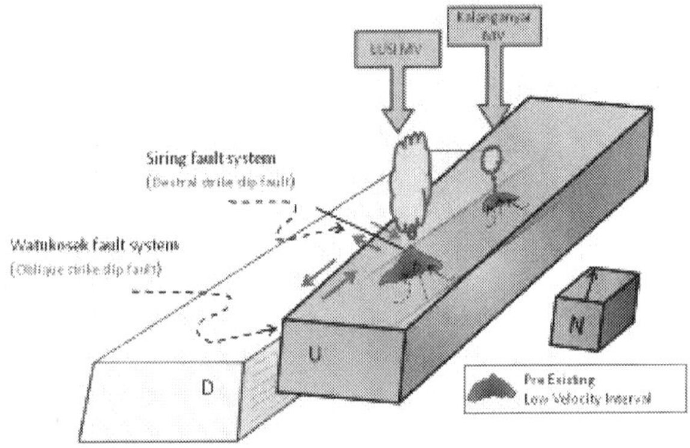

Fig. 12. Schematic cartoon (not to scale) of a mud volcano appearing along strike-slip faults. The shear zone along the Watukosek fault system and Siring fault that crosses LUSI where a low velocity interval existed before the eruption. Reactivation of the strike-slip fault after the earthquake caused the draining of fluids from the low density units towards the fault zone as the preferential pathway.

2.5 Response to earthquake

Due to its tectonic position at the front of the subducting Australian plate under the Sunda plate to the south, Java has been seismically active (see figure 13A). The compressional stresses, either due to subduction or its secondary effect that compresses the Sunda plate in a N-S direction, puts strain on local faults, especially those trending NE-SW. The latter caused a rupture on the NE-SW Opak fault, and had resulted in the magnitude 6.4 Yogyakarta earthquake, on 27 May 2006. This earthquake led to a new understanding of its effect on the volcanic plumbing system of Java Island. At the time of the earthquake, two Javanese volcanoes - Merapi and Semeru, were active; the distance of these volcanoes from the epicenter are around 50 km and 260 km respectively (see figure 13). It was observed that while there was no new volcanic eruption, the eruptive response of the heat and volume flux of these two volcanoes changed considerably by a factor of two-to-three starting on the third day after the earthquake (Harris and Ripepe, 2007). Their work revealed immediate eruptive response through processing of thermal data for volcanic hot spots detected by the Moderate Resolution Imaging Spectrometer (MODIS), (http://hotspot.higp.hawaii.edu). This implies that the earthquake triggered enhanced simultaneous output and identical trends in heat and volume flux at both volcanoes.

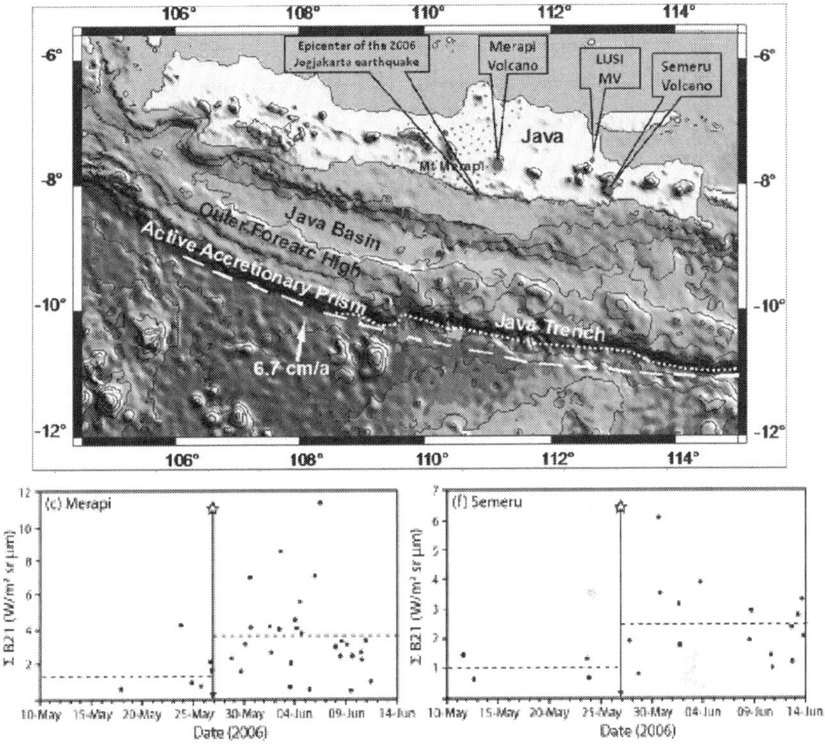

Fig. 13. Map of Java, showing the location of the Merapi and Semeru volcanoes. Increases in heat and volume flux occured 3 days after the Yogyakarta earthquake in the Merapi and Semeru Volcanoes. Thermally anomalous pixels detected by MODVOLC showing all band 21 pixel radiance. Source: Harris and Ripepe, 2007.

It was also reported that the magma extrusion rate and the number of pyroclastic flows from the volcano suddenly tripled [Walter et al., 2008]. This change did not last long, and everything was back to normal again after 12 days. This observation suggests that while this magnitude 6.4 earthquake may not able to trigger a new eruption, it is able to change the intensity of an erupting volcano at a long distance (260 km).

The May 2006 earthquake was one of the deadliest earthquakes in Java in historical times. Although it was as a magnitude 6.4, the scale of destruction was unprecedented in the region. The large scale destruction was concentrated in a 10 - 20 km distance along the Opak River Fault where the subsurface lithology consists mainly of soft volcaniclastic lahar deposit (Walters et al., 2007). Walters study suggests that such deposits have the property to amplify the ground motion such that even a relatively small magnitude earthquake could result in large scale destruction.

The two works of Harris and Ripepe, and Walters suggest the complex interdependency of the causes and effects in a seismically and volcanically active environment. The 27th May

2006 earthquake changed the static and/or dynamic stresses of the area. Their studies suggest a link between earthquake, changes in subsurface condition and its effect on the volcanic activity.

To monitor and record seismic waves around LUSI seismograph installation was carried out at several stations between April and July 2008 (see Figure 14). Seismic waves can be generated by the existing fault activity or by new cracks in the rock layers that had lost their cohesive strength as a result of subsidence around the main eruption vent of LUSI. The microseismic or seismic waves and energy released during crack formation in the rocks is relatively small compared to the energy released by earthquakes.

Microseismic activity recorded by the seismograph network installed around LUSI consists of 6 sensor units, of short period type and broadband seismographs. Each seismograph was

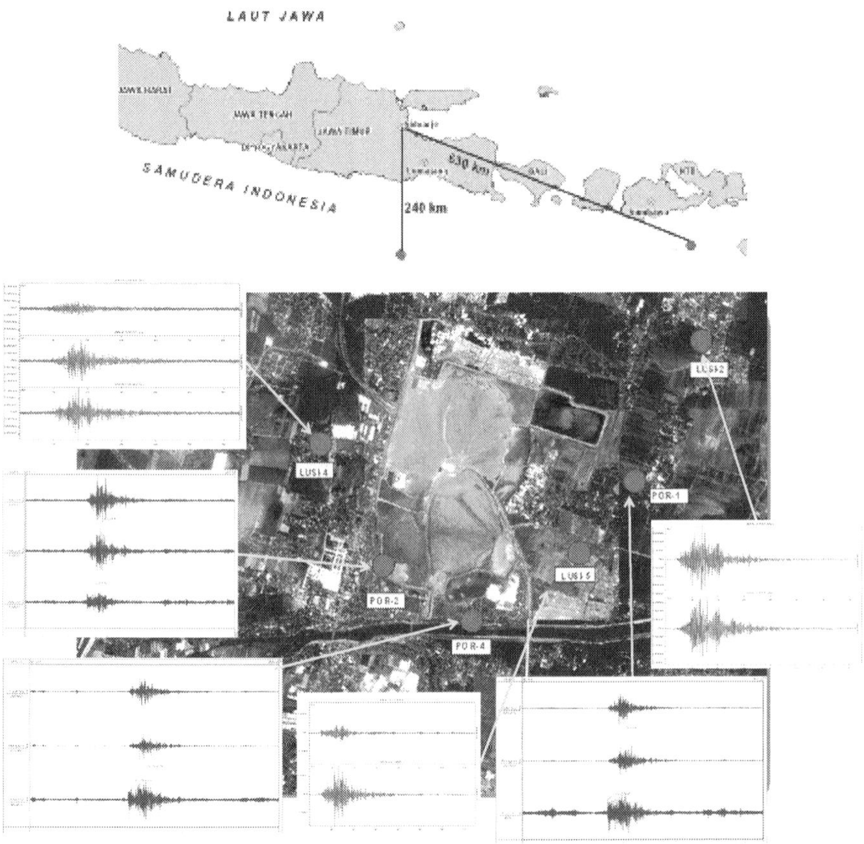

Fig. 14. (A) Epicenter locations of June 1st and 12th 2008 earthquakes located about 240 km and 630 km respectively from LUSI. (B) LUSI Microseismic monitoring network located around the center of the main crater. Seismographs show the June 12th 2008 earthquake with an epicenter located about 240 km South of LUSI.

equipped with a digital recorder system that records continuously for 24 hours, and GPS was used as timing marks on the seismic wave data. Data was processed by analyzing the arrival time of the P wave and S wave. The results of "picking" or "reading arrival rate" was analyzed with appropriate software, to determine the source of vibration.

To determine the location of the vibration source or microseismic hypocenter requires seismic wave velocity data at LUSI location. Wave velocity data was obtained from seismic surveys and wells logging data during drilling. Processed results in the form of coordinates of the location of the source of the wave system are plotted in three dimensions, so that the pattern of its occurence can be seen clearly. To facilitate processing, field data which is a mixture of different frequencies and microseismic noise are filtered, so as to identify microseismic events, arrival time, P wave and S wave, maximum amplitude and duration. All data was processed to determine the parameters of microseismic, namely: the timing, location coordinates, depth and magnitude. The results of the data processing are classified into two types of earthquakes, namely: the earthquake which occurred outside LUSI, and those that occurred around LUSI. In this case we will focus on earthquake data that occurred outside the LUSI area to determine earthquake response to changes in temperature, gas flux and behaviour that occur in the main vent.

The ability to detect an earthquake depends on the magnitude of the earthquake, the sensitivity of the sensors (seismometers), and the distance between the hypocenter and the location of the sensors. In general, earthquakes in Indonesia with magnitudes above 5.0 on the Richter scale, will be recorded by almost all seismograph networks in Indonesia. Like the two above mentioned tectonic earthquakes, wave energy can propagate from the source to the sensor around LUSI, with greater strength than the noise level around the sensors.

No.	Stations	Coordinates		Periods	Agency
		Latitude	Longitude		
1	POR 1	-7.53084	112.73086	29 April - 5 July 2008	BMKG
2	POR 2	-7.54043	112.70377	29 April - 5 July 2008	BMKG
3	POR 4	-7.54414	112.71470	29 April - 5 July 2008	BMKG
4	LUSI 2	-7.51485	112.74049	29 April - 5 July 2008	BMKG
5	LUSI 4	-7.52660	112.69772	29 April - 5 July 2008	BMKG
6	LUSI 5	-7.53700	112.72535	29 April - 5 July 2008	BMKG

Table 1. Coordinates of microseismic network stations in the area LUSI.

During the monitoring period two tectonic earthquake occurred outside LUSI. These are:

1. June 1, 2008, Time 15:59:50.2 GMT, the epicenter was located at latitude 9.53° South - longitude 118.04° East, at a depth of 90 km with a magnitude of 5.5 SR, about 630 km from LUSI
2. June 12, 2008 At 05:19:55 GMT, the epicenter was located at latitude 9.68° South - longitude 112.67° East, at a depth of 15 km and magnitude of 5.4 SR, about 240 km from LUSI.

In addition to microseismic monitoring, temperature, LEL (low explosive limit- in air where 20% LEL corresponds to 10000 ppm), and H_2S concentration monitoring was continuously performed using portable monitoring equipment by BPLS (Sidorajo Mud Mitigation Agency) officers in the field. Measurements from 1 to 20 June 2008 showed a fluctuation LEL, H2S, and temperature at the center of eruption. The peak value of the measurement period occurred on June 12 and 13, 2008, in which all measurement parameters rose sharply, particularly temperature and the concentration of H_2S (see figure 15).

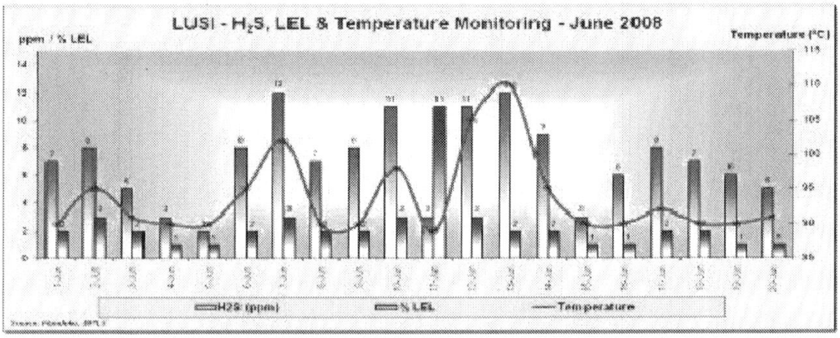

Fig. 15. Correlation between LUSI mud volcano activity and earthquakes. Increasing gas expulsion, temperature and mud eruption rates after earthquake are shown in the above graph after the 12th of June 2008 5.5 Mw earthquake. The epicenter was located some 240km South of LUSI.

The increase in temperature positively correlates with data from the installed seismograph network around LUSI which showed an earthquake occurred approximately 240 km south of LUSI on June 12, 2008. In The case of LUSI, the earthquakes have affected the rheology of fluid in term of permeability, changing the viscosity and the rate of mud eruption, consequently the increased concentration of expelled gases and temperature.

2.6 Horizontal displacement

Geodetic measurements were conducted at the LUSI site to quantify the ongoing deformation processes. The primary data sources were the GPS surveys periodically conducted at monitoring stations to measure vertical and horizontal movements relative to a more stable reference station. Seven GPS survey campaigns were conducted between June 2006 and April 2007. The GPS measurements were conducted at 33 locations using dual-frequency geodetic type receivers over various time intervals. Each measurement lasted from 5 to 7 h. (Istadi et al., 2009).

Areas within a 2-3 km radius of LUSI's main mud eruption vent are experiencing ongoing horizontal and vertical movement aligned to major faults. The horizontal displacements have spatial and temporal variations in magnitude and direction, but generally follows the two major trends, namely in the direction of NE - SW and NW - SE (see figure 16). Rates of horizontal displacement are about 0.5-2 cm/day, while vertical displacements are about 1-4 cm/day, with rate increasing towards the extrusion centre (Abidin et al, 2008).

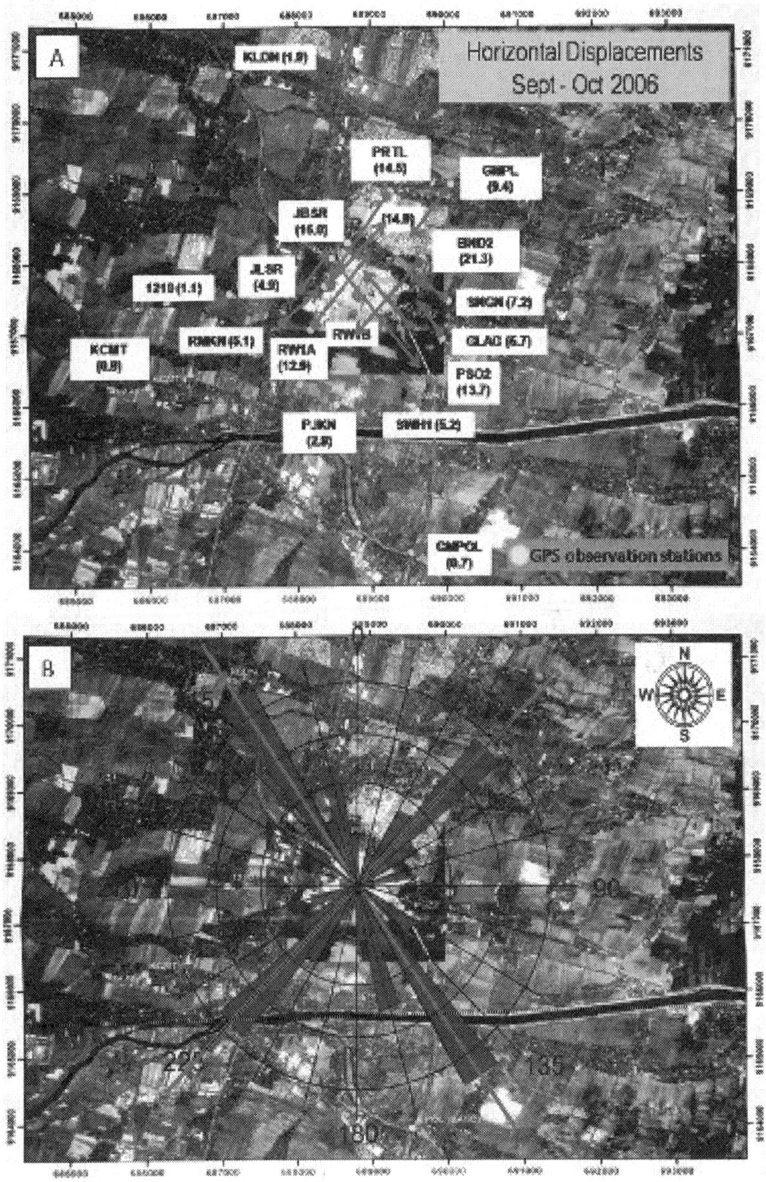

Fig. 16. (A). Horizontal displacement measurements in September - October 2006. Directions of the red arrows show the direction and magnitude of movement. (B). Measurements from June 2006 - March 2007 indicate the major trends are NW-SE and NE-SW as seen in the rose diagram.

2.7 Subsidence and uplift

Five years after the mud eruption, the area near LUSI has subsided at a considerable rate. Buildings and houses near the eruption site have completely disappeared under layers of mud. However, in the east and northeast uplift is occurring. To measure both the subsidence and uplift, four survey campaigns were conducted (Table 2):

Start	End	Points	Method
July 2006	March 2007	25	GPS
Dec. 2007	April 2009	30	Total Station
Dec. 2008	Feb. 2011	15	GPS
Dec. 2008	Feb. 2009	5	Level

Table 2. Four survey methods to measure elevation near LUSI MV

Data from these four surveys was used to show the changes in elevation, subsidence and uplift, as well as horizontal movement over time. Subsidence contour maps were created using GIS software by interpolating the measurement data. The results showed an almost concentric pattern shown in Figure 17.

The subsidence started as a crack in the ground that continued to grow and decrease its elevation. The existence of subsidence was evidenced by, among other things, the pattern of ground cracks, tilting of houses, cracking of flyover and bridges, as well as collapsing of buildings. The direction of the cracks varies depending on its location. In the Renokenongo area, southeast of LUSI, the cracks direction is NE- SW, whereas in West Siring area, west of LUSI, the cracks are North-South.

Subsidence and horizontal movements indicate the dynamic geological changes in the area. These movements have caused reactivation of pre-existing faults or newly formed faults. The continued movements along faults would likely result in the emergence of more fractures and gas bubbles (see figures 17 and 18).

Subsidence continues as the mud eruptions progress. The subsidence might result from any combination of ground relaxation due to mudflows, loading due to the weight of mud causing the area to compact, land settlement, geological structural transformation and tectonic activity (Abidin et al., 2007).

Based of field measurements, areas up to 3 km from the main eruption vent are experiencing subsidence to some degree. Presently however, due to much reduced volumes of mud eruption, the measured rate of subsidence on the West side of main eruption vent indicate a decrease from the original 25 cm/month when LUSI was very active in the first year, to less than 5 cm/month. If the decreasing trend continues, the affected subsidence area will likely decrease from earlier prediction of more than 3-4 km.

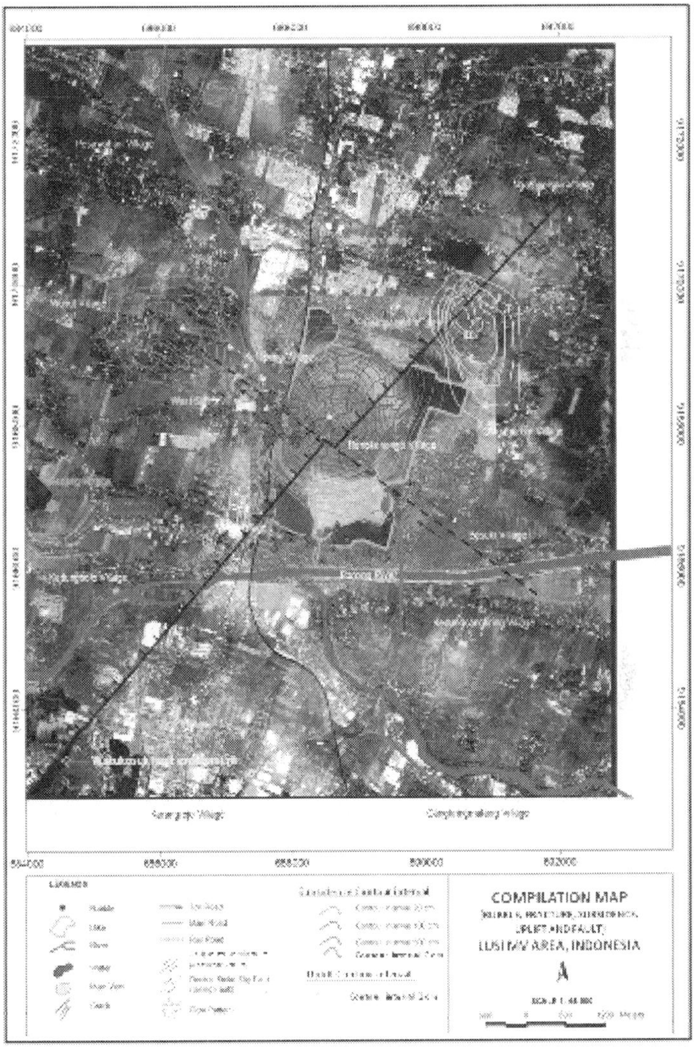

Fig. 17. LUSI post eruption map. The subsidence contour is status as of January 2010, constructed by interpolating the measurement data, and was created using GIS software. The contour showed an almost concentric pattern. The area West of the main vent was subsiding faster than other areas.
The map also shows fractures distribution around LUSI. East of the main vent, fractures trend NE - SW, whereas West of the main vent the fracture trend is North-South.
The Gas bubble distribution around LUSI status in May 2011 where more than 220 gas bubble locations have been recorded since the start of LUSI eruption in May 2006. Presently only a few are still active.

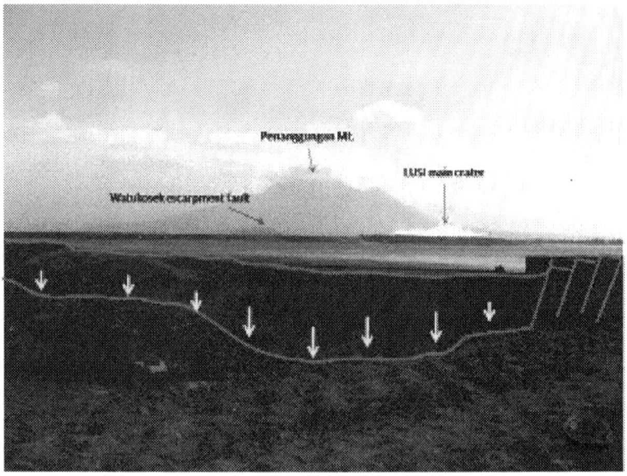

Fig. 18. Photo showing subsidence and collapse of the retaining mud dyke northeast of the LUSI main vent that occurred on 21 May 2008. In some parts, where slumping and subsidence occurred, local small scale faulting at the edge of subsiding wall occured. The continued subsidence proves very difficult to maintain the dyke.

2.8 InSAR data

InSAR (Interferometric Synthetic Aperture Radar) is a technique to map ground displacement with a high resolution of up to centimeter-level precision (e.g. Massonnet and Feigl, 1998; Hanssen, 2001). InSAR is effective tool to measure the amount of ground deformation caused by earthquake, volcanic activity has been useful for studying land subsidence associated with ground water movements (e.g. Amelung et al., 1999; Gourmelen et al., 2007), mining (e.g. Carnec and Delacourt, 2000; Deguchi et al., 2007a), and geothermal as well as oil exploitation (e.g. Massonnet et al., 1997; Fielding et al., 1998). The amount and pattern of deformation are shown by a range of colors in the spectrum from red to violet. The computed interferograms are interpreted using an inversion method that combines a boundary element method with a Monte-Carlo inversion algorithm (Fukushima et al., 2005). In LUSI, this technique was used to determine the surface deformation due to the mudflow starting from 19 June 2006 (three weeks after the mud eruption) to 19 February 2007. The measurement was done using PALSAR (Phased-Array L-band SAR) onboard the Japanese Earth observation satellite ALOS. Measurement of land subsidence is possible as the L-band microwave is less affected by vegetation (Deguchi et al., 2007a).
Deguchi et al. (2007a, 2007b) and Abidin et al. (2008) performed a study and measured the ground subsidence temporal changes of deformation obtained by applying time-series analysis to the deformation results extracted by InSAR.

- From 19 June 2006 to 4 July 2006 the subsidence showed an elliptical pattern, suggesting subsidence around the main vent and west of the main vent.
- From 4 July 2006 to 19 February 2007, the scale of subsidence and uplift became more significant. Both subsidence and uplift East of the main vent became more pronounced. In contrast to the high rate of mud eruption however, the InSAR results clearly showed that the ground deformation associated with mud eruption decreased after November 2006.

The results from the use of InSAR indicate subsidence has occurred in this area. Four different areas of deformation is suggested, these include areas centered around the main eruption vent; areas to the west-northwest of the main vent; areas to the northeast of main vent; and to the southwest of the main vent. Apart from the areas to the west-northwest which is associated with the deformation due to gas production in Wunut gas field, the other 3 deformation areas follow the regional fault pattern, contiguous to the Watukosek NE-SW fault trend.
The results also demonstrate the progressive subsidence evolution from time to time during the period of measurement. Subsidence in the main eruption area showed the most rapid subsidence rates. The 8-months measurements period showed ellipsoidal subsidence pattern covering an area of approximately 2 x 3 km2 with a long axis trending NE-SW.
Another area to the west-northwest of the main eruption area is also experiencing subsidence. This particular area is within the Wunut gas field which covers approximately 2 X 2.5 km2 with long axis trending NW-SE. This trend corresponds to the regional Siring NW-SE fault trend.

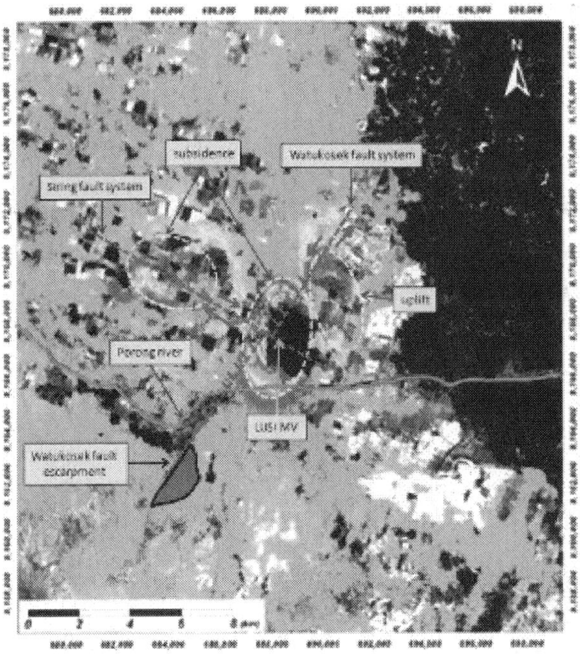

Fig. 19. The interpreted results of InSAR satellite imagery in February 2007 suggest an elliptical subsidence along the NW - SE long axis with a distance of 1-2 km from the main eruption vent, namely in the area around West Siring and Pamotan. In the vicinity of the main mudflow and the eastern regions about 2.5 km northeast of the main eruption, the subsidence occurred elliptical on the N-S long axis.
(figures modified from Deguchi et al, 2007)

Fault reactivation resulted in horizontal and vertical movement, which later manifested in the formation of uplift and subsidence or vertical and horizontal offset. An overlay of the ellipsoidal InSAR measurements with regional faults in these areas indicate a correlation between the two. Elipsoidal uplift suggest the long axis trending NNW - SSE is a restraining stepover to offset oblique strike slip fault of the reactivated Watukosek fault.

It is interesting to note that the InSAR measurements found that the deformation diminished after November 2006, only 6 months after the start of the eruption. Interpretation of interferogram for each periodic cycle for the period of May to July 2006 (beginning of eruption) showed more temporal change of deformation compared to the period of November 2006. In contrast, during the period of October - November 2006 field observations indicate increasing intensity of subsidence in the western side of the main vent, particularly in the village of Siring Barat. The main eruption vent and surrounding central area were experiencing most rapid rate of subsidence and continual collapse of the mud retaining dykes. Areas to the E-NE of the main vent were experiencing increases in uplift. The indication of contrasting InSAR measurements could be interpreted as lesser or diminishing effect of initial fault reactivation that triggered LUSI.

Interpretation of interferogram by Deguchi suggesting psudo anomaly in an area to the northeast of the main vent and does not indicate uplift based on conversion to rectangular coordinates (see Deguchi et al, 2007b). Field observation however suggest an uplift has occurred in areas to the east and northeast of the main eruption, in the Renokenongo village and surrounding areas. The uplifted area covers an area of approximately 1 X 1.5 km2 with a long axis trending NNW - SSE.

Fig. 20. The pattern of fractures trending NE -SW in the Village Renokenongo. A section of land on the right hand side of the picture is uplifted (east side) while the left is the downthrown block (west side). Note: The mineral water bottle is used as a comparison to indicate the amount of displacement (~20cm). In contrast to Degushi et al., 2007b psudo anomaly interpretation, the above photo taken 2 months after the eruption suggests displacement due to fault movement. Movement due to subsidence was unlikely as it was minimal at the early stages of the eruption.

2.9 Fracture orientation

Fractures appeared around LUSI area as a result of loss of cohesion due to ground movement, both vertical and horizontal movements. These fractures were concentrated mainly to the East of the main eruption (Renokenongo village), around the main vent and to the West (Siring Barat village), with displacements of varying degree and magnitude. The fractures follow the sinistral Watukosek NE - SW trend. Juxtaposed with the Watukosek fault reactivation, is the Siring fault movement that trends NW - SE which has dextral strike slip movement. These fractures were caused by reactivation of faults but their orientation pattern are often not apparent due to thick alluvial cover.

Fig. 21. (A). On June 2, 2008 the dyke on the East side of the main vent broke with an orientation NE-SW. Then on June 8, 2008 the 40 m long dyke collapsed as deep as 6 meters. (B). Fractures on the West Siring village west of the main vent showed an orientation trending North - South. (C)&(D) an active fault is located west of the main vent and trends North - South.

2.10 Gas bubbles

Gas bubbles of various sizes and pressures started to appear two days after the mud eruption. Those that appear from water wells generally have a higher pressure and high methane concentration than bubbles from surface fractures (see figure 22). The ejected materials from these gas bubbles typically had some water, mud with minor sand. A total of over 220 gas bubble locations have been identified since the start of the eruption, however the number that are still active continually decrease. Presently less than 20 gas bubbles are still active, suggesting LUSI is entering a more stable and less active phase.

Gas bubbles are not continuous; they may burst for several weeks or months then stop and reappear elsewhere. Some gas bubbles appear in straight lines that are contiguous with the fault trends. These gas bubbles are mainly concentrated on the West and South of the main eruption which reflect the existence of subsurface gas accumulation breached by deep fractures. The gas accumulation is believed to be a part of the Wunut gas field flanks with its sealing capacity breached by the reactivated Watukosek faults or newly formed fractures as a result of rapid subsidence in the area.

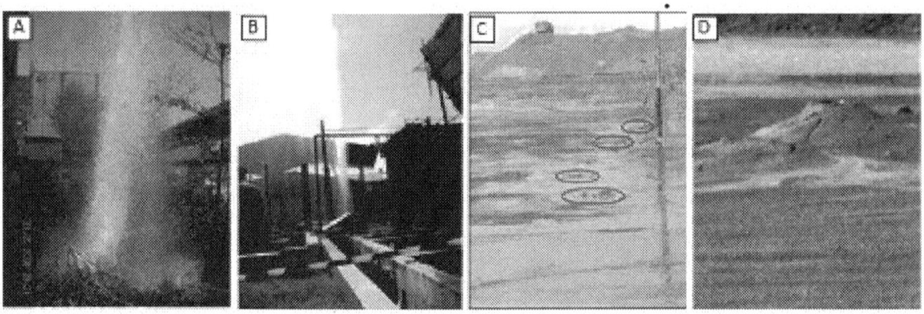

Fig. 22. (A) & (B) The gas bubble originating from water wells, with tremendous pressure and high content of methane gas. Besides removing water,the bubbles also ejected sand, shell fossils and a bit of mud from the swamp sediments. (C) Gas bubbles along the fracture to the west of the main vent, low pressure and in clusters. (D) a Gryphon located approximately 400 m west of the main vent.

Gas bubbles around the mud volcano have formed gryphons of around 30 cm in diameter and height of around 40 cm (see figure 22D). The ejected material was mainly methane gas and some water (see figure 22 A and B).

2.11 Source of mud, water, gas and heat

Mud material ejected from the mud volcanoes is believed to have originated from shale layers known as 'Bluish Gray' clay of the Upper Kalibeng Formation of Plio-Pleistocene in age. The similarity between the mud and the cutting samples from the nearby well Banjarpanji-1 from a depth of 1220 - 1828 meters is based on the following:

1. The similarity of foraminifera and nanno fossil collection, as well as index fossils containing Globorotalia truncatulinoides and Gephyrocapsa spp. that are Pleistocene in age. Benthos Foram collection shows that the sediment was deposited in the marine environment in the inner to middle neritic zones, ranging from shoreline to a depth of 100 meters.
2. Kerogen composition correlates with the side wall core from Banjarpanji -1 at a depth of 1707 m.
3. Thermal maturity based Vitrinite reflectance (Ro) correlates with cuttings and side wall core samples from Banjarpanji-1 at a depth of 1554-1920 meters.
4. Clay mineral composition has similarities with samples from the side wall core from Banjarpanji -1 at a depth of 1615-1828 meters where the illite content in illite-smectite mixture reached 65%.

LUSI muds contain various types of clay including smectite, kaolinite, illite and minor chlorite. It is known that illite minerals form at temperatures between 220 to 320 °C, smectite forms at surface temperatures of up to 180 °C and altered minerals chlorite forms at temperatures between 140 - 340 °C. The XRD analyses carried out on core samples from Banjarpanji-1 imply an intensive progressive smectite-illite transformation with the depth. This suggests that the intersected Upper Kalibeng Formation was exposed to a minimum temperature of 220 °C.

In the initial stages the volume of water in LUSI was very large reaching up to 70% of the total volume of mud with an average salinity of 14,151 ppm NaCl. The lower salinity than sea water suggests dilution. At the time of writing, the liquid composition made up 30% of the total volume. The source of water has been debated by various researchers. Davies et al (2007) states that the water originates from the carbonate Kujung formation, while Mazzini et al (2007) based on geochemical data concluded that the high-pressure water is derived from clay diagenetic dehydration of Upper Kalibeng Formation.

Indonesian Geological Agency, Ministry of Energy and Mineral Resources in 2008 conducted water analysis by using the Oxygen isotope method (σ18O) and Deuterium (σ D) to determine the magmatic origin of LUSI. Results showed deuterium concentration (σD) from -2.7‰ to -13.8 ‰ and Oxygen-18 (σ18O) from +7.59‰ up to +10.11‰, and high Chloride content of 12,000 - 17,000 ppm. Based on the above they concluded that the water source of LUSI is associated with igneous rock sourced from magma (Sutaningsih et al., 2010).

Perhaps it is quite impossible to determine the source of water as it could be a mixture of different sources, ie. clay diagenetic dehydration, carbonates, deeper source linked to geothermal, trapped water due to disequilibrium compaction and mixing with shallow meteoric waters. The importance of determining the source is for hydro-geological purposes, in handling the impact of the mud flow, its effect to the environment and contamination to the ground water. If the fluid is old (Tertiary), it is trapped water, whereas, if the water is young (Quaternary), it is likely to be recharged upslope. For the former, naturally, the eruption will stop after a certain time, whereas, for the later condition, the eruption will never stop (Hutasoit, 2007). However, Sunardi et al., 2007 suggested that LUSI will likely stop when hydrostatic pressure equilibrium is reached.

Groundwater samples from LUSI and its surrounding gas bubbles near the main vent have been chemically analyzed for major anions (Cl^-, HCO_{3-}, and SO_4^{2-}) and cations (Na^+, K^+, Ca^{2+}, and Mg^{2+}). The result shows that there is a significant difference in water chemistry between the main vent and the bubble. The concentration of Cl^-, Na^+, Ca^{2+}, and Mg^{2+} in the main vent water are much higher. This suggests that the water may be from different sources, or both are from the same source, but the gas bubble water has been diluted by shallow groundwater. The second case implies that the pressure in the gas bubble areas may be depleting so that shallow groundwater is mixed with deeper sourced water. If the pressure is still high, then flow from the eruption area will contaminate the shallow groundwater. In either case, the ongoing subsidence is also caused by the decreasing pore pressure as the water is discharged to the surface.

The composition of the erupted gas sampled in July in the proximity of the crater showed CO_2 contents between 9.9% and 11.3%, CH_4 between 83% and 85.4%, and traces of heavier hydrocarbons. In September, the steam collected from the crater showed a CO_2 content up to 74.3% in addition to CH_4. Simultaneously, the gas sampled from a 30.8 °C seep 500 m away from the crater had a lower CO_2 content (18.7%). The four gas samples collected during the September campaign were analysed for $\delta^{13}C$ in CO_2 and CH_4. The $\delta^{13}C$ values for CO_2 and CH_4 vary from − 14.3‰ to − 18.4‰ and from − 48.6‰ to − 51.8‰, respectively (Mazzini et al., 2007). The relatively low $\delta^{13}C_{CH4}$ (− 51.8‰) indicates input from biogenic gas mixed with a thermogenic contribution. The biogenic gas was derived from immature shale layers, probably from the overpressured shale at a depth of 1323-1871 meters, whilst the thermogenic gas was derived from shale layers that are more mature, probably of Eocene age. The CO_2 is postulated to come from the dissolved CO_2 in the water of the shale layer, at temperatures above 100 °C and low pressure. The constant presence of H_2S since the beginning of the eruption could also suggest a contribution of deep gas or, most likely, H_2S previously formed at shallow depth in layers rich in SO_4 and/or methane or organic matter. The rapidly varying composition of the erupted gas indicates a complex system of sources and reactions before and during the eruption (Mazzini, 2007).

Temperatures measured from a mud flow within 20 m of the LUSI crater revealed values as high as 97 °C (Mazzini, 2007, 2009). Given the visible water vapor and steam this suggests temperatures above 100 °C. The heat source of the erupted mud is believed to be from a formation at a depth of over 1.7 km where the temperature is over 100 °C. Geothermal gradients of c. 42 and 39 °C/km have been reported in the area. With such a high temperature gradient, LUSI can be viewed as a geo-pressured low temperature geothermal system that discharged hot liquid mud close to its boiling point the first four years of its life (Hochstein and Sudarman, 2010). Hochstein believed that the high temperature gradients are likely due to the low thermal conductivity of the highly porous, liquid saturated reservoir rocks. Mazzini, on the other hand, believed that the high geothermal gradient is due to the close proximity to Mount Arjuno-Welirang (about 40 km), which is part of the Java volcanic arc that formed since the Plio-Pleistocene (Mazzini, 2007, 2009).

Two shallow ground temperature surveys carried out in 2008 showed anomalously low temperatures at 1 m depth (possibly due to a Joule-Thompson effect of rising gases) and liquid mud temperature that varied between 88 and 110 °C with the highest temperatures occurring after a large, distant earthquake. The mud temperature of mud volcanoes is controlled by the gas flux (endothermic gas depressurizing induces a cooling effect), and by the mud flux (mud is a vector for convective heat transfer) Deville and Guerlais (2009).

2.12 Geomorphology of the area

In general, the geomorphology in Porong and the surrounding area is divided into 5 units: Under the volcanic slopes unit, Foot volcanic plateau unit, Cuesta unit, Alluvial plains unit, and Mud volcano unit. The geomorphological units division is based on morphology, the height difference and slope (Desaunettes, 1977).

2.12.1 Under the volcanic slopes unit

The unit is located at the northern foot of the Penanggungan mountain or in the Proximal facies. This unit is distributed mainly in the southern area of LUSI, adjacent to the mountain range. Lithologic constituents of the unit are generally in the form of volcanic breccia, tuff, lava, tuffaceous breccias, lava and agglomerates and the presence of shallow andesite intrusions in small dimensions. The dominant process in this unit is volcanism. Volcanism processes of Penanggungan Mountain produce volcanic cone morphology. The pattern of distribution in this area is a radial pattern.

2.12.2 Foot volcanic plateau unit

This unit has the morphology of the plains at the foot of Penanggungan mountain or in the medial facies. The unit was formed from the deposition of material surrounding the volcano eruption as laharic. Laharic deposits are found in the form of loose sand and gravel to boulder-sized fragments as products of volcanic eruptions. There is a wide variety of bedding igneous rock fragments to the level of weathering, colors and dimensions.
Lithologic constituents of this unit are fine tuff, sandy tuff, tuff and tuffaceous breccia. The dominant processes in this unit are erosion and sedimentation. The pattern of distribution in this area is a radial pattern.

2.12.3 Cuesta unit

The Cuesta unit is primarily distributed in the southern area of LUSI. The highest point is at an elevation of 150 m at the top of the Watukosek hill. The lowest point is at an elevation of 20 m on the valley of Watukosek. The dominant process in this unit is a tectonic process of faulting, which resulted in shear faults and the down thrown block to the West to form a steep escarpment in the area of Watukosek. This escarpment is known as the Watukosek Escarpment. Lithologic constituents are of andesite breccia, sandstone and tuff. Morphology in the region reflects the existence of Watukosek fault as indicated by the presence of steep slopes on the western escarpment while relatively gentle on the eastern slopes. The pattern of distribution in this unit is trellis pattern.

2.12.4 Alluvial plain unit

Alluvial plains unit make up most of the area and are widely distributed near LUSI. Geomorphological slope is approximately 0 -5%. Lithologic constituents are loose sand deposits, clay, sandy clay.

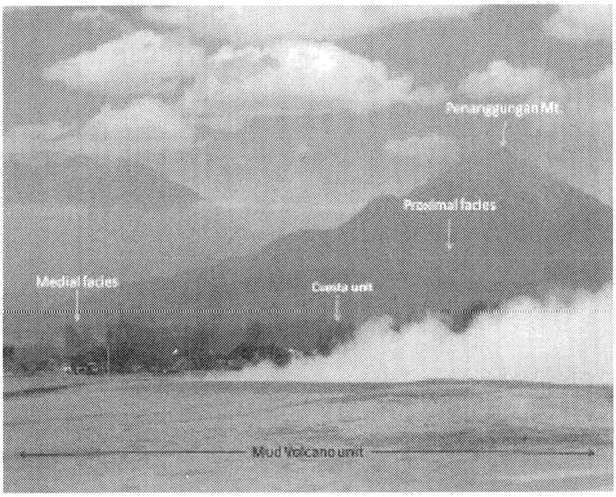

Fig. 23. LUSI area showing the division of volcanic facies. The central facies is located at the top Penanggungan mountain , proximal facies on the upper slopes and medial facies on the foot slope below the mountain. LUSI overlies the alluvial plains which are approximately10 km from Penanggugan mountain.

This geomorphological unit is controlled by alluvial rivers. Geologic processes that act on this unit are erosion, transport and deposition. Lateral erosion took place due to slopes of the mountains to the South causing lateral erosion to be more effective than vertical erosion. In this area there are large rivers namely Porong River which is flowing from West to East that ends up in the Madura Strait. Structural control is clearly visible on the morphology in this area evidenced by the abrupt deflections in the Porong River that follows the fault pattern.

2.12.5 Mud volcano unit

The unit was formed due to discharge of mud from formations below the surface. The morphology is like a low relief hill. The mud volcano Unit is limited by the retaining dykes so that the mud does not spill over into surrounding areas. This unit includes the Village of East Siring, Jatirejo, Tanggulangin Glagaharum, Ketapang and surrounding areas. Lithologically this unit is predominantly the mud itself that contain some fossils.

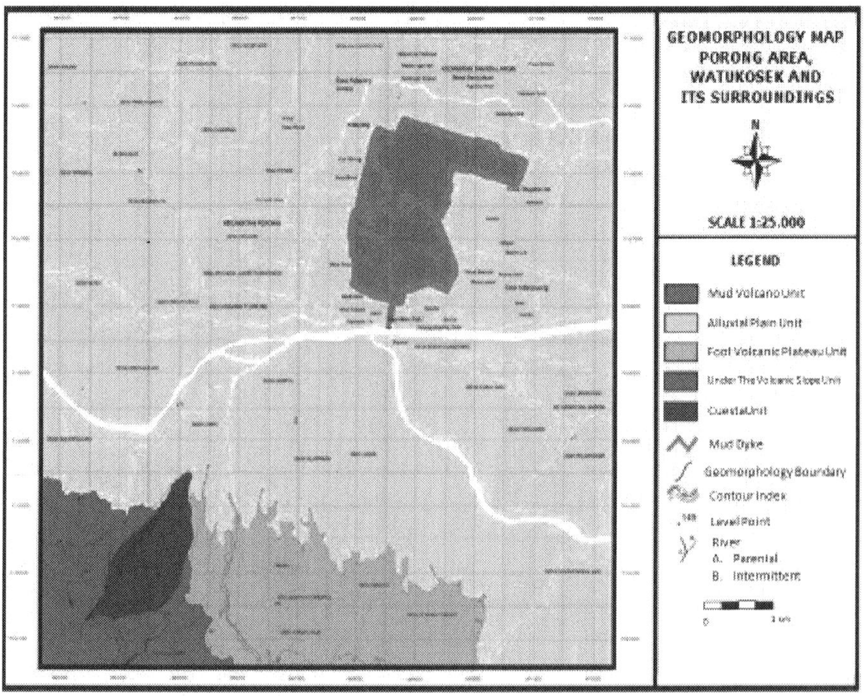

Fig. 24. Geomorphology map of the Watukosek area.

The morphological shape of LUSI is a semi-conical buildup with a peak around the main eruption vent. It is similar with the mud volcano models developed by Kholodov (1983) and Kopf (2002) where LUSI is classified as a swampy mud volcano type. The peak is not high due to the low viscosity of the extruding mud.

2.13 Forming of the Crater

The series of photographs in figure 25 below represent the changes through time at the main vent of LUSI Mud Volcano. In a time span of 5 years, LUSI has evolved from a small eruption of steam, hot water and mud to a destructive high rate mud flow, engulfing houses, schools, factories, neighboring villages and caused a large-scale ground deformation, damaged the highways, railroad, pipelines, electrical power lines and others; to presently a much more calm low rate ejection of mud and fluid and occasional intermittent stopping of steam eruption. LUSI evolved through time from a localized kilometre-scale fault zone in 2006 and expanded through pre-existing NE-SW Watukosek fault zone pathways in 2010.

May 2006
LUSI mud volcano on its first day, May 29th, 2006. The mud eruption is approximately 200 m from the Banjarpanji-1 well location. Initial eruption in the form of mud and hot water and clouds of steam with a discharge rate of less than 5,000 m^3/day.

June 2006
In June, the crater had swelled and the discharge has reached approximately 50,000 m3/day, with water temperatures as high as 97 °C.

August 2006
A semi conical structure is starting to form. The volume of water in LUSI is very large reaching up to 70% of the total volume of mud, shown in the picture as water reflection. The low viscosity of the mud results in mud spreads across, extending to large areas instead of building up vertically.

May 2007
In May 2007, retaining walls/dykes were built to prevent the mud from spilling over to the villages and major roads. The height of the dykes encircling the center of eruption reached approximately 15 m. The average mud flow rate at the time was around 100,000 m3/day.

May 2008
Ring levees/dyke were rebuilt and raised to prevent overflow of mud into the closest villages. Discharge rate was still around 100,000 m_3/day with surface water temperatures remain constant at 97 °C.

February 2009
By February 2009, the ring dyke on the south and north sides are rapidly sinking due to subsidence and are difficult to maintain despite efforts to continuously pile with soil and gravel.

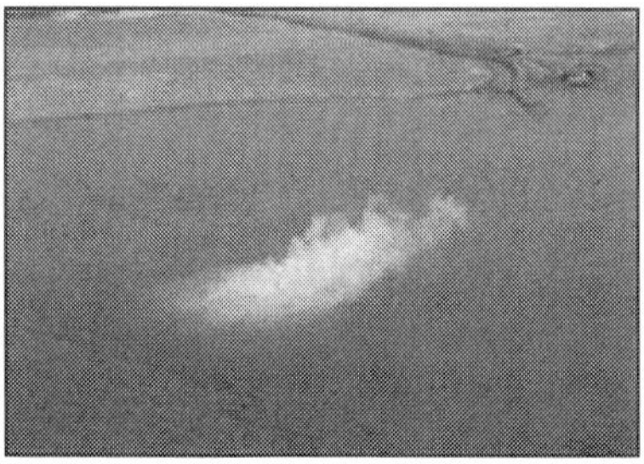

July 2009
The ring dyke around the main vent collapsed and sank in July 2009. The eruption discharge rate at this time is reduced by 60% to approximately 40,000 m^3/day.

January 2010
The diameter of the main crater at this time is approximately 120 m and flowing continuously 30-50,000 m3/day. At times instead of a single crater, it changes to two or three points aligned in the direction of the Watukosek fault. The flow is mainly liquid and hot steam. A gently sloping cone is starting to form. The mud covered area is mostly wet, covering 80% of the total area.

January 2011
At this time the mud flow rates and the scale of steam clouds are reduced. The eruption rate has decreased to less than 10,000 m3/day. LUSI is now entering a new phase, from an eruptive one to a mature and quiescence phase. The mud around the main vent is solidifying forming a dome.

May 2011
The mud volcano viewed from the west side. Note the reduced scale of the clouds of steam.

May 2011
The mud volcano viewed from the north side with the Watukosek escarpment hills and Mt. Penanggungan in the background.

Fig. 25. Changes from time to time at the main vent of LUSI Mud Volcano

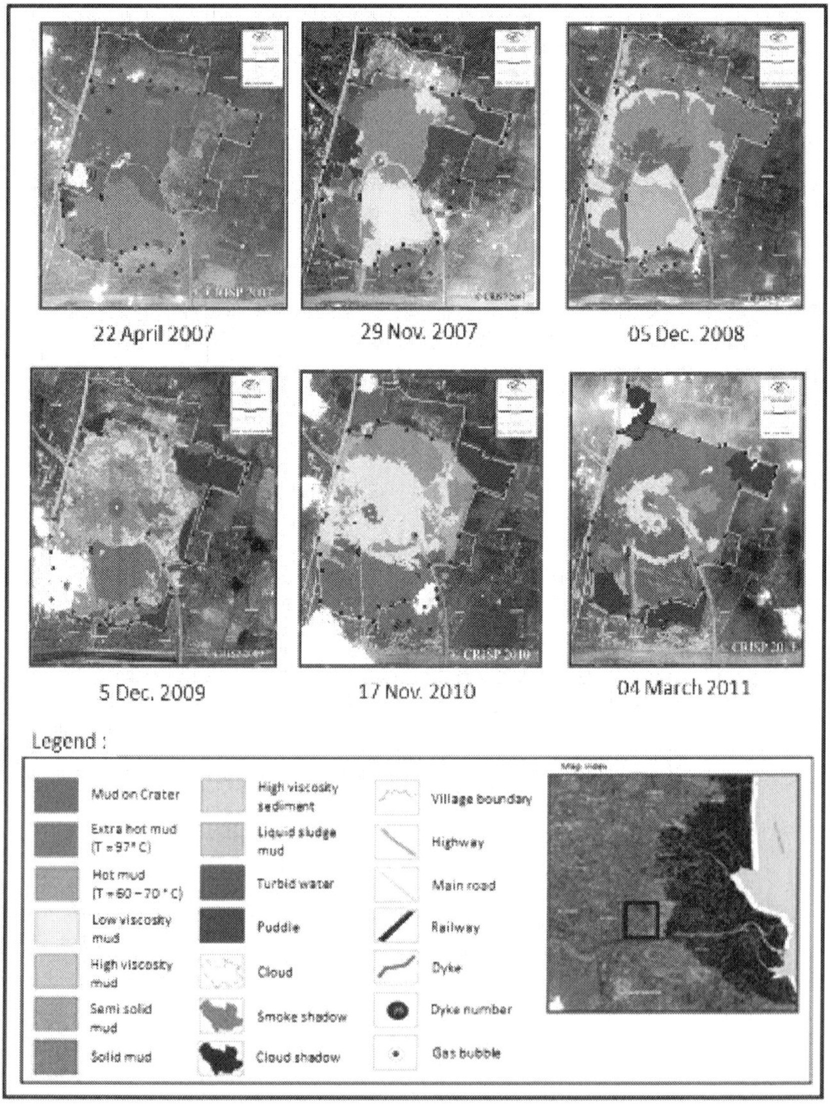

Fig. 26. Map interpretation of the results around the center of LUSI from IKONOS imagery using ERMAPPER Software from 2007 to 2011. The interpretation shows a decrease in the volume of hot mud around the main vent. In 2007, almost all the fluid inside the dyke is above 60°C. By November 2007, the rate of hot mud has begun to diminish. A year later, in December 2008, the LUSI morphology dome has begun to form. The next phase was the reduced production of hot mud in the main vent and the heightened dome coupled with the formation of the patterns of mud and water flow in the vicinity. (source: BPLS 2011)

2.14 Morphological changes

Spatial Aerial Photo Analysis was performed utilizing the CRISP satellite map regularly obtained from the Centre for Remote Imaging, Sensing and Processing, at the National University of Singapore (http://www.crisp.nus.edu.sg). Changes at the center of the eruption and the adjacent slopes can be observed.

The area of observation was between -7°27'04" / 112°40'27" and -7°35'52" / 112°49'35" an area of about 3.7 km x 4.0 km or 14.9 km^2, with the focus of the coverage area on the mud ponds.

The data processing stages were as follows:

1. Processed multitemporal IKONOS image data obtained from the CRISP in 2007 until 2010. ERMAPPER software was used for image enhancement, image correction, interpretation and image classification.
2. Spatial or geographical information analysis, information visualization, organization of information, combining thematic information using GIS software.
3. Classification of information based on the detection and identification of objects on the surface of the earth from a satellite image, to the next as the primary identifier elements and limits of an object is done by coloring.

3. Geohazard

The mud eruption in Sidoarjo has buried houses, villages, schools, factories, and displaced thousands of people and continues to pose a geohazard risks in a densely populated area with many activities and infrastructures. Studies of other mud volcanoes in East Java were used in the geohazard assessment.

Fig. 27. Dramatical sinking of a village gate of Siring Timur, located to the west of the main vent. The gate and the rubble was half buried but still visible in the 23 November 2007 photograph. Two months later just a part of the tile roof and walls remain visible on 20 January 2008.

LUSI initially had five mud eruption vents, but only one remains active. There is a possibility that inactive mud eruption vents may reactivate or new ones will emerge in other locations. The study suggests a possibility of mud erupting at one or more of the known gas bubble locations or at a new location along zones of weakness on reactivated pre-existing faults or on new fault zones. Methane gas bubbles have been identified in more than 220 locations, and are generally associated with fractures. Some are more active than others while some have died. In most cases, the methane is non-flammable because it is in such low concentrations from rapid dispersion in the air. However where bubbles are confined, the concentration of methane is high enough to burn. The gas leaks from these fractures suggest breach of seal and loss of sealing capacity of faults and the impervious shale overlying geological structures, in particular the flanks of Wunut anticline that contain gas accumulations.

The occurrence of gas bubbles also suggests that subsidence is not merely a shallow near-surface phenomenon as a result of surface loading by the weight of the mud or soft soil layer compaction, but instead also affecting deep horizons as the gas comes from a deep source. Gas chromatography of sampled gas bubbles near the main eruption vent in July 2006 indicates it primarily consists of methane but the presence of some heavier gases in small quantities including ethane, propane, butane, and pentane suggest a deep thermogenic origin and long extended fractures.

Fig. 28. (A)&(B). Photos before and after the collapse of levees west of the main vent. The levee has decreased in height by approximately 1 meter along a 150 m interval. (C)&(D). Photos before and after the collapse of the levee north of the main vent. The levee has decreased in height by 1.5 m along a 200 m interval.

Fig. 29. A mosque located in the Renokenongo village, to the east main of the vent is damaged due to fault reactivation. At this location a lot of fractures were oriented NE-SW, and these are patterns of the Watukosek fault syatem.

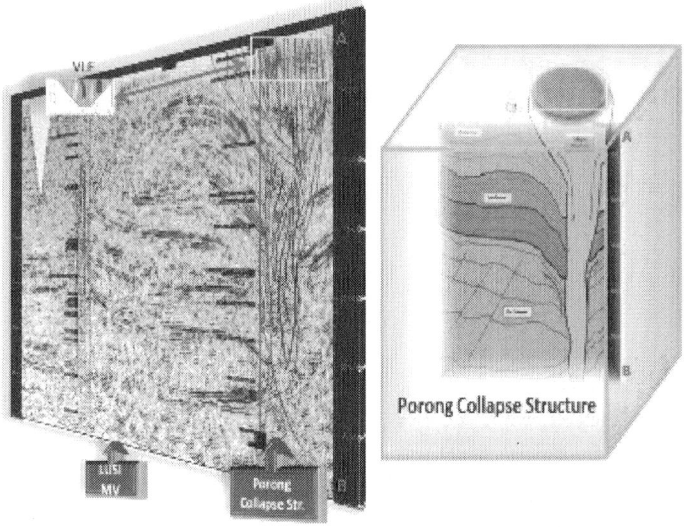

Fig. 30. The Porong paleo collapse structure was used as an analogy to predict the deformation pattern that will occur in LUSI. VLF measurement results indicate the existence of patterns of subsidence on the surface which is indicative of the onset of collapse

In terms of geohazard risks, the evidence and areas of concern include i) rupture of gas and water pipelines (shear and subsidence); ii) railroad bending (shear/faulting and subsidence); iii) road cracks (subsidence); iv) relief wells casing integrity (subsidence and shear); v) dyke collapse (subsidence); v) gas bubbles which appear along fractures and zones of weakness (shear/faulting and subsidence). (Istadi et al. 2009).

LUSI's continuing subsidence forms a depression bowl or funnel shaped structure. The subsidence forms an accommodation space, a natural basin to contain the mud. However, the high water content of the mud means it has a low viscosity and therefore cannot accumulate vertically to form a high and steep mountain-like structure. The mud, in particular the separated water tends to spread sideways which increases pressure on the mud retaining dykes that collapsed on a number of occasions and caused flooding. If the mud eruption continues with a high rate, then the potential flood prone areas will expand. The accumulated water at the peripheries away from the main eruption, which are held by the retaining dykes exert increasing hydrostatic pressure on the dyke walls. This increases the risk of retaining dyke failures.

Attempts to stop the mud flow should be implemented only after the likely causes of LUSI are studied and explained. The other alternative is to let nature find its own equilibrium and take care of itself, as LUSI mud volcano appears to be the result of a natural phenomena.

4. Other Mud volcanoes in East Java

Mud volcanoes are common in the northern part of Java and Madura Island (Satyana, 2008). Like elsewhere in the world, mud volcanoes in Java and Madura typically are located at the top of anticlines or along faults in the area. This phenomenon is demonstrated by the Sangiran mud volcano which is located at the top of the truncated dome on an up-thrown fault block, while the Bleduk Kuwu mud volcano is located on the top of the Purwodadi anticline, The Api Kayangan (means Fire of Heaven, or Eternal Fire) mud volcano is at the top of the Bojonegoro anticline. The Pengangson mud volcano located at the top of the Kedungwaru anticline, while the Pulungan and Kalang Anyar mud volcanoes are on the top of the Pulungan anticline. The Gunung Anyar (means New mountain) mud volcano is at the top of the Guyangan anticline, Bujel Tasek (Madura) and finally LUSI erupted on the extension of the Sekarputih anticlinal structure. Most of the mud volcanoes in East Java, with the exception of LUSI, are in the relative quiescence period and some can be considered in the dormant period with minimal activity. The existence of these mud volcanoes are described in the latter part of the paper

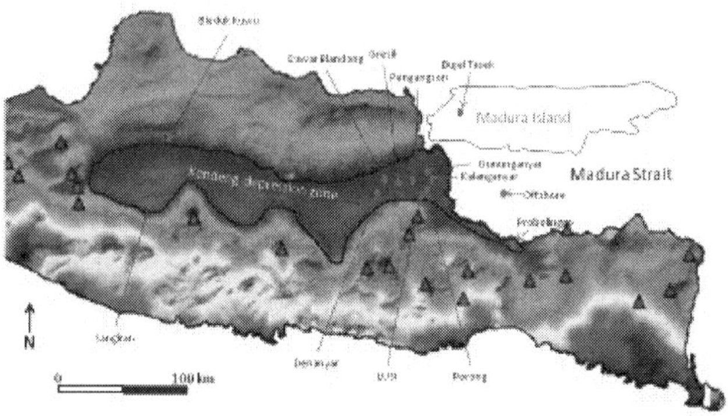

Fig. 31. Gravity map of East Java showing East Java Basin's depositional centers (blue) in the Kendeng depression zone. Red dots are the identified mud volcanoes, while the triangles are magmatic volcano locations.

The presence of mud volcanoes in the northern East Java Basin, especially along the Kendeng depression zone, is a common phenomenon. Their presence reflect the overpressure condition due to the very rapid deposition of the bluish-green mudstone and marlstone of the Sonde Formation during the Pliocene in a back arc basin setting that is folded and faulted (Dickinson, 1974).
In Central and East Java mud volcanoes are found within the Kendeng depression zone, except Bujel Tasek which appears in Rembang zone. The Kalang Anyar, Gunung Anyar, Pulungan, Bujel Tasek (Madura) and LUSI are in a straight line trending NE-SW contiguous with the regional fault trend, originating from the crater of Mt Penanggungan of the Arjuno-Welirang volcanic complex, following the Watukosek fault escarpment in the southern mountain ranges northward to Bujel Tasek in Madura island (see figure 32).

4.1 Kalanganyar

Kalanganyar mud volcano is located approximately 3 km south of Juanda Airport and approximately 15 km North-East of LUSI. The phenomenon of mud volcanoes in this area has been identified and mapped on the 1936 geological map of the Dutch era, suggesting it was formed long before the 1936 Duyfjes map. A temple known as Candi Tawangalun Majapahit (approximately 500 years ago) situated on the northern edge of the mud volcano suggest the significance of the mud volcano in the Majapahit era (see figure 33). Interestingly, some parts of the temple's material were made from material products of the Kalanganyar mud volcano.
The morphology of the Kalanganyar mud volcano forms a low relief hill overlying the alluvial plain deposit. The dimension of the main eruption cone is around 12x18 m. Some activities are still evident such as gas bubbles and some fresh mud around the main eruption vent (see figure 34) with a temperature of ± 38 °C. The unit is composed of silt-sized material and grains of fine sand and clay and saltwater that forms salt deposits.
The low relief of the mud volcano suggests that the mud has very low viscosity. The mud shrinks during dry season to form dessicated mud crack structures that are commonly found in the area. The mud material is derived from older rocks than the surrounding alluvial plains, correlatable with the mud at LUSI, the Upper Kalibeng Formation of Plio-Pleistocene age.
In the area around the active gas bubbles, rock fragments, such as siderite and salt deposits, are always found. The gas bursts are typically mild consisting of mixed gas, and formation fluids mainly connate water. Microbial activites are also found near the gas bubbles.
Mud breccias that are ejected from the Kalanganyar mud volcano are in the form of mud supported mudstone fragments which are light brown in color, composed of abundant carbonate mud, and quartz with opaque minerals in small quantities. Rudstone (Embry and Klovan, 1971) which is gray - brownish gray, grain supported, with fragment components consisting of shells of molluscs (Gastropoda and Pelecypoda, with a dominance of Ostrea shells) measuring 10-20 cm are abundant (> 10%) and bound by the matrix and carbonate cement (Figure 35 A&B). Balanus fossils were largely found freed from the carbonate (Figure 35 C) although some were still found to be bound by carbonate cement.
The mud eruption carried younger sediments and boulders which contain mud breccias and limestones. The stratigraphic position of the mud breccia based on Balanus fossil found on limestone within the mud breccia, indicate that this was sourced from the Sonde Formation of Pliocene age. The existence of mollusc and balanus contained in rudstone limestones suggest deposition in a shallow marine environment to the coastal littoral zone with strong

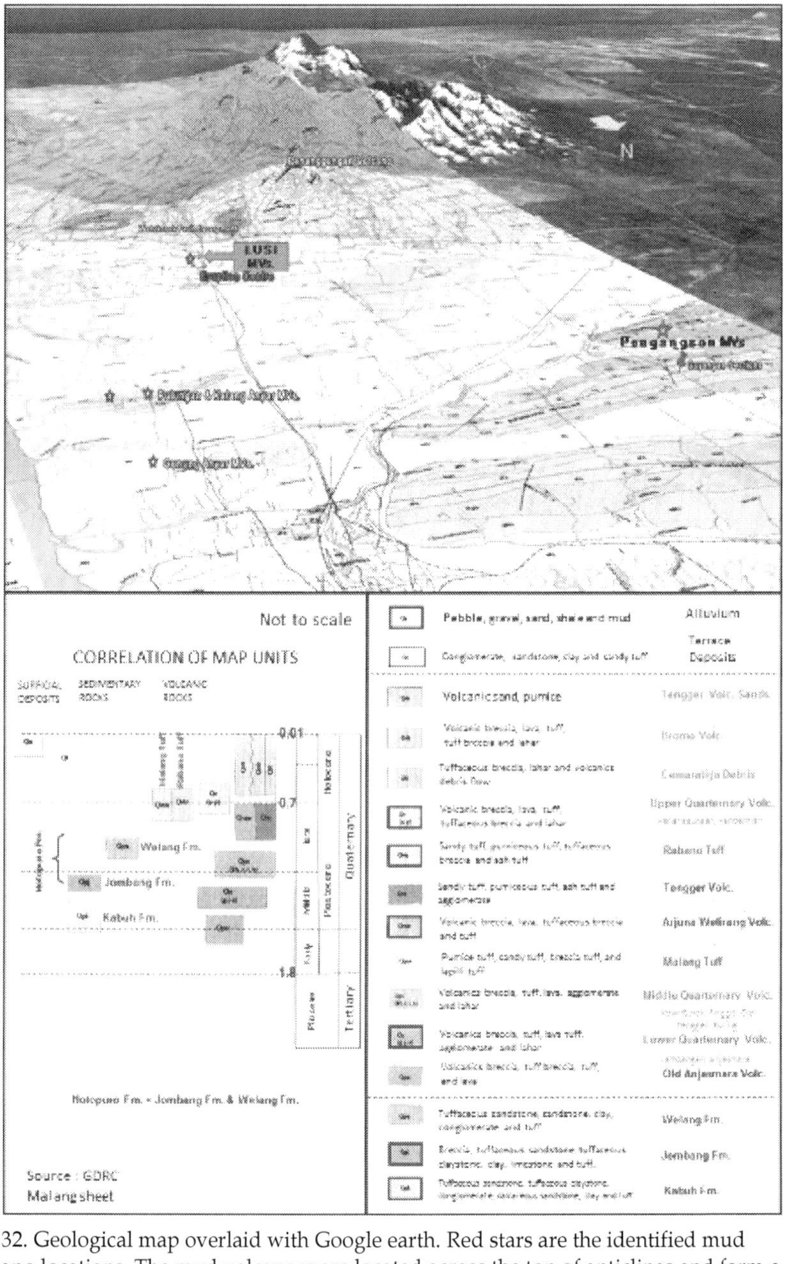

Fig. 32. Geological map overlaid with Google earth. Red stars are the identified mud volcano locations. The mud volcanoes are located across the top of anticlines and form a lineament.

Fig. 33. Tawangalun temple was built during the Hindu kingdom (500 years ago) situated in the Northern edge of Kalanganyar mud volcano. Some materials of the temple were made from fragments from the Kalanganyar mud volcano.

Fig. 34. Kalang Anyar mud volcano in Kalanganyar village, Sidoarjo, East Java

energy currents. The dominant grain supported matrix of rudstone further suggests a high energy depositional environment. A shallow marine environment is located on the continental shelf with the fore reef sea conditions that are less affected by the supply of silisiclastics. The existence of mudstone with mud supported textures indicate deposition below wave base conditions on the back reef, rocks consequently do not experience the washing process (winnowing) by wave activity. The older mud volcano material is thought to have been sourced from the deeper Upper Miocene Kalibeng Formation, suggesting a regressive sequence in this part of the Kendeng zone of the East Java Basin.

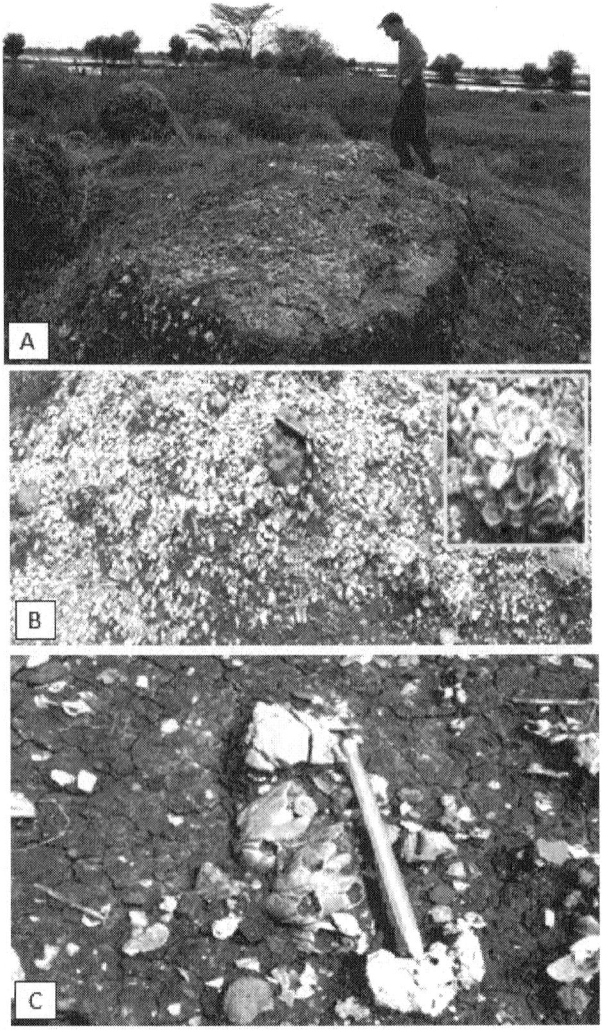

Fig. 35. (A)&(B) Sandstone with mollusc shells dominated by Ostrea , (C). Balanus fossils among carbonates and siderite.

4.2 Gunung Anyar

The Gunung Anyar mud volcano (means "new mountain") is located 8 km to the west-northwest of Kalang Anyar mud volcano and is surrounded by densely populated residential area in the Gununganyar village, Surabaya. The morphology of the Gunung Anyar mud volcano is a northeast orientation and elongated hill-shaped geometry on the

surrounding flat alluvial plains. The dimension of the still active main eruption vent is approximately 8x9m. The ejected material is composed of silt-sized grains of predominantly fine sand and saltwater. The temperature of the vent is ± 37.2 °C. More solid content than fluid has erupted from this mud volcano.
The lithology is similar to Kalang Anyar mud volcano where the composition is predominantly silt with the physical characteristics of brownish-grey, fine sand-sized to clay. In dry conditions, the structure shrinks, typically forming desiccated mud cracks. Seepage of crude oil of black color is also visible among the small bubbles that are still active.

Fig. 36. Gunung Anyar Mud Volcano morphology at Gunung Anyar Village in the Southern part of Surabaya, East Java.

Based on rocks and fragments carried by the mud eruption, the erupted materials were sourced from the following stratigraphic layers:

a. Marl

Brown marl with some weathered surfaces, white in color, with clay-sized fossil planktonic foraminifera and sand-sized bentonic fragments. Fresh outcrops of limestone fragments that appear to be newly ejected by the mud volcano. Stratigraphic position of the marl based on the physical properties and fossil content, indicate that this was derived from the Kalibeng and Sonde Formations.

b. Limestone

The limestone contains balanus fossils bounded by carbonate cement. The limestone based on fossil is thought to have been sourced from the Sonde Formation. The presence of balanus fossils typically suggest the coastal litoral zone depositional environment with the strong currents, while the texture of the limestone, suggest the low - moderate flow of energy environment with warm, calm and shallow waters possibly positioned on the back reef.

c. Calcareous sandstone

Weathered brown sandstone, poorly sorted, sub-rounded, fine sand-sized, composed of quartz, feldspar, biotite, and calcite. Black mudstone associated with these calcareous sandstones is also found in small quantities. The sandstone may have been sourced from the Pucangan Formation, Pleistocene in age.

d. Molluscs sandstone (grenzbank)

Sandstone containing molluscs of freshwater and seawater were found in this location. The freshwater molluscs which are characterized by the thin shell bi-valves are more dominant than the seawater molluscs. The presence of mixed seawater and freshwater molluscs suggest shallow to transitional marine deposition environment.

Fig. 37. Crude oil seepage coming out with salty connate water along with bubbles in Gunung Anyar Village.

e. Silt

Silt makes up most of the mud volcanic area. It is composed of silt-sized grains of predominantly silisiclastic brownish-grey colored clay materials and the salt water. The temperature is around 37 °C at the mud conduit. The mud contains more solid silt and clay materials than fluid. Silts were derived from underlying older rocks of possibly Upper Kalibeng Formation of late Miocene. The presence of rock fragments, siderite, and seepage of black crude oil are almost always found within the vicinity of active bubbles.

4.3 Pengangson

The location of Pengangson mud volcano is in the village of Kepuhklagen, Wringinanom, Gresik Regency. Compared to other mud volcanoes in East Java, the Pengangson mud volcano is the most ideal example of a mud volcano. The younger sediments outcropped and exposed at nearby excavated cliffs to the west of the mud volcano, and older rocks are found as fragments or clast carried by the mud volcano eruption. The geological structural components at this site are also clearly visible, such as folds, fractures, fault lines and sedimentary structures.

The morphology is formed of low hills between the alluvial plain surrounding. The unit is composed of silt sediment material with dominant clay-sized grains. The low temperature mud of ± 39.5 °C is typical of other mud volcanoes in other parts of East Java. The mud is thick and the liquid consists of a mix of formation fluid, crude oil seeps, salt deposits and gas bubbles.

Fig. 38. Pengangson Mud Volcano in Kepuhklagen village, Gresik, East Java

The materials ejected by the mud volcano are the following:

a. Sandy Mudstone - Sandy Siltstone

Rock outcrops are generally brownish-grey in fresh and partially weathered condition. Physical characteristics of the rock suggest that it is from the Sonde Formation. Based on the deposition enviroment, lithology, sedimentary structure, texture, mineralogical composition, and fossils suggest that this unit was deposited in a middle shelf environment - lower delta plain.

b. Tuffaceous Sandstone unit, Sonde Formation

The naming of the sandstone unit is based on the condition of the rock outcrop on the cliffs with tuffaceous sandstone lithology that shows the layered structure and planar crossbed. Rock samples in this unit do not contain foraminifera because of the dominance of volcanic material in the rocks.
The composition of the constituent material of plagioclase and the presence of volcanic glass which tends to be wacke, suggests that the source is not far from the sedimentation basin. The similarity of physical properties, and texture of rocks suggest this unit is part of the Sonde Formation.

c. Sandstone unit, Pucangan Formation

This unit is characterized by the presence of sedimentary structures such as grading, parallel lamination and slump structures. Outcrops of fresh rocks generally show somewhat weathered condition. The bottom of the unit is dominated by volcanic sandstone that contains calcareous mudstone layers. This unit is part of an eroded top of anticline.
Interpretation of the environment based on the lithology data, sedimentary structure, texture, mineralogical composition, and fossils indicates that this unit was deposited on the inner shelf environment - lower delta plain with traction and suspension flow mechanism. The sequence of lithologies indicates deposition in increasingly shallow conditions with the initial deposition on the inner shelf. The similarity of physical properties, texture, structure, age and environment of deposition of rocks suggests that this unit is part of the Pucangan Formation.

d. Tuffaceous Mudstone

The outcrop of rocks is generally fresh or slightly weathered. Overall this unit is dominated by massive mudstone and tuff. The bedding trends west - east and slopes to the north and south. Interpretation of the environment of deposition based on the lithology, texture, and mineralogical composition is that this unit was deposited in a braided stream environment or sub flood plain and stream sediment was transported through the mechanism of suspension. The similarity of physical properties and the texture of rocks suggests that this unit is part of the Kabuh Formation.

e. Volcanic sandstone

This unit is characterized by the presence of sedimentary structures such as grading, cross-lamination and parallel lamination. The outcrop of rocks generally show a somewhat weathered condition but rock structure is still visible. Interpretation of the environment of deposition based on the lithology, sedimentary structure, texture, and mineralogical composition suggests that this unit was deposited in a braided stream environment (minor channel) with traction flow mechanism. The similarity of physical properties, texture and structure of rocks suggests that this unit is part of the Kabuh Formation.

f. Silt unit

The naming of this rock unit is based upon the existence of a silt dominated mud volcano. The morphology of mud volcanoes forms a low hill between the alluvial plains. The unit is composed of silt sediment material dominant by clay-sized grains with a temperature 39.5 ° C. From the main vent fluid, gas bubbles and salty water are released.

Fig. 39. Crude oil seepage, brownish-black in color together with mud and gas that comes out of a gryphon.

In the vicinity, bubble-shaped Gryphons, siderite, and salt deposits are commonly found (see figure 39). Rock fragments are found in a limited number that consist of calcarenite (Sandy micrite), calcareous sandstone, calcareous mudstone (Micritic mudrock in the Mount, 1985) and sandstones with molluscs.
This mud deposit in Wringinanom contains foraminifera planktonic fossils suggesting a middle Pliocene age to late Pliocene (N20-N21) mud source, while the content of bentonic neritic foraminifera suggests the bathymetry position in the middle neritic.

4.4 Bujel Tasek

Bujel Tasek mud volcano is found in the Katol Barat village, Bangkalan district of Madura island. The morphology of the Bujel Tasek mud volcano is very different from other mud volcanoes in East Java. The shape is a cone edifice with a height of approximately 12 m with a diameter of approximately 5 m. The material that came out is a mixture of viscous mud, water and gas with a highly viscous mud. This cone shape is actually a giant gryphon formed by the high viscosity mud of a larger mud volcano that covers the area. Nearby this conic structure is a mud lake that represents an ancient mud caldera with mud conduits that are no longer active (see figure 40).

Fig. 40. Bujel Tasek mud volcano forms a cone morphology in the village of Katol Barat, Bangkalan, Madura island, East Java.

The mud sediment that came from the Lidah formation is characterized by its brownish-grey, fine clay. The mud breccia found is gravel to boulder sized, very abundant, with fragments ejected in the form of mudstone, calcareous sandstone, siderite and calcite.
The reddish-brown and gravel- pebble sized siderite mineral is found exposed. It is widely distributed, from the mud volcano to the valley. In addition, calcite is trapped in sandstones that fills the pores of the wood that look like silisified wood.

4.5 Bleduk Kuwu

This mud volcano is located in the Village Kuwu, Kradenan, Grobogan district, approximately 20 km south of Purwodadi in Central Java. The object of interest in Bleduk is the mud flow containing gas and salty water that takes place almost continuously in an area with a diameter of approximately 650 m (see figure 41). Etymologically, the name comes from Kuwu Bleduk. In the Javanese language 'Bleduk' means 'blast/burst' and 'kuwu' is derived from the word 'kuwur' which means 'run/scramble'.

Fig. 41. Bleduk Kuwu mud volcano during its almost continuous eruption. The gas is flammable and sometimes self-ignites. The expelled water is commercially used to extract salt. Eruptions generally occur four or five times a minute, as a burst of warm mud and gas.

The Kuwu mud volcano cluster covers about 45 hectares. The biggest vent can erupt materials as high as 5 meters with expelled mud in a diameter of about 9 meters. At the main Kuwu site the mud volcano usually erupts four or five times a minute consisting of mud accompanied by the release of gas and water (sometimes oil). Often the eruptions are accompanied by an explosion as the gas self-ignites. The temperature of the mud ranges from 28-30°C, while the smaller mud volcano is slightly cooler.

Bleduk Kuwu is surrounded by other mud volcanoes within a radius of approximately 1-2 km to the southwest, northeast and south with varying dimensional extents. To the southwest is Cangkring Bleduk mud volcano that occupies a larger area than the Bleduk Kuwu, while to the south is the Bleduk Banjarsari mud volcano, and to the East is Bleduk Crewek, and to the northeast is Medang Kamolan (figure 42). Other minor mud volcanoes in the area include Bledug Kesongo and Bledug Kropak. Geologically, these mud volcanoes are located at the boundary between North Serayu and Kendeng Depressions. Seismic sections across these mud volcanoes show disturbed zones from the top of the Kujung Formation, the top of Wonocolo Formation to the surface. The Bledug Kuwu disturbed zone is a chaotic mixture of upward convex and concave reflectors. Bledug Kesongo is characterized by a collapsed structure with upward concave horizons along the disturbed zone indicating a subsidence. The lower part of Late Miocene Wonocolo shales is believed to be the source of mud based on its fossil content. Seismic sections, however, show that the source of mud may also come from Early Miocene Tuban shales. Some diapirs also occur in this area and they are generally below the top of Wonocolo Formation. Folds in this area are considered to form diapirs as suggested by some seismic sections (Satyana and Asnidar, 2008).

Fig. 42. Medang Kamolan mud volcano, located approximately 3 km northeast of Bleduk Kuwu mud volcano.

4.6 Offshore mud volcanoes

The existence of submarine mud volcanoes and mud diapirs in the Madura Strait, offshore areas of East Java is visible in the seismic profiles. The cross sectional appearance looks like an upwards dipping strata around a venting system of seafloor-piercing shale diapir cutting the overlying sediment and forming a conic volcanic edifice (see figure 43).

The Madura Strait is an offshore extension of the Kendeng Depression. Thick Pliocene to Pleistocene sediments were deposited rapidly and compressed elisional system in the Madura Strait depression. Deepwater sedimentation is still taking place in this portion of the Kendeng zone, and it has not been uplifted. In the Madura Strait area, east-west trending left lateral wrench faulting triggered mobilization of Miocene basinal shales during the Plio-Pleistocene, resulting in a series of shale diapirs. Further south, the impact of on-going subduction along the Java Trench becomes increasingly significant and structures are dominated by north-directed thrusting, which may be independent of basement faulting (Satyana and Asnidar, 2008).

On the basis of structural style and the tectonic events, Widjonarko (1990) divided the Madura Strait block into five structural domains: wrench domain, slide domain, western basinal domain, eastern basinal domain, and southeastern fault block domain. Wrench and slide domains bound the Madura Strait to the Madura-Kangean High in the north. Southeastern fault block becomes the southern border of offshore Madura Strait. The main parts of the Madura Strait where mud diapirs and volcanoes exist are composed by western and eastern basinal domains. The Madura Strait Depression or Sub-Basin is one of the two deepest and thickest basins in Indonesia. In western basinal domain, very rapid sedimentation since the Late Miocene time resulted in the development of more than 3000 meters of Plio-Peistocene section. Eastern basinal domain is similar to western domain, the only difference is that the eastern basinal domain began to subside in the late Oligocene - early Miocene, much earlier than the western domain (Satyana and Asnidar, 2008).

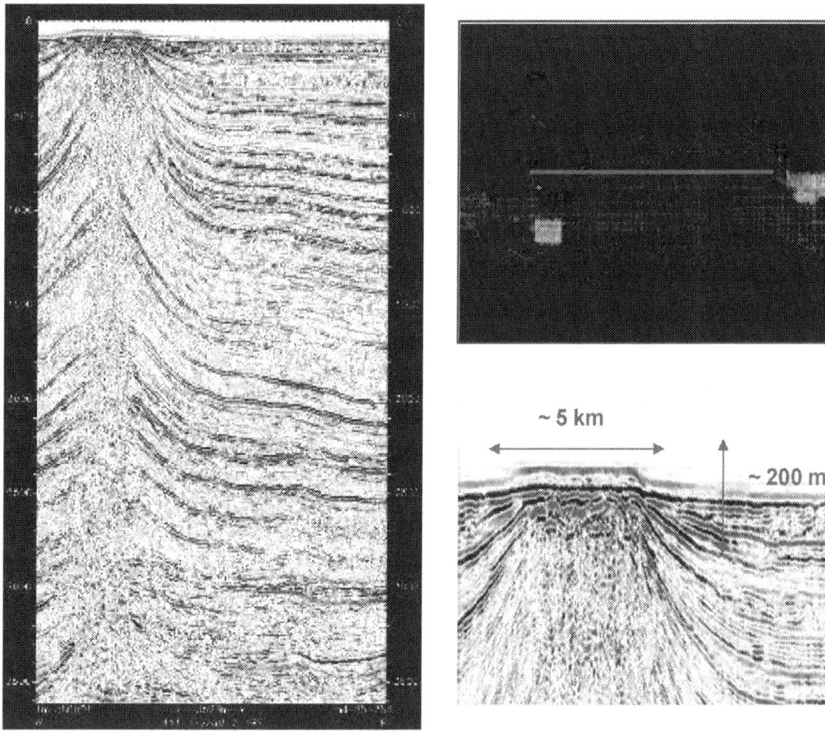

Fig. 43. Water Depth ~40 - 50 m in the offshore Madura Strait

Stratigraphy of the Madura Strait started in Middle Eocene time by deposition of transgressive clastics unconformably on top of pre-Tertiary basement. The deposition was terminated by a local uplift at the end of Eocene time. Subsidence during the Oligocene resulted in deposition of deep marine sediments. An uplift at the end of the Oligocene resulted in a regional unconformity throughout the basin. During the Early Miocene time the rapid subsidence resulted in deposition of deep marine sediments in the area. In the mid-Late Miocene time, the basin was filled and another uplift took place. After a short subsidence to the end of Late Miocene, sedimentation interrupted again by an uplift in Early Pliocene time. The rapid subsidence in the late Pliocene time is characterized by the deposition of overpressured thick clays. The area subsided again into a shallow marine environment after the Plio-Pleistocene regional uplift (Widjonarko, 1990; Satyana and Asnidar, 2008).

5. Conclusions

- Mud diapir and mud volcano are piercement structures showing the release of overpressured sediments piercing upward from subsurface to the Earth's surface due to buoyancy and differential pressure.

- The Kendeng-Madura Strait Zone is an ancient axial depression of Java to Madura Islands with elisional basin characteristics. The Mio-Pliocene and Pleistocene sediments were rapidly deposited into the depression and compressed as it was at the front of converging plate boundaries with high seismic activity. This resulted in numerous mud diapirs and mud volcanoes in the area. Sixteen mud volcanoes have so far been documented in East Jáva with the six mud volcanoes found along the Watukosek fault.
- LUSI, a new mud volcano, was born at the vicinity of the Watukosek fault. This geological phenomena as well as others occurring in the area such as the appearance of gas bubbles, cracks, subsidence, and vertical and horizontal displacement is believed to be due to a reactivation of existing faults in this area. The Watukosek fault system appeared to play a role in the existence of other six mud volcanoes at its vicinity.
- Early technical papers, such as Davies et al. (2007, 2008), Rubiandini et al. (2008) and Tingay et al. (2008) suggested a connection between the Banjarpanji-1 well and the mud volcano. These papers were based much on unverified and partial dataset. When the full dataset is integrated as in Sawolo et al. (2008, 2009 and 2010), it is evident that the well did not trigger LUSI mud volcano. Future analysis on the trigger of the mud volcano must consider this pitfall and integrate all available dataset. One must decipher and make use of the entire mud logger Real Time Data as it is the most reliable dataset, being automated, continuous and quantitative data that captures key operating parameters of the rig.
- The distance between the nearest volcanic complex, the Arjuno - Welirang volcanoes, to LUSI is about 10 km. This close proximity may have influenced the geothermal properties of LUSI and affected its temperature and geochemistry.
- Studies suggest that LUSI is likely to continue to flow for many years to come. Better understanding of its plumbing system and detailed subsurface studies must be conducted as part of the hazard mitigation effort. The more than 16 East Java mud volcanoes in the vicinity must be used as an analogy of mud volcano processes and its morphology.

6. Acknowledgement

The Authors wishes to express appreciation to the management of MIGAS, BPMIGAS, EMP, Lapindo Brantas Inc for the permission to publish the paper. Constructive discussions and inputs from our colleagues in particular Peter Adam and Awang H. Satyana are also appreciated.

7. References

Abidin, H.Z., Davies, R.J., Kusuma, M.A., Andreas, H., Deguchi, T., 2008., Subsidence and uplift of Sidoarjo (East Java) due to the eruption of the LUSI mud volcano (2006 present). Environmental Geology doi:10.1007/s00254-008-1363-4.

Akhmanov, G.G. and Mazzini, A., 2007, Mud volcanism in elisional basin, In: Proceedings of the International Geological Workshop on Sidoarjo Mud Volcano, Jakarta, IAGI-BPPT-LIPI, February 20–21, 2007. Indonesia Agency for the Assessment and Application of Technology, Jakarta.

Amelung, F., Galloway, D.L., Bell, J.W., Zebker, H.A., Laczniak, R.J., 1999. Sensing the ups and downs of Las Vegas: InSAR reveals structural control of land subsidence and aquifer-system deformation. Geology 27 (6), 483–486.

Davies, R.J., Brumm, M., Manga, M., Rubiandini, R., Swarbrick, R., Tingay, M., 2008. The East Java mud volcano (2006 to present): an earthquake or drilling trigger? Earth and Planetary Science Letters 272 (3–4), 627–638.

Davies, R.J., Swarbrick, R.E., Evans, R.J. and Huuse, M., 2007, Birth of a mud volcano: East Java, 29 May 2006: GSA Today, v. 17, p. 4-9.

De Genevraye, P., Samuel, L., 1972. Geology of the Kendeng zone (Central and East Java). In: Proceedings of the Indonesian Petroleum Association, 1st Annual Convention, pp. 17–30.

Deguchi, T., Maruyama, Y., Kato, M., Kobayashi, C., 2007a. Surface displacement around mud volcano, East Java captured by Insar using Palsar data. In: Proceedings of the 28th Asian Conference on Remote Sensing.

Deguchi, T., Maruyama, Y., Kato, M., 2007b. Measurement of long-term deformation by InSAR using ALOS/PALSAR data. In: Presented at FRINGE 2007 workshop, November 2007, Frascati, Italy. Workshop presentation available from: http://earth.esa.int/workshops/fringe07/presentations.html
Workshop proceedings are in press.

Desaunettes, J.R., 1977, *Catalogue of Landsform for Indonesia,* Trust Fund of The Goverment of Indonesia - FAO, Bogor.

Dimitrov, L.I., 2002. Mud volcanoes: the most important pathway for degassing deeply buried sediments. Earth Science Reviews 59 (1–4), 49–76.

Duyfjes, J., 1936, The geology and stratigraphy of the Kendeng area between Trinil and Surabaya (Java) (in German), De Mijningenieur, vol. 3, no. 8, August 1936, p. 136-149.

Embry, AF, and Klovan, JE, 1971, A Late Devonian reef tract on Northeastern Banks Island, NWT: Canadian Petroleum Geology Bulletin, v. 19, p. 730-781.

Etiope, G., Feyzullayev, A., Milkov, A.V., Waseda, A., Mizobe, K., Sun, C.H., 2009, Evidence of subsurface anaerobic biodegradation of hydrocarbons and potential secondary methanogenesis in terrestrial mud volcanoes, Marine and Petroleum Geology 26 (2009) pp. 1692–1703

Fielding, E.J., Blom, R.G., Goldstein, R.M., 1998. Rapid subsidence over oil fields measured by SAR interferometry. Geophysical Research Letters 25 (17), 3215–3218.

Fukushima, Y., Cayol, V., Durand, P., 2005. Finding realistic dike models from interferometric synthetic aperture radar data: the February 2000 eruption at Piton de la Fournaise. Journal of Geophysical Research 110 B03206. doi:10.1029/2004JB003268.

Fukushima, Y., Mori, J., Hashimoto, M., Kano, Y., 2009. Subsidence associated with the LUSI mud eruption, East Java, investigated by SAR interferometry. Journal Marine and Petroleum Geology 26, 1740–1750.

Gourmelen, N., Amelung, F., Casu, F., Manzo, M., Lanari, R., 2007. Mining-related ground deformation in Crescent Valley, Nevada: implications for sparse GPS networks. Geophysical Research Letters 34 L09309. doi:10.1029/2007GL029427.

Guliyiev, I.S., Feizullayev, A.A., 1998. All About Mud Volcanoes. Nafta Press, Azerbaijan Publishing House, 52 pp.

Hanssen, R.F., 2001. Radar Interferometry - Data Interpretation and Error Analysis. Kluwer Academic Publishers, Dordrecht, The Netherlands.

Higgins, G.E., Saunders, J.B., 1974. Mud volcanoes. Their nature and origin. Verhandlungen Naturforschenden Gesselschaft in Basel 84, 101-152.

Hochstein, M. P. and Sudarman, S., 2010, Monitoring of LUSI Mud-Volcano - a Geo-Pressured System, Java, Indonesia., Proceedings World Geothermal Congress 2010, Bali, Indonesia, 25-29 April 2010

Istadi, B., Pramono, G.H., Sumintadireja, P., Alam, S. Simulation on growth and potential Geohazard of East Java Mud Volcano, Indonesia. Marine & Petroleum Geology, Mud volcano special issue, doi: 10.1016/j.marpetgeo.2009.03.006.

Istadi, B.P., Kadar, A., Sawolo, N., 2008. Analysis & recent study results on East Java mud volcano. In: Subsurface Sediment Remobilization and Fluid Flow in Sedimentary Basin Conference, October 2008. The Geological Society, Burlington House, Piccadilly, London.

Kadar, A.P., Kadar, D., Aziz, F., 2007. Pleistocene stratigraphy of Banjarpanji#1 well and the surrounding area. In: Proceedings of the International Geological Workshop on Sidoarjo Mud Volcano, Jakarta, IAGI-BPPT-LIPI, February 20–21, 2007. Indonesia Agency for the Assessment and Application of Technology, Jakarta.

Kholodov, V.N., 1983, Postsedimentary Transformations in Elisional Basins (example from Eastern Pre-Caucasus) (in Russian), 150 ps.

Koesoemadinata, R.P., 1980. Geologi Minyak dan Gas Bumi. 2 jilid, ed ke-2. Penerbit ITB, Bandung.

Kopf, A., 2002. Significance of mud volcanism. Reviews of Geophysics 40 (2), 1005, 2.1–2.52. doi:10.1029/2000RG000093

Kumai, H. and Yamamoto, H., 2007, Earthquake, the major trigger of Mud Volcanism at Sidoarjo, East Java., In: Proceedings of the International Geological Workshop on Sidoarjo Mud Volcano, Jakarta, IAGI-BPPT-LIPI, February 20–21, 2007. Indonesia Agency for the Assessment and Application of Technology, Jakarta.

Link, W.K., 1952. Significance of oil and gas seeps in world oil exploration. Am. Assoc. Pet. Geol. Bull. 36, 1505–1540.

Manga, M. and Brodsky, E. (2006): Seismic triggering of eruptions in the far field: Volcanoes and Geysers. Annual Review of Earth and Planetary Sciences, 34, 263-291.

Manga, M., 2007. Did an earthquake trigger the May 2006 eruption of the LUSI mud volcano? EOS 88 (201).

Manga, M., 2007. Did an earthquake trigger the may 2006 eruption of the LUSI mud volcano? EOS 88 (18), 201.

Manga, M., Rudolph, M.L., Brumm, M., 2009. Earthquake triggering of mud volcanoes: a review 26, 1785–1798.

Massonnet, D., Feigl, K.L., 1998. Radar interferometry and its application to changes in the Earth's surface. Review of Geophysics 36, 441–500.

Mazzini, A., Nermoen, A., Krotkiewski, M., Podladchikov, Y.Y., Planke, S., Svensen, H., 2009. Strike-slip faulting as a trigger mechanism for overpressure release through piercement structures. Implications for the LUSI mud volcano, Indonesia. Marine and Petroleum Geology 26, 1751-1765.

Mazzini, A., Svensen, H., Akhmanov, G.G., Aloisi, G. Planke, S., Malthe-Serenssen, A and Istadi. B., 2007, Trigering and dynamic evolution of LUSI mud volcano, Indonesia: Earth and Planetary Sciences Letters, v. 261, p. 375-388.

Mellors, R., Kilb, D., Aliyev, A., Glasanov, A. and Yetirmishli, G. (2007): Correlations between earthquakes and large mud volcano eruptions. Journal of Geophysical Research, 112, doi:10.1029/2006JB004489.

Milkov, A.V., 2000, Worldwide distribution of submarine mud volcanoes and associated gas hydrates, Marine Geology, 167, p. 29–42.

Milkov, A.V., 2005. Global distribution of mud volcanoes and their significance in petroleum exploration, as a source of methane in the atmosphere and hydrosphere, and as geohazard. In: Martinelli, G., Panahi, B. (Eds.), Mud Volcanoes, Geodynamics and Seismicity. NATO Science Series, IV Earth and Environmental Sciences, vol. 51, Springer, pp. 29–34.

Mori, J. and Kano, Y., 2009, Is the 2006 Yogyakarta Earthquake Related to the Triggering of the Sidoarjo, Indonesia Mud Volcano? Journal of Geography 118 (3)492-498 2009

Nawangsidi, D., 2007. In search of theory on flow mechanism of wild mud blow. In: Proceedings of the International Geological Workshop on Sidoarjo Mud Volcano, Jakarta, IAGI-BPPT-LIPI, February 20–21, 2007, Indonesia Agency for the Assessment and Application of Technology, Jakarta.

Pitt, A.M., Hutchinson, R.A., 1982. Hydrothermal changes related to earthquake activity at mud volcano, Yellowstone National Park, Wyoming. Journal of Geophysical Research 87, 2762–2766.

Satyana, A.H. and Armandita, C., 2004, Deep-Water play of Java, Indonesia : regional evaluation on opportunities and risks, Proccedings International Geoscience Conference of Deepwater and Frontier Exploration in Asia and Australasia, Indonesian Petroleum Association (IPA) and American Association of Petroleum Geologists (AAPG), Jakarta, p. 293-320.

Satyana, A.H. and Asnidar, 2008, Mud Diapirs and Mud Volcanoes of Java to Madura: Origins, Natures, and Implications to Petroleum System, Proceedings Indonesian Petroleum Association (IPA), 32nd annual convention, Jakarta, 27-29 May 2008.

Satyana, A.H., 2007, Bencana Geologi dalam "Sandhyâkâla" Jenggala dan Majapahit : Hipotesis Erupsi Gununglumpur Historis Berdasarkan Kitab Pararaton, Serat Kanda, Babad Tanah Jawi; Folklor Timun Mas; Analogi Erupsi LUSI; dan Analisis Geologi Depresi Kendeng-Delta Brantas, Proceedings Joint Convention Bali 2007-HAGI, IAGI, and IATMI, 14-16 November 2007.

Satyana, A.H., 2008, Roles of Mud Volcanoes Eruptions in the Decline of the Jenggala and Majapahit Empires, East Java, Indonesia: Constraints from the Historical Chronicles, Folklore, and Geological Analysis of the Brantas Delta-Kendeng Depression, Majalah Geologi Indonesia, vol. 23, no. 1-2, p. 1-10.

Sawolo, N., Sutriono, E., Istadi, B.P., Darmoyo, A.B., 2008. East Java mud volcano (LUSI): drilling facts and analysis. In: Proceedings of the African Energy Global Impact, AAPG International Conference and Exhibition, October 2008, Cape Town, South Africa.

Sawolo, N., Sutriono, E., Istadi, B.P., Darmoyo, A.B., 2009. Mud volcano triggering controversy: was it caused by drilling? Journal Marine and Petroleum Geology 26, (2009) 1766–1784. doi:10.1016/j.marpetgeo.2009.04.002

Sawolo, N., Sutriono, E., Istadi, B.P., Darmoyo, A.B., 2010. Was LUSI caused by drilling? - Authors reply to discussion. Journal Marine and Petroleum Geology 27 (2010) 1658-1675. doi:10.1016/j.marpetgeo.2010.01.018

Schiller, D.M., Seubert, B.W., Musliki, S., Abdullah, M., 1994, The reservoir potential of globigerinid sands in Indonesia, *Indon. Petroleum Assoc. 23nd Ann. Conv.*

Sharaf, E.F., BouDagher-Fadel, M.K., Simo, J.A. (Toni) and Caroll, A.R., 2005, Biostratigraphy and strontium isotope dating of Oligocene-Miocene strata, East java, Indonesia., Stratigraphy, vol. 2, no. 3, pp. 1-19, text figure 1-4, tables 1, plate 1-5.

Sudarman and Hendrasto, 2007, Hot Mud Flow At Sidoarjo., In: Proceedings of the International Geölogical Workshop on Sidoarjo Mud Volcano, Jakarta, IAGI-BPPT-LIPI, February 20-21, 2007. Indonesia Agency for the Assessment and Application of Technology, Jakarta.

Sukarna, D., 2007, Mud Volcano In Indonesia: Distribution and its geological Phenomenon., In: Proceedings of the International Geological Workshop on Sidoarjo Mud Volcano, Jakarta, IAGI-BPPT-LIPI, February 20-21, 2007. Indonesia Agency for the Assessment and Application of Technology, Jakarta.

Sumintadireja, P., Purwaman, I., Istadi, B., Darmoyo, A.B., 2007. Geology and geophysics study in revealing subsurface condition of Banjarpanji mud extrusion, Sidoarjo, East Java, Indonesia. In: Proceedings of the International Union of Geodesy and Geophysics (IUGG) XXIV, Perugia, Italy.

Sunardi, E., Guntoro, A., Alam, S., Koesoema, M.A., Hadi, S., Budiman, A., 2007. Studi Geologi Dan Geofisika Semburan Lumpur Di Daerah Porong Sidoarjo, Jawa Timur., In: Proceedings of the International Geological Workshop on Sidoarjo Mud Volcano, Jakarta, IAGI-BPPT-LIPI, February 20-21, 2007. Indonesia Agency for the Assessment and Application of Technology, Jakarta.

Sutaningsih, N.E., Humaida, H., Zaennudin, A., Suryono, Primulyana, S., 2010. Indikasi Sistim Geotermal Pada Semburan Lumpur Sidoarjo Ditinjau Dari Karakteristik Kimia, Proceedings PIT IAGI Lombok 2010, The 39th IAGI Annual Convention and Exhibition

Tingay, M.R.P., Heidbach, O., Davies, R., Swarbrick, R., 2008. Triggering of the LUSI Mud Eruption: Earthquake Versus Drilling Initiation. Geology, vol. 36(8), pp. 639-642.

Tingay, M.R.P., 2010. Anatomy of the Lusi Mud Eruption, East Java. Australian Society of Exploration Geophysicists 21st International Conference and Exhibition, 23-26 August 2010, Sydney, Australia, doi:10.1071/ASEG2010ab241.

Walter, T.R., Wang, R., Luehr, B.G., Wassermann, J., Behr, Y., Parolai, S., Anggraini, A., Günther, E., Sobiesiak, M., Grosser, H., Wetzel, H.U., Milkereit, C., Sri Brotopuspito, P.J.K., Harjadi, P., Zschau, J., The 26 May 2006 magnitude 6.4 Yogyakarta earthquake south of Mt. Merapi volcano: did lahar deposits amplify ground shaking and thus lead to the disaster?. G-cubed: Geochemistry Geophysics Geosystems, 9, Q05006, doi: 10.1029/2007GC001810.

Watanabe, N., dan Kadar, D., 1985, Quaternary Geology of the Hominids Fósil Bearing Formations in Java : Report of the Indonesia - Japan Research Project, CTA - 41, 1976 - 1979, Geological Research and Development Centre Special Publication, Bandung,P. 378

Widjonarko, R., 1990, BD field - a case history, Proceedings Indonesian Petroleum Association (IPA), 19th Annu. Conv., p. 161-182.

Willumsen, P., Schiller, D.M., 1994. High quality volcaniclastic sandstone reservoirs in East Java, Indonesia. In: 23rd Annual Convention, vol. I. IPA, pp. 101-111.

18

Xujiaweizi Rift Lower Cretaceous Yingcheng Group Volcanic Sequence Stratigraphic Features[1]

Zhang Yuangao, Chen Shumin, Feng Zhiqiang, Jiang Chuanjin Zhang Erhua, Xin Zhaokun and Dai Shili
Daqing Oilfield Company Ltd.,Heilongjiang Daqing China

1. Introduction

Sequence stratigraphy is a discipline that is developed on the seismic stratigraphy (Vail, 1987). The concept of sequence stratigraphy is born Since 1950′(Sloss, 1959), which is composed of LST, TST and HST. The theoretical system of sequence stratigraphy has been widely used by geologists after half a century of development. It is also developing from the classical three kinds system tracts to four systems tracts(Catuneanu,2006). Although the schemes of sequence boundary division are different(Catuneanu,2006; Catuneanu,2009).but they have stressed the fact is that the inherent mechanism of sequence genesis and phase distribution is controlled by sea level change(Arimoto,1997; Saydam,2000; Caquineay,2008). The theoretical framework of Chinese modern sequence stratigraphy is mainly composed of continental sequence stratigraphy(Shanley,1994; Liu ZJ,2002;Ji YL,2002) and high resolution sequence stratigraphy (Deng HW,2002). at the moment, the application fields of sequence stratigraphy have been developed from the research of the whole basin layers to a intrabasinal measure detailed study(Xu YX,2001; Chen F,2010; Wang QC,2010). With the development of exploration technology, the division of the sequence is more and more detailed, the control factor research of sequence development is more in-depth. However, with the breakthrough of exploration field, a large number of oil and gas resources are found in the volcanic rocks. Sequence stratigraphy is facing new challenges, how to create a volcanic sequence is our new research hotspot (Fig.1).

Yingcheng formation is made up of volcanic and sedimentary rocks in Xujiaweizi depression. volcanic rocks include volcanic lava and pyroclastic rocks that is a direct result of volcanism (Wang PJ,2008). sedimentary rocks include two sections: the one is the normal sedimentary rocks that occurred in the intermittent period of volcanism; the other is composed of weather worn volcanic which transport by water. Scholars(Stewart A L,2006; Busby CJ,2007; Wang P J,2006) have carried out intensive research about the volcanic lithology and lithofacies. But the targeted research of volcanic action filling model has not been carried out. Volcanic eruption and accumulation can take place in the high position. so the volcanic occurrence are not controlled by the accommodation space of water bodies, which is the biggest difference between sedimentary strata. Lead to the base level of

[1]Foundation Project：National Basic Research Program of China (2009CB219307)

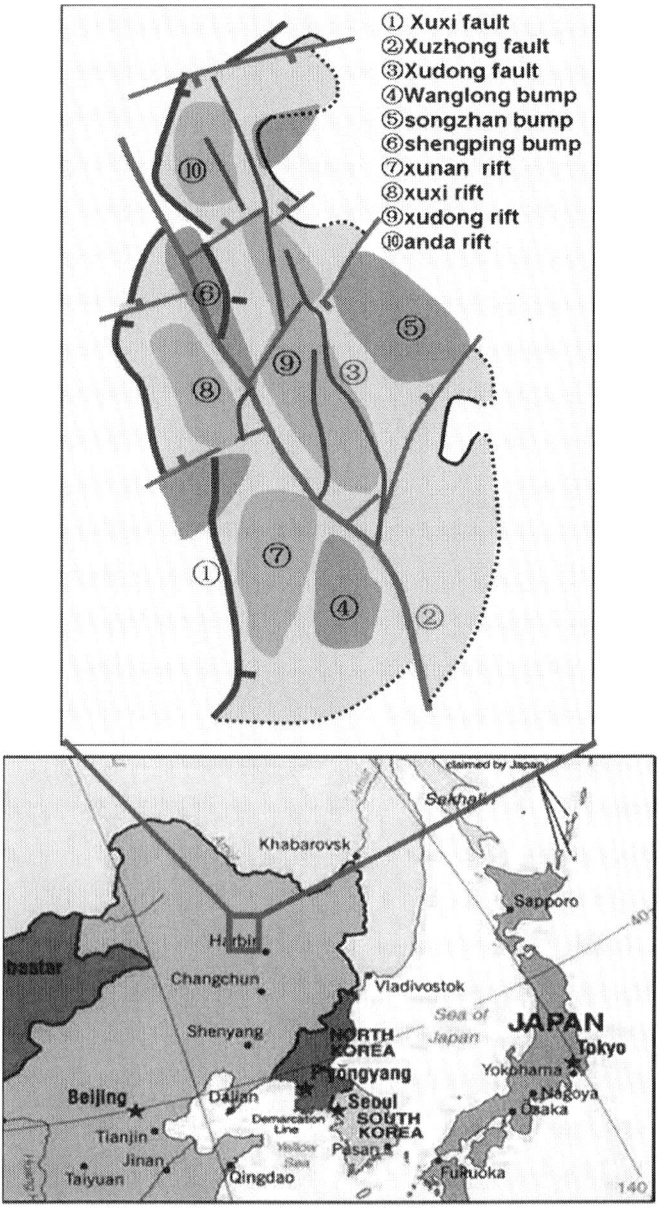

Map of the northeast of China

Fig. 1. Map of northeast of china ,in the upper shows locations of the fields mentioned in this article traditional sequence stratigraphy is difficult to find and compare. In particular, when the volcanic eruption is strong and multi-stage, the phase sequence changes further complicated in the horizontal and vertical. It is more difficulty to find a unified datum in volcanic rock formation.

As the features of the volcanic eruption localized, the volcanism product is not the same in different parts of the sequence boundaries. Volcanic strata exploration must to face this problem. It is also the bottleneck restricting the development of volcanic exploration. The stratigraphic sequence of volcanic rocks is still in the exploratory stage for global scientists. The scholars of home and abroad (Gamberi F,2001; Schmincke H U,2004; Wang P J,2010) focus on the type, characteristics, causes of volcanic ejecta, etc. The volcanic rocks is usually considered to be filling sequence part of the sedimentary basin for more reseach, that the volcanic sequence develop in the ultra-sequence sets, super-sequence and the top of the bottom of the third sequence(Meng QA,2005; Dong GC,2005; Qiu C G,2006). With further research, the time and location of volcanic development is controlled by tectonic activity has been recognized (Cheng R H,2005;Tang H F,2007). However, these studies mentioned above have ignored the fact that volcanic sequence has the specific formation mechanism and controlling factors relative to sedimentary sequence.

Volcanic formation is an important part of fault basin. The rapid accumulation volcanic rock has important implications for the formation and evolution of stratigraphic sequence with volcanic rocks. In order to obtain new ideas for volcanic sequence research and establish a interpretation contrast mode of volcanic sequence strata, volcanism and its products should be analyzed together in this paper. Based on the new research results of construction and volcanic eruption mechanisms, We propose a new development model of volcanic sequence strata, in order to find out the distribution of volcanic rocks and the formation mechanism of volcanic reservoir.

2. Volcanism features of Yingcheng formation

Because the multi-center multiphase volcanic eruption, there is widely developed volcanic rocks in Yingcheng formation of Xujiaweizi fault depression (Fig.2). Volcanic strata have the feature of multiphase superposition in the vertical and migration in the horizontal. These kinds of feature indicate volcanic eruption with cycling and direction.

2.1 Multiphase volcanic rock superposition in the vertical

This appearance is common that volcanic rocks is vertical superposition in the wholly Xujiaweizi fault depression. The formation process that alternating layers of different volcanic rocks is more complex, which is also a record of volcanism processes. There should be two reasons for different volcanic rocks superimposed. The one is that the eruption of magma from different craters at different times and different distances superimposed in the same region. The other is that the volcanic rocks come from the same crater, but the volcanic rocks are different in different parts of the same volcanic edifice (Fig.3). This appearance is asynchronous volcanic action with the different features of eruptive material, effusive activity and affected area. The characteristic of volcanic rocks vertical development, resulting in a single well by volcanic sequences in different parts of the sequence have different characteristics. So it is not good to divide the volcanic sequence by the traditional way, we also can not expect to find a unified interface feature in the regional area.

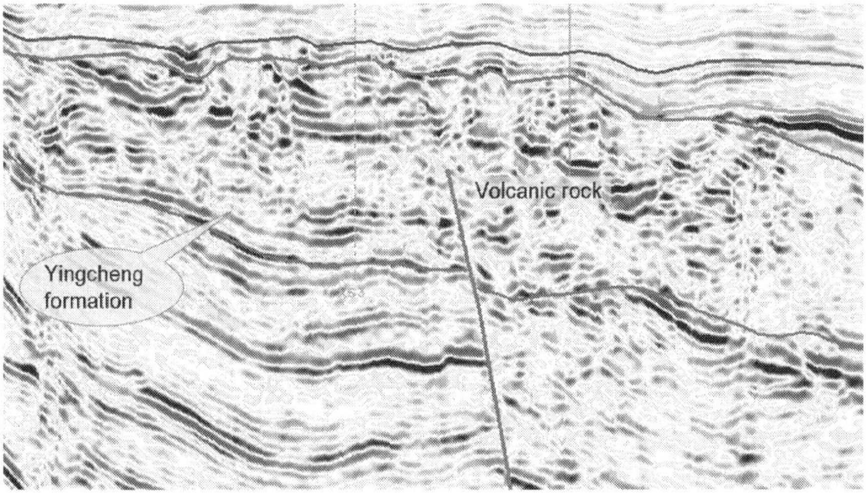

Fig. 2. Map of volcanic rocks development in Yingcheng formation

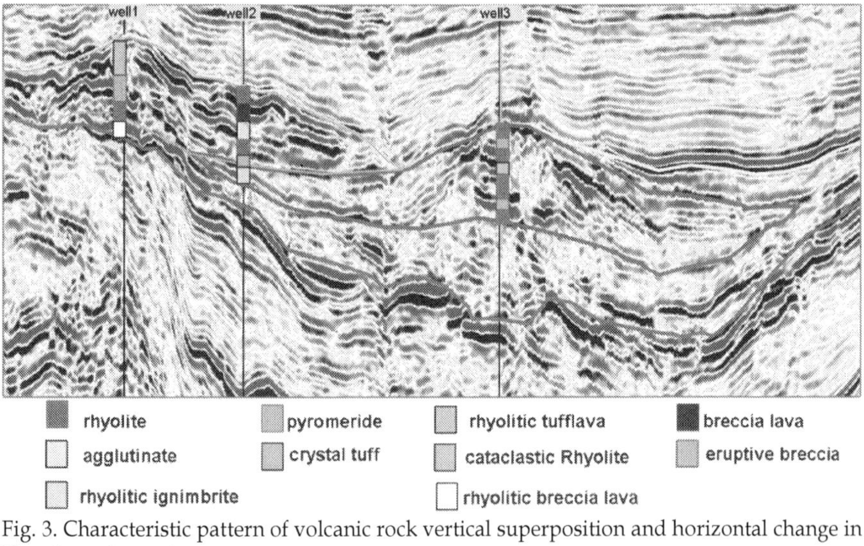

Fig. 3. Characteristic pattern of volcanic rock vertical superposition and horizontal change in Yingcheng formation

2.2 Volcanic eruption controlled by faults

Volcanic activity is related with the extensional movement of the mantle. When the unusual mantle bulge, causing supracrustal formation broken ground result in a volcanic eruption. The discordogenic fault that cut through the basement and earth's crust is the channel of magma upwelling. The Xuzhong and Xudong two major strike-slip fault systems (SSFS) and

theirs connected subsidiary fractures are the channels of magma eruption in Xujiaweizi fault depression. The thick volcanic rocks distribute along the strike-slip fault systems of Xuzhong and Xudong (Fig.4). According to drilling and geophysical data, the craters are mainly located in the transition location and intersection of fractures. Based on the correlation theory of rock fracture mechanics and structural geology, stress is concentrated in the end. However, the two endpoints of fracture are the stress concentration points, but also it is the weakest area. The volcanism is firstly happened in the weakest zone (Fig.5).
Because of the characteristics of SSFS, the strata are subjected to compressional and extensional coincident movement. There will be the phenomenon that magma is not eruption in various parts of the same fault at the same time along the strike-slip fault zone. The characteristics of the same fault activity are often different in different locations. There is a gradual process of change. The volcanic eruption first happened in the extensional location form the feature of multi-center fissure eruption. Because the alternating tension and compression effects. The characteristics of volcanic activity are different in different fault segments. The existence of such differences is the main reason for the volcanic superposition and migration in the horizontal (Fig.6).

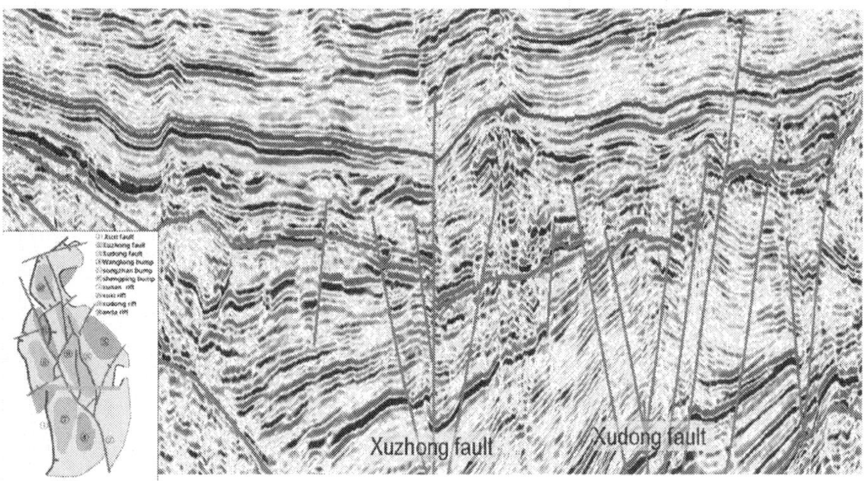

Fig. 4. The matched relationship of volcanic rocks thickness and fracture

2.3 The coexisting of volcanic and sedimentary rocks

The process of volcanic activity from start to finish is not continuous, but intermittent. The depositional interbedded stratum is the representative lithology for a quiet period of volcanic activity. Such as: there is about thickness of 45m deposition interlayer in the area of xs6 well volcanic rocks; there are also two sets of sedimentary sandwich thickness of 20m in the area of xs401 well. Regional deposition interlayer represents the interval of regional volcanic activity. There's a regional comparison deposition interlayer in the area of xs6-xs4-xs2-xs14-xs12 wells, which is the most obvious signs of volcanic cycle surface (Tang H F,2010).
The interbedded features of volcanic and sedimentary rocks, it is both on the evidence of intermittent volcanic activity, but also the evidence of underlying volcanic rocks below the

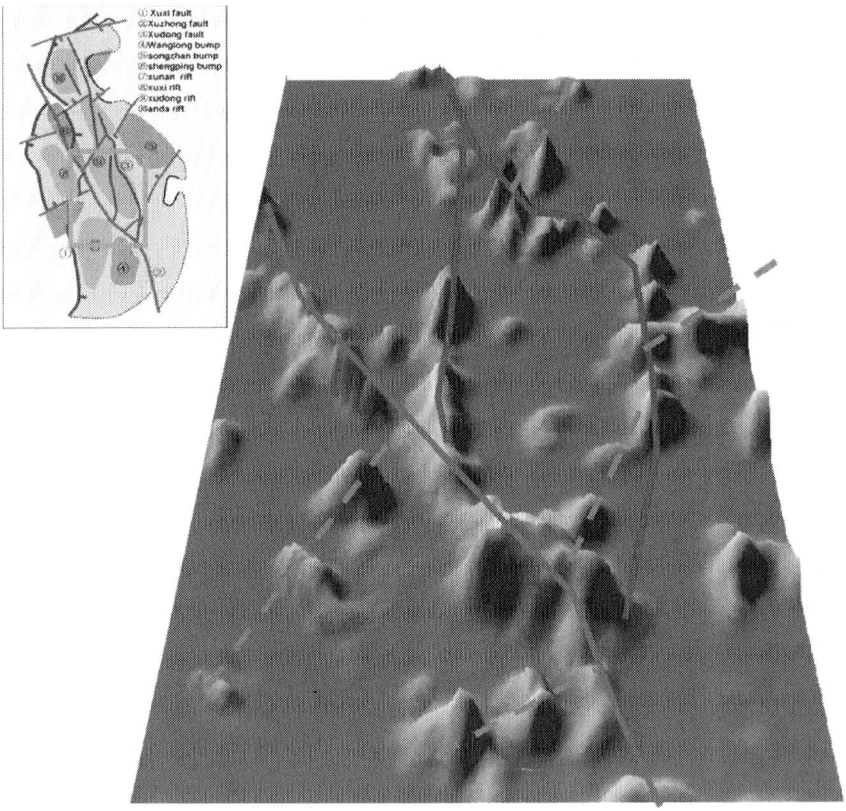

Fig. 5. The matched relationship of volcanic explosion vent and fracture

water surface a long period. This kind of sedimentary rocks indicates that the Yingcheng volcanic has the characteristics of subaquatic eruption. The pearlite of xs2, xs5 and deposit ash tuff of xs1-4 wells drilled are further indicates the presence of lacustrine environment. There is a complete set of clastic rock containing volcanic in the Yingcheng formation in the area of zs14 well, which confirms a certain amount of lake present at the same time of volcanic eruption. The contact relationship is diverse between sedimentary and volcanic rock, some sedimentary rock is located in the middle of volcanic rocks, some is also located at the bottom of volcanic rock. The difference of occurrence location is controlled by volcanic eruptive sequence and affected area for sedimentary rock.

3. The division of volcanic sequence for Yingcheng formation

The accurate division and attribution of layers is the basis to restore stratigraphic evolutionary sequence. It's also the basis for oil&gas exploration and development. Because the multiphase volcanic eruption in the same area, volcanic edifices occur reformed in the vertical and lateral migration. It's difficult to propose a simple evolution model of general

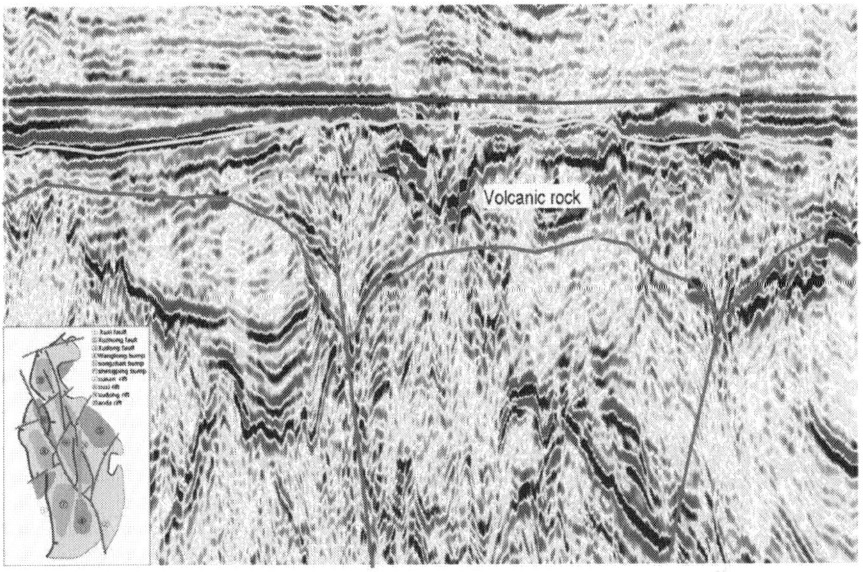

Fig. 6. The map of central vent eruption controlled by crack

application. It's also difficult to divide and compare volcanic sequence in accordance with this mode. Based on the summary of volcanic activity regular pattern, the authors analyze the volcanic sequence of the control factors and division and correlation in this article.

3.1 Distribution characteristics of Yingcheng strata

Based on the contrast from the regional stratigraphic features, there is main development of sedimentary rock strata containing volcanic material in the east and west sides of Xujiaweizi fault depression. In some areas of the rift center, such as xs801 well area, there is sedimentary rocks development in the lower part of Yingcheng formation. The set of sedimentary rocks is obviously different from the Shahezi sedimentary strata. The contact relationship of them is angle disconformity, this feature is very visible in the seismic profiles. It's further show that the volcanic rocks are not development in the whole Xujiaweizi rift in the period of volcanic eruption. It is jointly controlled by volcanic activities and sedimentary process for the temporal and spatial distribution of Yingcheng strata. It has the characteristics of multiphase volcanic eruption in the primary zone of volcanic eruption. Thus there is a large area of volcanic lava and athrogenic rocks development. There is the area of sedimentary rocks development in the outside of volcanic affected area in the Yingcheng formation. The construction feature of Yingcheng formation is associated with volcanic action and deposition (Fig.7).

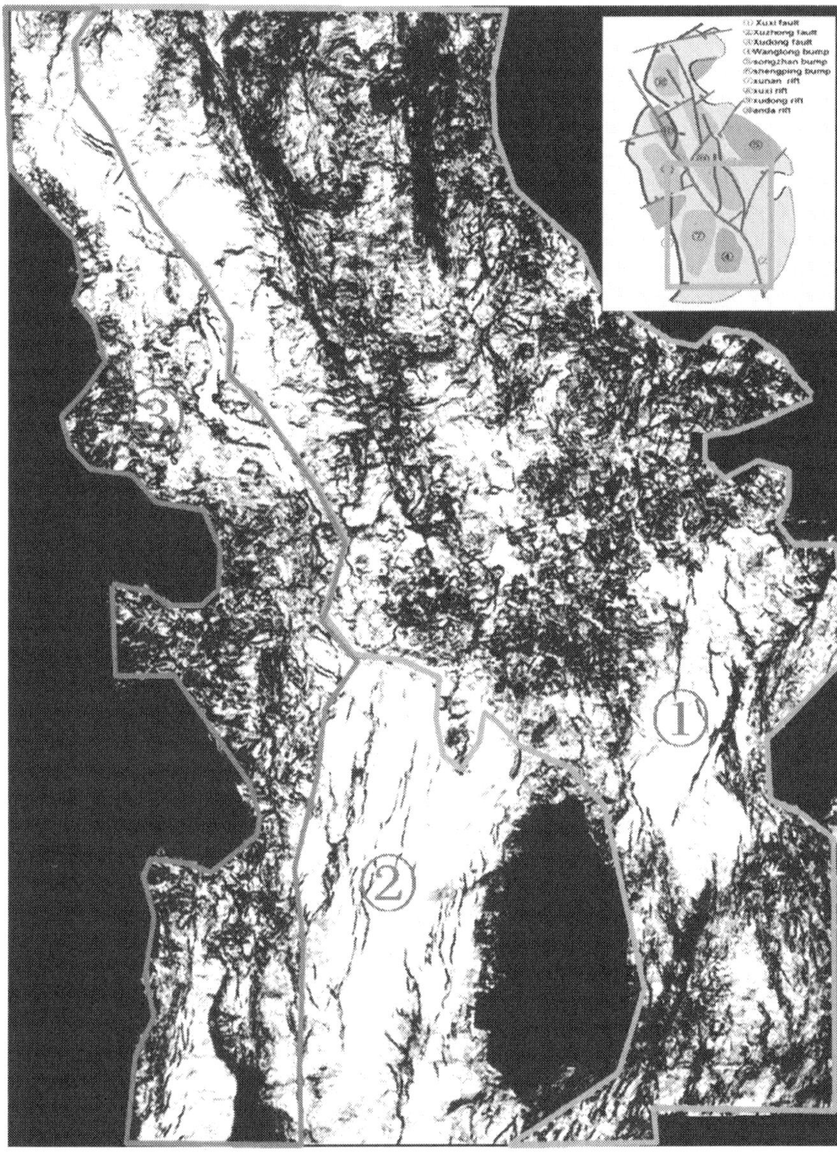

① Area of volcanic rock development
② Area of sedimentary rock development
③ Area of volcanic and sedimentary rock mixed development

Fig. 7. The forecasting map of lithology using coherence cube

3.2 Volcanic sequence classification foundation

It is to find the formation boundary that the upper and lower strata can be distinguished and correlatable in the horizontal for volcanic sequence division. For a concentrated continuous volcanic activity, the material composition, eruption methods and effusive activity will occur on a regular variation. which changes regularly must be inevitably form a set of genetic relationship of the volcanic sequence. Based on the single well detailed breakdown of volcanic rocks, Scholars (Wang PJ,etal, 2010) divided different numbers of volcanic rocks cycles in different regions in the Xujiaweizi rift. But these cycles explained is difficult to compare with others in the horizontal. Lithologic correlation is wrong correlation marker for volcanic sequence. It can reflects the variation of volcanic activity that superimposed array mode of volcanic apparatus(Chen SM,2011).Therefore, the directivity and superimposed mode of volcanic activity is different for diachronous volcano, volcanic sequence can be divided.

Based on comprehensive analysis of the volcanic rocks development characteristics of Yingcheng formation, the four correlation markers are proposed that is suited for volcanic rocks comparing in XJWZ rift. The first one is depositional interbedded stratum and ash tuff that can be regional compared. It is an important foundation for volcanic cycle dividing. The second one is the structural surface, in the upper and lower of the structural surface. There is distinguished difference for effusive activity, lithology, facies. So it is easy to form unconformity and the angular unconformity for volcanic formation, it is the good foundation for volcanic comparison (Fig.8). The third one is the superimposed mode of volcanic apparatus that diachronous volcano has different directivity of superimposition (Fig.9). The fourth one is depositional interface in the same period of volcano. Because the sedimentary formation records the evidence of volcanic event, classical infilling mode of fault depression has been changed when the volcanic goes into the rift. Each volcanic eruption correspond a tectonic movement, so the coupling of structure and sedimentary strata also recorded the volcanic activity cyclicity (Fig.10).

3.3 The division of volcanic sequence

Combination of the regional tectonic evolution, according to the macro-migration of volcanic activity, the volcanic rocks of Yingcheng formation can be divided into two three-stage sequences. The lower volcanic sequence is Y1 section, the upper volcanic sequence is Y3 section. The Y1 is mainly controlled by xuzhong strike-slip fault and its associated faults, the maximum thickness of volcanic rocks is along the fault zone. It is mainly distribution in the central and southern rift. The direction of volcanic eruption is mainly from north to south.As with the fault distance increases, the content of volcanic strata gradually reduced until it evolved into sedimentary strata. The Y3 is mainly controlled by xudong strike-slip fault and its associated faults. The direction of volcanic eruption is mainly from south to north. The xs22 well field is the starting point for the period of volcanic activity. Since the rapid release of volcanic energy, the xs22 Well field form a large-scale collapse crater, and form a thick pyroclastic filling. Because of the volcanic activity have the feature of direction and migration, there is a large area of debris deposition in the northern area that is the outside of volcanic activity sweep area. With the more large-scale violent volcanic eruptions and flooding, the larger and broader coverage of volcanic filling.

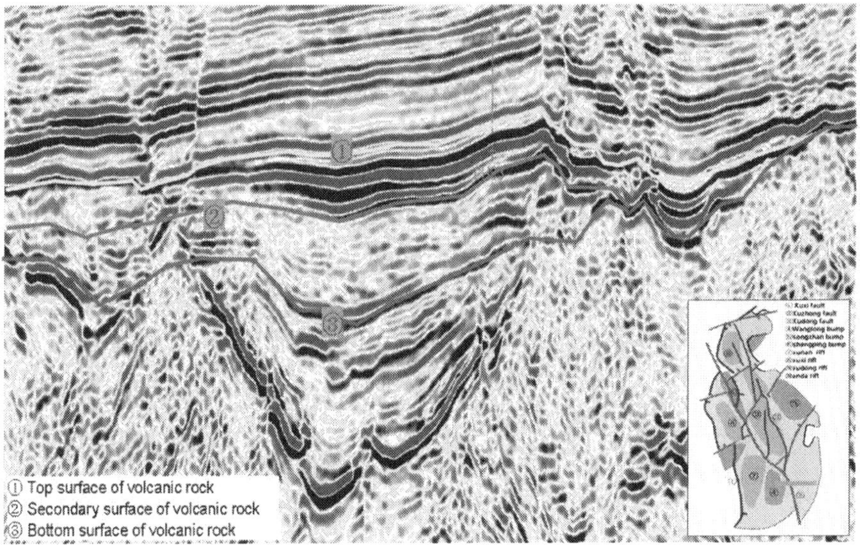

Fig. 8. The feature of sequence interface in volcanic rock

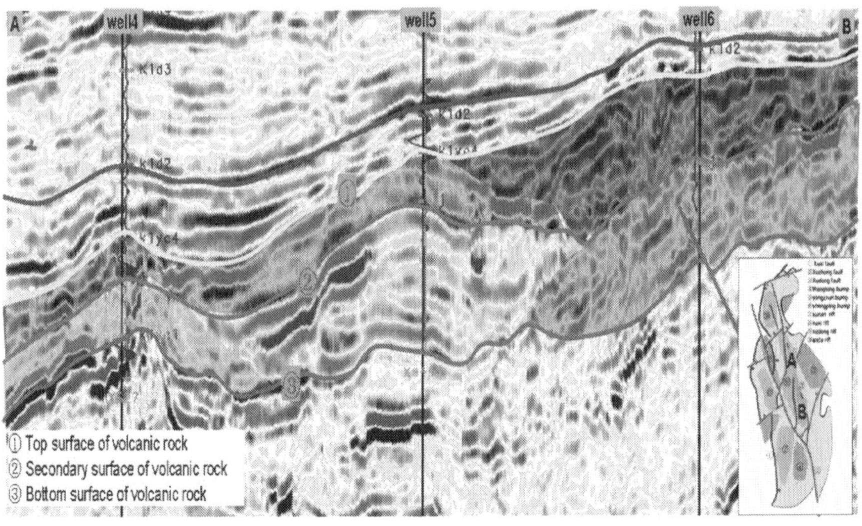

Fig. 9. The difference of volcanic edifice superimposed relationship nearby the volcanic sequence

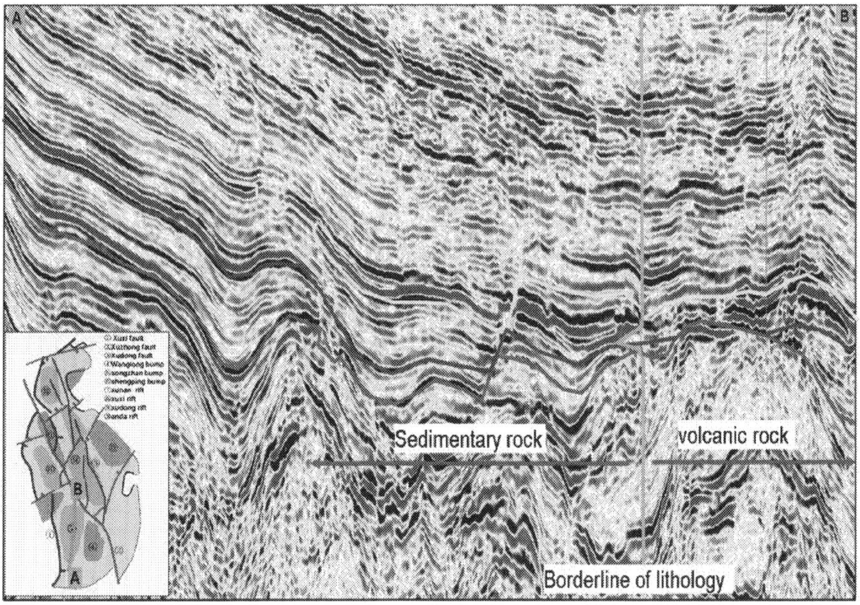

Fig. 10. The couple of sedimentary sequence and volcanic action

The volcanic sequence of Y1 and Y3 can be further identified three sub-sequences. they are the early volcanic eruption sequence(EVES), the strong volcanic eruption sequence(SVES) and the languid volcanic eruption sequence(LVES). Explosive eruption is primary in the EVES, there is also a certain amount of overflow phase developed. There is the area of sedimentary rocks development in the outside of volcanic affected area in the Yingcheng formation. The volcano shows the features of local central vent eruption along the cracks. Nowadays, the volcano of Wudalianchi has the same characteristics to EVES in the Heilongjiang province of China. The SVES has the typical feature that large-scale outbreak and the overflow happen simultaneously. There is the interbedded, and there is also some thick individual layer of lava flows and debris flows. The SVES is widely distributed in the large areas of sequence. It is not only the main volcanic sequence stratigraphy, but also the reservoir is the most development in the entire volcanic sequence. Local overflow and small-scale invasion is the main characterics of the LVES, because the size of LVES is small, the feature of sequence is no obvious. it is usually combined with the SVES as a sequence. But the LVES is the most favorable sequence for volcanic reservoir communication with deep fluid. it is also the favorable position for deep fluid easier to charging in the process of volcanic accumulation. It become the favorable accumulation area for carbon dioxide gas reservoir. The current exploration results have confirmed this characteristic in XJWZ rift (Fig.11).

4. Developmental patterns of Yingcheng sequence stratigraphy

The base level of volcanic sequence is unstable and not uniform in the region. Because of the existence of early volcanic eruption highland, the late sequence is not good overlay with the early. Therefore, the two-step volcanic sequence boundary is difficult to form a stable feature in the region. It is also hard to find the unified reflecting boundary for tracking.

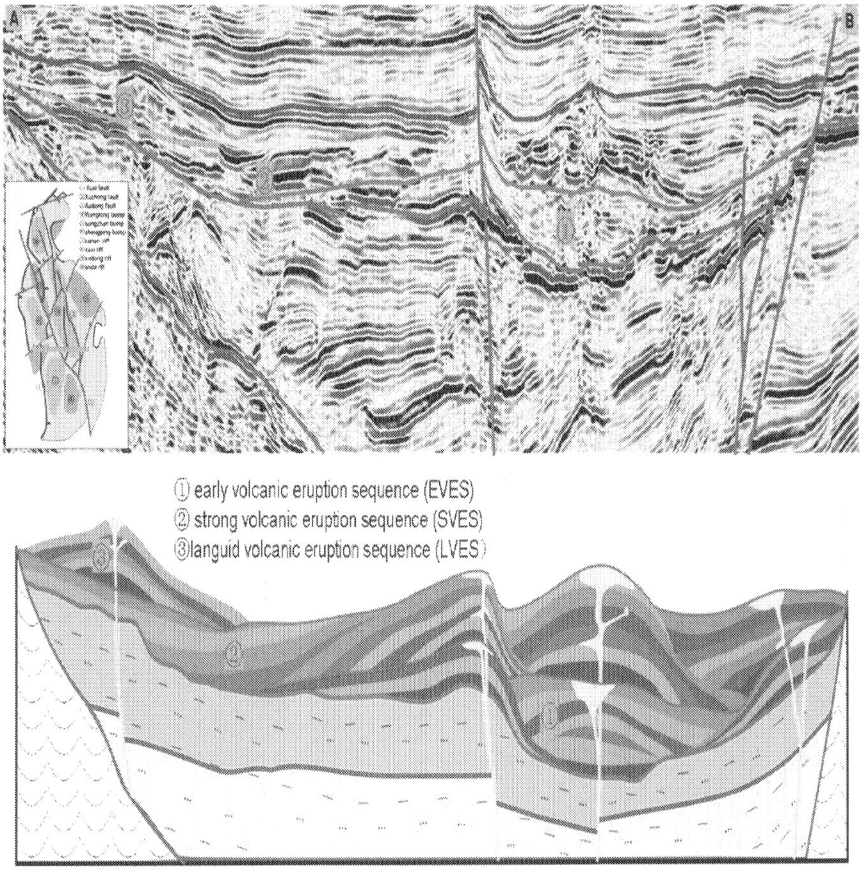

Fig. 11. The map of volcanic sequence division

4.1 The Y1 developmental patterns

The size of Y1 volcanic apparatus is certain scale, the general current occurrence height is above 300 meters. The early eruption of the volcano can be located above water level. Therefore, compared to the late period volcanic apparatus, the weather-worn extent of early is strong. such as: the transformation feature of volcano is relatively obvious in the north well field, there is thick layer weathering crust development; but the transformation feature of volcano is significantly decreased in the south well field. This change indicates the order of volcanic action that is from north to south.

Volcanic activity is also accompanied by tectonic movement, resulting in frequent lake level fluctuation. When volcanic material quickly build-in the rift basin, the accommodation space is changed, resulting in level surface rising. The volcanic eruption early is located under the level surface, received clastic sediments. With the thickness of cumulo-volcano continuous increase, the late crater outcrop above the water level. Therefore, the interbedded phenomenon of sedimentary and volcanic rocks significantly reduced, there is mostly volcanic with high purity development in the upper volcanic rocks. There is the area of sedimentary rocks development in the outside of volcanic affected area. These sediments containing volcanic material is high, because the highland formed by volcanic eruption suffered weather worn activity, provide a filling material for the basin (Fig.12). Thus the unique geological feature is formed for Yingcheng formation. where the volcanic eruption, volcanic intermittent, erosion, deposition is interaction. The ancient landscape of Y1 is not flat where the north is lower, the south is higher.

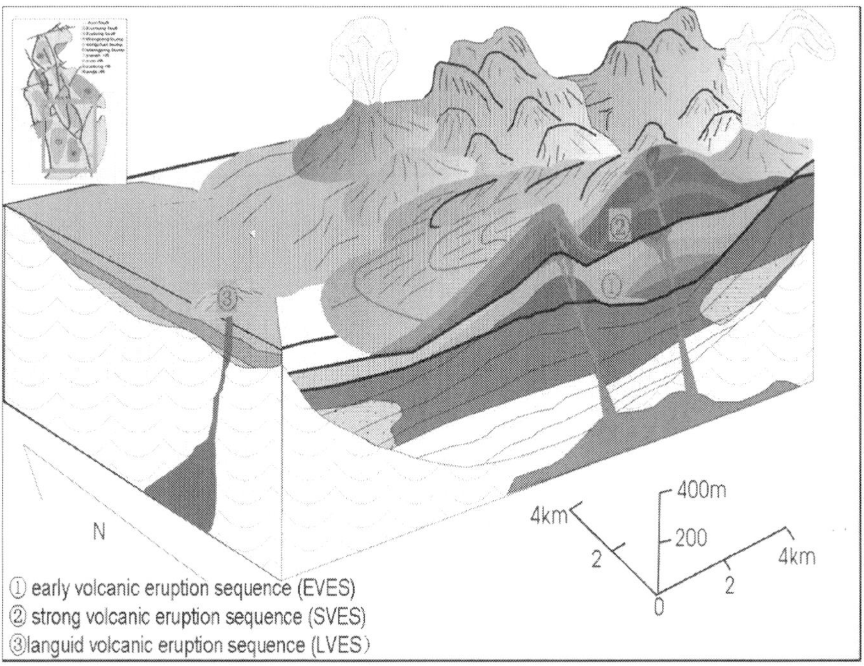

Fig. 12. The map of Y1 developmental patterns

4.2 The Y3 developmental patterns

Because of the different nature of the magma, the Y3 development pattern is completely different from Y1. The Y1 is mainly composed of acidic volcanic, but the basic volcanic rock is mainly development in Y3. Compared with the acid volcanic rocks, the basic volcanic rocks has different vents feature. Such as the volcanic vents is smaller, the plane position is relatively stable, the volcanic rocks is mainly volcanic lava. The local outbreaks characteristic of EVES is accompanied by fault activities. It is easy to form a barrier lake. Thus the formation of volcanic rocks and sedimentary rocks in symbiosis is the typical characteristic (Fig.13). When the volcano erupted violently, volcanic sub-sequence is in the stage of SVES. The rift basin is filled with volcanic material quickly, volcanic rock becomes the main rocks, and the sedimentary rock is little development. A large area of lava delta is development, in the vent, often accompanied by some small-scale outbreaks (Fig.14). The ancient landscape of Y3 is not flat where the north is higher, the south is lower.

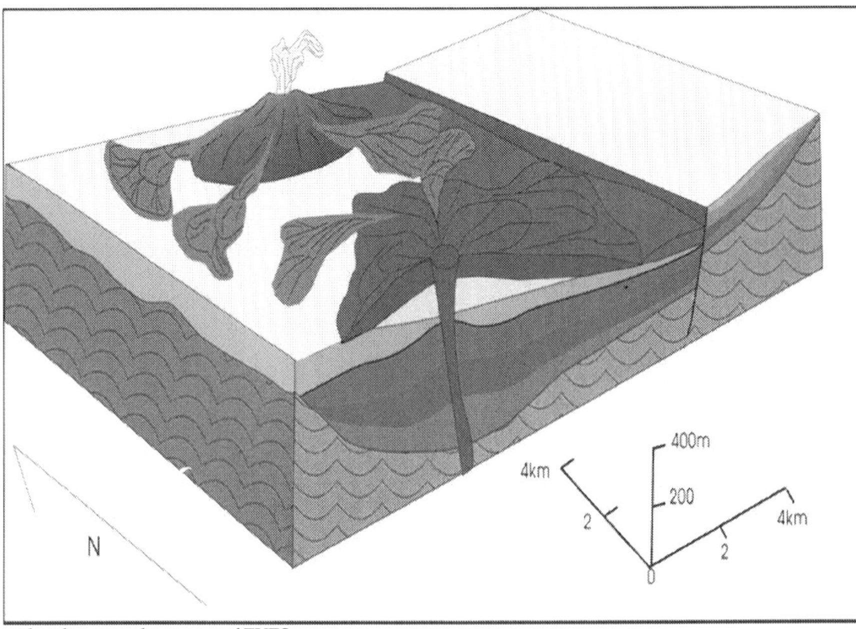

a. developmental patterns of EVES

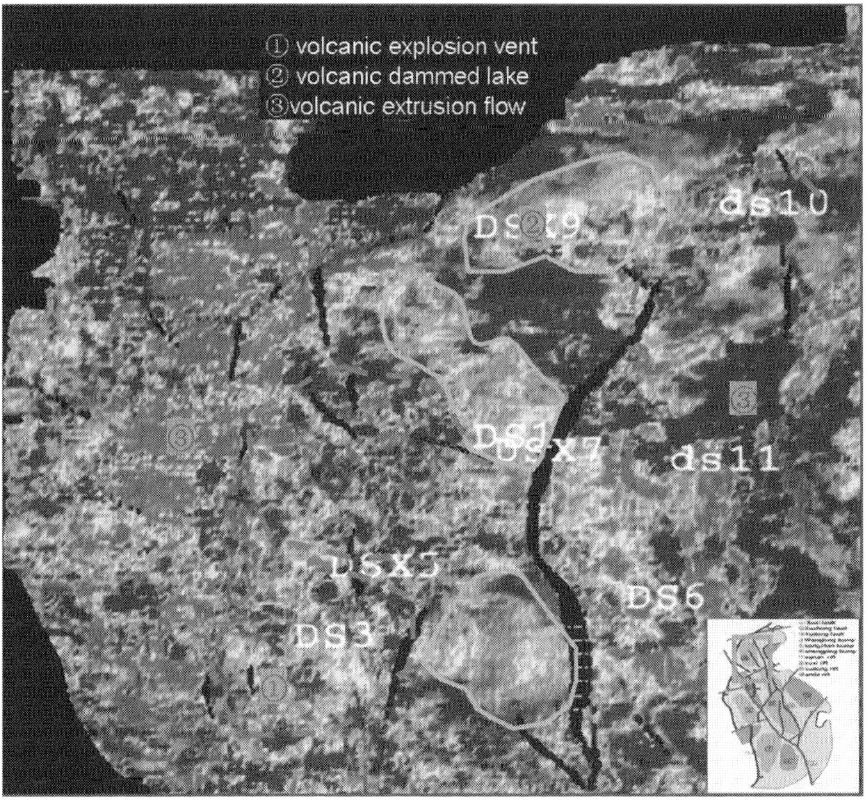

b. the example of EVES

Fig. 13. The map of Y3 developmental patterns(EVES)

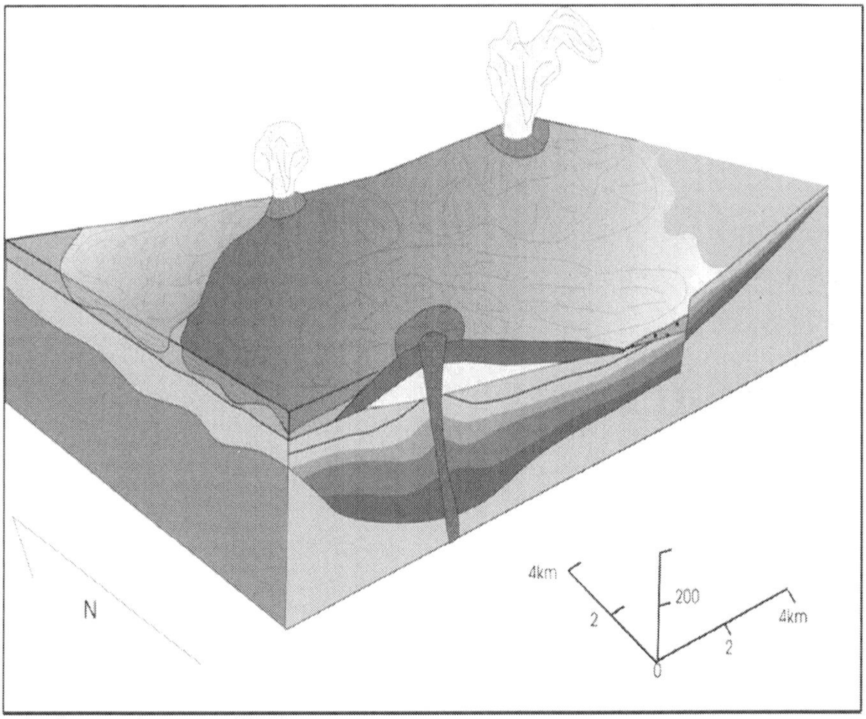

a. developmental patterns of SVES

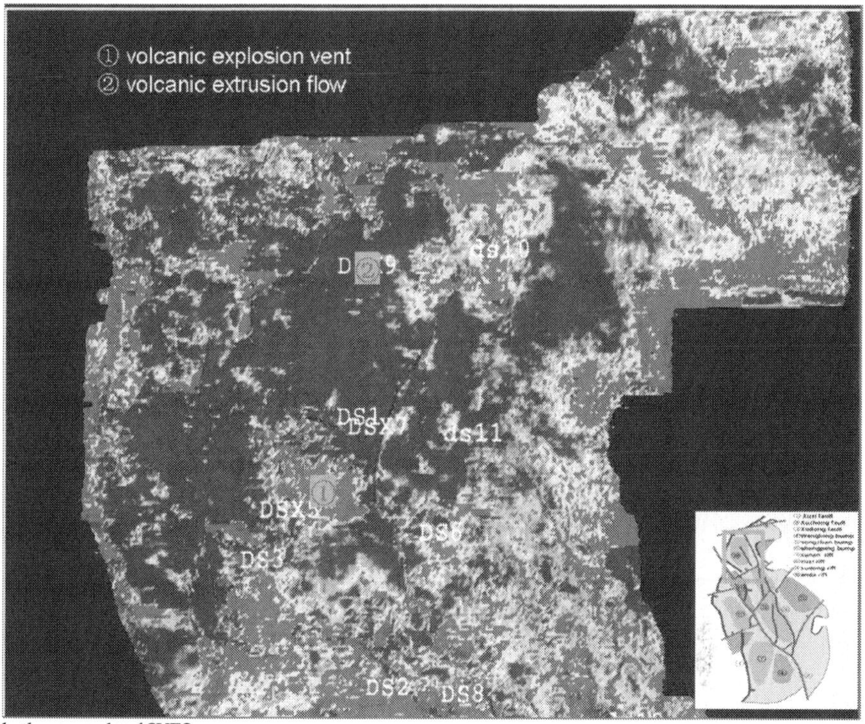

b. the example of SVES

Fig. 14. The map of Y3 developmental patterns(SVES)

5. Summary

1. The volcanic eruption is controlled by strike-slip fault systems(SSFS) and theirs connected subsidiary fractures. Because the order of fault activity is different, so that the volcanic rock of Yingcheng formation can be divided the upper and lower volcanic sequence. The the lower sequence(Y1) is mainly controlled by Xuzhong strike-slip fault and its associated faults, the upper sequence (Y3) is mainly controlled by strike-slip fault and its associated faults. So the research of volcanic sequence, it is very important to find the faults that control the volcanic eruption.
2. The regional sedimentary is present in volcanic rock strata. It is both on the evidence of intermittent volcanic activity, but also the evidence of underlying volcanic rocks below the water surface a long period. This kind of sedimentary rocks indicates that the Yingcheng volcanic has the characteristics of subaquatic eruption. The environment of volcanic eruption is shallow-water lacustrine facies. The interaction of fire and water, the volcanic eruption cycle can be found in sedimentary sequence.
3. Different parts of the same fault has different activities time, it also control the order of volcanic eruption. So it is the characteristic appearance that fissure flow multi-point

central vent eruption in XJWZ rift. The craters are mainly located in the transition location and intersection of fractures.

4. The difference of volcanic activity directional is the important basis of volcanic rock sequence division. It is also a reflection of plate movement. The direction change of volcanic activity is must be accompanied by the tectonic movement occurs.
5. Sedimentary and volcanic rocks are uniform in the strata, there is mainly volcanic sequence stratigraphy in the volcanic eruption area; There is the area of sedimentary rocks development in the outside of volcanic affected area . Sedimentary rock strata records the cyclicity of volcanism.
6. The volcanic sequence of Y1 and Y3 can be further identified three sub-sequences. They are the early volcanic eruption sequence(EVES), the strong volcanic eruption sequence(SVES) and the languid volcanic eruption sequence(LVES). It is mainly based on debris accumulation in EVES, it makes the reflection characteristics of chaotic and blank in seismic profile. The SVES usually have the reflection characteristics of strong energy in seismic profile. The LVES is local and sporadic development, it is difficult to find the regional characteristics compared in seismic profile. Therefore, it merged into the SVES. But it has the good indication for inorganic gas reservoir.
7. They are very important signs for volcanic sequence divided. Such as superimposed mode of volcanic apparatus,depositional interbedded stratum and ash tuff can be regional compared, structural surface, sedimentary sequence interface With the same period of volcanic activities.

6. References

[1] Vail P R.. 1987. Seismic stratigraphy interpretation using sequence stratigraphy. Part 1: Seismic stratigraphy interpretation procedure. American Association of Petroleum Geologists, Studies in Geology, 27:1-101

[2] Sloss L L. 1959. Sequences in the cratonic interior of North America, Part 2: Geological Society of America Bulletin, 70 (12) :1676-1677

[3] Catuneanu O, Khalifa M A, Wanas H A. Sequence stratigraphy of the Lower Cenomanian Bahariya Formation, Bahariya Oasis, Western Desert. Egypt Sedimentary Geology, 2006, 190: 121-137

[4] Catuneanu O.Principles of Sequence Stratigraphy. Elsevier, Amsterdam, 2006, 375

[5] Catuneanu O, Abreu V, Bhattacharya J P, et al. Towards the standardization of sequence stratigraphy. Earth-Science Reviews, 2009, 92: 1-33

[6] Arimoto R., B.J. Mass Particale Size Distribution of Atmospheric Dust and the Dry Deposition of Dust to the Remotr Ocean. Journal of Geopjyscical Research, 1997, 102: 15867-15874

[7] T., A.C., Saydam. Acidic and Alkaline Precipitation in the Cilician Basin, North-eastern Mediterranen Sea. Science of the Total Environment, 2000, 253: 93-109

[8] Caquineay, Gaudichet, Gomes. Saharan Dust: Clay Ratio as a Relevant Tracer to Assess the Origin of Soil-derived Aerosols. Geophysical Research Letters, 2008, 25: 983-986

[9] Shanley, K. W., McCabe, P. J., 1994. Perspectives on the sequence stratigraphy of continental strata. AAPG Bull., 78:544 - 568.

[10] Liu Z J, Dong Q S., Wang S. M., et al., 2002. Introduction and application to sequence stratigraphy of continental face. Petroleum Industry Publishing House, Beijing, 21 - 91 (in Chinese).

[11] Ji Y L. 2005. sequence stratigraphy. Tongji Univesity publishing house, Shanghai. (in Chinese)
[12] Deng H W, Wang H L, Zhu Y J, etal.2002. principle and application of high resolution sequence stratigraphy, Geological publishing house, Beijing (in Chinese).
[13] Hou M C, Chen H D, Tian J C. 2003. Sequence-filling dynamics--a new study direction on sequence on sequence stratigraphy. Journal of Stratigraphy, 27 (4) :358-364 (in Chinese)
[14] Xu Y X, Gao X L, Li Z Y, et al · Sequence stratigraphy and hydrocarbon distribution of the Eogene in the eastern slope of Chengdao area · Petroleum Exploration and Development, 2001, 28(4) :25-27 (in Chinese)
[15] Lin C S · Sequence and depositional architecture of sedimentary basin and process responses Acta Sedimentologica Sinica, 2009, 27(5) : 849-862 (in Chinese)
[16] Chen F, Luo P, Zhang X Y, et al. Stratigraphic architecture and sequence stratigraphy of upper Triassic Yanchang Formation in the eastern margin of Ordos Basin. Earth Science Frontiers, 2010, 17(1) :330-338 (in Chinese)
[17] Wang Q C, Bao Z D, He P. Sequence stratigraphic responses to the lacustrine basin deep-faulted period in the north area of the western sag, Liaohe Depression. Petroleum Exploration and Development, 2010, 37(1) :11-19 (in Chinese)
[18] Wang P J, Feng Z Q, Chen S M, et al. Basin Volcanic: Volcanic Rocks in petroliferous Basins: Lithology·Faces· Reservoir·Pool· Exploration. Science Press.2008, Beijing (in Chinese)
[19] Stewart A L, McPhie J. Facies architecture and LatePliocene-Pleistocene evolution of a felsic volcanic island, Milos, Greece. Bull Volcanol, 2006, 68:703-726.
[20] Busby C J, Bassett K N. Volcanic facies architecture of an intra-arc strike-slip basin, Santa Rita Mountains, Southern Arizona. Bull Volcanol, 2007, 69:85-103.
[21] Wang P J, Wu H Y, Pang Y M, et al. Volcanic facies of the Songliao Basin : sequence, model and the quantitative relationship with porosity& permeability of the volcanic reservoir. Journal of Jilin University (Earth Science Edition), 2006, 36(5) :805-812 (in Chinese)
[22] Gamberi F. Volcanic facies associations in a modern volcaniclastic apron (lipari and vulcano off shore, Aeolian island arc. Bull Volcanol, 2001, 63: 264-273.
[23] Schmincke H U. Volcanism. Berlin, Heidelberg, New York: Springer, 2004.
[24] Wang P J, Yin C H, Zhu R K, et al. Classification, description and interpretation of the volcanic products: ancient and modern examples from China. Journal of Jilin University(Earth Science Edition), 2010, 40(3) :469-481 (in Chinese)
[25] Meng Q A, Wang P J, Yang B J. Geological signatures of sequence boundary of the Songliao basin: new interpretation and their relation to gas accumulation. Geological Review, 2005, 51(1) :46-54 (in Chinese)
[26] Dong G C, Mo X X, Zhao Z D, et al. A new understanding of the stratigraphic successions of the Linzizong volcanic rocks in the Lhünzhub basin, northern Lhasa, Tibet, China. Regional Geology of China, 2005, 24(6) :549-557 (in Chinese)
[27] Qiu C G, Wang P J. Prelimimary study on sequence's location of volcanic stratigraphy in sedimentary basin-an example from Xujiaweizi depression of Songliao basion. Xinjiang Oil & Gas. 2006, 2(1):13-18 (in Chinese)

[28] Cheng R H, Wang P J, Liu W Z, et al. Sequence stratigraphy with fills of volcanic rocks in Xujiaweizi Faulted Depression of Songliao basin, Northeast China. Journal of Jilin University(Earth Science Edition), 2005, 35(4):469-474 (in Chinese)
[29] Tang H H, Wang P J, Jiang C J, et al. Seismic characters of volcanic facies and their distribution relation to deep faults in Songliao basin. Journal of Jilin University(Earth Science Edition), 2007, 37(1): 73-78 (in Chinese)
[30] Zhang Y G, Chen S M, Zhang E H. The new progress of Xujiaweizi fault depression characteristics of structural geology research. Acta Petrologica Sinica, 2010, 26(1):142-148 (in chinese)
[31] Zhang E H, Jiang C J, Zhang Y G, et al. Study on the formation and evolution of deep structure of Xujiaweizi fault depression. Acta Petrologica Sinica, 2010, 26(1):149-157 (in chinese)
[32] Jiang C J, Chen S M, Chu L L, et al. A new understanding about the volcanic distribution characteristics and eruption mechanism of Yingchen formation in Xujiaweizi fault depression. Acta Petrologica Sinica, 2010, 26(1):63-72 (in chinese)
[33] Tang H F, Bian W H, Wang P J, et al. Characteristics of volcanic eruption cycles of the Yingcheng Formation in the Songliao Basin. Natural Gas Industry, 2010, 30(3): 35-39 (in Chinese)
[34] Chen S M, Zhang Y G, Jiang C J. The analysis of volcanic edifice superimposition and its digital model parameters establishment. Chinese J. Geophys, 2011, 54(2): 499-507 (in Chinese)

Part 10

Remote Sensing

19

Laser Altimetry: What Can Be Learned About Geology and Surface Processes from Detailed Topography

Tim Webster
Applied Geomatics Research Group
Nova Scotia Community College, Middleton,
Canada

1. Introduction

Earth Science has utilized new remote sensing techniques for many years, weather it be airborne geophysics to sense the magnetic field or aerial photography and satellite imagery to obtain that ever important synoptic view that aids in our interpretation of the landscape and geology. The field of geomatics, which is the acquisition, analysis and mapping of the earth's surface, has emerged and drives the commonplace web applications like Google maps and Google earth. Geomatics is important in the earth science sector for many areas including: utilizing global positioning systems (GPS) for locating their property, infrastucture and geological samples, a geophysical-image analysis system for analyzing and display of their remote sensing data from geophysical (seismic, radiometric isotopes, electromagnetic, etc.) to imagery (airphotos, satellite) data, and a geographic information system (GIS) to house all of these data in addition to other geospatial data (points: wells, sample assays, etc.; lines: roads, streams, contours, etc.; and polygon: claim block, watershed, anomalies, etc.) and raster or grid cell based maps. Landscapes are influenced by several factors including geology, soils, climate, glaciations, topography, and vegetation cover, among others. In order to study geology and the influence on landscapes and their evolution, we attempt to map these different factors using geomatics.

2. Digital elevation models – lidar

One of the most critical layers to describe a landscape is the topography of the terrain, which is expressed as a digital elevation model (DEM) within a GIS environment. Most elevation models have been derived from stereo aerial photography, in which measurements of the ground are hampered by the tree canopy. The challenge to make accurate topographic measurements of the earth under the forest canopy has been a problem until recently. Airborne laser scanning has the ability to solve this problem and see through the vegetation, depending on the canopy density and closure. Light Detection and Ranging (lidar) is a technique that combines a laser ranging system with an inertial navigation system comprised of a survey grade GPS and an inertial measurement unit (IMU) in an aircraft (Fig. 1). Detailed technical overviews of lidar systems have been described by various authors (Flood et al., 1997; Gomes Pereira and Wicherson, 1999; When and Lohn, 1999; and Maune,

2001). Lidar has been used in a number of geoscience applications, including the analysis of river networks (Stock et al., 2005), the generation of cross-sections across rivers (Charlton et al., 2003), in general glaciology (Krabill et al., 1995, 2000), groundwater monitoring (Harding and Berghoff, 2000), investigation of landslides (McKean and Roering, 2003), and in the mapping of tectonic fault scarps and geomorphic features (Haugerud et al., 2003) and examining coastal processes (Brock et al., 2002). Lidar has been used to demonstrate improvements in mapping bedrock and surfical geology as well as landscape metrics such as stream incision, and to resolve and map the individual volcanic flow units of the North Mountain Basalt and the identification of crater structures within the lower flow unit (Webster et al., 2006, 2006 A). Lidar has been merged with geophysical data to revise the geological boundaries along the Avalon-Meguma terrain boundary in Nova Scotia, Canada (Webster, Murphy and Quinn, 2009). Webster et al. (2009) used lidar and drill logs to estimate the thickness of aggregate deposits in the Annapolis Valley, NS.

The detail and resolution of DEMs derived from lidar are ten times better than previous available data for these areas. Generally, DEMs derived from aerial photography or other remote sensing systems such as the Shuttle Radar Topography Mission (SRTM) have degraded accuracies in forested areas and have horizontal resolutions of ca. 20 - 30 m. The benefit of lidar is that a narrow laser beam is directed from the aircraft towards the earth's surface and reflected back in order to measure the range or distance from the aircraft to the ground. The beam divergence is typically very small (0.3 rmad), resulting in a laser footprint diameter of 30 cm on the ground at 1000 m flying height. The system is mounted in an aircraft and the laser fires hundreds of thousands of shots per second that are directed across a swath toward the earth's surface by an oscillating mirror (Fig. 1).

The laser pulse is reflected back to the sensor, which records the two-way travel time that is then converted into a range or distance based on the speed of light (Fig. 1). Since the laser pulse can partially hit several targets (top of canopy, branches, tree trunk, buildings, shrubs, and ground) the lidar sensor can record several returns. Earlier Lidar sensors, ca 2003, could record a single return, first or last. Today's sensors are capable of capturing multiple returns, for example the Optech ALTM-3100 model is capable of recording up to four returns per emitted pulse. For many surveys there is no requirement for these intermediate laser returns so only the first and last returns are recorded during the survey. The laser range distances are combined with the angular and trajectory data from the scan mirror, GPS and IMU to determine the three-dimensional location of the targets in the GPS World Geodetic System of 1984 (WGS84) mapping system. A local GPS base station is setup over a known monument to establish geodetic control for the survey (Fig. 1). Ground check points should also be acquired along road surfaces within the survey area in order to validate the lidar elevations as part of the vertical accuracy assessment process. The GPS from the aircraft (rover) is combined with that of the base station (reference) in order to obtain the position of the aircraft every second. The GPS information is combined with the angular measurements from the IMU that are 200 times per second (Fig 1). The lidar points are generally output to a map projection coordinate system, typically Universal Transverse Mercator (UTM) in meters east and north. The lidar elevations are referenced to the GRS80 ellipsoid and not above mean sea level or a local national vertical datum. A geoid-ellipsoid model can be used to convert the elevations from ellipsoidal to orthometric heights above the geoid. In Canada we currently use the HT_2 model supplied by the Canadian Geodetic Survey of Natural Resources Canada to relate ellipsoidal heights to the Canadian Geodetic Vertical Datum of 1928 (CGVD28). The lidar surveys are typically acquired in swaths along overlapping flight lines or strips (Fig. 2).

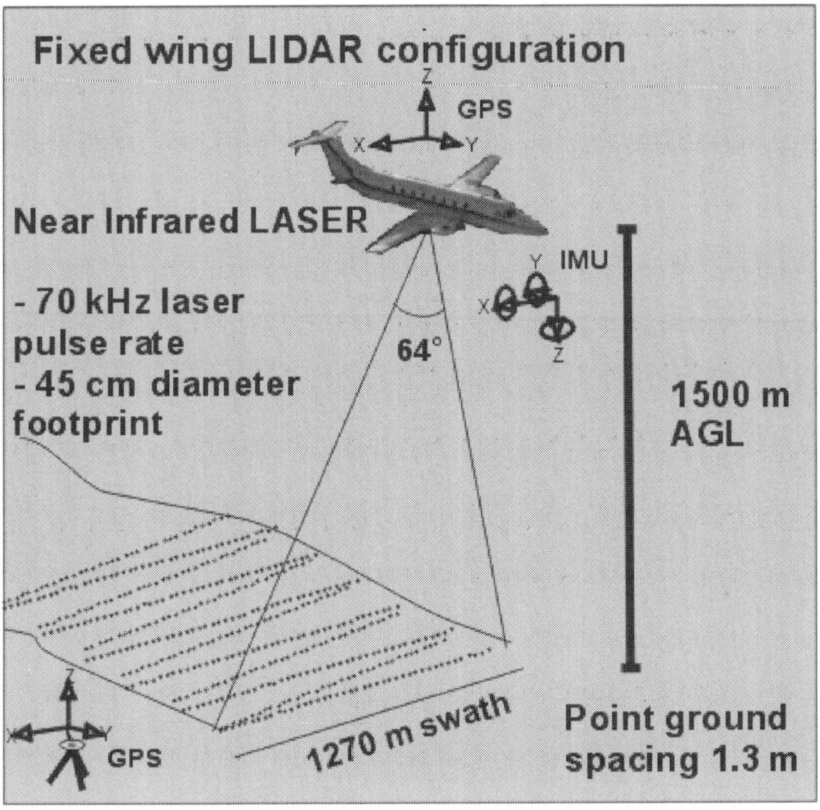

Fig. 1. Typical wide area lidar survey configuration. The laser firing at 70 kHz, with the pulses directed across the swath at 25 Hz at a height of 1500 m.

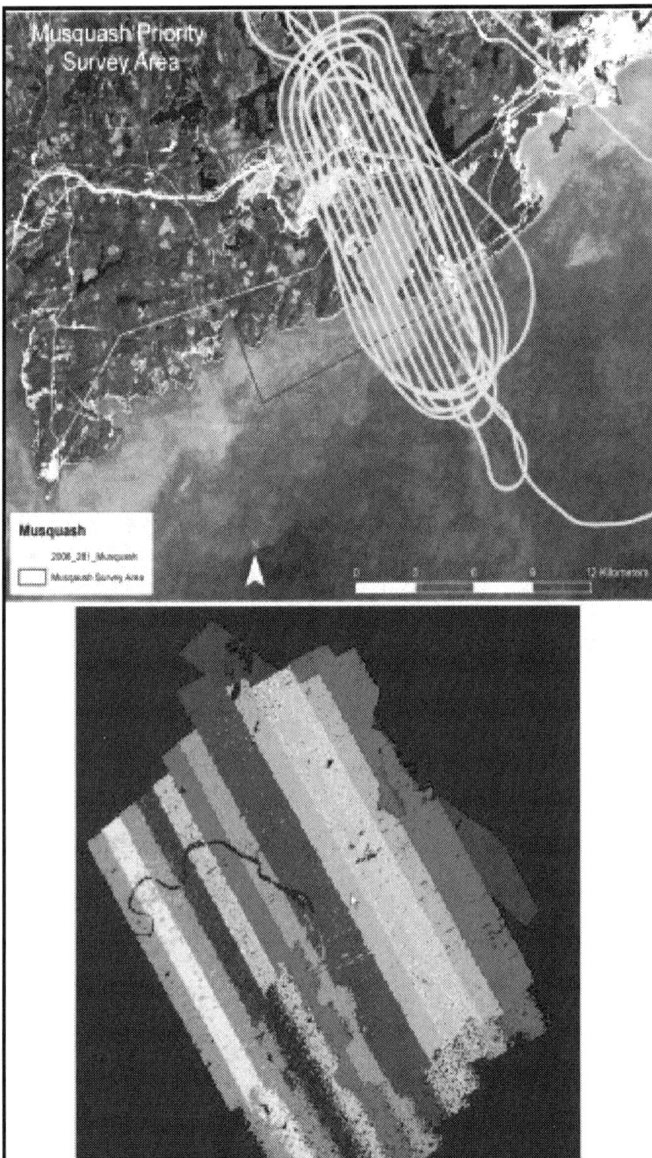

Fig. 2. Top map shows the flight trajectory of the aircraft during a lidar survey, yellow dots. The bottom map shows the lidar swaths associated with the above flight lines. Note the black outline of the river where no returns were detected because of the smooth mirror like surface of the water. Musquash, New Brunswick, Canada.

The trajectory is solved from the blend of GPS and IMU data to position the aircraft, then the laser ranges and scan mirror angles are used to compute the target position in space. The results are a set of high-density points known as a 'point cloud' that represent the ground and other targets, such as vegetation or anthropogenic features e.g. roads, buildings, bridges (Fig. 3).

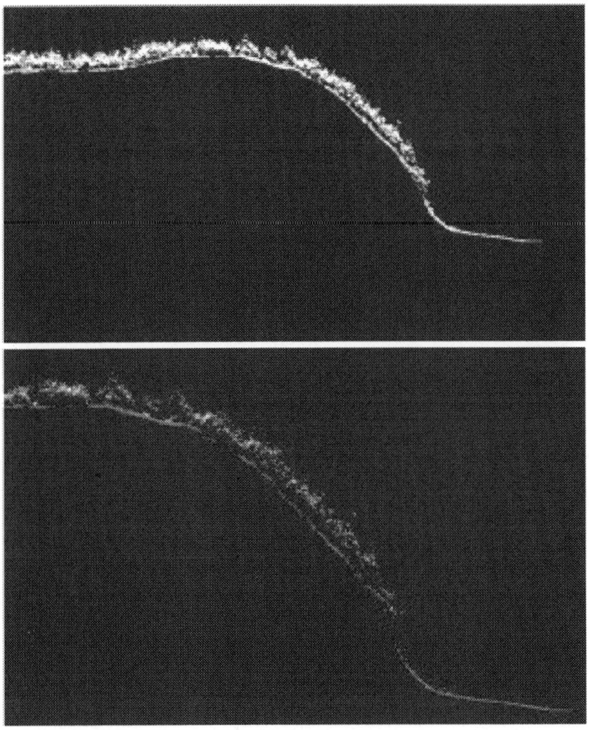

Fig. 3. Cross section of a lidar point cloud along a coastal area. Top cross-section of unclassified points. Bottom cross-section of classified ground points in orange. The remaining points are vegetation.

In order to derive an accurate DEM, the lidar points are classified or filtered into 'ground' and 'non-ground' target classes (Fig. 3). The point cloud is classified using specialized software where the points for each strip are merged together and broken down into a series of tiles based on a map projection grid system and processed individually. The classification algorithms can have problems producing accurate results in rough terrain or discontinuous slopes, dense forest areas where the beam cannot penetrate to the ground, and low vegetation being confused with the ground. Generally the lowest points are used to construct an initial surface from a Triangular Irregular Network (TIN). Then each additional point is added to the TIN if the parameters are below the threshold settings. The problem is that different thresholds are required for different terrain conditions. The two sets of lidar points, 'ground' and 'non-ground' are integrated into a GIS that can be used to interpolate different types of surfaces from the combination of points. Surfaces, such as a Digital Surface

Model (DSM) by using all of the lidar points (including those representing vegetation and ground) and the DEM using only the 'ground' points are constructed using the interpolation routines (Fig. 4). In addition to recording the time of the near-infrared 1064 nm laser pulse, the lidar system also records the amplitude of the returning pulse, known as the intensity. The intensity will vary depending on the material of the target; low reflective materials like asphalt will have very little energy returned and a low intensity compared to grass which will have a high intensity. The intensity of the points are interpolated to form a black and white type of photograph, since all of the points are used we refer to it as DSM intensity or DSMI (Fig. 4). Because the DSMI is grey-scale and is related to land cover rather than elevation, it can be combined with the DEM and DSM to form a hybrid image (Fig.4). The manipulation of lidar surface models in a GIS allows for the construction of maps that can preferentially highlight subtle geomorphic features (e.g. artificial sun illumination and vertical exaggeration). Such features are often not readily observed in traditional DEMs, or from stereoscopic inspection of aerial photographs, because of their low relief and obstructions from vegetation. Because of the scale of the many geological features being studied, regional-scale lidar surveys are required in order to assess its applicability to geomorphic research, such that features with a topographic expression can be detected and traced over long distances.

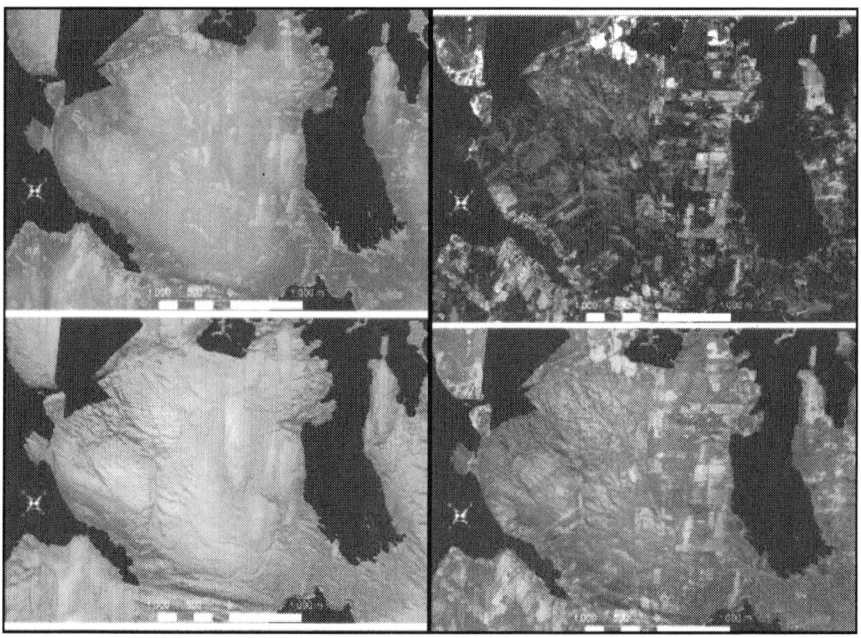

Fig. 4. Lidar surface models: Upper left map - Digital Surface model DSM using all the lidar points; Lower right map is a Digital Elevation Model DEM of ground only points; Upper Right map is the grey scale lidar intensity (land cover); Lower right map is the hybrid of the intensity plus the DEM. The folds and faults in the bedrock and drumlins are visible in the 2nd map of the DEM. Riverport, Nova Scotia, Canada.

3. Visualization

Grey-scale shaded relief maps can be constructed by illuminating the DEM from azimuth angles to highlight topographic features which trend in the perpendicular direction, for example if a topographic high trends east-west, then an illumination azimuth angle of 0 or 180 degrees would highlight the features. In addition to an azimuth illumination angle, a zenith angle can be expressed to denote where the light source is in the sky relative to the horizon. The cartographic convention for shaded relief maps is to illuminate the terrain (DEM) from the northwest or 315 degrees and with a zenith angle of 45 degrees. In order to enhance the terrain and have enough contrast between bright and darker surfaces, depending on the relief of the study area, a vertical exaggeration can be applied. A lidar DEM has been processed for a section of the Annapolis Valley, Nova Scotia and used to demonstrate various processing techniques. In the following example, the DEM has been illuminated from the eight cardinal directions 0, 45, 90, 135, 180, 225, 270 and 315 degrees at a constant zenith angle of 45 degrees and a 5 time vertical exaggeration applied (Fig. 5). The resultant grey-scale images reveal the texture or relief of the terrain, however they do not indicate the absolute elevation of features, for example a sloping surface will look the same regardless of the absolute elevation. By illuminating the terrain from different azimuth angles, different topographic features are revealed, depending on their orientation (Fig. 5). The ability to vary the sun angle and apply a vertical exaggeration is very useful in geology to enhance subtle topographic features such as lineaments, contact ridges or surficial deposits. These maps reveal two distinct morphological characteristics of the terrain with respect to the roughness of the topography. Some areas of the terrain are rough and ridges represent different volcanic flow units in contrast to smoother sections that represent areas that have a glacial till blanket covering the bedrock (Fig 5). This reflects differences in glacial history; areas to the west consist of glacially scoured bedrock with a thin till veneer, and areas to the east have a thick blanket of glacial till, known locally as the Lawrencetown Till (LT) (Stea and Kennedy, 1998). Since the shading of the terrain is limited to one direction, the resultant map only highlights features perpendicular to the shading direction and subdues features parallel to it. The application of principal components analysis, which is used to reduce data redundancy and compress multi-channel information into fewer components, has been used on multiple Radatsat images by Paganelli et al. (2003) for structural mapping. When PCA is applied to the 8 shaded relief maps, a new map is constructed where the top three principal components contain over 98% of the information, or the original variance of the 8 maps (Fig. 6). The three components PCA 1,2,3 are projected through the red, green and blue colour guns to form the new composite that highlights all of the topographic features, regardless of their orientation. The advantage of this map, in contrast to the grey-scale images, is that very few areas are in shadow and more features are highlighted (Fig. 6). The composite may be difficult to interpret since it reflects the dominant topographic features of the landscape in the 3 components. The dominant landforms are the North and South Mountains, highlighted in shades of red, separated by the Annapolis Valley in shades of green. The topographic feature which face southeast are in shades of blue (Fig. 6). In addition to the PCA composite the mapped surficial geology deposits of the Lawrencetown Till blanket (TL) have also been superimposed to compare to the terrain roughness. The arrow in figure 6 denotes a change in the glacial landforms visible in the valley. To the west of the arrow, the landforms resemble drumlins associated with an earlier ice movement. To the east of the arrow, streamline landforms are visible in the valley that represents the last movement of ice (Fig. 6). The black box in figure 6 denotes areas where raised beach terraces are visible on the DEM.

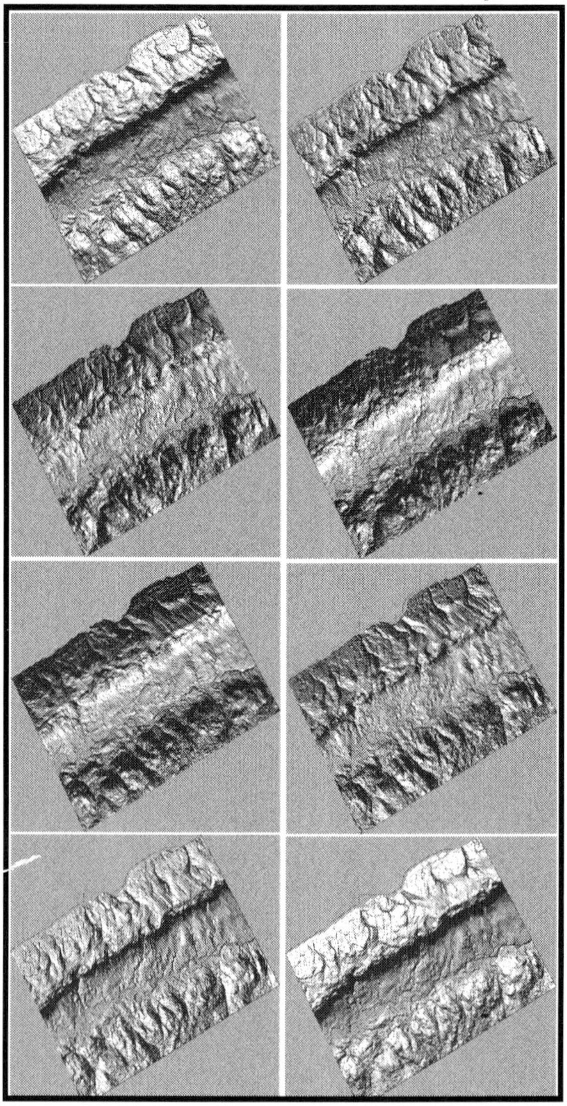

Fig. 5. Shaded relief maps on a lidar DEM for a section of the Annapolis Valley, Nova Scotia. All maps have had a zenith angle of 45 degrees and a vertical exaggeration of 5 times aplied. Top left azimuth of 0 degrees, top right azimuth of 45 degrees, second row left azimuth of 90 degrees, second row right azimuth of 135 degrees, third row left azimuth of 180 degrees, third row right azimuth of 225 degrees, bottom row left azimuth of 270 degrees, and bottom row right azimuth of 315 degrees. Images are approximately 15 km by 10 km. North is at the top of the page.

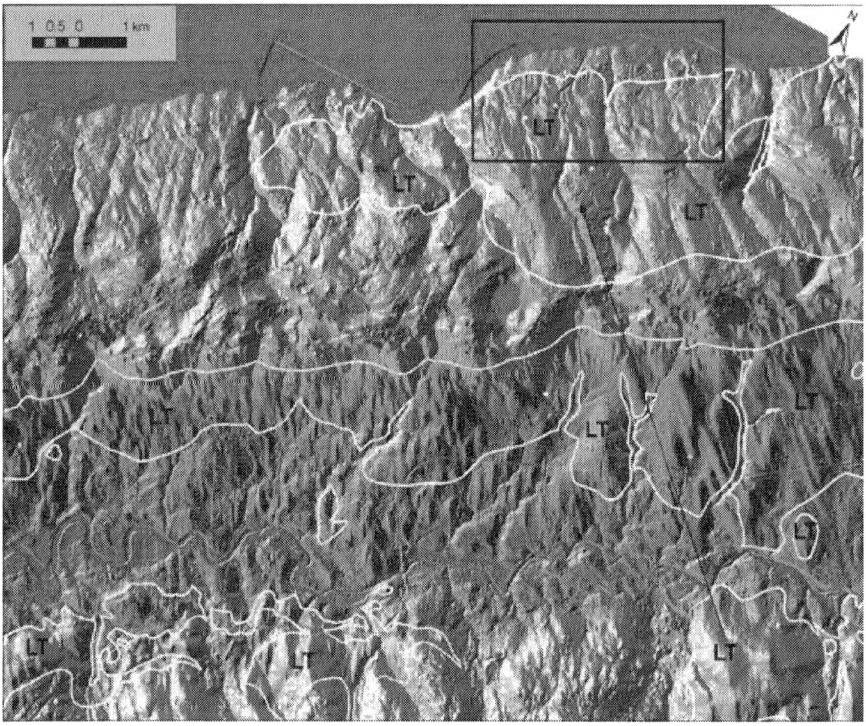

Fig. 6. Principal components 1,2,3 in red, green, blue respectively from the eight shaded relief maps (Fig. 5). The white outline represents the surficial geology boundaries of the Lawrencetown Till blanket (LT).

Another common method to enhance DEM for visualization is to construct a colour shaded relief model (CSR). This has the advantage of the shading enhancing the texture of the topography and the hypsometric colours denoting the absolute elevation of the terrain. An example of a CSR map for this area was constructed by first building a grey scale shaded relief map, with the sun illumination from the northwest (335°) at a zenith angle of 45° with a 5-times vertical exaggeration, then colour was applied to the DEM based on elevation, from below sea-level (hues of blue), to low lying land (green through yellow) to the highest point along the North Mountain (ca 265 m) (red) (Fig. 7). The colourized DEM is then merged with the grey scale shaded relief map in order to provide the texture of the terrain as a result of the shading effect. Since the colours have been applied to the terrain from the lowest elevation corresponding to the shortest visible wavelength (blue) to the highest elevation corresponding to the longest visible wavelength (red), the map appears in 3-D if viewed with Chromadepth™ glasses. Chroma stereoscopy is the technique of using colour to depict depth (Toutin and Rivard, 1995). The glasses are based on a diffraction grating which separates the incident light into different patterns depending on the wavelength. Since the map is coded from low to high elevation by low to high wavelengths of light we see depth when the brain is forced to fuse the multiple image patterns together to form a single image of the terrain. The benefit of this technique is that that map can be viewed and interpreted with or without the 3-D glasses. In comparison to the anaglyph method that requires the red and blue glasses to reveal a 3-D image in black and white. This technique offsets each image in red and blue proportional to the elevation; however, the map only can be easily interpreted when it is viewed with the glasses.

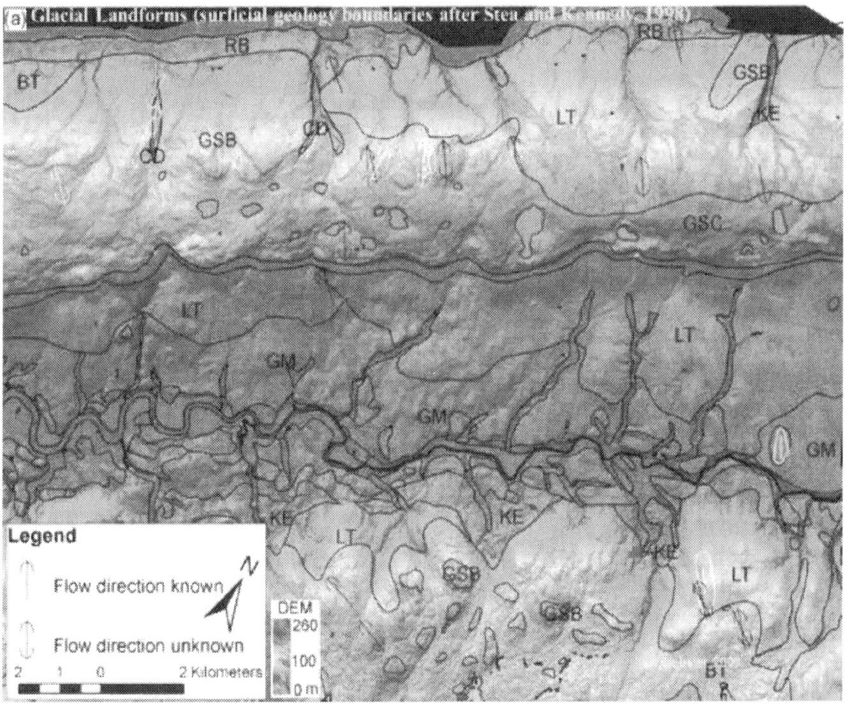

Fig. 7. Colour shaded relief map from lidar DEM. Surficial geological boundaries have been superimposed. The colour scheme is optimized for a Chroma-stereoscopic affect when viewed with 3-D Chromadepth™ glasses.

This technique of merging a grey-scale image with a colour image has been utilized in the past to integrate geophysical data with radar imagery or lidar shaded relief maps (Webster, Murphy and Quinn, 2009). It has also been used to generate "pan sharpened" images utilizing the new optical satellites where a panchromatic band at a 0.5 m resolution is combined with a multispectral (colour) set of bands at a courser resolution such as 2.5 m. The resultant hybrid image has the benefits of the 0.5 m panchromatic detail and the spectral colour information of the courser dataset. In geoscience, this method of data integration is especially useful when datasets that provide complimentary information that can be interpreted when they are integrated. For example, the lidar DEM highlights variations on the surface topography that reflects both bedrock and surficial geology features. Airborne radiometric surveys measure the amount of equivalent uranium, thorium and percent potassium near the surface and have been used in exploration. Since the gamma rays that the

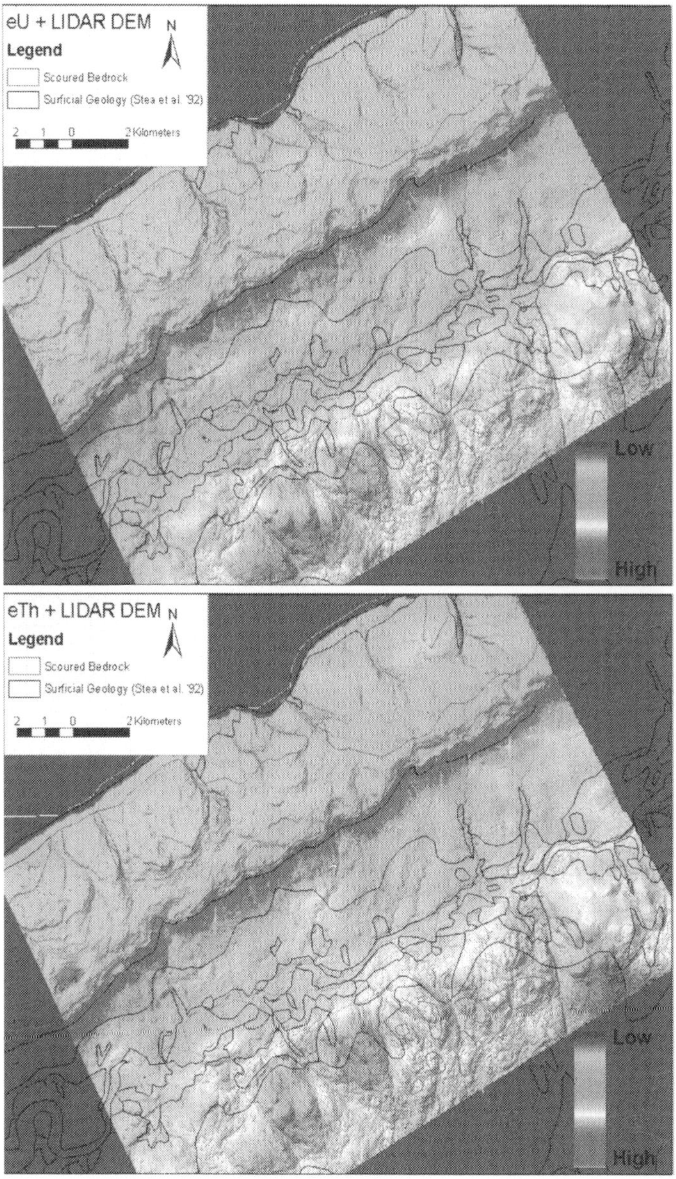

Fig. 8. Radiometric equivalent uranium (top) and thorium (bottom) have been colourized and merged with the lidar shaded relief maps. Low concentrations are colour coded blue through higher concentrations in red. The red areas on the south side of the valley correspond with granite bedrock.

sensor detects do not penetrate vary far through the soil, this geophysical measurement indicates what the concentration of the radioactive isotopes is near the surface. These airborne surveys are often gridded at a course resolution, ca. 250 m where the features appear blurry compared to the detail of lidar maps. The shaded relief lidar has been merged with the colourized equivalent uranium and thorium to produce hydrid maps (Fig. 8). The surficial geology boundaries have been superimposed which indicate the glacial till has a different radiometric signature that the underlying bedrock geology of the North Mountain (Fig. 8). The glacial till contains fragments of the South Mountain Batholith granite which occurs on the south side of the valley (MacDonald and Ham, 1994). The boundary between the scoured glacial bedrock (red outline) and the glacial Lawrencetown Till (Fig. 6-7) is highlighted by the contrast in radiometric element concentrations. The thorium values are anomalously low in the area of a crater within the basalt flow units and may reflect a difference in the chemistry of the basalt in that location (Fig 8 bottom). This technique allows two datasets to be interpreted at the same time and for the courser resolution dataset to be sharpened based on the detail of the grey-scale data. GIS systems allow the user to "fly through' the data or generate perspective views of the terrain and drape other GIS layers, either in the form of imagery or maps on the terrain. This technique further enables us to interpret the terrain and the relationship to lithology or glacial history. The ability to quickly visualize the terrain in a perspective view can often reveal relationships that are not readily visible from the standard top down map view (Fig. 9).

Fig. 9. Perspective view of shaded relief lidar map with the watershed boundaries superimposed.

4. Modern day processes: Watersheds and erosion

Understanding the relationship between stream incision and factors related to fluvial erosion such as rock-uplift, climate, base level changes, and bedrock resistance to erosion (e.g. Stock and Montgomery, 1999; Kirby and Whipple, 2001; Stock et al., 2005) is important for the analysis of landscape evolution (e.g. Kooi and Beaumont, 1996; Dietrich et al., 2003; Pazzaglia, 2003). The availability of high resolution lidar DEMs can facilitate quantitative analysis between incision and watershed morphometrics at sufficiently small scales to allow the examination of isolated influences on stream evolution. Previous studies have considered the relationship between the variations in the resistance of bedrock to erosion (Sklar and Dietrich, 2001) and stream or basin morphometry to the fluvial processes between regions (Belt and Paxton, 2005). However, the variations of bedrock resistance within a region (< 100 km^2) are less constrained, in part due to the scale of studies (Montgomery and Lopez-Blanco, 2003).

Fluvial processes in glaciated terrain are complex because glaciers and streams sequentially may occupy the same valleys but obey different laws of erosion, making the signatures of glacial and fluvial processes difficult to distinguish. Studies applying the stream power law often use the contributing drainage area as a surrogate parameter for stream discharge which, in addition to the local channel slope, controls the stream's ability to incise the underlying bed (e.g. Snyder et al., 2000). However, few studies examine the local hydrological effects of surface and groundwater interaction on discharge (Tague and Grant, 2004). At this scale, factors such as glacial till cover and the fracture density of bedrock can influence infiltration rates and affect peak annual stream discharge.

An example of utilizing a high-resolution lidar DEM to examine metrics of similarly-sized catchments that have been modified by glaciation is presented for the North Mountain within the Fundy Basin. The study area was selected because (i) the catchments are developed on three shallowly dipping volcanic flow units of the Jurassic North Mountain Basalt (NMB) which each have uniform resistance to erosion throughout the study area, (ii) the Bay of Fundy provides a uniform base level for all streams, (iii) there is a clear distinction in till cover thickness over the east and west portions of the study area, and (iv) the age of deglaciation and subsequent fluvial erosion is well documented and uniform throughout. The stream incision depths are related to the variability of the flow unit's resistance to erosion.

The land cover on the North Mountain is influenced by the occurrence of the till cover; farmland (pastures and hayfields) and mixed forest dominate in the east where the till is thickest, whereas the west has mostly mixed forest cover. There are more roads and anthropogenic influences in the east compared to the west where only one paved road occurs along the coast. The coastline varies between gently sloping bedrock platforms and ca. 25 m cliffs that occur in embayments. The streams on the Fundy side of NMB have evenly-spaced mainstems (1.5 km), similar catchment areas (ranging from 2 to 8 km^2) and are all consequent dendritic drainages with stream densities ranging from 0.9 to 2.9 km/km^2. The streambeds are typically 80% bedrock and 20% boulder-covered. Till is present in the streambed of some of the basins, attesting to the youthfulness of these catchments and to the inheritance of some low relief pre-glacial topography. Within the NMB study area, there are similar size basins (2 - 8 km^2) that drain scoured bedrock, and occur in the transition zone with scoured bedrock in their headwaters and glacial till near their outlets, and drain a glacial till blanket covering the basalt. The streams are ephemeral with their peak flows occurring in the spring and fall seasons. Their long profiles are ungraded and have several knick zones.

Watersheds are calculated for the main streams draining into the Bay of Fundy from the lidar DEM based on outlet locations identified on 1:10,000 scale topographic maps (Fig. 10). Most GIS systems can calculate the watershed draining into a stream based on the DEM. The standard D-8 algorithm (Jenson and Dominque, 1988; Costa-Cabral and Burges, 1994) is used to determine down-stream flow direction and sinks (depressions within the DEM treated as errors by the algorithm) are filled in the DEM to allow continuous down stream flow. However, when dealing with DEMs at high-resolution, other considerations must be made. Inspection of the drainage basin boundaries and stream longitudinal profiles indicates that most catchments have sinks. Many of these sinks are adjacent to the raised elevations of a roadbed captured by the high resolution of the lidar. As a culvert could not be represented on the DEM, a "notch" was cut across the roadbed and assigned an elevation of the nearest downstream cell to improve the accuracy of the flow direction algorithm and to prevent excessive erroneous sink filling operations in deriving the catchment basins and stream profiles. This modification improved accuracy of the flow direction algorithm, prevented excessive erroneous sink-filling operations in deriving the catchment basins and stream profiles, and allowed the stream to "pass through the roadbed". The overall result is the generation of a more accurate flow accumulation grid and basin boundary (Fig. 9-10). It

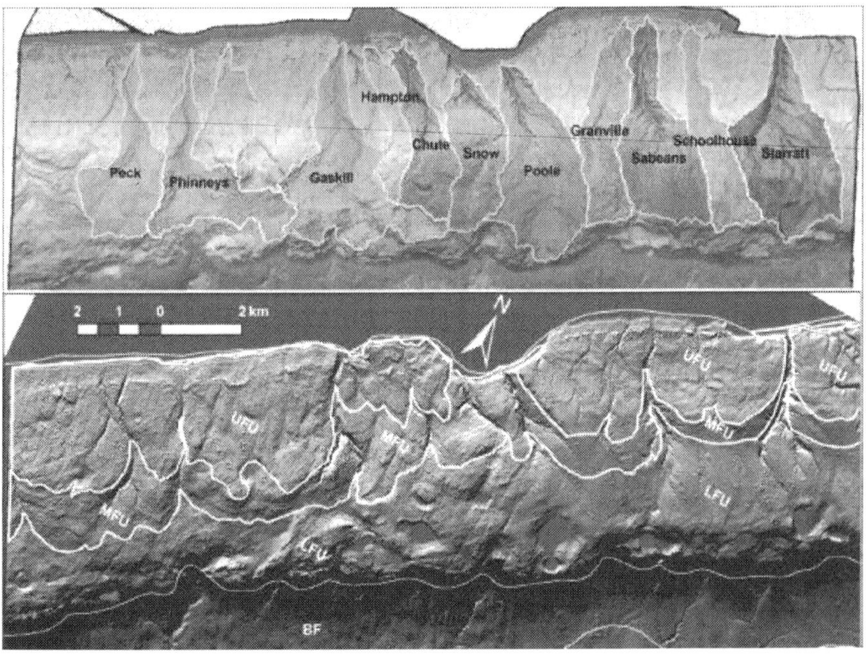

Fig. 10. Top: Lidar DEM with derived watershed boundaries for the streams draining the North Mountain Basalt. Bottom: Basalt flow units for the North Mountain over grey-scale shaded relief lidar DEM. UFU - Upper Flow Unit, MFU - Middle Flow Unit, LFU - Lower Flow Unit.

was determined that the streams from the topographic map and the longitudinal profiles obtained from the original DEM prior to sinks being filled are the most representative based on field observation and used for analysis. The flow units of the NMB have been subdivided into three distinct flow units: the lower flow unit (LFU) consists of a thick (40 - 150 m) massive single flow that is columnar jointed, the middle flow unit (MFU) conformably overlies the LFU, and consists of multiple thin flows that are highly vesicular and amygdaloidal, and the upper flow unit (UFU) conformably overlies the MFU, outcrops along the shore, and consists of 1-2 massive flows (Fig. 10 bottom).
The surface profiles of the drainage divides bordering each basin are averaged and the stream longitudinal profile is subtracted to compute the incision depth along the stream's entire length. The basalt flow units were intersected with the stream longitudinal profiles and the incision depth was summarized for each flow unit (Fig. 11).

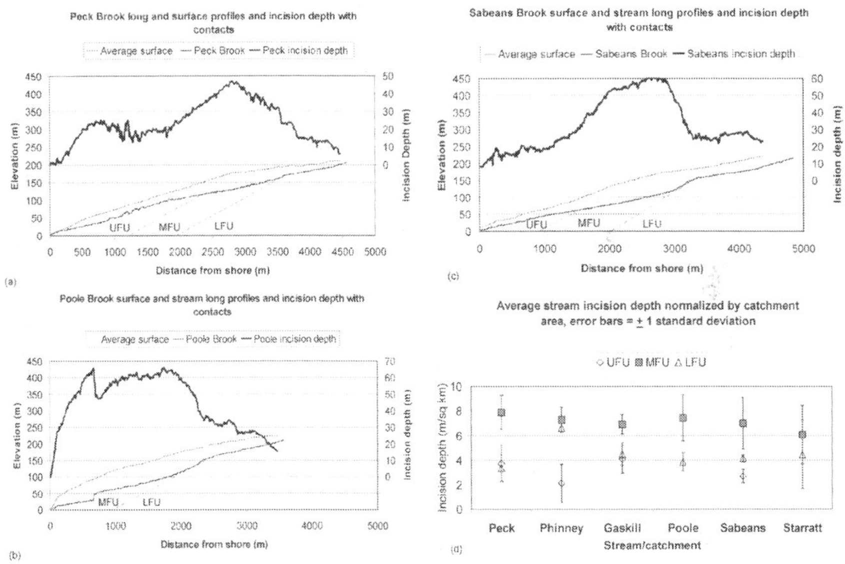

Fig. 11. Stream incision depth diagrams for the main drainage basins along the North Mountain. The surface profiles associated with the drainage divides and the stream long profile are plotted along with the depth of incision (difference between surface and stream profiles). The NMB flow unit (UFU, MFU, LFU) contacts have also been projected to intersection the streambed and related to the depth of incision. (A) Peck Brook profiles and incision. (B) Poole Brook profiles and incision. (C) Sabeans Brook profiles and incision. (D) Average incision depth for each flow unit of the NMB normalized by the drainage area for each basin and error bars indicates ± 1σ.

The stream profiles and incision depths were overlain on the flow unit map of the NMB in order to relate the incision depth to the basalt flow units. The flow units dip approximately 6° to the northwest and have been projected onto the stream profiles (Fig. 11). In general the stream incision depth reaches a maximum within the middle flow unit (MFU). Many knick zones occur either within the MFU or upstream of the contact between the MFU and lower flow unit (LFU). Incision in the upper flow unit (UFU) and LFU is similar in 3 of the 4 basins studied where both units outcrop in the streambed (Fig. 11, D). The average incision depth for the MFU is 45 m compared to 29 and 19 m for the LFU and UFU, respectively. The area percentage of each flow unit per basin and the length percentage of each flow unit per stream suggest that the percentage of flow unit per basin is a better indicator of stream incision depth than the percentage of stream length within a flow unit. The average incision depth is lowest in the catchments where the till cover is thinnest. However, the highest incision depths are associated with the catchments in the transition zone between the thin and thick till blanket areas. The valley cross-sections are used to compute the volume of material removed as described in Mather et al. (2002) for each basin. The elevations associated with the drainage divides were used to construct a paleosurface of the NMB following a similar method to that described by Brocklehurst and Whipple (2002) and Montgomery and Lopez-Blanco (2003). The lidar DEM was then subtracted from this surface in order to quantify the volume of material removed by glacial-fluvial processes and the patterns of erosion for each basin (Fig. 12).

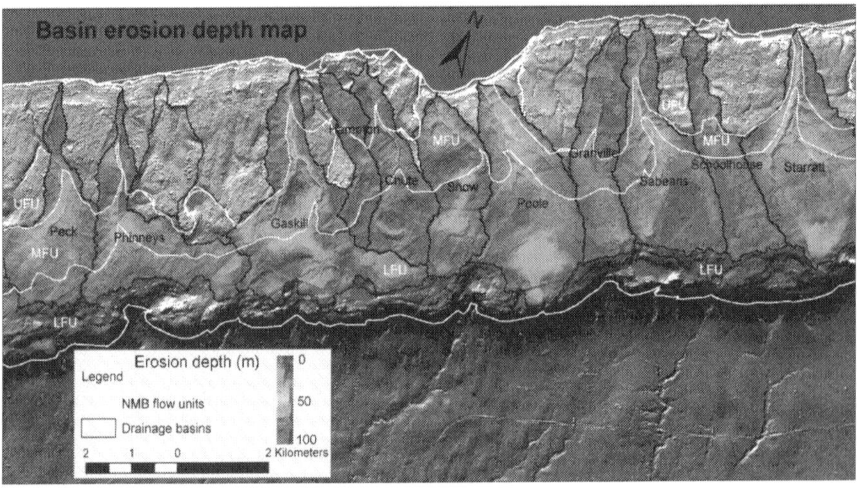

Fig. 12. North Mountain drainage basin erosion depth map with basalt flow unit boundaries. The western basins have incision depth maximums of approximately 50 m and the central and eastern basins have maximum incision depths approaching 100 m.

Erosion rates are calculated from the stream incision depth curves and sediment flux from the erosion depth map assuming erosion began after deglaciation at 12 ka ± 200 yr (1σ) (Stea and Mott, 1998), Table 1.

5. Examples of other lidar DEM geoscience applications

The improved resolution and accuracy under the forest canopy often reveals details that allow traditional geology maps to be improved and contacts between units better defined. Previously, geologists had to rely on sparse outcrop locations along the coast and along

Catchment	Till cover	Volume of sediment removed km³	Maximum sediment flux (km³/ka) assuming erosion started at 12 ka.
Peck	Thin veneer	38.3	3.2
Phinney	Thin veneer	37.4	3.1
Gaskill	Transition zone	81.8	6.8
Poole	Transition zone	91.7	7.6
Sabeans	Thick blanket	47.3	3.9
Starratt	Thick blanket	98.3	8.2

Table 1. Catchments grouped by the amount of till cover and sediment volume removed. Maximum sediment flux per catchment.

stream beds in combination with interpreting aerial photographs that did not penetrate the vegetation canopy. A lidar survey was flown to examine Piping Plover habitat, an endangered shore bird, along the south shore of Nova Scotia (Fig. 13). The area is completely forest covered except along the coast and had been mapped by the Geological Survey of Canada, Open File 1768 (Hope et al., 1988) (Fig. 13).

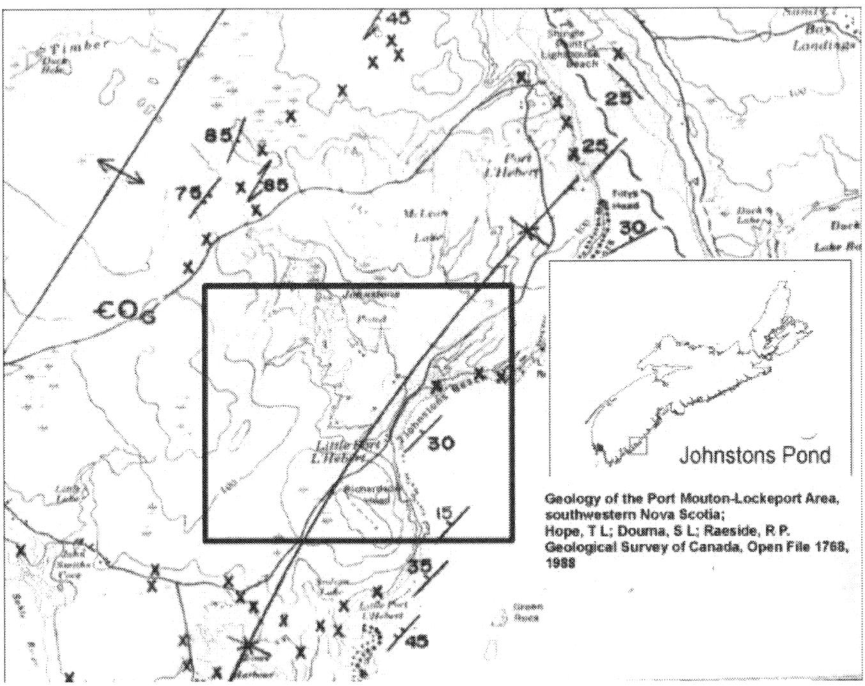

Fig. 13. Geological map of Johnston's Pond area. The black box outlines where the lidar survey was conducted. Note the only geological unit mapped is COg, indicating the Cambro-Ordivician Goldenville formation which is comprised of slates and greywacke. A syncline fold axis passes through the study area.

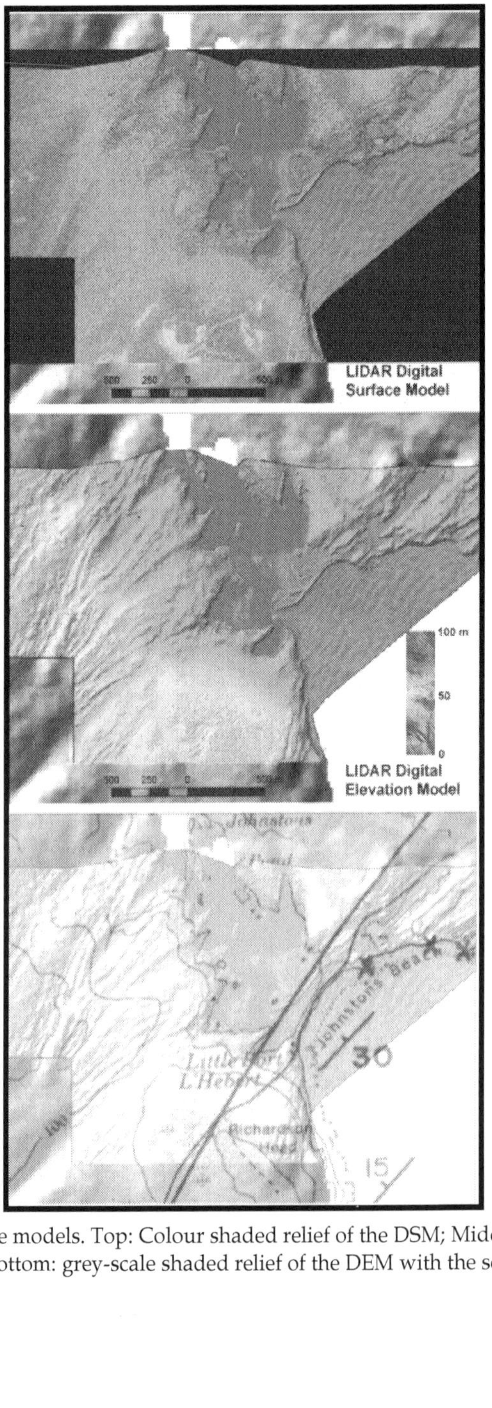

Fig. 14. Lidar surface models. Top: Colour shaded relief of the DSM; Middle: Colour shaded relief of the DEM; Bottom: grey-scale shaded relief of the DEM with the scanned geology (1988).

The lidar survey was conducted during full 'leaf-on' conditions, thus making penetration of the laser pulse to the ground more difficult. The lidar points were classified and surface models constructed, the DSM incorporating all of the lidar returns and the DEM utilizing only the ground points. Colour shaded relief maps of the surface models were constructed and interpreted (Fig. 14). The ability to remove the vegetation points reveals the bedding of the slates and a massive dome structure in the south (Fig. 14). The previous geology indicates that the entire area is made up of sedimentary rocks and is folded into a single syncline which passes directly through the dome structure (Fig. 14, bottom). Based on the visual interpretation of the terrain models, shaded relief and CSR, and a visit to the site for follow up field checks, a new geology map has been derived (Fig. 15). In addition to a new fault being mapped, a granite pluton has also been added to the map. Field evidence to support the occurrence of a fault at this location, where the bedding has been truncated on the CSR DEM, is based on the flat lying sedimentary rocks being tipped vertical in the area proximal to the fault (Fig. 15). A large granite boulder or possible outcrop was found in the field which further supports the interpretation that the topographic dome evident on the lidar is a granite pluton. The variable bed resistance to erosion allows the bedding planes to be traced over large distances even under the forest canopy in the lidar DEM.

Other examples of were the ability to penetrate the forest canopy has assisted geologist in identifying geohazards including sink holes and karst topography is presented next. The Windsor Group represents evaporates, gypsum and salt deposits of Carboniferous age in Nova Scotia. These deposits occur throughout Maritime Canada as sedimentary basins formed on the flanks of the highlands. The area of Oxford, Nova Scotia is used to demonstrate the ability of lidar to map karst topography (Fig. 16). This type of landscape can be a hazard as the bedrock is dissolved by the groundwater and the area can become undermined and local subsidence can occur. The lidar was processed to a DSM and DEM and colour shaded relief maps were constructed (Fig. 16). The bedrock geological boundaries are overlaid to highlight where the Windsor Formation occurs and contain rocks susceptible to the development of karst topography. As can be seen in figure 16, the karst topography crosses under the divided 100 series highway south of the town of Oxford and trends northeast-southwest.

In glaciated terrain, the topography reflects the glacial and fluvial deposits and often masks the bedrock structures. In these areas, the lidar surface models can be used to interpret the unconsolidated sediment deposits and better reconstruct the recent history of the area. If adequate control exists on the locations of bedrock, through outcrop locations or boreholes, a bedrock surface can be constructed and used to derive sediment thickness using the lidar surface model. Lidar was flown along a section of the North Mountain in May 2003 during 'leaf-off' conditions. As mentioned earlier, the North Mountain is underlain by basalt dipping northwest at 6 degrees. Webster et al. (2006) used field checks and the lidar to constrain the individual flow units, especially in areas covered by glacial till. In areas of thick glacial till, the morphology of the flow units is not evident in the lidar because of the smoothed till cover. In this case, planes representing the flow unit boundaries were projected through the DEM and used to define where the flow unit contacts intersected the surface topography (Fig. 17). A series of glacial deposits occur south of Port George along the North Mountain. As a result of the glacial till, the area supports local farms and the land cover is mixed between cleared and forest. The lidar DEM clearly highlights the mound of sediment on the North Mountain (Fig. 18 right). The glacial features evident on the CSR DEM represent a kame deposit with eskers to the south. The kame is formed by the glacier

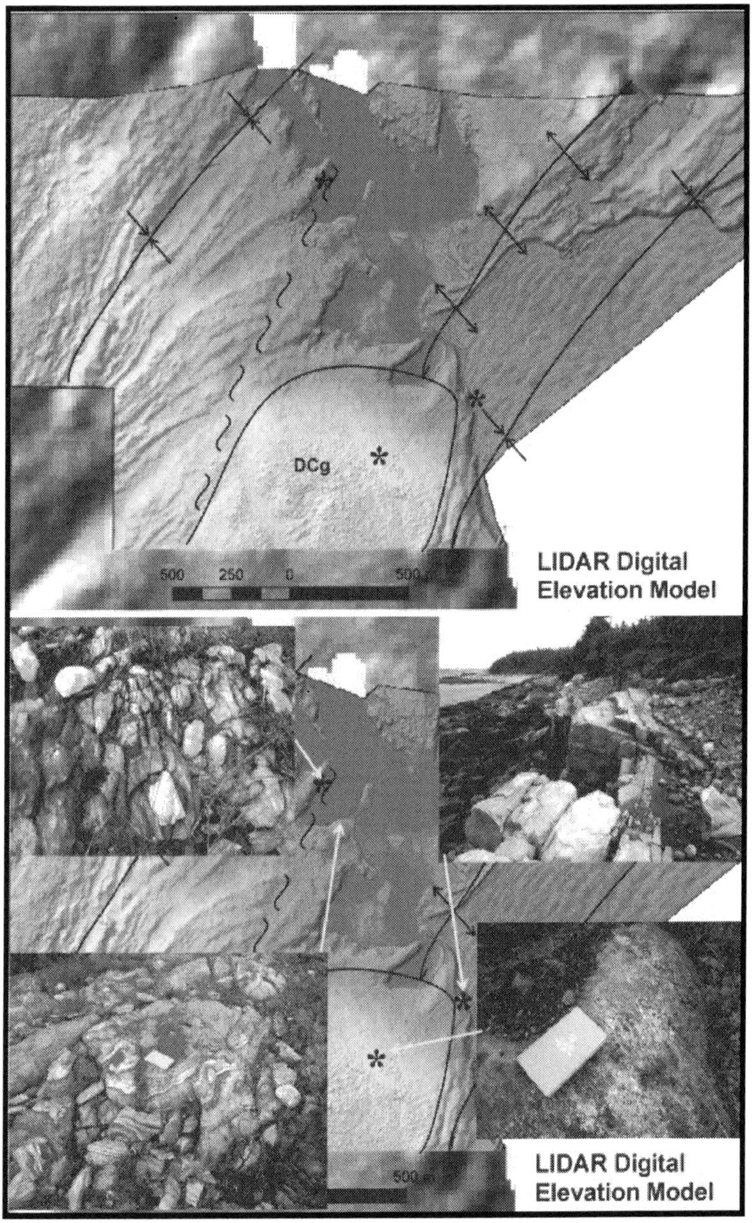

Fig. 15. Top: new interpretation of fold structures and contact between the Goldenville Formation (sedimentary rocks) and granite (DCg large dome to south). Bottom: field photos of rock outcrops that support the interpretation.

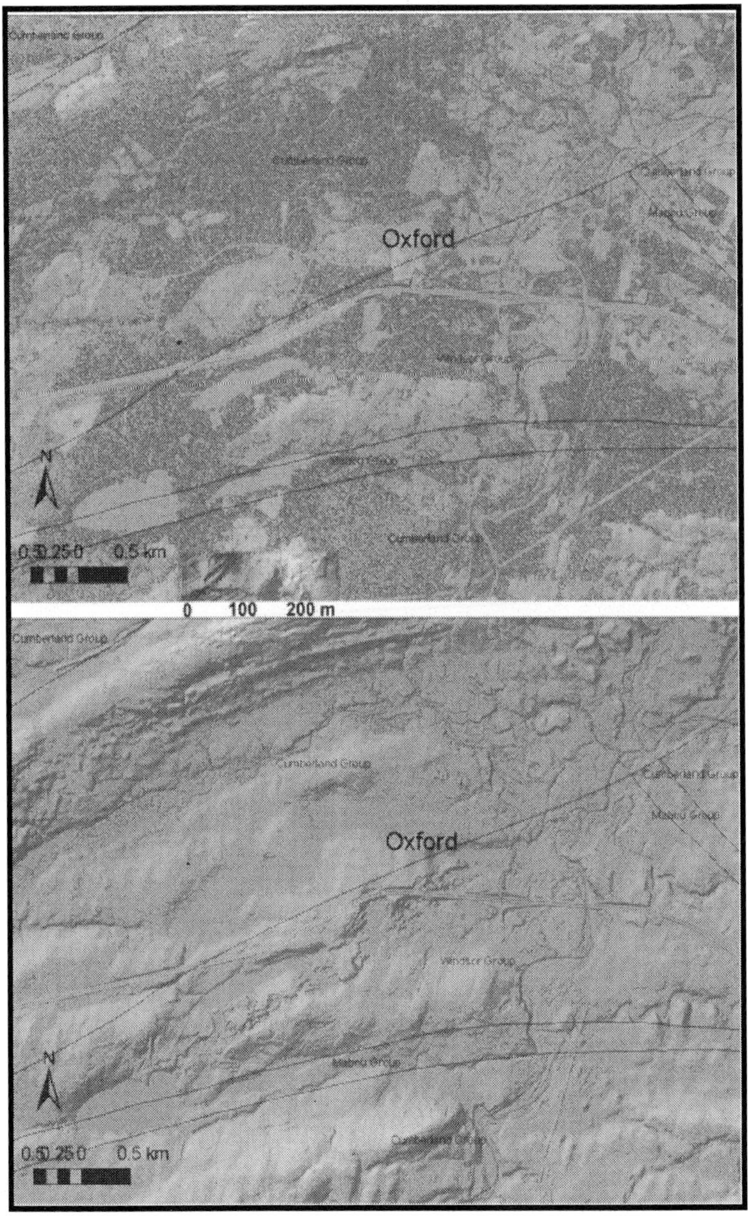

Fig. 16. Top Lidar DSM CSR of Oxford with geological boundaries. Bottom: Lidar DEM CSR with sink holes and karst topography developed below the town of Oxford. Geological boundaries (Keppie, 2000).

remaining stagnant and the trapped sediment dropping out of the ice. The mounds generally do not have well sorted sediment and show no fluvial bedding features. The eskers are the linear ridges running south of the kame and are formed by sediment collected in streams draining the melt-water within the glacier (Fig. 18). The kame and esker systems are comprised more of sand and gravel than the more clay rich glacial till and are a potential aggregate resource.

It appears that the kame and esker deposits are sitting on the bedrock surface or on a very thin glacial till veneer. GIS was used to calculate a surface representing the bedrock, which is dipping at 6 degree northwest (Fig. 19). This surface was constructed by placing a series of points around the perimeter of the kame and esker system and extracting the bedrock elevations from the lidar DEM. The points were then used to construct a TIN and a raster surface was extracted from the TIN using a linear interpolation method (Fig. 19).

Fig. 17. Basalt flow unit contact planes projected through the lidar DEM. The pink plane represents the contact between the Upper Flow Unit (UFU) and the Middle Flow Unit (MFU), the light green plane represents the boundary between the MFU and lower Flow Unit (LFU) and the dark green plane represents the base of the LFU. Adapted from Webster et al. 2006.

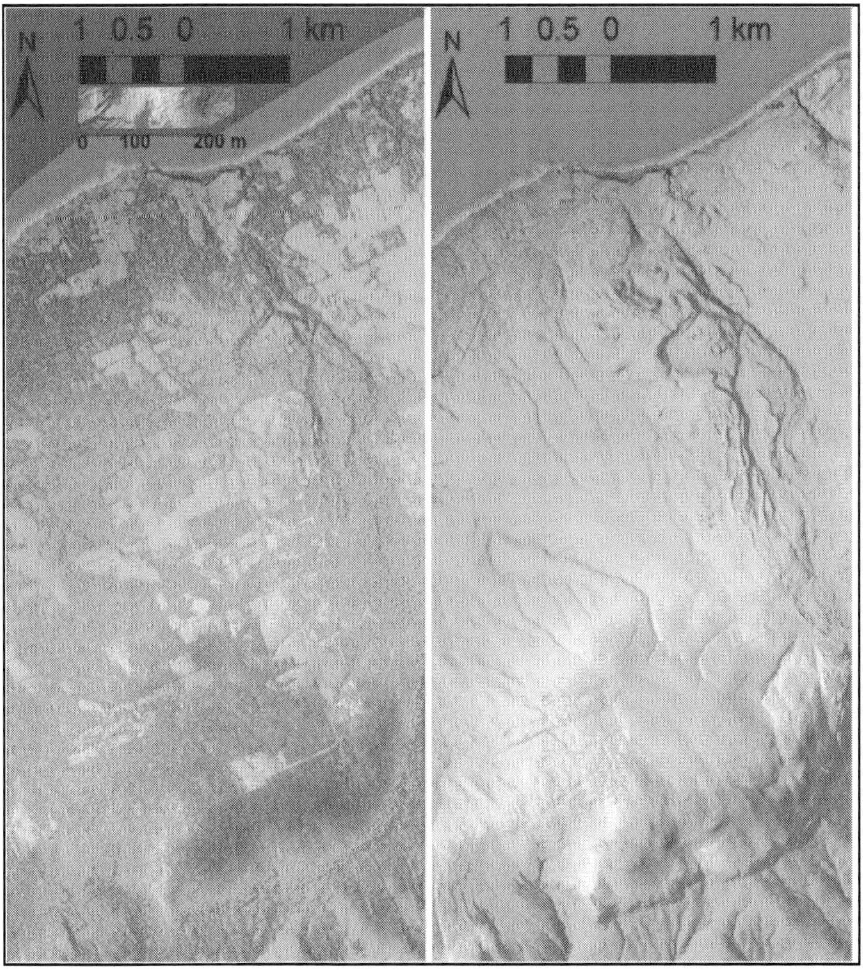

Fig. 18. Port George colour shaded relief lidar surface models. Left: DSM; Right: DEM with the kame and esker glacial deposits clearly visible.

Fig. 19. Perspective view of the lidar DEM along the North Mountain. The red plane represents the bedrock surface that intersects the DEM.

This new surface was used to calculate the thickness of the kame deposit by subtracting the lidar DEM from the bedrock planar surface (Fig. 20). Once the sediment thickness is calculate the volume of sediment can be easily derived. This allows the landowner to assess the potential value of the aggregate resource. As can be seen in figure 20 the thickness is up to 60 m in places and represents a significant amount of material that is available as an aggregate resource.

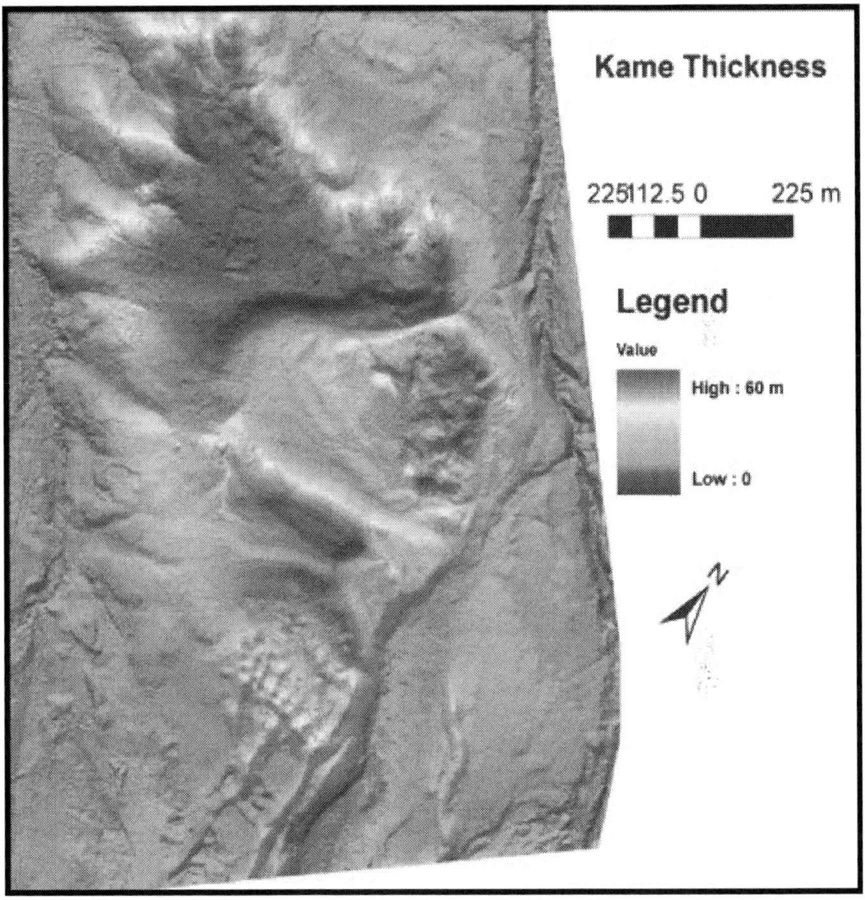

Fig. 20. Thickness of the sediment associated with the kame deposit along the North Mountain near Port George, Nova Scotia.

In some locations along the North Mountain the glacial deposits are thicker than others where ice moved into the Bay of Fundy. In areas of thicker glacial deposits along the coast, the raised beach terraces are more pronounced in the lidar DEM (Webster et al., 2006, A). These terraces represent sea-levels that where 35 m higher than present ca. 12,000 years ago. These higher sea-levels occurred after deglaciation when the melt-water caused the ocean to rise faster than the crust rebounded forming these terraces at the highest elevation.

As noted earlier the influence of glacial till over parts of the North Mountain have smoothed the topography which contrasts between the rough terrain where there is a thin veneer of glacial till and the basalt flows are evident and the smoothed surfaces where the glacial till is thickest (Figs. 6-8). Topography can exhibit a sense of being fractal, which means the measurements we make of the terrain surface are a function of the scale at which we make the measurements (Turcotte, 1992). In other words, the terrain will appear rougher and more undulating as you make observations at larger and larger scales (more detail and smaller areas), as compared to measurements of the terrain taken from data at smaller and smaller scales or courser resolution. Fractal roughness is the difference between the terrain undulations as measured from data at two different scales. To attempt to quantify the difference in roughness between the thin and thick glacial till areas along the North Mountain, a method has been devised that approaches a measure of fractal roughness (Fig. 21).

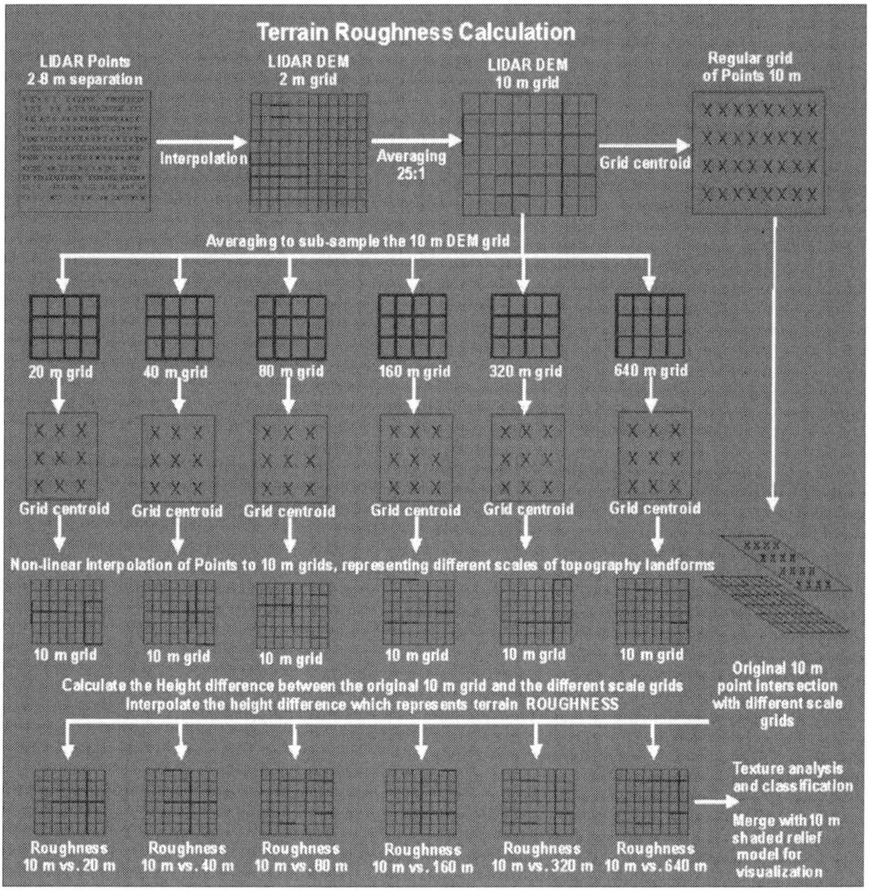

Fig. 21. Schematic of the data processing algorithm to calculate terrain roughness at different scales. Grids are represented by the square grid pattern and points are represented by "X" patterns.

The method involves starting with the irregular point spacing of the classified lidar 'ground' points and interpolating a DEM surface to a regular grid cell size of 2 m. This DEM grid is then averaged to a 10 m grid cell to facilitate processing and reduce the degree of noise of the surface. The 10 m grid is then converted back to points based on the grid centroid locations and assigned the elevation of that cell. The 10 m grid is also sampled up in resolution using an averaging technique to coarser and course DEMs of 20, 40, 80, 160, 320 and 640 m respectively. The choice of what grid cell resolution to resample to is dependent on the scale of the terrain features of interest. In this case, the grid cell resolution was increased by a factor of two each time for the purpose of demonstrating the technique (Fig. 21). These averaged DEMs of variable resolution are then converted to point centroids, where the point spacing would equal the grid cell spacing. These points are then non-linearly interpolated to 10 m grids, where each represents a different scale (decreasing scale with increasing point-grid cell spacing) of the terrain. The original 10 m grid cell points are then used to extract the elevation values from the variable scale grids and the difference between the 10 m scale elevation and the variable scale elevation is calculated. This difference in elevation, from different scale representations of the terrain, is then used to interpolate a new grid which represents the roughness or difference in roughness between the 10 m scale and the coarse scale grids (Fig. 21). The roughness grids have positive and negative values representing valleys and hills respectively on the 10 m DEM that are smoothed over as one moves to coarser and coarser scales of the topography. Profiles were extracted from the different roughness grids for a section of the North Mountain where the glacial till is thin and where it is thick (Fig. 22). The actual roughness profiles for the two

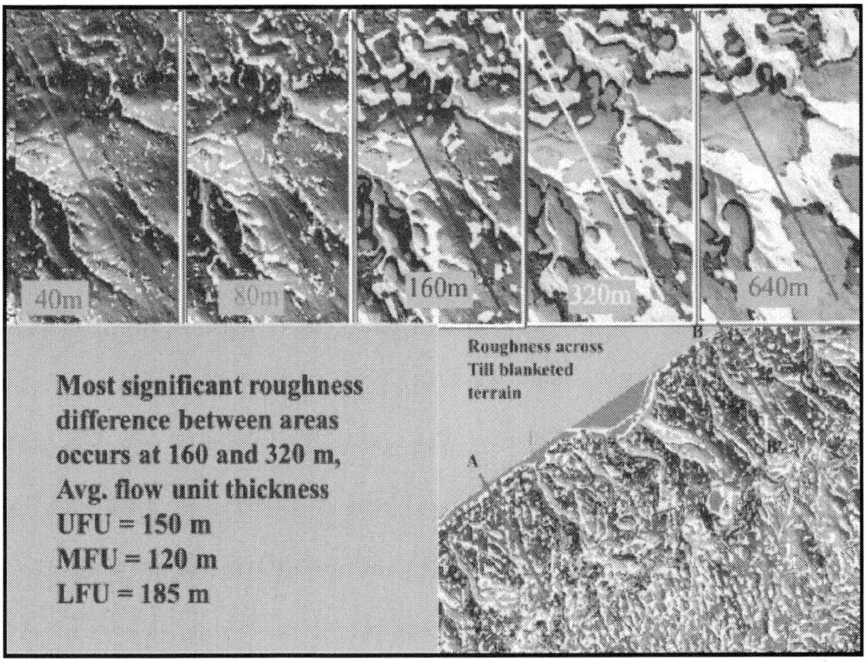

Fig. 22. Example of different terrain roughness grids along the North Mountain. Profile location A-A' is in the area of thin till and rough terrain, while profile B-B' is in the area of thick till and smoother terrain (See Fig. 23 for profiles). The top series of maps shows the location of profile B-B'.

locations are presented in figure 23. As can been seen in this figure the magnitude of the roughness difference between the two sites is significant with profile B-B′ being smoother. The differences in roughness at scales of topography closer to the 10 m grid show less differences between the two areas (e.g. 40 and 80 m grid differences) (Fig. 23). The most significant differences between the two profiles occurs once the scale of the DEMs are above 160 m (Blue line Fig. 23). This scale is interpreted to be related to the average volcanic flow unit thicknesses which varies between 150-185 m and are the dominant features that are causing the roughness in the areas of thin glacial till.

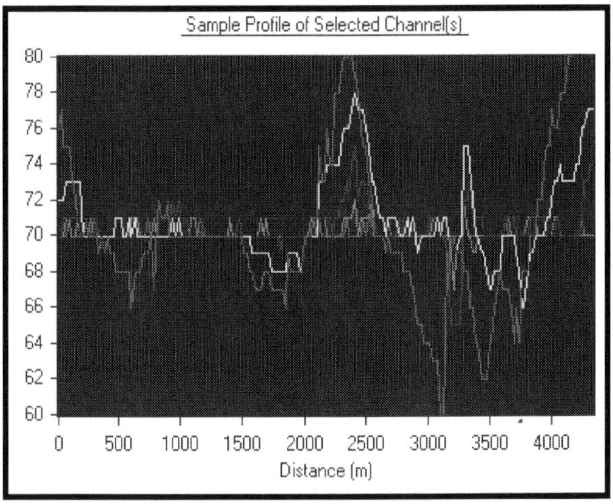

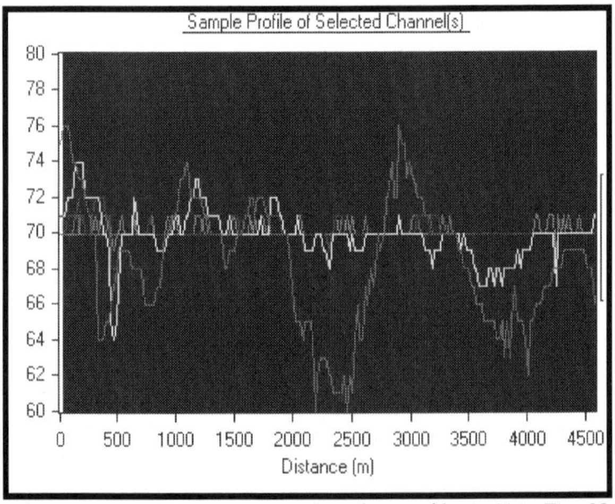

Fig. 23. Profiles of terrain roughness. Top graph is for profile A-A' in Fig. 22 thin glacial till, Top graph is for profile B B' in Fig. 22 thick glacial till. The colour of the profile line corresponds to the line and colour used for the different scales of topography (ie. 40 - red, 80 - green, 160 – blue, 320 - yellow and 640 m – magenta respectively Fig. 22).

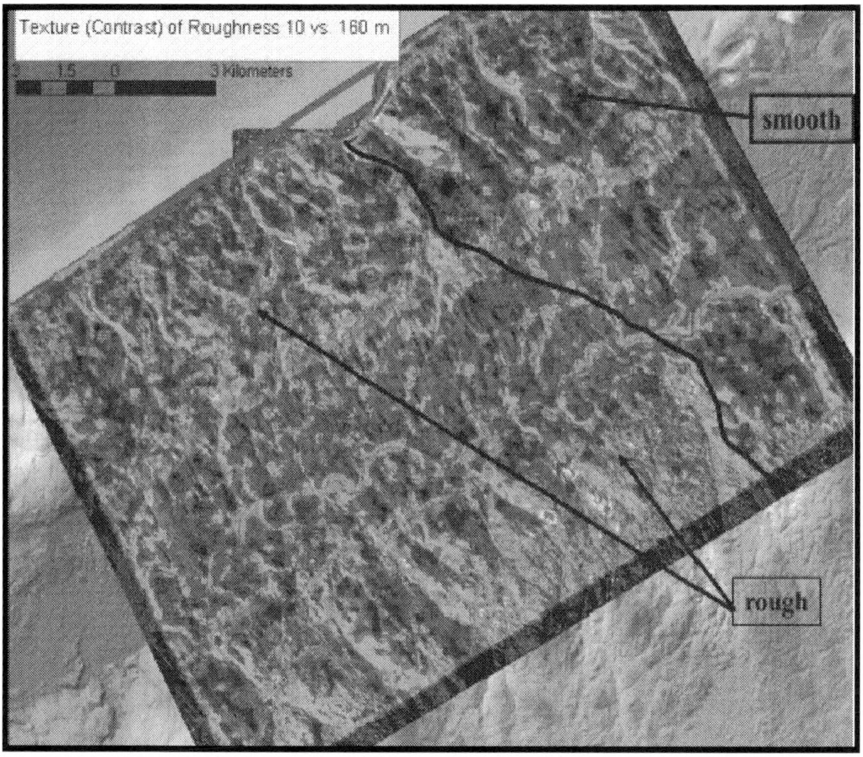

Fig. 24. Texture measure of the roughness grid derived from the difference of the 160 m grid cell terrain and the 10 m grid cell terrain map. Note the patches of dark grey represent a low texture value, or low roughness. The grey boarder around the map is no data.

A further metric was calculated from these difference roughness grids in the form of a texture measurement, which is typically related to the variance or standard deviation of the grid cell values within a moving window. The difference in roughness between the two profile locations begins to be significant at the 160 m scale; this roughness grid was used to calculate the texture in order to determine if the terrain is quantifiably different in these locations (Fig. 24). This map shows a significant contrast between the thick and thin glacial till cover, as marked by the heavy black line across the map. This difference in texture is more pronounced when applied to the 160 m scale roughness map than any of the original lidar DEMs at variable scales.

This approach of quantifying the degree of roughness of the terrain may help in eventually forming a fully automated landscape classification system. For example, if we examine the roughness difference grid at the 320 m scale, we can identify the streamlined landforms and drumlins within the valley floor (Fig. 25). The green areas represent topographic highs and the yellow areas represent topographic lows and the grey background represents smooth terrain (Fig. 25). This approach produces results that can complement other methods where the degree of curvature is estimated for the terrain, ie. concave or convex slopes, which are

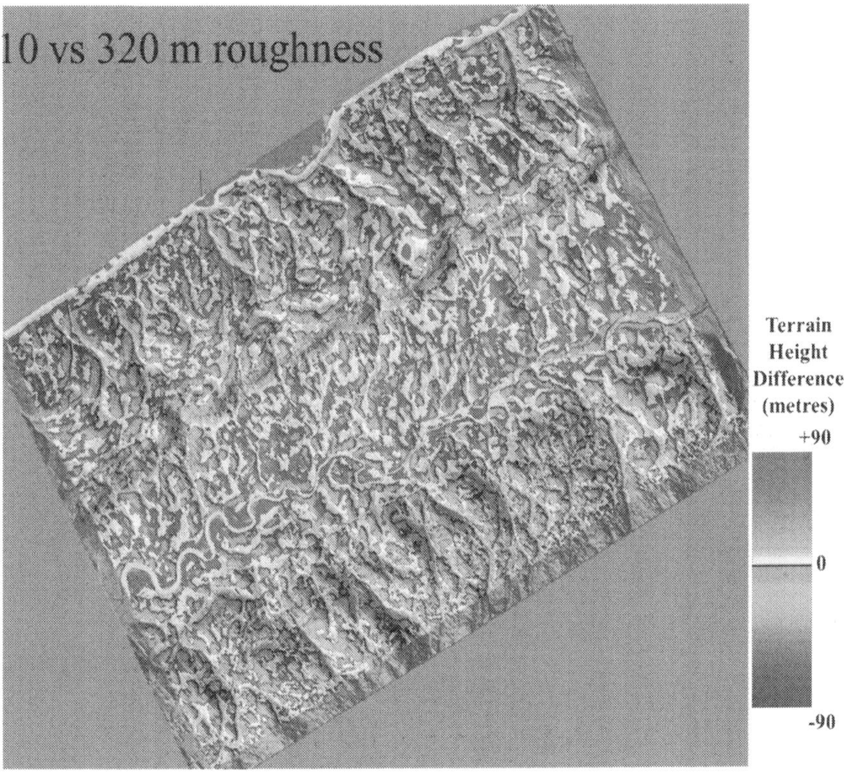

Fig. 25. Example of roughness grid difference between topographic scale 320 m and 10 m. Note the drumlins and streamline glacial landforms are highlighted in the valley floor as well as some of the basalt flow unit boundaries on the North Mountain.

calculated using the second derivative of the DEM. This method can be tailored to the scale of the topographic features of interest and through GIS processing, the features can be automatically extracted and quantified.

Recent advances in laser mapping technology have developed "mobile mapping" systems where laser scanners have been deployed on land and marine vehicles instead of aircraft. One disadvantage of an airborne lidar system is that it does not sample or measure the terrain of steep slopes very well and certainly not of any areas that are covered by an overhang because of the viewing geometry of the system. As a result of this limitation of airborne systems, steeper slopes and cliffs along the coast are not well resolved with airborne lidar DEMs. As with urban buildings and other structures, the cliff face is not resolved to the same level of detail as the rest of the terrain. This limitation affects the ability of airborne lidar to be used for detailed coastal change in these environments. In this case a laser scanner can be setup on the beach and the coastal cliff section can be imaged to capture all of the detail and merged with the airborne lidar to form a complete 3-D surface of the terrain. In urban landscapes, "mobile mapping" systems are being used to capture the sides of the buildings and all of the street furniture (e.g. signs, light poles, fire hydrants etc.). Joggins, Nova Scotia is a world UNESCO Heritage site because of the Carboniferous fossils that occur there in the outcrop exposed along the coast. Unfortunately the cliffs are actively eroding and expected to erode faster as sea-levels rise. An airborne lidar survey was conducted over the area in 2007 and a follow up ground-based lidar survey was conducted in 2009 to obtain details on the cliff face and monitor erosion (Fig. 26). Repeat ground-based surveys are planned in order to measure the change along the cliff face.

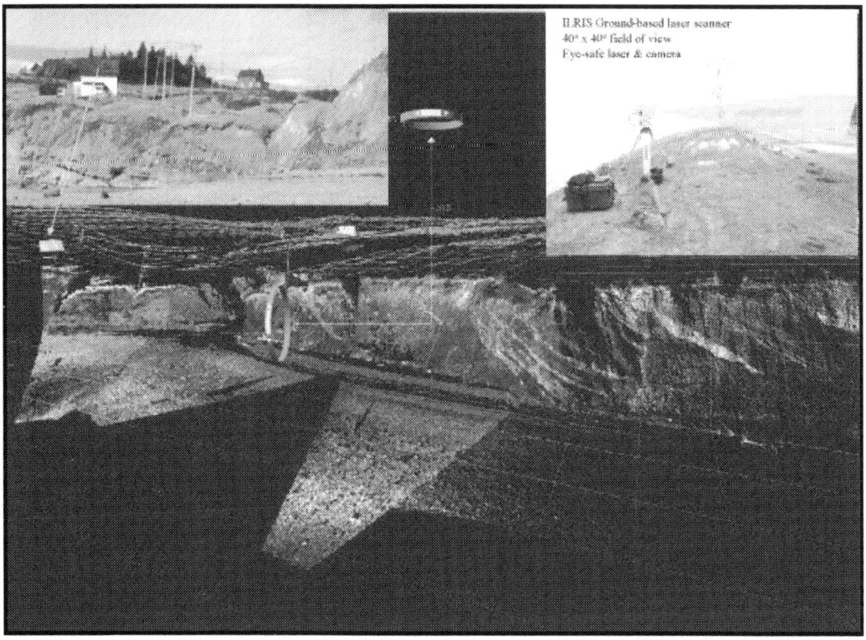

Fig. 26. Point clouds of ground-based lidar scans displayed as grey-scale intensity of the fossil cliffs at Joggins, NS. The wire frame diagram in the background is from an airborne lidar DEM. The top left photo shows the location of a house and power lines at the top of the cliff. The top right photo shows the lidar setup on the beach to image the cliff.

Other coastal areas in Nova Scotia are comprised of glacial till and are even more susceptible to erosion from the sea. Repeat ground-based lidar surveys were conducted at Cape John, Nova Scotia in 2010 and 2011 in order to measure the effects of winter storms on the coastline. Targets are placed within the landscape to be scanned and positioned using survey grade GPS. Once the terrain is scanned, a lidar point cloud can be georeferenced by identifying the targets and their coordinates to transform them into a map projection system so they can be integrated with other spatial data in a GIS. DEMs at 20 cm grid cells representing the bank were constructed from the georeferenced point clouds. Surveys were conducted in Dec. 2010 and Jan. 2011 after a major storm surge event on Dec 21 and 28th 2010 impacted the area. The DEMs from Dec. and Jan. were subtracted to map out the differences in the terrain and calculate the volume of sediment removed during the storm event (Fig. 27). Along a 150 m stretch of the coast, 771 cubic metres of sediment were removed based on the DEM analysis (Fig. 27). The storm surge associated with this event was 1.5 m above the usual water level and a local tide gauge measured the maximum water level to be 2.21 m above CGVD28 or approximate mean sea-level. This does not include breaking waves or wave run-up. A profile of the DEM before (Dec. 16) and after (Jan. 4) the storm indicates the erosion of the bank reached the 4.75 m elevation level. The erosion profile is typical of a coastal section where the tow of the sloped bank has been removed, causing the bank slope to steepen, and some of the material deposited in the near shore. The bank was frozen at the time of the second survey and could not maintain the steep slope and has since slumped during the spring thaw cycle to a stable slope causing the top of the bank to further retreat from the coast.

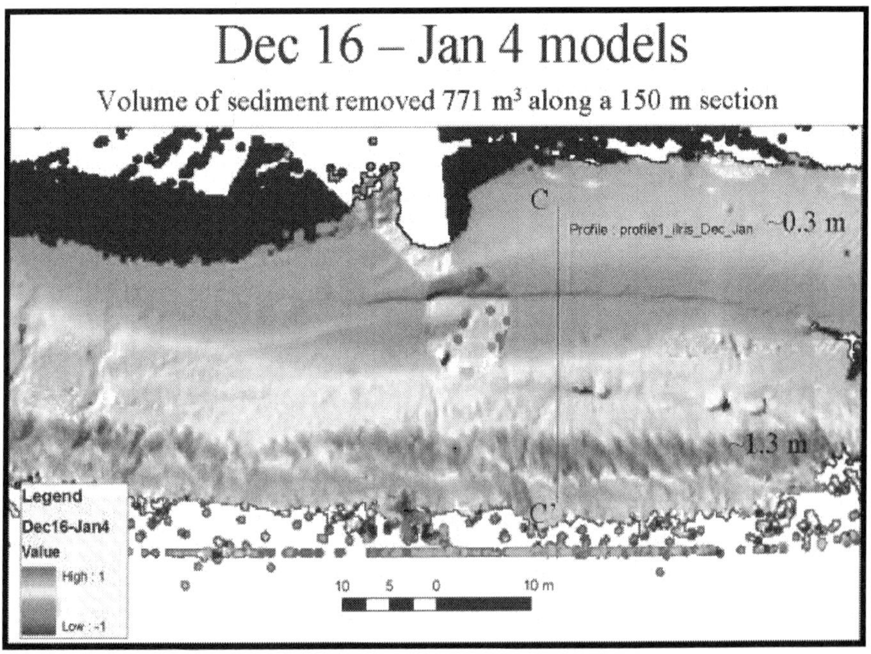

Fig. 27. Difference grid of Dec. and Jan. DEMs derived from ground-based lidar. Note the profile location C-C'.

This approach of utilizing a ground-based lidar allows for detailed analysis of changes of steep slopes that are not easily resolved with airborne techniques. The approach also has the advantage of allowing for quick deployment and is less expensive than an airborne survey. The method is fairly labour intensive and requires targets to be setup and precisely surveyed in order to georeference the scan. The latest "'mobile mapping" systems are equipped with a similar navigation system as the airborne lidars that provide a solution based on the GPS position of the sensor and the angular measurements of an IMU. As a result, the lidar scans are automatically georeferenced in a similar fashion as with the airborne systems, although some urban and cliffed environments present challenges to good GPS satellite reception.

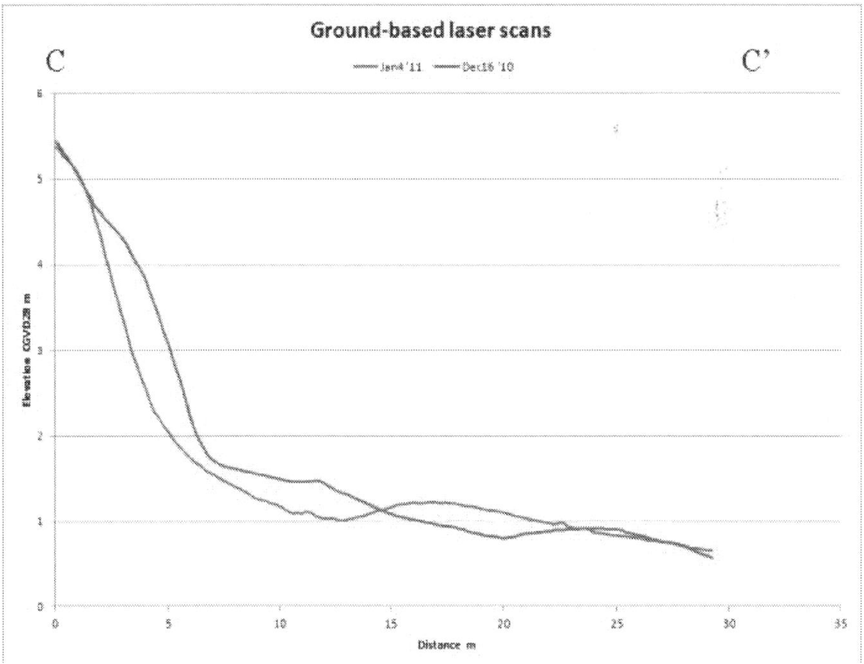

Fig. 28. Profile of coastal glacial till bank at Cape John, NS. before (Dec. 16 - red profile) and after (Jan. 4 – blue profile) a major winter storm event that occurred on Dec. 21, 2010. The upper erosion limit is at the 4.75 m elevation and the bank has been significantly steepened.

6. Conclusions

The objectives of this chapter were be to describe what terrain mapping lidar is and what map products can be derived from it. The ability of lidar to penetrate small openings in the forest canopy and sample the bare earth surface has revolutionized the accuracy and the way DEMs are constructed and used by Geoscientists. This review of lidar included the hardware and software involved in data collection and initial processing at a high level. The lidar point cloud must be processed and the ground points classified. Various surface models can be constructed from the classified point cloud including the Digital Surface Model, incorporating all of the lidar points (ground, buildings and vegetation), and of more importance to the geoscience community the DEM can be constructed by incorporating only the ground points into the model. An additional data product available from a lidar survey is based on the intensity of the reflected lidar pulse. All of the lidar point intensities are used to build the model which essentially resembles a near infrared photograph depending on the wavelength of the lidar laser system, typically 1064 nm. Once the lidar data are processed to map products, various GIS and image processing routines are applied to these data to allow visual and analytical interpretation (shaded relief maps for example).

Since the lidar only provides insights into the surface topography, the concepts of data integration and the generation of hybrid image maps were explained and demonstrated. For example, the grey scale shaded relief lidar was merged with geophysical data in the form of a radiometric survey of equivalent uranium and thorium. Different geophysical datasets can be integrated with the lidar to better understand the relationship between the surface topography and shallow structures and contacts (1st derivative magnetics), to deeper features (e.g. total field magnetics, Bouguer gravity). In glaciated terrains, the topography reflects the surficial deposits of unconsolidated material. The lidar DEM often reveals previously unseen details of the earth's surface that can be used to refine and revise geological maps. This was demonstrated in areas of folded and faulted sedimentary rocks in contact with a granitic pluton. In glaciated terrains, the lidar surface models were used to interpret the glacial and fluvial history of several areas in Nova Scotia. Lidar DEMs were used to measure modern surface processes such as fluvial incision and erosion. GIS and the lidar DEM were used to automatically extract watersheds and longitudinal stream profiles where knick points can be observed and related to erosion rates. Lidar surface models were also used to map geohazards in the form of sink holes and karst topography. In glaciated terrains, the thickness of the unconsolidated material associated with the glacial deposits was calculated using standard GIS techniques. The sediment thickness was calculated by constructing a bedrock surface and subtracting it from the lidar surface model. Other applications of lidar DEMs included research methods to quantify terrain roughness differences in areas of thin glacial till and areas of a thicker till blanket.

Lastly, the chapter concluded with the latest in lidar mapping which includes the use of ground-based scanners and "mobile mapping" which allows the lidar to be mounted on land and marine vehicles. This type of technology can be used to monitor the amount of material removed at open pit mines or gravel quarries. The application of a ground-based lidar was demonstrated to survey a bedrock cliff to establish a baseline of information. Repeat lidar surveys will be used to monitor the rate that the cliff is actively eroding since it is an important fossil heritage site. The impact of storm surge and erosion on a glacial till bank was demonstrated by comparing repeat ground-based lidar surveys. The volume of sediment eroded during a storm event was quantified and the vertical limit of erosion was measured from the lidar derived DEMs of the bank.

Geomatics offers the geoscientist a wide suite of data, tools and techniques to further our understanding of geology and earth surface processes.

7. Acknowledgements

There are several people who have contributed to the research presented in this chapter that I would like to thank. The lidar for the North Mountain was acquired with a grant from the Canada Foundation for Innovation. Some of the research related to the North Mountain was part of my PhD research that was funded through the NSCC and Brendan Murphy of St. Francis Xavier University. Also I would like to acknowledge my PhD supervisors: John Gosse of Dalhousie University, Ian Spooner of Acadia University and Brendan Murphy of St. Francis Xavier University. Lidar for the Oxford and Johnstons Pond areas was flown by AGRG and I would like to thank Bob Maher, Chris Hopkinson, Allyson Fox, David Colville, Ryan Goodale and Doug Stiff for their involvement in those projects. Other research interns and partners who contributed include Angela Templin, and Gordon Dickie and Matt Ferguson of Shaw Resources. The coastal fieldwork team for the repeat surveys of Cape John included Nathan Crowell, Kevin McGuigan and Candace MacDonald. Thanks to Grant Wach and Christian Rafuse for the use of Dalhousie Universities ILRIS laser scanner for the Dec. and Jan. surveys at Cape John. Funding for the repeat surveyes was provided by Will Green of the NS Department of Environment. The Joggins ground surveys were conducted with assistance from Nathan Crowell, Stephanie Rogers, Danik Bourdeau, and Kate Leblanc.

8. References

Belt, K. and Paxton, S.T. 2005. GIS as an aid to visualizing and mapping geology and rock properties in regions of subtle topography. Geological Society of America Bulletin, 117, no. ½: 149-160.

Brock, J.C., Wright, C.W., Sallenger, A.H., Krabill, W.B., Swift, R.N. 2002. Basis and methods of NASA airborne topographic mapper LIDAR surveys for coastal studies. Journal of Coastal Research, 18: 1-13.

Brocklehurst, S.H. and Whipple, K.X. 2004. Hypsometry of Glaciated Landscapes. Earth Surface Processes and Landforms, 29: 907-926.

Costa-Cabral, M.C. and Burges, S.J. 1994. Digital Elevation Model Networks (DEMON): A model of flow over hill slopes for computation of contributing and dispersal areas. Water Resource Research, 30, no. 6: 1681-1692.

Charlton, M.E., Large, A.R., and Fuller, I.C. 2003. Application of airborne LIDAR in River Environments: The River Coquet, Northumberland, UK. Earth Surface Processes and Landforms, 28: 299-306.

ChromaDepth 3-D Glasses. www.Chromatek.com

Dietrich, W.E., Bellugi, D.G., Sklar, L.S., Stock, J.D., Heimsath, A.M., and Roering, J.J. 2003. Geomorphic Transport Laws for Predicting Landscape Form and Dynamics. *In* Prediction in Geomorphology. Geophysical Monograph 135, pp. 1-30.

Dostal, J., and Dupuy, C. 1984. Geochemistry of the North Mountain Basalts (Nova Scotia, Canada). Chemical Geology, 45: 245-261.

Flood, M. and Gutelius, B. 1997. Commercial implications of Topographic Terrain Mapping Using Scanning Airborne Laser Radar. Photogrammetric Engineering and Remote Sensing, 4: 327-366.

Gomes Pereira, L.M. and Wicherson, R. J. 1999. Suitability of laser data for deriving geographic information - a case study in the context of management of fluvial zones. International Journal of Photogrammetry and Remote Sensing, 54, no. 2-3: 105-114.

Harding, D.L., and Berghoff, G.S. 2000. Fault scarp detection beneath dense vegetation cover: Airborne lidar mapping of the Seatle fault zone, Bainbridge Island, Washington State. *In* Proceedings of the American Society of Photogrammetry and Remote Sensing Annual Conference, Washington, D.C., pp. 9.

Haugerud, R.A., Harding, D.J., Johnson, S.Y., Harless, J.L., Weaver, C.S., and Sherrod, B.L. 2003. High-resolution lidar topography of the Puget Lowland-A bonanza for earth science. Geological Society of America Today, 13, no. 6: 4-10.

Hope , T.L., Douma, S.L., Raeside, R.P. 1988. Geology of Port Mouton-Lockeport Area, southwestern Nova Scotia. Geological Survey of Canada Open File 1768.

Jenson, S.K., and Dominque, J.O. 1988. Extracting Topographic Structure from Digital Elevation Data for Geographic Information Systems Analysis. Photogrammetric Engineering and Remote Sensing, 54, no. 11: 1593-1600.

Kirby, E. and Whipple, K. 2001. Quantifying differential rock-uplift rates via stream profile analysis. Geology, 29, no. 5: 415-418.

Kooi, H. and Beaumont, C. 1996. Large-scale geomorphologt: classical concepts reconciled and integrated with contemporary ideas via a surface-process model. Journal of Geophysical Research, 101: 3361-3386.

Krabill, W.B., Thomas, R.H., Martin, C.F., Swift, R.N., and Frederick, E.B. 1995. Accuracy of airborne laser altimetry over the Greenland ice sheet. International Journal of Remote Sensing, 16: 1211-1222.

Krabill, W., Abdalati, W., Frederick, E., Manizade, S., Martin, C., Sonntag, J., Swift, R., Thomas, R., Wright, W., and Yungel, J. 2000. Greenland Ice Sheet: high-elevation balance and peripheral thinning, Science 289: 428-430.

MacDonald, M. A. and Ham, J.A. 1994. Geological Map of Bridgetown, South Mountain Batholith Project. Nova Scotia Department of Natural Resources, Halifax N.S. Map 94-08.

Mather, A.E., Stokes, M., and Griffiths, J.S. 2002. Quaternary Landscape Evolution: A Framework for understanding contemporary erosion, southeast Spain. Land Degradation & Development, 13: 89-109.

Maune, D. F. 2001. Digital Elevation Model Technologies and Applications: The DEM Users Manual. *Edited by* D.F. Maune, American Society of Photogrammetry and Remote Sensing. pp. 1-250.

McKean, J., and Roering, J. 2003. Objective landslide detection and surface morphology mapping using high-resolution airborne laser altimetry. Geomorphology, 1412: 1-21.

Montgomery, D.R., and Lopez-Blanco, J. 2003. Post-Oligocene river incision, southern Sierra Madre Occidental, Mexico. Geomorphology, 55: 235-247.

Paganelli, F., Grunsky, E.C., Richards, J.P., and Pryde, R. 2003. Use of RADARSAT-1 principal component imagery for structural mapping. A case study in the Buffalo Head Hills, northern central Alberta, Canada. Canadian Journal of Remote Sensing, 29, no. 1: 111-140.

Pazzaglia, F.J. 2003. Landscape evolution models. Developments in Quaternary Science. Vol. 1. pp. 247-274.

Sklar, L.S. and Dietrich, W.E. 2001. Sediment and rock strength controls on river incision into bedrock. Geology, 29, no. 12: 1087-1090.

Snyder, N. P., Whipple, K.X., Tucker, G.E., and Merritts, D.J. 2000. Landscape response to tectonic forcing: Digital elevation model analysis of stream profiles in the Mendocino triple junction region, northern California. Geological Society of America Bulletin, 112, no. 8: 1250-1263.

Stea, R.R., and Kennedy, C.M. 1998. Surficial Geology of Bridgetown (NTS sheet 21A/14). Nova Scotia Department of Natural Resources Minerals and Energy Branch, Halifax, N.S. OFM 1998-002.

Stea, R.R. and Mott, R.J. 1998. Deglaciation of Nova Scotia: Stratigraphy and Chronology of Lake Sediment Cores and Buried Organic Sections. Geographie physique et Quaternaire, 50, no. 1: 3-21.

Stock, J.D., Montgomery, D. R, Collins, B.D., Dietrich, W.E., and Sklar, L. 2005. Field measurements on incision rates following bedrock exposure: Implications for process controls on the long profiles of valleys cut by rivers and debris flows. Geological Society of America Bulletin. 117: 174-194.

Stock, J. and Montgomery, D. R. 1999. Geologic constraints on bedrock river incision using the stream power law. Journal of Geophysical Research, 104, no. B3: 4983-4993.

Tague, C. and Grant, G.E. 2004. A geological framework for interpreting the low-flow regimes of Cascade streams, Willamette River basin, Oregon. Water resources Research, 40: W04303.

Toutin, T, Rivard, B. 1995. A New Tool for Depth Perception of Multi-Source Data. Photogrammetric Engineering & Remote Sensing, 61, no. 10: 1209-1211.

Turcotte, D. 1992. Fractals and Chaos in geology and geophysics. Cambridge Press, pp. 199.

Webster, T., Templin, A., Ferguson, M., Dickie, G. 2009. Remote Predictive Mapping of Aggregate Deposits using LiDAR. Canadian Journal of Remote Sensing. Vol. 35, Suppl. 1 (Special Issue on Remote Predictive Mapping), pp. S154-S166.

Webster, T., Murphy, J.B., Quinn, D. 2009. Remote Predictive Mapping of a Potential Volcanic Vent Complex in the Southern Antigonish Highlands using LiDAR, Magnetics, & Field Mapping. Canadian Journal of Remote Sensing. Vol. 35, No. 5, pp. 486-495.

Webster, T.L., Murphy, J.B., Gosse, J.C., and Spooner, I. 2006. The Application of LIDAR-derived DEM analysis for geological mapping: An Example from the Fundy Basin, Nova Scotia, Canada. Canadian Journal of Remote Sensing, Vol. 32. No. 2, pp. 173-193.

Webster, T.L., Murphy, J.B., and Gosse, J.C. 2006A. Mapping Subtle Structures with LIDAR: Flow Units and Phreomagmatic Rootless Cones in the North Mountain Basalt, Nova Scotia. Canadian Journal of Earth Sciences. Vol. 43, pp. 157-176.

Wehr, A., and Lohr, U. 1999. Airborne laser scanning—an introduction and overview, ISPRS Journal of Photogrammetry and Remote Sensing, 54, no. 2-3: 68-82.

20

Remote Predictive Mapping: An Approach for the Geological Mapping of Canada's Arctic

J. R. Harris, E. Schetselaar and P. Behnia
Geological Survey of Canada, Ottawa
Canada

1. Introduction

Due to its vast territory and world-class mineral and energy potential, efficient methods are required for upgrading the geoscience knowledge base of Canada's North. An important part of this endeavour involves updating geological map coverage. In the past, the coverage and publication of traditional geological maps of a limited region demanded multiple years of fieldwork. Presently more efficient approaches for mapping larger regions within shorter time spans are required. As a result, an approach termed Remote Predictive Mapping (RPM) has been implemented since 2004 in pilot projects by the Geological Survey of Canada. This project falls under the larger Geo-mapping for Energy and Minerals (GEM) program initiated by Natural Resources Canada.

Remote predictive mapping comprises the compilation and interpretation (visual or computer-assisted) of a variety of geoscience data to produce predictive maps containing structural, lithological, geophysical, and surficial information to support field mapping. Predictive geological maps may be iteratively revised and upgraded to publishable geological maps on the basis of evolving insight by repeatedly integrating newly acquired field and laboratory data in the interpretation process. The predictive map(s) can also serve as a first-order geological map in areas where field mapping is not feasible or in areas that are poorly mapped. The fundamental difference between RPM and traditional ground-based mapping is that in the latter, the compilation of units away from field control (current and legacy field observations) is largely based on geological inference while in RPM this geological inference is repeatedly tested and calibrated against remote sensing imagery.

Remote predictive mapping is of course not an entirely new philosophy for geological mapping. Geologists have long assembled diverse layers (primarily aerial photographs and aeromagnetic contour maps) of geoscience data to study the relationships between the spatial patterns for resource exploration and mapping endeavours. In the past this has been accomplished using an 'analog' approach, forcing maps printed on mylar to be portrayed on a uniform map scale on a light table. However, with the increasing availability of digital data sets and the routine use of geographic information systems (GIS), the task of studying relationships between data and producing innovative maps to assist field mapping has become easier and more versatile. Contrary to the 'light table' approach, GIS allow maps and image data to be combined, overlaid, and manipulated at any scale with any combination of layers and subjected to any integrated enhancement.

A predictive map does not represent geological *truth* but rather a best estimate of what that area may represent on the ground based on the signatures derived from the interpreted data (geophysical, geochemical, remotely sensed). For that matter, even a traditionally produced geological map may not represent the geological *truth*, as all maps, no matter how they are produced, may contain spatial and classification errors. Thus, geological features of a predictive map do not necessarily correspond to how these features would be classified on the ground by a field geologist. At the categorization level, the geological term attached to a unit or structural feature may even prove to be wrong; yet at the detection level, the identified feature may correspond to a hitherto unrecognized mappable unit or structure that can be targeted for follow-up fieldwork.

The amount, variety, and quality of data used are obviously key factors in how closely the predictive map matches the geological patterns obtained by field mapping. Another factor is the nature of the geological terrain being mapped as the data sets and associated processing and enhancement techniques being employed will vary depending on the bedrock, surficial, and topographic environment. A remote predictive map can assist the geologist in a number of ways: (1) by predicting map units that would tentatively be assigned to rock types and/or geological formations (bedrock and surficial). This is based on establishing critical relationships between imaged physical properties (magnetic intensity, gravity, gamma-ray spectrometry, spectral reflectance, radar backscatter) and patterns obtained from available geological maps and field data, (2) by predicting areas that appear to be characterized on remotely sensed images by more complex and spatially heterogeneous geological patterns, thus focusing and prioritizing field work in these areas; likewise, areas with more homogenous signatures and simpler patterns can also be identified as possibly requiring less field work to geologically calibrate, (3) by predicting a variety of structures (foliation traces, faults, dykes, lineaments, glacial flow directions, etc.). The structural information can be used in advance of field work, to supplement field observations or as stand-alone geological information and, (4) by predicting the distribution of bedrock outcrop and other physiographic features such as wetlands, areas of forest fire burns, vegetation cover, and infrastructure to support fieldwork planning.

Predictive maps can also result in a different paradigm for planning field traverses. Instead of regularly spaced traverse lines, more detailed traverses can be set up that are focused on more complex areas and on areas where bedrock outcrop has been identified. This is especially advantageous in Northern mapping campaigns where the territory is vast and mapping expenses are high.

The mechanics of producing interpretations from various geoscience data sets can be greatly facilitated by GIS and image analysis technology. For example, image interpretation can be accomplished directly on a computer touch -sensitive screen as opposed to interpreting on mylar overlays. The advantage of this screen digitization process (i.e. heads-up digitization or interpretation) is that various enhanced images can be displayed quickly to facilitate interpretation by virtually real-time comparison between different data types at any scale. Multiple iterations can be undertaken and each digital interpretation can be stored as a different GIS layer. This by-passes the cumbersome procedure of scanning and digitizing hard-copy interpretations followed by georeferencing, which can introduce spatial errors. Similar to field mapping, the successful recognition and extraction of geological information is a learning process based on experience in interpreting image data in a variety of physiographic and geological settings.

2. RPM approach

Remote predictive mapping involves the acquisition, processing, and geological interpretation of available remotely sensed data sets as well as legacy geological data. The results are predictive maps (or GIS layers) of interpreted bedrock and surficial units as well as geologic structures. Remote Predictive Mapping can be either completed in isolation from field-based mapping or can be intimately integrated with it in order to ground truth the interpretation as field mapping proceeds. **Figure 1** shows a summary of the RPM

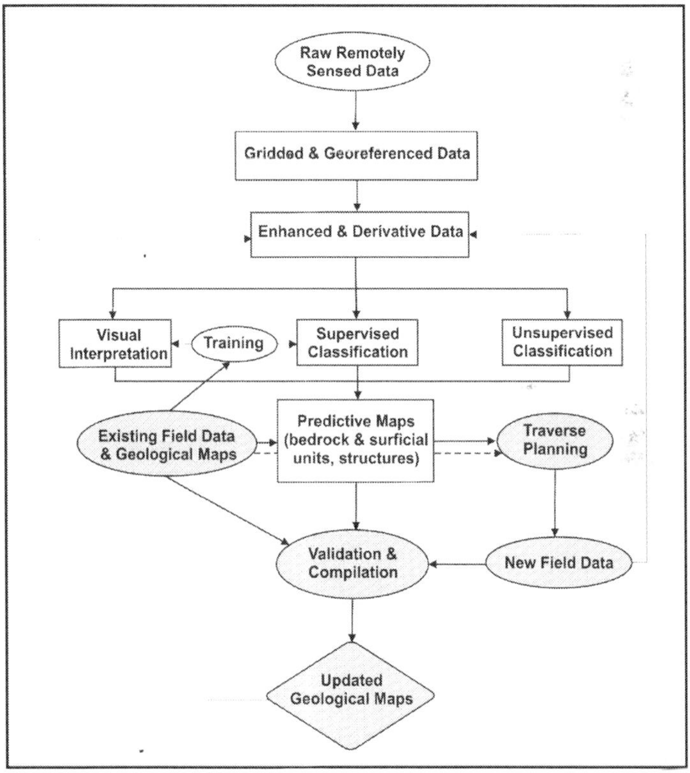

Fig. 1. Flow chart showing how RPM methods can be integrated in a geological mapping project. The grey area represents traditional field mapping methods whereas the white area represents remote predictive mapping methods. Predictive maps can be produced by enhancing and fusing various remotely sensed data and visually extracting geologic information from these products. Alternatively, a computer can be employed to automatically produce a predictive map (unsupervised approach) or by utilizing the geologist's expertise in concert with computer analysis (supervised approach). The geological interpretations are constrained or 'trained' by existing geological field data and existing geological maps. The arrow that loops back from *Updated Geological Maps* to *Enhanced and Derivative Data* emphasizes that the interpretation and map compilation process can be integrated over multiple iterations of field mapping.

process integrated into the work flow of a geological mapping project. The shaded portion represents the activities common to the *traditional* geological mapping process, whereas the portion that is not shaded represents the additional activities of the RPM approach. Regardless of whether the interpretation of remotely sensed data is fully integrated into a geological mapping project or not, the following provides a systematic outline of RPM work flow.

2.1 Mapping objectives

The first step in a RPM project is to define the mapping context, which includes the following:

- mapping focus (bedrock, surficial),
- nature of the geological terrain,
- surficial conditions and degree of exposure, physiography,
- data availability, quantity, and quality.

These factors will determine the data that will be most useful for bedrock mapping. Bedrock mapping projects that are planned in well exposed terrain and have thin residual till cover will benefit from the integration of magnetic, gamma-ray spectrometry, optical, and radar image data. In areas where sparse outcrops alternate with thick overburden, bedrock mapping will primarily profit from the interpretation of magnetic data.

In surficial mapping, optical and radar remote sensing techniques, together with gamma-ray spectrometry and digital terrain data, will contribute to distinguishing various types of surficial materials, identifying and mapping geomorphic features, and mapping streamlined glacial landforms that provide information on glacial movement. Geological setting and physiography of the terrain in combination with the spatial and spectral resolution, penetration depth, season of image acquisition, and aerial coverage of the remote sensing system (including airborne geophysics) are all important factors when choosing data sets for geological interpretation.

2.2 Data selection

Governments and private-sector contractors and/or vendors now provide much of the geoscience data in digital format that increasingly can be accessed through the internet. The core data types that are generally acquired and interpreted for RPM projects are listed in **Table 1** along with references to a sample list of websites to obtain them. In Canada, most of these data sets cover the complete landmass with the exception of gamma-ray spectrometry data. Nonproprietary, medium to low-resolution geophysical data, including magnetic and gamma-ray data were obtained from the Geological Survey of Canada's Geophysical Data Centre. LANDSAT 7 enhanced thematic mapper scenes of 180 x 180 km optical remotely sensed data with one 60 metre resolution thermal band, six 30 metre multispectral bands in the visible to mid-infrared range, and one 15 metre panchromatic band in the visible range can be obtained, free of charge, from the Geogratis website (http://www.geobase.ca). Radarsat data is obtained from the Canadian Space Agency (CSA). Digital elevation data (DEM) (CDED at 1:50,000 and/or 1:250,000 scale) can be downloaded from the Canadian Council on Geomatics (CCOG) website (http://www.geobase.ca). The internet providers of optical remotely sensed data often include a quick-look download service that allows for the inspection of cloud cover of the scenes before downloading.

Data	Data provider	Cost
LANDSAT TM	Geogratis web-site http://geogratis.cgdi.gc.ca/ MDA Geospatial Services	Free to download from Geogratis Cdn$720 per scene from MDA
RADARSAT	Geogratis web-site (100m pixel mosaic of Canada) MDA Geospatial Services (http://www.rsi.ca/)- for individual scenes from the archive or acquire new data - commercial users Canadian Space Agency (CSA) - as above - for government users	mosaic -free to download MDA - Cdn$3,000 - 4,000 /scene CSA - Cdn$300/ scene
Magnetic data	Geophysical Data center (GSC) http:// gdcinfo.agg.nrcan.gc.ca	Free to download
Gamma-ray spectrometry data	Geophysical data centre http:// gdcinfo.agg.nrcan.gc.ca	Free to download
DEM - CDED	Geobase http://www.geobase.ca/	Free to download
ASTER	USGS (http://edcdaac.usgs.gov/main.asp) Information can be found at: http://asterweb.jpl.nasa.gov/ http://asterweb.jpl.nasa.gov/gallery.asp	$40.0 US per scene
SPOT	IUNCTUS Geomatics Corp. http://www.terraengine.com/	$1200.0 per scene for SPOT 4 $1.00 - $6.00 per sq km - SPOT 5
IKONOS	MDA Geospatial Services (http://www.rsi.ca/) Information can be found at: http://www.infoterraglobal.com/ikonos.htm http://www.satimagingcorp.com/gallery-ikonos.html	$15.0 -$30.0 per sq km
QIUCKBIRD	MDA Geospatial Services (http://www.rsi.ca/) Information can be found at: http://www.satimagingcorp.com/gallery-quickbird.html http://www.ballaerospace.com/quickbird.html	See MDA web-site
ENVISAT	MDA Geospatial Services (http://www.rsi.ca/)	See MDA web-site
RESOURCESAT	MDA Geospatial Services (http://www.rsi.ca/)	$2750.0 per scene
IRS	MDA Geospatial Services (http://www.rsi.ca/)	$900.0 - $2,500.0
ERS-1 Radar	MDA Geospatial Services (http://www.rsi.ca/)	$660.0 per scene
Airborne hyperspectral data	- selected coverage of PROBE data - Baffin Island, Sudbury - Canada Centre for Remote Sensing (CCRS) and Geological Survey of Canada - Geophysical Data Centre	Selected scenes free to download

Table 1. Data sets used for the RPM projects discussed in this paper

There are also a number of other specialized remote sensing systems included in **Table 1** that do not yet provide complete coverage of the Canadian landmass. Optical sensors, including ASTER, SPOT, IKONOS, QUICKBIRD, WORLDVIEW I and II and airborne hyperspectral, can provide a wealth of geological information but these data are not available for all of Canada. However, these data can be acquired and when available their use should be considered, since they offer imagery with either higher spectral resolution (ASTER, 14 spectral bands) or higher spatial resolution (SPOT 5, IKONOS, QUICKBIRD, WORLDVIEW). The higher spatial resolution of the latter sensor systems with 4.0 to 2.4 metre multispectral and 1.0 to 0.4 metre panchromatic data acquisition is not only useful for mapping and logistical planning but also as a navigational guide in hand-held field computers.

Existing field and laboratory data and published geological maps can be integrated into the RPM process to guide, calibrate, and test interpretations (see Case Studies 5 and 9 in Harris, 2008 and Schetselaar et al., 2000). This can be accomplished by overlaying the field observations (lithological unit, strike and dip measurements) on the predictive map(s) in a GIS environment to calibrate the interpretation of geological units and structures. Field data can also be used in training computer classification algorithms. The statistical relationships between the numerical values of image data (representing spectral reflectance, magnetic field intensity, radar backscatter, etc) and lithological units can be computed at field stations and then used to predict other areas with similar signatures. Geological mapping is increasingly being supported by digital field-data capture technology using hand-held computers and global positioning systems (GPS). This is a revolutionary development in RPM as it allows the validation of remote predictive maps on the outcrop. Simultaneous display of remote predictive maps and GPS position in real time may lead the mapping geologist to make small deviations from planned traverses to inspect subtle anomalous patterns that appear to be geologically significant when analyzed in the context of the immediate surroundings of an outcrop. This may apply, for example, to confirming the presence of a dyke, when short-wavelength linear magnetic anomalies from near surface magnetic bodies appear to be in close proximity to the field site.

2.3 Data processing and enhancement

A wide range of processing and enhancement methods can be used to facilitate extraction of geological information from RPM data sets (**Table 2**). Harris (2008) provides many examples of enhanced image data (mainly from Canada's North). Generally the methods employed depend on the data type to be enhanced. Derivatives of potential field data include vertical derivatives, upward continuation, analytic signal, magnetic susceptibility, and pseudogravity, among others (Pilkington et al., 2008). Grids of measured magnetic and gravity data, as well as their derivatives, are improved by applying contrast-enhancement and relief-shading algorithms or both in combination (Milligan and Gunn, 1997). Spatial convolution-filters and colour-enhancement techniques, such as decorrelation stretch (Gillespie et al., 1986) and saturation enhancement (Kruse and Raines, 1994) may be applied to enhance optical remotely sensed (Chapter 5 in Harris, 2008), multibeam radar (Chapter 6 - Harris, 2008) and gamma-ray spectrometry data (Chapter 4 - Harris, 2008) , while band ratios or pairwise principal component analysis (Jensen, 1995; Jolliffe, 2004; Richards and Jia, 2006) are useful to enhance geological information on multispectral or multibeam radar imagery. Most of these enhancements can be generated semi-automatically using computer algorithms available with GIS and/or image analysis systems. User input, however, is always important to fine-tune the enhancement, since this is guided by insight on how the dynamic range and spatial frequency distribution of the imaged physical properties are associated to geology. In addition to the enhancement of individual data types, image fusion

(Harris et al., 1999) combines image data into single images to highlight features of interest and assist in the analysis of complementary geological information.

Data Source (including various enhancements)	RPM Product
Magnetics	Map of magnetic units (domains)
	Map of structures (faults (ductile, brittle), dykes, lineaments, foliation/ bedding traces, folds, potential lithologic contacts)
Gamma ray	Map of radioelement units (domains) that can provide insight into lithologies, different granitic phases and regional metamorphic conditions
Digital elevation data (DEM)	Map of terrain units (based on relief)
	Glacial landforms
	Map of structures (based on topographic expression) - bedrock or glacial (ice-flow features)
	Map of drainage basins (watersheds)
LANDSAT	Map of structures (faults (ductile, brittle), dykes, lineaments, foliation/ bedding traces, folds, potential lithologic contacts)
	Map of spectral units (spectral absorption features due to white mica, clay minerals (potentially associated to hydrothermal alteration) and carbonates) especially carbonates - may represent a combination of bedrock lithology and surficial units
	Fe -oxide map (3/1 - ratio)
	Clay-alteration map, Carbonate, white mica and other OH-group minerals (5/7 - ratio)
	Map of vegetation (4/3 ratio)
	Outcrop map (1+7/4 or 7/4 ratio)
	Map of wetlands (band 4)
	Map showing forest fire burns
	Map of snow and ice
	Drainage map (can provide more detail than topographic maps depending on scale)
Radarsat data	Map of terrain units that may represent surficial or lithologic units
	Map of structures (faults (ductile, brittle), dykes, lineaments, foliation/ bedding traces, folds, potential lithologic contacts)
Hyperspectral	Map of spectral units (can be calibrated to actual lithologic units or specific minerals in certain environments)
	Map of structures (as above)
	Alteration map (if good exposure)

Table 2. RPM data types and products (maps)

2.4 Data analysis

Interpretation can be undertaken visually, on various enhanced and fused images using the well-known principals of photo-geologic interpretation or by employing computer-assisted techniques that can lead to automatically generated maps or products that require some geologic interpretation and calibration by the geologist (**Fig.1**).

2.4.1 Visual interpretation

Visual interpretation of the enhanced and or fused remotely sensed data can be based either on making hard-copy images or by digitizing on a touch-sensitive computer screen. The latter method is more flexible as it allows for instantaneous display of different data sets, thus facilitating the extraction of complementary information while weighing the geological significance of image patterns in each of the data layers. It can provide interpretations of units, unit contacts, or faults that are automatically georeferenced to the database, can be virtually overlain on other data for comparison, and serve as a basis for geological map compilation once new field data are acquired.

Regardless of the data type being rendered, visual interpretation is based on recognizing geological features using seven diagnostic elements. These include tone and/or colour, texture, patterns, shape, size, shadow, and association (Lillesand and Kieffer, 2000; Drury, 2001). Depending on the type of geoscience data used for predictive mapping (including remotely sensed and geophysical) data one or more of these photo-geologic elements can be captured. *Tone* and/or *colour* refers to the relative brightness or colour of objects in an image. It is the most fundamental element of image interpretation, as its variation also allows appreciating other elements, such as texture, pattern, and shape. Tonal and/or colour response can be captured from optical sensors (i.e. LANDSAT and may others) sensitive to reflectance properties of the Earth's surface and entail the use of spectral signatures to characterize various earth materials. Magnetic data captures tonal response due to variations in magnetic susceptibility and these tonal variations often reflect underlying lithology and geologic structure. Gamma ray spectrometer tonal variations reflect radioelement emissions (eU, eTh and %K) from the surface and are useful for mapping geochemical variations at the surface. *Size*, *shape* and surface *texture* can be captured by both optical and microwave remote sensors as well as digital elevation models. Radar is particularly useful for capturing textural responses from the Earth's surface due to variations in surface roughness and moisture. *Pattern* refers to the repetitive arrangement of discernable features in an image and different patterns can be captured based on what each sensor responds to, as discussed above. *Shadow* refers to the part of an object that is obstructed from incoming radiation from a natural, active, or artificial energy source. Shadow provides a perception of the profile or relative height of a target. It, however, may also hamper the identification of an object since it lowers or completely obstructs the reflectance from that object. *Association* refers to the relationship of an object with other recognizable objects in the vicinity. The identification of features that one would expect to associate with other features may provide information to facilitate identification. Typical geological examples include radial drainage patterns around circular objects, such as those associated with impact structures, and intrusive and tectonic domes and volcanoes.

2.4.2 Computer-assisted (numerical methods)

In addition, or as a compliment to visual interpretation, numerical interpretation methods can be used to produce remote predictive maps (**Fig. 1**). Automated numerical methods can include supervised and unsupervised classification and image segmentation algorithms (Lillesand and Kieffer, 2000; Richards and Jia, 2006). These methods provide alternatives for extracting geological information in a systematic and unbiased manner, although visual interpretation is commonly judged to outperform methods of automated pattern recognition. However, numerical methods are superior to visual methods at simultaneously manipulating and interpreting multiple data sets having a large number of image variables.

Supervised classification methods allow geologists to have input into the map-making process by using geological field data during the training stage of the classification

(Schetselaar and de Kemp, 2000; Schetselaar et al., 2000; also see Case Studies 2, 5, 6, and 7 in Harris, 2008). In supervised classification, decision rules for class allocation are derived from multivariate statistics computed from the relationships between classes and image variables at the sample sites (i.e. field sites considered representative for bedrock or surficial units). The decision rules are used in the classification stage to allocate all pixels or grid cells to particular classes. The available classification algorithms differ in the way probability density functions for each class are modelled and estimated from the training data. The classification algorithms can be broadly categorized into (1) parametric classifiers that model the class probability density functions with the estimated parameters of a multivariate normal distribution or (2) nonparametric classifiers that directly estimate the class probability density functions from the data.

2.5 Data Integration (making a predictive map)

Various aspects of the surface can be emphasized and enhanced on various geoscience datasets. The difficulty comes in how all this information can be integrated into a final geologic map. Firstly the concept of what constitutes a map has changed with the explosion of digital data and tools (i.e. GIS and image analysis systems) to manipulate, enhance, combine and analyse data. A map now can be defined on demand by extracting themes of interest from a geodatabase housed within a GIS comprising a series of geo-referenced layers. These layers can then be combined to create a customized, or in fact a *virtual* geologic map representing different aspects of a geologic terrain. Two examples, one dealing with bedrock geology and the other with surficial materials (surficial geology) are presented below to illustrate this concept.

2.6 Validation

All maps whether predictive or based on field measurements and observations are a generalized model of the Earth's surface. Both approaches (remote and ground-based) are complimentary. There are obviously geological features that can only be observed and mapped in the field, complimented by various laboratory analysis. However, *the view from above* using a variety of geoscience datasets offers a different geologic perspective of the terrain to be mapped, highlighting features and patterns not easily seen or evident when on the ground. Both methods of producing a geologic map, are characterized by different types of uncertainties and these should be (but not always are!) indicated on the map. These include uncertainties in what feature is being mapped, and the spatial location of these features. Capturing these uncertainties is an integral part of the map-making process and example 2, discussed below, illustrates how statistical and spatial uncertainty was quantified when producing a predictive surficial materials map.

3. Examples of predictive maps

Two examples are discussed demonstrating how the concepts discussed above can be applied to make a predictive geological map. The first example deals with the creation of a bedrock geology map which includes spectral/lithologic units as well as structural features over a small portion of the Hall peninsula, Baffin Island, in Canada's Arctic. Both visual and computer-assisted techniques will be presented, compared and contrasted. The second example deals with the creation of a predictive surficial materials map using computer-assisted techniques over a much broader region of the Hall peninsula, Baffin Island. Data used to create these predictive maps include freely available Canadian geoscience datasets including LANDSAT 7 TM, CDED, 1:50,000 DEMS, airborne magnetic geophysical data, hydrographic and geographic GIS layers and legacy field data (digital maps and GIS databases). Image processing software (ENVI™) in concert with GIS software (ArcGIS™) were used to produce the maps using touch-screen display technology.

The study areas for these two examples (Fig. 2) are from the Hall peninsula of south-central Baffin Island, Canada. This area has not been systematically mapped since the 1960's and thus requires updating for both bedrock and surficial information. The geology of the Hall Peninsula corridor can be divided into three principal lithological domains. An eastern domain of Archean tonalitic gneisses, monzogranite and minor metasedimentary rocks, a central domain of Paleoproterozoic siliciclastic metasedimentary rocks and subordinate Paleoproterozoic metaplutonic rocks, and a western domain dominated by orthopyroxene- and garnet bearing monzogranites of the Paleoproterozoic Cumberland batholith (Scott, 1997). The terrain is rough and rocky, with hills near the coast. The Hall peninsula has permanent ice; the Grinnel glacier calves icebergs into Frobisher Bay. The Hall Peninsula is part of the Arctic Tundra biome—the world's coldest and driest biome.

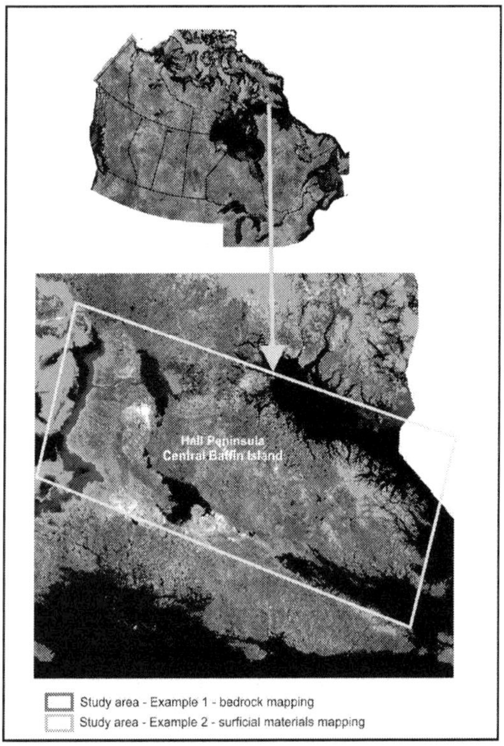

Fig. 2. Study areas for the two examples (bedrock and surficial) of predictive mapping - Hall Peninsula, south-central Baffin Island, Nunavut, Canada.

3.1 Example 1 – Bedrock mapping

3.1.1 Visual assessment

Figure 3 presents a generalized flow-chart summarizing the RPM protocol for producing a bedrock geology map by visual interpreting the enhanced LANDSAT (**Fig. 4**) and magnetic data (**Fig. 5**). Both structural form lines comprising potential lithological contacts, bedding and foliation trends, faults and lineaments (no dykes were evident in the area) and spectral units and magnetic domains were identified. A *heads-up* digitization (interpretation) process was utilized in which interpretations were undertaken directly on a touch-sensitive display

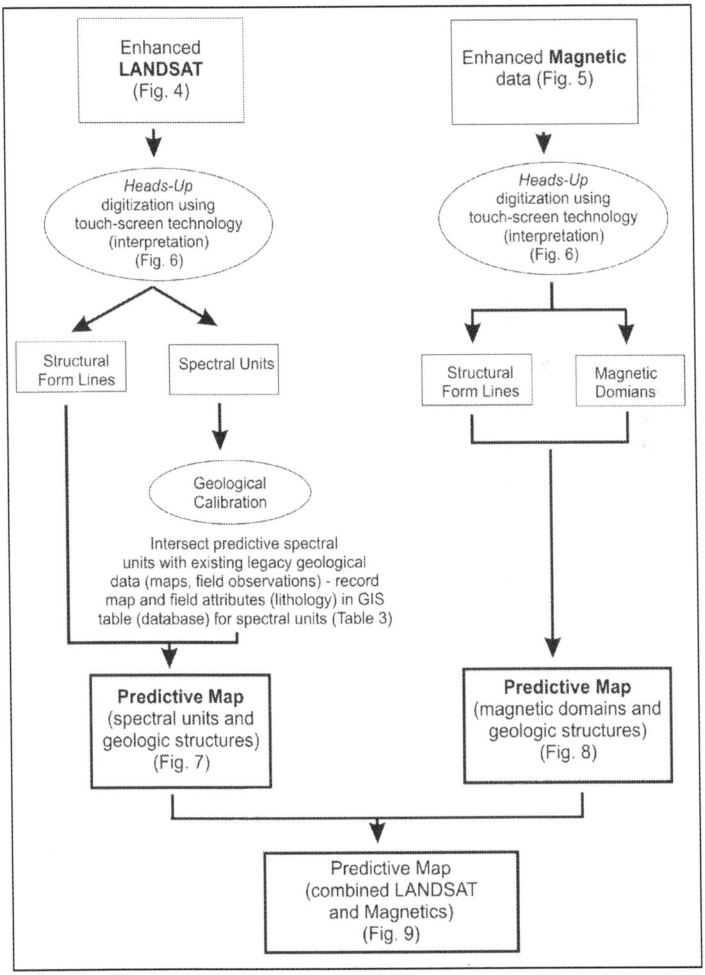

Fig. 3. Flow chart outlining the steps for producing a bedrock predictive map from LANDSAT and airborne magnetic data using visual interpretation techniques and the final integration of the two predictive maps.

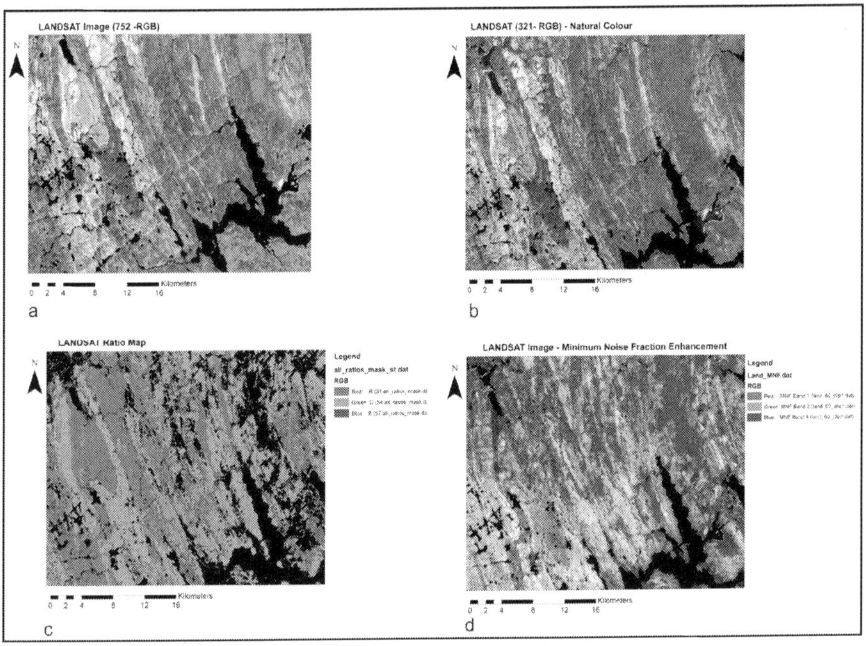

Fig. 4. Enhanced LANDSAT data used for predictive mapping (visual and computer-assisted) (a) band 7,5,2 (RGB) ternary composite image (contrast enhanced), (b) band 3,2,1 (RGB) ternary natural colour composite image (contrast enhanced), (c) LANDSAT ratio ternary composite image (R = ferric iron ratio - red/ blue wavelengths (bands 3/1); G = ferrous iron ratio - SWIR / NIR wavelengths (bands 5/4); B = clay ratio - SWIR / SWIR (bands 7/5). Red areas are higher in ferric iron content, green higher ferrous iron and blue, higher clay (possible sericite), (d) Minimum Noise Enhancement (transform), R = MNF component image 1, G = MNF component image 2, R = MNF component image 3. In these images, note the good spectral separation leading to the identification of distinct spectral units.

screen (Cintiq screen) using a stylus pen (**Fig 6**). An ArcGIS geodatabase was first defined with pre-selected structural and lithological attributes and using the touch screen, all interpretations were immediately incorporated and attributed within feature classes of the geodatabase. The *heads-up* interpretation process is akin to overlaying transparent paper over a hard-copy image and conducting photo-geologic interpretation. However it offers the advantage of flexibility and efficiency as the enhanced image data displayed on the background can be interactively changed while interpretations are fully geo-referenced and are immediately incorporated within the geodatabase. **Table 3** shows the results (GIS attribute table) of geologically calibrating the spectral units by intersecting the polygon map of spectral units with legacy geological data (maps, field stations) thus assisting in assigning a lithological name to each spectral unit. This was accomplished within the GIS by comparing the interpretation of the spectral interpretations with lithological units displayed as polygons on the digital geology maps and field stations in which rock type was recorded

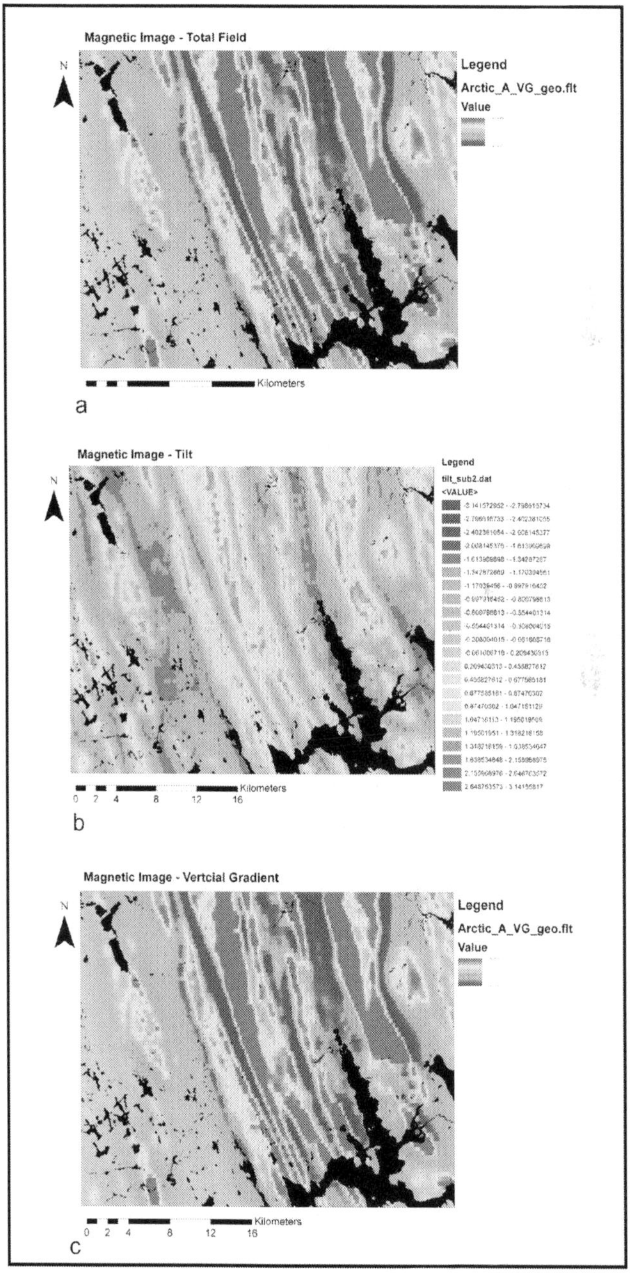

Fig. 5. Enhanced airborne magnetic data (a) total field, (b) tilt, (c) vertical gradient

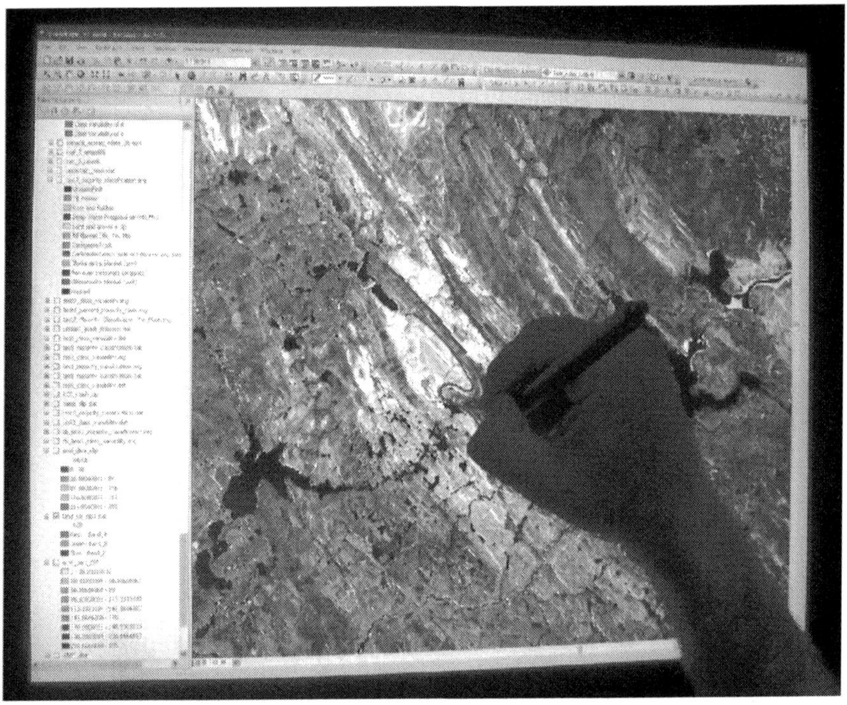

Fig. 6. Example of the *heads-up* digitization (interpretation) process using a touch-sensitive screen - geologist is drawing boundaries on an enhanced LANDSAT image.

in a point database. Note that initially a spectral unit was assigned based on interpretation of the LANDSAT data and after comparing these to the geological data (maps and field stations) a tentative rock unit was assigned. The tentative rock name of course requires field validation. The final predictive map produced by visually interpreting the LANDSAT data, which combines spectral units and the associated database with structural form lines, is shown in **Figure 7** whereas **Figure 8** shows the predictive map produced by visually interpreting the enhanced magnetic data. Five divisions (RPM units 1 - 1d) of the sedimentary rock assemblage (Lake Harbour Group -St- Onge et al., 1998)) , four intrusive units (RPM units 2a,b, comprising the Ramsey River orthogneiss assemblage and 4, 6 comprising the Cumberland Batholith (St- Onge et al., 1998)) and one gneissic unit (RPM unit 5), have been identified by differing spectral responses (**Fig.7**)
Components of these two predictive bedrock maps are combined in the final predictive map, shown in **Figure 9**. The process of overlay the interpretations is a crucial decision process in RPM that is often difficult as this requires the conflicts between interpretations from different image types to be resolved. One approach is to combine the interpretations after all are complete. An alternative approach is to combine the interpretations *on the fly* by dynamically changing the imagery on the computer screen during the interpretation process.

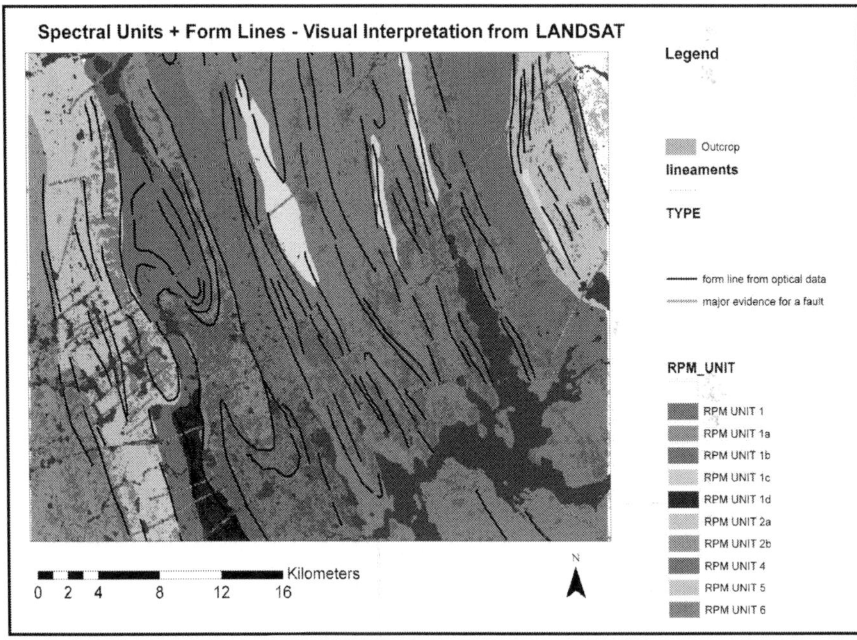

Fig. 7. Predictive bedrock geology map produced by visually interpreting enhanced LANDSAT data (Fig. 4) using a *head-up* digitization process (Fig. 6). The steps for producing such a map are outlined in Fig. 3. Note hat the grey shaded areas within each spectral unit are areas of bedrock outcrop identified on the LANDSAT data. This was accomplished by producing a Blue / NIR wavelength (1/4) ratio as exposed outcrop reflects blue energy and absorbs NIR energy. An upper threshold on the histogram of this ratio image was identified creating a binary raster map of outcrop and non outcrop areas that were included as part of the predictive map.

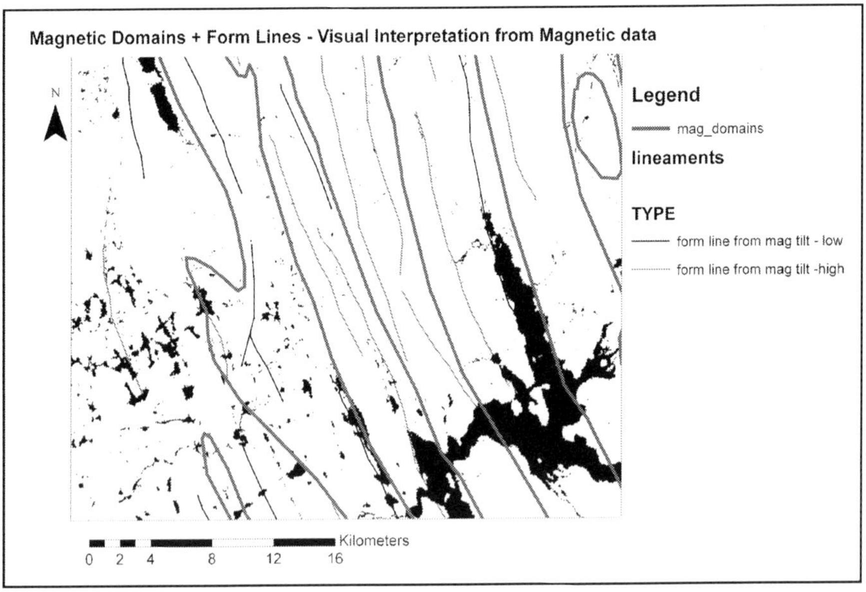

Fig. 8. Predictive bedrock geology map produced by visually interpreting enhanced airborne magnetic data (Fig. 5) using a *head-up* digitization process (Fig. 6). The steps for producing such a map are outlined in Fig. 3. The boundaries of each magnetic domains (which have not been polygonized and thus are not coloured as are the spectral units in Fig. 7) are shown in purple the structural form lines, interpreted largely form the tilt image (Fig.5) in black and red.

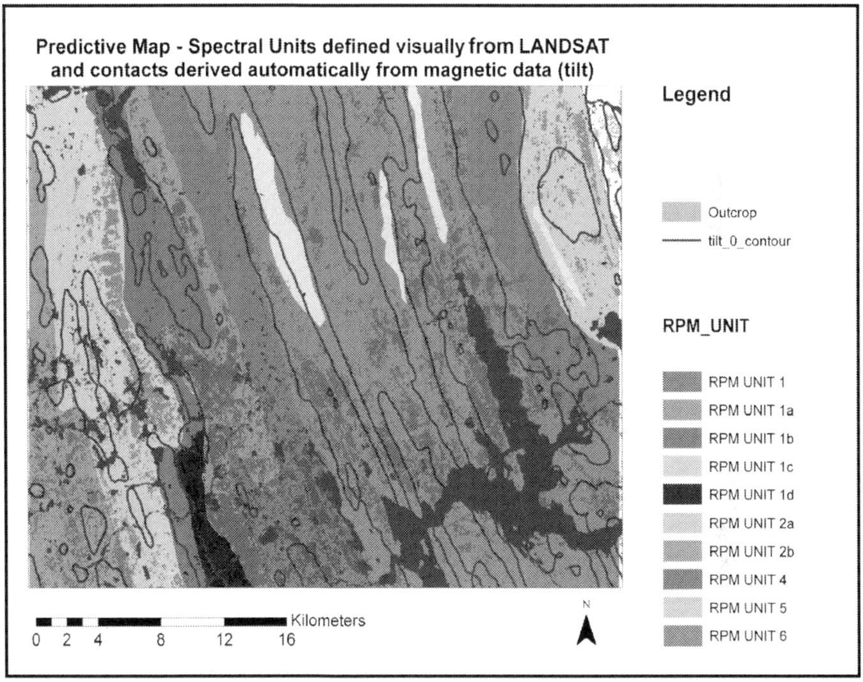

Fig. 9. Predictive map which combines spectral units (geologically calibrated - see Table 3) visually interpreted from the enhanced LANDSAT imagery and magnetic contacts extracted automatically from the magnetic tilt data (0 contour - see description in the text). Areas of bedrock, as described on Fig. 7 have been overlaid in grey. Note that there is good correspondence between the magnetic contacts and the boundaries of the spectral units. However, certain spectral units (RPM 6 for example) are characterized with more frequent and apparent magnetic contacts, perhaps representing significant differences in magnetic susceptibility contrast within each spectral unit, which may be due to metamorphic and /or tectonic processes (e.g. new growth and retrograde destruction of magnetite). This would, of course, benefit from field follow-up work.

SPECTRAL UNIT	MAP UNIT 1 (International Polar Map -IPY- not shown) (Harrison et al., 2011)	DESCRIPTION	FIELD UNIT (Fig 14)	MAP UNIT 2 (Fig. 14)	RPM UNIT
RPM5	orthogneiss	monzogranite-tonalite	gneiss (mafic enclaves)	drift	Orthogneiss - monzogranite-tonalite
RPM4	Igneous intrusive	monzogranite-tonalite orthogneiss	quartz feldspar gneiss	drift	Intrusive - orthogneiss
RPM6	Intrusive	charnockite - monzogranite to syenogranite	quartz feldspar gneiss	quartz-feldspar gneiss	Intrusive - charnokite
RPM1	Sedimentary	psammite - semipelite	gneiss -buff , grey	garnet biotite quartz feldspar gneiss	Meta-sediment 1- psammite - semipelite
RPM1a	Sedimentary	psammite -garnet-biotite-quartz-feldspar	granite, rusty gneiss, gneiss	rusty paragneiss	Meta-sediment 2 - psammite
RPM1b	Sedimentary	psammite, semipelite	rusty gneiss, gneiss, granite	rusty paragneiss - gneiss	Meta-sediment 3 - psammite - semipelite - (rusty - high Fe content))
RPM1c	Sedimentary	psammite garnet-biotite-quartz-feldspar	gneiss (buff)-granite	garnet-biotite -quartz-feldspar Gneiss + rusty paragneiss	Mea-sediment 4 - psammite (less rusty)
RPM1d	Sedimentary	psammite - semipelite	quartz feldspar gneiss	quartz-feldspar gneiss	Meta-sediment 5 -psammite - semipelite
RPM2a	Intrusive	monzogranite-tonalite orthogneiss	quartz feldspar gneiss - buff gneiss	quartz feldspar gneiss	Gneiss 1 - quartz feldspar
RPM2b	Intrusive	monzogranite-tonalite orthogneiss	quartz feldspar gneiss	quartz feldspar gneiss	Gneiss 2 - quartz feldspar

Table 3. Attribute table produced by intersecting the spectral (RPM) units visually interpreted from the LANDSAT data (see Fig. 7) with 2 legacy geological maps (note the column labeled *Map Unit 2* was derived from the geological map shown in Fig. 14 –) *Map Unit* 1 was derived from the International Polar Year Map (Harrison et al., 2011), the field data was derived from field stations shown on Fig. 14

3.1.2 Computer-assisted

The numerical power of an image analysis system in concert with a GIS can be leveraged to extract geological features automatically from remotely sensed imagery producing a stand-alone interpretive, GIS layer and/or a product that will facilitate visual photo-geologic interpretation. **Figure 10** presents a generalized flow-chart summarizing the RPM protocol for producing a bedrock geology map utilizing computer-assisted techniques. Spectral units that may or may not relate to underlying lithologic patterns can be extracted from optical data such as LANDSAT using unsupervised and/or supervised classification techniques in which the geologist provides *a priori* information on the spectral /lithologic features to be classified. Training areas, representing distinct spectral units, were identified on the

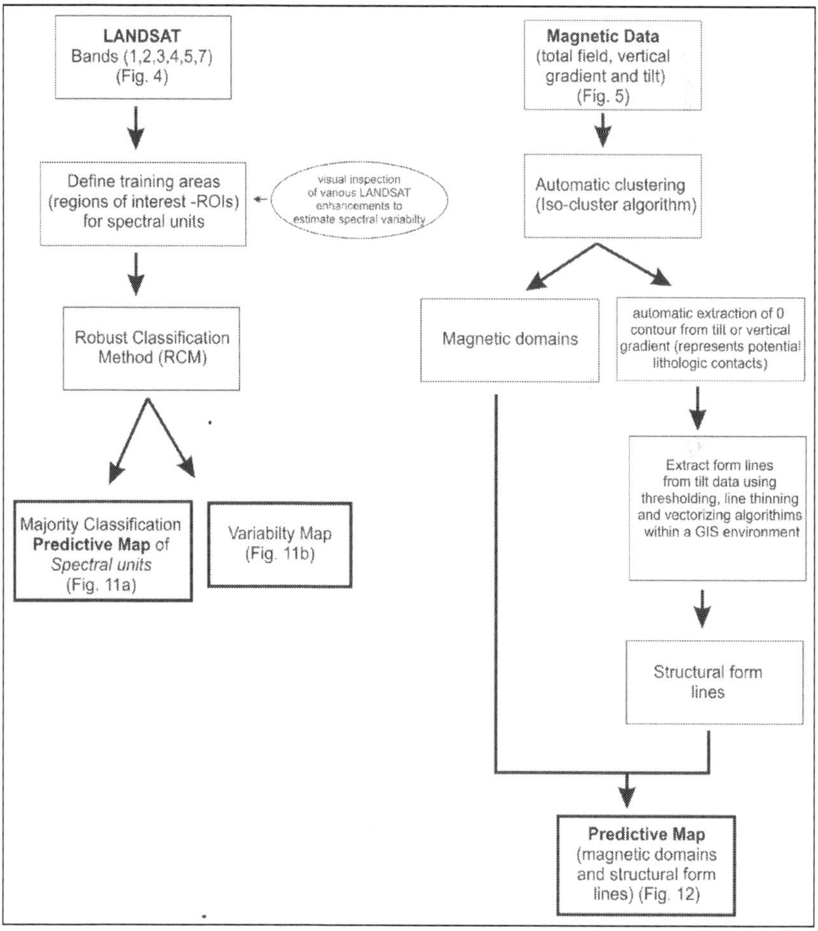

Fig. 10. Flow chart outlining the steps for producing a bedrock predictive map from LANDSAT and airborne magnetic data user computer-assisted (semi-automatic to automatic) techniques.

enhanced LANDSAT data (**Fig. 4**) and used to classify the entire image. The Robust Classification Method (RCM) was employed using the maximum likelihood algorithm to classify the data into spectral units. The RCM method involves a repetitive sampling of a training dataset in concert with cross validation to produce a user-specified number of predictions (classified maps) of spectral units. The RCM process provides a better classification result as the final map comprises a majority classification whereby each pixel is assigned the class that occurred most frequently over the user-specified number of repetitions and the spatial uncertainty of the process is captured by a variability map (cross-validation process). A majority classification map (**Fig. 11a**) for the 10 repetitions of RCM as well as a map that shows the spatial variability (uncertainty) (**Fig. 11b**) over the 10 repetitions are produced as part of the outputs from RCM. Interested readers can find more details on RCM in Harris et al., (2011).

A fair degree of correspondence between the automatically derived and visual derived spectral boundaries exist (**Fig. 7a vs. 11a**). The main difference is that the spectral map derived through supervised classification techniques provides more potential detail within the main visually derived spectral units, perhaps reflecting slightly different lithologic compositions and/ or weathering conditions. With respect to the classification variability map (**Fig. 11b**) no large areally extensive zones of classification uncertainty (variability) exist. However, a few NNW-SSE trending linear zones in the central portion of the study area (green and yellow) have been identified as uncertain using RCM.

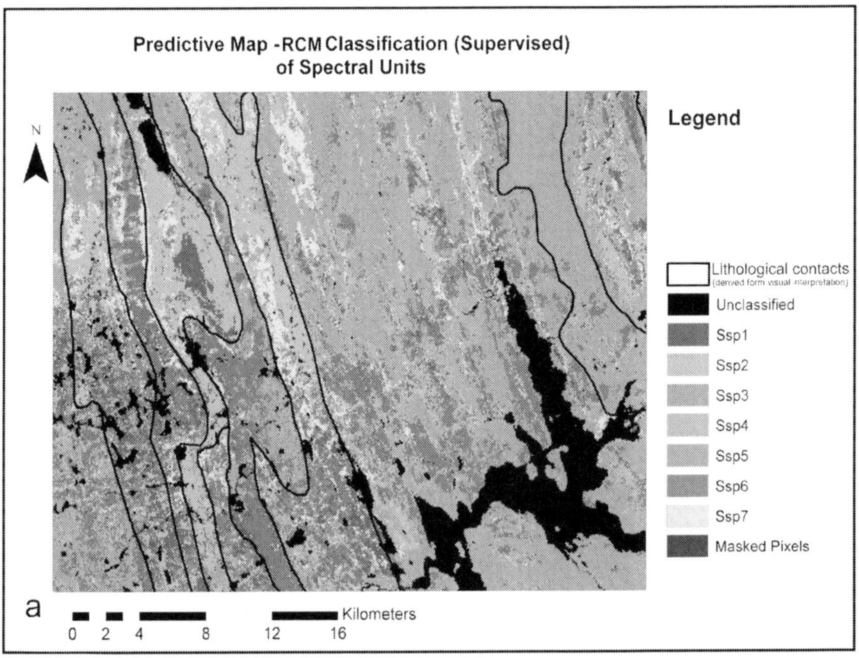

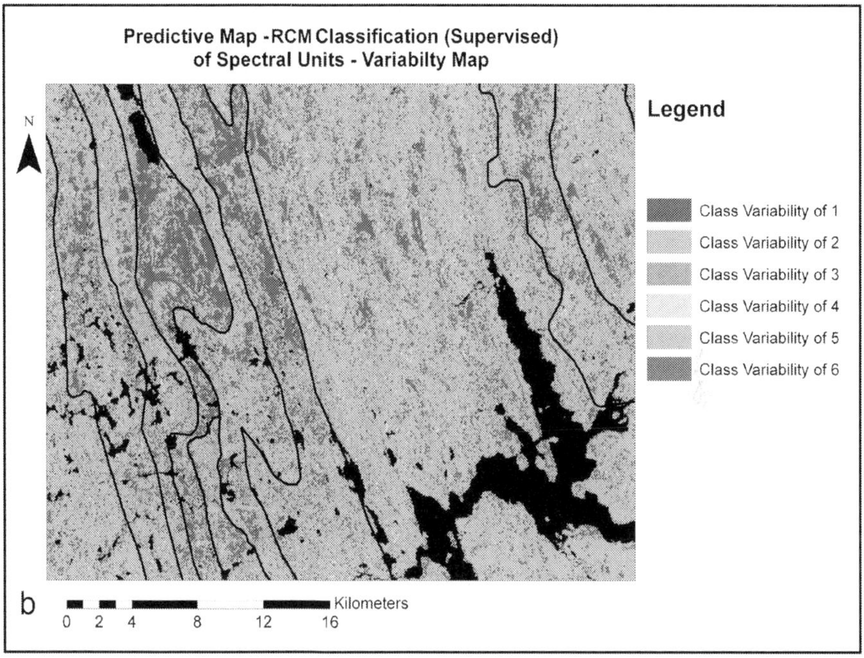

Fig. 11. Predictive bedrock maps of spectral units identified using a supervised classification technique referred to as the Robust Classification Method (RCM) (see description in text and Harris et. al., 2011 for more details on this algorithm). (a) Majority classification predictive map of spectral units. The main spectral boundaries identified through visual interpretation (Fig. 7) have been overlaid for comparison purposes. (b) associated map produced from RCM showing the spatial uncertainty in the spectral classification (i.e. spectral variability map).

Magnetic domains can be automatically produced from the multi-band magnetic dataset (total field, tilt and vertical gradient) by employing unsupervised clustering techniques. This processing involves identifying similar statistical clusters in N-dimensional space (in this example - 3 dimensions (i.e. 3 magnetic images)) based on magnetic susceptibility and then plotting these spatially creating a magnetic domain map (**Fig. 12**).

Potentially meaningful geologic structural features can be automatically extracted from magnetic data forming the basis of a structural map comprising form lines (**Fig 12**). Mapping the locations of lateral magnetization contrasts (i.e. the edges of magnetic bodies or sources) is one of the most useful applications of magnetic data for geological mapping (Pilkington et al., 2009). Contacts can be automatically extracted from magnet tilt data by selecting zero values (which exist over potential edges) and then contoured in the GIS environment creating a vector map of potential lithologic contacts. Furthermore, the linear high and low areas from a vertical gradient or tilt image can be extracted by simple density (thresholding) slicing, followed by thinning the binary map produced from thresholding to a single pixel and then vectorizing producing a vector map of structural form lines (**Fig. 12**).

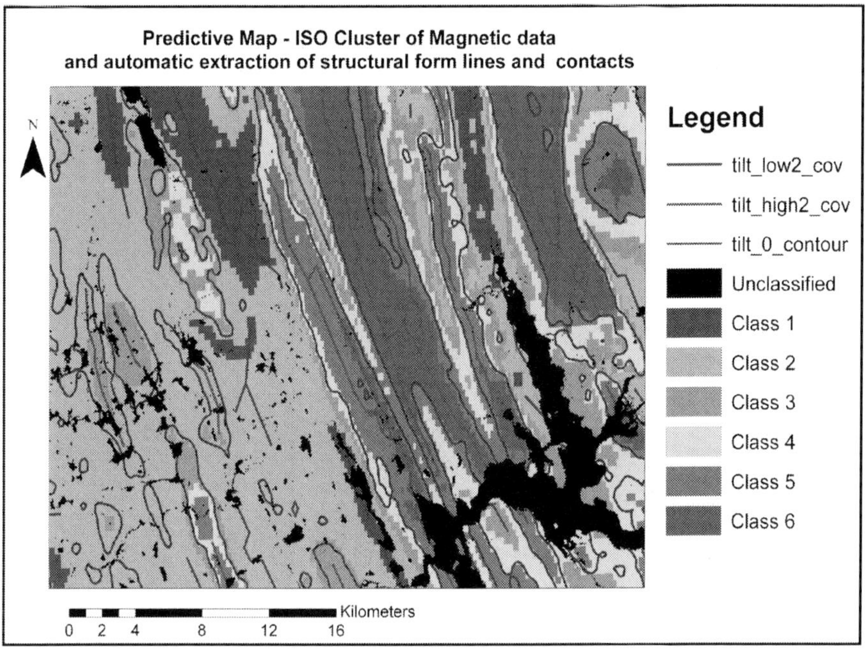

Fig. 12. Predictive bedrock map produced from automatic and semi-automatic processing of the airborne magnetic data. Magnetic domains have been identified and mapped by automatically clustering the total field, tilt and vertical gradient data and contacts and structural form lines have been extracted from the tilt data using semi-automatic methods (see Fig. 10 and descriptions in the text).

3.1.3 Evaluation of predictive bedrock maps

Selected components from the predictive bedrock maps produced from visual and computer-assisted techniques can be combined creating a predictive map which is a hybrid of both interpretation techniques (**Fig. 13**). Although this is a somewhat busy bedrock map it illustrates the power of using the GIS to compile and integrate various layers from the LANDSAT and magnetic data contained within a geodatabase. The various layers can then be combined producing a custom geologic map determined by the geologist and to meet the requirements of what the map is designed to highlight and display (i.e. be it for mapping, exploration etc). Thus the concept of a geologic map now is the geodatabse containing the various geological and geoscience information as points, lines, polygons and rasters as opposed to the traditional static paper map. This new paradigm of a geologic map now allows customization depending on the geological application and fully supports a print-on-demand concept.

There are some similarities in the patterns between the predictive and legacy geological map and in fact the legacy map (**Fig. 14**) was used to geologically calibrate the spectral RPM units as discussed above (see **Table 3**). However, on the legacy map the entire central-north area has been mapped as Quaternary cover. This is clearly not the case as evidenced (and mapped) on the LANDSAT in concert with the magnetic data, both of which offer more detailed geological information in this area. Of course the predictive map would benefit from field follow-up especially with respect to verifying and assigning rock names to each RPM unit.

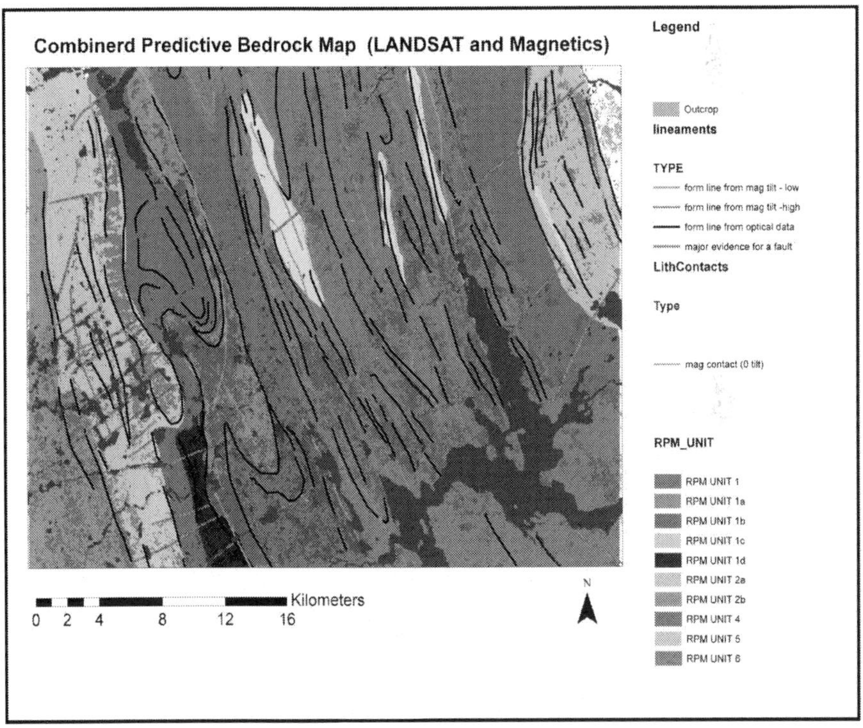

Fig. 13. Predictive bedrock map combining spectral units, bedrock outcrop and form lines derived from visual interpretation of the enhanced LANDSAT imagery with form lines and contacts extracted from semi-automatic interpretation of the magnetic (tilt) data.

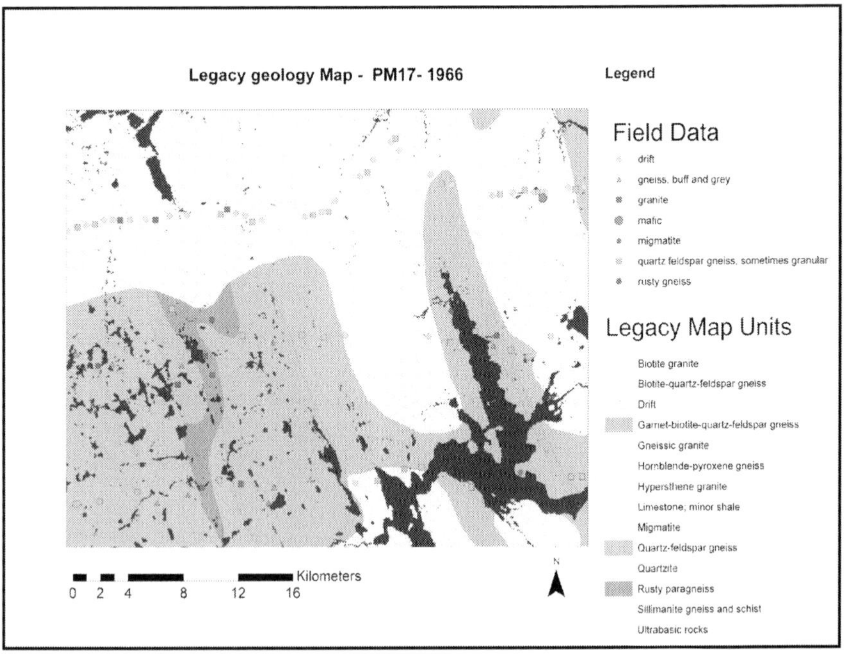

Fig. 14. Legacy geological map (Blackadar, 1966)

3.2 Example 2 – Surficial materials map

3.2.1 Computer-assisted (supervised classification)

The RPM protocol for producing a predictive map of surficial materials is presented as a processing flow-chart in **Figure 15**. This process involves selecting representative training areas (regions of interest) by an expert surficial geologist, knowledgeable about the area to be mapped, selection of geoscience and remotely sensed data to use and selection of an algorithm to perform the classification. In this example, the Robust Classification Method (RCM), discussed and used for bedrock mapping in example 1, was again employed. The data used to produce the predictive surficial materials map included LANDSAT, to capture spectral reflectance characteristics of surficial materials, derived textural derivatives of the LANDSAT bands (entropy and homogeneity) to capture spatial variations in surface texture and finally derivatives from a digital elevation model (DEM) designed to capture topographic characteristics of the terrain. The derivatives of the DEM were based on a 16 by 16 pixel neighbourhood filter which was passed over the DEM and at each pixel the difference from the mean, standard deviation and percent difference were calculated based on the total number of pixels in the neighbourhood. The difference from the mean was used as a measure of topographic position, the standard deviation as a measure of local relief and percent as the range in elevation (Wilson, 2000). Thus in this case both surface reflectance, textural and topographic properties were used to classify surficial materials.

The majority classification map (**Fig. 16**), as described above in example 1, shows the class that was most frequently assigned on a pixel-to-pixel basis over 10 repetitions of RCM whereas

Figure 17 shows a variability map in which the warmer colours represents pixels (areas) that showed much variability in the class each was assigned to through the repetitive classification process. In fact, these variable pixels could be excluded from the majority classification map, as they represent a high degree of uncertainty in the classification process.

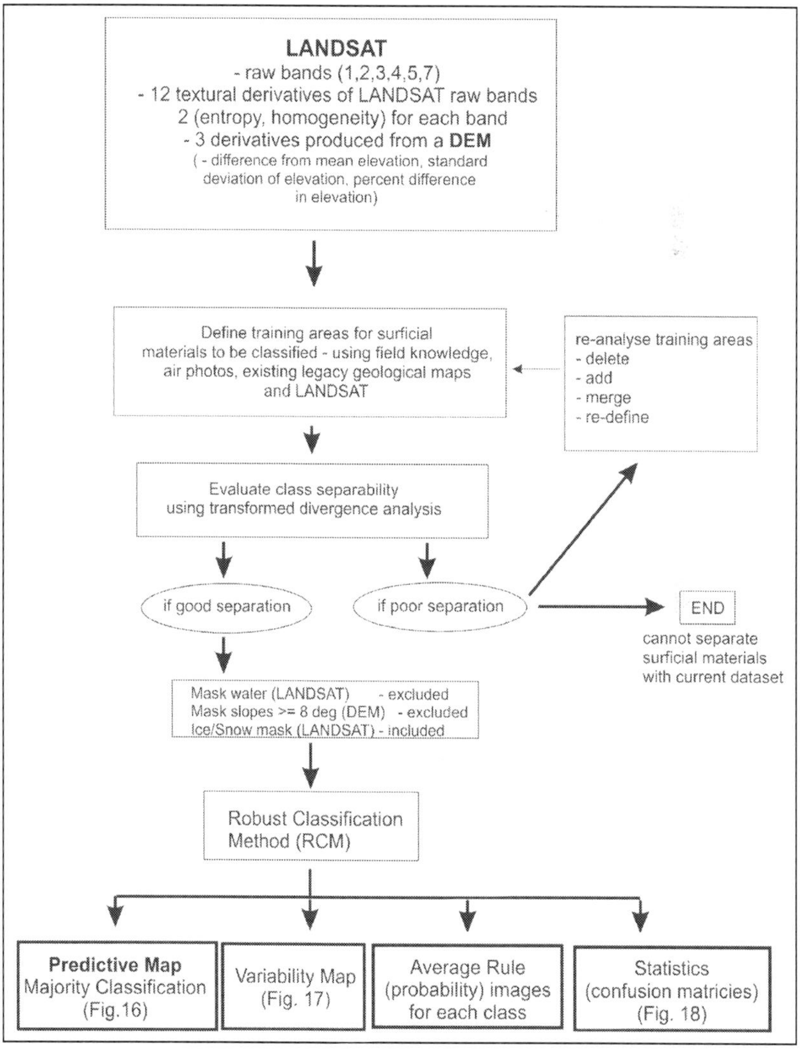

Fig. 15. Flow chart outlining the steps involved in producing a predictive map of surficial materials using a supervised classification technique referred to as the Robust Classification Method (RCM) (see description in text and Harris et. al, 2011 for more details on this algorithm).

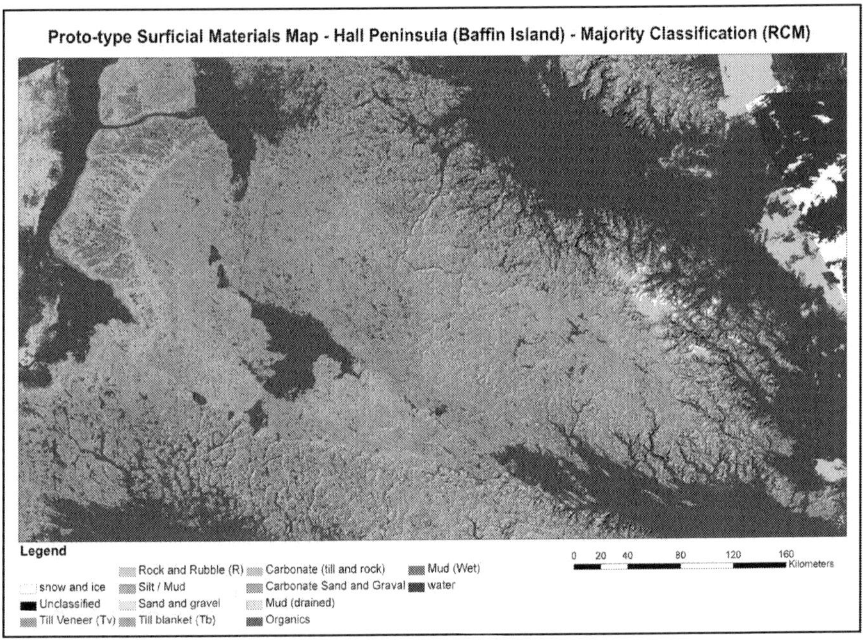

Fig. 16. Predictive surficial materials map - This map produced by RCM shows the majority classification of surficial material on a pixel-to-pixel basis for 10 iterations of the classification algorithm. This map was produced in the same manner as the predictive map of spectral units (Fig. 11). The classification has been combined with a shaded DEM (CDED data) to enhance topographic and geomorphologic variations in the landscape as they relate to the distribution of surficial materials.

The overall average classification accuracy of the majority classification map (**Fig. 16**) is 75.9% whereas the mean accuracy (based on the average of the producer's accuracy) is somewhat lower at 64%. These accuracies do not reveal whether the classification errors are evenly distributed over all classes. Thus, **Figure 18** shows plots of both user's and producer's accuracy for each surficial class which gives a better representation of error as a function of each class. Although the overall accuracy is good some classes are characterized by very poor user's accuracy yet good producer's accuracy and vice versa. Specifically, surface materials with poor producer's accuracy (errors of exclusion - pixels on the classified map that do not match the reference data (training pixels)) yet good user's accuracy (errors of inclusion - pixels on the map that are not the class specified or pixels incorrectly excluded from a particular class.) are : silt/ mud, till veneer and sand and gravel. Thus, pixels in these classes have a much lower probability of being classified correctly on the image, yet on the map they have a higher probability of being correct. Materials that have an opposite relationship (i.e. high producer's but low user's accuracy - pixels incorrectly assigned to a particular class that actually belong in other classes.) are carbonate (till and rock) and organics. Thus the materials that have the least uncertainty of being misclassified are rock and rubble, carbonate sand and gravel, both dry and wet mud and to a lesser extent, till blanket.

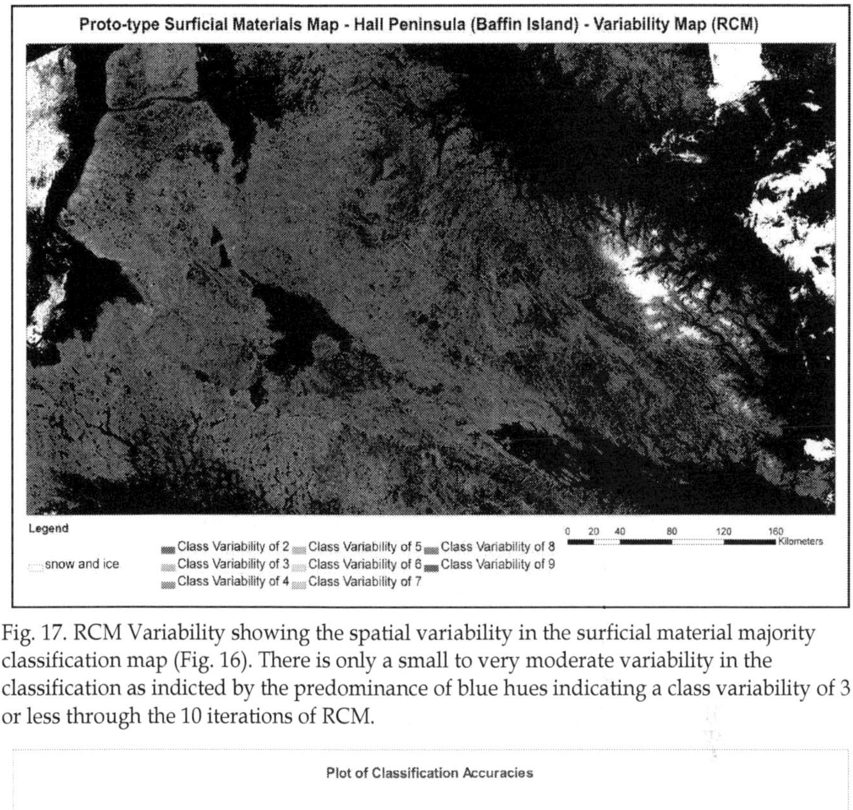

Fig. 17. RCM Variability showing the spatial variability in the surficial material majority classification map (Fig. 16). There is only a small to very moderate variability in the classification as indicted by the predominance of blue hues indicating a class variability of 3 or less through the 10 iterations of RCM.

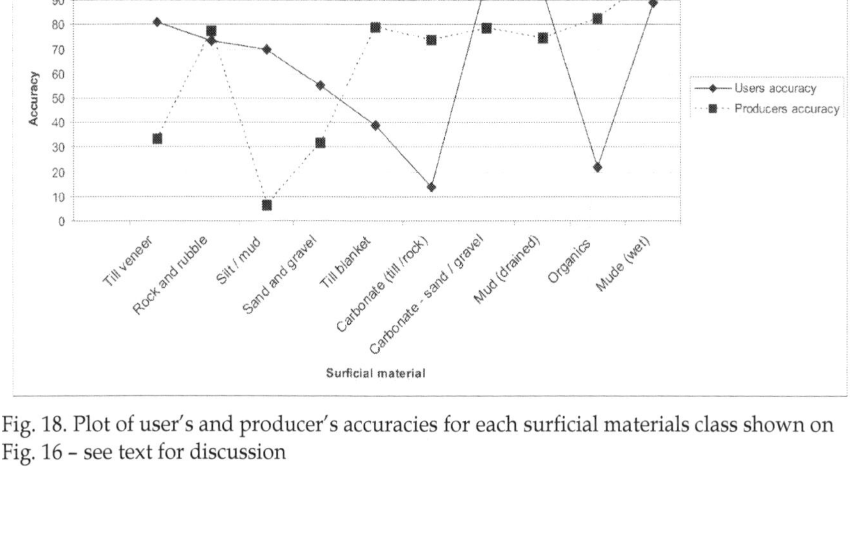

Fig. 18. Plot of user's and producer's accuracies for each surficial materials class shown on Fig. 16 - see text for discussion

Thus, with respect to a user of this map, a high percentage of silt/mud, till veneer and sand and gravel are classified as these materials on the ground. However, the producer's accuracy of these categories are quite low indicating much misclassification of the original training (reference) data. The opposite situation exists for carbonate (till and rock) and organics.

4. Discussion and conclusion

With respect to the best method for producing a predictive geological map, a number of factors, discussed in the introduction section, are important. Mapping bedrock geology is generally more difficult than mapping surficial materials as most remotely acquired data, with the exception of magnetic data, respond to surface parameters (spectral reflectance, backscatter, radioelement emission, topography) only. Capturing all the factors that comprise a bedrock map arguably is more easily done visually as a decision to draw a geological boundary often requires the geologist to integrate all the photo-geologic parameters in the interpretation process. This is difficult to do using computer assisted algorithms unless these photo-geologic parameters can be readily transformed into numerical variables that yield complementary discrimination potential in using multivariate image classification. Furthermore, even in Arctic terrains, the target (bedrock) is often covered by glacial deposits and lichen which can obscure important spectral, radar, backscatter and radioelement characteristics of the underlying bedrock. It is critical to note that the nature of the glacial overburden and whether it is residual or transported is an important factor in determining the effectiveness of remotely sensed data for mapping bedrock patterns. For example, if the glacial material is largely residual, the overburden often reflects the underlying bedrock composition and thus the bedrock can be mapped in part remotely using spectral reflectance, backscatter and radioelement characteristics of the surface. Glacial and vegetative cover, of course, is not a severe limitation with magnetic data. The Canadian Arctic islands and coastal areas are better environments for predictive bedrock mapping using optical remote sensors due to less lichen and vegetation cover whereas inland areas, even though bedrock outcrop is plentiful, are largely covered by lichen which suppresses spectral reflectance variations. This, however, does not apply to structural mapping as several types of geologic and glacial structures, regardless of whether the mapping area is inland, island or coastal, are often clearly expressed on optical, radar and topographic data. The only issue is separating glacial from bedrock structures. It is suggested the best method for producing a predictive bedrock map is to combine both visual and computer-assisted approaches. Automatic or semi-automatic methods can be employed and the results incorporated in the GIS database. The geologist is then free to screen, geologically calibrate and use these automatically derived results in whole or in part on a predictive bedrock map as shown on **Figure 13** which combines distinct spectral boundaries and units, derived through classification of optical data and automatically derived form lines from the magnetic data. Furthermore, the structural data can be screened based on attributes such as orientation, length and correlation with structural features interpreted from optical, topographic and microwave data.

Mapping of surficial materials is a somewhat easier endeavour than bedrock mapping using remotely sensed data as it is the surface material (which may be noise for bedrock mapping!) that forms the target for surficial mapping. Furthermore surficial materials mapping, as demonstrated in example 2 above, is more amenable to computer-assisted techniques for producing a predictive map. The key to producing meaningful predictive surficial material maps lies in the identification of representative training areas. The protocol being followed by RPM efforts in Canada is to establish a database of representative training areas by eco-region which are regions defined based on similar terrain, geologic and biophysical characteristics.

Validation of predictive maps is certainly a key issue. Statistical and spatial uncertainties can be quantified when using computer-assisted algorithms (i.e. classification) as demonstrated by both examples presented in this paper (variability maps, confusion analysis). However, the process of characterizing uncertainty is more subjective when creating a predictive map using visual interpretation techniques. This has traditionally been done by the geologist making the map by adding symbologies such as inferred contacts, extrapolated boundaries etc. as demonstrated in example 1. However, these types of uncertainties are not always included in the final map product and are dependent on the geologist making the map. Part of the Canadian RPM project is to develop these standard mapping protocols.

Canada's Arctic region (north of 60°) comprises a vast territory that is difficult to access and is extremely expensive to map by a traditional *"boots on the ground"* approach characterized by evenly spaced traverses (3- 5 km) that transect all rock and surficial material types, regardless of complexity and variability. This traditional approach often leads to under sampling areas of complex geology and oversampling areas that are characterized by less complex geology. It is often the more complex areas that are of interest from a mineral exploration point of view. Field work is an integral and absolute essential part of geological mapping and of course this will always be the case. No geologist would disagree with this! Remote Predictive Mapping protocols are not meant as a replacement for traditional mapping methods but as a compliment. In many case the view from above captures different geological information than that observed on the ground. The integration of the two approaches is essential in order to provide systematic geological data over large tracts of Canada's North. This combined style of mapping utilizing RPM protocols (and variations of) presented in this paper will provide consistent, efficient and broad coverage of Canada's North. Associated with predictive mapping is a different form of field work which relies on focused traverses in areas of complex geology, as indicated by the predictive map, and less dense field checks in areas characterized by more homogeneous signatures and patterns. Ultimately this will lead to a more complete geoscience database of Canada's northern territory.

5. References

Blackadar, R.G., 1966. Geology, Cumberland Sound, District of Franklin, Geological Survey of Canada, Preliminary Map 17-1966.

Drury, S.A., 2001. Image Interpretation in Geology, 3rd edition Cheltenham, UK: Nelson Thornes; Malden, MA : Blackwell Science, 304 p.

Gillespie, A.R., Kahle, A.B., and Walker, R.E. 1986. Colour enhancement of highly correlated images. I. Decorrelation and HSI contrast stretches; Remote Sensing of the Environment, v. 20, p. 209-235.

Harris, J.R. (ed), 2008. Remote Predictive Mapping: An Aid for Northern Mapping, geological Survey of Canada Open File 5643, DVD.

Harris, J.R., Viljoen, D., and Rencz, **A.** 1999. Integration and visualization of geoscience data, Chapter 6 *in* Manual of Remote Sensing, Volume 3: Remote Sensing for the Earth

Sciences, 3rd edition, (ed.) A. Rencz; John Wiley and Sons Inc., New York, v. 3, p. 307-354.

Harris, J.R., He, J., Grunsky, E. Gorodetsky and Brown, N., 2011. A Robust, Cross Validation Classification Method (RCM) for Improved Mapping Accuracy and Confidence Metrics- Canadian Journal of Remote Sensing -in press)

Harrison, J C; St-Onge, M R; Petrov, O V; Strelnikov, S I; Lopatin, B G; Wilson, F H; Tella, S; Paul, D; Lynds, T; Shokalsky, S P; Hults, C K; Bergman, S; Jepsen, H F; Solli, A., 2011. Geological map of the Arctic / Carte géologique de l'Arctique, Geological Survey of Canada, "A" Series Map 2159A, 9 sheets 1 DVD; Natural Resources Canada / Ressources naturelles Canada; 1:5,000,000.

Jensen J.R. 1995. Introductory Digital Image Processing: A Remote Sensing Perspective, 2rd edition; Prentice Hall, 316 p.

Jolliffe, I.T. 2004. Principal Component Analysis, 2nd edition; Springer-Verlag, New York, Springer Series in Statistics, 486 p.

Kruse, F. and Raines, G. 1994. A technique for enhancing digital color images by contrast stretching in Munsell color space, *in* Proceedings of the ERIM Third Thematic Conference, Environmental Research Institute of Michigan, Ann Arbor, MI, p. 755-760.

Lillesand, T.M. and Kieffer, R.W. 2000. Remote Sensing and Image Interpretation, 4th edition; John Wiley and Sons Inc., New York, 724 p.

Milligan, P. R. and Gunn, P. J., 1997. Enhancement and presentation of airborne geophysical data; AGSO Journal of Australian Geology and Geophysics, v. 17, p. 63-75.

Pilkington M., Keating, P.B., and Thomas, M.D., 2008. Chapter 3 -Geophysics, in Harris, J.R. (ed), Remote predictive Mapping: An Aid for Northern Mapping, Geological Survey of Canada Open File 5643, DVD.

Richards, J.A. and Jia, X. 2006. Remote Sensing Digital Image Analysis: An Introduction; Springer-Verlag, New York, 4th edition, 439 p.

Schetselaar, E.M., Chung, C.F, and Kim, K.E., 2000. Integration of Landsat TM, gamma-ray, magnetic, and field data to discriminate lithological units in vegetated granite-gneiss terrain, Remote Sensing of the Environment, v. 71, pp. 89-105.

Schetselaar, E.M., and deKemp, E.A., 2000. Image classification from Landsat TM, airborne magnetics and DEM data for mapping Paleoproterozoic bedrock units, Baffin Island, Nunavut, Canada. ISPRSS Amsterdam, July 2000.

Schetselaar, E. M., Harris, J.R., Lynds, T. and de Kemp, E. A. 2007. Remote Predictive Mapping (RPM): A strategy for geological mapping of Canada's North, Geoscience Canada, v. 34, no. 3/4, pp. 93 -111.

Scott, D. J., 1997. U-Pb geochronology of the eastern Hall Peninsula, southern Baffin Island, Canada: a northern link between the Archean of West Greenland and the Paleoproterozoic Torngat Orogen of northern Labrador. Precambrian Research, 93: 5-26.

St-Onge, M.R., Scott D J., and Corrigran, D., 1998. Geology, Central Baffin Island area, Nunavut, Geological Survey of Canada Open File Reports 3536 and 3537.

Wilson, J. P., amd Gallant, J.G., 2000. Terrain Analysis: Principles and Applications, John C Wiley and Sons Inc. New York, 479 p.

Part 11

Environmental Sciences

21

Monitoring of Heavy Metal Concentration in Groundwater of Mamundiyar Basin, India

Imran Ahmad Dar[1], K. Sankar[1], Dimitris Alexakis[2] and Mithas Ahmad Dar[1]
[1]Department of Industries and Earth Sciences, Tamil University- Thanjavur
[2]Centre for the Assessment of Natural Hazards and Proactive Planning, Laboratory of Reclamation Works and Water Resources Management
National Technical University of Athens, Athens
[2]Greece
[1]India

1. Introduction

Heavy metals designate a group of elements that occur in natural system in minute concentration and when present in sufficient quantities and are toxic to living organisms. The behavior of trace metals in groundwater is complicated and is related to source of group water and the bio-geochemical process in elemental conditions.

It is often assumed that natural, uncontaminated waters from deep (bedrock) wells are clean and healthy (Banks et al., 1998b). This is usually true with regards to bacteriological composition. The inorganic chemical quality of these waters is, however, rarely adequately tested before the wells are put into production. Due to variations in the regional geology and water rock interactions, high concentrations of many chemical elements can occur in such waters. During the last 5–10 years several studies have shown that wells in areas with particular geological features yield water that does not meet established drinking water norms (e.g. Varsanyi et al., 1991; Bjorvatn et al., 1992, 1994; Edmunds and Trafford, 1993; Banks et al., 1995a,b, 1998a; Sæther et al., 1995; Reimann et al., 1996; Edmunds and Smedley, 1996; Smedley et al., 1996; Williams et al., 1996; Morland et al., 1997, 1998; Midtga˚rd et al., 1998; Misund et al., 1999; Frengstad et al., 2000) without any influence from anthropogenic contamination. These studies also document that quite a number of elements for which no drinking water guideline values (GL) or maximum acceptable concentration limits (MAC) have been established can occur at unpleasantly high levels in natural well waters (e.g. Be, Th, Tl). In Norway, F and radon (Rn) are the most problematic elements (see Frengstad et al., 2000) in terms of possible health effects. In Hungary, Bangladesh and India, arsenic represents one of the most drastic examples of unwanted natural chemical 'contamination' of groundwater. Several 100 000 people in these regions suffer skin cancer due to high As concentrations in drinking water from drilled wells (Chatterjee et al., 1995; Das et al., 1995; Smith et al., 2000; Smedley and Kinniburgh, 2002).

It has been established that various trace elements have certain health on living organisms (WHO, 1984). But the extent to which these elements affect health of living organisms depends on the chemical characteristics and the concentration of the element in the water consumed. Furthermore, the time of exposure will also determine the level of the element on the organism. Some elements are biocumulative and therefore get increased with time in the body. The present paper reports analytical results for 6 chemical elements (trace elements) from 50 sampling stations of Mamundiyar basin, India.

2. Geography and geology of the study area

Mamundiyar basin, India lies in hard rock terrain. Groundwater is available only in weathered and fractured zones. In this area assured surface water supplies are nominal and most of the farmers depend on groundwater for drinking and irrigation purposes. Average annual rainfall is around 464 mm which is mostly lost as surface runoff and evaporation. Only one-fifth of it is recharging to groundwater. Therefore, groundwater development assumes great significance in improving the quality of life of the most deprived and vulnerable people of this basin by improving their access to safe drinking water.

The Mamundiyar basin extends over approximately 720 km^2 and lies between 10^0 25` and $10^0$ 40`N latitudes and 78^0 10` and $78^0$ 30` E longitudes in the southern part of Tamilnadu, India (Fig. 1). Mamundiyar River originates at an altitude of 315 m above Irungadu group of hills and joins Ariyavur River near Maravanur about 25 Km south-west of Tiruchirapalli. The western, north-western and south-western parts are characterized by the presence of residual hills. The basin is generally hot and dry except during winter season. The mean maximum monthly temperature varies from 37^0C in May to 29^0C in December. While as mean minimum monthly temperature ranges from 27^0C in June and 20^0C in January. The area receives an average annual rainfall of about 464 mm. The surface runoff goes to stream as instant flow. Rainfall is the direct recharge source and the irrigation return flow is the indirect source of groundwater in the Mamundiyar hydrographic basin. The study area depends mainly on the North-east monsoon rains which are brought by the troughs of low pressure established in the South Bay of Bengal.

Several digital image processing techniques, including standard color composites, intensity-hue-saturation (IHS) transformation and decorrelation stretch (DS) were applied to map rock types. The statistical technique adopted by Sheffield (1985) was employed to select the most effective Three-band color composite image. The band combination 1, 4 and 5 is the best triplet and was used to create color composites with Landsat TM bands 5, 4 and 1 in red, green and blue, respectively. IHS transformation and DS were also applied to the selected band combination in order to enhance the difference between rock types. Better contrast was obtained due to color enhancement and this facilitated visual discrimination of various rock types. Eleven lithologic units were mapped and could be distinguished by distinct colors in the processed images. These are: Ultramafics, Hornblende biotite gneiss, Basic rocks, Charnockite, Pyroxene granulite, Pink magmatite, Quartzite, Pegmatite vein, Quartz vein, Granite, and Calc granulite and limestone. Fig. 2 is a map of the interpreted distribution of rock types Mamundiyar basin (Dar et. el, 2010).

3. Sampling

Most samples reported here were taken from drinking water wells in small villages and settlements scattered throughout the Mamundiyar basin, India. Factory new, unwashed 100-ml high-density polyethylene (HDPE) bottles were used for sampling. Different brands of plastic bottles had previously been thoroughly checked for possible contamination (Reimann et al., 1999a). No risk of contamination from such bottles was found for the

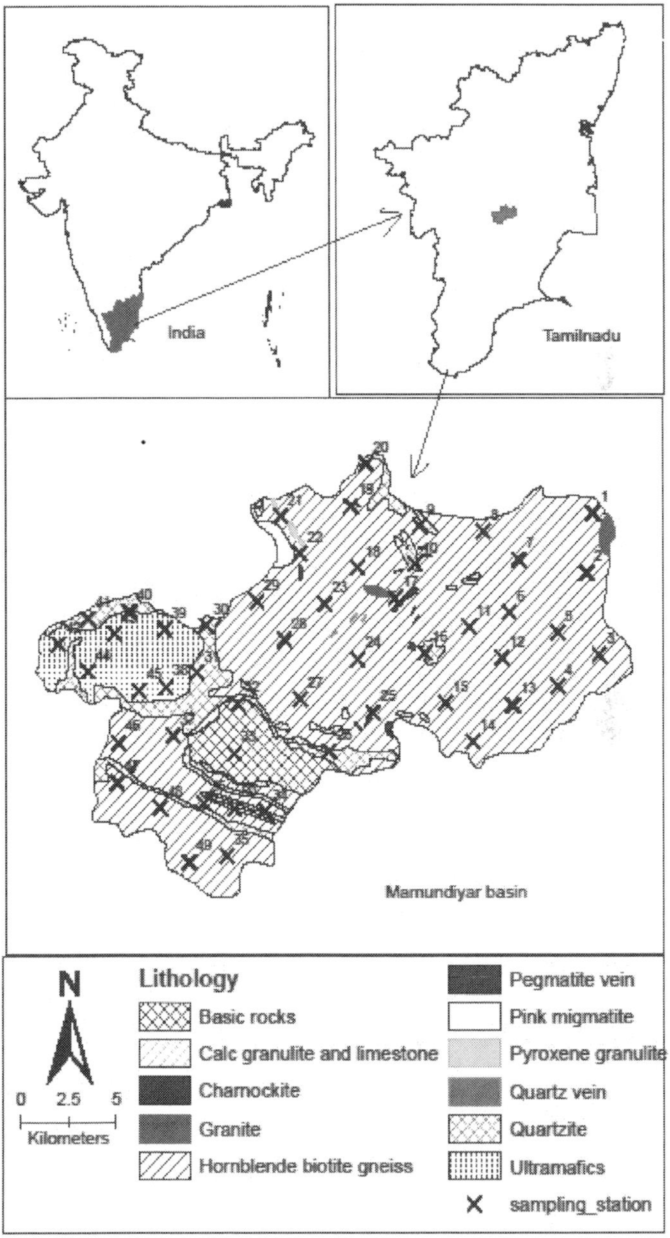
India
Tamilnadu
Mamundiyar basin
N
0 2.5 5
Kilometers
Lithology
Basic rocks
Calc granulite and limestone
Charnockite
Granite
Hornblende biotite gneiss
Pegmatite vein
Pink migmatite
Pyroxene granulite
Quartz vein
Quartzite
Ultramafics
sampling_station

Fig. 1.

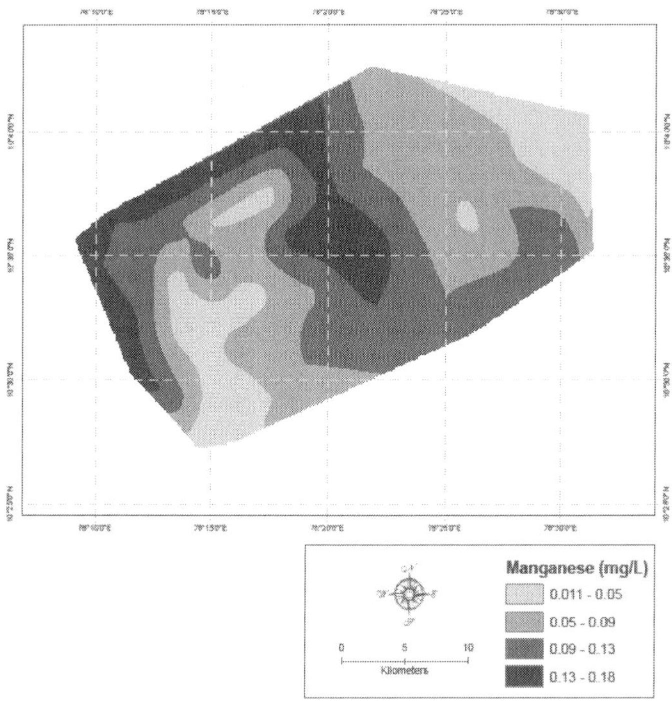

Fig. 2.

parameters reported here, as long as the bottles are thoroughly rinsed with water prior to sampling. In the field the bottles were rinsed three times with running water and then filled to the top.

Sampling took place directly at the tap or the wellhead. In order to collect fresh well water, the water was left running for at least 5 min or until temperature and conductivity remained stable. In most cases, each of these wells supplies more than 100 people with their daily drinking water. The water, therefore, never accumulates over longer periods in the well.

Two 100-ml bottles were collected at each site. The first sample, which was intended for anion analyses, was left unfiltered and unacidified. The unfiltered water of the second sample was acidified with 2 ml of concentrated nitric acid (Merck, Ultrapure). This second sample was used later for cation analysis. The acid was tested for its trace element content using the same analytical procedure as for the water samples. In the field, the samples were stored in a cool box and in the evening transferred to a refrigerator, where they were stored until shipment to the laboratory.

4. Analysis

The trace elements analyzed included manganese (Mn), iron (Fe), chromium (Cr), copper (Cu), zinc (Zn) and boron (B).

5. Results and discussion

The result of the analysis of 50 groundwater sampling stations is shown in tables 1 and 2.

Sample station	Zinc	Copper	Iron	Manganese	Chromium	Boron
1	0.01	0.02	0.03	0.011	0.001	0.38
2	0.01	0.2	0.02	0.012	0.001	0.35
3	0.04	0.03	0.03	0.07	0.002	0.31
4	0.03	0.03	0.04	0.13	0.001	0.31
5	0.04	0.03	0.03	0.12	0.001	0.4
6	0.01	0.02	0.01	0.09	0.001	0.55
7	0.07	0.03	0.02	0.05	0.002	0.56
8	0.07	0.02	0.02	0.06	0.001	0.25
9	0.006	0.03	0.04	0.08	0.001	0.38
10	0.07	0.02	0.06	0.05	0.001	0.55
11	0.03	0.03	0.06	0.04	0.001	0.31
12	0.04	0.04	0.02	0.07	0.001	0.31
13	0.02	0.03	0.2	0.13	0.002	0.4
14	0.05	0.01	0.03	0.12	0.002	0.41
15	0.002	0.02	0.03	0.09	0.002	0.32
16	0.1	0.02	0.04	0.09	0.001	0.48
17	0.1	0.02	0.04	0.09	0.001	0.55
18	0.1	0.02	0.04	0.09	0.001	0.41
19	0.1	0.02	0.04	0.09	0.001	0.31
20	0.01	0.01	0.03	0.06	0.001	0.4
21	0.07	0.02	0.04	0.18	0.001	0.12
22	0.08	0.03	0.03	0.15	0.002	0.18
23	0.002	0.03	0.03	0.14	0.001	0.38
24	0.07	0.01	0.02	0.17	0.001	0.55
25	0.08	0.01	0.03	0.13	0.001	0.31
26	0.05	0.02	0.01	0.12	0.002	0.31
27	0.002	0.03	0.03	0.08	0.001	0.4
28	0.07	0.01	0.02	0.15	0.001	0.23
29	0.05	0.03	0.03	0.02	0.002	0.31
30	0.08	0.01	0.04	0.04	0.001	0.38
31	0.07	0.02	0.06	0.13	0.001	0.55
32	0.08	0.01	0.01	0.02	0.001	0.41
33	0.07	0.03	0.01	0.07	0.002	0.31
34	0.01	0.02	0.02	0.05	0.001	0.4
35	0.07	0.02	0.02	0.04	0.002	0.32
36	0.02	0.03	0.03	0.03	0.001	0.38
37	0.08	0.01	0.12	0.02	0.002	0.42
38	0.1	0.02	0.02	0.02	0.001	0.44
39	0.1	0.01	0.03	0.1	0.001	0.47
40	0.09	0.03	0.21	0.15	0.001	0.38
41	0.07	0.01	0.02	0.14	0.002	0.55
42	0.09	0.02	0.21	0.17	0.001	0.31
43	0.1	0.01	0.03	0.12	0.001	0.31
44	0.09	0.03	0.03	0.13	0.001	0.4
45	0.07	0.03	0.04	0.1	0.002	0.38
46	0.05	0.01	0.02	0.15	0.001	0.55
47	0.02	0.02	0.02	0.17	0.001	0.31
48	0.07	0.01	0.03	0.12	0.001	0.31
49	0.08	0.03	0.03	0.02	0.001	0.4
50	0.05	0.03	0.02	0.02	0.001	0.42

Table 1.

Parameters	z	Cu	Fe	Mn	Cr	B
Minimum	0	0.01	0.01	0.01	0	0.12
Maximum	0.1	0.2	0.21	0.18	0	0.56
Range	0.1	0.19	0.2	0.17	0	0.44
Mean	0.06	0.03	0.04	0.09	0	0.38
Median	0.07	0.02	0.03	0.09	0	0.38
First quartile	0.03	0.01	0.02	0.05	0	0.31
Third quartile	0.08	0.03	0.04	0.13	0	0.42
Standard error	0	0	0.01	0.01	0	0.01
95% confidence interval	0.01	0.01	0.01	0.01	0	0.03
99% confidence interval	0.01	0.01	0.02	0.02	0	0.04
Variance	0	0	0	0	0	0.01
Average deviation	0.03	0.01	0.03	0.04	0	0.07
Standard deviation	0.03	0.03	0.05	0.05	0	0.1
Coefficient of variation	0.57	1.06	1.09	0.55	0.35	0.26
Skew	-0.4	6.01	3.09	0	1.13	0.06
Kurtosis	-1.2	40.1	9.06	-1.2	-0.8	0.2
Kolmogorov-Smirnov stat	0.22	0.39	0.38	0.13	0.46	0.15
Critical K-S stat, alpha=.10	0.17	0.17	0.17	0.17	0.17	0.17
Critical K-S stat, alpha=.05	0.19	0.19	0.19	0.19	0.19	0.19
Critical K-S stat, alpha=.01	0.23	0.23	0.23	0.23	0.23	0.23

Table 2.

6. Manganese

The U.S. Public Health Service Drinking Water Standards of 1925, 1942, and 1946 included manganese with iron for a combined maximum level of 0.30 mg/L, but in 1962 the regulations included, in addition, a maximum concentration of 0.05 mg/L for manganese. The USEPA adopted the 0.05 mg/L of the USPHS as recommendation, and issued a secondary standard in 1989. The WHO recommended 0.05 mg/L (maximum acceptable) and 0.50 mg/L as maximum allowable. The European Community (1980) used a guide value of 0.02 mg/L and a maximum of 0.05 mg/L.
Figure 2 shows the spatial distribution of manganese through Natural Nearst Neighbor interpolation technique. It's quite obvious from the map that manganese tends to dominate the central part of the Mamundiyar basin, and its concentration diminishes radially outwards from the centre. The highest concentration of manganese found was 0.18; which means all the samples fall within the permissible limit set by WHO.

7. Iron

Since the standards for iron have been set for less than 0.3 mg/L, acceptability of water sources was a condition for meeting this concentration. Groundwater exceeding this limit may need a treatment to meet the standard at the distribution system. Groundwater containing soluble iron may remain clear when pumped out, but exposure to air will cause precipitation of iron due to oxidation, with a consequence of rusty color. The presence of iron bacteria may clog well screens particularly when sulphate compounds in addition to iron may be subjected to chemical reduction. Solubility of iron is increased by a low pH (<5).

High turbidity may help to keep acid- soluble iron in suspension. Iron in raw or potable water may be either ferrous or ferric or both and categorized as in solution, in colloidal state, in organic or inorganic compounds or in the form of coarse suspended or settled particles. The 1925 , 1942, 1946 and 1962 regulations of the U.S Public Health Service always reported the maximum concentration for iron as 0.30 mg/L. the USEPA did not include iron in the National Drinking Water Quality Regulations, but maintained in the secondary Drinking Water Regulations of 1989 the limit of 0.3 mg/L based on aesthetic and taste consideration. WHO (1963) also adopted a 0.3 mg/L as a maximum acceptable level and 1.0 as maximum allowable. The European Community adopted in 1980 a guide of 0.05 mg/L and a maximum of 0.20 mg/L; WHO (1984 and 1993) recorded a guideline of 0.30 mg/L. USEPA (1979 and 1991) confirmed the original ruling for iron as a contaminant to be included in the Secondary Drinking Water Standards with a level of 0.3 mg/L as the final rule.
Figure 3 shows the spatial distribution of iron using interpolation method in GIS environment. It's quite evident from the map that iron contaminations occur at few locations, mainly around the Kadavur (western) region of the Mamundiyar basin. In each patch, the concentration of iron was found to decrease/diminish radially outwards from centre. That means, at the point of rock (iron bearing) - water interaction, the concentration of Iron is maximum; and as the distance increases from the interface of rock (bearing iron element) and water, the concentration also gets decreased because of dilution factor. The values of iron were within the permissible limit of drinking water standards.

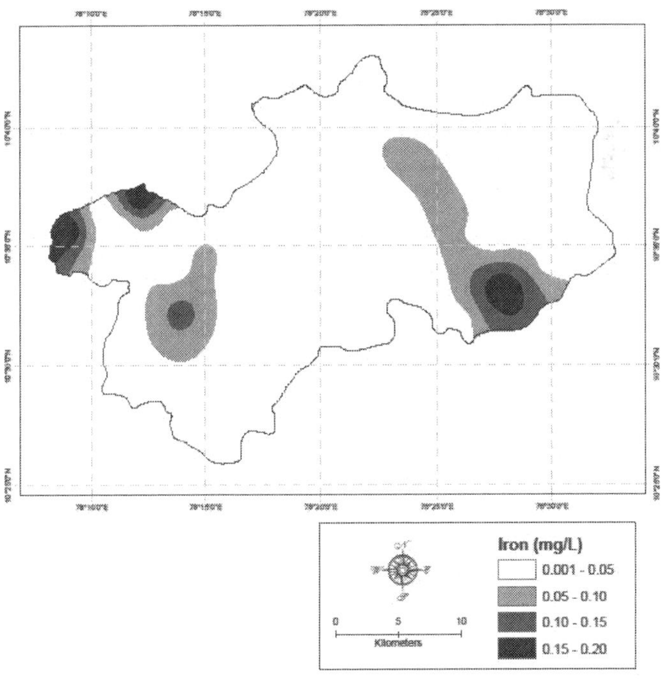

Fig. 3.

8. Chromium

It's a naturally occurring metal in drinking water.

USPHS 1925	= not stated
USPHS 1942	= 0 mg/L as hexavalent
USPHS 1946	= 0.05 mg/L
USPHS 1962	= 0.05 as hexavalent
WHO guidelines	= 0.05 mg/L (as Cr^{6+} and total chromium)
European Community	=0.05 mg/L (as Cr^{6+} and total chromium)
MCLG and MCL (USEPA, 1989)	= 0.1 mg/L (proposed)
MCLG and MCL (USEPA, 1991)	= 0.1 mg/L (final; effective 7/30/1992)

The spatial distribution map of chromium is shown in figure 4; which is created using Nearest Neighbors interpolation technique. It's clear from the figure that the maximum concentration (0.002 mg/L) of chromium occurs in the southern region of Mamundiyar basin; and its concentration decreases towards the north-west, due to dilution. The concentration of chromium wherever recorded is well within the limits of drinking water standards prescribed by WHO.

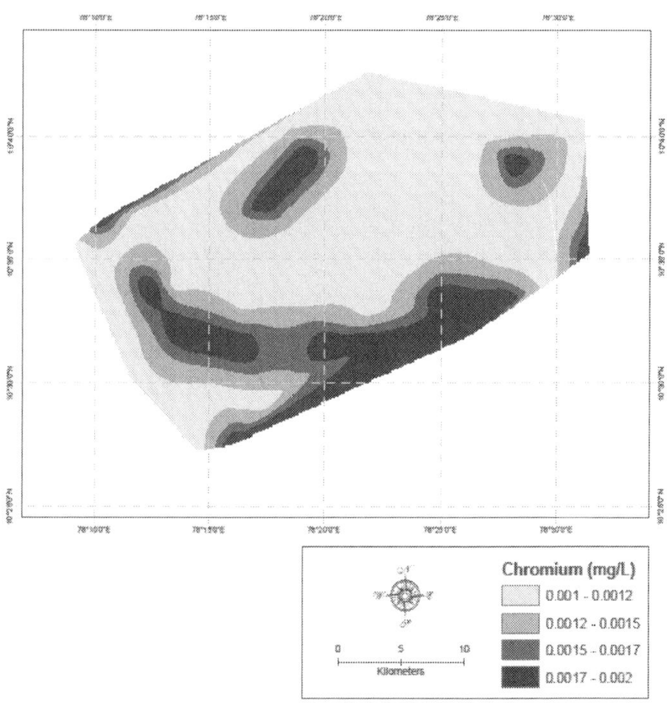

Fig. 4.

9. Copper

USPHS 1925 = 0.2 mg/L
USPHS 1942 = 3 mg/L
USPHS 1962 = 1 mg/L
WHO guidelines = 1 mg/L (1.5 mg/L excessive)
European Community =0.1 mg/L
MCLG and MCL
(USEPA, 1988) = 1.3 mg/L (proposed)
MCLG and MCL
(USEPA, 1991) = 1.3 mg/L at the consumer`s tap (final revised regulations for lead and copper according to the New Lead and Copper Rule as requested by the Safe Drinking Water Act- Revision of 1986).

Copper can exist in aquatic environment in three forms namely soluble, colloidal and particulate. It is found in less quantity as an essential element for organisms. Excess of copper in human body is toxic and causes hypertension and produces pathological changes in brain tissues. Excessive ingestion of copper is responsible for specific disease of the bone (Krishnamurthy, C.R. and V. Pushpa. 1995). The spatial distribution map of copper (figure 5) is prepared using interpolation technique in GIS environment. It's quite obvious from the map that the maximum concentration of copper is present at eastern edge of Mamundiyar basin. In the present study, the values of copper are showed within the limit of drinking water standards.

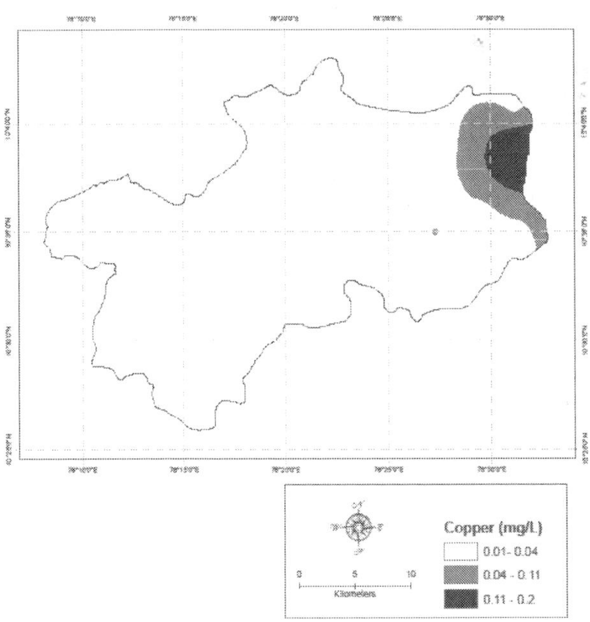

Fig. 5.

10. Zinc

The USPHS recommended a maximum zinc concentration of 15.0 mg/L in 1942 and 1946 standards, and 5.0 mg/L in the 1962 standards.USEPA recommended 5 mg/L in 1980 and a SMCL of 5 mg/L in 1989. WHO (1971) recommended 5 mg/L with a maximum of 15mg/L. the European Community advised 0.1 mg/L, with a maximum of 1.5 mg/L. the WHO (1984) adopted a guideline of 5 mg/L based on the taste consideration. USEPA (1991) issued a final status for Zinc as a Secondary Drinking Water Standard (SDWS) of 5 mg/L, confirming the final rule of 5 mg/L issued in 1980.
The spatial distribution map of zinc (figure 6) is prepared using interpolation technique in GIS environment. It's quite obvious from the map that the maximum concentration of zinc (0.5 MG/L) is present at the Kadavur (western) and central part of Mamundiyar basin. In the present study the values of zinc are showed within the limit of drinking water standard.

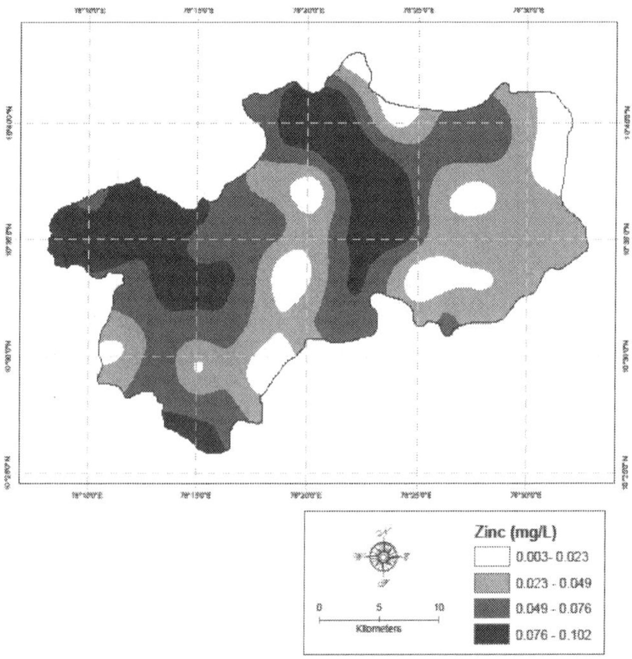

Fig. 6.

11. Boron

Spatial distribution map of boron (figure 7) depicts that maximum concentration of boron in patches at Central, Eastern and western part of Mamundiyar basin. Boron concentration varied between 0.11 to 0.56, indicating that the samples fall within the permissible limit set by WHO. Overall, boron dominates the trace metal pool of the Mamundiyar basin as shown graphically in figure 8.

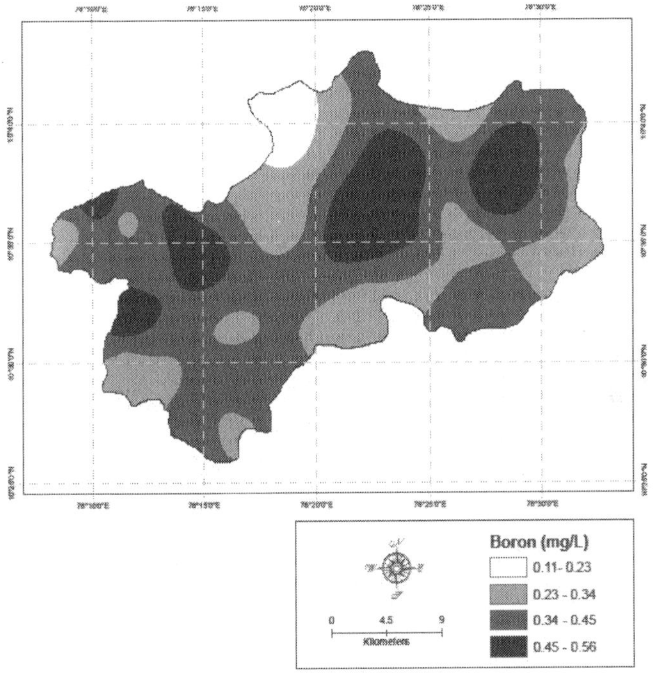
Boron (mg/L)
0.11- 0.23
0.23 - 0.34
0.34 - 0.45
0.45 - 0.56
0
4.5
9
Kilometers

Fig. 7.

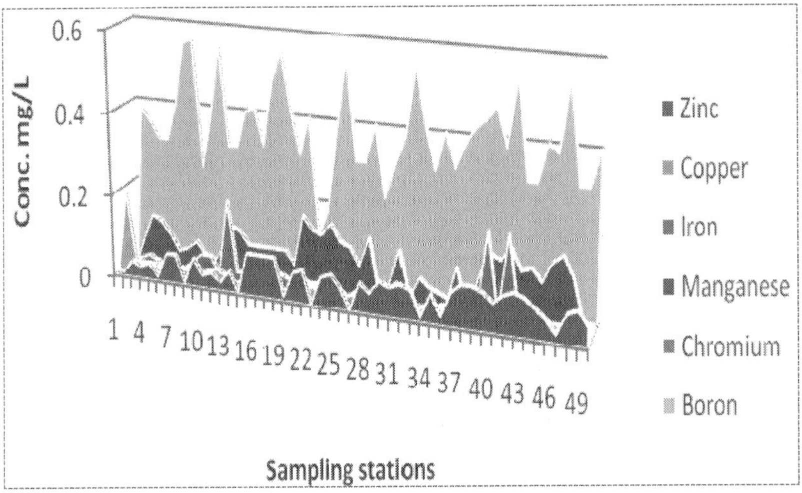
0.6
0.4
0.2
0
Conc. mg/L
1 4 7 10 13 16 19 22 25 28 31 34 37 40 43 46 49
Sampling stations
Zinc
Copper
Iron
Manganese
Chromium
Boron

Fig. 8.

The correlation between these parameters is shown in table 3. The highest positive correlation (0.244) was found between iron and manganese, followed by iron and zinc (0.138). The lowest positive correlation (0.047) was found between iron and chromium. While as the highest negative correlation (- 0.293) was found between zinc and copper. Moreover, boron was found to show the negative correlation with all the parameters except zinc.

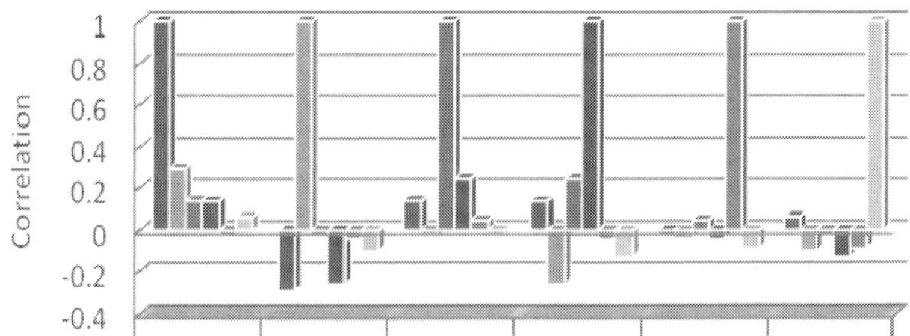

	Z	Cu	Fe	Mn	Cr	B
Z	1	-0.29338	0.138309	0.136238	-0.0242	0.062718
Cu	0.29338	1	-0.02784	-0.26347	-0.0433	-0.10313
Fe	0.138309	-0.02784	1	0.244276	0.047197	-0.01913
Mn	0.136238	-0.26347	0.244276	1	-0.0448	-0.1304
Cr	-0.0242	-0.0433	0.047197	-0.0448	1	-0.08985
B	0.062718	-0.10313	-0.01913	-0.1304	-0.08985	1

Table 3.

12. Conclusion

The concentrations of the investigated heavy metals (Mn, Fe, Cr, Cu, Zn and B) in the drinking water samples from Mamundiyar basin, India were found below the guidelines for drinking waters given by the WHO (World Health Organization), EC (Europe Community), EPA (Environment Protection Agency). It was concluded that drinking waters in Mamundiyar contain low heavy metal levels.

13. References

Banks D, Frengstad B, Midtga°rd AK, Krog JR, Strand T. The chemistry of Norwegian groundwaters: I. The distribution of radon, major and minor elements in 1604 crystalline bedrock groundwaters. Sci Total Environ 1998;222:71 -91.

Banks D, Midtga˚rd AK, Morland G, Reimann C, Strand T, Bjorvatn K, Siewers U. Is pure groundwater safe to drink? Natural 'contamination' of groundwater in Norway. Geol Today 1998;14(3):104 -113.

Banks D, Reimann C, Røyset O, Skarphagen H. Natural concentrations of major and trace elements in some Norwegian bedrock groundwaters. Appl Geochem 1995;10:1 - 16.

Banks D, Røyset O, Strand T, Skarphagen H. Radioelement (U,Th, Rn) concentrations in Norwegian bedrock groundwaters. Environ Geol 1995;25:165 -180.

Bjorvatn K, Ba˚rdsen A˚ , Thorkildsen AH, Sand K. Fluorid I norsk grunnvann—en ukjent helsefaktor wFluoride in Norwegian drinking water—an unknown health factorx. Vann 1994;2:120 -128. in Norwegian.

Bjorvatn K, Thorkildsen AH, Holteberg S. Sesongmessige variasjoner i fluoridinholdet i sør og vestnorsk grunnvann [Seasonal variations of the fluoride content in south and west Norwegian groundwaters]. Den norske tannlegeforenings tidende 1992;102:128 -133. in Norwegian

Chatterjee A, Das D, Mandal BK, Chowdhurry TR, Samanta G, Chakraborti D. Arsenic in groundwater in six districts of West Bengal, India: the biggest arsenic calamity in the world. Part 1: arsenic species in drinking water and urine of affected people. Analyst 1995;120:643 -650.

Dar IA, Sankar K, Dar MA, Remote sensing technology and geographic information system modeling: An integrated approach towards the mapping of groundwater potential zones in Hardrock terrain, Mamundiyar basin. Journal of Hydrology (2010) 394: 85-295

Das D, Chatterjee A, Mandal BK, Samanta G, Chakraborti D. Arsenic in groundwater in six districts of West Bengal, India: the biggest arsenic calamity in the world. Part 2: arsenic concentration in drinking water, hair, nails, urine, skin scale and liver tissue (biopsy) of the affected people. Analyst 1995;120:917 -924.

Edmunds WM, Smedley PL. Groundwater geochemistry and health: an overview. In: Appleton JD, Fuge R, mccall GJH, editors. Environmental geochemistry and health. Geological Society Special Publication 113 1996. P. 91 -105.

Edmunds WM, Trafford JM. Beryllium in river baseflow, shallow groundwaters and major aquifers of the UK. Appl Geochem 1993;2(Suppl):223 -233.

European Union. 80y778yeec Council Directive of 15 July 1980 relating to the quality of water intended for human consumption. Official Journal of the European Community 1980. P. L229y11 -L229y29.

European Union. Council Directive 98y83yec of 3 November 1998 on the quality of water intended for human consumption. Official Journal of the European Community 1998. P. L330y32 -L330y54.

Frengstad B, Midtga˚rd AK, Banks D, Krog JR, Siewers U. The chemistry of Norwegian groundwaters. III. The distribution of trace elements in 476 crystalline bedrock groundwaters, as analysed by ICP-MS techniques. Sci Total Environ 2000;246: 21 -40.

Krishnamurthy, C.R. and V. Pushpa. 1995. Toxic metals in the Indian Environment. Tata McGraw Hill Publishing Co. Ltd., New Delhi. pp 280.

Midtga˚rd AK, Frengstad B, Banks D, Krog JR, Strand T, Siewers U. Drinking water from crystalline bedrock aquifers— not just H2O. Min Soc Bull 1998;121:9 -16.

Misund A, Frengstad B, Siewers U, Reimann C. Natural variation of 66 elements in European mineral waters. Sci Total Environ 1999;243y244:21 -41.

Morland G, Reimann C, Strand T, Skarphagen H, Banks D, Bjorvatn K, Hall GEM, Siewers U. The hydrogeochemistry of Norwegian bedrock groundwater-selected parameters (ph, Fy, Rn, U, Th, B, Na, Ca) in samples from Vestfold and Hordaland, Norway. NGU Bull 1997;432:103 -117.

Morland G, Strand T, Furuhaug L, Skarphagen H, Banks D. Radon concentrations in groundwater from Quaternary sedimentary aquifers in relation to underlying bedrock geology. Ground Water 1998;36:143 -146.

Reimann C, Hall GEM, Siewers U, Bjorvatn K, Morland G, Skarphagen H, Strand T. Radon, fluoride and 62 elements as determined by ICP-MS in 145 Norwegian hardrock groundwaters. Sci Total Environ 1996;192:1 -19

Reimann C, Siewers U, Skarphagen H, Banks D. Does bottle type and acid washing influence trace element analyses by ICP-MS on water samples? A test covering 62 elements and four bottle types: high-density polyethene (HDPE), polypropene (PP), fluorinated ethene propene copolymer (FEP) and perfluoroalkoxy polymer (PFA). Sci Total Environ 1999;239:111 -130.

Sæther O, Reimann C, Hilmo BO, Taushani E. Chemical composition of hard- and softrock groundwaters from central Norway with special consideration of fluoride and Norwegian drinking water limits. Environ Geol 1995;26(3):147 - 156.

Smedley PL, Edmunds WM, Pelig-Ba KB. Mobility of arsenic in groundwater in the Obuasi gold-mining area of Ghana: some implications for human health. In: Appleton JD, fuger, mccall GJH, editors. Environmental geochemistry and health. Geological Society Special Publication 113 1996. P. 163 -181.

Smedley PL, Kinniburgh DG. A review of the source, behavior and distribution of arsenic in natural waters. Appl Geochem 2002;17(5):517 -568.

Smith AH, Lingas EO, Rahman M. Contamination of drinking water by arsenic in Bangladesh: a public health emergency. Bull WHO 2000;78(9):1093 -1101.

USEPA, 1989. National Primary and Secondary Drinking Water Regulations, Proposed Rule, Fed. Reg. (Vol. 54, No. 97).

USEPA, 1991. Safe Drinking Water Act, 1991 Amendments, EPA 570/9-86-002. Washington, D.C.

USPHS, Drinking water standards, Unitede States Public Health Services, 1987, Washington DC.

Varsanyi I, Fodre Z, Bartha A. Arsenic in drinking water and mortality in the Southern Great Plain, Hungary. Environ Geochem Health 1991;13:14 -22

WHO. Guidelines for drinking water quality. Geneva: World Health Organisation, 1993.

WHO. Guidelines for drinking-water quality. Addendum to vol. 1. Recommendations, 2nd ed. Geneva: World Health Organisation, 1998. P. 10 -11.

WHO: 1984, *Guidelines for Drinking Water Quality*, Geneva.

Williams M, Fordyce F, Paijiprapapon A, Charoenchaisri P. Arsenic contamination in surface drainage and groundwater in part of the Southeast Asian tin belt, Nakhon Si Thamarat Province, southern Thailand. Environ Geol 1996;27:16 -33.

22

Age Dating of Middle-Distillate Fuels Released to the Subsurface Environment

Gil Oudijk
Triassic Technology, Inc.
USA

1. Introduction

The term "age dating" is defined as: estimating the time frame of a contaminant release to the environment. Because of the high costs of environmental cleanups, age-dating studies have now become an integral part of environmental investigations. Knowledge of the local geology, hydrology and geochemistry are required to perform these studies and, therefore, geologists are commonly involved.

The "middle distillates" include products such as diesel fuel, heating oils, kerosene and jet fuels. Middle-distillate fuels are used throughout the world to power motors, heat residences, fuel jet engines and propel ships, among many other uses. Middle-distillate fuels are commonly stored in aboveground or underground tanks and these tanks are often unprotected and exposed to the elements. Because of corrosion, leaks from storage tanks are a severe environmental problem, especially in locations where groundwater is used for potable supplies. Numerous underground storage tanks (USTs) were installed in North America during the "boom" years following World War II and impacts from leakage are now being found in the subsurface.

An understanding of the problems associated with leaking petroleum USTs has been known since the 1950s (Kehoe, 1960). However, action was not undertaken until the late 1970s and in some places, even much later. For example, the US state of New Jersey did not pass UST regulations until 1986 (State of New Jersey, 1986).

The average non-leaking lifespan of unprotected steel USTs may be as little as 15 years (Robinson et al., 1988). The Canadian province of Nova Scotia requires that USTs older than 25 years be removed (Hankey-Masui, 1998). Thus, numerous leaking USTs existed over the years and many probably continue today. Because of costs, the number of people impacted and the large number of cases, releases of middle-distillate fuels from USTs are a serious problem in North America (Oudijk et al., 1999). In the US states of New Jersey and Maine, several leaks are reported daily to regulatory agencies (Pearson & Oudijk, 1993; McCaskill, 1999). Similar problems exist in Europe (Bennet, 1997).

Remediation costs can be high and cases exist where buildings were removed, razed or structurally supported to complete a cleanup. It is not uncommon for costs to exceed US$500,000 and many cases costing over US$1 million exist. Costs are often borne by insurance policies, although carriers may subrogate and obtain contribution from previous carriers or others responsible. For this reason, carriers and law firms commonly request information on the time frames of releases. Because of costs, many cases are litigated and, consequently, a legally defensible method to age date releases is needed. Kanner (2007) provided the legal criteria needed to defend such methods.

Many methods exist to assess contaminant-release ages, such as UST corrosion models (Morrison, 2000), groundwater flow calculations (Morrison, 2000; Lee et al., 2007), isotope surveys (Oudijk, 2005), tree-ring investigations (Balouet et al., 2007), petroleum-weathering studies (Christensen & Larsen, 1993; Wade, 2001; Douglas et al., 2004; Hurst & Schmidt, 2005; Galperin & Kaplan, 2008; Oudijk, 2009a,b). The most common technique now in use, and normally the least expensive, is the Christensen & Larsen (C&L) method, a procedure employing petroleum-weathering rates. Hurst & Schmidt (2005), with additional data, expanded on the C&L method to date diesel-fuel releases, reporting a best-case precision of ±1.5 years.

2. Fuel composition

2.1 Diesel fuels and heating oils

In North America, diesel fuel and no. 2 heating oil have a similar composition. Diesel fuel may contain some additives during the winter and heating oil often contains a dye for tax purposes. These fuels normally have a density of 0.87 to 0.95 grams per cubic centimeter (g cm^{-3}) at 20°C and are lighter but more viscous than water (Schmidt, 1985; Wang et al., 2003) (Table 1). Heating oils and diesel fuels are composed of hydrocarbons, which are chains or rings of hydrogen and carbon. Hydrocarbons are classified by the number of carbon atoms present. For example, benzene is a C_6 molecule because it contains 6 carbon atoms.

Diesel fuels and no. 2 heating oils are composed predominantly of hydrocarbons in the range of about C_6 to C_{24} (but most of the hydrocarbons are heavier than C_8); sometimes hydrocarbons can be found up to C_{28}. The boiling range is from 150°C to 380°C (Song, 2000; Owen & Coley, 1995)(Figure 1). They contain aromatics (benzene, toluene, *o,m,p*-xylenes, naphthalenes, phenanthrenes), *n*-alkanes (such as *n*-heptadecane), *iso*-alkanes (such as the isoprenoids: pristane, phytane or norpristane), *cyclo*-alkanes and poly-aromatics plus sulfur-containing compounds such as dibenzothiophenes (Kramer & Hayes, 1987; Potter & Simmons, 1998; Bruya, 2001). The dominant hydrocarbons are the *n*–alkanes (straight-chain alkanes) and isoprenoids (methyl-substituted "*iso*-alkanes"). The aromatics include: mono-aromatics (such as benzene and toluene), alkyl-benzenes, naphthalenes, tetralins, biphenyls, acenaphthenes, phenanthrenes, chrysenes and pyrenes (Song, 2000). The predominant poly-aromatic hydrocarbons (PAHs) in no. 2 heating oil and diesel fuel are the napthalenes and phenanthrenes, whereas pyrogenic PAHs, such as chrysene and pyrene, may exist at reduced concentrations.

There can be variations in the composition of diesel fuels. For example, diesel fuels in colder climates tend to contain lighter hydrocarbons to prevent freezing problems (Figure 2).

2.2 Kerosene and jet fuels

Kerosenes are complex mixtures of hydrocarbons generally within a range of C_6 to C_{16} and a boiling range of about 145°C to 300°C (Table 1, Figures 3 & 4). Jet fuels are quite similar in composition to kerosene. The major components of kerosenes are *n*-alkanes, *iso*-alkanes and *cyclo*-alkanes. Aromatic hydrocarbons, predominantly alkyl-benzenes and alkyl-naphthalenes, normally comprise less than 25% of the volume (CONCAWE, 1995).

	Kerosene	Jet fuel (Jet-A)	Jet fuel (Jet-B)	Jet fuel (JP-5)	Jet fuel (JP-8)
Color	colorless, multiple dyes	colorless multiple dyes	colorless, multiple dyes	Clear and bright	Clear and bright
Carbon range	C_9 to C_{16}	C_9 to C_{16}	C_9 to C_{16}	C_9 to C_{16}	C_9 to C_{16}
Density	0.81@15°C	0.775 to 0.840 @15°C	0.751 to 0.802 @15°C	0.788 to 0.845 @15°C	0.775 to 0.840 @15°C
Boiling range (°C)	145 to 300	205 to 300	145 to 245	150 to 290	150 to 290
Flash point (°C)	62[b]	38	0	60	38
Kinematic viscosity[a] ($mm^2\ s^{-1}$)	1.5 to 2.5 @20°C	8 @-20°C	-	1.28 to 1.60 @40°C	1.05 to 1.58 @40°C
CAS No.	8008-20-6	8008-20-6	8008-20-6	70892-10-3/ 8008-20-6	70892-10-3/ 8008-20-6

	No.2 heating oil/Motor diesel fuel	No. 4 heating oil	No.6 heating oil/Bunker oil
Color	colorless to brown, often dyed red	colorless to brown, can be dyed red	colorless to brown
Carbon range	C_6 to C_{24}	C_{15} to C_{40}	C_{20} to C_{50}
Density	0.87 to 0.95 @ 20°C	0.876 to 0.979 @20°C	0.95 to 1.01 @15°C
Boiling range (°C)	160 to 360	177 to 371	350 to 650
Flash point (°C)	58	>60	>60
Kinematic viscosity ($mm^2\ s^{-1}$)	20 to 30 @ 20°C	4 to 50 @20°C	6 to 55 @ 20°C
CAS No.	68476-30-2	68476-31-3	68553-00-4

a One millimetre squared per second ($mm^2\ s^{-1}$) equals one centistoke.
b Flash point is for US kerosene. Flash points for European kerosene are normally 40° to 45°C.

Table 1. General characteristics of middle distillate fuels. Sources: Bowden et al., 1988; CONCAWE, 1995;

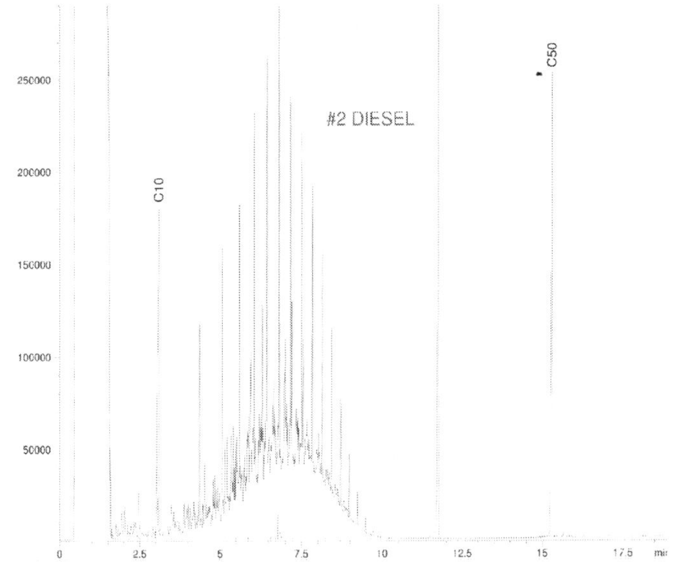

Fig. 1. A GC/FID chromatogram of a motor diesel fuel (2010). Source: Maxxam Analytics (Mississauga, Ontario, Canada).

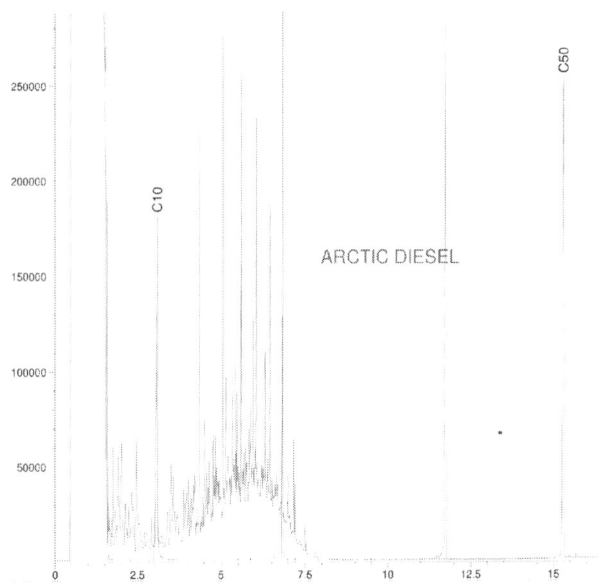

Fig. 2. A GC/FID chromatogram of an Arctic diesel fuel (2010). Source: Maxxam Analytics (Mississauga, Ontario, Canada).

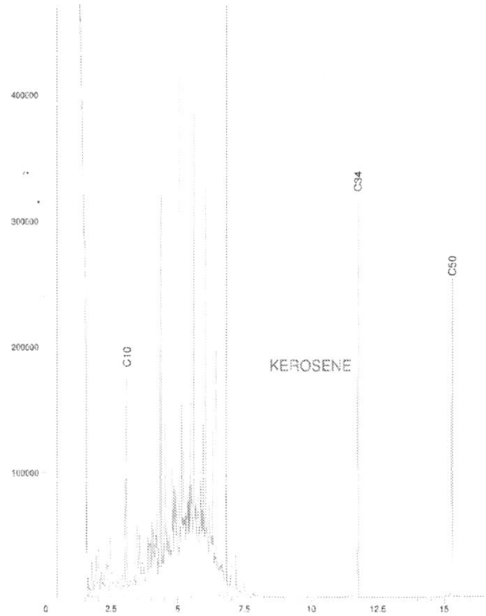

Fig. 3. A GC/FID chromatogram of a kerosene (2010). Source: Maxxam Analytics (Mississauga, Ontario, Canada).

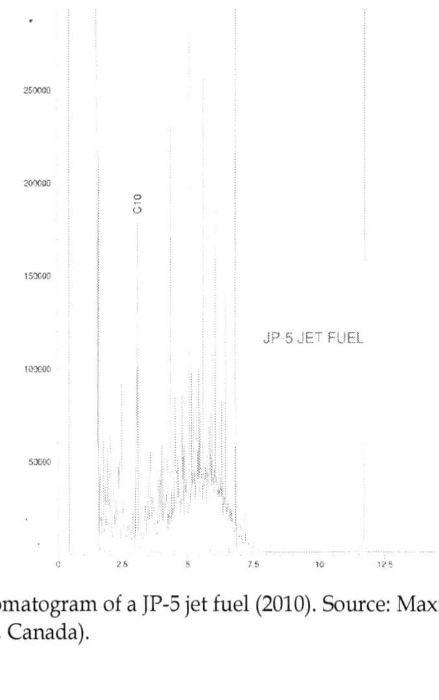

Fig. 4. A GC/FID chromatogram of a JP-5 jet fuel (2010). Source: Maxxam Analytics (Mississauga, Ontario, Canada).

2.3 No. 6 oils and bunker oils

No. 6 oil is a heating fuel, whereas bunker oil fuels ship engines. Both fuels are complex mixtures of hydrocarbons normally within a range of C_{15} to more than C_{30} (CONCAWE, 1998; Stout et al., 2002)(Table 1 & Figure 5). However, these fuels often differ greatly in composition. Furthermore, to prevent freezing problems during the winter, no. 6 oil is often mixed with lighter fuels such as kerosene or no. 2 heating oil.

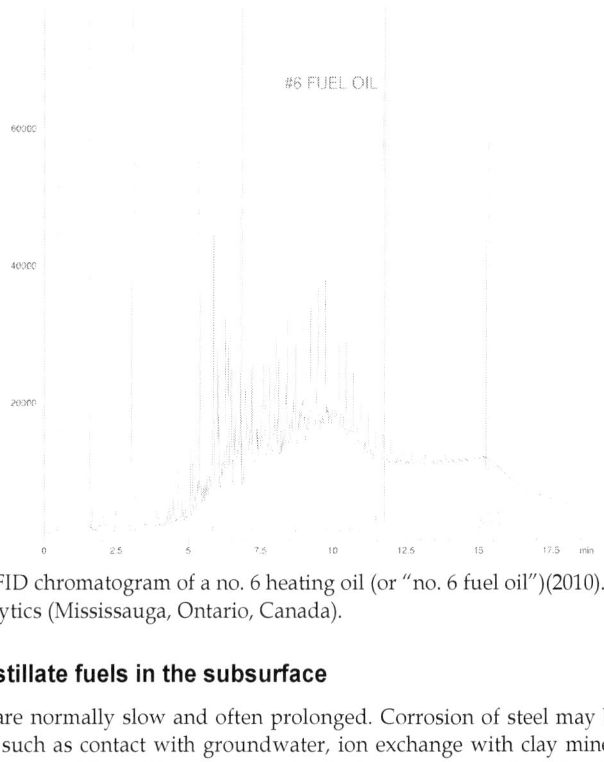

Fig. 5. A GC/FID chromatogram of a no. 6 heating oil (or "no. 6 fuel oil")(2010). Source: Maxxam Analytics (Mississauga, Ontario, Canada).

3. Middle distillate fuels in the subsurface

UST releases are normally slow and often prolonged. Corrosion of steel may be caused by many factors, such as contact with groundwater, ion exchange with clay minerals or stray electrical currents. Holes will begin as pin-sized openings and, with time, expand. Accordingly, petroleum in soil or groundwater is a mixture of ages and the ages will be skewed younger because leakage rates increase with time. The most downgradient portions of a middle-distillate plume are commonly the oldest and most age discrete. Moving closer to the source, for example towards an UST, the oil becomes progressively less age discrete (or less of a mix of ages)(Oudijk et al., 2006). To assess the maximum release age, sampling is needed within these downgradient areas.

4. Assessing a middle-distillate release

Over time, the chemistry of middle distillates placed into an UST may change because the crude-oil source for different refiners can be dissimilar. If the owner changed distributors or the distributor obtained its supply from different refineries, the initial composition of the product will not be identical over the leakage time frame. Accordingly, to properly conduct an age-dating study, it is important to assess if the plume all originates from the same source and if the initial petroleum chemistry differed.

A test to assess different compositions is through isoprenoids ratios. These compounds are relatively resistant to degradation and their ratios may be used to determine if the initial chemistry varied across the plume (Wade, 2005). There are numerous other methods to fingerprint spilled hydrocarbons (Bruce & Schmidt, 1994; Douglas et al., 1996; Galperin & Camp, 2002; Wang & Fingas, 1995a; Wang et al., 2005; Galperin & Kaplan, 2008a). These methods include a comparison of compounds such as PAHs, dibenzothiophenes or bicyclic sesquiterpanes. However, use of pristane/phytane (pr/ph) ratios seems to be the easiest and least expensive.

The pr/ph ratio is dependent on the crude-oil source and may reflect the depositional environment during oil formation (Illich, 1983; ten Haven et al., 1987; Paul et al., 1994; Sun et al., 2004; Peters et al., 2005; Osuji et al., 2009). The refining process, whereby middle distillate are produced from crude oil, normally does not alter pr/ph ratios (Stout & Wang, 2007).

Between 2004 and 2007, pr/ph ratios were calculated from 141 petroleum-saturated soil samples collected from 48 sites in the US state of New Jersey (Oudijk, 2009a). The ratios were then compared to the apparent weathering for each sample, grouped into fresh, moderate, degraded and very-degraded classes (Table 2). Apparent weathering was based on review of chromatograms and dependent on depletion of specific hydrocarbon classes, such as *n*-alkanes, aromatics or *iso*-alkanes (Kaplan et al., 1996; Senn & Johnson, 1987). The data revealed that weathering did not alter pr/ph ratios until significant degradation occurred. Hence, pr/ph values in middle distillates are dependent on their crude-oil source and consequently, they can be a simple and effective fingerprint to assess the origin of relatively unweathered, middle distillates. Other helpful isoprenoid ratios include pristane/norpristane and pristane/farnesane. However, norpristane and farnesane are less resistant to weathering and ratios can be altered in the environment.

Magnitude of petroleum weathering in sample	Number of samples	pr/ph average	Standard deviation
Fresh	41	1.61	0.19
Moderate	39	1.73	0.21
Degraded	49	1.59	0.23
Very degraded	12	1.18	0.40

Table 2. Mean and standard deviation for pristane/phytane values from soil samples (contaminated with no. 2 heating oil or motor diesel fuel) collected in the US states of New Jersey, Pennsylvania and New York, 2002-2007. Source: Oudijk (2009a).

5. Factors influencing petroleum weathering

Petroleum degradation in soil is predominantly controlled by (Stout et al., 2002b):

- evaporation, occurring when petroleum is in contact with air, causing constituents to volatilize;

- dissolution, occurring when petroleum is in contact with water, causing constituents to dissolve, and
- biodegradation, the digestion of petroleum constituents by microbes.

Researchers found that normally, but not always, the predominant weathering process in the subsurface is biodegradation (de Jonge et al., 1997; Kaplan, 2003). With a subsurface leak, evaporation is often not a factor and the remaining processes dominate. Because hydrocarbons in middle distillates are relatively insoluble in water, especially the heavier ones, biodegradation often predominates over dissolution (Christensen & Larsen, 1993).

Many microbes can use hydrocarbons as a sole energy source in their metabolism (Zobell, 1946). Energy is obtained through transfer of electrons between donors such as organic carbon, although some reduced forms of nitrogen, iron and sulfur also play a role. Dissolved oxygen (O_2) produces the most energy per mole of organic carbon oxidized than any other commonly-available electron acceptor and it is preferred by subsurface microbes (McMahon & Chapelle, 2008).

Atlas & Bartha (1992) concluded that, in one environment, spilled petroleum could persist almost indefinitely, whereas under other conditions, the same hydrocarbons might be completely removed within a few hours or days. Therefore, each environmental setting is specific and significant differences could exist in the rates and types of biodegradation. In general, biodegradation depends on:

- presence of microbes with the metabolic capacity to degrade the petroleum;
- recalcitrance of compounds (in the middle-distillate mixture);
- growth and activity factors, such as temperature, nutrients, electron acceptors and pH, influencing the microbial-population dynamics; and
- bioavailability (de Jonge et al., 1997). Bioavailability is the amount of contaminant present in the soil or groundwater that is available for uptake by microbes (Harmsen et al., 2005).

Microbes with the potential to degrade hydrocarbons in soil and ground water include bacteria, fungi, and yeasts, although bacteria are normally the most plentiful, followed by fungi (Markovetz et al., 1968; Leahy & Colwell, 1990). The byproducts of microbial degradation of, for example *n*-alkanes, are normally alcohols, aldehydes, and then fatty acids and possibly ketones of similar chain length plus water, CO_2 or CH_4 (Klug & Markovetz, 1967; Atlas & Bartha, 1992). Dashti et al. (2008) found that bacteria prefer *n*-alkanes, whereas fungi prefer the oxidized byproducts; however, the same consortium of microbes could degrade both the original alkanes and the degradation byproducts.

For biodegradation to occur, electron acceptors, such as O_2 and NO_3^-, and nutrients, such as NH_4^+ and PO_4^{3-}, are needed. Aromatics can be mineralized in the absence of O_2 under denitrifying, iron-reducing, methanogenic and/or sulfate-reducing conditions, whereas *n*-alkanes can mineralize under sulfate-reducing or denitrifying conditions (Bregnard et al., 1996; Ehrenreich et al., 2000). However, biodegradation rates under anaerobic conditions may be slow. Under aerobic conditions, the *n*-alkanes and mono-aromatics are the first hydrocarbons to be depleted. They are usually followed by alkyl-benzenes, alkyl-naphthalenes, alkyl-*cyclo*-hexanes and then the isoprenoids, thiophenes and PAHs (Cerniglia, 1984; Singer & Finnerty, 1984; Hostettler & Kvenvolden, 2002). The lighter hydrocarbons in each series are commonly removed earliest. For example, the naphthalenes often degrade in series of methylnaphthalene to dimethylnaphthalene to trimethylnaphthalene (Garrett et al., 2003). According to Kaplan et al. (1996), the most resistant isoprenoid is pristane, although our field data do not confirm this conclusion (Oudijk, 2009a)(Table 2).

In uncontaminated soils, hydrocarbon-degrading bacteria often constitute less than 1% of the microbial community. In polluted soils, hydrocarbon-degrading bacteria are often 10% of the community and, in some cases, may comprise 100% (Atlas, 1981; Atlas & Bartha, 1992). Furthermore, previously polluted environments, although remediated, tend to contain elevated percentages of hydrocarbon-degrading microbes (Leahy & Colwell, 1990). Therefore, microbial weathering may be accelerated in urban soils in comparison to pristine soils.

Environmental factors that may influence petroleum biodegradation, but are not always investigated or quantified in environmental investigations, include (Atlas, 1981; Leahy & Colwell, 1990):

5.1 Hydrocarbon physical state

The physical state of the spilled hydrocarbons, such as separate, dissolved, vapor or adsorbed phases, impacts biodegradation rates. Dissolved or vapor phases are often more susceptible to weathering processes.

The addition of large quantities of separate phase can suppress or completely stop bacterial growth. However, some researchers found that addition of large quantities to previously polluted environments increases bacterial growth (Colwell, 1978). At a previously pristine site, Hostettler & Kvenvolden (2002) found unweathered separate-phase crude oil almost 20 years after a spill. De Jonge et al. (1997) found that *n*-alkane biodegradation decreased significantly when petroleum concentrations exceeded 4,000 milligrams per kilogram (mg/kg). Lapinskiene et al. (2005) found that diesel concentrations in excess of 30,000 mg/kg were generally toxic to microbes in an aerated soil. Swindell & Reid (2006) found that, at total diesel concentrations of 20 mg/kg, 200 mg/kg and 2,000 mg/kg, phenanthrene was biologically removed from soil. However, at 20,000 mg/kg, phenanthrene removal was retarded.

The degree of separate-phase spreading will also influence biodegradation. A thick pool of separate phase will biodegrade slower compared to a pool that spread across the water table (Atlas & Bartha, 1992). The thinner pool will have more surface area in contact with ground water or the unsaturated zone, allowing increased dissolution and volatilization and hence more biological activity.

Concentrations of hydrocarbons in dissolved or vapor phases can have a strong influence on the petroleum-degrading microbes. Highly elevated dissolved petroleum concentrations may also limit biological alteration (Atlas, 1981).

5.2 Soil and groundwater chemistry

Factors such as pH or salinity can reduce biological activity. For example, elevated salt concentrations can prevent microbes from consuming petroleum. However, in some instances, salt-tolerant microbes exist and can accelerate biodegradation (Atlas, 1981; Diaz et al., 2002). The optimum pH for microbial activity is normally about 8 and soil water or groundwater commonly exhibits lower pH values (Atlas & Bartha, 1992). In some instances, soil with pH values greater than 6 can exhibit accelerated biodegradation. However, biodegradation often produces acids, allowing the pH to lower, often to as low as 3 or 4. A pH value in this range can often inhibit microbial degradation (Zaidi & Imam, 1999). The presence of certain elements in the soil or groundwater, in particular heavy metals, is toxic to certain microbes and can reduce or prevent biodegradation.

5.2.1 Redox conditions

Under aerobic conditions, *n*-alkanes commonly degrade readily, whereas isoprenoids are generally recalcitrant. Bouchard et al. (2008) found that, based on isotopic studies, biological degradation of *n*-alkanes in aerobic, unsaturated sand was dependent on chain length with smaller molecules degrading quicker. Isoprenoids, such as pristane, can weather under anaerobic conditions (Bregnard et al., 1997), whereas light *n*-alkanes may become recalcitrant compared to heavier *n*-alkanes (Hostettler & Kvenvolden, 2002; Siddique et al., 2006; Hostettler et al., 2008). In particular, Bregnard et al. (1997) found that pristane can weather under nitrate-reducing conditions. Hostettler & Kvenvolden (2002) found that under anaerobic conditions the degradation order is the same compared to aerobic conditions: *n*-alkanes are removed first followed by alkyl-*cyclo*-hexanes and *iso*-alkanes. However, anaerobic conditions can cause the order to reverse within each homologous series. Heavier *n*-alkanes may be removed first and the same is true for alkyl-*cyclo*-hexanes. Other researchers finding similar reversals include Setti et al. (1995)(and references therein). However, Davidova et al. (2005) did not find a reversal in the degradation order, at least under sulfate-reducing conditions, and Stout & Uhler (2006) and Galperin & Kaplan (2008b) contend that reversals are caused by other means. Also, *n*-alkane degradation up to C_{28} was observed under sulfate-reducing conditions (Caldwell et al., 1998). Therefore, use of *n*-alkane/isoprenoid ratios, as a measure of weathering under anoxic or sub-anoxic conditions, may be problematic.

Under nitrate-reducing or methanogenic conditions, nitrogen gas (N_2) or methane (CH_4) can form through degradation of aromatics. If the gas accumulates, it can limit groundwater flow and retard biological processes (Reinhard et al., 2000).

Fungi degrade long-chain *n*-alkanes (*n*-nonane to *n*-octadecane) in preference to shorter-chain varieties (Merdinger & Merdinger, 1970; Teh & Lee, 1973). Because fungi are dependent on oxygen for growth, depletion of long-chain *n*-alkanes may be indicative of fungi, instead of low redox. However, Jovancicević et al. (2003) found that an accumulation of heavier, even-numbered *n*-alkanes, such as *n*-C_{16} and *n*-C_{18}, may occur during biodegradation because of the presence of algae.

5.2.2 Temperature

Near-ground-surface temperatures fluctuate greatly, whereas underground temperatures remain somewhat constant. Biological alteration of spilled petroleum generally increases with temperature. Furthermore, volatilization of lighter *n*-alkanes at colder temperatures may decrease.

Atlas (1981) found that degradation was an order of magnitude greater at 25°C compared with 5°C, whereas Sexstone et al. (1978) found diesel contamination in Arctic soils 28 years after a spill. Ludzack & Kinhead (1956) found that motor oil rapidly oxidized at 20°C, but not at 5°C. Margesin & Schinner (2001) found that diesel degradation at a cold, high-altitude location occurred mostly during the summer and at a reduced rate. Man (1998) found that *n*-alkane depletion was similar regardless of temperature if the range was between 10°C and 22°C. Bonroy et al. (2007) found that heating-oil biodegradation rates in shallow soil almost doubled during the summer months compared to the winter.

Ground cover can impact the temperature of surface soils and consequently the temperature of percolating rainwater (Huang et al., 2008). Paved surfaces, such as asphalt or concrete, retain heat, whereas grass-covered or forested areas cool quicker during summer months. Increased temperature will decrease petroleum viscosity, allowing increased spreading, additional surface area in contact with groundwater, and enhanced biodegradation (Atlas & Bartha, 1992).

5.2.3 Contact with water

Many constituents of middle distillates exhibit low aqueous solubilities. Aromatics are more soluble than aliphatics of the same carbon number, whereas *cyclo*-alkanes tend to be slightly more soluble than *n*-alkanes (Bobra, 1992). Two compounds often used to represent petroleum weathering are the *n*-C_{17} alkane (*n*-heptadecane: $C_{17}H_{36}$) and pristane (2,6,10,14-tetramethylpentadecane: $C_{19}H_{40}$)(or "*n*-C_{17}/pr"). Bregnard et al. (1997) reported that pristane's aqueous solubility is less than 0.1 microgram per litre (µg/l), whereas Ritter (2003) found that solubility differences (in petroleum) between *n*-C_{17}, *n*-C_{18}, pristane and phytane are small. Middleditch et al. (1978) reported *n*-heptadecane concentrations in seawater ranging from 2 to 747 µg/l. Leahy & Colwell (1990) report that microbial degradation of long-chain *n*-alkanes (≥C_{12}) occurs at rates that exceed the rates of hydrocarbon dissolution.

LaFargue & Barker (1988) found that *n*-alkanes lighter than C_{14} in crude oils were susceptible to dissolution, whereas the heavier *n*-alkanes were not. Isoprenoids heavier than C_{16} were not susceptible to dissolution, whereas the C_{13} through C_{15} isoprenoids were somewhat vulnerable.

For a given carbon number, ring formation, unsaturation, and branching cause an increase in aqueous solubility. Therefore, one could expect that when dissolution occurs, aromatics of a given carbon number would decrease first, followed by *cyclo*-alkanes, *iso*-alkanes and *n*-alkanes (Palmer, 1991).

Dissolution of hydrocarbons into groundwater or soil water may be impacted by:

- the surface area of hydrocarbons in contact with water, also known as the oil-water ratio. A higher ratio may impart greater dissolution; accordingly, geologic materials with a greater porosity may allow greater dissolution (Bobra, 1992);
- ambient groundwater chemistry and, in particular, temperature, pH and oxidation-reduction potential (ORP). The aqueous solubility of hydrocarbons often increases with temperature; however, the relationship between variables such as pH or ORP and solubility is often compound specific and possibly site-specific;
- the magnitude of precipitation and recharge. Recharge commonly increases dissolution, and
- the groundwater migration rate. Slow-moving groundwater will lessen transfer of hydrocarbons to a dissolved state, whereas the opposite occurs with rapidly migrating groundwater (Fried et al., 1979). In column experiments, Miller et al. (1990) found that the rate of mass transfer between a toluene separate phase and the aqueous phase was directly related to the groundwater migration rate.

As a result of mass transfer, dissolution and biodegradation are coupled processes because contact with water stimulates biological activity. Addition of petroleum to groundwater or soil water can allow indigenous bacteria to multiply and preferentially attack *n*-alkanes (Solević et al.,2003). Therefore, contact with groundwater may cause dissolution of lighter *n*-alkanes and isoprenoids and induce microbial degradation of lighter and heavier *n*-alkanes

and isoprenoids. Degradation can also begin inside an UST if sufficient water infiltration occurs (Gaylarde et al., 1999).
A rapidly fluctuating water table will foster emulsification and can enhance biological activity because of greater contact between the separate phase and water. Therefore, production of an emulsification can increase biodegradation rates (Atlas & Bartha, 1992).

5.2.4 Light

The rate of photochemical reactions is directly proportional to the number of photons absorbed by a chemical. Nearness to the Equator or an increase in altitude will accelerate the reactions (Sukol et al., 1988). Photodecomposition is not a significant process in the subsurface, although immediately adjacent to the ground surface, it may be important.

5.2.5 Oxygen and nutrients

Aerobic microbes need electron acceptors and nutrients to degrade petroleum. Lack of oxygen and nutrients may limit biological activity. Even though anaerobic microbes exist, anaerobic degradation is normally slower. For example, Bonin & Betrand (2000) found lowering oxygen contents could stop *n*-heptadecane mineralization. Numerous researchers found that oxygen availability is the most important factor in petroleum degradation (Raymond et al., 1976; Song et al., 1990). Factors affecting oxygen availability in soil include (Atlas & Bartha, 1992):

- *Drainage*: in water-logged soils, oxygen diffusion can be slow and bacterial movement restricted;
- *Soil texture*: coarse-grained soils have higher permeabilities and oxygen can be quickly replenished. Furthermore, coarser textures allow greater contact area between water and petroleum, increasing dissolution. However, for reasons stated earlier, medium-grained soils may exhibit the most biodegradation potential;
- *Proximity to the ground surface*: in laboratory column experiments, degradation was 3 to 5 times greater at the top versus the base (Atlas, 1981). This observation is related to proximity to greater oxygen abundance, temperature and recharge. Biological degradation can vary significantly over short distances in the horizontal and vertical directions. Variations will be dependent on nutrient and oxygen content and microbial diversity of geologic layers (Maila et al., 2005), and
- *Quantity of hydrocarbons*: Areas saturated with hydrocarbons may exhaust oxygen faster than it can be resupplied. Oxidation of 1 litre (L) of hydrocarbons can exhaust the dissolved oxygen in close to 400,000 L of water (Atlas & Bartha, 1992). Furthermore, large quantities of separate phase may decrease soil permeability with respect to water.

5.2.6 Bacteriocides

For biodegradation to occur, toxic concentrations of bacteriocides must not exist. Bacteriocides are elements or compounds toxic to bacteria. For example, H_2S may be toxic to some microbes (Prince & Walters, 2007). Under sulfate-reducing conditions, H_2S may form through biodegradation of aromatics.

5.3 Soil composition: Chemistry, lithology and texture

Coarser-grained soils permit freer movement of liquids such as soil gas, soil water and groundwater, allowing replenishment of oxygen, nutrients and microbes. Pore diameters of less than 3 micrometres are an obstacle to bacteria, thereby limiting biodegradation (Aichberger et al., 2006). Zibiske & Risser (1986) found that medium-grained soil might have the most biodegradation potential: a combination of sufficient permeability and soil-surface area is the cause for increased biological activity. Increased surface area allows attachment of a greater number of microbes.

One cause for the persistence of spilled petroleum in the subsurface is a concept known as burial (Owens et al., 2008). If petroleum migrates into an enclosed area, for example, a sand layer sandwiched between clay, replenishment of nutrients and oxygen may be limited and petroleum could last for many years or decades.

5.3.1 Soil chemistry

The chemical composition of soil will impact conditions such as pH, redox and cation/anion exchange capacities (McVay et al., 2004). For example, soil derived from or overlying carbonate-type rocks will tend to exhibit higher pH values, whereas sandier soil (derived from sandstones, quartzites, etc.) will be less buffered and impacted more readily by acid rain. Higher organic carbon content tends to induce more biological activity in the soil. The organic carbon content commonly lessens in older soil and is often high in glacial sediments (Jobbágy & Jackson, 2000).

5.3.2 Soil moisture

Soils lacking moisture normally exhibit decreased biodegradation rates. The lack of moisture prevents influx of oxygen and nutrients and reduces contact between microbes and spilled petroleum. Waterlogged soils may retard biological processes. Laboratory studies performed by Schroll et al. (2006) showed a linear relationship between soil moisture and pesticide biodegradation. Bekins et al. (2005) reported on a crude-oil release where the shallowest soil samples exhibited the least petroleum degradation. The lack of degradation was attributed to reduced moisture within the shallow soil.

5.4 Petroleum chemistry

The chemical composition of petroleum products can influence weathering rates. Distillates derived from certain crudes can weather at varying rates, despite similar compositions (Atlas, 1981). Eganhouse et al. (1996) reports that certain petroleum constituents may inhibit degradation of others. For example, degradation rates of heavier *n*-alkanes may increase once lighter *n*-alkanes are removed.

Contaminant mixtures also impact biodegradation. In one study, *iso*-alkanes degraded individually, but when introduced with other hydrocarbons, degradation proceeded slowly. This finding suggests a competition effect (Kampbell & Wilson, 1991). However, there is evidence to the contrary, suggesting that degradation for some compounds is more rapid when in a mixture (Smith, 1990).

5.5 Distance from source

Distance from the source of the release will impact petroleum weathering. Because of the effects of source-area sequestration, increased surface area, and decreased contaminant mass, peripheral portions of the middle-distillate plume often weather at a faster rate than the core area (Parsons, 2003). It is unlikely that petroleum will weather at a uniform rate across the plume (Landon & Hult, 1991).

5.6 Hydrologic conditions

In areas with fluctuating water tables, separate phase can become engulfed by groundwater, forming an emulsion and enhancing biodegradation. Bekins et al. (2005) reports that in areas of significant recharge, enhanced degradation can occur because of increased contact with nutrient-rich water. At sites exhibiting rapid groundwater migration rates, mass transfer to the aqueous phase may increase, thereby enhancing hydrocarbon degradation.

5.7 Vegetation

Nearby plants and associated microbes can metabolize petroleum and convert it to harmless byproducts through a process known as phytoremediation. Microbial populations can be 5 to 100 times greater in the vicinity of roots, an area called the rhizosphere (Frick et al., 1999; Kechavarzi et al., 2007). McPherson et al. (2007) found that diesel removal in soil can be up to 40% greater when poplar trees exist. Hence, heavily vegetated areas may increase weathering of spilled petroleum.

Increased vegetation will also increase the number and density of roots in the subsurface. Because of transpiration, increased vegetation will lessen recharge and possibly decrease petroleum dissolution.

6. Sequence of biodegradation

The *n*-alkanes and aromatics (benzene, toluene, ethylbenzene and *o, m, p*-xylenes) are commonly the first compounds to be removed through biological processes (Chapelle, 2001). The *n*-alkanes are more readily converted to long-chain fatty acids (for subsequent beta-oxidation) compared to unsaturated or branched-chain hydrocarbons.

Because it has the highest solubility, benzene is commonly the first mono-aromatic to be depleted from a middle-distillate separate phase (Kaplan et al., 1996). However, Barker et al. (1987) found benzene to be the most persistent aromatic in ground water. Depletion is then normally followed by alkyl-benzenes and alkyl-naphthalenes. Alkyl-naphthalenes appear more resistant than alkyl-benzenes. Furthermore, homologues with longer alkyl chains will be more resistant to biodegradation (Kaplan et al., 1996). For example, a C_1-naphthalene (such as 1-methylnaphthalene) is normally less resistant than a C_4-naphthalene (such as diethylnaphthalene). Alkyl-*cyclo*-hexanes are commonly more resistant than *n*-alkanes and alkyl-benzenes and may be found in the environment much later in the life of a spill. In general, compound classes in order of decreasing susceptibility to biodegradation are *n*-alkanes > *iso*-alkanes (except isoprenoids) > low-molecular-weight aromatics > *cyclo*-alkanes (Leahy & Colwell, 1990).

Kaplan et al. (1997) found that weathering of petroleum products could be divided into seven progressive stages, which we term the *Kaplan Stages*. Similar weathering stages have been presented by Philp & Lewis (1987), Peters et al. (2005), Zytner et al. (2006) and Prince & Walters (2007). The *Kaplan Stages* are depicted on Table 4. Biodegradation including and beyond Stage 5 indicates substantial alteration and normally implies residence times greater than 20 years (Kaplan, 2003; Peters et., 2005).

7. Christensen & Larsen method

Microbes preferentially digest some hydrocarbons, leaving behind a biomarker (Christensen & Larsen, 1993). A biomarker is an organic compound that can be structurally related to its precursor molecule, which occurs as a natural product in a plant, animal, bacteria, spore, fungi or petroleum (Philp & Lewis, 1987). Biomarkers are often resistant to degradation. For example, the isoprenoids: pristane, phytane, norpristane and farnesane, are resistant to microbial alteration, and their relative concentrations compared to *n*-alkanes, can be used as a proxy for weathering (Schaeffer et al, 1979). Therefore, ratios, such as n-C_{17} alkane to pristane (n-C_{17}/pr) or n-C_{18} alkane to phytane (n-C_{18}/ph) have been used as a measure of biodegradation. These *n*-alkanes and isoprenoids have similar solubilities and partitioning coefficients and the absence of *n*-alkanes is a result of biological activity and not transport or sorption (Bregnard et al., 1996).

Biodegradation of *n*-alkanes with molecular weights of up to n-C_{44} is known (Atlas, 1981). However, under aggressive conditions, isoprenoids may be susceptible to microbial oxidation; farnesane and norpristane are the most vulnerable (Pirnik et al., 1974; Pirnik, 1977; Nakajima et al., 1985).

The Christensen & Larsen (C&L) study reported a linear correlation between the n-C_{17}/pr ratio and the diesel-fuel age in soil from numerous spills where release dates were known. The n-C_{17}/pr ratio has been used as a measure of biodegradation for several decades (Atlas, 1981; Swannell et al., 1996), especially with marine spills. Christensen & Larsen (1993) report that statistical analysis of the correlation between the n-C_{17}/pr ratio and known spill ages can provide an age estimate to ± 2 years at a 95% confidence level, with some slight variability for releases <5 and >20 years old. Kaplan et al. (1996) provided an equation to calculate the C&L age where,

$$T(\text{year}) = -8.4(n\text{-}C_{17}/\text{pr}) + 19.8$$

According to Christensen & Larsen (1993), their method may be valid if several conditions are met:

- samples are collected from below an impervious cover such as asphalt or concrete;
- samples are obtained from at least 1 m below the ground surface;
- samples are acquired from at least 1 m above the water table;
- petroleum concentrations in the samples are at least 100 mg/kg, and
- the release is sudden.

Christensen & Larsen (1993) do not define a sudden release, but it can be assumed that a discharge lasting 1 year or less is implied. Most UST releases are slow and prolonged.

The C&L method dealt solely with contaminated soil samples. It did not apply to groundwater or separate-phase samples.

There has been much discussion on the validity of the C&L method (Alimi, 2002; Kaplan, 2002; Stout et al., 2002a; 2002b; Wade, 2002; Galperin & Kaplan, 2008c). Several claim that the method is invalid (Bruya, 2001; Smith et al., 2001; Shepperd & Crawford, 2003; Zemo, 2007). For example, Hostettler & Kvenvolden (2002) found weathered products (crude oils and distillates) with n-C_{17}/pr ratios in excess of 3.0. Stout & Douglas (2007) presented a case study where the C&L method failed to accurately predict the age of a known and sudden release of diesel fuel. However, several recent studies conclude that the method is viable, although with limitations; for example, more than one sample is recommended and knowledge of the original n-C_{17}/pr ratio is needed (Wade, 2001; Hurst, 2003; Hurst & Schmidt, 2005; Oudijk et al., 2006; Hurst & Schmidt, 2007; Oudijk, 2007; Hurst & Schmidt, 2008). Galperin & Kaplan (2008d) recently provided a model based on different initial n-C_{17}/pr values.

As discussed earlier, de Jonge et al. (1997) found that biodegradation rates decreased significantly when petroleum concentrations exceeded 4,000 mg/kg. Accordingly, one might argue that a window exists, only between 100 mg/kg and 4,000 mg/kg, where the C&L method might be valid.

To assess the validity of the assumption the C&L, nine samples of heating oil and motor diesel were collected from residential tanks and commercial service stations in the northeast United States in 2007. The samples were analyzed with a GC/FID to evaluate n-C_{17}/pr ratios. Furthermore, a literature review was conducted to establish n-C_{17}/pr ratios in middle distillates and crude oils (Palacas et al., 1982; Collins et al., 1994; Buruss & Ryder, 1998; Porter & Simmons, 1998; Wang et al., 2003; Chung et al., 2004; Environment Canada, 2004; Hurst & Schmidt, 2005; Blanco et al., 2006; Hwang et al., 2006; Stout et al., 2006; Røberg et al., 2007).

Christensen & Larsen (1993) claim that n-C_{17}/pr ratios for fresh diesel fuel range from around 2.0 to 2.4 (based on Figure 4 of their article). Based on 11 samples, they obtained an average n-C_{17}/pr value of 1.98 with a standard deviation (σ) of 0.83. Hurst & Schmidt (2005) conducted a search of n-C_{17}/pr ratios in fresh distillates and crude oil and found a mean value of 2.3±0.7. However, our samples revealed n-C_{17}/pr ratios ranging from only 0.95 to 1.54 with a mean of 1.15 and σ of 0.18 (Table 3). There are several potential reasons for the discrepancy between our findings and the others:

- n-C_{17}/pr ratios were previously around 2.0, but more recently lowered to the 0.95-to-1.54 range because of changes in crude-oil sources;
- lower n-C_{17}/pr ratios are an artifact of only northeast-US refineries, and
- C&L reveal a mean value of around 2.0, but data are highly variable. Assuming the cited σ value, a 95% confidence interval would be between 1.15 and 2.81.

Type	Town	State	n-C_{17}/pr	pr/ph
Heating oil	Frenchtown	New Jersey	1.25	1.87
Heating oil	North Bellemore	New York	1.10	1.69
Heating oil	Toms River	New Jersey	1.54	1.46
Diesel fuel	Morrisville	Pennsylvania	1.05	1.89
Diesel fuel	Millstone	New Jersey	1.11	1.74
Diesel fuel	North Brunswick	New Jersey	1.29	1.56
Diesel fuel	South Plainfield	New Jersey	0.95	1.62
Diesel fuel (1)	Trenton	New Jersey	1.00	1.66
Diesel fuel (2)	Trenton	New Jersey	1.04	1.85
		Average:	1.15	1.70
		Standard deviation:	0.18	

NOTES: Laboratory analyses performed by Precision Testing Labs, Inc., Toms River, New Jersey. Based on Hurst and Schmidt (2005), the origin of these heating oils and diesel fuels may be Venezuelan and Canadian crude oils, which have average n-C_{17}/pr ratios of 1.4 and 1.0, respectively. Because much of New Jersey's heating oil originates from the Hess Corporation refinery in Port Reading, New Jersey, and Hess obtains crude oil from Petroleo de Venezuela, SA (PDVSA), this conclusion seems probable. The Venezuelan crude oil is fairly immature and exhibits low n-C_{17}/pr values. Furthermore, as of 2008, much of the United States' East Coast crude oil comes from the oil sands of Alberta, Canada (Oudijk, 2009a), which also exhibit much low n-C_{17}/pr values.

Table 3. n-C_{17}/pristane (n-C_{17}/pr) and pristane/phytane (pr/ph) ratios in samples of fresh no. 2 heating oil and motor diesel fuel collected in the US states of New Jersey, Pennsylvania and New York in 2007. Source: Oudijk (2009a).

Stage	Description
1.	Abundant *n*-alkanes, dye still present
2.	Light-end *n*-alkanes removed (such as *n*-C_8 in diesel fuels)
3.	Middle-range *n*-alkanes (*n*-C_9 through *n*-C_{14} in diesel fuels), benzene, toluene removed
4.	More than 90% of *n*-alkanes removed
5.	Alkyl-*cyclo*-hexanes & alkyl-benzenes removed
6.	Isoprenoids, C_1-naphthalenes, benzothiophenes and alkyl-benzo-thiophenes removed, C_2-naphthalenes selectively reduced
7.	Phenanthrenes, dibenzothiophenes and other PAHs reduced

NOTE: In many North American localities, a dye is placed into no. 2 heating oil to distinguish it from motor diesel fuel for taxing purposes. Accordingly, fresh motor diesel fuel would not have this dye.

Table 4. Stages of biodegradation of no. 2 heating oil or motor diesel fuel, known as the Kaplan Stages. Based in part on Kaplan et al. (1997) and Peters et al. (2005).

Our literature review showed that *n*-C_{17}/pr ratios for crude oil worldwide range from <1.0 to about 7.0. The *n*-C_{17}/pr ratio in diesel fuel or heating oil would not be significantly different from its crude source, although Stout & Wang (2007) report that if the fuel is blended with cracked components during refining, *n*-C_{17}/pr ratios may be altered.

Based on the crude-oil data and our findings, C&L ages for today's fresh diesel fuel are unreliable. Therefore, it is unlikely that *n*-C_{17}/pr ratios can presently assist in age-dating studies, especially if litigation ensues. Because original *n*-C_{17}/pr ratios have changed, the C&L method may no longer be appropriate for age dating, at a minimum in North America, and a new method is needed.

8. Age-dating methodology

Significant laboratory studies and/or field investigations have not been performed to determine specific weathering rates of spilled middle distillates. Furthermore, Chapelle & Lovely (1990) report that laboratory studies tend to overestimate biodegradation rates. Field studies with known spill time frames are not plentiful. Therefore, specific data on subsurface weathering rates are generally not available. To obtain such data may be an extremely cumbersome endeavor because of the numerous variables involved. Studies of this type would need to address all the different geological, hydrological and biological conditions, which are numerous.

Previous age-dating methods for spilled middle distillates have been based, for the most part, on the chemistry of the petroleum. These methods have, in general, used weathering or biodegradation rates as a proxy for age. Because weathering at and within each spill site could be different, such a method can be problematic. Cherry et al. (1984) found that "Because the proportion of each [microbial] species present at any point in space and time is environmentally dependent, predictions of actual organic transformation pathways and rates are all but impossible (p. 57)". In their study of a crude-oil spill, Bekins et al. (2005) concluded that ". . . techniques for dating the time of a spill on the basis of the degree of degradation may yield very different results. . . (p. 140)". Accordingly, the use of only degradation rates for age dating is not sound and a technique is needed that considers many parameters, such as weathering, geology, site history and the numerous site-specific environmental factors.

Because a mix of historical and scientific data will be used for our age estimates, each with possibly a large error range, a purely quantitative method, such as the equation used by Kaplan et al. (1997) (equation 1), is not practical. For that reason, a semi-quantitative method is proposed. This technique is based on an evaluation of five major factors and 15+ sub-factors, some of which are used to select a site-specific, weathering-potential regime (Atlas, 1981; Atlas & Bartha, 1992; Providenti et al., 1993) (Tables 4 and 5).

With the technique described here, five site-specific weathering-potential regimes are proposed to describe each release site (Table 6). The regimes are: very weak, weak, moderate, aggressive and very aggressive, and they are based on site-specific environmental factors. To obtain the age-date range, the weathering regimes are compared through a matrix to the Kaplan Stages, as described in Oudijk (2009a) and Table 7.

Environmental conditions	Chemical conditions
Soil permeability & effective porosity	*n*-alkane depletion
Water-table depth/hydrologic setting	*n*-alkane distribution
Soil cover	aromatic content/depletion
Organic-matter content of soil	unresolved complex mixture (UCM) size and location
Groundwater and soil-water salinity & pH	Carbon range
Lithology/chemical composition of geologic formation	Ratio of heavy versus light hydrocarbons
Sulfide content of soil & groundwater	Depletion of:
Vegetation	*iso*-alkanes
Dissolved oxygen content of ground water/oxidation- reduction potential (aerobic v. anaerobic or oxic v. anoxic)	PAHs
	alkyl-*cyclo*-hexanes
	alkyl-benzenes
Moisture content of soil	thiophenes
Temperature	
Indigenous microbial community	
Co-presence of other contaminants	
Presence of electron acceptors	
Presence of emulsifying agents	
Distance from source	

Table 5. Examples of environmental factors impacting the weathering of middle-distillate fuels and resulting chemical responses

Regime	Examples and description
Very aggressive	Water-logged soils; areas prone to flooding; surficial or very shallow releases; moderate to high permeability and effective porosity; stray electrical currents, heavily vegetated; nutrient-rich soils and/or groundwater, moderate pH (7–8); high DO content (at or near saturation) and ORP in groundwater; lack of soil cover; high recharge rates, and history of environmental pollution (acclimated microbes).
Aggressive	High water table; highly-permeable and porous soils; lack of soil cover; shallow release; moderate pH (7–8); high oxygen content in soil and groundwater (>5 mg/L); high organic-matter content; stray electrical currents; high salinity content of groundwater; high soil sulfide content; high soil moisture; heavily vegetated, and urban environment (not pristine).
Moderate	Moderate water content in soil; moderate depth to groundwater; moderate permeability and porosity; moderate pH (5–9); moderate oxygen content in soil and groundwater (>2 mg/L); lack of stray electrical currents, and moderate vegetation.
Weak	Low moisture content in soils; water-logged soil (as per biodegradation); very high or very low pH; low oxygen content in soil and groundwater (<1 mg/L); deep water table; no stray electrical currents; low organic-matter content in soil; non-existent or sparse vegetation, and pristine environment.
Very weak	Extremely cold, hot or harsh environment; extremely high or extremely low pH; total lack of oxygen; pristine; no soil moisture, and sterilized environment (elevated bacteriocides).

Table 6. Site-specific weathering-potential regimes. Source: Oudijk (2009a)

9. Assessing petroleum weathering with chromatograms

Petroleum weathering may be assessed through collection of soil or separate-phase samples and laboratory analysis with a gas chromatograph (GC) equipped with flame-ionization (GC/FID) or mass spectrometry (GC/MS) detectors (Senn & Johnson, 1987).

To assess the magnitude of weathering in each sample, either the peak height or area for the *n*-alkanes and *iso*-alkanes (in particular, the isoprenoids) must be calculated. Calculation of the peak areas is preferred; however, peak heights are acceptable if there is a linear relationship between heights and areas (Wade, 2001). There are two methods to calculate the peak height: either directly from the base of the chromatogram, or from the base of the UCM. The UCM method is preferred (Hostettler et al., 1999).

Assessment of petroleum weathering is needed to determine into which *Kaplan Stage* a sample is placed. There are several factors to consider:

- *Compound depletion*: Specific compounds are more resistant to biodegradation and their presence or depletion can be used to assess weathering;
- *Carbon range*: No. 2 heating oil and diesel fuel are normally within a range of C_9 through C_{24}. Lighter hydrocarbons (less than C_9, but not the mono-aromatics) may be evidence of the presence or mixture with gasoline or kerosene. A heavier fraction may be evidence of increased weathering (Wang & Fingas, 1995b). In addition, heavier constituents (greater than C_{24}) may be evidence of a mixture with no. 6, lubricating or motor oils;
- *The n-alkane distribution*: The *n*-alkanes in middle distillates, such as diesel fuel, heating oils or kerosene, normally show an even distribution, evidenced by a bell-shaped

Weathering regime:	Very aggressive	Aggres-sive	Moderate	Weak	Very Weak
Fresh fuel	0	0	0	0	0
Kaplan Stages:					
1. Abundant *n*-alkanes	<0.25	0-2	0-4	0-8	0-10
2. Light *n*-alkanes removed, benzene & toluene removed	<0.5	2-4	4-8	8-16	10-20
3. Middle-range *n*-alkanes removed, ethylbenzene & xylenes removed	<1	4-6	8-12	16-24	20-30
4. More than 50% of the *n*−alkanes removed	<2	6-8	12-16	24-32	30-40
5. More than 90% of *n*-alkanes removed, alkyl-benzenes and alkyl-*cyclo*-hexanes begin to degrade	<3	8-10	16-20	32-40	40-50
6. All n-alkanes removed, alkyl-benzenes	<4	10-12	20-24	40-48	50-60
7. Isoprenoid removal significant	<5	>12	>24	>48	>60

NOTE: The age ranges cited above must be compared to site-specific information, such as underground storage tank (UST) age, UST condition and the extent of contamination, to assess their accuracy The age ranges provided in this table should be used solely as a guide. Accordingly, additional information is needed to estimate the actual age as described in the text herein. In some situations, however, these age ranges may not apply and should not be used at all. Such situations, for which an age estimate cannot be done with the method described herein, include but are not limited to the following: multiple releases as well as changes of environmental conditions since the release has occurred. Such conditions should be carefully evaluated and excluded before applying this age-dating method.

Table 7. Matrix of *Kaplan Stages* and weathering-potential regimes providing potential age ranges in years for a release of a middle-distillate fuel. Source: Oudijk (2009b).

envelope. In diesel and no. 2 heating oils, the envelope reaches a maximum at C_{14} to C_{17} (Kaplan et al., 1996). In kerosene and jet fuels, the maximum is normally between C_{10} and C_{12}. An uneven or jagged distribution is often evidence of weathering (Figure 6a through 6c);

- *Unresolved complex mixture (UCM)*: The UCM is the hump at the base of a GC/FID trace (Figures 6a through 6c) and a mixture of complex *cyclo*- and *iso*-alkanes that are unresolvable through gas chromatography (McGovern, 1999). UCMs are a typical appearance on chromatograms for crude oil and crude-oil distillates (Frysinger et al., 2003). The UCM normally increases in relative height and width as biodegradation proceeds (Wang & Fingas, 1995b). The presence of multiple UCMs is commonly evidence that more than one distillate is present, for example, a mixture of no. 2 and no. 6 heating oil, and
- *Heavy versus light n-alkanes*. Under aerobic conditions, lighter *n*-alkanes are normally removed quicker compared to the heavier *n*-alkanes (Mohantya & Mukherji, 2008). A comparison of heavy *n*-alkanes, such as *n*-C_{20} through *n*-C_{22}, versus lighter *n*-alkanes, such as *n*-C_8 through *n*-C_{10}, can demonstrate the magnitude of evaporation. Lighter *n*-alkanes are often more volatile (Wang & Fingas, 1995b). Experiments have shown that

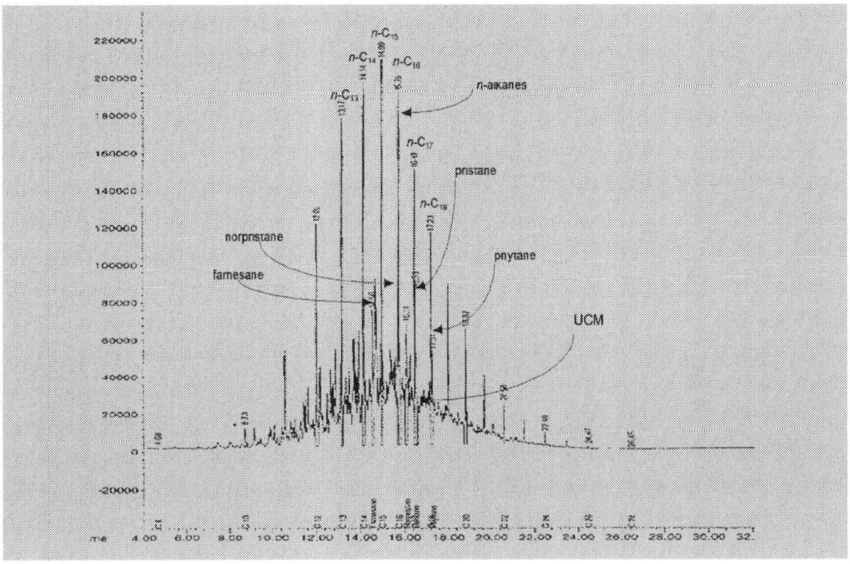

Fig. 6a. GC/FID chromatogram for a 2007 fresh motor diesel fuel from New Jersey (USA) showing the n-alkane peak envelope and the unresolved complex mixture (UCM). Source: Precision Testing Labs, Inc., Toms River, New Jersey (USA).

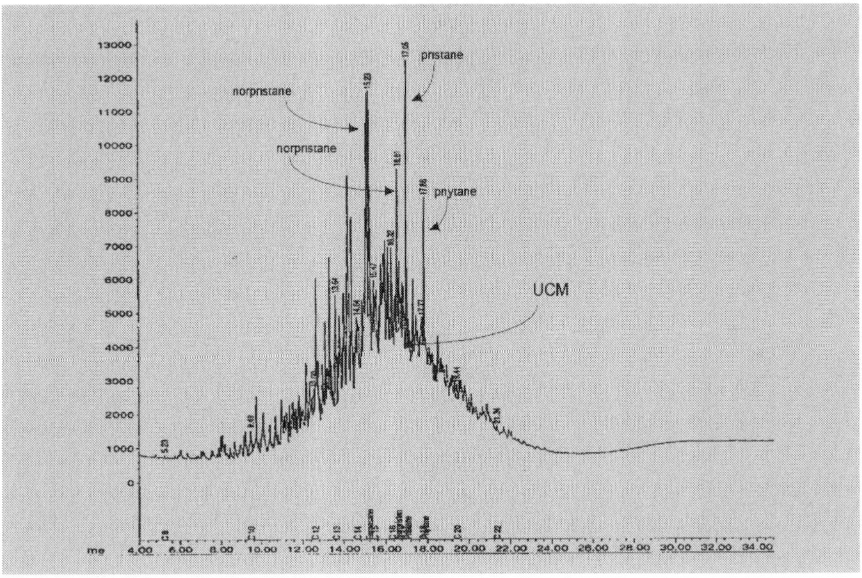

Fig. 6b. GC/FID chromatogram for a weathered motor diesel fuel obtained in 2007 from New Jersey (USA). Source: Precision Testing Labs, Inc., Toms River, New Jersey (USA).

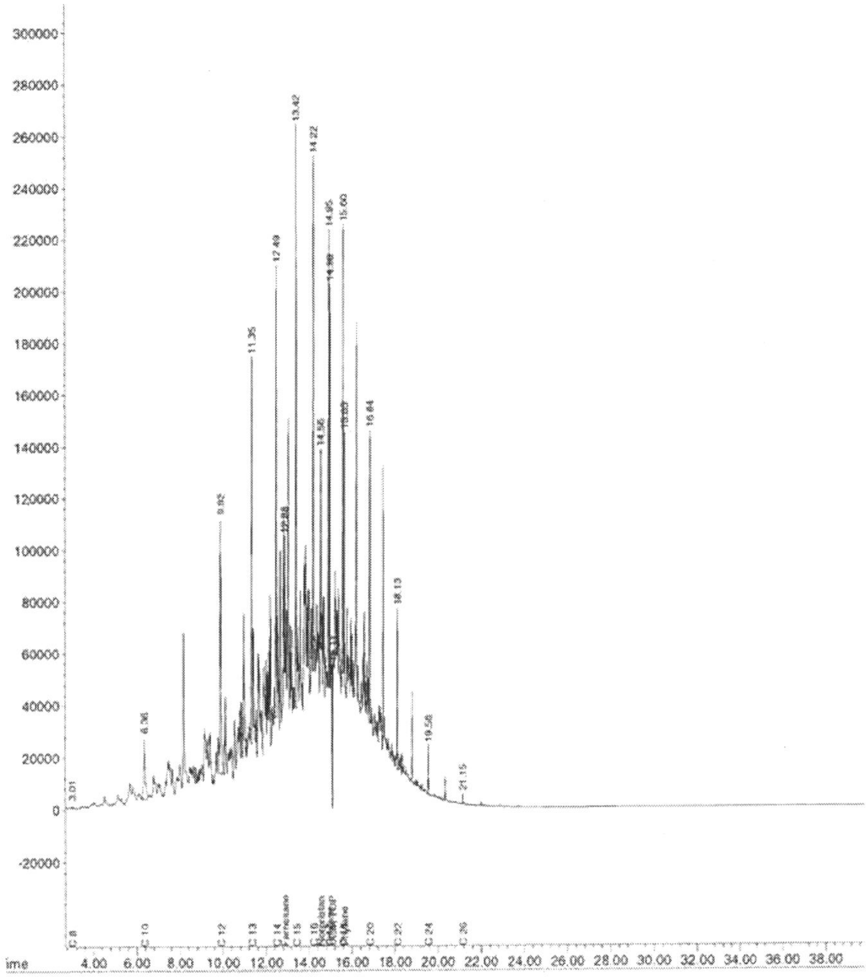

Fig. 6c. GC/FID chromatogram for a mixture of weathered and fresh motor diesel fuel obtained in 2011 from New Jersey (USA). Note the even distribution of *n*-alkane peaks and the relatively large unresolved complex mixture (UCM). Source: Precision Testing Labs, Inc., Toms River, New Jersey (USA).

n-alkanes lighter than *n*-C_{11} can be lost within 9 days in a surface spill (Payne et al., 1991). In some cases, elevated salinity can increase evaporation rates and decrease dissolution (Oyewo, 1988). However, hydrocarbons heavier than C_{14} are only slightly impacted by evaporation or dissolution (Blumer et al., 1970). The ratio of *n*-C_{10} to *n*-C_{20} (*n*-C_{10}/*n*-C_{20}) is normally 0.5 to 1.5 in an unweathered diesel fuel, whereas evaporated diesel fuel often exhibits lower *n*-C_{10}/*n*-C_{20} values. Therefore, *n*-C_{10}/*n*-C_{20} values can help to assess the magnitude of evaporation and dissolution. Furthermore, a formula, similar to the weathering index (WI*) suggested by Wang & Fingas (1995b), used to assess weathering for diesel fuels or no. 2 heating oil, is:

$$WI^* = (n\text{-}C_8 + n\text{-}C_{10} + n\text{-}C_{12} + n\text{-}C_{14})/(n\text{-}C_{16} + n\text{-}C_{18} + n\text{-}C_{20} + n\text{-}C_{22})$$

Under aerobic conditions, lower WI* values are indicative of weathering, whereas higher values are evidence of less degradation.

GC/FID traces can show evidence of a mixture of fresh and weathered middle distillates. Evidence for this phenomenon is normally an enlarged UCM overlain with an envelope of evenly distributed *n*-alkanes. A mixture of highly weathered and fresh product is often evidence of two (or more) releases, although it does not necessarily reveal more than one source. Furthermore, a subsurface release could be superimposed by a surficial spill, overfill or piping failure (Figure _).

The above methods can be employed to assess the weathering characteristics of middle-distillate fuels such as kerosene, the jet fuels, diesel fuels (such as motor diesel or railroad diesel), heating oils (such as no. 2, no. 4 and no. 6) and bunker oil. In some cases, such as with no. 6 oil/bunker oil, quantifying the *n*-alkane/isoprenoid ratios may be difficult because of low concentration of the marker compounds. Furthermore, evaluation of ratios such as $n\text{-}C_{10}/n\text{-}C_{20}$ in the heavier oils may not be possible.

10. Site-specific environmental and non-environmental factors

A proper age-dating study will include information on the following five factors and several associated sub-factors: 1. site history; 2. on-site environmental conditions; 3. extent and magnitude of known impact; 4. condition (and age) of the UST (or other types of sources), and 5. other conditions. The environmental factors can be used to select a site-specific, weathering-potential regime (Table 5).

10.1 Site history

Historical factors include: 1) first date when petroleum was observed in the environment or when problems first began, such as a malfunctioning furnace; 2) age of the UST (or other types of sources, such as aboveground storage tanks or pipelines). Quite commonly, the UST age is not known and it must be assumed that it is the same as the site, service station or residence (although not always correct), and 3) known or calculated petroleum quantity in the subsurface.

It is assumed that the petroleum age will be less than the UST age, but older than the date of its first environmental appearance. Furthermore, trouble with the furnace, because of water or lack of oil, may be a clue surrounding the onset of UST failure in a residential case. The average lifespan of an UST may be as low as 15 years, whereas the commonly used "rule-of-thumb" is 25 years. The 15- to 25-year average lifespan should be kept in mind when estimating a release age. However, we have seen USTs develop leakage within days (because of improper installation) and USTs older than 75 years in close-to-perfect condition.

A calculation of the petroleum quantity in the subsurface may be helpful, although this result is often fraught with error. The calculation may be performed by: 1) computing the petroleum quantity through soil-sampling results and separate-phase-thickness measurements in wells, or 2) comparing the amount of fuel delivered versus average usage (with the use of "degree days" to estimate fuel usage). The calculated value may then be divided by an estimated leakage rate to obtain the time frame. A minor leakage rate could be 0.01 L/day, whereas a high rate might be greater than 0.5 L/day.

10.2 Environmental conditions

Environmental factors impacting age-dating evaluations include: 1) depth to groundwater, 2) lithology and texture of geologic materials; 3) geochemical conditions of soil and groundwater, such as: a) pH; b) salinity; c) redox potential, and d) dissolved oxygen content; 4) biological conditions; 5) overlying soil cover, and 6) other factors.

10.2.1 Depth to groundwater

Petroleum in soil samples collected beneath the water table may experience increased dissolution, in particular, lighter *n*-alkanes. The hydrologic locality must also be considered. For example, in recharge zones, the water table may fluctuate several meters seasonally. Therefore, soil-sampling locations may have previously been within groundwater. To assess this problem, governmental agencies often have nearby observation wells with water-level data spanning decades and these data should always be consulted.

The amount of recharge is dependent on soil cover, vertical permeability and topographic location (such as within a hill or valley). Soil samples collected beneath a building may be subjected to less water contact. However, petroleum in samples collected beneath covers such as grass or bare ground may experience increased dissolution and, consequently, additional weathering.

10.2.2 Lithology and texture of geologic materials

Soil texture will impact drainage and permeability. Poor drainage and low permeability prevent oxygen and nutrient replenishment. Fine-grained soils, such as clays, more commonly exhibit anaerobic conditions. Soils exhibiting high cation-exchange capacities, such as silts or clays, or contain large organic-carbon contents, may adsorb petroleum readily. Adsorption decreases with increasing temperature and soil moisture and adsorbed petroleum is less available to microbes (Providenti et al., 1993). Hence, biological activity may be subdued and weathering minimized in fine-grained soils; however, large organic carbon contents could also induce greater microbial activity.

Inadequate soil hydration depresses microbial metabolism and movement. Furthermore, lack of moisture decreases nutrient replenishment (Providenti et al., 1993). However, waterlogged soils can also limit oxygen concentrations. Accordingly, soil-moisture extremes may decrease biological activity and weathering.

10.2.3 Geochemical and biochemical conditions of soil and groundwater

Subsurface geochemical conditions impact petroleum weathering. Extreme pH limits biological activity, whereas redox dictates if aerobes or anaerobes exist. Both microbes use middle distillates as substrates, but aerobic degradation is often quicker. The dissolved oxygen content and oxidation-reduction potential (ORP) of groundwater (if the water table is shallow) can help to identify these conditions.

10.2.4 Biological conditions

Microbes are a part of the geochemical framework of saturated and unsaturated zones. Populations may increase in response to petroleum releases and alter geochemical conditions.

Certain bacteria may increase or decrease mineralization rates of hydrocarbons. Furthermore, microbes can act as emulsifying agents, suspending petroleum in the aqueous phase and enhancing dissolution and biodegradation (Zajic et al., 1974). Additionally, vegetation impacts weathering, especially in the rhizosphere where microbial action is plentiful.

10.2.5 Overlying soil cover

The type and extent of soil cover will impact infiltration and potentially the magnitude of petroleum dissolution. This factor could have an impact on *n*-alkane depletion. Christensen & Larsen (1993) recommend that samples be collected at least 1 m below the ground surface. Petroleum located at shallow depths may be subject to increased volatilization or photodecomposition. Furthermore, shallow locations face greater temperature changes and increased weathering.

10.2.6 UST-related factors

In addition to affecting petroleum weathering, environmental conditions will also impact UST corrosion. These environmental factors include:

1. soil exhibiting low resistivity increases ion exchange between steel and soil minerals. Highly-corrosive soil is often composed of clay with elevated sulfide concentrations and low pH;
2. coarse-grained and/or angular-grained soil. The periodic filling of the UST can cause angular stones to puncture the UST;
3. increased recharge, such as an adjacent roof leader or a location within a low area, will increase UST contact with water, and
4. periodic or constant immersion in water, in particular acidic groundwater.

Anthropogenic conditions that can impact UST corrosion include:

- quality of the UST materials, such as the steel thickness. Based on our field observations, North American USTs installed prior to the 1970s are often constructed of a thicker gage steel and, therefore, less susceptible to corrosion. In coastal areas, where the water table is shallow, aboveground tanks are often used underground. These tanks are commonly constructed with thin-gage steel and more susceptible to corrosion;
- UST size. Larger diameter USTs will corrode at an accelerated rate, possibly because of the greater load on the steel (Holt, 1997);
- improper installation procedures, such as use of jagged backfill;
- lack of cathodic protection;
- use of dissimilar materials in UST construction, such as a mix of galvanized- and stainless-steel, and
- stray electrical currents from nearby buried power lines.

Many of these factors also apply to aboveground tanks (ASTs). Furthermore, ASTs are particularly susceptible to lightning strikes, which can cause immediate failure.

10.3 Extent and magnitude of impact

The extent and magnitude of impact can be measured by: 1) area of impacted soil and/or groundwater and the quantity of separate-phase-saturated soil; 2) vertical extent of impact, and 3) extent and thickness of separate phase on the water table.

It is assumed that a middle-distillate plume with a significant distance will also exhibit a significant age. If sufficient data are available, it may be possible to calculate the migration rate and back-calculate the time frame.

The volume of petroleum in the environment will have an impact on weathering. Samples collected within a large pool of separate phase may not exhibit any weathering many years after a release. Accordingly, samples collected within a highly contaminated location may not provide productive evidence.

10.4 UST condition

The UST condition can be assessed by: the number, size and location of corrosion holes. Corrosion often takes considerable time to develop. Metallurgists are commonly consulted to provide opinions on the UST release ages. However, soil conditions should be evaluated to determine their connectivity.

Stray electrical currents may have a significant impact on unprotected steel USTs. In particular, central air-conditioning units, which normally run on underground 220-volt currents, are often the culprits with newer buildings or homes. Unfortunately, USTs are commonly installed adjacent to these units and leakage can often initiate within 5 years.

10.5 Other considerations

There may be additional site-specific factors in addition to the five listed. For example, USTs are often abandoned in-place, possibly by a previous owner. The abandonment date might represent when a previous owner suspected leakage. Ecosystem responses may also need to be evaluated. For example, contamination may induce stressed vegetation or impacts to water bodies. The time needed to produce such impacts may be significant. Investigators need to evaluate these factors and determine if they are sufficiently important to consider in the matrix.

11. Recommended sampling and laboratory analyses

Oudijk et al. (2006) and Oudijk (2009b) provided guidelines for age-date sampling. Samples can be either impacted soil or separate phase, although soil samples are preferred. Samples as distant as possible from the source are needed to assess the maximum release age; however, a sufficient quantity of petroleum must be present to perform laboratory analyses. A hydrocarbon concentration of greater than 1,000 mg/kg is recommended.

An important decision is the sampling locations. It is assumed that locations distant from the source represent older ages. However, downgradient locations may be more susceptible to excessive weathering. Furthermore, petroleum in samples collected from within separate-phase pools may not weather as much as locations proximate or outside the pool. Accordingly, an understanding of sampling locations with respect to accumulations of separate phase is needed. As discussed by Wade (2001) and Oudijk et al. (2006), age dating based on one sample is unwise.

Field analyses of groundwater samples are needed (unless the water table is deep and inaccessible). The analyses should include pH, dissolved oxygen (DO), ORP, specific conductance, temperature and salinity. The samples should be laboratory analyzed by GC/FID. We have found that analyses for *n*-alkanes by GC/MS are often inaccurate. The GC/FID analyses must be conducted so that sufficient separation exists between peaks. For example, the *n*-C_{17} alkane and pristane elute very close to each other. To enhance peak separation, a run time of about 40 minutes is recommended. In some cases, some of the samples should be analyzed for aromatics such as benzene, toluene, ethylbenzene and *o, m, p-*

xylenes and base/neutral extractable compounds (B/Ns), also targeting C_1- and C_2-naphthalenes, alkyl-benzenes, alkyl-*cyclo*-hexanes and dibenzothiophenes. It is further recommended that a sample of fresh fuel be collected from each site for comparison purposes. However, it is possible that the fresh oil may be significantly different from the spilled oil.

12. Evaluating the age range

To evaluate the age, a Kaplan Stage is selected for each sample, whether soil or separate phase. Based on known environmental conditions, a weathering-potential regime is then chosen for the site. A matrix, comparing the Kaplan Stages to the weathering-potential regimes, is provided detailing potential release ages (Table 7). Compared to Kaplan et al. (1997), the Kaplan Stages on Table 7 were modified to include additional parameters. These potential age ranges are based on:

- under very aggressive conditions, *n*-alkanes can be completely removed in less than 5 years (Hurst, 2003). For example, in marine environments, which are very aggressive, *n*-alkanes can be removed in a matter of days (Colwell, 1978). In the 2002 *Prestige* tanker spill off the coast of Spain, the *n*-C_{17}/pr ratios in nearby sediments were cut in half after less than one year (Blanco et al., 2006). With marine spills, processes in addition to biodegradation, such as volatilization and dissolution, may cause the *n*-alkane depletion. Conversely, under these same aggressive conditions, PAHs may still last decades (DeLaune et al., 1990);
- under very-weak conditions, such as a Arctic environments, *n*-alkanes can persist in soil for decades (Sexstone et al., 1978a; Collins et al., 1994);
- high concentrations of nutrients and oxygen, indicative of aggressive environments, can allow complete middle-distillate degradation in soil within less than one decade (Bregnard et al., 1996). However, removal or lessening of oxygen and nutrients can effectively retard hydrocarbon removal (Bonin & Bertrand, 2000). Unless extreme conditions exist, such as permafrost or drought, complete or near-complete removal of hydrocarbons is normally accomplished within 20 to 30 years;
- benzene and toluene often biodegrade and dissolve quicker than ethylbenzene and *o,m,p*-xylenes (although not always)(Kaplan et al., 1996). Based on our field observations, these aromatics are removed rapidly at most sites, and under moderate weathering conditions, *iso*-alkanes, such as pristane and phytane, and the alkyl-naphthalenes are commonly the predominant compounds after about 20 years (Caredda et al., 2007).

The matrix was constructed with the following assumptions:

- the very-aggressive column represents a marine-spill situation. The *n*-alkanes degrade quickly in such environments. There are reports that *n*-alkanes may persist for two or three years (Colwell, 1978), but they will, in general, be gone within 4 years (and often much earlier). With regard to *n*-alkane depletion, time frames of weeks or months are more common than years (de Souza & Triguis, 2004). Accordingly, <4 years can be placed into the matrix for Stage 6 in an aggressive environment.

the very-weak column represents spills in an Arctic or Antarctic climate. The *n*-alkanes degrade slowly here (Sexstone et al., 1978a; Collins et al., 1994). Sexstone et al. (1978b) reported that biodegradation in tundra soil could be slow "with no major preferential utilization of classes of hydrocarbons during the period of exposure". The longer-chain *n*-alkanes (>C_{10}) are commonly solid at temperatures less than 10° C (Whyte et al., 1998). Kershaw & Kershaw (1986) found significant depletion at surface locations from a 35-year old spill in the Canadian Northwest Territories, but with depth, up to 80% of the oil persisted. Collins et al. (1994) found only marginal depletion of *n*-alkanes in subsurface soils from a 12-year-old crude-oil spill in a permafrost region of Alaska. Gore et al. (1999) and Kerry (1993) also found minimal subsurface biodegradation in a similar Antarctic environment. A very weak environment is where microbes are dormant, cannot come in contact with hydrocarbons or have been removed because of toxicity. Accordingly, >60 years can be placed into the matrix for Stage 6 in a very-weak environment;

- under moderate conditions, *n*-alkanes are normally removed from subsurface soils in about 20 years (Christensen & Larsen, 1993; Kaplan, 2003). Accordingly, 20 to 24 years can be placed into the matrix for Stage 6 in a moderate environment;
- degradation follows a clear sequential pattern as depicted by Kaplan et al. (1996) and Table 4. This sequential pattern is normally the case. However, there are cases where different compounds of the same class degrade at significantly different rates. For example, Olson et al. (1999) found that components within the aliphatic fraction of diesel fuel degraded at different rates, although the aliphatic fraction, as a whole, degraded quicker than the aromatic fraction;
- removal of *n*-alkanes tends to be linear, instead of exponential (Christensen & Larsen, 1993; Hurst & Schmidt, 2005; Galperin & Kaplan, 2008c). Therefore, age ranges are extrapolated in a linear manner from Stage 6 to Stage 1 (zero-order kinetics). Chapelle (2001) explains that there is often a lag time between introduction of a contaminant to a soil or groundwater system and acclimation of microbes. However, in the first days or weeks after release and acclimation, once that acclimation has occurred, biodegradation rates may be significant. Díez et al. (2007) found that biodegradation rates of heavy oils can be slow, even in a marine environment. Accordingly, it can be assumed that as oil ages and becomes more viscous, the rate of biological consumption decreases. Colwell (1978) found that rates could initially be logarithmic for marine spills and later linear. Bonroy et al. (2007) reports that biodegradation rates will vary seasonally. Walker et al. (1976) found that the degradation rate of the alkane fraction of a crude oil was linear, whereas the rate for aromatics varied. Ostendorf et al. (2007) found that *n*-alkane degradation rates in unsaturated soil followed zero-order kinetics, whereas aromatics followed first-order. Bjorklof et al. (2008) found that petroleum degradation rates can be linear, but they are mass-transfer dependent. Therefore, rates will more likely be linear in a permeable soil where the petroleum can dissolve more easily. It is assumed that rates averaged across the years are linear, although it is understood that this assumption may not apply everywhere, and
- age ranges can then be extrapolated between the very-aggressive and very-weak environments and the moderate environments.

The matrix should only be used to provide potential ages and should not be the sole factor for an age-date opinion. The matrix is a method to lead the investigator towards the correct age, but it is not the final say.

13. Critiquing the matrix age range

Once a potential age range is obtained from the matrix, it must be compared to several factors to assess its reliability. Other factors that may impact the age are:

- age of the UST;

- time frames of known or possible pollution events, such as appearance of petroleum in waterways, furnace failures, excessive oil usage or abandonments. This process may entail a review of fuel bills to determine when excessive usage began;
- state of the UST, such as the number and sizes of corrosion holes and pitting;
- extent and magnitude of petroleum-impacted soil and groundwater, and
- miscellaneous factors such as high-voltage underground electrical lines.

With the date of petroleum appearance, such as when oil began seeping into a creek, the upper and lower ages can be constrained. For example, if it is known that an UST was installed in 1980 and the furnace began to take in water in 2004, we can conclude that UST failure occurred between 1980 and 2004.

If heating oil was released, the history of the heating system should be considered. The dates of repairs and service calls should be reviewed. The fuel company is often called when a furnace fails and causes for any failures should be determined. The date of any on-site work that included excavating should also be identified. Gardening is a common cause for severing feed and return lines.

Reports should be obtained on the UST condition. If one pin-sized hole is present, it is probable that the leak began within the last 1 to 2 years. However, if 25 coin-sized holes are found, it is more likely that the leak began 10 years ago or later.

Calculations of plume extent are important in assessing the reasonableness of estimated age ranges. A large and extensive plume is more indicative of an older release. However, care is needed in making such conclusions because aggressive biodegradation can greatly reduce plume sizes. If monitoring wells are available, groundwater migration rates may be calculated and time frames for plume movement assessed and compared to geochemical data.

Lastly, underground power lines should be assessed. Residential USTs are often adjacent to air-conditioning units and associated power lines can quickly instigate leakage. At a site in New Jersey, a power line produced a sequence of holes on the top and bottom of an underlying UST. The power line had only existed for 2 years.

Table 7 provides an age range, for example, of 4 to 8 years. This range could be interpreted as 6 years ±2 years. It is stressed that the values on Table 7 are only potential ranges and might contain significant error, if they are not verified through alternative means. Each release is different and on-site petroleum weathering may not necessarily fall within the confines of this technique.

An additional geochemical method to critique the age-date estimate is to establish the sulphur content of the oil. In many countries, the government limits sulphur contents in on-road diesel fuel (Table 8). By determining the sulphur content, the age of the diesel fuel can often be constrained.

14. Range of error

Error is the difference between an observed or calculated value and the true value. Usually, we do not know the true value, but can approximate error ranges from earlier experiments or theoretical predictions (Bevington & Robinson, 1992). The purpose of calculating error ranges is to evaluate the confidence of our results.

All geochemical data collected to assess some type of natural phenomenon contain error caused by sampling and analysis (Miesch, 1967). Subsurface modeling is plagued by uncertainties that are both epistemic (reducible through observations) and aleatory (irreducible because of inherent stochasticity) (Srinivasan et al., 2007). The amount of error

	Maximum sulphur concentration allowable (mg/l)	Regulated as of
Australia	10	2009
Canada	15	1997
	500	1994
China	2,000	2002
European Union	50	2005
New Zealand	10	2009
Singapore	50	2005
Taiwan	50	2007
USA	15	2007[a]
	500	1993

[a] Mandated year was 2006 in California and 2010 in Alaska.

Table 8. Maximum allowable sulphur concentrations allowable in motor diesel fuel in selected countries

often cannot be found in textbooks or with a formula. It is commonly based on a 'rule-of-thumb' or experience. Nevertheless, some calculations can be performed to provide insight into potential error ranges.

To assess error ranges in our calculations, we can use the C&L method as an example. C&L uses the n–C_{17}/pristane ratio as its basis and, to calculate the ratio, an analysis with a gas chromatograph is needed. This analysis generally has a precision of about 5%. Furthermore, the method assumes that n–C_{17}/pristane ratios for unweathered oil range from 2.0 to 2.2. Hence, the starting point has an error of about 10%. Therefore, errors are summed and the resulting error is about 15%. This exercise assumes that the C&L method perfectly reflects weathering processes in the subsurface. This, we know to be untrue. Consequently, we can assume that the method's error is at least 20% or ±2 years

In their article, Christensen & Larsen (1993) cite an error range of ±1 year or about 10%. Hurst & Schmidt (2005), employing a similar method, cite a minimum error of 15% or ±1.5 years. These cited error ranges seem optimistic and there are several references in the scientific literature arguing that the range is greater (Stout et al., 2002; Oudijk, 2007).

Because the age-dating method described herein is semi-quantitative, it may be difficult to assign an error range. Furthermore, there have not yet been experiments to test these techniques and evaluate what these error ranges may be. Nevertheless, one should assume that the range, under the best of scenarios, may be ±2 years and, under many circumstances, it may only provide a 5-year age window. In other circumstances, the method may only be able to constrain the age, wherein an example of an opinion might be 'less than 5 years' or 'greater than 10 years.'

One important limitation to this method is the presence of multiple, overlapping releases. It may be possible to distinguish the releases, but applying an age to each may be exceedingly difficult.

The limitations of the methods described herein need to be fully understood by its users and should be explained whenever an age-dating opinion is provided. As stated by Morgan & Bull (2007, p. 56),

"Fundamental differences exist between forensic geoscience and its sister disciplines, fundamental enough to make the unwary geoscientist succumb to philosophical and practical pitfalls which will not only endanger the outline of their report, but may well indeed provide false-negative or false-positive results leading to contrary or inaccurate conclusions. In the law, such outcomes have devastating and untenable consequences".

The need to apply age-dating methods cautiously cannot be understated. It must be noted that the methods described herein may not be applicable to all spill sites. It is certainly possibly that site-specific circumstances could exist to preclude these methods from consideration. The investigator will need to carefully consider each site individually and the possibility exists that other methods, such as tree-ring studies, isotopic analyses or groundwater migration rate calculations, may be more applicable or financially viable.

15. Conclusions

Because of costs associated with cleanups, many cases come to litigation. Age dating is often central to the litigation. For a plaintiff or defendant to be successful, a legally defensible age-dating method is needed. The C&L method, which is solely dependent on chemistry, has been strongly attacked inside and outside the courtroom. Hence, a revised method is needed and proposed here.

Each pollution site has specific characteristics that must be evaluated. Investigators need a thorough understanding of the geologic and hydrologic conditions, in addition to the nature of the release. Investigators also need to combine knowledge of chemistry, microbiology, and site-specific history to provide an opinion on the release time frame. Because scientific processes associated with releases are too complex to model through a simple formula, a qualitative or semi-quantitative technique is needed. The methods described herein are an effort to develop such an approach.

16. References

Aichberger, H., Loibner, A. P., Celis, R., Braun, R., Ottner, F. & Rost, H. 2006. Assessment of factors governing biodegradability of PAHs in three soils aged under field conditions. *Soil Sediment Contam.* 15:73-85.

Alimi, H. 2002. Invited commentary of the Christensen and Larsen technique. *Environ. Forensics* 3:5.

Atlas, R. M. 1981. Microbial degradation of petroleum hydrocarbons: An environmental perspective. *Microbiol. Rev.* 45:180-209.

Atlas, R. M. & Bartha, R. 1992. Hydrocarbon biodegradation and oil spill bioremediation. *Adv. Microbial Ecol.* 12:287-315.

Balouet, J-C., Oudijk, G., Smith, K. T., Petrisor, I., Grudd, H. & Stocklassa, B. 2007. Applied dendroecology and environmental forensics. Characterizing and age dating environmental releases: Fundamentals and case studies. *Environ. Forensics* 8:1-17.

Barker, J. F., Patrick, G. C. & Major, D. 1987. Natural attenuation of aromatic hydrocarbons in a shallow sand aquifer. *Ground Water Monitor Remed.* 7:64-71.

Bekins, B. A., Hostettler, F. D., Herkelrath, W. N., Delin, G. N., Warren, E. & Essaid, H. I. 2005. Progression of methanogenic degradation of crude oil in the subsurface. *Environ. Geosc.* 12:139-152.

Bennet, S. M. 1997. Groundwater contamination from leaking home heating oil systems. *J. Environ. Hydrology* 5:1-6.

Bevington, P. R. & Robinson, D. K. 1992. *Data Reduction and Error Analysis for the Physical Sciences, 2nd edition*: New York, NY: McGraw Hill, Inc.

Bjorklöf, K., Salminen, J., Sainio, P. & Jørgensen, K. 2008. Degradation rates of aged petroleum hydrocarbons are likely to be mass transfer dependent in the field. *Environ. Geochem. Health* 30:101-107.

Blanco, C. G., Prego, R., Azpíroz, M. D. G. & Fernandez-Domínguez, I. 2006. Characterization of hydrocarbons in sediments from Laxe Ria and their relationship with the Prestige oil spill (NW Iberian Peninsula): *Ciencias Marina* 32:429-437.

Blumer, M., Sass, J., Souza, G., Sanders, H. Grassle, F. & Hampson, G. 1970. *The West Falmouth Oil Spill: Persistence of the Pollution Eight Months After the Accident.* Woods Hole Oceanographic Institution Reference No. 70-44.

Bobra, M. 1992. Solubility Behaviour of Petroleum Oils in Water. Ottawa, ON: Environment Canada, Environmental Protection Directorate.

Bonin, P. & Betrand, J. C. 2000. Influence of oxygen supply on heptadecane mineralization by Pseudomonas nautical. *Chemosphere* 41:1321- 1326.

Bonroy, J., van de Steene, J., van de Velde, R., van Eetvelde, G., Verplancke, H., Verfaillie, F. & Boucneau, G. 2007. Monitored natural attenuation of domestic fuel oil in the unsaturated zone. In *In Situ and On-Site Bioremediation – 2007.* Proceedings of the Ninth International In Situ and On-Site Bioremediation Symposium, eds., Gavaskar, A. R. & Silver, C. F. Columbus, OH: Battelle Press.

Bouchard,D., Hunkeler, D. & Hohener, P. 2008. Carbon isotope fractionation during aerobic biodegradation of n-alkanes and aromatic compounds in unsaturated sand. *Org. Geochem.* 39:23-33.

Bowden, J. N., Westbrook, S. R. & LePera, M. E. 1988. A survey of JP-8 and JP-5 properties. U. S. Army Belvoir Research, Development and Engineering Center, Contract No. DAAK70-87-C-0043.

Bregnard, T. P-A., Hohener, P., Haner, A. & Zeyer, J. 1996. Degradation of weathered diesel fuel by microorganisms from a contaminated aquifer in aerobic and anaerobic microcosms. *Environ. Toxicol. Chem.* 15:299-307.

Bregnard, T. P-A., Häner, A., Höhener, P., & Zeyer, J. 1997. Anaerobic degradation of pristane in nitrate-reducing microcosms and enrichment cultures. *Appl. Environ. Microbiol.* 63:2077-2081.

Bruce, L. G. & Schmidt, G. W. 1994. Hydrocarbon fingerprinting for application in forensic geology: Review with case studies. *AAPG Bulletin* 78:1692-1710.

Bruya, J. 2001. Chemical fingerprinting. In *Practical Environmental Forensics,* eds. J. P. Sullivan, F. J. Agardy & R. K. Traub, New York, NY: John Wiley & Sons.

Buruss, R. C. & Ryder, R. T. 1998. Composition of crude oil and natural gas produced from 10 wells in the Lower Silurian "Clinton" sands, Trumbull County, Ohio. U.S. Geological Survey Open-File Report 98-799.

Caldwell, M. E., Garrett, R. M., Prince, R. C. & Suflita, J. M. 1998. Anaerobic biodegradation of long-chain n-alkanes under sulfate-reducing conditions. *Environ. Sci. Technol.* 32:2191–2195.

Caredda, P., Pintus, M., Ruggeri, C., Viola, A., Tamburni, E., Suardi, E., Franzetti, A. and Bestetti, G. 2007. Laboratory tests and bioremediation of a site chronically contaminated by diesel. In *In Situ and On-Site Bioremediation – 2007.* Proceedings of the Ninth International In Situ and On-Site Bioremediation Symposium, eds., Gavaskar, A. R., and Silver, C. F. Columbus, OH: Battelle Press.

Cerniglia, C. E. 1984. Microbial transformation of aromatic hydrocarbons. In *Petroleum Microbiology,* ed. R. M. Atlas. New York, NY: MacMillan Publishing Company.

Chapelle, F. H. 2001. *Ground-water microbiology and geochemistry, Second Edition.* New York: John Wiley & Sons.

Chapelle, F. H. & Lovely, D. R. 1990. Rates of microbial metabolism in deep coastal plain aquifers. *Appl. Environ. Microbiol.* 56:1865–1874.

Cherry, J. A., Gillham, R. W. & Barker, J. F. 1984. Contaminants in groundwater: Chemical processes. In *Groundwater Contamination (Studies in Geophysics).* Washington, DC: National Academy Press, 46– 64.

Christensen, L. B. & Larsen, T. H. 1993. Method for determining the age of diesel spills in the soil. *Ground Water Monitor Remed.* 13: 142– 149.

Chung, J. W., Lee, W. S., Yoon, J. Y. & Kim, H. G. 2004. The study for identification of waterborne spilled oil by fast gas chromatography [in Korean]. *J. Korean Soc. Marine Environ. Eng.* 7:122–130.

Collins, C. M., Racine, C. H. & Walsh, M. E. 1994. The physical, chemical, and biological effects of crude oil spills after 15 years on a Black Spruce forest, interior Alaska. *Arctic* 47:164–175.

Colwell, R. R. 1978. Toxic effects of pollutants on microorganisms. In *Principles of Ecotoxicology,* ed. G. C. Butler, New York, NY: John Wiley & Sons.

CONCAWE. 1995. Kerosines/jet fuels . CONCAWE product dossier no. 94/106, Brussels.

CONCAWE. 1998. Heavy fuel oils. CONCAWE product dossier no. 98/109, Brussels.

deCourcy Hinds, M. 1979. Homeowners buying larger oil tanks. *New York Times,* July 5. http://www.nytimes.com.

Dashti, N., Al-Awadhi, H., Khanafer, M., Abdelghany, S. & Radwan, S. 2008. Potential of hexadecane-utilizing soil-microorganisms for growth on hexadecanol, hexadecanal and hexadecanoic acid as sole sources of carbon and energy. *Chemosphere* 70:475–479.

Davidova, I. A., Gieg, L. M., Nanny, M., Kropp, K. G. & Sulfita, J. M. 2005. Stable isotopic studies of n-alkane metabolism by a sulfate-reducing bacterial enrichment culture. *Appl. Environ. Microbiol.* 71:8174–8182.

DeLaune, R. D., Gambrell, R. P., Pardue, J. H. & Patrick, W. H. 1990. Fate of petroleum hydrocarbons and toxic organics in Louisiana coastal environments. *Estuaries* 13:72–80.

Diaz, M. P., Boyd, K. G., Grigson, S. J. W. & Burgess, J. G. 2002. Biodegradation of crude oil across a wide range of salinities by an extremely halotolerant bacterial consortium MPD-M, immobilized onto polypropylene fibers. *Biotech. Bioeng.* 79:145– 153.

Dıez, S., Jover, E., Bayona, J. M. & Albaigés, J. 2007. Prestige oil spill. III. Fate of a heavy oil in the marine environment. *Environ. Sci. Technol.* 41:3075–3082.

Douglas, G. S., Bence, A. E., Prince, R. C., McMillen, S. J. & Butler. E. L. 1996. Environmental stability of selected petroleum hydrocarbons source and weathering ratios. *Environ. Sci. Technol.* 30:2332- 2339.

Douglas, G. S., Stout, S. A., Uhler, A. D., McCarthy, K. J. & Emsbo-Mattingly, S. D. 2007. Advantages of quantitative chemical fingerprinting in oil spill source identification. In *Oil Spill Environmental Forensics*, eds. Z. Wang & S. A. Stout, San Diego, CA: Academic Press, 267- 299.

Douglas, G. S., Ziegler, S., Pinzone, C., Hardenstine, J. & McCarthy, K. 2004. The application of the Federal On Road Diesel Fuel Sulfur Reduction Act of 1993 to the age dating of diesel fuels: A case study (Abs.). In *The Annual International Conference on Soil, Sediment & Water.* Amherst, MA: Association of Environmental Health and Soils.

Eganhouse, R. P., Dorsey, T. F., Phinney, C. S. & Westcott, A. M. 1996. Processes affecting the fate of monoaromatic hydrocarbons in an aquifer contaminated by crude oil. *Environ. Sci. Technol.* 30:3304–3312.

Ehrenreich, P., Behrends, A. & Harder, J. 2000. Anaerobic oxidation of alkanes by newly isolated denitrifying bacteria. *Arch. Microbiol.*173:58–64.

Environment Canada. 2004. Oil composition and properties for oil spill modeling. USEPA 3D-6152-NAFX.

Frick, C. M., Farrell, R. E. & Germida, J. J. 1999. Assessment of phytoremediation as an in-situ technique for cleaning oil-contaminated sites: Report prepared for the Petroleum Technology Alliance of Canada by the Department of Soil Science, University of Saskatchewan, Saskatoon, SK. Saskatoon, SK: University of Saskatchewan.

Fried, J. J., Muntzer, P. & Zilliox, L. 1979. Ground-water pollution by transfer of oil hydrocarbons: *Ground Water* 17:586–594.

Frysinger, G. S., Gaines, R. B., Xu, L. & Reddy, C. M. 2003. Resolving the unresolved complex mixture in petroleum-contaminated sediments. *Environ. Sci. Technol.* 37:1653–1662.

Galperin, Y. & Camp, H. 2002. Petroleum product identification in environmental samples: Distribution patterns in fuel-specific homologous series. *Soil Sediment Contam.* April/May: 27–29.

Galperin, Y. & Kaplan, I. R. 2008a. Zero-order kinetics model for the Christensen-Larsen method for fugitive fuel age estimates. *Ground Water Monit. Remed.* 28 (2): 94-97.

Galperin, Y. & Kaplan, I. R. 2008b. Forensic environmental geochemistry in dispute resolution—Case history 2: Differentiating sources of diesel fuel in a plume at a fueling station. *Environ. Forensics* 9:55–62.

Galperin, Y. & Kaplan, I. R. 2008c. Comments on the reported unusual progression of petroleum hydrocarbon distribution patterns during environmental weathering: *Environ. Forensics* 9:117–120.

Galperin, Y. & Kaplan, I. R. 2008d. Age significance of nC_{17}/pr ratios in forensic investigations of refined product and crude oil releases: Discussion. *Environ. Geosc.* 1585–86.

Galperin, Y. & Kaplan, I. R. 2008e. Zero-order kinetics model for the Christensen-Larsen method for fugitive fuel age estimates. *Ground Water Monitoring Rev.* 28:94–97.

Garrett, R. M., Rothernburger, S. J. & Prince, R. C. 2003. Biodegradation of fuel oil under laboratory and Arctic marine conditions. *Spill Sci. Technol. Bull.* 8 (3): 297-302.

Gaylarde, C. C., Bento, F. M. & Kelly, J. 1999. Microbial contamination of stored hydrocarbon fuels and its control. *Revista Microbiol.* 30:1–10.

Gore, D. B., Revill, A. T. & Guille, D. 1999. Petroleum hydrocarbons ten years after spillage at a helipad in Bunger Hills, East Antarctica. *Antarctic Science 11*:427–429.

Hankey-Masui, K. M. 1998. Bioremediation of Nova Scotia tills contaminated with weathered heating oil. Master's thesis, Dalhousie University, Halifax, Nova Scotia, Canada.

Harmsen, J., Rulkens, W. & Eijsackers, H. 2005. Bioavailability: concept for understanding or tool for predicting? *Land Contam. Reclam.* 13 (2): 161-171.

Harper, R. G. 2000. Comparisons of independent petroleum supply statistics. Petroleum Supply Monthly. Washington, DC: Energy Information Administration www.eia.doe.gov.

ten Haven, H. L., de Leeuw, J. W., Rullkotter, J. & Sinninghe Damste´, J. S. 1987. Restricted utility of the pristane/phytane ratio as a palaeoenvironmental indicator: *Nature* 330:641-643.

Holt, G. S. 1997. Risk assessment program of underground storage tank systems. In IIR Corrosion Conference. Johannesburg, South Africa: South African (SA) Oil Industry Corrosion Control Group.

Hostettler, F. D. & Kvenvolden, K. A. 2002. Alkylcyclohexanes in environmental geochemistry. *Environ. Forensics* 3:293–301.

Hostettler, F. D., Pereira, W. E., Kvenvolden, K. A., van Geen, A., Luoma, S. N., Fuller, C. C. & Anima, R. 1999. A record of hydrocarbon input to San Francisco Bay as traced by biomarker profiles in surface sediment and sediment cores. *Marine Chem.* 64: 115-127.

Hostettler, F. D., Bekins, B. A., Rostad, C. E. & Herkelrath, W. N. 2008. Response to commentary on observed methanogenic biodegradation progressions. *Environ. Forensics* 9:121–126.

Huang, L., Li, J., Zhao, D., and Zhu, J. 2008. A fieldwork study on the diurnal changes of urban microclimate in four types of ground cover and urban heat island of Nanjing, China. *Building and Environment* 43:7–17.

Hurst, R. W. 2003. Invited commentary on Dr. Isaac Kaplan's paper "Age dating of environmental organic residues". *Environ. Forensics* 4:145–152.

Hurst, R. W. & Schmidt, G. W. 2005. Age significance of nC17/pr ratios in forensic investigations of refined product and crude oil releases. *Environ. Geosc.* 12:177–192.

Hurst, R. W. & Schmidt, G. W. 2007. Age significance of nC17/pr ratios in forensic investigations of refined product and crude oil releases: Reply. *Environ. Geosci.* 14:111–112.

Hurst, R. W. & Schmidt, G. W. 2008. Age significance of nC17/pr ratios in forensic investigations of refined product and crude oil releases: Reply. *Environ. Geosc.* 15:85–86.

Hwang, E-Y., Park, J-S., Kim, J-D. & Namkoong, W. 2006. Effects of aeration mode on the composting of diesel-contaminated soil. *Ind. Eng. Chem. Res.* 12:694–701.

Illich, H. A. 1983. Pristane, phytane, and lower molecular weight isoprenoid distributions in oils. *AAPG Bulletin* 67:385–393.

Jobbágy, E. G. & Jackson, R. B. 2000. The vertical distribution of soil organic carbon and its relation to climate and vegetation. *Ecol. Applications* 10 (2): 423-436.

de Jonge, H., Freijer, J. I., Verstraten, J. M., Westerveld, J. & van der Wielen, F. W. M. 1997. Relation between bioavailability and fuel oil hydrocarbon composition in contaminated soils. *Environ. Sci. Technol.* 31:771–775.

Jovancicevic, B., Polic ́, P., Vrvic ́, M., Scheeder, G., Teschner, M. & Wehner, H. 2003. Transformations of n-alkanes from petroleum pollutants in alluvial groundwaters. *Environ. Chem. Letters* 1:73–81.

Kampbell, D. H. & Wilson, J.T. 1991. Bioventing to treat fuel spills from underground storage tanks. *J. Haz. Materials* 28:75–80.

Kanner, A. 2007. Meeting the burden for admissibility of environmental forensic testimony. *Environ. Forensics* 8:19–23.

Kaplan, I. R. 2002. Invited commentary of the Christensen and Larsen technique. *Environ. Forensics* 3:7.

Kaplan, I. R. 2003. Age dating of environmental organic residues. *Environ. Forensics* 4:95–141.

Kaplan, I. R., Galperin, Y., Alimi, H., Lee, R-P. & Lu, S-T. 1996. Patterns of chemical changes during environmental alteration of hydrocarbon fuels. *Ground Water Monitoring Rev.* 16:113–125.

Kaplan, I. R., Galperin, Y., Lu, S-H. & Lee, R. P. 1997. Forensic environmental geochemistry: Differentiation of fuel-types, their sources and release time. *Org. Geochem.* 27:289–317.

Kechavarzi, C., Petterson, K., Leeds-Harrison, P., Ritchie, L. & Ledin, S. 2007. Root establishment of perennial ryegrass (L. perenne) in diesel contaminated subsurface soil layers. *Environ. Pollution* 145: 68–74.

Kehoe, R. A. 1960. Impacts of pollution on health. In *Proceedings of the National Conference on Water Pollution.* Washington, DC: US Department of Health, Education and Welfare, Public Health Service, 60–84.

Kerry, E. 1993. Bioremediation of experimental petroleum spills on mineral soils in the Vestfold Hills, Antarctica. *Polar Biology* 13:163–170.

Kershaw, G. P. & Kershaw, L. J. 1986. Ecological characteristics of 35-year-old crude-oil spills in tundra plant communities of the Mackenzie Mountains, N.W.T. *Can. J. Botany* 64:2935- 2947.

Klug, M. J. & Markovetz, A. J. 1967. Degradation of hydrocarbons by members of the Genus Candida. *J. Bacteriology* 93:1847–1852.

Kramer, W. H. & Hayes, T. J. 1987. Water soluble phase of number 2 fuel oil: Results of a laboratory mixing experiment. New Jersey Geological Survey Technical Memorandum 87:3.

LaFargue, E. & Barker, C. 1988. Effect of water washing on crude oil compositions. *AAPG Bulletin* 72:263–276.

Landon, M. K. & Hult, M.F. 1991. Evolution of physical properties and composition of a crude oil spill, in Mallard, G. E. & Aronson, D.A. (Eds.) *U.S. Geological Survey Toxic*

Substances Hydrology Program. Proceedings of the Technical Meeting, Monterey, California. U.S. Geological Survey Water Resources Investigations Report 91-4034, 641-645.

Lapinskiene, A., Martinkus,P. & Rebzdaite,V. 2005. Eco-toxicological studies of diesel and biodiesel fuels in aerated soils. *Environ. Pollution* 142:432–437.

Leahy, J. G. & Colwell, R. R. 1990. Microbial degradation of hydrocarbons in the environment. *Microbiol. Rev.* 54:305–315.

Lee, Y-K., Ryu, H-W., Lee, J-Y., Han, J-G., Chang, K-H. & Lee, J-H. 2007. The occurrence time estimation of oil contaminated source considering hydrogeological features. *Geosc. J.* 11:95–103.

Ludzack, F. L. & Kinkead, D. 1956. Persistence of oily wastes in polluted water under aerobic conditions. *Ind. Eng. Chem. Res.* 48:263–267.

McGovern, E. 1999. Bioremediation of Unresolved Complex Mixtures in Marine Oil Spills: Final Report. Dublin, Ireland: Marine Institute Fisheries Research Centre:. EU Project B$-3040/95/928/jnb/C4.

McMahon, P. B. & Chapelle, F. H. 2008. Redox processes and water quality of selected principal aquifer systems. *Ground Water* 46:259–271.

McPherson, A., Fleming, I., Farrell, R. & Headley, J. 2007. Monitoring phytoremediation of petroleum hydrocarbon-contaminated soils in a closed and controlled environment. In *In Situ and On-Site Bioremediation – 2007.* Proceedings of the Ninth International In Situ and On-Site Bioremediation Symposium, eds., Gavaskar, A. R., and Silver, C. F. Columbus, OH: Battelle Press.

Maila, M. P., Randima, P., Surridge, K., Drønen, K. & Cloete, T. E. 2005. Evaluation of microbial diversity of different soil layers at a contaminated diesel site. *Int. Biodeterior. Biodegrad.* 55:39–44.

Man, A. G. 1998. Temperature effect on bioremediation of diesel fuel contaminated Winnipeg Clay. Master's thesis, University of Manitoba, Winnipeg, Manitoba, Canada.

Margesin, R. & Schinner, F. 2001. Bioremediation (natural attenuation and biostimulation) of diesel-oil-contaminated soil in an alpine glacier skiing area. *Appl. Environ. Microbiol.* 67:3127–3133.

Markovetz, A. J., Cazon, J. & Allen, J. E. 1968. Assimilation of alkanes and alkenes by fungi. *Appl. Microbiol.* 16:487–489.

McCaskill, W. D. 1999. A report from Maine on the trials and tribulations of leaking aboveground home heating oil tanks. *LUSTLine Bull.* 33:20–23.

McVay, K. A., Radcliffe, D. E., West, L. T. & Cabrera, M. L. 2004. Anion exchange in saprolite. *Vadose Zone J.* 3:668–675

Merdinger, E. & Merdinger, R. P. 1970. Utilization of n-alkanes by Pullularia pullulans. *Appl. Microbiol.* 20:651–652.

Middleditch, B. S., Basile, B. & Chang, E. S. 1978. Discharge of alkanes during offshore oil production in the Buccaneer Oilfield. *Bull. Environ. Contam. Toxicol.* 20:59–65.

Miesch, A. T. 1967. Theory of error in geochemical data: U. S. Geological Survey Professional Paper 574–A.

Miller, C. T., Poirier-McNeill, M. M. & Mayer, A. S. 1990. Dissolution of trapped nonaqueous phase liquids: Mass transfer characteristics. *Water Resour. Res.* 26(11): 2783-2796.

Mohantya, G. & Mukherji, S. 2008. Biodegradation rate of diesel range *n*-alkanes by bacterial cultures *Exiguobacterium aurantiacum* and *Burkholderia cepacia*. *Int. Biodet. Biodegrad.* 61 (3): 240-250.

Morgan, R. M. & Bull, P. A. 2007. The philosophy, nature and practice of forensic sediment analysis. *Prog. Phys. Geog.* 31: 43–58.

Morrison, R. D. 2000. Application of forensic techniques for age dating and source identification in environmental litigation. *Environ. Forensics* 1:131–153.

Nakajima, K., Sato, A., Takahara, Y. & Iida, T. 1985. Microbial oxidation of isoprenoid alkanes, phytane, norpristane and farnesane. *Agri. Biol. Chem.* 49:1993-2002.

Olsen, J. J., Mills, G. L., Herbert, B. E., and Morris, P. J. 1999. Biodegradation rates of separated diesel components. *Environ. Toxicol. Chem.* 18:2448–2453.

Ostendorf, D. W., Schoenberg, T. H., Hinlein, E. S. & Long, S. S. 2007. Monod kinetics for aerobic biodegradation of petroleum hydrocarbons in unsaturated soil microcosms. *Environ. Sci. Technol.* 41:2343–2349.

Osuji, L. C., Ogali, R. E. & Kalu, A. U. 2009. The use of pristane and phytane biomarkers: A rethink of Cognoscenti. *Scientia Africana* 8 (2): 42-52.

Oudijk, G. 2005. The use of atmospheric contaminants to estimate the minimum age of environmental releases impacting groundwater. *Environ. Forensics* 6:345–354.

Oudijk, G. 2007. Age significance of nC17/pr ratios in forensic investigations of refined product and crude oil releases: Discussion. *Environ. Geosc.* 14:110–111.

Oudijk, G. 2009a. Age dating heating-oil releases. Part 1. Heating-oil composition and subsurface weathering. *Environ. Forensics* 10(2): 107-119.

Oudijk, G. 2009b. Age dating heating-oil releases, Part 2: Assessing weathering and release time frames through chemistry, geology and site history. *Environ. Forensics* 10(2): 120-131.

Oudijk, G., Duffy, B. & Ochs, L. D. 1999. Aerobic bioremediation of residential petroleum releases. In *In Situ and On-Site Bioremediation of Petroleum Hydrocarbons and Other Organic Compounds*, eds. Alleman, B. C. & Leeson, A., 271- 276.

Oudijk, G., Obolensky, M. & Polidoro, K. 2006. The use of the Christensen-Larsen model to age date residential heating oil releases: Conditions, limitations, and recommended practices. *Environ. Claims J.* 18:257–273.

Owens, E. H., Taylor, E. & Humphrey, B. 2008. The persistence and character of stranded oil on coarse-sediment beaches. *Marine Poll. Bull.* 56:14–26.

Owen, K. & Coley, T. 1995. *Automotive Fuels Reference Book, Second Edition.* Warrendale, PA: Society of Automotive Engineers.

Oyewo, E. O. 1988. *Preliminary Studies on the Initial Weathering of Oil on Water.* Nigerian Institute for Oceanography and Marine Research Technical Paper No. 26.

Palacas, J. G., Snyder, R. P., Baysinger, J. P. & Threlkeld, C. N. 1982. Geochemical analysis of potash mine seep oils, collapsed breccia pipe oil shows and selected crude oils, Eddy County, New Mexico. U. S. Geological Survey Open-File Report 82–421.

Palmer, S. E. 1991. Effect of Biodegradation and Water Washing on Crude Oil Composition: Chapter 4: *Petroleum Generation and Migration.* AAPG Special Volumes, Volume TR: Source and Migration Processes and Evaluation Techniques, 47–54.

Parsons Engineering Science. 2003. *Final, Light nonaqueous-phase liquid weathering at various fuel release sites, 2003 Update*. U. S. Air Force Center for Environmental Excellence Science and Engineering Division, Brooks City-Base San Antonio, Texas.

Payne, J. R., McNabb, G. D. & Clayton, J. R. 1991. Oil-weathering behavior in Arctic environments. *Polar Research* 10: 631-662.

Pearson, G. & Oudijk, G. 1993. Investigation and remediation of petroleum product releases from residential storage tanks. *Ground Water Monitoring Rev*. 13:124-128.

Peters, K. E., Walters, C. C. & Moldowan, J. M. 2005. *The Biomarker Guide: Volume 1, Biomarkers and Isotopes in the Environment and Human History*. Cambridge, UK: Cambridge University Press.

Philp, R. P. & Lewis, C. A. 1987. Organic geochemistry of biomarkers. *Ann. Rev. Earth Plan. Sci*. 15:363-395.

Pirnik, M. P. 1977. Microbial oxidation of methyl branched alkanes. *CRC Critical Rev. Microbiol*. 5:413-422.

Pirnik, M. P., Atlas, R. M. & Bartha, R. 1974. Hydrocarbon metabolism by Brevibacterium erythrogenes: Normal and branched alkanes. *J. Bacteriol*. 119:868-878.

Potter, T. L. & Simmons, K. E. 1998. Composition of Petroleum Mixtures, vol. 2. Amherst, MA: Amherst Scientific Publishers.

Prince, R. C. & Walters, C. C. 2007. Biodegradation of oil hydrocarbons and its implications for source identification. In *Oil Spill Environmental Forensics*, eds. Z. Wang & S. A. Stout. San Diego, CA: Academic Press, 349-379.

Providenti, M. A., Lee, H. & Trevors, J. T. 1993. Selected factors limiting microbial degradation of recalcitrant compounds. *J. Ind. Microbiol. Biotechnol*. 12:379-395.

Raymond, R. L., Hudson, J. O. & Jamison, V. W. 1976. Oil degradation in soil. *Appl. Environ. Microbiol*. 31:522-535.

Reinhard, M., Hopkins, G., Cunningham, J. & Lebron, C. A., 2000. Enhanced in situ anaerobic bioremediation of fuel-contaminated ground water: Naval Facilities Engineering Command, Washington, DC. Contract Report CR 00-005-ENV.

Ritter, U. 2003. Solubility of petroleum compounds in kerogen— implications for petroleum expulsion. *Org. Geochem*. 34:319- 326.

Røberg, S., Stormo. S. K. & Landfald, B. 2007. Persistence and biodegradation of kerosene in high arctic intertidal sediment. *Marine Environ. Res*. 64:417-428.

Robinson, J. E., Scott, D. W., Knocke, W. R. & Conn, W. D. 1988. Underground storage tank disposal: Alternatives, economics, and environmental costs. Virginia Polytechnic Institute and State University, Center for Environmental and Hazardous Materials Studies, Bulletin 160.

Schaeffer, T., Cantwell, S. G., Brown, J. L., Watt, D. S. & Fall, R. R. 1979. Microbial growth on hydrocarbons: Terminal branching inhibits biodegradation. *Appl. Environ. Microbiol*. 38 (4): 742-746.

Schmidt, P. F. 1985. *Fuel Oil Manual, 4th edition*: New York, NY: Industrial Press, Inc.

Schroll, R., Becher, H. H. & Dorfler, U. 2006. Quantifying the effect of soil moisture on the aerobic microbial mineralization of selected pesticides in different soil. *Environ. Sci. Technol*. 40:3305- 3312

Senn, R. B. & Johnson, M. S. 1987. Interpretation of gas chromatographic data in subsurface hydrocarbon investigations. *Ground Water Monitoring Rev*. 7:58-63.

Setti, L., Pifferi, P. G. & Lanzarini, G. 1995. Surface tension as a limiting factor for aerobic n-alkane biodegradation. *J. Chem. Technol. Biotechnol.* 64:41-48.

Sexstone, A., Everett, K., Jenkins, T. & Atlas, R. M. 1978a. Fate of crude and refined oils in North Slope soils. *Arctic* 31:339-347.

Sexstone, A., Gustin, A. & Atlas, R. M. 1978b. Long term interactions of microorganisms and Prudhoe Bay crude oil in tundra soils at Barrow, Alaska. *Arctic* 31:348-354.

Shepperd, D. A. & Crawford, T. R. 2003. Age-dating diesel fuel: Facts and fallacies: *Proceedings of the American Academy of Forensic Sciences* IX:101-102.

Siddique, T., Fedorak, P. M. & Foght, J. M. 2006. Biodegradation of short-chain n-alkanes in oil sands tailings under methanogenic conditions. *Environ. Sci. Technol.* 40:5459-5464.

Singer, M. E. & Finnerty, W. R. 1984. Microbial metabolism of straight- chain and branched alkanes. In *Petroleum Microbiology*, ed. R. M. Atlas. New York, NY: MacMillan Publishing Company.

Smith, J. S., Eng, L. & Shepperd, D. A. 2001. Age-dating oil: Is Christensen and Larsen applicable? *The Chemist* March/ April Issue, 9- 12.

Smith, M. R. 1990. The biodegradation of aromatic hydrocarbons by bacteria. *Biodegradation* 1:191-206.

Solevic, T., Jovancićević, B., Vrvić, M. & Wehner, H. 2003. Oil pollutants in alluvial sediments—influence of the intensity of contact with ground waters on the effect of microorganisms. *J. Serbian Chem. Soc.* 68:227-234.

Song, C. 2000. Introduction to chemistry of diesel fuels. In *Chemistry of Diesel Fuels*, eds. Song, C., Hsu, C. S. & Mochida, I. New York, NY: Taylor & Francis.

Song, H-G., Wang, X. & Bartha, R. 1990. Bioremediation potential of terrestrial fuel spills. *Appl. Environ. Microbiol.* 56:652- 656.

de Souza, E. S. & Triguïs, J. A. 2006. Degradação do petróleo em derrames no mat—intemperismo e biorremediacão [in Portuguese]. In 3° *Congresso Brasileiro de P&D em Petroleo e Gas.*

Srinivasan, G., Tartakovsky, D. M., Robinson, B. A. & Aceves, A. B. 2007. Quantification of uncertainty in geochemical reactions. *Water Resour. Res.* 43:W12415.

State of New Jersey. 1986. Underground storage tanks containing hazardous substances; legislative findings and declarations. N.J.A.C 58:10A- 21, effective September 3, 1986.

Stout, S. A., Uhler, A. D., McCarthy, K. J. & Emsbo-Mattingly, S. D. 2002a. Invited commentary of the Christensen and Larsen technique. *Environ. Forensics* 3:9-11.

Stout, S. A., Uhler, A. D., McCarthy, K. J. & Emsbo-Mattingly, S. 2002b. Chemical fingerprinting of hydrocarbons. In *Introduction to Environmental Forensics*, eds. Morrison, R. D. & Murphy, B. M. San Diego, CA: Academic Press.

Stout, S. A., Uhler, A. D., and McCarthy, K. J. 2006. Chemical characterization and sources of distillate fuels in the subsurface of Mandan, North Dakota. *Environ. Forensics* 7:267-282.

Stout, S. A. & Douglas, G. S. 2007. Age-dating diesel fuel: A case study averse to the Christensen and Larsen method. In *The Annual International Conference on Soil, Sediment & Water.* Amherst, MA: Association of Environmental Health and Soils.

Stout, S. A. & Uhler, A. D. 2006. Causation of variable n-alkylcyclohexane distributions in distillate nonaqueous phase liquids from Mandan, North Dakota. *Environ. Forensics* 7:283–287.

Stout, S. A. & Wang, Z. 2007. Chemical fingerprinting of spilled or discharge petroleum—Methods and factors affecting petroleum fingerprints in the environment. In *Oil Spill Environmental Forensics*, eds. Z. Wang & S. A. Stout. San Diego, CA: Academic Press, 1-72.

Sun, P., Gao, Z., Zhou, Q., Zhao, Y., Wang, X., Cao, X. & Li, G. 2009. Evaluation of the oil spill accident in Bohai Sea, China. *Environ. Forensics* 10 (4): 308-316.

Sukol, R., Woolson, E. & Thompson, W. 1988. Fate and Persistence in Soil of Selected Toxic Organic Chemicals. U. S. Environmental Protection Agency Project Summary EPA/600/S6-87/003.

Swannell, R. P. J., Lee, K. & McDonaugh, M. 1996. Field evaluations of marine oil spill bioremediation. Microbiology Reviews 60:342–365.

Swindell, A. L. & Reid, B. J. 2006. Influence of diesel concentration on the fate of phenanthrene in soil. *Environ. Poll.* 140:79–86.

Teh, J. S. & Lee, K. H. 1973. Utilization of n-alkanes by Cladosporium resinae. *Appl. Microbiol.* 25:454–457.

Wade, M. J. 2001. Age-dating diesel spills: Using the European empirical time-based model in the USA. *Environ. Forensics* 2:347- 358.

Wade, M. J. 2002. Invited commentary of the Christensen and Larsen technique *Environ. Forensics* 3:13.

Wade, M. J. 2005. The use of isoprenoid ratios to calculate percentage mixing of different distillate fuels released to the environment. *Environ. Forensics* 6:187–196.

Walker, J. D., Colwell, R. R. & Petrakis, L. 1976. Biodegradation rates of components of petroleum. *Can. J. Microbiol.* 22:1209- 1213.

Wang, Z. & Fingas, M. 1995a. Use of methyldibenzothiophenes as markers for differentiation and source identification of crude and weathered oils. *Environ. Forensics* 29:2842–2849.

Wang, Z. & Fingas, M. 1995b. Study of the effects of weathering on the chemical composition of a light crude oil using GC/MS GC/FID. *J. Microcolumn Sep.* 7 (6): 617-639.

Wang, Z., Hollebone, B. P., Fingas, M., Fieldhouse, B., Sigouin, L., Landriault, M., Smith, P., Noonan, J. & Thouin, G. 2003. Characteristics of spilled oils, fuels, and petroleum products: 1. Composition and properties of selected oils. EPA/600/R-03/072.

Wang, Z., Yang, C. & Fingas, M. 2005. Characterization, weathering, and application of sesquiterpanes to source identification of spilled lighter petroleum products. *Environ. Sci. Technol.* 39:8700–8707.

Whyte, L. G., Hawari, J., Zhou, E., Bourbonniere, L., Inniss, W. E. & Greer, C. W. 1998. Biodegradation of variable-chain-length alkanes at low temperatures by a psychrotrophic Rhodococcus sp. *Appl. Environ. Microbiol.* 64:2578–2584.

Zaidi, B. R. & Imam, S. H. 1999. Factors affecting microbial degradation of polycyclic aromatic hydrocarbon phenanthrene in the Caribbean coastal water. *Marine Poll. Bull.* 38 (8): 737-742.

Zajic, J. E., Supplisson, B. & Volesky, B. 1974. Bacterial degradation and emulsification of no. 6 fuel oil. *Environ. Sci. Technol. 8:664–668.*

Zemo, D. 2007. Forensic tools for petroleum hydrocarbon releases. *SW Hydrol.* July/August:26–35.

Zibiske, L. M., and Risser, J. A. 1986. Effects of soil texture on respiration and metal solubility in heating oil-amended soils. *Bull. Environ. Contam. Toxicol.* 36:540–547.

Zobell, C. E. 1946. Action of microorganisms on hydrocarbons. *Bacteriol. Rev.* 10:1–49.

Zytner, R G, Salb, A C. & Stiver, W H. 2006. Bioremediation of diesel fuel contaminated soil: Comparison of individual compounds to complex mixtures. *Soil Sed. Contam.* 15:277–297.

23

Geology and Geomorphology in Landscape Ecological Analysis for Forest Conservation and Hazard and Risk Assessment, Illustrated with Mexican Case Histories

María Concepción García-Aguirre[1], Román Álvarez[2] and Fernando Aceves[3]
[1]*Centro de Ciencias de la Complejidad(C3), Departamento de Ecología y Recursos, Naturales, Facultad de Ciencias. Universidad Nacional Autónoma de México (UNAM) Ciudad Universitaria, C.P. Coyoacán, D.F.*
[2]*Instituto de Matemáticas Aplicadas y Sistemas, Universidad Nacional Autónoma de México (UNAM)*
[3]*Instituto de Geografía, Universidad Nacional Autónoma de México (UNAM)*
México

1. Introduction

The aim of landscape ecology is to understand both the effects of spatial patterns on ecological processes, and the development of those spatial patterns. It is considered holistic since it regards nature as a whole and it deals also with all human environment interactions. Application of landscape ecological principles for prioritizing rich species sites has the advantage of integrating spatial information, non-spatial information, and horizontal relationships in space and time.

Landscapes are complex systems constituted by a large number of heterogeneous components (with different geology, geomorphology, vegetation cover, ecological communities, land uses and so on) interacting in a non-linear way, that are hierarchically structured and scale-dependent (Wiens, 2009, Hall *et al.*, 2004). Landscape description and structure has traditionally been done on the basis of landscape metrics, that is, basic measures of the amount of habitat and core habitat, the number of discrete patches and the perimeter to area ratio. However many of these metrics are generally poorly tested and require of rigorous validation if they are to serve as reliable indicators of habitat loss and fragmentation (McAlpine and Eyrie, 2002). There are strategies which can help to improve the reliability of landscape pattern analysis (Shao and Gu, 2008). Since landscape pattern is spatially correlated and scale dependent, often multiscale information is required (Wu, 2004).

The study of causes, processes and consequences of land/cover change is one of the main research topics of landscape ecology. It is important to study processes and not merely spatial patterns considering cultures as a drivers of landscape change. Knowledge of temporal changes in landscape composition and structure, and their driving processes, can provide insight into regional landscape dynamics (Luke, 2000; Burgui *et al*, 2004).

Landscape information is of the out most importance to develop appropriate policies for environmental planning and nature conservation (Gulink *et al.*, 2000). An example of recognition of landscape ecology as an essential field of science for territorial planning is the project of the metropolitan region of Barcelona (Forman, 2004), in which environmental principles based on landscape ecology and sustainable use of resources and basic spatial models are applied, even when lacking quantitative regional analysis.

2. Geology and geomorphology as fundamental elements in landscape analysis

Several terrain characteristics are important for soil scientists, geologists, and geographers, because of their strong influence in the capability of the land to support various plant or animal species, or for terrain evaluation. Geologic origin and structure can be estimated by air photo interpretation and satellite image analysis. Sedimentary (sandstone, shale, limestone) or igneous rocks can be differentiated using digital analysis (remote sensing and GIS). The recognition of strike and dip attitudes, land form types, drainage patterns and the orientation of highlights and shadows, as well as susceptibility to flooding, are based on geomorphology (Sabins, 1978, Lillesand and Keiffer, 1979, Verstappen, 1988). Geomorphological analysis is greatly improved by the use of aerial photographs and satellite images, since they provide a synoptic view of terrain and a relatively rapid description of geographic distribution of major landforms and dominant land cover. Terrain classification based on landforms, lithology and genesis (historical processes) can be further specified into biogeomorphic land units on the basis of geomorphologic processes, relative age, sediment, drainage, and land cover/use (García-Aguirre *et al.*, 2010).

Ecological research provides ample evidence that topography can exert a significant influence on the processes shaping broad-scale landscape vegetation patterns. Unfortunately, the standard methods for landscape pattern analysis are not designed to include topography as a pattern shaping factor. Topography features may be derived from the digital elevation models (DEM) to obtain slope and aspect maps (Dorner *et al.*, 2002, Peiffer *et al.* 2003). A DEM and Landsat images were used to assess topographical complexity and evaluate changes in landscape composition and structure after fire (Viedma, 2008). Simultaneous analysis of maps of non-biotic elements (such as geology, geomorphology and topography) and biotic elements (land use/cover) allow to generate synthetic and systematic information of landscape in the form of biogeomorphic land unit maps (Zonneved, 1995).

3. Remote sensing and GIS in landscape analysis

Remote sensing and GIS are essential tools for generation of landscape thematic information (Gulink *et al.*, 2000) even when it is common to face problems during integration of different data sources in the GIS (Tinker *et al.*, 1998). Advances in remote sensing technologies have provided practical means for land use/cover mapping, which is particularly important for landscape ecological studies. These tools can also efficiently identify and assess areas of landscape damage at different scales and help land managers to solve specific problems. However, it is a key consideration to evaluate the remote sensing data and methods used as well the scale and information needed, for instance, to correctly define the best resolution to use (Ludwing *et al.*, 2007), as well as to find the best procedures to follow when linking data of different qualities (Falcucci, *et al.*, 2007). Land cover change information through time, combined with thematic information can be stored and managed efficiently in a GIS since it relates different layers of spatial information. In addition, it is a powerful analysis tool that allows identification of spatial relationships among different maps, through connection of spatial data with its attributes (Belda and Melia, 2000, Baysnat *et al.*, 2000). Time series remote sensing provide researchers with a valuable tool for the dynamic analysis of landscape (Staus *et al.*, 2002).

Environmental models implemented in computers have become important tools for designing management plans towards ecological and economic sustainability. Computers help to deal with the tremendous complexity reflected in the extensive temporal and spatial scales at which human and natural processes occur. Iverson and Prasad (2007) evaluated tsunami damage and built empirical vulnerability models of damage/no damage based on elevation, distance from shore, vegetation, and exposure.

4. An outline of the geology of Mexico

The geology of Mexico is the result of multiple tectonic processes that have taken place along its geologic history. Current geologic configuration of Mexico is the consequence of continental block interaction with surrounding oceanic provinces. As a result, young sedimentary and volcanic outcrops are dominant. 80% of exposed units are placed on Cenozoic and Mesozoic eras (less than 250 million years), 13% correspond to the Paleozoic and only 7 % belong to the Precambrian, belonging to the Proterozoic (up to 2500 million years ago (Ortega *et al.* 1992).

Precambrian

The oldest metamorphic Rocks in Mexico were found in Sonora State and belong to the Bamori Complex (Figure 1), it is conformed by muscovite schist, hornblende-amphibolites schist and quartzite, dated at 1755 ± 20 million years (myr) by Anderson and Silver (1981). In Chihuahua State a metamorphic complex outcrops (metagranite, metadiorite, amphibolite, gneiss, metalimestone and quartzite), these rocks have an age between 1025 ± 21 myr and 948 ± 14 myr (Blount 1983). In southeast Mexico one finds dispersed outcrops from the Proterozoic composed of metamorphic rocks (augengneiss, orthogneis, marble, amphibolite and migmatite) that belong to the Oaxaqueño Complex, with an age between 1300 to 700 myr (SGM, 2007). Along the southern coast from Zihuatanejo, Guerrero to Puerto Ángel, Oaxaca emerges a group of paragneiss, pelitic schist, boitite schist, quartzite, marble, orthogneiss, amphibolite and migmatites that has been grouped inside the Xolapa complex (between 980 and 1300 myr; SGM, 2007). Intrusive rocks of the Paleo and Mesoproterozoic, emerge in Sonora state, showing small outcrops of granite, granodiorite, and in less proportion, diorite. Their ages vary between 1440 and 1140 myr (Anderson and Silver, 1981).

Sedimentary rocks of the Proterozoic outcrop in Sonora state like small patches of dolomite, limestone and sandstone. These deposits were dated as Neoproterozoic owing to the presence of conic stromatolite fossils of the Conophyton genus (SGM, 2007).

Paleozoic

Metamorphic rocks from this period are schist, marble and quartzite. They are located in the states of Baja California, Sinaloa, Sonora and Chihuahua, their ages fluctuate between

Cambric and Carboniferous (Figure 1). Phyllites and schist with quartzite can be found at the southeast of the State of Chiapas (Late Mississippian). In Tamaulipas state metamorphic rocks appear in Huizachal-Peregrina Structure presenting mica-schist interstratified with green rocks, metaflints, serpentine, and metalimestones. In the Huizachal-Peregrina structure, near Ciudad Victoria, Tamaulipas, there are outcrops formed by mica-schist of low grade. The age for this unit was 330 myr (Stewart *et al.*, 1999). Low grade metavolcanic sedimentary rocks from the Permic (SGM, 2007) exist north of Durango city. Outcrop schists, phyllites, quartzite's and metalavas are found northeast of Puebla. These rocks are from Early Permic (280 myr; Iriondo *et al.*, 2003). There are outcrops of schists and quartzites in southeast Oaxaca. The age of these rocks ranges between 289±5 Ma and 219±6 Ma. (Grajales-Nishimura *et al.*, 1999). In Zacatecas outcrops are found of a metamorphosed sedimentary succession, from Late Paleozoic, 260.2 ± 3 Myr (Díaz-Salgado, 2004).

Fig. 1. Distribution of metamorphic rocks of Precambrian and Paleozoic

The metamorphic rocks from Mesozoic (Figure 2) in the northeast of the country are phyllites and chlorite and biotite schist, amphibolite gneiss, metatonalite and metadiorite (De Cserna *et al.*, 1962; 220 myr). A meta-volcanosedimentary succession was deposited from Late Jurassic to Early Cretaceous along the west of Mexico.
States of Michoacán, México and Guerrero present outcrops of metamorphic rocks of low grade, there are schist's, slates and quartzites. (Tejupilco Schist, Taxco Schist and Green Rock Taxco Viejo formations (De Cserna 1982) and probably belong to the Jurassic.

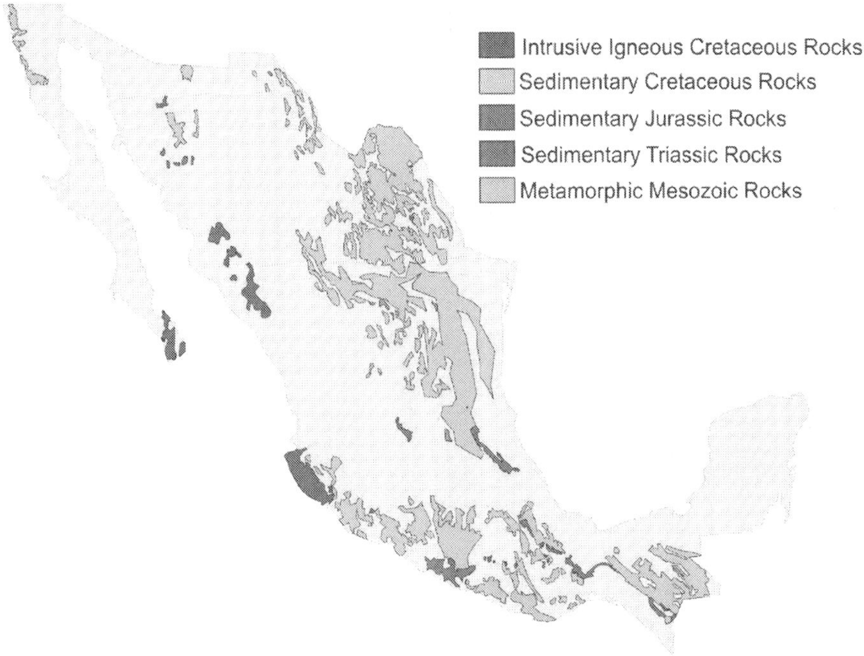

Fig. 2. Mesozoic Lithologic Units

Outcrops from Triassic are scarce. In southwest Sonora there is calcareous sandstone, alternated with limonite. This rock contains fossils of gastropods, coral, bryozoans, sponges and ammonites from this period (González-León, 1980), and the same in northeast Mexico were deposited at the Huizachal Formation conglomerate (SGM, 2007).
Along the Lower Jurassic an alternate succession of shales and limestones are deposited in the states of San Luis Potosí, Querétaro and Hidalgo. Jurassic sediments in southern Mexico (Oaxaca and Guerrero states) is composed of a conglomerate and sandstone succession with quartz clasts (Cualac Formation). The Upper Cretaceous was identified with ammonites and radiolarian fossils. In south Baja California the lower Cretaceous is represented by sandstones, limonites, shales and conglomerate successions.
To the east, in Chihuahua and Nuevo León states the great sea transgression deposited thick beds of calcareous and siliceous rocks (Formations Taraises, Cupido, La Peña, La Virgen.). At the same time calcareous anhydrite and clayish successions were deposited in Tamaulipas, San Luis Potosí, Hidalgo, Queretaro, Puebla and Veracruz states. The most important formations are Lower and Upper Tamaulipas, Otates, El Abra, Tamabra, Tamasopo and Cuesta del Cura. The last Formation is composed of limestone and chert. In south and central Mexico sedimentary rocks appear in the states of Jalisco, Michoacán, Guerrero, México, Morelos and Oaxaca, where calcareous successions settled, and clayish components are constituted by conglomerates, sandstones and limonites, with interstratified limestone, marls, and gypsum in different facies. The most important formations are Zicapa, Tepexi de Rodríguez, Xochicalco, Morelos, Cuautla y Mexcala. These formations contain a wide variety of mollusca, gasteropods, ammonites, and milliolids.

A change in sedimentation takes place in the Upper Cretaceous marked by the suffocation of the platforms with a series of terrigenous successions that show the evolution of deltas and basins. These successions form big bundles of sandstone, conglomerate, limonite, marl, and shales.

The outcrops of intrusive rocks are scarce and dispersed along the territory. The most important deposits are found in southern Baja California (granites, Peridotite, utramafic sequence, and an ophiolite sequence (Kimbrough and Moore, 2003). A sequence of granite and diorite from Middle Jurassic is located in Puebla (Macizo de Teziutlán), dated between 163 ± 13 and 134 ± 11 Myr (Manjarrez and Hernández, 1989). In Guanajuato state outcrops a sequence of granite, diorite and tonalite from the Upper Jurassic outcrops, these deposits are related to the evolution of Mesozoic insular arcs from this period (SGM, 2007). Plutonic magmatism appears in southern Mexico, probably from Late Jurassic to Early Cretaceous, with a variable composition: granite, granodiorite and diorite. Some localities in which this plutonic intrusive appears are Tumbiscatio, near Zitácuaro city.

Cenozoic

A continental sedimentation of the Early Cenozoic marks the change of sedimentary rocks to volcanic sequences (Figure 3). The Red Conglomerates with intercalation of sandstones and limonites represent them in the Balsas, Red Conglomerate of Guanajuato, and Tehuacán formations. Transgressive events are reflected by the horizons of sandstone, lutite and conglomerates. In the Eocene-Oligocene, deposits of conglomerate and sandstone, limonite,

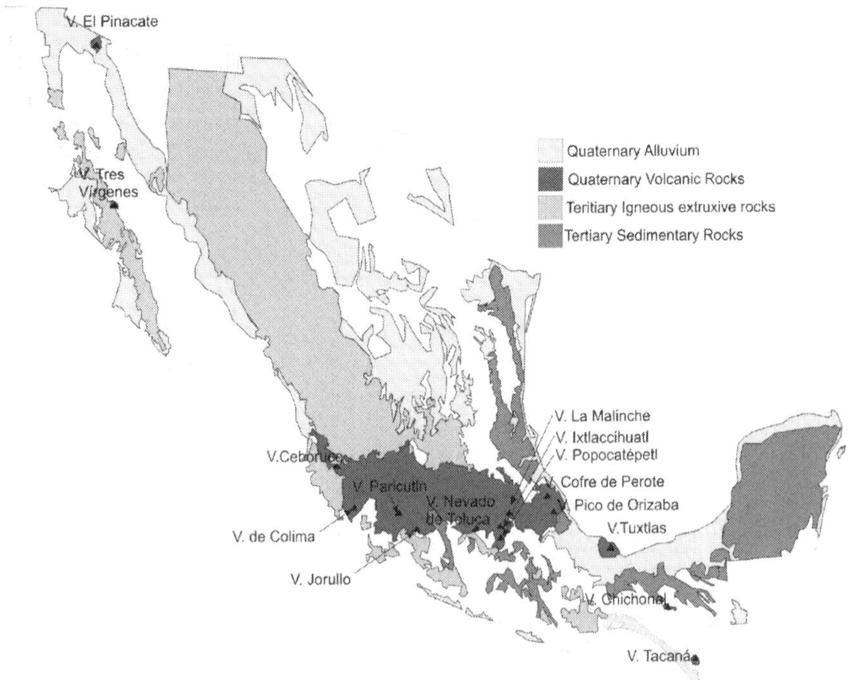

Fig. 3. Cenozoic Lithologic Units

and sandstone and limestone are distributed in grabens and synclinal valleys in the Sierra Madre Oriental and Sierra Madre Occidental. In the Sierra de Chiapas, the Miocene differed in the deposition of the clay-calcareous successions, as well as in the thin horizons of the conglomerates.

On the other hand, in the regions of the rim of the Yucatan Platform the bar reefs and lagoons keep on developing carbonate sediments of limestone and dolomites. The Holocene deposits of coastal environments of the coast of the Gulf of Mexico in the states of Tamaulipas, Veracruz, Tabasco, Campeche, Yucatan, and Quintana Roo, are still in the process of sedimentation of silts, clays and marshy sand, flat dunes of coastal sand, and continental shelf carbonate sediments.

The volcanic units of the Cenozoic are widely distributed in the Mexican territory, including the ignimbrites of the Sierra Madre Occidental and the Pliocenic-Quaternary sequences of the Transmexican Volcanic Belt (TMVB). The Sierra Madre Occidental is formed by an extensive volcanic plateau affected by grabens and normal faults. It spreads from Sonora to Guerrero states, although in the states of Jalisco, Michoacán and Guerrero it is fragmented and mixed with the Transmexican Volcanic Belt and the rocks of the Sierra Madre del Sur. The TransMexican Volcanic Belt (TMVB) extends from the Pacific Ocean up to the coast of the Gulf of Mexico along 920 km between parallels 19 and 20. It is formed by a large variety of volcanic rocks produced by a number of volcanic buildings, some of which constitute the main elevations in the country. Likewise, this activity has caused a big number of endorreic basins with the consistent development of lacustrine landscapes. The principal volcanoes of the TMVB are stratovolcanoes of variable dimensions, such as Pico de Orizaba, Popocatépetl, Iztaccíhuatl, La Malinche, Nevado de Toluca and Nevado de Colima.

5. Case studies

Landscape studies in Mexico are numerous: Ochoa (2001) undertook the integration of geologic and geomorphic units, incorporating afterwards variables of climate, hydrology, vegetation and soil in the Tehuacán-Cuicatlán, Puebla, area. Casals-Carrasco *et al.*, (2000) performed a geomorphologic analysis in order to establish relationships among land cover, landforms and soils using interpretation of a stereoscopic pair of SPOT-PAN, and a TM false color composite image. Martínez (2002) elaborated environmental land unit maps of a sub watershed in Morelos State. García (1991) studied the influence of relief dynamics in landscape structure on the vegetation in Zapotitlán, Puebla watershed. García (1998) analized the east slope of Sierra de la Cruces, Monte Alto and Monte Bajo while, Garcia-Aguirre *et al.* (2007) related geology, landform and vegetation in the Ajusco volcano area in central Mexico. Aguilar (2007) performed an environmental diagnostic of the Parque Nacional Nevado de Toluca from the biogeomorphic land units of the region.

Two case studies will be described in detail. The first is related to the use of biogeomorphic land units of a region located nearby Mexico City as a basis for hydrology and vegetation analysis. The second case refers to the evaluation of risks and hazards of the fourth highest summit in México, the Nevado de Toluca.

5.1 Sierra de las Cruces

The objectives of the first study are two-fold: to identify the most degraded areas of the region through the land unit analysis, and to study the relationship between forest loss and runoff in the region (scale 1:250,000) through a conceptual and cartographic model.

Landscape analysis was performed to describe regional characteristics in an integral form; land cover features were overlaid with other landscape elements (geology, geomorphology, soils and climate) to obtain a land unit map (García-Aguirre, 2008).
Mountainous relief and flat plain may be appreciated in the shadow relief model derived from a DEM (figure 4). Chichinautzin region, towards the south of the zone, being an infiltration zone is very important from the hydrological view point. Refief of the east slope of the Sierra de las Cruces, Monte Alto and Monte Bajo is constituted by four geomorphic units: mountain, upper and lower piedmont, and hills. Notice the N-S orientation of this mountain range.

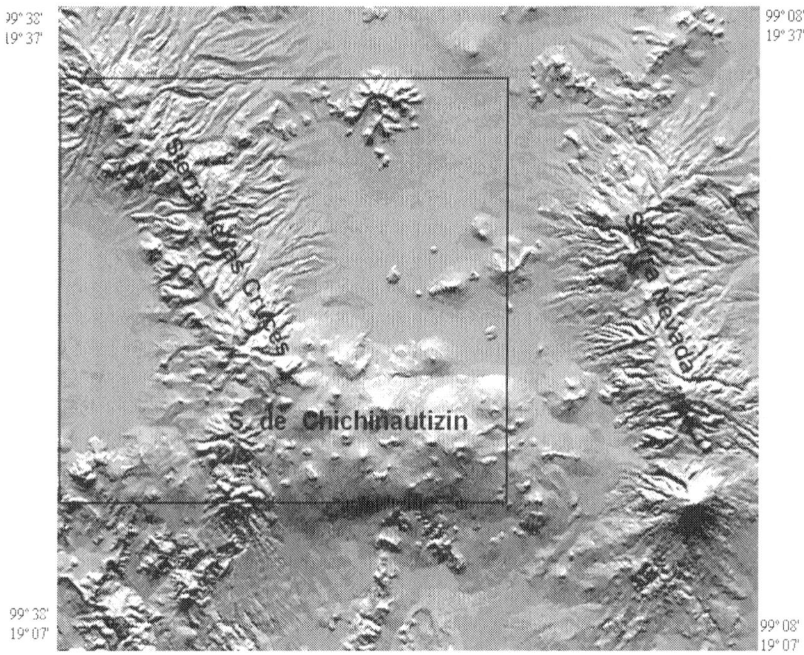

Fig. 4. Shadow relief model derived from the DEM. Study area is located inside the square. Sierra de Chichinautzin is located to the south, and Sierra Nevada toward the east

Land unit map

Remote sensing and GIS were linked to integrate geomorphological and geological information to find mayor associations among variables. Then, biogeomorphic land units were delineated on the basis of homogeneity of a dominant factor.
Figures 5 and 6 show the geology and geomorphology maps of the region. These maps were obtained by digitizing hardcopy maps of INEGI (1993) and reclassified using IDRISI (Eastman, 1997). Andesite and basalt are dominant in the area and in turn, andosols and lithosols (Figure 5). Lugo (1984) points the south of Cuenca de México as one of the zones of the country with higher concentration of young volcanoes, from the late Pleistocene and Holocene (Figure 6).

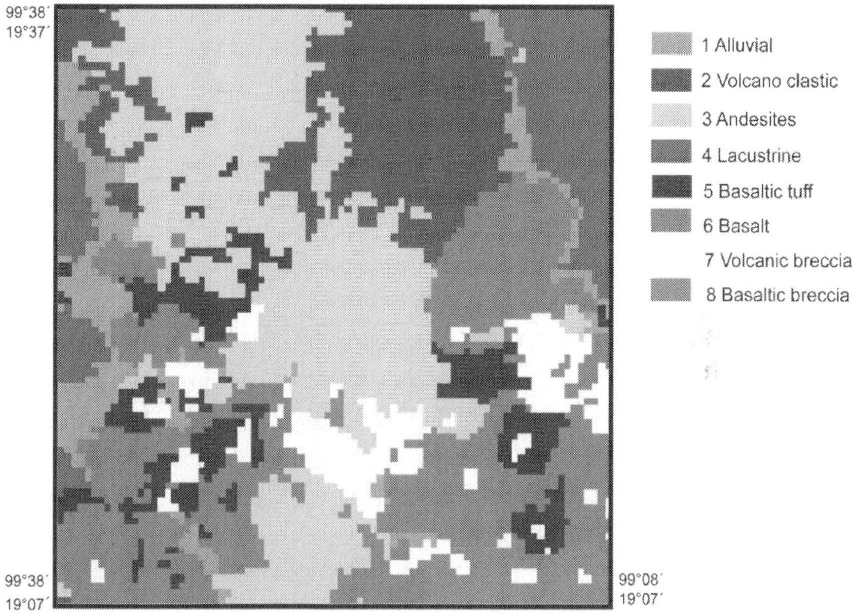

Fig. 5. Geologic map. Andesites and Basalt are dominant in this region

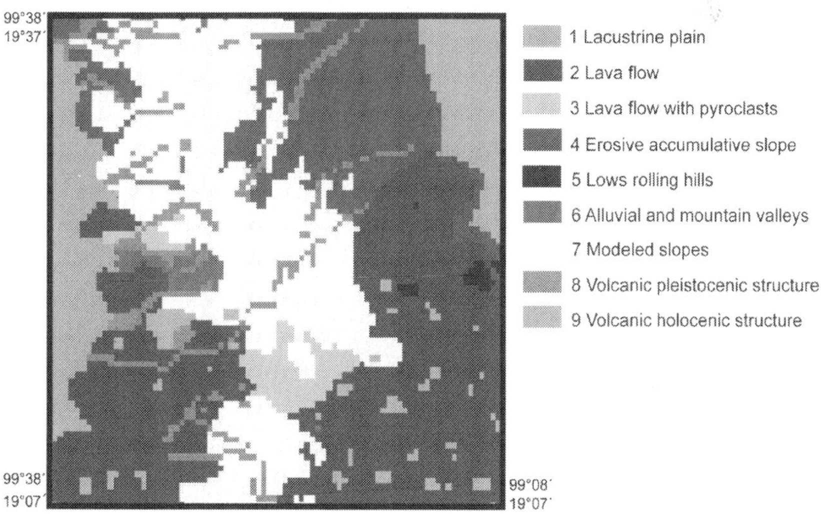

Fig. 6. Geomorphologic map. Dominant geoforms are modeled slopes and lava flows

Geomorphologic map (Figure 6) shows modeled slopes and lava flows as dominant geoforms. The region has an extense footslope, in which half the area slopes down to the piedemont and the other half is mountainous terrain. Also extensive lava flows show the active quaternary volcanism in this zone.
Combination of biotic and non biotic features generated more than 100 units, that were reclassified into 48 units on the basis of its surface (only units with more than 500 ha), for map legibility (Figure 7). Alluvial and lacustrine units are mainly covered with pheozem and hystosol, agriculture and grasslands. The modeled slopes of andesites with andosol are subdivided into those with Abies and those with Pine forest. Towards the east (BGU39, BGU40, and BGU2), there are abundant lacustrine forms with agriculture, grasslands and human settlements. Basalt is abundant to the south with forest cover, agriculture and grassland (BGU29, BGU33). In the footslopes, mainly towards the north, there are units constituted by andesites, modeled slopes and cambisols, that are covered by oak forests, crops and grasslands. There are many holocenic volcanic structures over the mountainous area, with andosol and litosol, covered mainly by forests.
The regional vision provided by this study allowed to have a rapid overview of sites that should be preserved, such as basalt zones, that are nevertheless continuously invaded by human settlements. Results indicate that Sierra de Chichinautzin is the main recharging area, but the foot slopes of the Sierra de Las Cruces are also important infiltration and recharging zones as a result of the abundance of clastic volcanic forms therein.

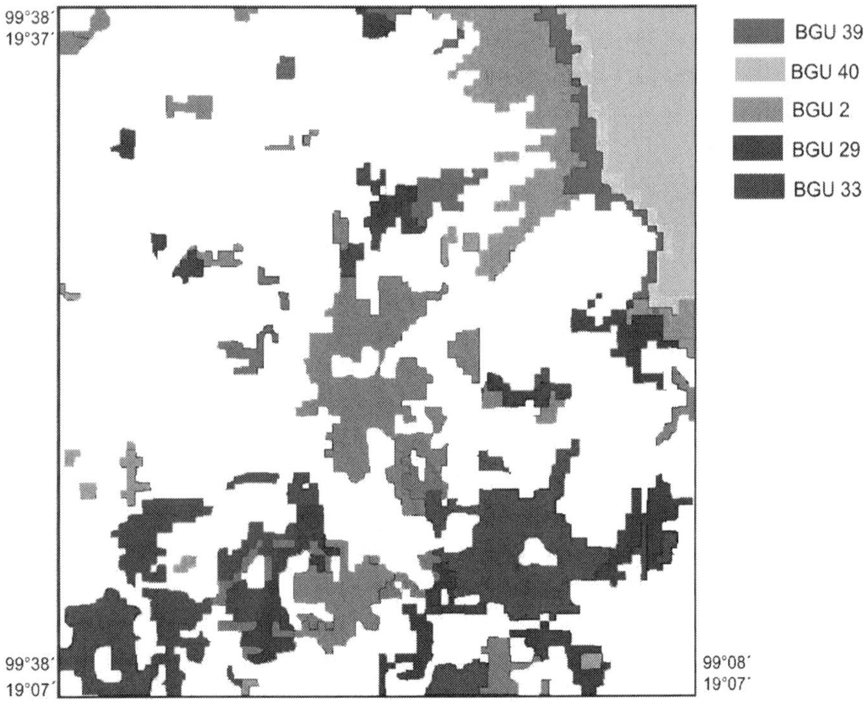

Fig. 7. Land unit map (BGU=biogeomorphic land units)

5.2 Nevado de Toluca geologic history and hazards

Nevado de Toluca Volcano (NTV), located in central Mexico, is a large stratovolcano, with an explosive history. The area is one of the most important human developing centers (>2 million people) in Mexico and in the last 30 years large population growth and urban expansion have increased the potential risk in case of a reactivation of the volcano. NTV is the fourth highest summit in Mexico (4,665 masl) and it is a potentially dangerous large stratovolcano, that lies in the southeastern part of the Toluca Basin, some 70 km east of Mexico City (Figure 8). The NTV has been characterized by very explosive eruptions with long periods of dormancy; the periods between eruptions are discussed below.

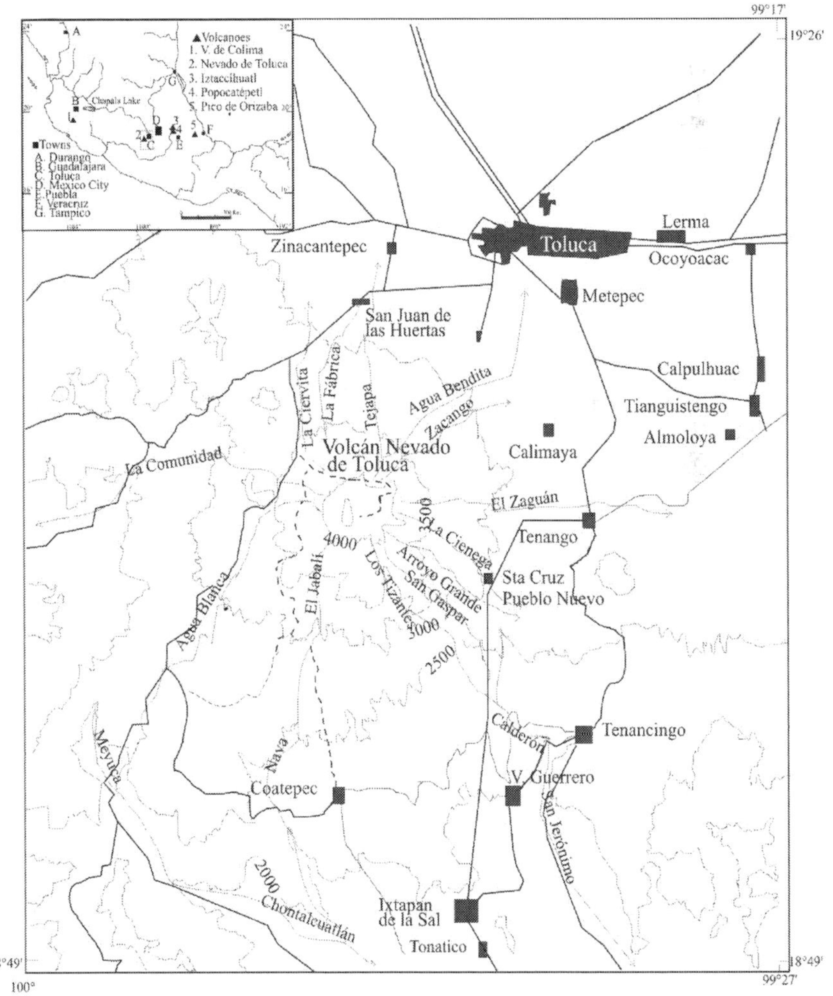

Fig. 8. Nevado de Toluca, location map

Nevado de Toluca Volcanic hazards

Pyroclastic flow Hazards: These deposits are widely spread around the volcano, filling the stream valleys where several settlements are located (Figure 8). The pyroclastic flow deposits cover a minimum area of 630 km^2 and assuming that they have an average thickness of 5 m, the approximate volume is 3.15 km^3 (Macias *et al.* 1997). The maximum distance reached by these deposits is 32 km from the crater towards the south, in the Tizantes and Calderón stream valleys. The block and ash flows form massive units interstratified with surge horizons. The first dome collapse deposited the Zacango Block and Ash flow (37 kyr BP) composed of three massive units with associated surge horizons. The second dome collapse (28 kyr BP) deposited El Capulín Block and Ash Flow. These deposits are distributed around the volcano and cover approximately 630 km^2 with a volume of 2.6 km^3 (Aceves *et al.*, 2007).

The pumice flows erupted by the NTV are: the Pink Pumice Flow (43 kyr BP); the White Pumice Flow (26 kyr BP). The MF2 (13.4 kyr BP) pumice flow is a gray ash flow enriched in pumice clasts (<2 cm) and charcoal (Aceves, *et al.* 2007). The Intermediate White Pumice (12.1 kyr BP) is composed of white ribbon pumice clasts and gray to reddish (altered) dacitic lithic clasts, interbedded with a surge (Cervantes 2001). The pumice flows are distributed around the volcano and cover more than 200 km^2, with a volume of 0.2 km^3. The pumice flows are related to the Plinian eruptions, many of which result in mudflows and include charcoal and variable amounts of pumice. Some are altered to paleosoils. One of the most pumice-rich flows belongs to the Upper Toluca Pumice Formation (UTP), dated 10,500 yr BP (Macias *et al.* 1997).

Debris avalanches hazards: The debris avalanches have been located towards the south of the NTV in the Meyuca, Calderón Chontalcuatlán and San Jerónimo river valleys. Two units compose the debris avalanches. The oldest unit (DAD1) is massive, with 35% blocks up to 2.5 m in diameter, in a heavy pink partially hardened sand sized matrix. The lithological composition is heterogeneous dacitic, andesitic and schist lithics, with"Jig saw" blocks (Capra and Macias, 2000). DAD1 is around 10 m thick in Coatepec and continues towards the south to the Valley of the Chontalcuatlan River. The youngest avalanche (DAD2) is composed of two large cohesive debris flows: the Pilcaya (PDF) and the El Mogote (MDF) (Capra 2000). Thickness varies from 6 m in the proximal section to 40 m in the intermediate zone, extending out to a distance of 75 km from the crater. It covered an area of 220 km^2 with a volume of 2.8 km^3 (Aceves *et al.*, 2007, Capra 2000).

Lahar hazards: Lahars at NTV are wide spread around the volcano filling new and old valleys. Lahars have rounded and subrounded dacite lithics (15-25 cm), small pumice fragments (<5 cm) fixed in a muddy-sand sized matrix. To the south, the lahar thickness is more than 30 m. The oldest lahars are made up of rounded and subrounded gray and red andesite blocks fixed in red clay sized matrix with scarce pumice fragments. The recent deposits contain subangular and subrounded blocks of gray and red dacite, with pumice fragments, some of which are hydrated, fixed in a siltclay sized matrix. The flow direction of these lahars was controlled by the topography, principally deep tectonic, glacial and fluvial valleys. In distal areas, the lahar deposits were transformed into fluvial mixing with the stream and river waters. To the east, the lahars contain more pumice fragments in a pale brown silt-clay matrix. There is a lahar with large hydrated pumice fragments (20-30 cm), in the Arroyo Grande channel. In this area, many secondary lahars exist, such as the one deposited in 1952 in the Ciénaga channel.

These secondary lahars are not related to volcanic activity and represent an increased hazard, because these materials are not consolidated. In torrential periods these materials can be removed by rain, triggering these secondary lahares, which can come up to the low zones affecting the populations who are settled in the mouths of the valleys as it happened to the people of Santa Cruz Pueblo Nuevo. To the north, the ravines are shallower (<30 m). The origin of the secondary lahars are uncompacted volcanic products (ash fall and, pyroclastic flows) mixed with water from the glacial melt and torrential rains (Aceves *et al.*, 2007).

Ash fall hazards: Ash and pumice fall deposits cover wide zones around the volcano. The most important deposits are the Upper and Lower Toluca Pumice. For the ash fall hazards map, the distribution and thickness of these deposits were considered as well as the present wind direction for the UTP. These isopach and isopleth maps show dispersal to the northeast. In the Toluca Basin, the maximum thickness measured for the UTP was 40 cm. In order to plot the isopach of 10 cm, map scales of 1:100,000, and 1:250,000 scale were used. The dominant wind direction was obtained from the National Meteorological Service and calculated for a height between 20 and 30 km (Fonseca 2003). The results were: east-northeast from November to March, west-northwest in April and west from May to October (Aceves et al., 2007).

In conclusion, in the last 50,000 yr, NTV had eight vulcanian, and four Plinian eruptions, three large dome collapses, and one ultraPlinian eruption. Block and ash and pumice flows are the most common deposits. The eruptive history of NTV shows cataclysmic events of Plinian and utraPlinian type, beside Vulcanian eruptions, which represent a large hazard for the Toluca Basin. The regions that would be most affected by pyroclastic flows and lahars are: (1) Toluca, Lerma, Metepec, San Mateo Atenco, Santiago Tianguistenco and Capulhuac, located to the northeast of the volcano. These areas concentrate the major population and industrial centers. (2) Calimaya, Zacango, Tenango among several others located to the east of NTV are highly populated and important agricultural areas. (3) To the south, the flower producing centers: Coatepec and Villa Guerrero, and the tourist towns of Ixtapan de la Sal and Tonatico.

Four volcanic hazards types were identified: pyroclastic flows (block and ash flows and pumice flows), lahars, debris avalanches and ash fall. The most destructive (based on energy and frequency) in the NTV are the block and ash flows and the pumice flows, both of them have reached distance of up to 35 km from the volcano summit. The principal affected areas are the northeast and south of the volcano, because these areas have major differences in altitude and present the major development of ravines. Lahars have been present in most of the eruptions. The deep and large valleys located to the east and south of the volcano are the most hazardous areas. The most active valleys are San Jerónimo, Chontalcuatlan, Grande, and El Zaguán Rivers. Debris avalanches present a hazard for areas to the east and south of the volcano, because of the active faults in the area and the instability caused by the difference in altitude, favoring the gravitational collapse of NTV. This type of event has occurred twice in the last 100,000 yrs, the deposits are located to the south of the volcano. The hazards from ash fall, arising from the dominant winds, are: from November to March, they would affect mainly the east and north-east sectors of the volcano, in April affectation would be to the north-west, and from May to October to the west. In case of small and medium eruptions (VEI = 1-3), the affected zone would be the Toluca Basin, but in case of large eruptions (VEI > 4) the affected zone would even include Mexico City.

6. Acknowledgment

Authors wishes to thank Verónica Aguilar for help in drawing of Figures 5, 6 and 7

7. References

Aceves-Quesada, F., Martin del Pozzo, A.L., López-Blanco J. 2007. Volcanic hazards zonation of the Nevado de Toluca Volcano, Center of Mexico" *Natural Hazards*. 41(1):159-180.

Aguilar, V. 2007. Diagnóstico del Parque Nacional Nevado de Toluca con base en unidades de Paisaje. Tesis de Maestría en Ciencias Biológicas (Biología Ambiental). Facultad de Ciencias. Universidad Nacional Autónoma de México. 94 p.

Anderson, T.H., Silver, L.T. 1981. An overview of precambrian rocks in Sonora: *Revista del Instituto de Geología*, 5(2): 131-139.

Basnyat, P., Teeter, L.D. , Lockaby, B.G., Flynn K.M. 2000. The use of remote sensing and GIS in watershed level analyses of non-point source pollution problems. *Forest Ecology and Management*. 128: 65-73.

Belda, F., Melia,J. 2000. Relationhips between climatic parameters and forest vegetation: application to burned area in Alicante (Spain). *Forest Ecology and Management*. 135: 195-204.

Blount, J.G., 1983, The geology of Rancho Los Filtros, Chihuahua, Mexico, in Clark, K.F., Goodel, P.C., (eds.), Geology and mineral resources of north-central Chihuahua: El Paso, Texas, El Paso Geological Society Guidebook, p. 157-164.

Burgui, M., A.M. Hersperger and N. Schneeberger. 2004. Driving forest of landscape change. Current and New directions. *Landscape Ecology*, 19:857-868.

Capra PL (2000) Colapsos de Edificios Volcánicos: transformación de Avalanchas de Escombros en flujos de escombros cohesivos. Tesis de Doctorado, UNAM, p176

Capra, L., Macías, J.L., 2000. Pleistocene cohesive debris fl ow at Nevado de Toluca volcano, central Mexico: Journal of Volcanology and Geothermal Research, 102, 149-168.

Casals-Carrasco, P., Kubo S., Madhavan, B. 2000. Application of spectral mixture analysis for terrain evaluation studies. *Int. J. Remote Sensing*, 21(16):3039-3055.

Cervante, K.E. 2001. La Pómez Blanca Intermedia: depósito producido por una erupción pliniana-subpliniana del Volcán Nevado de Toluca hace 12,100 años. Tesis de Mestría en Ciencias de la Tierra, UNAM.

De Cserna, Z., Schmitter, E., Damon, P.E., Livingston, D.E., Kulp, J.L., 1962, Edades isotópicas de rocas metamórficas del centro y sur de Guerrero y de una monzonita cuarcífera del norte de Sinaloa: *Boletín del Instituto de Geología*, 64(número), p. 71-84.

De Cserna, Z., 1982, Hoja Tejupilco 14Q-h (7), Geología de los Estados de Guerrero, México y Morelos: México, D.F., México, Universidad Nacional Autónoma de México, Instituto de Geología, Cartas Geológicas de México serie 1:100 000, 1 mapa con texto.

Díaz-Salgado, C., 2004, Caracterización Tectónica y Procedencia de la formación Taray, Región de Pico de Teyra, Estado de Zacatecas: México, D.F., México, Universidad Nacional Autónoma de México, Instituto de Geología, Tesis de Maestría, 95 p.

Dorner, B., Lertzman, K, Fall, J. 2002. Landscape in topographically complex landscapes: issues and techniques for analysis. *Landscape Ecology*, 17:729-743.

Eastman, J.R. 1997. IDRISI for windows user's guide, version 3.2, Clark laboratories for cartographic technologies and geographic analysis. Clark University, Worcester, MA.

Falcucci, A., Maiorano , L., Boitani, L. 2007. Changes in land-use/land-cover patterns in Italy and their implications for biodiversity conservation. *Landscape Ecol* 22:617–631

Fonseca R (2003) Efectos Ambientales de la Erupcio´n de 1913, del Volcán de Colima, Proyecciones a Futuro. Tesis de Maestría, Colegio de Geografía, Universidad Nacional Autónoma de México, p 90

Forman, R.T.T. 2004. Mosaico territorial para la región metropolitana de Barcelona. Barcelona regional/Editorial Gustavo Gili, S.A. Barcelona, España.

García, F. 1991. Influencia de la dinámica del paisaje en la distribución de las comunidades vegetales en la Cuenca del Río Zapotitlán, Puebla. *Investigaciones Geográficas. Boletín del Instituto de Geografía*. No. 23: 53-70.

García, A. 1998. Análisis integrado de paisajes en el occidente de la Cuenca de México (La vertiente oriental de la Sierra de las Cruces, Monte Alto y Monte Bajo) Tesis de Doctorado. Facultad de Geografía e Historia. Universidad Complutense de Madrid. 543 p.

García-Aguirre, M.C., M.A, Ortiz, J.J. Zamorano y Y. Reyes. 2007. Vegetation and landform relationships at Ajusco volcano Mexico, using a geographic information system (GIS). *Forest Ecology and Management* 239: 1-12.

García-Aguirre M.C., Álvarez, R., Dirzo, R., Ortiz, M.A., Mah Eng, M. 2010. Delineation of biogeomorphic land units across a tropical natural and humanized terrain in Los Tuxtlas, Veracruz, México. *Geomorphology*. 121:245-256.

González-León, C.M., 1980, La Formación Antimonio (Triásico Superior-Jurásico Inferior) en la Sierra del Alamo, Estado de Sonora: *Revista del Instituto de Geología*, 4(1), p. 13-18.

Grajales-Nishimura, J.M., Centeno-García, E., Keppie, J.D., Dostal, J., 1999, Geochemistry of Paleozoic basalts from the Juchatengo complex of southern Mexico: tectonic implications: *Journal of South American Earth Sciences*, 12 (6), p. 537-544.

Gulinck, H., Dufourmont, H., Coppin, P., Hermy, M.. 2000. Landscape research, landscape policy and Earth observation. *Int. J. Remote Sensing*. 21(13&14): 2541-2554.

Hall, O., Hay, G.J., Bouchard, A., Marceau, D.J. 2004. Detecting dominant landscape objects through multiple scales: An integration of object-specific methods and watershed segmentation. *Landscape Ecol* 19:59–76.

INEGI (Instituto Nacional de Estadística, Geografía e Informática). 1993. Mapas de uso del suelo y vegetación escala 1:250,000. México.

Iriondo, A., Punk, M.J., Wininck, J.A., 2003, 40Ar- 39Ar Dating studies of minerals and rocks in various areas in Mexico: USGS/CRM scientific collaboration (Part I), Open file report 03-020 on-line edition: USGS, Consejo de Recursos Minerales(COREMI), http://www.coremisgm.gob.mx/productos/novedades/estudios/USGS%202003_Texto01.pdf.

Iverson, L. R., Prasad, A.M. 2007. Using landscape analysis to assess and model tsunami damage in Aceh province, Sumatra. *Landscape Ecol.* 22:323–331.

Kimbrough, D.L., Moore, T.E., 2003, Ophiolite and volcanic arc assemblages on the Vizcaíno Peninsula and Cedros Island, Baja California Sur, México, in Johnson, S.E., Paterson, S.R., Fletcher, J.M., Girty, G.H., Kimbrought, D.L., Martín-Barajas, A. (eds.), Mesozoic forearc lithosphere of the Cordilleran magmatic arc, in Tectonic evolution of northwestern Mexico and the southwestern: Boulder, Colorado, EE.UU., Geological Society of America, Special Paper 374, p. 43-71.

Lillesand, T.M. and R.W. Keifer, 1979. Remote sensing and Image Interpretation. Second Edition. John Wiley and Sons. 612 p.

Ludwig, J. A., Bastin, G. N., Wallace, J.F., McVicar, T.R. 2007. Assessing landscape health by scaling with remote sensing: ¿when is it not enough? *Landscape Ecol* 22:163–169.

Lugo, H. J. 1984. Geomorfología del Sur de la Cuenca de México. *Serie Varia* T.1. Núm. 8. Instituto de Geografía, Universidad Nacional Autónoma de México. 95 p.

Luque, S. 2000. The challenge to manage biological integrity of nature reserves: a landscape ecology perspective. *Int. J. Remote Sensing*. 2000, 21(13&14): 2613-2643.

Manjarrez, P.P., Hernández, R., 1989, Informe geológico final Prospecto Cardel: Petróleos Mexicanos, Superintendencia General de Exploración Geológica, IGPR-278

McAlpine, C.A. and T.J. Eyre. 2002. Testing landscape metrics as indicators of habitat loss and fragmentation in continuous *Eucaliptus* forest (Queensland, Australia). *Landscape Ecology*, 17:711-728.

Macias JL, García A, Arce JL, Siebe C, Espíndola JM, Komorowski JC, Scott K., 1997. Late Pleistocene-Holocene cataclysmic eruptions at Nevado de Toluca and Jocotitlan volcanoes, Central Mexico. BYU Geology Studies, vol 42, Part I, pp 493–528

Martínez, F. 2002. Síntesis de las unidades ambientales biofísicas de la subcuenca del río Colotepec, Edo. de Morelos, mediante la aplicación del enfoque geomorfológico y un GIS. Tesis de Maestría en Ciencias (Ecología y Ciencias Ambientales). Fac. de Ciencias, División de Estudios de Posgrado. UNAM. 91p.

Ochoa, V. 2001. Geomorfología, clima y vegetación del valle de Tehuacan-Cuicatlán, Pue-Oax. México. Tesis de licenciatura en Biología. Facultad de Ciencias, UNAM. 80 p.

Ortega-Gutiérrez, F., Mitre-Salazar L.M., Roldán-Quintana J., Aranda-Gómez J.J., Morán-Zenteno D., Alaniz-Álvarez S.A., Nieto-Samaniego A.F. 1992. Texto Explicativo de la Quinta Edición de la Carta Geológica de la Republica Mexica a Escala 1:2,000,000. Instiuto de Geología, Universidad Nacional Autónoma de México, Consejo de Recursos Minerales. 74 p.1 Mapa.

Peiffer, K., Pebesma, E.J., Burrough, P.A. 2003. Mapping alpine vegetation using vegetation observations and topographic attributes. *Landscape Ecology*, 18:759-776.

Servicio Geológico Mexicano (SGM). 2007, Texto Explicativo de la Sexta Edición de la Carta Geológica de la República Mexicana Escala 1:2,000,000. Servicio Geológico Mexicano. Secretaria de Economía. 30 p. 1 Mapa.

Shao, G., Gu, J. 2008. On the accuracy of landscape pattern analysis using remote sensing data. *Landscape Ecology*, 23:505-511.

Sabins, F.F 1978. Remote Sensing. Principles and Interpretation. Second Edition. W.H. Freeman and Company. 449 p.

Staus,N.L., Striitholt, J.R., Delta, D.A., Robinson, R. 2002. Rate and pattern of forest disturbance in the Klamath Siskiyoa ecoregion, USA, between 1972 and 1992. *Landscape Ecology*, 17:455-470.

Stewart, J.H., Blodgett, R.B., Boicot, A.J., Carter, J.L., López, R., 1999, Exotic Paleozoic strata of Gondwanan provenance near Ciudad Victoria, Tamaulipas, México, in Ramos, V.A., Keppie, J.D. (eds.), Laurentia Gondwanan connections befote Pangea: Geological Society of America, Special Paper 336, p. 227-252.

Tinker, D.B., C.A.C. Resor, G. P. Beauvais, K.F. Kipfmueller, Ch. I. Fernandes and W.L. Baker. 1998. Watershed analysis of forest fragmention by clearcuts and roads in a Wyoming forest. *Landscape Ecology* 13:149-165.

Verstappen, H.Th. 1988. Remote Sensing in geomorphology. Elsevier Scientific Publishing Company. Amsterdam. 214 p.

Viedma, O., 2008. The influence of topography and fire in controlling landscape composition and structure in Sierra de Gredos (Central Spain). *Landscape Ecology*, 23: 657-672.

Wiens, J.A. 2009. Landscape ecology as a foundation for sustainable conservation. *Landscape Ecology*, 24:1053-1065.

Wu, J. 2004. Effects of changing scale on landscape pattern analysis: scaling relation. *Landscape Ecology*, 19:125-138.

Zonneveld, I.S. 1995. Land Ecology. SPB Academic Publishing, Amsterdam. 90 p.

Radiolarian Age Constraints of Mid-Cretaceous Black Shales in Northern Tunisia

Ben Fadhel Moez[1], Soua Mohamed[2], Zouaghi Taher[1], Layeb, Mohsen[3], Amri Ahlem[1] and Ben Youssef Mohamed[1]
[1]*CERTE, Technopole de Borj Cédria,*
[2]*Entreprise Tunisienne d'Activités Pétrolières, ETAP-CRDP 4 Rue des Entrepreneurs, 2035 la Charguia II,*
[3]*ISMP, Tunis,*
Tunisia

1. Introduction

Mid-Cretaceous pelagic deposits outcropping in Northern Tunisia include organic-rich beds locally associated with high abundance of radiolarian microfauna, which are interpreted as the signature of the two global oceanic anoxic events OAE1 and OAE2 (Talbi, 1991; Saïdi & Belayouni; 1994; Caron et al., 1999; Amédro et al., 2005, Heldt et al., 2008; Khazri et al., 2009; Soua et al., 2009; Robascynski et al., 2010; Ben Fadhel et al., 2011). Several studies have stated the close association between organic-rich sediments and radiolarian in the Atlantic and Tethyan realms (Marcucci-Passerini et al, 1991; O'Dogherty, 1994; Erbacher & Thurow, 1998; Danelian et al., 2004, 2007).

In North African margins, the radiolarian biostratigraphy have focused upon radiolarian-bearing Jedidi Formation which has been thoroughly discussed by Cordey et al, (2005) and Boughdiri et al, (2007). The first attempts at dating radiolarian series in Northern Tunisia show that radiolarian associated with carbonate-siliceous beds, have yielded useful diagnostic radiolarian assemblages (Cordey et al., 2005; Soua et al., 2006; Ben Fadhel et al., 2010).

Albian and Cenomanian-Turonian black shales of Northern Tunisia were considered to have good generative oil source rock (Layeb, 1990; Saidi & Belayouni, 1994; Bechtel et al., 1998; Ben Fadhel et al., 2011). In this overall context, the restudy and high-resolution biostratigraphy of Albian black shale beds of Lower Fahdene Formation and C/T cherty beds of organic-rich Bahloul Formation outcropping in Northern Tunisia domain have yielded well-preserved and age-diagnostic radiolarians species.

The aim of this paper is to: 1) give new illustrations of radiolarian taxa recovered from albian pelagic deposits of north african margins 2) establish a direct age of black shales using radiolarian assemblages 3) compare the radiolarian assemblages with time equivalent investigated in tethyan and east Pacific domains.

2. Geological setting

The area of investigation is located in Northern Tunisia (Fig. 1). Three sections are selected in this study on the basis of occurrence of organic and radiolarian-rich layers:

1. The Jebel Srassif area (Fig. 1a) is located in the northwestern extremity of the 'Dome Belt', a complex structure linked to Triassic extrusions and strike-slip faults. According to Chikhaoui et al., (1991) and Chikhaoui & Turki (1996), the observed structural complexity was the result of the extensive tectonic movement, which led to the extrusion of Triassic evaporites during the Albian–Aptian period. Consequently, halokinetic and tilted blocks movements are responsible for the horst and- graben architecture. The so-called 'tectonic corners', described by the previous authors, are induced by the reactivation of strike slip faults during the Tertiary compressive phase.
   The Jebel Srassif section belongs to the subsiding basin of the Mellegue 'paleograben' (Chikhaoui et al., 1991) bordered by two structural highs: Koumine to the west and Nebeur to the east.
   Cretaceous successions are characterized by a thick pile of Aptian to Campanian pelagic sequences, which are affected by multiple non-depositional unconformities and condensed layers (Chikhaoui, 1988)
2. The Fadeloun-Garci-Mdeker structure (fig.1b) in which belongs the Jebel Garci section, is composed of three anticlines, trending North South and considered as the northern prolongation of the N-S axis (Saadi, 1990). The anticlines are separated from the Atlasic domain by the Zaghouan thrust, which its North-eastern part becomes south-verging, commonly defined as the Chérichira-Kondar thrust (Khomsi et al., 2004).
   The Cretaceous sedimentation was under the control of syn-sedimentary faults trending N140-160 reflected by chaotic and gravitational deposits (Saadi, 1991). Early Cretaceous successions show northward, reduced thickness and affected by hiatus and extreme condensations in Hammam Zriba (Saadi, 1990). The motion of a corridor trending north-south by N140-160 faults has led to the compartmentalization of the seafloor in losangic basins (Saadi, 1991).
   During the Valanginian - Barremian time span, theses basins were supplied by siliciclastic deposits while condensed sedimentation occupied uplifted horsts (Biely et al., 1973; Saadi et al., 1994)
3. The Oued Kharroub section (Fig. 1c) is located in the Atlas domain (Northern and Central Tunisia), characterized by various facies of Cenomanian-Turonian transition (C-T) deposits, including benthic fauna-rich carbonates e.g. Zebbag Formation by Burollet (1956) and Gattar Formation by Boltenhagen and Mahjoub (1974) and organic-rich black shales with pelagic fauna (e.g. Bahloul Formation by Burollet, 1956) respectively of shelf and slope in the southern margin of the Tethyan realm. During this period and since the Jurassic, this domain has been influenced by the opening of the Tethyan palaeosea, its deepening as well as its southern margin migration. Generally speaking, the Bargou area, connected palaeogeographically to central Tunisia, is characterized by (1) emerged palaeohighs displaying gaps and discontinuities (Turki, 1985) and (2) subsiding zones affected by deep-water sedimentation. This area is dominated by N140° and N70° trend faults limiting several blocks. Cretaceous sedimentation varies on both sides. Its structural evolution may be summarized as follow : (1) during the late Jurassic to early Cretaceous, the area was subjected to a major extensional phase that delimited horst and graben systems (Martinez & Truillet, 1987) (2) In the uppermost Aptian, a regional compressional pulsation affecting the north-African platform had resulted from a transpressional scheme (Ben Ayed & Viguier, 1981) (3) New NNE-SSW trend anticline structures appeared attested by the Albian Fahdene Formation onlap features on the reefal aptian Serj deposits in subsurface (Messaoudi & Hammouda, 1994) or upper

Aptian - Albian unconformity in outcrops (Ouahchi et al., 1998). (4) During the Albian, the geodynamic evolution is marked by the sealing of lower Cretaceous structures during an extensional phase that persisted to form graben systems promoting organic-rich and siliceous strata deposition throughout upper Cenomanian to Lower Turonian times (Soua et al., 2009). The major faults in this area are represented by N140° and N70° trend features. The Bahloul thickness is significantly variable in this area. It may varies from 10m to 40m in thickness (Layeb & Belayouni, 1989; Soua & Tribovillard, 2007). Uniquely, in this area, the top of the Bahloul represents many cenomanian olistolith levels (Soua et al., 2006) marking syndepositional tectonic activities (Turki, 1985).

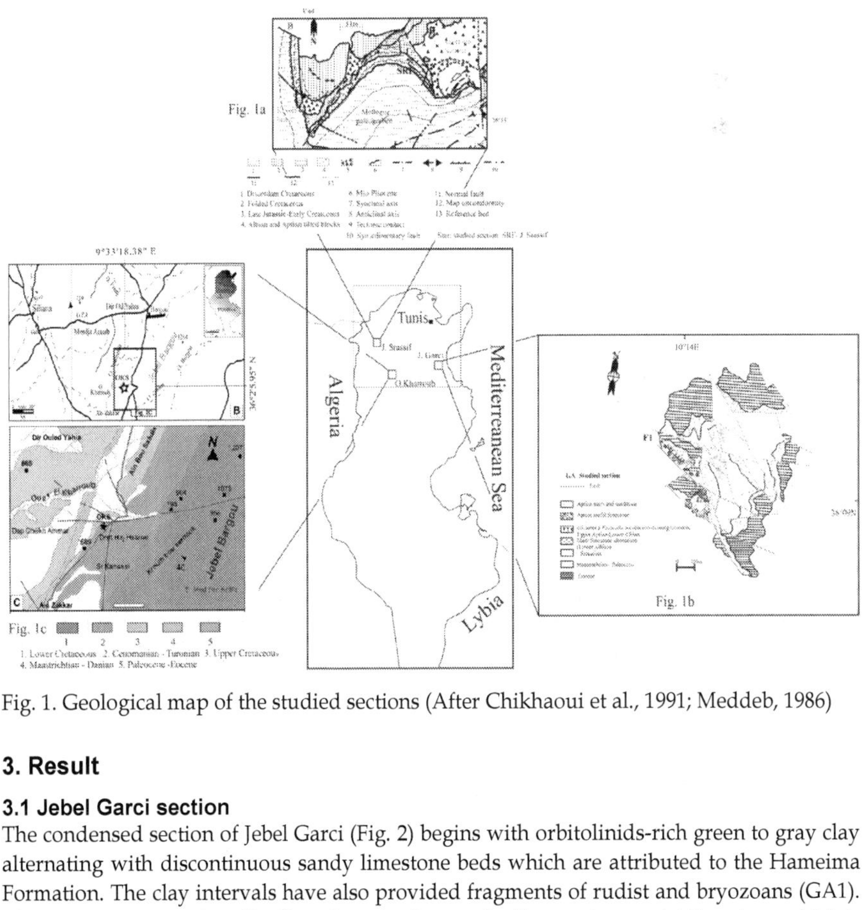

Fig. 1. Geological map of the studied sections (After Chikhaoui et al., 1991; Meddeb, 1986)

3. Result

3.1 Jebel Garci section

The condensed section of Jebel Garci (Fig. 2) begins with orbitolinids-rich green to gray clay alternating with discontinuous sandy limestone beds which are attributed to the Hameima Formation. The clay intervals have also provided fragments of rudist and bryozoans (GA1). The upper part contains olistolites deposits that gradually pass to a reefal limestone which is outlined at the top by burrowed hardground.

The next successions (GA9 - GA23) which correspond to the "Allam" Member consist of centimeter-thick grey to dark laminated limestone bed and organic-rich black marl intervals.

Upwards, the succession becomes rhythmic and the marly intervals increase in thickness in opposition to limestone beds. The microfauna content yields depauperate planktic foraminiferal assemblages and radiolarian rich microfauna.

Fractures related to a strike slip fault outlining the black hales unit are onlapped by a marly intervals and gray limestone beds alternation (GA24-GA27).

The organic-rich beds (GA17 - GA 23) have released a moderately to well-preserved and age-diagnostic radiolarian species. Twenty nine species were recorded in the studied section.

The radiolarians appear with few discrete taxa within GA6 level. It provides an assemblage composed of *Holocryptocanium barbui* Dumitrica, *Spongostichomitra elatica* (Aliev), *Pseudoeucyrtis hanni* (Tan), *Archeodictyomitra vulgaris* Pessagno, *Thanarla brouweri* (Tan), *Stichomitra simplex* (Smirnova and Aliev), *Angulobracchia portmanni* Baumgartner, *Thanarla pacifica* Nakaseko and Nishimura. They become diversified and abundant within GA7. It yields an association of *Dictyomitra* aff. *gracilis* (Squinabol), *Dictyomitra communis* Squinabol, *Dictyomitra montisserei* (Squinabol), *Pseudodictyomitra lodogaensis* Pessagno, *Thanarla praeveneta* Pessagno, *Archaeodictyomitra* aff. *A. vulgaris* Pessagno, *Hiscocapsa sp., Thanarla* aff.*pulchra* (Squinabol), *Spongostichomitra elatica* (Aliev), *Thanarla brouweri* (Tan), *Angulobracchia portmanni* Baumgartner, *Stichomitra simplex* (Smirnova and Aliev), *Stichomitra communis* Squinabol

GA15 sample provided very diversified and abundant radiolarian population. It is composed by *Dictyomitra montisserei* (Squinabol), *Holocryptocanium barbui* Dumitrica *Pseudodictyomitra lodogaensis* Pessagno, *Stichomitra simplex* (Smirnova and Aliev), *Pseudoeucyrtis hanni* (Tan), *Diacanhocapsa sp., Hiscocapsa grutterinki* (Tan) *Angulobracchia portmanni* Baumgartner, *Stichomitra communis* Squinabol, *Pseudodictyomitra paronai* (Aliev), *Cryptamphorella conara* (Foreman)

GA18 provided *Dictyomitra gracilis* (Squinabol), *Thanarla conica* (Squinabol)

The upper part of black shales (G17-23), composed by rhythmic bundles of limestone and marl beds, is characterized by a decrease of radiolarian abundance. The sample GA20 has released a radiolarian assemblages composed of *Thanarla brouweri* (Tan), *Spongostichomitra phalanga* O' Dogherty, *Pseudodictyomitra paronai* (Aliev), *Dictyomitra communis* Squinabol, *Holocryptocanium barbui* Dumitrica, *Pseudodictyomitra lodogaensis* Pessagno, *Dictyomitra gracilis* (Squinabol), *Dictyomitra montisserei* (Squinabol), *Spongostichomitra elatica* (Aliev), *Pseudodictyomitra paronai* (Aliev)

Although the uppermost beds have yielded (GA24-27) benthic foraminiferal-rich assemblages, we identified well-preserved radiolarian population (GA24) composed of *Pessagnobrachia rara* (Squinabol), *Stichomitra communis* Squinabol, *Dictyomitra montisserei* (Squinabol), *Cryptamphorella conara* (Foreman), *Holocryptocanium barbui* Dumitrica, *Pseudodictyomitra lodogaensis* Pessagno, *Torculum coronatum* (Squinabol), *Xitus spicularius* (Aliev), *Obeliscoites vinassai* (Sqinabol), *Hiscocapsa asseni* (Tan), *Thanarla pulchra* (Squinabol), *Dictyomitra communis* Squinabol, *Hiscocapsa grutterinki* (Tan)

Marly interval of the top GA27 have released an assemblage of *Holocryptocanium barbui* Dumitrica, *Stichomitra simplex* (Smirnova and Aliev), *Pseudodictyomitra paronai* (Aliev), *Dactyliosphaera maxima* (Pessagno)

3.2 Jebel Srassif section

The base of Jebel Srassif section (Fig. 3) which constitutes the "Marnes Moyennes" Member, consists of 130 meter-thick alternations of grey marl and limestone, which become dark and laminated at the top. A cyclic marl/limestone bundles (10m) can be distinguished having an

Legend

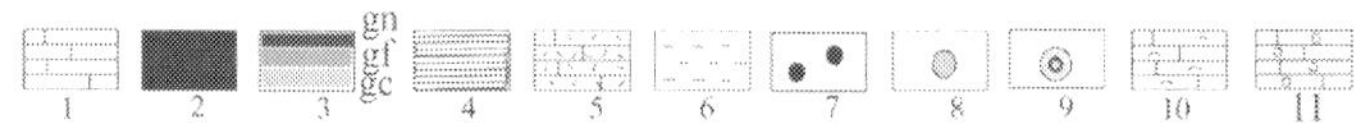

1. Limestone 2. Black shales 3. Marl (gc: light gray, gf:dark to gray, gn:gray to black) 4 Thin laminated limestone
5. Siliceous limestone 6. Shales 7. Iron oxide level 8. Olistolith layers 9. Ammonites horizon 10. Reefal limestone 11. Sandy limestone

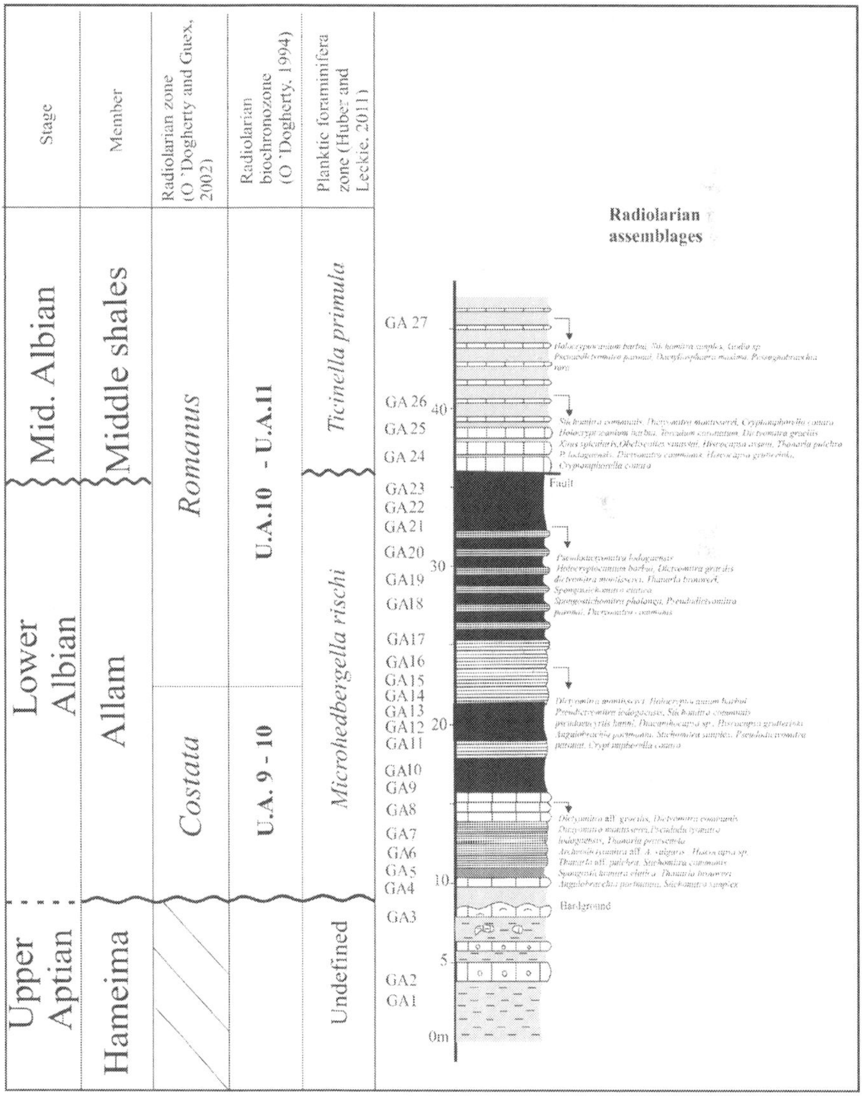

Fig. 2. Jebel Garci section

organic and radiolarian-rich mudstone texture. It is capped by a thick organic-rich limestone bed (20 m) characterized by bituminous odor and yellowish color in patina. This level corresponds to the Mouelha member (Burollet, 1956). The 40 meters of the top consist of an alternation of grey limestones and dark grey-ochre marls yielding septarian nodules characterizing the Defla member. They are overlain by a succession of lenticular limestone beds and grey marl of Azreg member (50 m).

Three samples were selected, based on the good preservation of the faunal assemblages. Among 35 radiolarian morphotypes, only 23 species were figured. Biostratigraphic analysis of the fossil record and planktic foraminifer's zones (Fig. 3, and Plates 1 and 2) correlation allow us to distinguish the following three radiolarian assemblages:

1. The sample 37 has provided a diversified radiolarian fauna with the co-occurrence of *Dictyomitra montisserei* (Squinabol), *Obeliscoites perspicuus* (Squinabol), *Tubilustrium transmontanum* O'Dogherty, *Dictyomitra gracilis* (Squinabol), *Holocryptocanium barbui* Dumitrica, *Stichomitra* aff. *navalis* O'Dogherty, *Cryptamphorella conara* (Foreman), *Torculum dengoi* (Schmidt-Effing), *Stichomitra communis* Squinabol, *Torculum coronatum* (Squinabol), *Distylocapsa micropora* (Squinabol), *Patellula verteroensis* (Pessagno), *Godia concava* (Li &Wu).
2. Radiolarian assemblage recovered from sample 62 is highly diversified at the top of Mouelha blackshales. Likewise, it records an acme of species belonging to Hagiastridae and Cavaspongiidae taxa. This interval shows the co-occurrence of *Dispongotripus acutispinus* Squinabol, *Dactyliosphaera maxima* (Pessagno), *Pessagnobrachia* sp., *Cavaspongia euganea* (Squinabol), *Cryptamphorella conara* (Foreman), *Pessagnobrachia rara* (Squinabol), *Dorypyle communis* (Squinabol), *Pseudodictyomitra paronai* (Aliev), *Pseudodictymitra* sp., *Torculum coronatum* (Squinabol), *Holocryptocanium tuberculatum* Dumitrica, *Distylocapsa micropora* (Squinabol), *Obeliscoites perspicuus* (Squinabol), *Dactyliosphaera acutispina* Squinabol, *Dictyomitra gracilis* (Squinabol), *Thanarla spoletoensis* O'Dogherty, *Dactyliosphaera lepta* (Foreman), *Patellula verteroensis* (Pessagno), *Savaryella novalensis* (Squinabol), *Savaryella quadra* (Foreman), *Pessagnobrachia fabianii* (Squinabol), *Stichomitra communis* Squinabol, *Holocryptocanium barbui* Dumitrica, *Xitus* aff. *spicularius* (Aliev), *Torculum coronatum* (Squinabol), *Crolanium* aff. *spineum* (Pessagno),
3. 3. Sample 68 is characterized by the abundance of cryptocephalic nassellaria (Holocryptocanium). Moreover, we notice the first occurrence and bloom of *Mallanites triquetrus*. This interval shows the co-occurrence of *Xitus mclaughlini* (Pessagno), *Hexapyramis pantanelli* Squinabol, *Mallanites triquetrus* (Squinabol), *Thanarla spoletoensis* O'Dogherty, *Dictyomitra montisserei* (Squinabol), *Godia concava* (Li & Wu), *Cryptamphorella conara* (Foreman), *Torculum coronatum* (Squinabol), *Cavaspongia euganea* (Squinabol), *Distylocapsa micropora* (Squinabol), *Dactyliosphaera maxima* (Pessagno), *Holocryptocanium barbui* Dumitrica, *Dactyliodiscus longispinus* (Squinabol), *Dispongotripus acutispinus* Squinabol.

3.3 Oued Kharroub section

The outcrop (Fig. 4) is composed mainly by dark clayey limestone and organic-rich black shales with abundant planktic foraminifera. These organic-rich deposits include siliceous beds with abundant radiolarians, an equivalent to "Livello Bonarelli" bed marker (Marcucci Passerini et al, 1991; Salvini and Marcucci Passerini, 1998; Premoli-Silva et al, 1999; Scopelliti et al, 2004; Musavu-Moussavou et al, 2007)

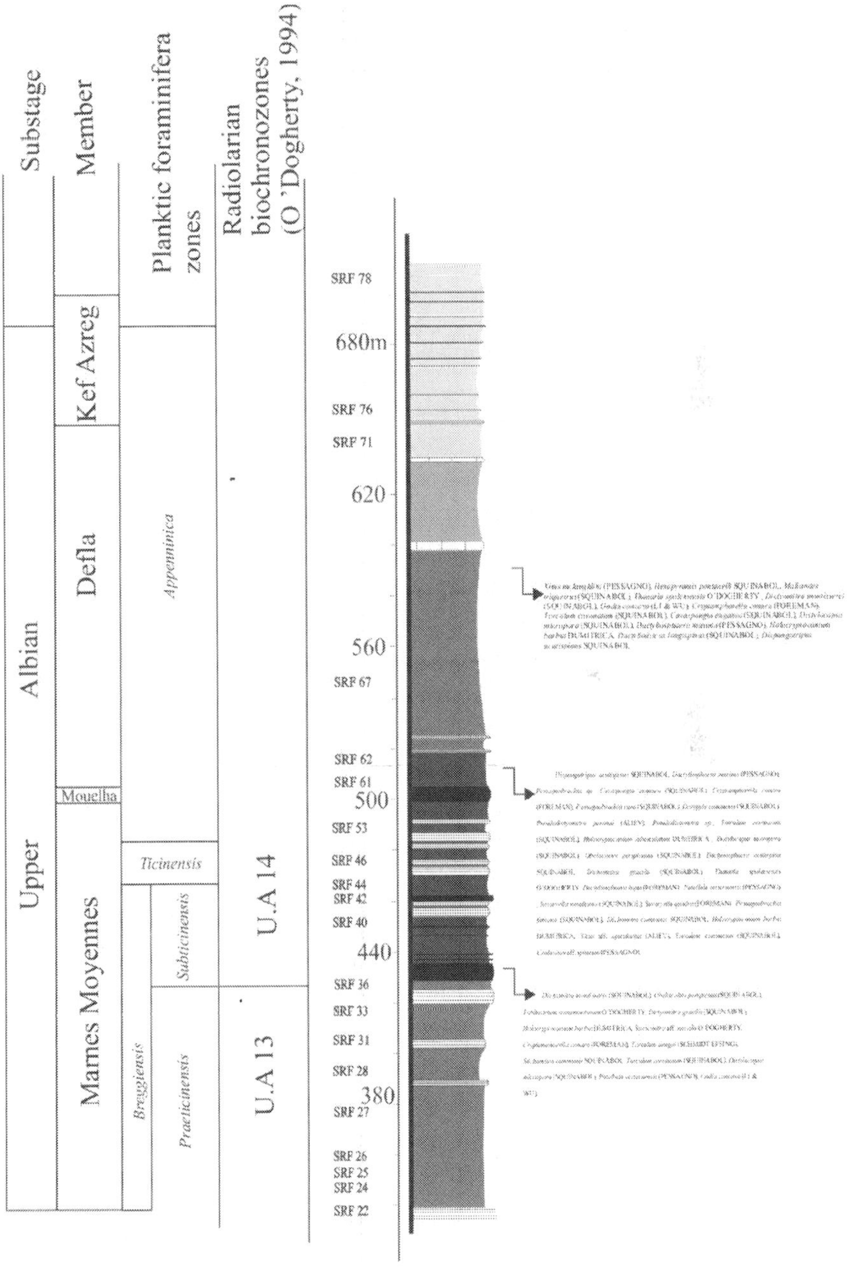

Fig. 3. Jebel Srassif section

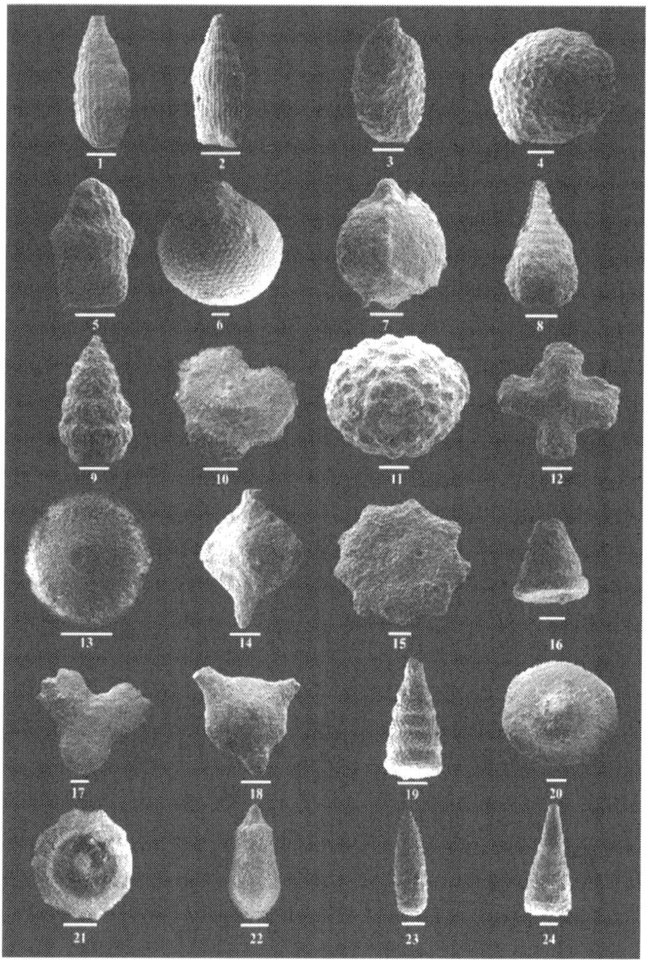

Plate 1. 1 - *Dictyomitra gracilis* (SQUINABOL), scale bar: 50m, sample 62. 2 - *Dictyomitra montisserei* (SQUINABOL), scale bar: 50m, sample 68. 3 - *Tubilustrium transmontanum* O'DOGHERTY, scale bar: 50m, sample 37. 4 - *Holocryptocanium barbui* DUMITRICA, scale bar: 50m, sample 37. 5 - *Stichomitra* aff. *navalis* O'DOGHERTY, scale bar: 50m, sample 37. 6 - *Cryptamphorella conara* (FOREMAN), scale bar: 50m, sample 37. 7 - *Mallanites triquetrus* (SQUINABOL), scale bar: 100m, sample 68. 8 - *Dictyomitra gracilis* (SQUINABOL), scale bar: 100m, sample 62. 9 - *Xitus* aff. *spicularius* (ALIEV), scale bar: 100m, sample 62. 10 - *Dispongotripus acutispinus* SQUINABOL, scale bar: 100m, sample 62. 11 - *Holocryptocanium tuberculatum* DUMITRICA, scale bar: 50m, sample 62. 12 - *Savaryella quadra* DUMITRICA, scale bar: 100m, sample 62. 13 - *Patellula verteroensis* (PESSAGNO), scale bar: 150m, sample 37. 14 - *Dactyliosphaera lepta* (FOREMAN), scale bar: 50m, sample 62. 15 - *Dactyliodiscus longispinus* (SQUINABOL), scale bar: 50m, sample 68. 16 - *Torculum dengoi* (SCHMIDT-EFFING) Scale bar:

50m, sample: 37. 17 - *Pessagnobrachia* sp., scale bar: 100m, sample 62. 18 - *Cavaspongia euganea* (SQUINABOL), scale bar: 100m, sample 62. 19 - *Stichomitra communis* SQUINABOL, scale bar: 50m, sample 62. 20 - *Dactyliosphaera maxima* (PESSAGNO), scale bar: 100m, sample 62. 21 - *Godia concava* (LI &WU), scale bar: 100m, sample 37. 22 - *Torculum coronatum* (SQUINABOL), scale bar: 150m, sample 62. 23 - *Crolanium* aff. *spineum* (PESSAGNO), scale bar: 100m, sample 62. 24 - *Obeliscoites perspicuus* (SQUINABOL), scale bar: 100m, sample 62.

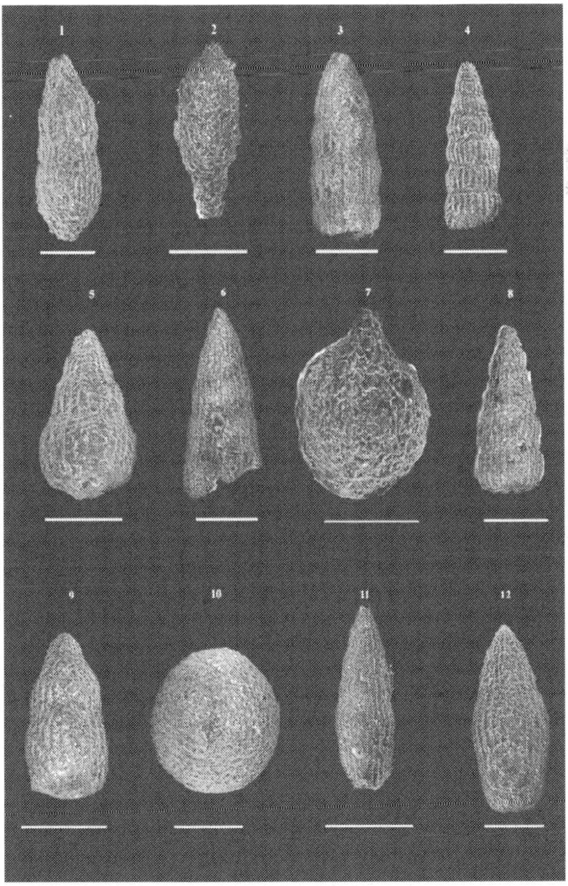

Plate 2. 1 - *Dictyomitra gracilis* (SQUINABOL), 100µm, GA18. 2 - *Pseudoeucyrtis hanni* (TAN), 100µm, GA15. 3 - *Dictyomitra montisserei* (SQUINABOL), 100µm, GA21. 4 - *Pseudodictyomitra lodogaensis* PESSAGNO, 100µm, GA21. 5 - *Thanarla praeveneta* PESSAGNO, 100µm, GA8. 6 - *Archeodictyomitra* aff. *vulgaris* PESSAGNO, 100µm, GA8. 7 - *Hiscocapsa sp.* 100µm, GA8. 8 - *Dictyomitra communis* (SQUINABOL), 100µm. GA8. 9 – *Thanarla* aff.*pulchra*,(SQUINABOL), 100µm, GA8. 10 - *Holocryptocanium barbui* DUMITRICA, 100µm, GA20. 11 - *Dictyomitra gracilis* (SQUINABOL), 200µm, GA9. 12 - *Thanarla brouweri* (TAN), 100µm, GA20.

A total of twenty five of radiolarian species are recognized belonging to nassellarians and spumellarians with maximum of eighteen (18) species in sample OKS 11. Their differential stratigraphical range and relative abundance allow to distinct two successive assemblages (R_I and R_{II}) through the C-T transition.

Although, the studied radiolarian species do not exhibit a good potential for biostratigraphic dating, the section is calibrated either by foraminifers and ammonites.

The R_{II} assemblage spans the upper part of the OAE-2 interval and the organic-poor deposits overlying this interval. It is characterized by a decrease trend of the nassellarian relative abundances (from 87% to 42%). Therefore, maybe dissolution or bad preservation conducted to the absence of this group close to the base of the upper half of the section, across the OKS40-OKS45 samples interval. Many species show rapid and gradual disappearing following a stepwise-like pattern (e.g. *Guttacapsa* sp., *Spongostichomitra elatica, Novixitus* sp., *Stichomitra stocki, Mita gracilis, Pseudodictyomitra pseudomacrocephala, Thanarla pacifica D. montisserei*).

About the associated spumellarians, several species from the R_I assemblage persisted more or less long time (e.g. *Archaeocenosphera* aff. *vitalis, Crucella messinae, Praeconocaryomma lipmanae, Rhopalosyringium hispidum Pyramispongia glascockensis* Pessagno., *Cavaspongia euganea* (Squinabol),, *C. Californiaensis* Campbell and Nishimura, *Pseudoeucyrtis spinosa* (Squinabol), *Archaeocenosphaera ? mellifera* O'Dogherty,). Nevertheless, very few species of nassellarians first occurred across the upper half part of the studied section. All these species are represented by dwarf and poorly preserved specimens.

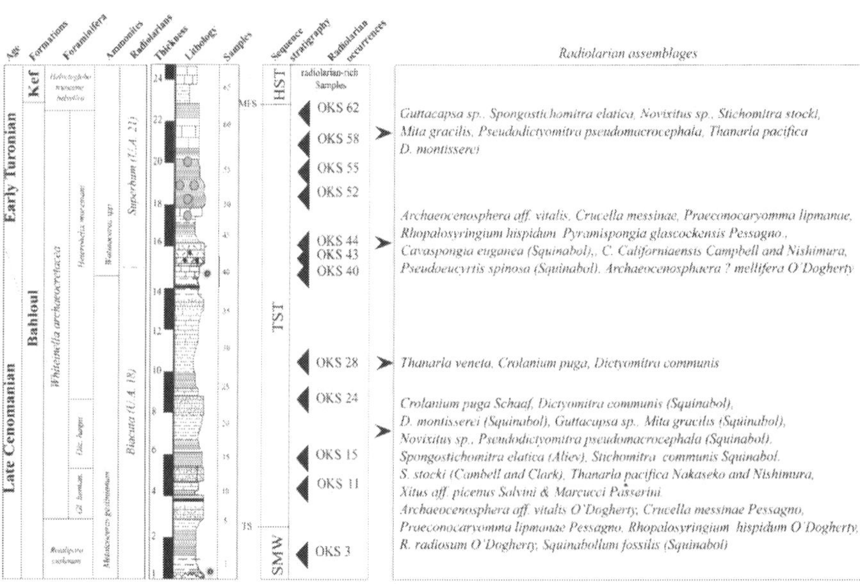

Fig. 4. Oued Kharroub section

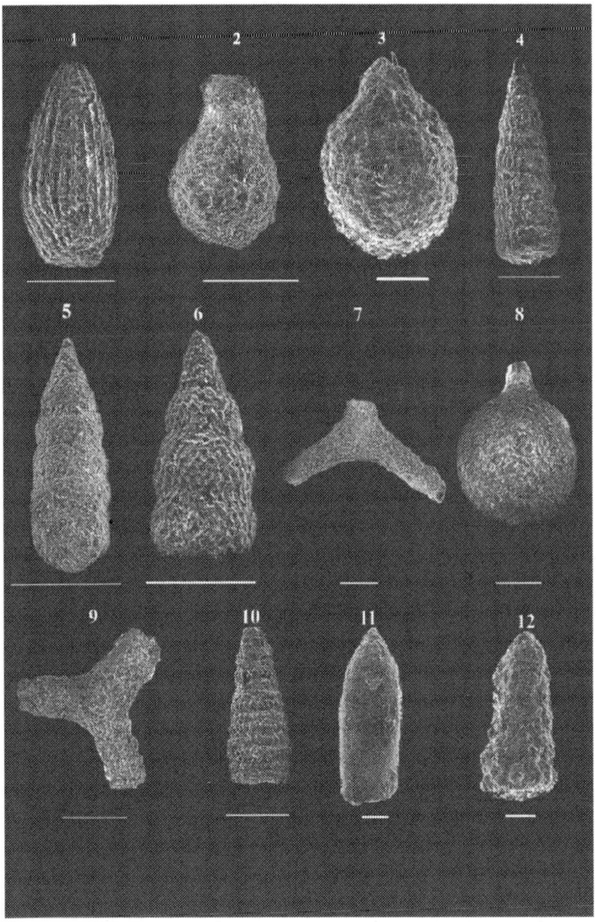

Plate 3. 1 - *Thanarla conica* (SQUINABOL), 100µm, GA18. 2 - *Diacanthocapsa sp.* 200µm, GA15. 3 - *Cryptamphorella conara**(FOREMAN), 50µm, GA22. 4 - 8 – *Pseudodictyomitra paronai* (ALIEV), 100µm, GA20. 5 – *Torculum coronatum* (SQUINABOL), 200µm, GA21. 6 - *Stichomitra simplex* (SMIRNOVA et ALIEV), 100µm, GA15. 7 - *Angulobracchia portmanni* BAUMGARTNER, 100µm, GA15. 8 - *Hiscocapsa* aff. *grutterinki* (TAN) 50µm, GA23. 9 – *Pessagnobrachia rara* (SQUINABOL), 100µm, GA27. 10 - *Xitus spicularius* (ALIEV), 100µm, GA25. 11 - *Spongostichomitra elatica* (ALIEV), 50µm, GA20. 12 - *Stichomitra communis* SQUINABOL, 50µm, GA8.

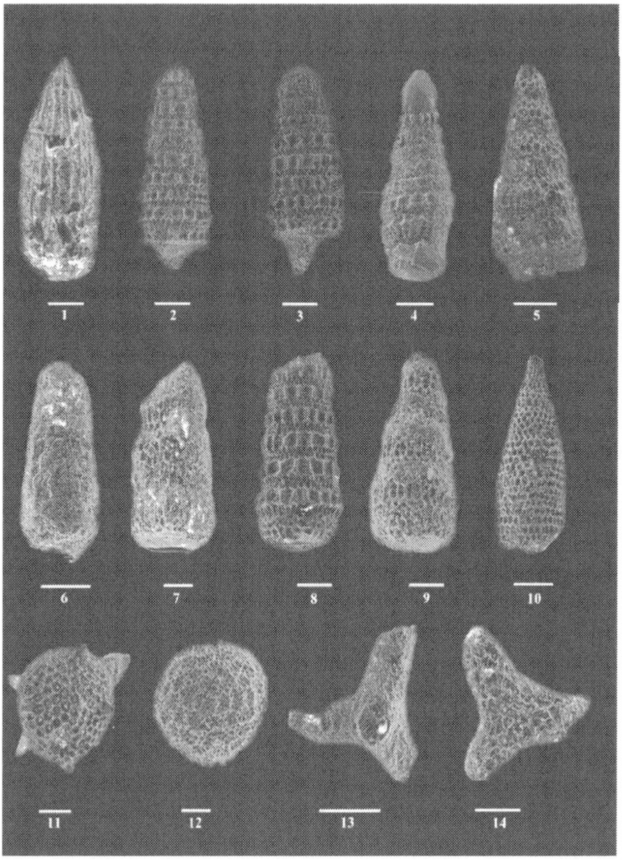

Plate 4. 1- *Mita gracilis* (?) (Squinabol), scale bar: 100, OKS-40. 2- *Novixitus* (?) sp., scale bar: 100, OKS-43;3-*Novixitus* sp. Scale bar: 100, OKS-42. 4- *Pseudodictyomitra pseudomacrocephala* (Squinabol), scale bar: 100, sample OKS-11. 5- *Guttacapsa* sp. Scale bar: 50, OKS-24. 6- indetermined species, scale bar: 100, OKS-24. 7- indetermined species, scale bar: 50, sample OKS-58. 8- *Pseudodictyomitra* sp. Scale bar: 100, OKS-15. 9- *Xitus* aff. *picenus* Salvini and Marcucci Passerini, scale bar: 50, OKS-40. 10- *Phalangites* (?) sp. Scale bar: 50, OKS-58; 11- *Squinabollum fossile* (Squinabol), scale bar: 50, OKS-28. 12- *Archaeocenosphaera? mellifera* O'Dogherty, scale bar: 50, OKS-62. 13- *Cavaspongia* sp., scale bar : 50, OKS-3. 14- *Cavaspongia californiaensis* Pessagno, scale bar: 50, OKS-58.

4. Discussion

Detailed analysis of radiolarian assemblages allows us to attribute a biostratigraphic framework for the organic-rich beds. In the following section, we used zonal scheme proposed by O'Dogherty (1994) for tethyan realms. The age-diagnostic assemblages are discussed and compared with time equivalent investigated in adjacent tethyan domains.

O'Dogherty (1994) proposed a radiolarian zonation for the Albian based on Unitary Associations. He described for the Upper Albian to the base of the Cenomanian the Spoletoensis zone divided into three radiolarian subzones: the Romanus, Missilis and Anisa subzones. Bak (1995) established a radiolarian zonation (H. barbui - H. geysersensis) for the Albian-Cenomanian of northern Tethyan domains, based on the co-occurrence of *Holocryptocanium barbui* DUMITRICA, *Holocryptocanium geysersensis* PESSAGNO, *Novixitus weyli* SCHMIDT-EFFING, *Squinabollum fossile* (SQUINABOL), *Crymptamphorella macropora* DUMITRICA, *Hemicryptocapsa tuberosa* DUMITRICA

The first appearance of *B. breggiensis*, recorded within Upper Albian basal intervals, coincides with first appearance of radiolarian species *Tubilustrium transmontanum* O'DOGHERTY, which is confined with the upper part of the Romanus subzone (O'Dogherty, 1994). An assemblage containing *Stichomitra navalis* and *Torculum coronatum* was recorded also within this subzone.

Babazadeh & de Wever (2004) described a radiolarian assemblage yielding the co-occurrence of *Dictyomitra gracilis, Holocryptocanium barbui* and *Dictyomitra montisserei* and assigned it to Middle-Late Albian age. Nevertheless, the presence of *T. dengoi*, whose first appearance coincides with the Missilis - Anisa subzones boundary (O'Dogherty, 1994), allows rejuvenating the assemblage age.

Samples recovered from the succession overlying the Mouelha Member blackshales show an assemblage composed of *Cryptamphorellla conara* Dumitrica, *Pessagnobrachia* sp., and *Thanarla spoletoensis* O'DOGHERTY, which correspond to the lower part of the *Appenninica* zone and the middle part of the Anisa subzone of O'Dogherty (1994). Although the coexistence of *D. lepta, Stichomitra communis* and *Patellula verteroensis* is assigned to early Late Cenomanian age (Erbacher, 1998), this assemblage possibly characterize the Late Albian taking into account the presence of *D. maxima* whose last occurrence is coeval with the base of Anisa subzone (O'Dogherty, 1994).

Samples recovered from basal beds (GA2-GA6) show high abundance of *Pseudodictyomitra lodogaensis* and contain some early Cretaceous taxa from *Turbocapsula* Zone such as *A.portmanni* and *Th. pacifica* (O' Dogherty, 1994; Erbacher and Thurow, 1998; Danelian et al., 2007; Michalik et al., 2008). Thus, a late Aptian age of these beds could not hitherto be ruled out.

According to Erbacher & Thurow (1998), the first occurrence of *Pseudodictyomitra lodogaensis* coincides with the upper part of *G. algerianus* Zone. Its last occurrence coincides with the Aptian-Albian boundary and the first occurrence of *Mita gracilis* (= *Dictyomitra gracilis*). This taxon is also reported from the Albian to Cenomanian deposits of the Atlantic domain, California and Pacific realms (Thurow, 1988; Karminia, 2006; Palechek et al., 2010).

It is possible that black shale unit of Jebel Garci could underlines the Aptian-Albian boundary. In fact, Danelian (2008) have reported the presence of *Thanarla praeveneta* from the Upper Aptian - Lower Albian bed which occurs in GA7 beds underlying the black shale successions.

On the other side, Slazcka et al., (2009) described an assemblage containing *Angulobracchia portmanni* Baumgartner, *Dictyomitra communis* (Squinabol), *Hiscocapsa asseni* (Tan), *Pseudodictyomitra lodogaensis* Pessagno, *Pseudoeucyrtis hanni* (Tan), almost similar to GA7 taxa. These authors attributed the assemblage to *Costata* zone that is confined to UA6-9 biochronozones of mid to late Aptian age (O'Dogherty, 1994).

It is noteworthy to point the coexistence of Albian species in all samples such as *D.montisserei* and *D.gracilis* with Aptian taxa particularly in GA7, GA 15 and GA26.

In that score, an assemblage recovered from Mid Cretaceous outcrops of Northern Tethys margins was described by Danelian et al., (2007), shows the co-occurrence of *P. lodogaensis, Dictyomitra gracilis, Thanarla brouweri, Archaeodictyomitra* aff.*vulgaris* assigning it to the early Albian UA10-11 biochronozone. Danelian et al (2004) consider that an early Albian age of Dercourt Member cannot be ruled out despite the presence of *Angulobracchia portmanni* and *pseudoeucyrtis hanni* characteristic of U.A.9. These species are observed hitherto within assemblage from GA15, associated with *Dictyomitra montisserei.*

Kurilov & Vishnevskaya (2011) described an assemblage extracted from Early Cretaceous outcrops of Pacific domain that does not differ from GA21. It contains *Thanarla brouweri, Pseudodictyomitra paronai, Pseudodictyomitra lodogaensis, Holocryptocanium barbui, Dictyomitra cf. montisserei, Dictyomitra communis, and Dictyomitra gracilis* indicating an early Albian age.

The sample GA26 has provided an assemblage characterized by high abundance of *Hiscocapsa asseni,* co-occurring with *D.gracilis* and *D.montisserei.* It lies with the UA10 biochronozone of Romanus zone (O'Dogherty, 1994; Danelian et al., 2004).

We suggest that lower part of black shale intervals could be assigned to the upper part of *Costata* zone (GA5 - GA14) based on the presence of Aptian taxa (i,e. *Angulobracchia portmanni, Pseudoeucyrtis hanni*). The lower part of this zone coincide with the first occurrence of *Microhedbergella praeplanispira* planktic foraminifera. Whereas the top coincide with the last occurrence of *Angulobracchia portmanni* and *Pseudoeucyrtis hanni* associated with a relative increase in abudance of Archaedictyomitrae and Williriedellidae families.

The Romanus zone (GA14 - GA27) show the dominance of high diversified nassellarian species. The assemblage recovered from GA17 is composed of *Thanarla brouweri, Archaeodictyomitra montisserei, Thanarla conica* which is attributed to the middle Albian *Mallanites* romanus subzone (U.A. 10 -11 biochronozone) (O'Dogherty, 1994; Danelian et al, 2004). However, the first occurrence of *Ticinella primula* planktic foraminifera is recorded 24 m above GA17 bed. Thus, we suggest that lower part of Romanus zone may be attributed to the Early Albian.

Studies on Cenomanian - Turonian boundary interval show that deposition of radiolarian, organic-rich sediment and large positive carbon isotopic excursion are coeval with extreme fertility conditions and correspond to a large-scale proxy that indicate a hypersiliceous period (Premoli Silva et al, 1999; Racki & Cordey, 2000)

The Bonarelli equivalent in Tunisia is commonly known by the Bahloul Formation (Burollet, 1956). In the Bargou area, the Bahloul Formation shows organic-rich intervals interbedding cherty and radiolarian limestone layers (Layeb and Belayouni 1999, Soua and Tribovillard, 2007)

Although the C/T boundary interval outcropping in the Tunisian realm was extensively studied by planktic foraminifera and ammonite biostratigraphy (Maamouri et al, 1994; Nederbragt and Fiorentino, 1999; Abdallah et al., 2000; Amédro et al, 2005), radiolaria assemblages have provided a useful tool for age calibration and subdivision of C/T organic-rich beds in this study.

Two black shale levels were identified in Oued Kharroub section:

1. The first lies with the lower part of *Withinella archaeocretacea* planktic foraminifera zone, above the highest occurrence of *Rotalipora cushmani*
2. The second coincides with the middle part of *Heterohelix moremani* zone

The calibration of these levels is based on age-diagnostic radiolarian recovered from biosiliceous limestone beds (Fig. 4).

The OSK 24 yields an assemblage composed of *Rhopalosyringium radiosum* O'Dogherty, *Praeconocaryomma lipmanae* Pessagno, *Acaeniotyle vitalis* O'Dogherty *Rhopalosyringium hispidum* O'Dogherty. The three first taxa have been described by Bak et al (2005) and attributed them to the late Cenomanian - early Turonian. Erbacher (1998) attribute *Rhopalosyringium radiosum* to the early Turonian, but later Musavu-Moussavou and Danelian (2006) expand its range to late Cenomanian. The assemblage contains *Xitus picenus* Salvini and Marcucci - Passerini which its range do not exceed the Silviae Zone of Bonarelli (O'Dogherty, 1994; Salvini & Marcucci-Passerini, 1998). Consequently, we assign the lower black shale beds (OKS 11) to the late Cenomanian and to upper part of *Silviae* Zone [U.A 18 biochronozone of O'Dogherty (1994)]

Many authors have stated the occurrence of *Archaeocenosphaera mellifera* O'Dogherty within Turonian strata of Boreal and northern Tethyan domains (Bandini et al., 2006; Smreckova, 2011). In East Pacific domain, this taxon, associated with *C. californaensis* and *Pyramispongia glascockensis* PESSAGNO, is recorded within the Silviae Zone of late Cenomanian age (Bragina, 2009). Salvini & Marcucci-Passerini (1998) stated that *C. californiaensis* occurs only in the base of upper assemblage C of Bonarelli Level which lies with the base Superbum Zone defined by O'Dogherty (1994). In the Atlantic domain, the last occurrence of *C. californiaensis* is recorded in the late Cenomanian just beneath the organic-rich beds related to the OAE2 (Musavu-Moussavou and Danelian, 2006). Taking into account the paleogeographic similarities between northern and southern Mediterranean Tethys margins, the radiolarian assemblage recovered from OKS44 level could be correlated with upper assemblage (Superbum Zone) of Bonarelli level in Central Italy. Thus, the second black shale lie with the upper part of *Biacuta* subzone of late Cenomanian age, if we take into consideration the position of turonian *Watinoceras* spp. ammonite (Amédro et al., 2005)

5. Conclusion

Biostratigraphic investigations of Albian and C/T boundary intervals in Northern Tunisia show that organic-rich beds are generally associated with high abundance of radiolarian fauna.

Age constraint of organic-rich sediments is established and correlated with biochronozones of O'Dogherty (1994). In the light of these results, we deduce that:

1. Black shale interval of Jebel Garci which is embedded within the "Allam" Member is assigned to the early Albian U.A.10 biochronozone. However, the latest Aptian could not be excluded for the lower part.
2. Late Albian organic-rich beds of Jebel Srassif including cyclic limestone/marl beds of "Marnes Moyennes" and Mouelha Members lie with the boundary interval between U.A. 13 and U.A. 14 biochronozones.
3. Two black shale levels embedded within Bahloul Formation are probably of late Cenomanian age and confined with the U.A.18 biochronozone. The first occurrence of turonian *Watinoceras* spp. ammonite is recorded 70 cm above the second black shale bed (OSK40)

It seems that distribution of radiolarian assemblages of albian and cenomanian-turonian boundary intervals shows some difference from those of Atlantic and east Pacific domains. Preservation index and range discrepancies of some radiolarian species could affect the subdivision resolution. Further studies on radiolarian distribution assemblages and relationships with environmental changes during Mid-Cretaceous time are needed to establish paleogeographic reconstructions of southern tethyan margins.

6. Acknowledgments

We thank Drs Moncef Saidi and Hedia Bessaies from ETAP center research for giving all facilities needed for SEM photographs. Authors gratefully acknowledge Dr. Imran Ahmad Dar, Editor-in-chief, for accepting the publication of this work.

7. References

Abdallah, H., Sassi, S., Meister, C., & Souissi, R. (2000). Stratigraphie séquentielle et paléogéographie à la limite Cénomanien-Turonien dans la région de Gafsa-Chotts (Tunisie centrale). *Cretac. Res.*, vol.21, No. 1, pp. 35-106, ISSN 01956671

Amédro F., Accarie H., & Robaszynski F. (2005). Position de la limite Cénomanien-Turonien dans la Formation Bahloul de Tunisie centrale : apports intégrés des ammonites et des isotopes du carbone (δ13 C). *Eclogae Geologicae Helvetiae,* vol. 98, No. 2, pp. 151-167, ISSN 0012-9402

Babazadeh, S. A. & De Wever, P. (2004a). Radiolarian Cretaceous age of Soulabest radiolarites in ophiolite suite of eastern Iran. *Bulletin de la Société Géologique de France,* Vol. 175, No.2, pp. 121-129, ISSN 0037-9409

Bak, M. (1995). Mid Cretaceous radiolarians from the Pieniny Klippen Belt, Carpathians. Poland. *Cretaceous Research,* vol. 16, No. 1, pp. 1-23, ISSN 01956671

Bak, M., Bak, K. & Ciurej, A. (2005). Mid-Cretaceous spicule-rich turbidites in the Silesian Nappe of the Polish Outer Carpathians: radiolarian and foraminiferal biostratigraphy. *Geological Quarterly,* vol, 49, No. 3, pp. 275-290, ISSN 1641-7291

Bandini, A.N, Baumgartner, P.O. & Caron, M. (2006). Turonian Radiolarians from Karnezeika, Argolis Peninsula, Peloponnesus (Greece). *Eclogae geol. Helv.* 99 Supplement, Vol.2, pp.1-20, ISSN 0012-9402

Biely, A., Memmi, L., Salaj, J. & (1973). Le Crétacé inferieur de la région d'Enfidaville. Découverte d'Aptien condense, *Livr. Jub. M. Solignac, Ann. Min. Geol.,* vol. 26, pp.169-1 78, ISSN: 03300013

Bechtel, A., Savin, S.M. & Hoernes, S. (1999). Oxygen and hydrogen isotopic composition of clay minerals of the Bahloul Formation in the region of the Bou Grine zinc-lead ore deposit (Tunisia): evidence for fluid-rock interaction in the vicinity of salt dome cap rock. *Chemical Geology,* Vol. 156, Issues 1-4, pp. 191-207, ISSN 0009-2541

Ben Ayed, N. & Viguier, C. (1981). Interprétation structurale de la Tunisie atlasique. *CRAS* Paris, t 292 série II pp. 1445-1448, ISSN 16310713

Ben Fadhel, M., Layeb, M. & Ben Youssef, M. (2010). Upper Albian planktic foraminifera and radiolarian biostratigraphy (Nebeur - northern Tunisia). *Comptes Rendus Palevol,* vol. 9, No. 3, pp. 73 - 81, ISSN 16310713

Ben Fadhel, M., Layeb, M., Hedfi, A. & Ben Youssef, M. (2011). Albian Oceanic Anoxic Events in northern Tunisia: Biostratigraphic and geochemical insights. *Cretaceous Research* vol. 32, No. 6, pp. 685-699, ISSN 01956671

Boltenhagen, C. & Mahjoub, M. N. (1974-79). Divers rapports inédits sur la géologie du Cretacé moyen de Tunisie centrale. Archives SEREPT.

Boughdiri, M., Cordey, M., Sallouhi, H., Maalaoui, K., Masrouhi, M. & Soussi, M. (2007). Jurassic radiolarian-bearing series of Tunisia: biostratigraphy and significance to western Tethys correlations. *Swiss Journal of Geoscience,* vol. 100, No.3, pp. 431-441, ISSN 16618726.

Bragina, L.G. (2009). Radiolarians and Stratigraphy of Cenomanian-Coniacian Deposits in the Crimean and West Sakhalin Mountains, Pt. 1: Biostratigraphic Subdivision and Correlation. *Stratigraphy and Geological Correlation,* vol. 17, No. 3, pp. 316–330, ISSN: 08695938

Burollet, P.F. (1956). Contribution a l'étude stratigraphique de la Tunisie centrale. *Ann. Mines Geol.,* Tunis, n° 18, 350p. IVpl, ISSN 03300013

Caron, M., Robaszynski, F., Amédro, F., Baudin, F., Deconinck, J.F., Hochuli, P., Salis-perch nielsen, K. & Tribovillard, N. (1999). Estimation de la durée de l'événement anoxique global au passage Cénomanien/Turonien. Approche cyclostratigraphique dans la formation Bahloul en Tunisie centrale. Bull. Soc. géol. France, vol. 170, No.2, pp. 145–160, ISSN 0037-9409

Chikhaoui, M. & Turki, M. (1996). Rôle et importance de la fracturation méridienne dans les déformations crétacée de la zone des diapirs (Tunisie septentrionale). *J. Afr. Earth Sci.* vol. 21, No. 2, pp. 271–280, ISSN 1464343X

Chikhaoui, M., Turki, M. & Delteil, M. (1991). Témoignage de la structurogenèse de la marge téthysienne en Tunisie, en Jurassique terminal - Crétacé (région d'El Kef Tunisie septentrional). *Geol. Mediterraneenne,* vol.XVIII, No. 3, pp. 125–133, ISSN 0397-2844

Cordey, F., Boughdiri, M. & Sallouhi, H. (2005). First direct age determination from the Jurassic radiolarian-bearing siliceous series (Jedidi Formation) of north-western Tunisia. *Comptes Rendus Geoscience,* vol. 337, No. 8, pp. 777–785, ISSN 16310713

Danelian, T., Tsikos, H., Gardin, S., Baudin, F., Bellier, J.P.& Emmanuel, L. (2004). Global and regional palaeoceanographic changes as recorded in the mid-Cretaceous (Aptian-Albian) sequence of the Ionian zone (northwestern Greece). *Journal of the Geological Society,* vol. 161, No. 4; pp. 703-709, ISSN 00167649

Danelian, T., Baudin, F., Gardin, S., Masure, E., Ricordel, C., Fili, I., Mecaj, T. & Muska, K. (2007). The record of mid Cretaceous oceanic anoxic events from the Ionian zone of southern Albania. *Revue de micropaléontologie,* vol. 50, No. 3, pp. 225–237, ISSN 00351598

Danelian, T. (2008). Diversity and biotic changes of Archaeodictyomitrid Radiolaria from the Aptian/Albian transition (OAE1b) of southern Albania. *Micropaleontology,* 54, No. 1, pp. 3-12, ISSN 00262803

Erbacher, J. (1998). Mid-Cretaceous radiolarians from the Eastern equatorial Atlantic and their paleoceanography. In: *Proceedings of the Ocean Program, Scientific Results* Mascle, J., Lohmann, G.P., Moullade, M. (Ed.), vol. 159, pp.363–373, ISSN 1096-7451 College Station, TX (Ocean Drilling Program)

Erbacher, J. & Thurow, J. (1998). Mid-Cretaceous radiolarian zonation for the North Atlantic: an example of oceanographically controlled evolutionary processes in the marine biosphere? *In: Geological Evolution of Ocean Basins: Results from Ocean Drilling Program,* Cramp, A., Macleod, C.J., Lee, S.V., Jones, E.J.W. (Ed.),. Geological Society, London Special Publications 131, 71–82, ISBN 0585235759 9780585235752

Heldt, M., Bachmann, M. & Lehmann, J. (2008). Microfacies, biostratigraphy, and geochemistry of the hemipelagic Barremian - Aptian: Influence of the OAE 1a on the southern Tethys margin. *Palaeogeography, Palaeoclimatology, Palaeoecology,* vol. 261, No. 3-4, pp. 246–260, ISSN 00310182

Kariminia, S.M. (2006). Upper Jurassic and Lower Cretaceous radiolaria biostratigraphy of California coast ranges. Ph. D. Thesis, University of Texas, Dallas, 143 pp.

Kurilov, D. V. & Vishnevskaya, V. S. (2011). Early cretaceous radiolarian assemblages from the East Sakhalin Mountains. *Stratigraphy and Geological Correlation,* vol. 19, No. 1, pp. 44-62, ISSN: 08695938.

Layeb M. & Belayouni, H. (1989). La formation Bahloul au Centre et au Nord de la Tunisie un exemple de bonne Roche mère de pétrole à fort intérêt pétrolier. Mémoires de l'ETAP, n°3, *Actes des II ème journées de géologie Tunisienne appliquée à la recherche des hydrocarbures* , pp, 489-503, OCLC 32815633, Tunis, Tunisie, Nov - 1989

Layeb, M., 1990. Etude geologique, geochemique et mineralogique, regionale, des facies riches en matiere organique de la Formation Bahloul d'age Cenomano-Turonien dans le domaine de la Tunisie Nord-Centrale. PhD Thesis, Univ. de Tunis II, Tunis, 209 pp.

Layeb, M., and Belayouni, H., 1999. Paléogéographie de la Formation Bahloul (passage Cénomanien - Turonien), *Annales des mines et de la géologie,* vol. 40, pp. 21–44, ISSN 03300013

Maamouri, A.L., Zaghbib-Turki, D., Matmati, M.F., Chikhaoui, M. & Salaj, J. (1994). La Formation Bahloul en Tunisie centro-septrentionale: variations latérales, nouvelle datation et nouvelle interprétation en termes de stratigraphie séquentielle. *J. Afr. Earth Sci.*, vol. 18,No.1, pp. 37–50, ISSN 1464343X

Marcucci Passerini, M., Bettini, P., Dainelli, J. & Sirugo, A. (1991). The "Bonarelli Horizon" in the central Apennines (Italy): radiolarian biostratigraphy. *Cretaceous Research,* vol. 12, No. 3, pp. 321–331, ISSN 01956671

Martinez C. & Truillet R. (1987). Évolution structurale et paléogéographie de la Tunisie. *Memoria de la Societa Italiana de Geologia*, vol. 38, No. 35-45, ISSN 2038-1727

Messaoudi F. & Hammouda F. (1994). Evènement structuraux et types de pièges dans l'offshore Nord-Est de la Tunisie. *Proceedings of the 4th tunisian petroleum exploration conference,* pp. 55-64, OCLC 502635466, Tunis, Tunisia, may 1994).

Michalík, J.,Soták, J., Lintnerová, O., Halásová, E., Bak, M., Skupien, P. & Boorová, D. (2008). The stratigraphic and paleoenvironmental setting of Aptian OAE black shale deposits in the Pieniny Klippen Belt, Slovak Western Carpathians. *Cretaceous research* vol. 29, No. 5-6, pp. 871-892, ISSN 01956671

Musavu-Moussavou, B. & Danelian, T. (2006). The radiolarian biotic response to Oceanic Anoxic Event 2 in the southern part of the Northern proto-Atlantic (Demerara Rise, ODP Leg 207). *Revue de Micropaléontologie,* vol. 49, No. 3, pp. 141-163, ISSN 00351598

Musavu-Moussavou, B., Danelian, T., Baudin, F., Coccioni, R. & Frohlich, F. (2007). The radiolarian biotic response during OAE2. A high-resolution study across the Bonarelli level at Bottaccione (Gubbio, Italy). *Revue de Micropaléontologie*, vol. 50, No. 3, pp. 253–287, ISSN 00351598.

Nederbragt, A. J. & Fiorentino, A. (1999). Stratigraphy and paleoceanography of the Cenomanian-Turonian Boundary Event in Oued Mellegue, north-western Tunisia. *Cretaceous Research,* vol.20, No. 1, pp. 47–62, ISSN 01956671

O'Dogherty, L. (1994). Biochronology and paleoecology of Mid-Cretaceous radiolarians from Northern Apennines (Italy) and Betic Cordillera (Spain). *Mémoires de Géologie* (Lausanne), 21, 415pp, ISSN 1015-3578

Ouahchi, A., M'Rabet, A., Lazreg, J. Mesaoudi, F. & Ouazaa, S. 1998. Early structuring, paleoemersion and porosity development: a key for exploration of the aptian serdj carbonate reservoir in Tunisia. *Proceedings of the 6th Tunisian petroleum exploration and production conference*, pp.267-284, Tunis, Tunisia , May 5th - 9th, 1998, OCLC 704163368

Palechek, T. N., Savel'ev, D. P. & Savel'eva, O. L. (2010). Albian–Cenomanian Radiolarian Assemblage from the Smaginsk Formation, the Kamchatskii Mys Peninsula of Eastern Kamchatka. *Stratigraphy and Geological Correlation,* vol.18, No. 1, pp. 63–82, ISSN: 08695938

Premoli Silva, I., Erba, E., Salvini, G., Locatelli, C. & Verga, D. (1999). Biotic changes in Cretaceous oceanic anoxic events of the Tethys. *Journal of Foraminiferal Research,* vol.29, No.4, pp. 352–370, ISSN: 00961191.

Racki, G. & Cordey, F. (2000). Radiolarian palaeoecology and radiolarites: is the present the key to the past? *Earth-Science Reviews,* vol.52, No.1-3, pp. 83–120, ISSN 00128252

Robaszynski, F., Zagrarni, M.F., Caron, M. & Amédro, F. (2010). The global bio-events at the Cenomanian-Turonian transition in the reduced Bahloul Formation of Bou Ghanem (central Tunisia). *Cretaceous Research,* vol.31, No.1, pp. 1-15, ISSN 01956671

Saadi, J., (1990). Exemple de sédimentation syntectonique au Crétacé inférieur le long d'une zone de décrochement NS. Les structures d'Enfidha (Tunisie nord-orientale). *Géodynamique,* vol.5, No. 1, pp. 17-33, ISSN 0766-5105

Saadi, J., (1991). Sedimentation en zone mobile coulissante, l'exemple du Cretace inferieur des structures submeridiennes de la region d'Enfidha. (Prolongement septentrional de l'Axe N.S-Tunisie nord-orientale). PhD Thesis, Université de Pau,244 pp.

Saïdi, M. & Belayouni, H. (1994). Etude géologique et géochimique des roche mères albovraconienne dans le domaine de la Tunisie centro-septentrional. *Proceeding of the 4th Petroleum Exploration Conference*, No. 7, pp. 91-116, OCLC 502635466, Tunis, Tunisia, may 1994

Salvini, G. & Marcucci Pesserini, M. (1998). The radiolarian assemblages of the Bonarelli Horizon in the Umbria-Marches Apennines and Southern Alps, Italy. *Cretaceous Research,* vol. 19, No. 6, pp. 777–804, ISSN01956671

Scopelliti, G., Bellanca, A., Coccioni, R., Luciani, V., Neri, R., Baudin, F., Chiari, M. & Marcucci, M. (2004). High-resolution geochemical and biotic records of the Tethyan "Bonarelli Level" (OAE2, Latest Cenomanian) from the Calabianca-Guidaloca composite section, northwestern Sicily, Italy. *Palaeogeography, Palaeoclimatology, Palaeoecology,* vol. 208, No. 3-4, pp.293–317, ISSN 00310182

Ślączka, A., Gasiński, M. A., Bąk, M. & Wessely, G. (2009). The clasts of Cretaceous marls in the conglomerates of the Konradsheim Formation (Pöchlau quarry, Gresten Klippen Zone, Austria). *Geologica Carpathica,* vol. 60, No. 2, pp. 151-164, ISSN 13350552

Smrečková, M. (2011). Lower turonian radiolarians from the locality Červená Skala section. Mineralia Slovaca 43, 31-38, ISSN 0369-2086

Soua, M., Zaghbib-Turki, D. & O'Dogherty, L. (2006). Les réponses biotiques des radiolaires à l'événement anoxique du cénomanien supérieur dans la marge sud téthysienne (tunisie). *Proceedings of the 10th tunisian petroleum exploration & production conference.* pp. 195-216.

Soua M. & Tribovillard, N. (2007). Modèle de sédimentation au passage Cénomanien /Turonien pour la formation Bahloul en Tunisie. *Comptes Rendus Geoscience,* vol. 339, No. 10, pp. 692-701, ISSN 1631-0713

Soua, M., Echihi, O. Herkat, M., Zaghbib-Turki, D., Smaoui, J., Fakhfakh-Ben Jemia, H. & Belghaji, H. (2009). Structural context of the paleogeography of the Cenomanian - Turonian anoxic event in the eastern Atlas basins of the Maghreb. *C. R. Geoscience,* 341, No. 12, pp. 1029-1037, ISSN 1631-0713

Talbi, R. (1991). Etude géologique et géochimique des faciès riches en matière organique d'âge Albien du bassin de Bir M'Cherga (NE de Tunisie) : déterminisme de leur genèse et intérêt pétrolier de la région. PhD Thesis, University de Tunis, 223 pp.

Thurow, J. (1988). Cretaceous Radiolarians of the North Atlantic Ocean: ODP Leg 103 (Sites 638, 640 and 641) and DSDP Legs 93 (Site 603) and 47B (Site 398). In: *Proceedings of the Ocean Drilling Program, Scientific Results*, Boillot, G., Wintere, E.L., et al. (Ed.). Proceedings of the Ocean Drilling Program, Scientific Results, 103, 379 -418, ISSN 0884-5891.

Turki, M.M. (1985). Polycinematique et contrôle sédimentaire associé sur la cicatrice Zaghouan-Nebhana. Ph.D. thesis, Univ. Tunis, 252 pp.

Miogypsinid Foraminiferal Biostratigraphy from the Oligocene to Miocene Sedimentary Rocks in the Tethys Region

Kuniteru Matsumaru
Innovation Research Organization, Saitama University, Saitama Japan

1. Introduction

Tan Sin Hok (1936, 1937) has done the anatomical and morphometrical analysis of the family Miogypsinidae at the first time. This is regarded as the important contribution for the Micropaleontology on foraminiferal studies. His studies have developed from the phylogenetic history of the genus *Cycloclypeus* and their relative species of the family Nummulitidae de Blainville, 1827 (Tan Sin Hok, 1932). The basic materials of these two families have been gathered based on detailed geological fieldworks of many geologists for a long time from the East Indies (Indonesia and its surroundings) as eastern Tethys region or Indo - Pacific region. Therefore the research results of Miogypsinid foraminiferal Biostratigraphy through Tan Sin Hok's morphogenetic method could compare easily with the results of Miogypsinid foraminiferal biostratigraphic research from many areas (Drooger, 1993).

The purpose of this study is to describe the introduction of the Miogypsinid foraminiferal Biostratigraphy and its evolutional lineage based on the author's research and other colleagues results, and research of materials from three areas (Maraş, Palu, and Muş) of Menderes - Taurus Platform, Turkey, respectively.

2. Method of study

All microscopical studies were conducted by examination of all sectioned foraminiferal specimens from sample materials collected from the biostratigraphial columnar sections or spot samplings in order to reinforce the space and time distribution of species. Concerning to the observation of outer and inner structure of the foraminiferal test, the microscope used had the lens combination from x 20 to x 200. The biometrical measurement of the equatorial sectioned nucleoconch and peri-nucleoconchal chambers were made by means of a curvimeter and/or scale protractor from a drawing or direct thin section of the nucleoconch and peri-nucleoconchal chambers at magnification of x 200. The present study is based on the sectioned specimens and free specimens, which were collected from various localities and/or drill core in Japan, Taiwan, and Turkey.

In the present paper, the morphological terms used are given in the glossary and the important criteria for detailed measurements (Figure 1). The measurements were taken from the equatorial and axial (= vertical) sections of megalospheric specimens which exhibited considerable variations in measurements. Also those of microspheric specimens are used supplementary for the measurements and for observation of structure. Since all characters are not measurable, the statistical analysis and consideration from measurements are not perfectly alternative to traditional description, but merely supplementary to it, but provides a more objective basis for comparison between the measurable characters of important morphology and/or structure. The measured parameters are explained for the following terminology as defined by many authors (Figure 1).

On nucleoconch (= embryonic chambers), as showing the development of embryonic chambers is explained as below.

DI: diameter of protoconch, the first chamber of embryonic chambers, consisting protoconch and deuteroconch (Figure 1a). Generally inner protoconch is measured at right angle for the center line of both protoconch and deuteroconch. DI is Drooger's (1952) symbol or definition. The next two is also his symbol and definition.

DII: diameter of duteroconch, the second chamber of embryonic chambers (Figure 1a).

DII/DI: ratio of diameter between protoconch and deuteroconch.

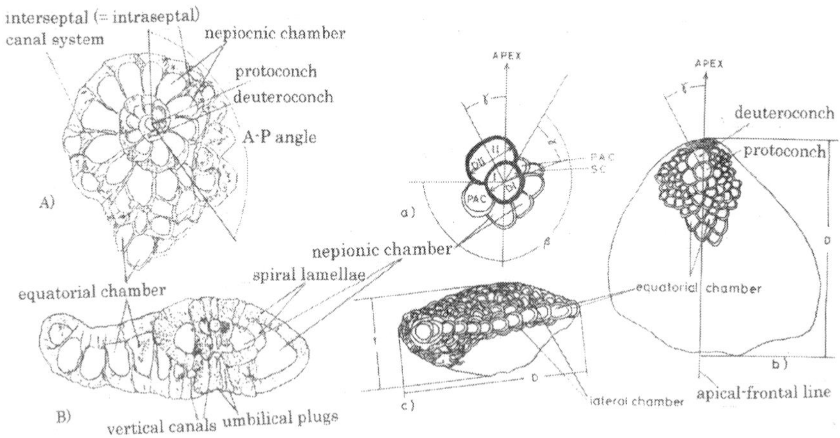

Fig. 1. Terminology of the Miogypsinid foraminifera. A. Equatorial section of *Miogypsinella boninensis* Matsumaru, and B. Axial (= Vertical) section of *Miogypsinella boninensis* Matsumaru. a. Enlargement of the apical portion of equatorial section of *Miogypsina globulina* (Michelotti). b. Equatorial section of *Miogypsina globulina* (Micgelotti). c. Axial (=Vertical) section of *Miogypsina globulina* (Michelotti). I = protoconch. II = deuteroconch. PAC = principal auxiliary chamber. SC = symmetrical (= closing) nepionic chamber. DI = diameter of protoconch. DII = diameter of deuteroconch. α = angle between a line joining the center of the protoconch and the junction of both walls of the protoconch and deuteroconch, and another line connecting the said center and a mid-point of the posterior wall of the symmetrical chamber. ß = angle representing the whole development area of both nepionic spirals of the protoconch surround with both spirals starting from both PAC. V = 200 α/ß = indicate from the absence of the second PAC representing by the value 0 to the protoconchal spirals of equal length by the value 100. γ = angle between the apical - frontal line and line joining the center of the protoconch and deuteroconch.

On nepionic chambers, as base character of the principle of nepionic acceleration (including nepionic retardation, but with additional development of new nepionic spirals) by Tan Sin Hok's (1936, 1937) theory is explained as below.

Parameter X: number of spiral nepionic chambers developed in peri-nucleoconch (= peri-embryonic chambers). This is Tan Sin Hok's (1936) "*Rotalia*-Anfang und mit intraseptalen Spalten". Drooger (1952) counts and used as one of important characters as symbol or definition, parameter X. The value X is progressed until the new nepionic spirals of parameter α as stated below. In the genus *Miogypsinella* with trochoid spirals (Figure 2, top left) and genus *Miogypsinoides* with planispiral (Figure 2, top right), the nepionic chambers are counted in number as parameter X in a spiral arrangement. This is continued until the presence of *Miogypsina primitiva* (Tan Sin Hok) (Figure 2, second left from top), which is considered to be the genus *Miogypsinopsis* Hanzawa, 1940, and *Miogypsina borneensis* Tan Sin Hok, 1936. The number of parameter X is getting decrease from *Miogypsinella boninensis* Matsumaru, 1996 (Figure 1A, X = 25) to *Miogypsina primitiva* (Figure 2, X = 10) and *Miogypsina borneensis* (Plate 2, figures 1-3, X = 6), and it is regarded as nepionic retardation. The next step is beginning at the development of secondary nepionic spirals from the second prinicipal auxiliary chamber and situated in the opposite side of the primary principal auxiliary chamber (Figure 1a; parameter α). Therefore this evolutionary lineage is generally regarded as Tan Sin Hok's (1936, 1937) nepionic acceleration from the genus *Miogypsinella* to genus *Miogypsina*. The following four parameters are based on Drooger (1952, 1963).

Parameter α: small nepionic spiral developed from the second principal auxiliary chambers (Figure 1a). This is the secondary or short nepionic spirals, situated under the outer wall of protoconch. This spiral is arc length and measured by the angle between the line connecting the center of protoconch and rough inscribed line touchrd between deuteroconch and second principal auxiliary chamber, and the line connecting the center of protoconch and center line of closing chamber, which is situated between large and small nepionic spirals developed from opposite direction of two principal auxiliary chambers (Figure 1a). Parameter α in Figure 1a = 40°.

Parameter ß: total nepionic spirals including a closing chamber (= symmetrical chamber, Figure 1a, sc) under the outer wall of protoconch, developed from two primary and secondary principal auxiliary chambers. This spiral is also arc length and measured by the angle between two rough inscribed lines from both two principal auxiliary chambers and deuteroconch, connecting with the center of protoconch (Figure 1a). Parameter ß in Figure 1a = 240°. Generally the primary or long nepionic spirals from the primary principal auxiliary chamber is larger arc than the secondary or short nepionic spirals.

Parameter V (= 200 α/ß): ratio of small or short nepionic spirals (α) for total short and long nepionic spirals (ß), and parameter α will be stopped at the midpoint of parameter ß. Then the ratio times 200 are expressed as a continuous scale with units from 0 to 100. When a closing or symmetry chamber is situated at the midpoint of both protoconchal nepionic spirals, V value is indicated as 100. When a short nepionic spiral or a closing chamber isn't present, V value is indicated as 0. Parameter V in Figure 1a = 40°/240° x 200 = 33.332. When measuring these parameters in numerous specimens in a sample, there is sometimes exceed over 100 in V value, but it is few case.

Parameter γ: angle between the apical-frontal line of test through the center of protoconch and the line connecting centers of embryonic chambers (Figure 1b). If the primary principal auxiliary chamber is situated below the line connecting of centers of embryonic chambers, parameter γ is positive, and reverse is negative (Plate 2, figure 2). Parameter γ in Figure 1a = positive 30°.

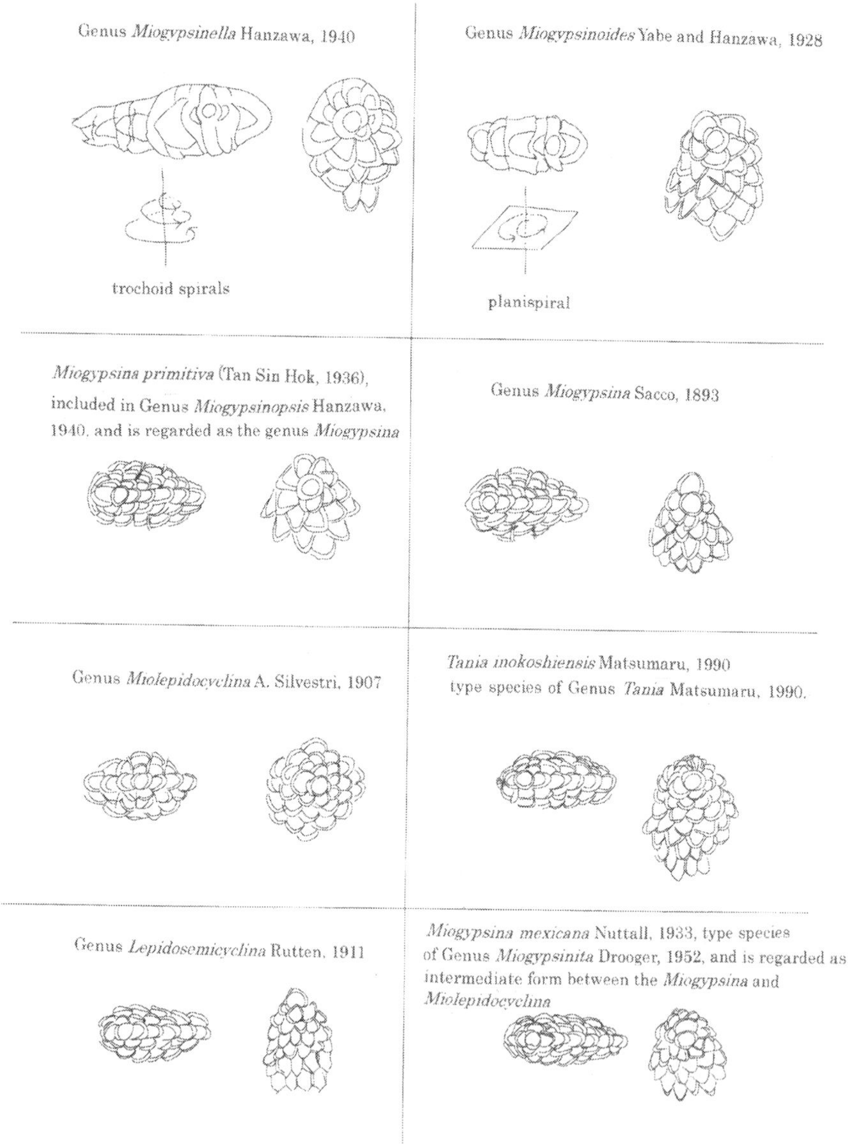

Fig. 2. Sketch of several genera and species of the family Miogypsinidae Vaughan, 1928 in the Tethys region. Two nepionic spirals are shown as trochoid spirals and planispirals.

Parameter A-P angle: arc length of nepionic spirals starting from embryonic chambers, and ending to the apical point of test (Hanzawa, 1957, p. 91; table 6), and A-P angle in Figure 1A , AP = 360° + 145° = 505°.

Designation of the species: In the previous investigation by many authors, the species units of miogypsinid foraminifera have mainly been established by applying from the mean X value to mean V value (mean X - mean V value scale), in addition to traditional main observation, i.e. presence and/or absence of lateral chambers, shape and arrangement of equatorial chambers, arrangement of stolons and canal system in chamber walls, and development of pillars and/or sometimes spines. Species of Figure 1A, B is *Miogypsinella boninensis* Matsumaru, carrying Parameter X (X = 25) and A-P angle (AP = 505°), and that of Figure 1 a-c is *Miogypsina globulina* (Michelotti), carrying Parameter V (V = 33.3) and Parameter γ (γ = + 30°). Parameter X exists until 5, and doesn't exist 4.

3. Stratigraphy, faunal succession, and correlation

In this chapter, the author describes the introduction of the fundamental Miogypsinid foraminiferal Biostratigraphy, faunal succession and phylogenetic lineage, and correlation.

1. Ogasawara Islands, Japan

The basal Oligocene carbonate sedimentary rocks in Japan has been known in Chichi-Jima (island) and Minami-Jima, Ogasawara Islands, Japan (Matsumaru, 1996) (Figures 4-7). In there, six stratigraphic sections in the Minamizaki cape, SW of Chichi-Jima, and two stratigraphic sections and several spot samplings in the Minami-Jima, Ogasawara Islands were examined for the larger foraminiferal biostratigraphy of the Minamizaki Limestone (Formation) with maximum 244 m thick, overlying the basement volcanic rocks (boninite, andesite, dacite, and others) (Figures 5-6). Two larger foraminiferal assemblages (Assemblage IV and Assemblage V) during Oligocene age were recognized in the respective sections of biostratigraphic sequence, based on the stratigraphic range of larger and smaller foraminifera in association with planktonic foraminifera. The Assemblage IV is the *Eulepidina dilatata* (Michelotti) - *E. ephippioides* (Jones and Chapman) - *Heterostegina borneensis* van der Vlerk Assemblage and the Assemblage V is the *Miogypsinella boninensis* - *Spiroclypeus margaritatus* (Schlumberger) - *Austrotrillina howchini* (Schlumberger) Assemblage. Both assemblages were correlative with Tertiary c and/or Tertiary d, and Tertiary e1-2 to Tertiary e4 of the East Indies Letter Stages (Leupold and van dr Vlerk, 1931), respectively, and were also correlative with Zone P 18?-21 or *Globigerina sellii* (Borsetti) Zone - *Globorotalia opima opima* Bolli Zone, and Zone P 21? or P 22 of planktonic foraminiferal zonations. The Assemblage IV is correlated with the fauna of the Tertiary beds of 1629 to 2687 feet, in Eniwetok Atoll Drill Holes (Cole, 1957), and 1723.5 to 2359.5 feet, in Bikini Atoll Drill Holes (Cole, 1954), respectively, because of the coexistence and range of *Eulepidina ephippioides* (Jones and Chapman), *Heterostegina borneensis* van der Vlerk, *H. duplicamera* Cole, and *Halkyardia minima* (Liebus) (Figure 4). The Assemblage V is also correlated with Tertiary e limestones in bore-holes at Eniwetok Atoll Drill Holes at depth from 1210 to 1599 feet, and at Bikini Atoll Drill Holes at depth from 1597.5 to 1671 feet, respectively, where *Miogypsinella grandipustula* (Cole) and *Miogypsinella ubaghsi* (Tan Sin Hok) were reported by Cole (1954, 1957) (Figure 4).

Two assemblages (IV and V), Ogasawara Islands are referable in the geological age to Early to late Early Oligocene, and early Late Oligocene, respectively (Figures 6-7). According to Kaneoka et al. (1970) and Tsunakawa (1983), K-Ar radiometric ages on boninite, andesite, dacite and quartz dacite of basal volcanic rocks of Chichi-Jima, Ogasawara Islands is regarded as 43.0 to 29.4 Ma, and the most young age of volcanics is 26.7 Ma. The Minamizaki Limestone is regarded as submerged karst topography, with summits sticking up from the sea as peninsulas of Minamizaki Cape, Chichi-Jima and Minami-Jima (island) and a lot of islets. The largely submerged Minamizaki Limestone is estimated to be more than 244 m thick and overlies the basement volcanic rocks of lavas and pyroclastics of boninite and other rocks as stated above.

The Assemblage IV is at least regarded as Tertiary d, in this study, due to occurrence of *Heterostegina borneensis* van der Vlerk, *H. duplicamera* Cole, *Eulepidina dilatata* (Michelotti), *E. ephippioides* (Jones and Chapman), *Pararotalia mecatepecensis* (Nuttall), *Paleomiogypsina boninensis* Matsumaru, *Borelis pygmaeus* (Hanzawa) and *Nephrolepidina marginata* (Michelotti), with associated planktonic foraminifera of Zone P 21(Blow, 1969) such as *Globorotalia opima nana* Bolli, *G.* cf. *opima opima* Bolli, and *G.* gr. *opima* Bolli. They are correlated with the Late Eocene to Neogene time scale (official website of ICS, 2004; Berggren et al., 1995) (Figures 6-7, 9). The Assemblage V is assigned to Tertiary e1-2 to Tertiary e3, in this study, due to occurrence of *Miogypsinella boninensis* Matsumaru, *Spiroclypeus margaritatus* (Schlumberger), *Cycloclypeus eidae* Tan Sin Hok, which is junior synonym of *C. koolhoveni* Tan Sin Hok and/or *C. oppenoorthi* Tan Sin Hok, *Paleomiogypsina boninensis, Boninella boninensis* Matsumaru, *Flosculinella reicheli* Mohler, which is a synonym of *Flosculinella globulosa* Rutten, and *Austrotrillina howchini* (Schlumberger). This fauna didn't associate with diagnostic planktonic foraminifera, but it should be assumed to be Zone P 22 from the biostratigraphical occurrence (Figure 6). Although the basal part of the Minamizaki Limestone is obscure due to subsidence under the sea, *Pararotalia mecatepecensis* may evolve into *Paleomiogypsina boninensis* due to nepionic acceleration and well-developed subsidiary chambers during early Oligocene (Rupelian) and/or latest Eocene (Priabonian?) due to basal volcanic radiometric age (Figures 3-4, 7). Also *Paleomiogypsina boninensis* evolved into *Miogypsinella boninensis* due to biostratigraphical occurrence, Tan Sin Hok's nepionic acceleration, and development of equatorial chambers (Figures 3-4, 7). *Miogypsinella boninensis* has the character of number of nepionic chambers (mean X = 27) and A-P angle (mean AP = 578°) (Figure 4).

2. Komahashi-Daini Seamount, Japan

The larger foraminiferal assemblage has been discovered from limestone blocks dredged at two sites on the Komahashi-Daini Seamount of the Kyushu-Palau Ridge, Japan (sample DG-04-01; 30°02.98′N. lat., 133°19.88′E. long.; sample DG-05-02; 29°53.98′N. lat., 133°22.66′E. long.; Mohiuddin et al., 2000) (Figure 5). The assemblage is dominated by the occurrence of *Miogypsinella ubaghsi* (Tan Sin Hok), *Spiroclypeus margaritatus, Heterostegina borneensis, Eulepidina dilatata, E. ephippioides, Nephrolepudina marginata,* and *Austrotrillina howchini,* and was correlated with the top part of the Minamizaki Limestone of Ogasawara Islands. In this study, the Komahashi-Daini larger foraminiferal fauna may be regarded as the fauna from the covering limestone of the Minamizaki Limestone, Ogasawara Islands, because *Miogypsinella ubaghsi* didn't occur from the top member of the Minamizaki Limestone (Figures 4, 6-7). *Miogypsinella ubaghsi* has the character such as number of nepionic chambers (X = 21) and A-P angle (AP = 395°) (Mouhiddin et al, 2000, fig.8-3). Judging from the stratigraphic correlation and nepionic acceleration, *Miogypsinella boninensis* evolved into *Miogypsinella ubaghsi* as the author's consideration (Matsumaru, 1996, p. 39, fig. 24) (Figure 4).

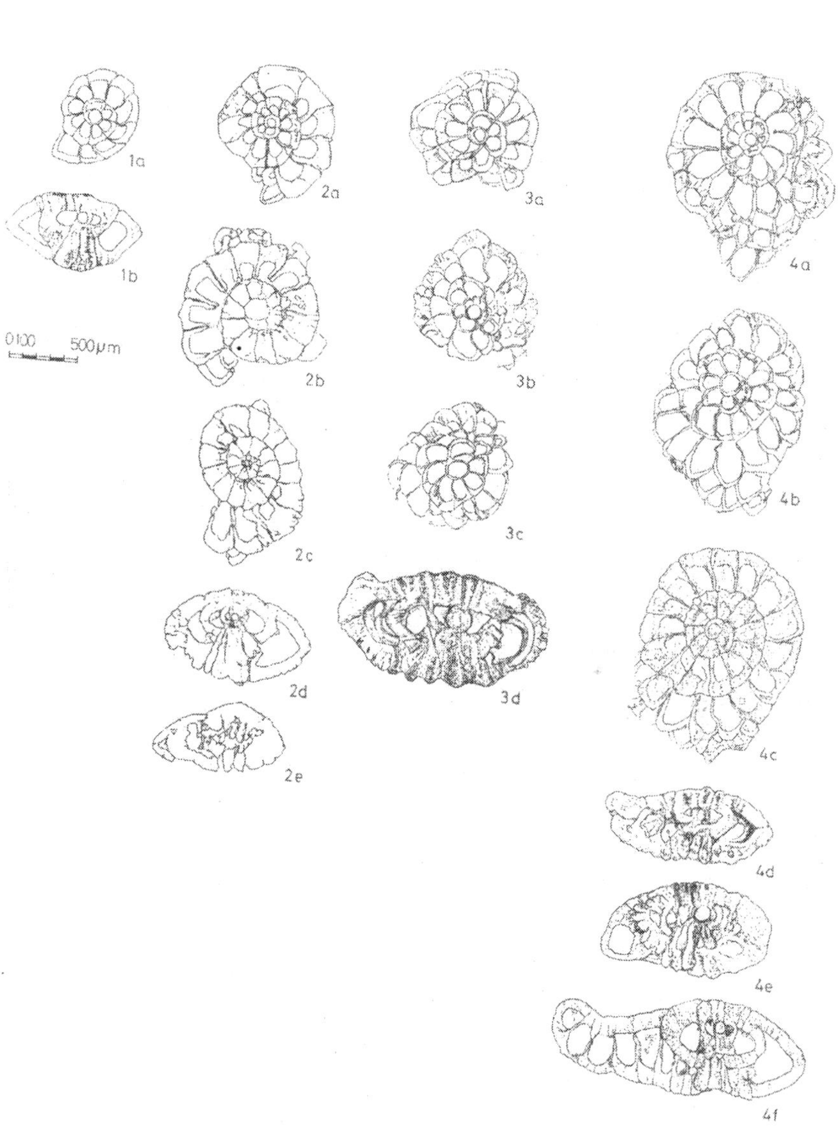

Fig. 3. Drawings of the embryonic, nepionic, and neanic stages in the equatorial and axial sections of species of: 1. *Pararotalia mecatepecensis* (Nuttall), 2. *Paleomiogypsina boninensis* Matsumaru, 3. *Boninella boninensis* Matsumaru, 4. *Miogypsinella boninensis* Matsumaru (Matsumaru, 1996, fig. 23).

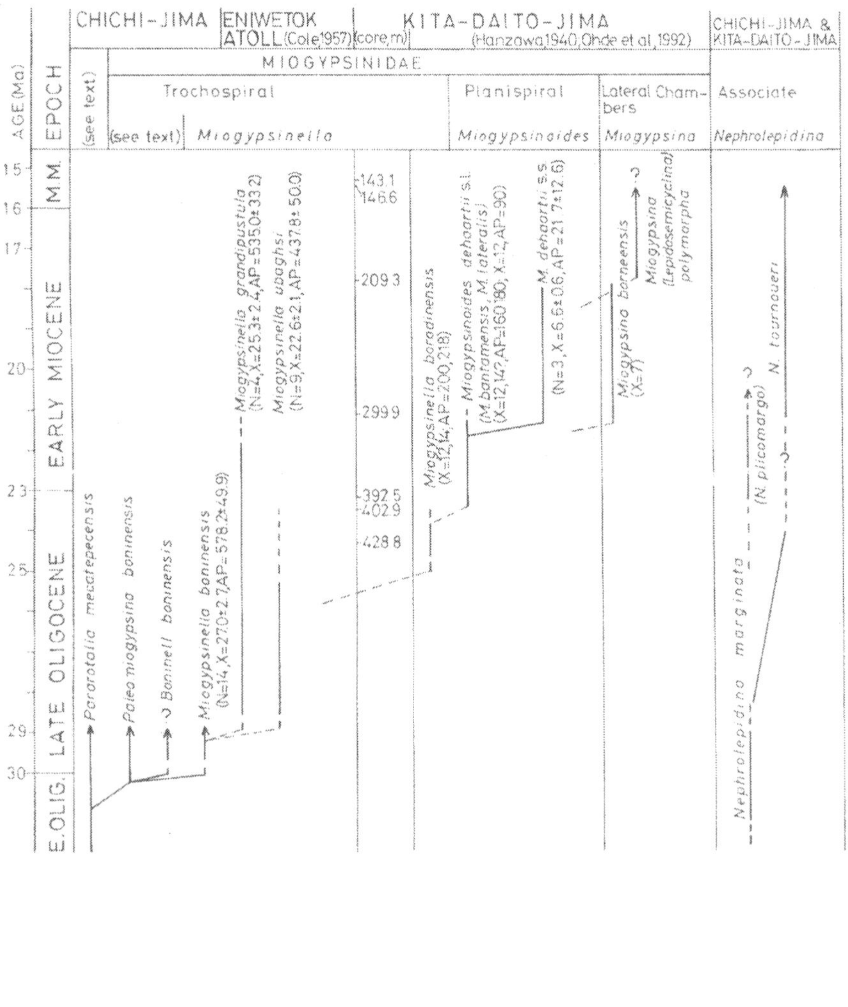

Fig. 4. Evolution of the western Pacific Miogypsinids from Chichi-Jima, Eniwetok Atoll, and Kita-Daito-Jima, and the stratigraphic position of associated *Nephrolepidina* species from Chichi-Jima and Kita-Daito-Jima (Matsumaru, 1996, fig. 24).

Fig. 5. Geographical locations from Japan and Taiwan treated in this study (Retouch to Matsumaru, 1996, fig. 1).

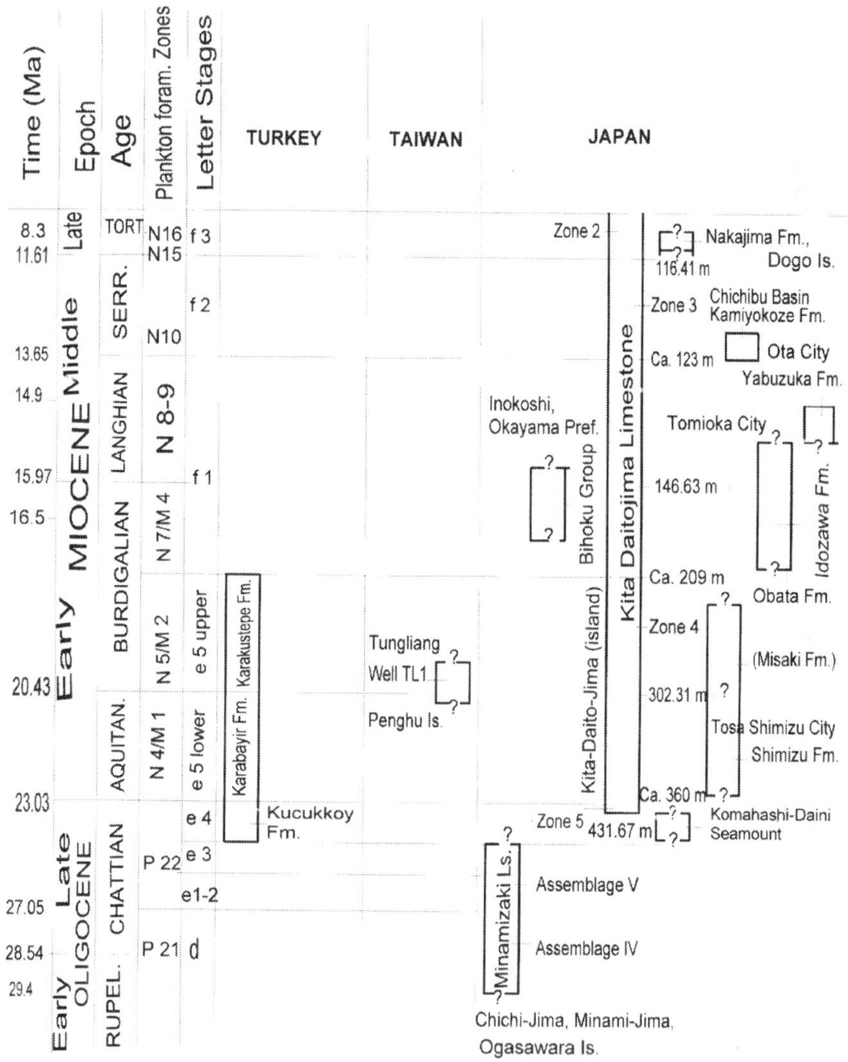

Fig. 6. Correlation chart between the stratigraphic columnar sections treated in Turkey (Matsumaru et al., 2010); Taiwan (Matsumaru, 1968); and Japan (Hanzawa, 1940; Matsumaru, 1967, 1971, 1972, 1977, 1980, 1982, 1996; Matsumaru et al., 1993; Mohiuddin et al., 2000; Nomura et al., 2003).

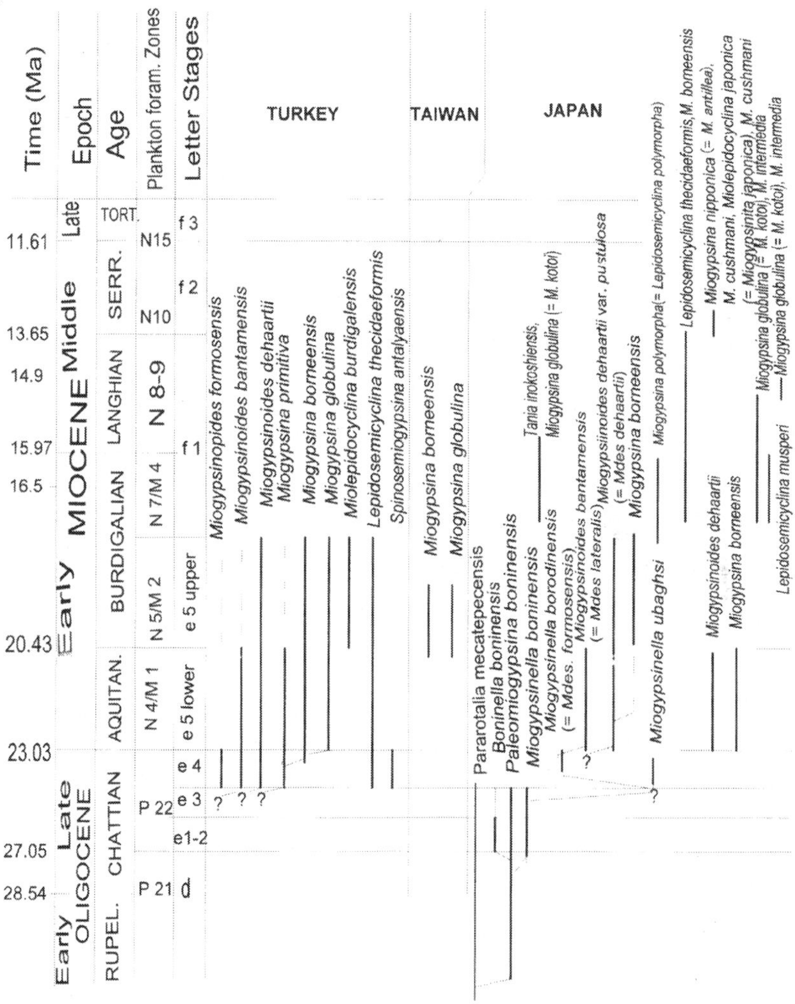

Fig. 7. Biostratigraphic occurrence of Miogypsinid foraminifera from Turkey, Taiwan and Japan.

3. Kita-Daito-Jima, Okinawa Prefecture, Japan

Five foraminiferal fauna have been established into depth zones of the drill cores (431.67-2.68 m thick) of the Kita Daitojima Limestone, at Kita-Daito-Jima (North Borodino Island, 25°56′47″N. lat., 131°17′30″E. long.), Okinawa Prefecture, Japan (Hanzawa, 1940) (Figures 4-7). Hanzawa's Zone 5 (431.67-394.98 m) is characterized by the occurrence of *Miogypsinella borodinensis* Hanzawa, which is later assigned to *Miogypsinoides borodinensis* (Hanzawa) without lateral chambers or with incipient lateral chambers by Hanzawa (1964). This species evolved from *Miogypsinoides formosensis* Yabe and Hanzawa (Hanzawa, 1964, p. 309, 311) due to decrease of number of nepionic chambers. In this study, *Miogypsinella borodinensis* of probable holotype specimen has the character of both number of nepionic chambers (X = 13) and A-P angle (AP = 220°) (Hanzawa, 1940, pl. 39, fig.6), and *Miogypsinoides formosensis* of probable holotype specimen has the character of both number of nepionic chambers (X = 16) and A-P angle (AP = 240°) (Yabe and Hanzawa, 1928, fig. 1a). Hanzawa's consideration is right, but there is unknown on species variation of both species. Both forms could fortunately be found from the Küçükkoy Formation in Korkuteli area, Bey Dağlari Autochton, Menderes-Taurus Platform, SW Turkey (Matsumartu et sl., 2010) (Figures 6-9). The author in Matsumaru et al. (2010) described *Miogypsinoides formosensis* (Yabe and Hanzawa) and regarded their all specimens of schizont (A1 form) and gamont (A2 form) of sexual reproduction, rather planispiral, and carrying rudimentary lateral chambers (Matsumaru et al., 2010, pl. 2, fig. 1) from the Küçükkoy Formation. A specimen (Matsumaru et al., 2010, pl. 1, fig. 8) has the character of number of nepionic chambers (X = 13) and A-P angle (AP = 210°), while a specimen (Matsumaru et al., 2010, pl. 1, fig. 9) has the character of number of nepionic chambers (X = 16) and A-P angle (AP = 250°). Another specimen (Matsumaru et al., 2010, pl. 1, fig. 10) has the character of number of nepionic chambers (X = 13) and A-P angle (AP = 260°). As such the Küçükkyoy Formation carrying *Miogypsinoides formosensis* was correlated with the Zone 5 drill cores (431.67-ca. 360 m, as stated below) of the Kita Daitojima Limestone due to occurrence of *Miogypsinella borodinensis* (= *Miogypsinoides formosensis*) (Matsumaru et al., 2010).

Sr isotope age of Hanzawa's Zone 5 is regarded as 24.3 to 23.5 Ma (Ohde and Elderfield, 1992), and then Hanzawa's Zone 5 is applied for drill cores from 431.67 to ca. 360 m from their age assignment (Figures 6-7). Judging from the Tan Sin Hok's nepionic acceleration, *Miogypsinella ubaghsi* occurred from the limestone of Komahashi-Daini Seamount evolved into *Miogypsinella borodinensis* (= *Miogypsinopides formosensis*) occurred from Zone 5 drill cores of Kita Daitojima Limestone due to reduction of number of nepionic chambers and low value of A-P angle (Figure 7).

4. Tosa-Shimizu City, Shikoku, Japan

The Shimizu Formation, Ashizuri Cape, Tosa Shimizu City, Shikoku, Japan occupies the southernmost part of the Shimanto Belt, one of Japanese Tectonic Zones, and consists of calcareous sandstone and volcanic conglomerate into the coherent rock facies and chaotic rock facies (Figures 5-7). The calcareous sandstone of the Shimizu Formation occurred *Miogypsinoides dehaartii* (van der Vlerk) (Plate 2, figure 12), *Miogypsina* sp, which is assigned to *M. borneensis, Nephrolepidina praejaponica* Matsumaru, *Spiroclypeus margaritatus* (Schlumberger) and *Victoriella conoidea* (Rutten) in the location of Ashizuri Cape, Tosa Shimizu City, Kochi Prefecture, Japan (32°47′15.4 ″N. lat., 132°57′34.6″E. long., Matsumaru et al., 1993). The Misaki Formation of Tosa Shimizu City is composed on alternation of sandstone and mudstone, and crops out in 4 km NW of Ashizuri Cape and there is no contact with the Shimizu Formation. The lower member of the Misaki Formation occurs *Nephrolepidina praejaponica* Matsumaru, *Amphistegina radiata* (Fichtel and Moll), *Sphaerogypsina globulus* (Reuss) and *Rotalia* spp. and also occurrs planktonic foraminifera such as *Catapsydrax stainforthi* Bolli, *Globigerina altiapertura* Bolli, *G. immaturus* Leroy, *G. subquadratus* Brönnimann, *Globorotalia zealandica* Hornibrook, *Globorotaloides suteri* Bolli and *Praeorbulina sicana* (de Stefani) (Matsumaru and Kimura, 1989). At least *Nephrolepidina praejaponica*-bearing Misaki Formation is assumed to be correlated with Zone N5 to lower Zone N7 of planktonic foraminiferal Zonations (Blow, 1969). Therefore the *Miogypsinoides dehaartii* and *Miogypsina borneensis* bearing Shimizu Formation is underlain the Misaki Formation, and its age is regarded as Zone N4 or Zone N 5? (Blow, 1969) (Figures 6-7). Both embryonic and nepionic stages of *Miogypsinoides dehaartii* and *Miogypsina borneensis* from the Shimizu Formation are insufficient in oblique and vertical sections. The Shimizu Formation carrying *Miogypsinoides dehaartii* and *Miogypsina borneensis* may mostly be correlated with the upper Zone 4 drill cores (302.31 to ca, 209 m thick) of the Kita Daitojima Limestone due to the upper occurrence of *Miogypsinoides dehaartii* var. *pustulosa* (= *M. dehaartii,* s. l.) and *Miogypsina borneensis*, and without *Miogypsinoides bantamensis* (Hanzawa, 1940) (Figures 4-7). *Miogypsinoides dehaartii* s. l. of the Kita Daitojima Limestone has the character of number of nepionic chambers (X = 7, 7 and 7) and A-P angle (AP = 20°, 10° and 20°) in three specimens (Hanzawa, 1940, pl. 40, figs. 29, 27 and 26), and *Miogypsina borneensis* of the Kita Daitojima Limestone has the character of number of nepionic chambers (X = 7 and 7) and A-P angle (AP = 10° and 30°) in two specimens (Hanzawa, 1940, pl. 41, figs. 19-20). However *Miogypsinoides bantamensis* (Tan Sin Hok) in the lower Zone 4 drill cores (ca. 360 to 302.31 m thick) has the character of number of nepionic chambers (X = 12, 12 and 13) in three specimens and A-P angle (AP = 165°, 180° and 195°) (Hanzawa, 1040, pl. 39, figs. 16-17, 19). Then *Miogypsinoides bantamensis* evolved into *Miogypsinoides dehaartii* s. l. due to biostratigraphic occurrence and Tan Sin Hok's nepionic acceleration, with reduction of number of nepionic chambers as explained above (Figure 7). Moreover *Miogypsinoides dehaartii* without lateral chambers evolved into *Miogypsina borneensis* with lateral chambers (Figure 7).

5. Tungliang Well TL1, Paisa Island, Penghu Islands, Taiwan

Miogypsina globulina (B form, but not microspheric form; Matsumaru, 1968, pl. 36, figs. 1-6) with nepionic chambers arranged single type (= *Miogypsina borneensis* Tan Sin Hok) and *Miogypsina globulina* (A form; Matsumaru, 1968, pl. 35, figs. 1-6) with two unequal protoconchal nepionic spirals (= *Miogypsina globulina* (Michelotti)) have been found in *Miogypsina* bearing calcareous sandstone at about 500 m depth in the Tungliang Well TL1, located at about 800 m NE of Tungliang Village, Paisa Island, Penghu Islands, Taiwan (Figure 5). *Miogypsina borneensis* has the character of number of nepionic chambers (X = 6, 5, 6, 6, 7, and 6; mean X = 6) in 6 specimens (n = 6) and A-P angle (AP = 10°, 15°, 6°, 6°, 15° and 25°) (Matsumaru, 1968, pl. 36, figs. 1-6). *Miogypsina globulina* has the characters of ratio of two nepionic spirals (V = 32.56, 18.18, 15.20 and 28.36; mean V = 23.58) in 4 specimens (n = 4) (Matsumaru, 1968, pl. 35, figs. 1-4). Therefore the *Miogypsina* bearing sandstone of Well TL1, carrying *Miogypsina borneensis* and *M. globulina,* but not *Miogypsinoides dehaartii,* is stratigraphically younger than the Shimizu Formation and upper Zone 4 of the Kita Daitojima Limestone, which carry *Miogypsinoides dehaartii* and *Miogypsina borneensis* (Figures 6-7). As such *Miogypsina borneensis* carrying number of nepionic chambers (mean X = 6) from the *Miogypsina* sandstone, Well TL1, Paisa Island, Penghu Islands is necessarily fewer number of nepionic chambers than *M. borneensis* carrying number of nepionic chambers (mean X = 7) from the upper Zone 4 drill cores of Kita Daitojima Limestone, Kita-Daito-Jima.

6. Early to Middle Miocene *Miogypsina* from Honshu, Japan

The Obata and Idozawa Formations, Tomioka Group, Honshu, Central Japan are known as representative sedimentary rocks of late Early to early Middle Miocene age in Japan (Matsumaru, 1967, 1977) (Figures 5-7). Japanese *Miogypsina* has been known to occur from the Lower/Middle Miocene sedimentary rocks in Honshu, Japan as *Miogypsina kotoi* Hanzawa, 1931, *Miolepidocyclina* (= *Miogypsinita*) *japonica* Matsumaru, 1972, *Miogypsina japonica* Ujiie, 1973 (= *M. globulina* (Michelotti)), *M. nipponica* Matsumaru, 1980 (= *M. antillea* (Cushman) and *M. cushmani* Vaughan steps of nepionic acceleration), and *Tania inokoshiensis* Matsumaru, 1990, in addition to *Miogypsina borneensis* Tan Sin Hok, *Lepidosemicyclina thecidaeformis* (Rutten), and *Lepidosemicyclina musperi* (Tan Sin Hok) (Plates 1-2). According to Matsumaru and Takahashi (2004), Japanese *Miogypsina* is discussed as the followings: The measurement data of topotype specimen of *Miogypsina kotoi* Hanzawa is as follows: V = 20, DI = 120 x 70 micron, DII/DI = 1.0, and γ = 35°, and *Miogypsina kotoi* Hanzawa is junior synonym of *Miogypsina globulina* (Michelotti). *Miogypsina japonica* Ujiie from type locality and other three stations has the following data: V = 40.9 and DII/DI = 1.34, V = 39.2 and DII/DI = 1.28, V = 36.1 and DII/DI = 1.32, and V = 47.2 and DII/DI = 1.24. Then *Miogypsina japonica* Ujiie doesn't represent *Miogypsina cushmani* Vaughan, 1924 of V scale of Drooger (1963), but represent *M. globulina* due to having of V value (mean V = 40.85) (n = 4). The *Miogypsina* population at Nogami locality found from the Obata Formation, Tomioka Group, Tomioka City, Gunma Prefecture is known as V value (mean V = 43.93 in 20 specimens), which is assigned to *Miogypsina globulina* (Matsumaru, 1967, 1977) (Figures 5-6). However, critical viewing Miogypsinid population, *Lepidosemicyclina musperi* (Rutten) and *Miogypsina cushmani* Vaughan with V value (V = 77) can be found from the Obata Formation (Figure 7). The Obata Formation is conformably overlain by the basal tuff (T6 Tuff or Wagoubashi Tuff, Matsumaru, 1967) of the Idozawa Formation carrying *Miogypsina globulina,* and the fission truck age of the T6 Tuff is 16. 5 ± 1.9 Ma by Nomura and Ohira (1998). The Idozawa Formation is conformably overlain by the basal tuff (T5 Tuff, Matsumaru, 1967) of the Haratajino Formation, which yields *Orbulina suturalis* Brönnimann, *O. universa* d'Orbigny, *Globorotalia birnagea* Blow, *Globigerinoides sicanus* de Stefani and others (Matsumaru, 1977). The fision truck age of T5 Tuff is 15.2 ± 0.5 Ma (Nomura and Ohira, 2002). Then the geological age of the Idozawa Formation is roughly Langhian of early Middle Miocene based on Zone N8 (Blow, 1969; Berggren et al., 1995), and the age of T5 Tuff is regarded to be the age of the *Orbulina* datum-plane. The *Miogypsina* population at Kanayama locality found from the Yabuzuka Formation at Ota City, Gunma Prefecture is known as V value (mean V = 47.38 in 24 specimens), which is assigned to *Miogypsina intermedia* of Drooger's mean V scale (Figures 5-7). The Kanayama *Miogypsina* population is found from the medium sandstone below the pumice tuff of the Yabuzuka Formation, and fission truck age of this tuff is 14.9 ± 0.5 Ma (Nomura et al., 2003). As such Miogypsinid foraminifera from the Obata, Idozawa and Yabuzuka Formations are known as *Miogypsina globulina* (Michelotti) and *Miogypsina intermedia* Drooger Assemblage. Raju (1974) and Mishra (1996) regarded *Miogypsine globulina* as population with mean V value between zero and 45 and positive γ, but *Miogypsina intermedia* couldn't find from the study of Indian *Miogypsina*. These criteria are arbitrary, and why they cannot find *Miogypsina intermedia* between *Miogypsina globulina* and *Miogypsina cushmani* or *M. antillea* in a series of mean V value scale? (Figure 7).

Miogypsina nipponica Matsumaru is found from the middle member of the Kamiyokoze Formation, at sample location UN-1 (35°58′31″ N. lat., 139°5′42″E. long.), Chichibu Basin, Saitama Prefecture (Matsumaru, 1980) (Figures 5-6). Also the planktonic foraminifera such as *Globigerinoides immaturus* (Leroy), *Globigerinoides subquadratus* Brönnimann, *Globigerina praebulloides* Blow, *Globorotalia* (*Turborotalia*) *peripheroacuta* Blow and Banner, and *Globorotalia* (*Turborotalia*) *birnagea* Blow are found from the upper member of the Kamiyokoze Formation at sample location NG-1 (35°58′25″ N. lat., 139°6′20″ E. long.), Chichibu Basin, and these fauna is shown in the lower part of Zone N10 of Blow (1969) (Matsumaru, 1980). *Miogypsina nipponica* has the character of V value in 23 specimens (mean V = 88.38 ±5.20) and this mean V value is regarded as *Miogypsina antillea* (Cushman) step of Drooger (1952). V value of *Miogypsina nipponica* varies from 78 to 100, and more than 42 % of specimens having V value larger than 90. Then *Miogypsina nipponica* Matsumaru has both V value of *Miogypsina cushmani* (Cushman) step and *M. antillea* step of Drooger (1952). Raju (1974) regarded *Miogypsina cushmani* as miogypsinid population with mean V value between 70 and 100, usually less than 90, and more than 50 % carrying less than 90. Also *Miogypsina antillea* has mean V value between 70 and 100, and more than 50 % carrying greater than 90.

Miogypsina nipponica resembles the topotype of *Miogypsina antillea* according to Cole (1957, pl. 29, fig. 1), but *M. nipponica* frequently possess small nepionic chambers on the deuteroconch (Matsumaru, 1980, pl. 25, figs. 4-5) (Plate 1, figure 10). The ogival to lozengic equatorial chambers and very short hexagonal equatorial chambers are sometimes distributed in *Miogypsina nipponica*, but the elongate hexagonal equatorial chambers are distributed in the frontal margins in *Miogypsina antillea* by Raju (1974, pl. 2, figs. 25-26). *Miogypsina nipponica* is distinguished from *Miogypsina antillea* by Frost and Langenheim (1974) in view of the description and illustration of Mexican specimens. In this study, *Miogypsina nipponica* is regarded as *Miogypsina antillea* step of V value, and associated with *Miogypsina cushmani* Vaughan and *Miolepidocyclina* (= *Miogypsinita*) *japonica* Matsumaru (Figure 7).

Miogypsina kotoi, which is junior synonym of *Miogypsina globulina*, is found from the Nakajima Formation, Dogo Island, Oki Islands (Matsumaru, 1982) (Figures 5-6). The sample location is Kumi at left side of the Kumi River about 1.1 km NW of Kumi Tunnel, Goka-Mura (village), Oki-Gun, Shimane Prefecture, Japan (36°18′11″ N. lat., 133°15′5″ E. long.). *Miogypsina kotoi* has the character of V value of 34 specimens (mean V = 31.29 ± 11.0), and is in association with planktonic foraminifera such as *Globorotalia acostaensis acostaensis* Blow, *G. continuosa* Blow, *G. quinifalcata* Saito and Maiya, *G. scitula* (Brady), *Globigerinoides quadrilobaturus* Leroy and *Globigerina* sp. Then the geological age of the Nakajima Formation carrying *Miogypsina kotoi* is early Late Miocene based on Zone N16 (Blow, 1969). Therefore it is inferred from the planktonic foraminiferal zones that *Miogypsina kotoi* with low mean V value from the Nakajima Formation were reworked from the pre-Nakajima Formation. Moreover *Miogypsina kotoi* (= *M. globulina*) from the Nakajima Formation is associated with *Miogypsina borneensis* Tan Sin Hok, which is known to occur from the Lower/Middle Miocene sedimentary rocks in Japan in the Yatsuo Formation, Toyama Prefecture; and Hirashio Formation, Ibaraki Prefecture (Matsumaru and Takahashi, 2004).

Tania inokoshiensis Matsumaru is found from the sandstone of the Lower Formation ("Yamaga" Formation) of the Bihoku Group, Okayama prefecture (Matsumaru, 1990) (Figures 2, 5-7). The sample location is the same place as Hanzawa's (1935) and Tan Sin Hok's (1937) Inokoshi, Koyamaichi Village, Kawakami-Gun, Okayama Prefecture (34°45′ N. lat., 133°24′ E. long.) (Figures 5-6). *Tania inokoshiensis* Matsumaru is characterized by having

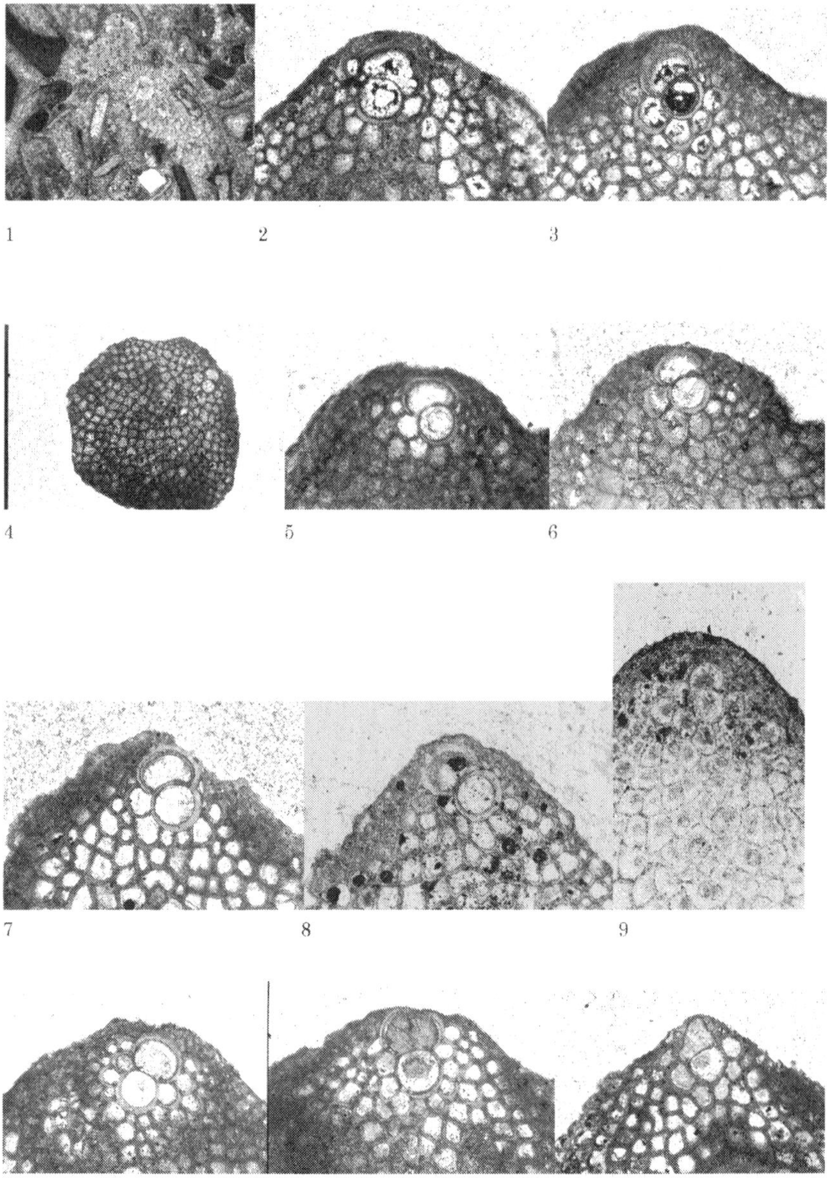

Plate 1.
Figures 1-5. *Miogypsina globulina* (Michelotti)
1. Centered oblique section. Topotype specimen of *Miogypsina kotoi* Hanzawa, 1931.

Otsuki Limestone at Otsuki locality, Yamanashi Prefecture, Japan. x 19. 2-5. Equatorial sections. 2-3, 5. Miyato Formation, correlated with Obata Formation, at Komori locality, Saitama Prefecture, Japan, x 53. 2: V = 25, γ = 20°. 3; V = 29, γ = 25°; 5: V = 40, γ = 20°; 4. Shiomizaki Formation at Todoroki locality, Aomori Prefecture, Japan. x 19. V = 35, γ = 35°.
Figures 6-8. *Miogypsina intermedia* Drooger.
Equatorial sections. 6-7. Ichinokawa Formation, correlated with Idozawa Formation, at Kawabata locality, Saitama Prefecture, Japan. x 53. 6: V = 46, γ = 25°; 7: V = 48, γ = 20°. 8. Nakahara Formation at Hota locality, Chiba Prefecture, Japan. x 53. V = 62, γ = 40°.
Figure 9. *Miogypsina cushmani* Vaughan
Equatorial section. Yabuzuka Formation, correlated with Idozawa and/or Haratajino Formations, at Kanayama locality, Gunma Prefecture, Japan. x 53. V = 70, γ = 10°.
Figures 10-12. *Miogypsina nipponica* Matsumaru
Equatorial sections. 10-12. Kamiyokoze Formation, at Une locality, Saitama Prefecture, Japan. x 53. 10. V = 84, γ = 30° (This specimen possess *Miogypsina cushmani* step of V value, and small nepionic spirals on the outer wall of deuteroconch). 11. Holotype, Saitama Univ. coll. no. 800301, V = 93, γ = 5°; 12. V = 90, γ = 20°.

the peculiar structure of two unequal sets of spiral nepionic chambers, situated along the outer wall of deuteroconch, but not outer wall of protoconch, and having lozengic and short hexagonal shaped equatorial chambers and rectangular shaped lateral chambers (Plate 2, figures 8-9). Then *Tania inokoshiensis* represents more primitive arrangement of embryonic chambers and more advanced hexagonal equatorial chambers than each one of *Miogypsina globulina*, and is associated with *Miogypsina globulina*, which carry the character of V value (mean V = 44.31 in 37 specimens) at Inokoshi (Matsumaru and Takahashi, 2004). *Tania inokoshiensis* is similar to *Lepidosemicyclina thecidaeformis* due to having short hexagonal shaped equatorial chambers. But *Tania inokoshiensis* is different from *Lepidosemicyclina thecidaeformis* in having characteristic structure of two sets of spiral nepionic chambers developed along the outer wall of deuteroconch, but not along the outer wall of protoconch. *Tania inokoshiensis* is similar to *Miogypsina primitiva* Tan Sin Hok due to having deuteroconch situated on the frontal side of test and/or situated beside protoconch along the outer wall of protoconch. However*Tania inokoshiensis* is different from *Miogypsona primitiva* in having two sets of nepionic spirals along the outer wall of deuteroconch. The *Miogypsina* Sandstone of the Lower Formation, Bihoku Group is correlated with the Obata and Idozawa Formations, Tomioka Group due to similar mean value of parameter V of *Miogypsina globulina*. Moreover the *Miogypsina* Sandstone of the Lower Formation, Bihoku Group occurs *Miolepidocyclina japonica* Matsumaru, and is at lest correlated with the lower Zone 3 drill cores (ca. 209-146.63 m) of the Kita Daitojima Limestone due to occurrence of *Miogypsina* (= *Lepidosemicyclina*) *polymorpha* (Rutten) with hexagonal equatorial chambers (Hanzawa, 1940) (Figures 6-7). *Miolepidocyclina japonica* is known to occur from the Lower/Middle Miocene Yatsuo Formation, Toyama Prefecture; Shiomizaki Formation, Aomori Prefecture; Saigo Formation, Shizuoka Prefecture; Naeshiroda Formation, Ibaraki Prefecture; Gassanzawa Sandstone, Yamagata Prefecture; Saginosu Formation, Saitama Prefecture; and Nakahara Formation, Chiba Prefecture, respectively. Their locations are shown in Honshu, Japan (Matsumaru and Takahashi, 2000, fig. 1).

7. Korkuteli Area, Bey Dağlari Autochton, Menderes - Taurus Platform, Turkey

The Oligocene - Miocene succession of the Küçükkoy, Karabayir and Karakuştepe Formations is known to occur in the Bey Dağlari Autochton in the Menderes - Taurus Platform of main

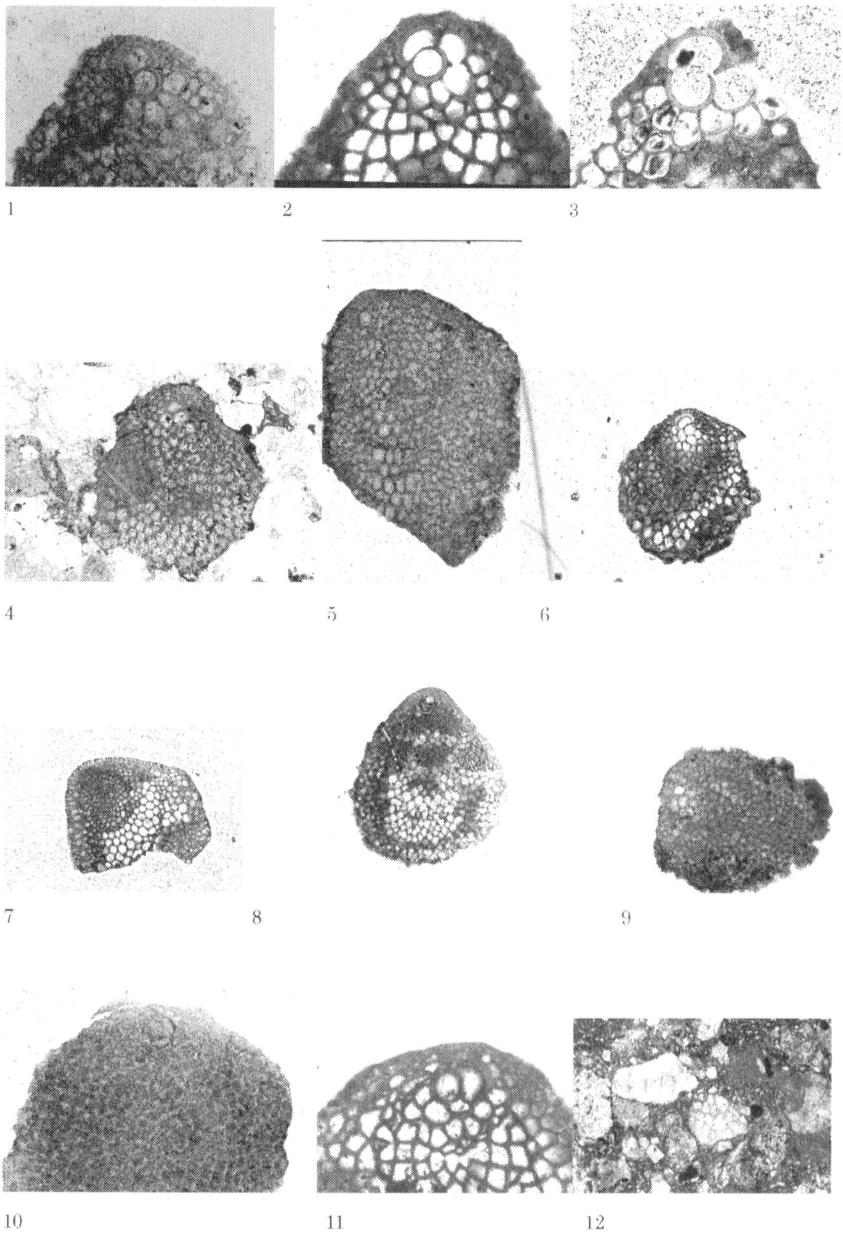

Plate 2.
Figures 1-3. *Miogypsina borneensis* Tan Sin Hok

Equatorial sections. 1. Naeshiroda Formation at Tsukiori-Toge locality, Ibaraki Prefecture, Japan. x = 6, γ = 25°; 2. Nakajima Formation at Dogo locality, Shimane prefecture, Japan. x = 6, γ = - 30°; 3. Hirashio Formation at Tanagura locality, Ibaraki Prefecture, Japan. x = 6, γ = 20°. x 53.
Figure 4. *Lepidosemicyclina musperi* (Tan Sin Hok)
Equatorial section. Obata Formation at Nogami locality, Gunma Prefecture, Japan. x 19.
Figures 5-7. *Lepidosemicyclina thecidaeformis* Rutten
Equatorial sections. 5-6. Megalospheric specimens, 5. Koguchi Formation at Kushimoto, Wakayama Prefecture, Japan, 6. Nakahara Formation at Hota locality, Chiba Prefecture, Japan. 7. Microspheric specimen. Nakahara Formation at Hota locality, Chiba prefecture, Japan. x 19.
Figures 8-9. *Tania inokoshiensis* Matsumaru
Equatorial sections. 8-9. Lower ("Yamaga") Formation, Bihoku Group, at Inokoshi, Okayama Prefecture, Japan. 8. Holotype, Saitama Univ. coll. no. 8803. x 19.
Figures 10-11. *Miolepidocyclina japonica* Matsumaru
Equatorial sections. 10. Gassanzawa Sandstone at Gassanzawa, Yamagata Prefecture, Japan. Holotype, Saitama Univ. coll. no. 720301. 11. Saigo Formation at Shinzaike locality, Shizuoka Prefecture, Japan. 10. x 19, 11. x 53.
Figure 12. *Miogypsinoides dehaartii* (van der Vlerk)
12 left. Axial section. 12 right. Oblique section. Shimizu Formation at Ashizuri Cape, Kochi Prefecture, Japan. x 19.

tectonic units, 40 km NW Antalya City, Turkey (Figure 8). 13 columnar sections from Korkuteli to Karabayir Villages in Korkuteri area, Bey Dağlari Autochton are examined for Miogypsinid biostratigraphy (Matsumaru et al., 2010, figs. 1-3). As results, three larger foraminiferal assemblages are established as the following: *Miogypsinoides formosensis - Miogypsinoides bantamensis - Miogypsinoides dehaartii - Miogypsina primitiva - Spiroclypeus margaritatus* Assemblage (Assemblage 1), *Miogypsinoides bantamensis - Miogypsinoides dehaartii - Miogypsina primitiva - Miogypsina borneensis – Miogypsina globulina – Spiroclypeus margaritatus* Assemblage (Assemblage 2), and *Miogypsinoides dehaartii – Miogypsina borneensis – Miogypsina globulina – Miolepidocyclina burdigalensis* Assemblage (Assemblage 3). The Assemblage 1 is known from the Küçükkoy Formation, and 7 species of *Miogypsinoides formosensis, Miogypsinoides bantamensis, Miogypsinoides dehaartii, Miogypsina primitiva Miogypsina borneensis, Lepidosemicyclina thecidaeformis,* and *Spinosemiogypsina antalyaensis* Matsumaru, Özer and Sari are found in this Assemblage 1 (Figure 7). However two species of *Paleomiogypsina boninensis* Matsumaru and *Miogypsinella complanata* (Schlumberger) in the Assemblage 1 are considered to be reworked. The Assemblage 1 is a younger assemblage than *Miogypsinella boninensis – Spiroclypeus margaritatus – Austrotrillina howchini* Assemblage (Assemblage V) from the upper Minamizaki Limestone, Ogasawara Islands, Japan (Matsumaru, 1996). Because the Ogasawara assemblage (V) has the occurrence of *Miogypsinella boninensis* carrying primitive nepionic spirals and probable planktonic foraminifera belonging to Zone P22 than Zone P21 of Blow (1969) (Matsumaru, 1996) (Figures 6-7). Moreover the Assemblage 1 of Turkey is correlated with Zone 5 drill cores (431.67-ca.360 m) of the Kita Daitojima Limestone (Hanzawa, 1940) due to occurrence of *Miogypsinella borodinensis* (= *Miogypsinoides formosensis*) (Matsumaru et al., 2010). The measurement data of Miogypsinid foraminifera in the Assemblage 1 is as follows: A schizont specimen (A1 form; DI = 88 x 92 micron, DII = 96 x 40 micron)) of *Miogypsinoides formosensis* (Matsumaru et al., 2010, pl. 1, fig. 8) from locality 97-95 in Section 7 has the character of number of nepionic chambers (X = 13) and A-P angle (AP = 210°), while a schizont specimen (DI = 88 x 96 micron, DII = 96 x 48 micron) of *Miogypsinoides formosensis* (Matsumaru et al., 2010, pl. 1, fig. 9) from locality 97-96 in Section 7 has the character of number of nepionic chambers (X = 16) and A-P angle (AP = 250°). A gamont specimen (A2 form; DI = 160 x 160 micron, DII = 128 x 48 micron) of *Miogypsinoides formosensis* (Matsumaru e al., 2010, pl. 1, fig. 10) from locality 96-136 in Section 11 has the character of nepionic chambers (X = 13) and A-P angle (AP = 260°), and also a schizont specimen (DI = 104 x 112 micron, DII = 84 x 44 micron) of *Miogypsinoides bantamensis* (Matsumaru et al., 2010, pl. 2, fig. 3) from locality 97-95 in Section 7, in associated with *Miogypsinoides formosensis,* has the character of number of nepionic chambers (X = 13) and A-P angle (AP = 180°). A gamont specimen (DI = 184 x 168 micron, DII = 192 x 136 micron) of *Miogypsinoides bantamensis* (Matsumaru et al., 2010, pl. 2, fig. 4) from locality 96-137 in Section 11, associated with *Miogypsinoides formosensis,* has the character of number of nepionic chambers (X = 11) and A-P angle (AP = 150°). A gamont specimen (DI = 136 x 120 micron, DII = 128 x 80 micron) of *Miogypsinoides bantamensis* (Matsumaru et al., 2010, pl. 2, fig. 5) from locality 96-137 in Section 11 has the character of number of nepionic chambers (X = 10) and A-P angle (AP = 150°). Moreover on *Miogypsinoides dehaartii,* associated with *Miogypsinoides formosensis* and *Miogypsinoides bantamensis,* a gamont specimen (DI = 152 x 116 micron, DII = 168 x 88 micron; Matsumaru et al., 2010, pl. 2, fig. 7) from locality 97-95 in Section 7 has the character of number of nepionic chambers (X = 9) and A-P angle (AP = 60°), while a gamont specimen (DI = 176 x 168 micron, DII = 208 x 160 micron; Matsumaru et al., 2010, pl. 2, fig. 8) from locality 96-119 in Section5 has the character of nepionic chambers (X = 7) and A-P angle (AP = 50°). A gamont specimen (DI = 224 x 200 micron, DII = 244 x 112 micron; Matsumaru et al., 2010, pl. 3, fig. 6) from locality 97-95 in Section 7 has the character of number of nepionic chambers (X = 10) and A-P angle (AP = 60°).

On *Miogypsina primitiva,* associated with *Miogypsinoides formosensis, Miogypsinoides bantamensis* and *Miogypsinoides dehaartii,* a schizont specimen (DI = 84 x 88 micron, DII = 72 x 56 micron) from locality 97-95 in Section 7 has the character of nepionic chambers (X = 12) and A-P angle (AP = obscure due to twist) (Matsumaru et al., 2010, pl. 3, fig. 4), while a gamont specimen (DI = 200 x 176 micron, DII = 176 x 128 micron) from locality 97-95 in Section 7 has the character of nepionic chambers (X = 10) and A-P angle (AP = 110°) (Matsumaru et al., 2010, pl. 3, fig. 5).

The Assemblage 2 is known from the Karabayir Formation, and 6 species of *Miogypsinoides bantamensis, Miogypsinoides dehaartii, Miogypsina primitiva, Miogypsina borneensis, Miogypsina globulina* and *Lepidosemicyclina thecidaeformis* are found from the Assemblage 2 (Figur 7). The Assemblage 2 is correlated with lower Zone 4 drill cores (ca. 360-302.31 m) of the Kita Daitojima Limestone (Hanzawa, 1940) due to occurrence of *Miogypsinoides bantamensis, Miogypsinoides lateralis,* and *Miogypsinoides dehaartii* var. *pustulosa* (= *M. dehaartii*) (Matsumaru et al., 2010). The measurement data of Miogypsinid foraminifera of the Assemblage 2 is described as follows: a schizont specimen (DI = 112 x 96 micron, DII = 120 x 72 micron) of *Miogypsinoides bantamensis* (Matsumaru et al., 2010, pl. 2, fig. 2) from locality 96-121 in Section 5 has the character of number of nepionic chambers (X = 12) and A-P angle (AP = 180°). On *Miogypsina primitiva,* two specimens are measured: a schizont specimen (DI = 96 x 72 micron, DII = 88 x 40 micron) from locality 97-90 in Section 8 has the character of nepinic chambers (X = 11) and A-P angle (AP = 145°) (Matsumaru, 2010, pl. 3, fig. 2), and a schizont specimen (DI = 96 x 90 micron, DII = 88 x 64 micron) from locality 97-153 in Section 4 has the character of nepionic chambers (X = 9) and A-P angle (AP = 110°) (Matsumaru et al., 2010, pl. 3, fig. 3). On *Miogypsiona borneensis*, three specimens are measured: a gamont specimen (DI = 120 x 112 micron, DII = 128 x 84 micron) from locality 97-90 in Section 8 has the character of nepionic chambers (X = 8) and A-P angle (AP = 50°) (Matsumaru et al., 2010, pl. 3, fig. 9). A gamont specimen (DI = 120 x 116, DII = 140 x 83 micron) from locality 97-152 in Section 4 has the character of nepionic chambers (X = 7) and A-P angle (AP = 25°) (Matsumaru et al., 2010, pl. 3, fig. 10), while a schizont specimen (DI = 96 x 72, DII = 88 x 40 micron) from locality 97-90 in Section 8 has the character of nepionic chambers (X = 7) and A-P angle (AP = 20°) (Matsumaru et al., 2010, pl. 4, fig. 1). Also a schizont specimen (DI = 128 x 104, DII = 144 x 80 micron) of *Miogypsina globulina* from locality 97-90 in Section 8 shows the character of V value (V = 30) and γ value (γ = 10°) (Matsumaru et al., 2010, pl. 4, fig. 7).

The Assemblage 3 is known from the Karakuştepe Formation, and 5 species of *Miogypsinoides dehaartii, Miogypsinopides borneensis, Miogypsina globulina, Miolepidocyclina burdigalensis* and *Lepidosemicyclina thecidaeformis* are found from the Assemblage 3 (Figure 7). The Assemblage 3 is correlated with upper Zone 4 drill cores (302.31 - ca. 209 m) of the Kita Daitojima Limestone (Hanzawa, 1940) due to occurrence of *Miogypsinoides dehaartii* var. *pustulosa* (= *M. dehaartii*) and *Miogypsina borneenisis* (Matsumaru et al., 2010). In the Assemblage 3, the following Miogypsinid foraminifera are measured: On *Miogypsina globulina,* two specimens are measured; a gamont specimen (DI = 200 x 136, DII = 216 x 120 micron) from locality 97-142 in Section 4 has the character of V value (V = 35) and γ value (γ = 40°) (Matsumaru et al., 2010, pl.4, fig. 5), and a gamont specimen (DI = 176 x 152, DII = 248 x 152 micron) from locality 97-125 in Section 4 has the character of V value (V = 25) and γ value (γ = 34°) (Matsumaru et al., 2010, pl. 4, fig. 6).

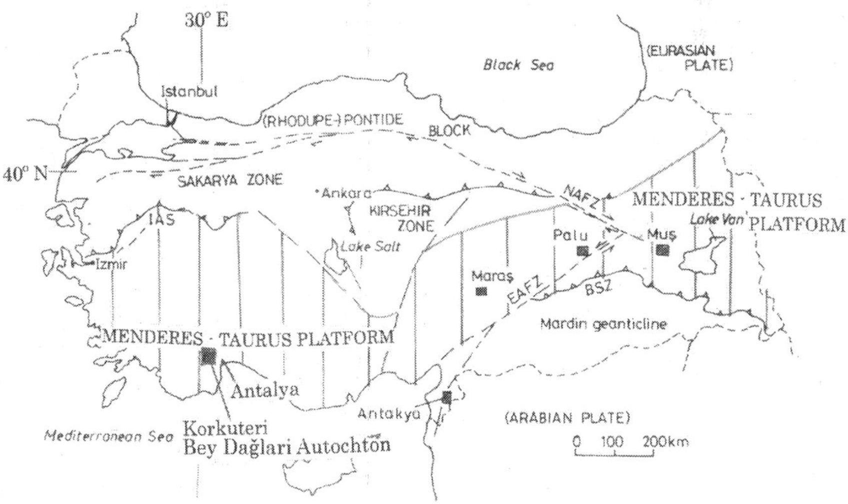

Fig. 8. Locations of research areas (Maraş Palu and Muş) in Turkey treated in this study, in addition to Korkuteri area (Matsumaru et al., 2010). Antakya area without Miogypsinid samples is described in the text.

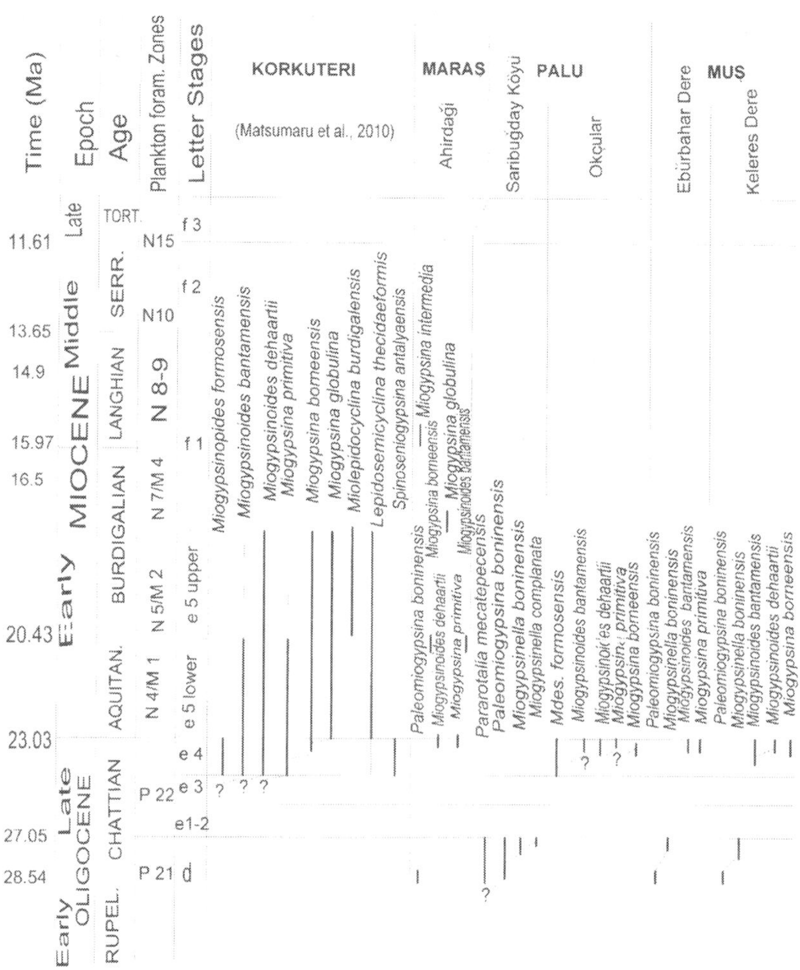

Fig. 9. Biostratigraphic occurrence of Miogypsinid foraminifera from Korkuteri, Maraş, Palu, and Muş areas in the Menderes - Taurus Platform, Turkey.

4. Biostratigraphic occurrence of Miogypsinid foraminifera from Maraş, Palu and Muş areas in the Menderes - Taurus Platform, Turkey

While the author has visited the General Directorate of Mineral Resaerch and Exploration (MTA; T.C. Maden Tetkik Ve Arma Genel Müdürlüğü), Ankara as the Japan Society for the Promotion of Science (JSPS) fellowship researcher for a half year in 1992, he has investigated the foraminifers from the upper Cretaceous to middle Miocene sedimentary rocks in Turkey. This study is to give a note on Miogypsinid foraminiferal Biostratigraphy of Maraş, Palu, and Muş Areas in the eastern Turkey after Uysal et al. (1985), except Antakya Area due to lack of

Miogypsinid foraminifera-bearing samples, where *Nummulites fabianii* (Prever), *N. perforatus* (Montfort), *Pellatispira orbitoidea* (Provale) and others are found from samples A-1 to A-8. Here the author describes useful scientific contribution for Miogypsinid foraminifera (Figures 8-9).

1. Ahirdaği Section, Maraş Area

According to Uysal et al (1985), there are four columnar sections in Maraş Area. In Ahirdaği section in this study, there are five Miogypsinid horizons. Basal samples 77/59 to 77/57 yield *Paleomiogypsina boninensis* Matsumaru, *Lepidocyclina boetonensis* van der Vlerk, carrying the character (DI = 300 x 188 micron, DII = 390 x 223 micron, and DI = 305 x 200 micron, DII = 335 x 215 micron, and thickness of embryonic wall (T = 12 micron)) in two specimens, *Nephrolepidina marginata* (Michelotti), *Heterostegina borneensis* van der Vlerk, carrying the character (1 or 2? operculine chamber(s) and 19 to 20 nepionic septa), and *Cycloclypeus koolhoveni* Tan Sin Hok, carrying the character (more than 23 and 25 heterostegine septa, and DI = 118 x 118 micron), and *Eulepidina dilatata* (Michelotti). They are regarded as the age of late Early to early Late Oligocene. Also they are assigned to probable Tertiary d of the Letter Satges (Leupold and van der Vlerk, 1931; Matsumaru, 1996), because of the occurrence of *Paleomiogypsina boninensis, Heterostegina borneensis,* and *Eulepidina ephippioides,* which are dominated in the Minamizaki Limestone, Ogasawara Islands, Japan (Matsumaru, 1996) (Figures 6-7). Moreover *Paleomiogypsina boninensis* in the Assemblage 1 was found in samples 97-486, 97-502, and 97-503 from the Küçükkoy Formation in the Korkuteri Area, Bey Dağlari Autochton, but *Paleomiogypsina boninensis* was regarded as the reworked species in those samples as stated before (Matsumaru et al., 2010, pl. 1, figs. 1-4) (Figure 9). Sample 77/60, 10 m below from Sample 77/59, yields *Eulepidina dilatata, Lepidocyclina boetonensis* van der Vlerk, carrying the character (DI = 320 x 220 micron, DII = 325 x 125 micron, and 6 nepionic spirals), *Nephrolepidina marginata,* and *Heterostegina borneensis,* carrying the character (1 operculine chambers, 8 chambers in 1 whorl, and 21 chambers in 2 whorls), although there is no Miogypsinid foraminifera, but it is worth to describe the fauna.

More than 40 m above from sample 77/57, sample E479 in Ahirdağgi Section yields *Miogypsinoides dehaartii* (van der Vlerk), carrying number of nepionic chambers (X = 7), *Miogypsina primitiva* Tan Sin Hok, carrying the character (X = 9), *Elphidium* spp., and *Planorbulinella larvata* (Parker and Jones). This horizon is probable assigned to the boundary between the Oligocene and Miocene, due to occurrence of *Miogypsinoides dehaartii* and *Miogypsina primitiva* based on foraminiferal biostratigraphic occurrences between the Küçükkoy and Karabayir Formations, Korkuteri Area (Matsumaru et al., 2010) (Figures 6-7, 9).

Above 90 m from sample E479, Sample E477 occurs *Miogypsina borneensis* Tan Sin Hok, carrying the character (X = 6 to 8), and *Miogypsinoides bantamensis* (Tan Sin Hok), carrying the character (X = 14), and *Elphidium* spp. This horizon is correlated with the Karabayir Formation due to occurrence of *Miogypsinoides bantamensis* and *Miogypsina borneensis,* and then the age of the horizon is assigned to the Early Miocene (Aquitanian) (Figure 9).

Sample E475, at about 50 m above from sample E477, yields *Miogypsina globulina* and *Operculina complanata* (Defrance). *Miogypsina globulina* with 11 specimens has been measured, and they are the character of the followings: DI = 108 x 108 micron, DII = 150 x 125 micron, DII/DI = 1.39, V = 40.7, γ = 40°; DI = 200 x 163 micron, DII = 238 x 175 micron, DII/DI = 1.29, V = 37.0, γ = 10°; DI = 165 x 140 micron, DII = 173 x 103 micron, DII/DI = 1.05, V = 43.5, γ = 15°; DI = 148 x 105 micron, DII = 198 x 118 micron, DII/DI = 1.34, V = 29.6, γ = 5°; DI = 165 x 135 micron, DII = 207 x 116 micron, DII/DI = 1.25, V = 38.5, γ = 30°; DI = 140 x 130 micron, DII = 210 x 125 micron, DII/DI = 1.50, V = 44.4, γ = 15°; DI = 125 x 123 micron, DII = 163 x 125 micron, DII/DI = 1.30, V = 40.0, γ = 25°; DI = 168 x 166 micron, DII = 235 x

123 micron, DII/DI = 1.39, V = 37.0, γ = 20°; DI = 175 x 140 micron, DII = 240 x 150 micron, DII/DI = 1.37, V = 34.8, γ = 20°; DI = 118 x 108 micron, DII = 125 x 116 micron, DII/DI = 1.06, V = 33.3, γ = 15°, and DI = 145 x 165 micron, DII = 213 x 116 micron, DII/DI = 1.47, V = 30.4, γ = 25°. *Miogypsina globulina* is generally characterized by the data of mean V of 37.2 and mean γ of 20° (n = 11), and is regarded as the form of the stratigraphic position between *M. globulina* of Tungliang Well TL1, Taiwan (mean V = 23.58, mean γ = 15°, n = 4) and *M. globulina* of Nogami Area, Obata Formation, Tomioka Group, Japan (mean V = 43.93, mean γ = 40°, n = 20)(Figures 6-7). Therefore the age of *Miogypsina globulina* from sample E475 is probably situated in Early Miocene (Burdigalian) (Figure 9).

Sample E471, about 20 m above from sample E475, yields *Miogypsina intermedia* Drooger, which has the character such as DI = 173 x 148, DII = 175 x 75 micron, DII/DI = 1.01, V = 49, and γ = 30°. As such *Miogypsina intermedia* from sample E475 is correlated with *M. intermedia,* associated with *M. globulina* from the Obata and Idozawa Formations, Tomioka Group, and other Japanese Miocene sedimentary rocks, i.e. Yabuzuka Formation in Ota City (Matsumaru and Takahashi, 2000) (Figures 6-7). The age of *M. intermedia* bearing sample E471 horizon is assigned to the Middle Miocene (Langhian) (Figure 9).

2. Saribuğday Köyü Section, Palu Area

Sample M230 in Saribuğday Köyü Section, Palu Area (Uysal et al., 1985) yields *Pararotalia mecatepecensis* (Nuttall), *Paleomiogypsina boninensis* Matsumaru, *Nummulites fichteli* Michelotti, *Lepidocyclina isolepidinoides* van der Vlerk, *Eulepidina dilatata* (Michelotti), *Borelis pygmaeus* (Hanzawa), and *Austrotrillina* spp. As such this fauna is correlated with the fauna of samples 77/59 to 77/57 in Ahirdaği Section, Maraş Area, as stated above, due to occurrence of *Paleomiogypsina boninensis* (Figure 9). As the author has described the evolution from *Pararotalia mecatepecensis* (Nuttall) to *Paleomiogypsina boninensis* Matsumaru, both species could be found in sample M230 (Matsumaru, 1996, p. 56, fig. 24) (Figure 4). The basal Sample M224, about 740 m below from Sample M230, yields *Lepidocyclina isolepidinoides* van der Vlerk, carrying the character (DI = 163 x 110 micron, DII = 175 x 105 micron, and 6 nepionic spirals), *Eulepidina dilatata, Nummulites fichteli, N. vascus* Joly and Leymerie, and *Borelis pygmaeus* (Hanzawa), and is regarded as the basal Tertiary d of the Letter Stages, although there is no miogypsinid foraminifera. However it is worth to describe the basal Oligocene in this area.

Sample M232, about 40 m above from sample M230, yields *Miogypsinella boninensis* carrying the character (X = 23, DI = 110 x 90 micron, and DII = 90 x 60 micron), *Nummulites fichteli, Nephrolepidina marginata, Eulepidina dilatata, Borelis pygmaeus, Heterostegina* spp., *Halkyardia minima* (Liebus), and *Operculina* spp. Top sample M233, about 130 m above from sample M232, yields *Miogypsinella boninensis,* carrying the character (X = 26, DI = 110 x 95 micron, and DII = 85 x 40 micron), *Miogypsinella complanata* (Schlumberger), carrying the character (X = more than 18, DI = 110 x 103 micron, and DII = 88 x 30? micron), *Paleomiogypsina boninensis, Pararotalia mecatepecensis,* and rarely *Nummulites fichteli.* Therefore these fauna from three samples (M230, M232, and M233) are correlated with the Assemblage IV from the Minamizaki Limestone due to occurrence of *Paleomiogypsina boninensis, Miogypsinella boninensis, Eulepidina dilatata, Nephrolepidina marginata, Borelis pygmaeus,* and *Halkyardia minima* (Matsumaru, 1996). Then these fauna from Saribuğday Köyü is assigned to Tertiary d stage of the Letter Stages. *Nummulites fichteli* is known in the fauna from Saribuğday Köyü, but isn't known in the Assemblage IV from Ogasawara Islands. Also *Miogypsinella complanata* carrying the number of nepionic chambers (X = more than 18), is known to occur from the fauna of Saribuğday Köyü, but this species isn't known in association with *Nummulites fichteli* in the Tethys region as far as the author knows. Then *Nummulites fichteli* in sample M233 is considered to be reworked, and the fauna of sample M233 may be partly correlated with the Assemblage V from the uppermost Minamizaki Limestone, Ogasawara Islands, Japan (Matsumaru, 1996).

3. Okçülar Section, Palu Area

Sample 77/27B below the basalt layer in the Okçülar Section, Palu Area (Uysal et al., 1985) yield *Miogypsinoides formosensis* (Yabe and Hanzawa), carrying the character (X = 15, DI = 116 x 110 micron, DII = 95 x 63 micron, and AP = 210°), *Eulepidina dilatata, Nephrolepidina marginata* (Michelotti), *Cycloclypeus* spp. and *Operculina complanata* (Defrance). The fauna of sample 77/27B is correlated with the Assembalge 1 of the Küçükkoy Formation in the Korkuteri Area due to occurrence of *Miogypsinoides formosensis* (Matsumaru et al., 2010). In sample 77/27B, *Miogypsinella complanata,* carrying the character (X = more than 18, DI = 100 x 88 micron, and DII = 75 x 50 micron) is, however, found in association with *Miogypsinoides formosensis,* but *Miogypsinella complanata* is considered to be reworked.

Sample 77/26 above the same basalt layer yield *Miogypsinoides formosensis, Spiroclypeus* spp., *Heterostegina* spp., and *Operculina complanata,* in addition to *Nummulites vascus* Joly and Leymerie, which is characterized by having the character (DI = 100 x 98 to 263 x 193 micron, DII = 83 x 40 to 208 x 108 micron, distance and number of chambers in 1/2 whorl = 360 to 400 micron and 3, those in 1 whorl =825 to 875 micron and 8, those of 1 1/2 whorl = 1125 to 1200 and 13 to 17, and those in 2 whorl = 1405 to 1475 micron and 20 to 24), *Cycloclypeus koolhoveni* Tan Sin Hok, carrying the character (DI = 125 x 125 to 120 x 118 micron, DII = 168 x 73 to 175 x 58 micron, number of opeculine chamber = 3, and number of nepionic septa = more than 16), and *Miogypsinella complanata,* carrying the character (DI = 88 x 78 micron, DII = 72 x 38 micron, and X = 19), respectively. The latter three species of *Nummulites vascus, Cycloclypeus koolhoveni,* and *Miogypsinella complanata* are considered to be reworked due to non coexistence.

Sample 77/20, about 50m above from Sample 77/26, yield *Miogypsinoides dehaartii,* carrying the character (X = 6, DI = 145 micron, DII/DI = 1.14; X = 6, DI = 153 x 125 micron, DII = 150 x 123 micron, DII/DI = 0.98; and X = 7, DI = 175 micron, DII/DI = 0.74) in three specimens, *Miogypsina borneensis,* carrying the character (X = 6, DI = 145 x 125 micron, γ = 40°; and X = 7, DI = 175 x 175 micron, DII = 130 x 63 micron, γ = 40°) in two specimens, *Eulepidina dilatata, Nephrolepidina marginata,* and *Operculina complanata.* Sample 77/19, 20 m above from Sample 77/20 yields *Miogypsinoides dehaartii, Operculina complanata,* and *Cycloclypeus* spp., in addition to *Paleomiogypsina boninensis,* and *Miogypsinella ubaghsi* (Tan Sin Hok), carrying the character (X = 24, DI = 63 x 58 micron, DII = 50 x 43 micron, AP = 425°, and diameter of spiral chambers = 650 x 763 micron). The latter two species are considered to be reworked, but the discovery of *Miogypsinella ubaghsi* is important to consider the evolutional lineage from *Miogypsinella boninensis* to *Miogypsinella ubaghsi* based on Tan Sin Hok's nepionic acceleration (Figure 4). This lineage has been considered as the evolution from *Miogypsinella boninensis* in the Assemblage V of the Minamizaki Limestone, Ogasawara Islands to *Miogypsinella ubaghsi* in the dredge limestones of the Komahashi-Daini Seamount, Kyushu - Palau Ridge as stated before (Figure 6-7).

Sample 77/18, about 20 m above from Sample 77/19 yields *Miogypsinoides formosensis,* carrying the character (X = 16, DI = 65 x 45 micron, DII = 112 x 80 micron, and AP = 270°), *Miogypsinoides bantamensis* (Axial section), *Miogypsina primitiva,* carrying the character (X = 11, DI = 125 x 116 micron, DII = 138 x 78 micron, and AP = 160°), *Heterostegina* spp., carrying the character (DI = 230 x 200 micron, and number of operculine chambers =3), and *Planorbulinella larvata* (Parker and Jones). Moreover Sample 77/16, about 40 m above from Sample 77/18, yields probable *Miogypsinoides formosensis, Heterostegina* spp., *Cycloclypeus* spp, and *Operculina complanata,* and is overlain by the basalt. In this section, all samples treated in the study belong to upper Oligocene and can be correlated with the Küçükkoy Formation in Korkuteri Area (Figure 9).

4. Ebürbahar Dere Section, Muş Area

Sample M27 in the Ebürbahar Dere, Muş Area (Uysal et al., 1985) yields *Paleomiogypsina boninensis,* carrying the character (DI = 125 x 125 micron, DII = 130 x 83 micron, 17 to 20 spiral chambers in 2 whorls, and diameter of nepionic spirals = 900 to 1050 micron), *Nephrolepidina marginata* (Michelotti), *Borelis pygmaeus, Operculina* spp., and *Peneroplis* spp. This sample is a horizon about 800 m above from the boundary between the Eocene and Oligocene sedimentary rocks, and is correlated with Samples 77/59 to 77/ 57 in the Ahirdaği Section, Maraş Area, and Sample M230 in the Saribuğday Köyü Section, Palu Area, due to occurrence of *Paleomiogypsina boninensis* (Figure 9). Sample M5, about 722 m above from Sample M27 yields *Miogypsinella boninensis,* carrying the character (DI = 70 x 50 micron, and diameter of nepionic spirals =745 micron), *Eulepidina dilatata, Nephrolepidina* spp., and *Spiroclypeus* spp. This horizon is correlated with Samples M 232 and M233 in Saribuğday Köyü Section, Palu Area, due to occurrence of *Miogypsinella boninensis* (Figure 9). Moreover Sample M1, about 450 m above from Sample M5 yields *Miogypsinoides bantamensis,* carrying the character (dimension of protoconch (diam. x height.) of 223 x 208 micron, and dimension of deuteroconch (diam. x height.) of 175 x 118 micron in axial section, and form ratio of diameter/thickness (F. R. = 1.5 mm/ 0.58 mm = 2.61)), *Miogypsina primitiva,* carrying the character (X = more than 10, DI = 183 x 175 micron, and DII = 173 x 123 micron), *Eulepidina dilatata, Heterostegina* spp., carrying the character (DI = 318 x 283 micron, DII = 365 x 188 micron, and 7 nepionic chambers in 1 whorl), and *Spiroclypeus* spp. This fauna from Sample M1 is correlated with the fauna of Sample 77/18 in the Okçülar Section, Palu Area, due to occurrence of *Miogypsinoides bantamensis* and *Miogypsina primitiva* (Figure 9).

5. Keleres Dere Section, Muş Area

Sample O155 in the Keleres Dere Section, Mus Area (Uysal et al., 1985) yields *Paleomiogypsina boninensis,* carrying the character (X = more than 20, DI = 110 x 108 micron, and DII = 112 x 58 micron), *Heterostegina* spp., *Borelis pygmaeus, Peneroplis* spp. and *Austrotrillina* spp. This fauna is correlated with the fauna of Sample M27 in the Ebürbahar Dere Section, Muş Area; Sample M230 in the Saribuğday Köyü Section, Palu Area; and Samples 77/59 to 77/57 in the Ahirdaği Section in Maraş Area, due to occurrence of *Paleomiogypsina boninensis,* respectively. Sample O148, placed about 310 m thick above from Sample O155 yields *Miogypsinella boninensis, Heterostegina* spp. and *Planorbulinella larvata.* Sample 142, placed more than 1000m above from Sample O148, yields *Miogypsenella boninensis,* carrying the character (X = more than 23, DI = 110 x 105 micron, DII = 112 x 73 micron, and diameter of nepionic spirals = 865 micron), and *Operculina complanata,* carrying the character (DI = 250 x 208 micron, DII = 238 x 135 micron, and distance and number of chambers in 1/2 whorl = 700 micron and 3, those in 1 whorl = 1125 micron and 8, those in 1 1/2 whorl = 1400 micron and 18, and those in 2 whorl = 3700 micron and 28). The fauna from Samples O148 to M142 is correlated with the fauna of Samples M232 to M233 in the Saribuğday Köyü Section, Palu Area, due to occurrence of *Miogypsinella boninensis,* respectively.

Sample M139, placed about 650 m above from Sample M142, yields *Miogypsinoides bantamensis,* carrying the character (X = 12, DI = 118 x 120 micron, DII = 100 x 60 micron, and diameter of nepionic spirals = 600 micron), *Spiroclypeus* spp., *Eulepidina dilatata,* and *Lepidocyclina boetonensis* van der Vlerk. Sample M133, placed about 300 m above from Sample M139, yields *Miogypsina borneensis,* carrying the character (X = 7, DI = 171 x 170 micron, DII = 190 x 100 micron, γ = 30°, and diameter of nepionic spirals = 625 micron),

Heterostegina spp., and *Operculina complanata*. Sample M131, placed obscure rightly, but about 60 m thick above from Sample M133, yields *Miogypsinoides bantamensis*, carrying the character (X = 11, DI = 113 x 105 micron, DII = 113 x 75 micron, γ = 20°, and diameter of nepionic spirals = 525 micron, and AP = 170°), *Miogypsinoides dehaartii*, carrying the character (X = 6, DI = 158 x 113 micron, DII = 158 x 105 micron, and γ = 20°), and *Miogypsina borneensis*, carrying the character (X = 7, DI = 135 x 120 micron, DII = 150 x 85 micron, and γ = 20°). As such the fauna of Samples M139 to M131 is correlated with the fauna of Sample M1 in the Ebürbahar Dere Section, Muş Area; and Samples 77/20 to 77/18 in the Okçülar Section, Palu Area, due to occurrence of *Miogypsinoides bantamensis*, *Miogypsinoides dehaartii*, *Miogypsina primitiva*, and/or *Miogypsina borneensis*, respectively. Moreover these fauna are correlated with the fauna of the Assemblage 1 in the Küçükkoy Formation in Korkuteri Area (Figure 9).

5. Conclusion

The Miogypsinid foraminifera (Order Foraminiferida) in the Tethys Region are known to occur from the Early Oligocene (Rupelian) to Middle Miocene (Serravallian) age. Characteristic faunal assemblages from the Miogypsinid foraminiferal Biostratigraphy in Japan, Taiwan and Turkey have been known and correlated each other, respectively (Figures 6-7, 9). Judging from the correlation and analysis of faunal assemblages, the following evolution is established: *Paleomiogypsina boninensis* was proved to be a diagnostic species for the basal assemblage of the Early Oligocene (Rupelian), and *Paleomiogypsina boninensis* evolved from *Pararotalia mecatepecensis* due to having co-existence, and trochoid nepionic spirals in the Minamizaki Limestone, Ogasawara Islands, Japan (Matsumaru, 1996) (Figures 6-7). *Miogypsinella boninensis* evolved from *Paleomiogypsina boninensis*, and evolved into *Miogypsinella ubaghsi* during Late Oligocene (Chattian), based on the biostratigraphic relationship between the Minamizaki Limestone and limestones of Komahashi-Daini Seamount, Kyushu - Palau Ridge, Japan (Figures 6-7). *Miogypsinella ubaghsi* may evolve into *Miogypsinella complanata* due to nepionic acceleration, but there is no discovery on direct evidences in the field. However there is evidence of the evolution from *Miogypsinella ubaghsi* to *Miogypsinella borodinensis* (= *Miogypsinoides formosensis*) during the Late Oligocene (Chattian), based on the biostratigraphic relationship between limestones of Komahashi-Daini Seamount and basal Zone 5 drill cores of the Kita-Daitojima Limestone, Kita-Daito-Jima, Okinawa Prefecture, Japan (Figures 6-7). However *Miogypsinella complanata* is missing in both limestones as stated above, but probably has been existed as co-existence. *Miogypsinella complanata* and *Miogypsinoides formosensis* are found together, but both species are associated with *Paleomiogypsina boninensis* and *Miogypsinoides bantamensis* in Sample 97-486 in Section 4 and Sample 97-502 in Section 6 in the Küçükkoy Formation, Bey Dağlari Autochton, Menderes–Taurus Platform, Turkey (Matsumaru et al., 2010). Then *Paleomiogypsina boninensis* and *Miogypsinella complanata* are regarded as the reworking. During Late Oligocene (Chattian), *Miogypsinoides formosensis* evolved into *Miogypsinoides dehaartii* through *Miogypsinoides bantamensis* due to the nepionic acceleration, and *Miogypsinoides dehaartii* evolved into *Miogypsina primitiva* due to having the lateral chambers during the depositional age of the Küçükkoy Formation (Figures 6-7, 9). Moreover, *Miogypsina primitiva* evolved into *Miogypsina borneensis* due to the nepionic acceleration in the Küçükkoy Formation during Late Oligocene (Chattian) (Figures 6-7, 9). In the Early Miocene (Aquitanian), *Miogypsina borneensis* from the Küçükkyoy Formation evolved into *Miogypsina globulina* from the Karabayir Formation, Bey Dağlari Autochton (Figures 6-7, 9). Further *Miogypsina globulina* evolved into *Miogypsina intermedia* due to occurrence and nepionic acceleration (Drooger, 1952, 1963). *Miogypsina intermedia* evolved into *Miogypsina cushmani* due to the nepionic acceleration during Early Miocene (Burdigalian)/ Middle Miocene (Langhian) age, and these are shown in Indian and Japanese *Miogypsina* (Raju, 1974; Matsumaru, 1967, 1977; Matsumaru and Takahashi, 2004). Also their evolution is shown in the biostratigraphical correlation between the Karakuştepe Formation carrying *Miogypsina globulina* in Korkuteri Area, Bey Dağlari Autochton, Menders - Taurus Platform, and Sample E471 beds carrying *Miogypsina intermedia* in Ahirdaği Section in Maraş Area, Menderes - Taurus Platform, Turkey (Figure 9). *Miogypsina nipponica* (= *M. antillea* and *M. cushmani* steps of nepionic acceleration) was found from the Kamiyokoze Formation of the Middle Miocene (Serravallian) age, and this species evolved from *Miogypsinid cushmani* of nepionic acceleration by Indian and Japanese Miogupsinid researches (Raju, 1974; Matsumaru, 1980; Matsumaru and Takahashi, 2004). The ancestor of *Miolepidocyclina burdigalensis* (Gümbel), *Lepidosemicyclina thecidaeformis* (Rutten), *Tania inokoshiensis* Matsumaru, *Boninella boninensis* Matsumaru and *Spinosemiogypsina antalyaensis* Matsumaru, Özer and Sari isn't known, although some are considered, and it will be solved from further Miogypsinid foraminiferal Biostratigraphy. Some new genera by the author's research have been known from Miogypsinid foraminifera from the Philippines Archipelago, eastern Tethys region, and they will contact the unknown lineage soon.

6. Acknowledgments

The author thanks the Japan Society for the Promotion of Science (JSPS) for his fellowship researcher in 1992; and the General Directorate of Mineral Research and Exploration (MTA), Ankara, Turkey and their colleagues (Drs. and Messrs. E. Yazgan, E. Sirel, S. Acar, G. Tunay, S. Örcen, and K. Erdoğan) for their kind facility and comment. The author thanks Mr. M. Matsuo, Emeritus Professor of Saitama University, for his kind facility, and Mr. V. Grebro, and Ms. D. Duric, InTech, for their kind managing.

7. References

Berggren, W. A., Kent, D. V., Swisher III, C. C., and Aubry, M. P., 1995: A revised Cenozoic Geochronology and Chronostratigraphy. *In,* Berggren, W. A., Kent, D. V., Aubry, M. P. and Hardenbol, J. eds., *Geochronology, Time Scale and Global Stratigraphic Correlation.* SEPM Special Publication, Tulsa, 54: 129-212.

Blainville, H. M. Ducrotay de, 1827: *Manuel de malacologie et de conchyliologie* (1825). Paris: F. G. Levrault.

Blow, W. H., 1969: Late Middle Eocene to Recent planktonic foraminiferal biostratigraphy. *In,* Brönnimann, P. and Renz, H. H., eds, *Proceedings of the First International Conference on Planktonic Microfossils.* 1. Geneva: E. Brill, 199-422.

Cole, W. S., 1954: Larger foraminifera and smaller diagnostic foraminifera from Bikini Drill Holes. *U. S. Geological Survey Professional Paper* 260-0: 569-608.

________, 1957: Larger foraminifera from Eniwetok Atoll. *U. S. Geological Survey Professional Paper* 260-V: 743-784.

Cushman, J. A., 1918: The larger fossil Foraminifera of the Panama Canal Zone. *Bulletin of the U. S. National Museum,* 103: 89-102.

Drooger, C. W., 1952: *Study of American Miogypsinidae.* Utrecht University, Thesis, 80 pp.

________, 1963: Evolutionary trends in the Miogypsinidae. *In,* Von Koenigswald, G.H. R., Emeis, J. D., Buning, W. L. and Wagner, C. W. eds., *Evolutionary Trends in Foraminifera.* Elsevier, Amsterdam, 315-349.

________, 1993: *Radial Foraminifera, Morphometrics and Evolution.* Verhandelingen der Koninklijke Nederlandse Akademie van Wetenschappen, Afd. Natuurkunde, Eerste Reeks 41:242 pp.

Ehrenberg, C. G., 1839: Uber die Bildung dr Kreidefelsen und des Kreidemergels durch unsichtbare Organismen. *Physikalische Abhandlungen der Koniglichen Akademie der Wissenschaften zu Berlin,* 1838: 59-147.

Frost, S. H., and Langenheim, R. L., 1974: Tertiary larger foraminifera and scleractinian corals from Mexico. *In, Cenozoic reef biofacies.* 388 pp. Northern Ilinois University press, Dekalb.

Hanzawa, S., 1940: Micropaleontological studies of drill cores from a deep well in Kita-Daito-Zima (North Borodino Island). *Jubilee Publication of Professor H. Yabe's 60th birthday,* 755-802.

________, 1957: Cenozoic foraminifera of Micronesia. *Geological Society of America, Memoir,* 66: 163 pp.

________, 1964: The phylomorphogeneses of the Tertiary foraminiferal families, Lepidocyclinidae and Miogypsinidae. *Science Reports of the Tohoku University, second ser. (Geology),* 35: 295-313.

Kaneoka, I., Issiki, N., and Zashu, S., 1970: K-Ar ages of the Izu-Bonin Islands. *Geochemical Journal,* 4: 53-60.

Le Calvez, Y., 1949: Révision des foraminifères Lutetiens du Bassin de Paris. II. Rotaliidae et familles affines. *Mémoires du Service de la Carte Géologique Détaillée de la France,* 1-54.

Leupold, W., and Vlerk. I. M. van der, 1931: The Tertiary. *Leidsche Geolofische Mededeelingen, Leiden,* 5: 611-648.

Matsumaru, K., 1967: Geology of the Tomioka area, Gunma Prefecture with a note on "*Lepidocyclina*" from the Abuta Limestone Member. *Science Reports of the Tohoku University, second ser. (Geology),* 39: 113-147.

________, 1968: Miogypsinid population from the Tungliang Well TL-1 of the Penghu Islands, China. *Transactions and Proceedings of the Paleontological Society of Japan, New Ser.* 72: 340-344.

________, 1971: Studies on the genus *Nephrolepidina* in Japan. *Science Reports of the Tohoku University, second ser. (Geology),* 42: 97-185.

________, 1972: The genus *Miolepidocyclina* from Japan. *In, Prof. Jun-Ichi Iwai Memorial Volume,* 679-681. Sasaki Pub. Co., Sendai.

________, 1977: Neogene stratigraphy of the northern to northeastern marginal areas of the Kanto Mountainland, Central Japan. *Journal of the Geological Society of Japan,* 83: 213-225. (in Japanese with English abstract)

________, 1980: Note on a new species of *Miogypsina* from Japan. *In, Professor Saburo Kanno Memorial Volume,* 213-219. Sasaki Pub. Co., Sendai.

________, 1982. On *Miogypsina* (*Miogypsina*) *kotoi* Hanzawa from Zone N. 16 on Dogo Island, Oki Islands,Japan. *Proceedings of the Japan Academy,* 58, ser. B: 52-55.

________, 1990: A new genus of the Miogypsinid foraminifera from Southwest Japan. *Transactions and Proceedings of the Paleontological Society of Japan, New Ser.* 158: 535-539.

________, 1996: Tertiary larger foraminifera (Foraminiferida) from the Ogasawara Islands, Japan. *Paleontological Society of Japan, Special Papers,* 36: 239 pp.

________, Myint Thein, and Ogawa, Y., 1993: Early Miocene (Aquitanian) larger foraminifera from the Shimizu Formation, Ashizuri Cape, Kochi Prefecture, Shikoku, Japan. *Transactions and Proceedings of the Paleontological Society of Japan, New Ser.* 169: 1-14.

_______, and Takahashi, M., 2004: Studies on the genus *Miogypsina* (Foraminiferida) in Japan. *Journal of Saitama University, Faculty of Education (Mathmatics and Natural Sciences)*, 53: 17-39.

_______, Sari, B., and Özer, S., 2010: Larger foraminiferal biostratigraphy of the middle Tertiary of Bey Dağlari Autochton, Menderes-Taurus Platform, Turkey. *Micropaleontology*, 56: 439-463.

Mishra, P. K., 1996: Study of Miogypsinidae and associated planktonics from Cauvery, Krishna-Godavari and Andaman Basins of India. *Geoscience Journal*, 17: 123-251.

Mohiuddin, M. M., Ogawa, Y., and Matsumaru, K., 2000: Late Oligocene larger foraminifera from the Komahashi-Daini Seamount, Kyushu-Palau Ridge and their tectonic significance. *Paleontological Research*, 4: 191-204.

Nomura, M., and Ohira, Y., 1998: Fission track age of the Miocene tuff in the Tomioka area, Gunma Prefecture. *Journal of Gunma Museum of Natural History*, 2: 35-42. (in Japanese)

_______, and _______, 2002: Fission track age of the Miocene Haratajino and Shiohatado tuffs in the Tomioka Group, Gunma Prefecture, Central Japan. *Journal of Gunma Museum of Natural History*, 6: 75-80. (in Japanese)

_____, Ishikawa, H., Kaneko, M., and Matsumaru, K., 2003; The discovery of *Miogypsina* from the Miocene Series of Kanayama Hill, Ota City, Gunma Prefecture, Central Japan. *Journal of the Geological Society of Japan*, 109: 611-614. (in Japanese)

Ohde, S., and Elderfield, H., 1992: Strontium isotope stratigraphy of Kita-Daito-Jima Atoll, North Philippine Sea: implications for Neogene sea-level change and tectonic history. *Earth and Planetary Science Letters*, 113: 473-486.

Raju, D. S. N., 1974: Study of Indian Miogypsinidae. *Utrecht Micropaleontological Bulletins*, 9: 148 pp.

Rutten, L. M. R., 1911: Over Orbitoiden uit de omgeving or Balik Papan-baai (Oostkust van Borneo). *Verhandlingen van het Geologisch Mijnbouwkundig Genootschap voor Nederlandsch Kolonien, Amsterdam, 25 Febr., 1911*. 1/1122-17/1139.

Sacco, F., 1893: Sur quelques Tinoporinae du Miocène de Turin. *Bulletin de la Sociéte Belge de Géologie, de Paléontologie, et d'Hydrologie* (1893-1894) 7: 204-207.

Tan Sin Hok, 1932: Over *Cycloclypeus* voorloopige resultaten einer biostratigrafische studies. *Mijngenieur Bandoeng*, 11: 233-242.

_______, 1936: Zur Kenntniss der Miogypsiniden. *Ingenieur in nederlandsch-Indie, IV, Mijnbouw en Geologie*, 3 (3): 45-61; 3 (5); 84-98; 3 (7): 109-123.

_______, 1937: Weitere Untersuchungen uber die Miogypsiniden I,und II. *Ingenieur in nederlandsch-Indie, Mijnbouw en Geologie*, 4 (3): 35-45; 4 (6): 87-113.

Tsunakawa, H., 1983: K-Ar dating on volcanic rocks in the Bonin Islands and its tectonic implication. *Tectonophysics*, 95: 221-232.

Ujiie, H., 1973: Distribution of the Japanese *Miogypsina*, with description of a new species. Restudy of the Japanese Miogypsinids, Part 3. *Bulletin of the National Science Museum, Tokyo*, 16: 99-114.

Uysal, S., Sirel, E., and Gündüz, H., 1985: Guneydogu anadolu boyunca (Muş-Palu-Maraş-Hatay) bazi Tersiyer Kesitleri. *Maden Tetkik Ve Aramą Genel Müdürlüğü (Jeoloji Etudleri Dairesi) Jeol. Elid. Ar.*, No. 236 (in Turkish)

Vaughan, T. W., 1928: Subfamily Miogypsininae Vaughan. *In*, Cushman, J. A., 1928, *Foraminifera their classification and Economic Use, Special Publications Cushman Laboratory for Foraminiferal Research*, 1: 401 pp.

Yabe, H., and Hanzawa, S., 1928: Tertiary foraminiferous of Taiwan. *Proceedings of Imperial Academy of Tokyo*, 4: 222-225.